WELLINGTON SEARS HANDBOOK OF INDUSTRIAL TEXTILES

Sabit Adanur, B.S., M.S., Ph.D.

Assistant Professor, Department of Textile Engineering
Auburn University, Alabama, U.S.A.

TECHNOMIC
PUBLISHING CO., INC.
LANCASTER · BASEL

Published in the Western Hemisphere by
Technomic Publishing Company, Inc.
851 New Holland Avenue, Box 3535
Lancaster, Pennsylvania 17604 U.S.A.

Distributed in the Rest of the World by
Technomic Publishing AG
Missionsstrasse 44
CH-4055 Basel, Switzerland

Printed in the United States of America
10 9 8 7 6 5 4 3 2 1

Main entry under title:
  Wellington Sears Handbook of Industrial Textiles

Library of Congress Catalog Card No. 95-61229
ISBN No. 1-56676-340-1

Table of Contents

16.0 GENERAL INDUSTRIAL TEXTILES

Foreword

Wellington Sears Company had its beginning in 1845 as N. Boynton & Company, an outfitter of sailing ships. By the late 1800s, a textile mill had been constructed in Langdale, Alabama, and was selling large quantities of fabric to N. Boynton & Company. Unfortunately a fire destroyed the mill in 1887 leaving N. Boynton & Company with no source of fabric. This event led to N. Boynton & Company providing needed capital to rebuild the mill and become exclusive selling agent for the mill that had been renamed West Point Manufacturing Company. N. Boynton & Company's name was changed at the turn of the century to Wellington Sears Company and the former Ship's Chandlery continued as an exclusive selling agent until 1966 when it was renamed the Industrial Fabrics Division of West Point Pepperell Inc. The Wellington Sears name was retired at that time only to be revived in November of 1992 when Jupiter National, Inc.* of Rockville, Maryland, purchased the division from West Point Pepperell Inc. A great tradition had been reborn in the Wellington Sears Company.

Another great tradition of Wellington Sears Company from 1934 through 1963 was the *Wellington Sears Company Handbook of Industrial Fabrics*. During these years the book was updated and republished five times. Over thirty years have passed since the 1963 edition and the

textile industry has seen dramatic technological and scientific changes. Today, there is little in our lives that is not touched by a textile product. The application areas of industrial textiles cover a very broad range, from deep inside the human body to outer space. Development of man-made fibers with tailored properties opened up new opportunities for the application of industrial textiles. Application of new chemicals increased the performance of industrial textiles. With the ever advancing technology, new manufacturing methods offered possibilities of producing more complex and sophisticated products. As a result, the use of industrial textiles is on the rise.

The management of Wellington Sears Company is proud to present this revised edition of the Handbook and is thankful for the opportunity to do so provided by the following people:

David L. Chandler
Chairman, Johnston Industries
Chairman, Jupiter National, Inc.
Chairman, Wellington Sears Company

Gerald B. Andrews
President & C.O.O., Johnston Industries

This new edition of the *Wellington Sears Handbook of Industrial Textiles* is designed to be a comprehensive resource for professionals and students in textiles and other industries that use textiles. It contains the latest technology. It is not a revision but a completely new book.

We are fortunate to have as author of the handbook, Dr. Sabit Adanur who currently is a professor in the Textile Engineering Department of Auburn University. Dr. Adanur holds a Ph.D.

---

*Jupiter National, Inc. is a public company (symbol, JPI—American Stock Exchange) which is majority owned by Johnston Industries, Inc. (symbol, JII—New York Stock Exchange). Johnston Industries and Jupiter National announced on August 16, 1995 a merger agreement whereby Johnston would acquire the remaining outstanding shares of Jupiter National, Inc.

in Fiber and Polymer Science and a Master of Science in Textile Engineering and Science from North Carolina State University as well as a B.S. degree in Mechanical Engineering from Istanbul Technical University. Before joining Auburn University, he had three years of industry experience as R&D manager with a major industrial fabric manufacturer.

L. ALLEN HINKLE
President and C.O.O.
Wellington Sears Company

Preface

The 21st century will demand a new kind of textile leader, and one of the key attributes will be the ability to adapt to change. Change is the process by which the future invades our lives. Those individuals and corporations that are unable to make the transition will be overwhelmed by what they encounter. Those companies that are smart, quick, flexible and adaptable will grow and prosper.

In the past few years, the world has become a very different place. Monumental events – such as the fall of the Berlin Wall, the collapse of Communism, end of apartheid in South Africa, NAFTA and GATT – have ushered in an era few of us would have dreamed about just a few years ago. The velocity of change in this small global village in which we live, will impact our quality of life, and shape the decisions of business leaders, educators, and government officials far beyond the end of this century.

Today, we stand on the threshold of the greatest economic revolution and challenge in history. The rock solid foundation upon which we must build our future is education – which leads to innovation and creativity, the keys to a successful and competitive nation. These are essential ingredients in a corporation's recipe to compete in the rapidly emerging NAFTA and GATT trade environment.

Trade is always about jobs – not economics. We can not look in our rear view mirror, we have to look out in the headlights – our front – to the future. The challenge of the next century will not be shutting down factories and replacing men with machines, it will be much more profound and rewarding than that. The ultimate challenge of textile executives will be the uniting of the individual, machines, computers and innovation

in the best possible working environment, so they can manufacture the highest-quality products at the most economical cost. Competitiveness is a function of the ability of companies to continuously improve and innovate their products and processes. The firms, in all major industries, that can do this are and will continue to be world leaders.

Textile companies have traditionally viewed investment in terms of hard assets: tangible investments that can be measured, such as plants and equipment. Yet it is investment in the intangible assets of the company – such as training, product development, R&D and supplier/customer relationships – that will help determine success in tomorrow's global economy. We have a tendency to over-invest in some areas and under-invest in those areas most required for competitiveness. In the 21st century, greater investment must be made in innovation, creativity, product development and product improvement. This will mandate a major change in management's perspective of the textile industry.

Corporations will have to adopt less stringent strategies and structures to remain competitive. Organizations of the future will be less bureaucratic, more entrepreneurial, more flexible and creative. These will not be commendable characteristics; they will become essential. The organization will be leaner with a simplified structure, but also have increased channels for action. Leaders will have to work synergistically with other divisions, other companies and even competitors to optimize performance. Corporate strategy will be based not just on products or markets, but on competencies that give a company access to several markets and are difficult for competitors to imitate.

Change itself can really be exciting and invigorating. Our consistent aim has to be to create products today that anticipate our customer's needs for tomorrow. We believe this adage not only captures our determination to anticipate the needs of a changing world, but also our resolve to embrace change as an active participant in future growth and prosperity. We have to be more concerned with inventing the future than conserving the past. We should be more interested in creating new, competitive space, than positioning ourselves in existing markets. Creativity and innovation will go just as far as we take it.

The *Wellington Sears Handbook of Industrial Textiles* is one of our contributions to the textile industry. It is our recognition of the significant role that this important industry has played for mankind in promoting free trade and democracy throughout this shrinking world community. Johnston Industries is very pleased to be able to support this progressive effort.

GERALD B. ANDREWS
President and C.O.O.

DAVID L. CHANDLER
Chairman and C.E.O.

Acknowledgements

This handbook has been prepared with the help of many people. First of all, I would like to thank my colleagues who contributed to the handbook by writing some sections indicated below. Their help is very much appreciated. Some of these sections are co-authored.

Dr. Royal M. Broughton and Mr. Paul B. Brady
- Section 2.1 – Fiber Forming Polymers
- Section 2.2 – Natural and Man-Made Fibers
- Section 3.1 – Manuafacture of Man-Made Fibers
- Section 4.8 – Nonwoven Fabrics

Dr. Yehia E. El-Mogahzy
- Section 3.3 – Manufacture of Staple Yarns

Dr. David M. Hall
- Section 2.2 – Natural and Man-Made Fibers
- Section 4.3 – Slashing
- Section 20.1 – Textile Testing

Prof. Warren S. Perkins
- Section 5.1 – Dyeing, Printing and Finishing

Dr. Lewis B. Slaten
- Section 20.1 – Textile Testing

Mr. Chettoor G. Namboodri
- Section 5.2 – Coating and Laminating

Prof. Robert P. Walker
- Section 4.1 – Classification of Fabrics
- Section 4.2 – Yarn Preparation
- Section 4.4 – Weaving
- Section 4.5 – Knitting
- Section 4.7 – Tufting

The author requested and received considerable assistance from industry professionals, professors, students, organizations, companies and federal agencies.

I would like to thank Mr. Michael J. Ravnitzky of IFAI who spent a considerable amount of his time and provided excellent data, information and suggestions for the book. His help made a difference. Special thanks are extended to other associates of the IFAI including Tim Arens, Juli Case, Jean M. Cook, Joseph A. Dieltz, Danette R. Fettig, Joel R. Hoiland, Maike Liekweg, Frank McGinty, Robert Meany, Karen Musech, Del Swanson and Suzanne M. Zorichak, for their help.

Many associates from Wellington Sears Company have provided industry experience and insight to the handbook. I would like to thank Owen J. Hodges III, Bobby G. Crutchfield, Gayron N. Davis, Donald W. Lauderdale, Harold L. Lauderdale, Alan Lightholder, Darwin L. Sears, M. Wayne Smith and Tommy B. Strength for their help.

The chapters of the handbook are reviewed and edited by many professionals in industry and academia. We are most grateful to the following people who reviewed the sections in their specialty areas for their time and valuable suggestions: Peter Alling, The Haartz Corp.; Hilton Barrett, Red Kap Industries; Jerry Bauerle, Ontario Research Foundation; Lutz Bergman, Filter Media Consulting; Royal Broughton, Auburn University; Ann Brooks, Hamby Textile Research Laboratories; Danny Campbell, Hoechst Celanese; Elliot Chaikof, Emory University Hospital; Tsu-Wei Chou, University of Delaware; Patricia Christian, Meadox Medical, Inc.; Barry Christopher, Polyfelt, Inc.; Jerry Cogan, Milliken Research Center; James Connelly, Highland Industries; Jim Conner,

American Yarn Spinners Association; Dave Crawford, J. A. King & Company, Inc.; Payton Crosby, Asten Dryer Fabrics; Stephen De-Berardino, LINQ Industrial Fabrics, Inc.; Marcel Dery, Chemfab; Joseph Dieltz, IFAI; Martin Dodenhoff, Dodenhoff Industrial Text.; Numan Dogan, Tuskegee University; Steve Duerk, JPS Automotives; David Elton, Auburn University; Carole Faria, U.S. Army Natick R, D & E. Center; Marsha Feldstein, AlliedSignal Inc.; Chris Gardner, Inside Automotives; Milton Gilbert, Darlington Fabrics Inc.; Cheryl Gomes, ILC Dover; Charles Hamermesh, SAMPE; Bill Hawkins, Reemay Inc.; Heidi Jameson, Sulzer Ruti; Charlie Jones, Hoechst Celanese; Carla Kalogeridis, ANCAR Publications; Bill Kennedy, Highland Industries; Andy Kistler, Bridgestone-Firestone; Alan Lightholder, Wellington Sears Textest; Jean Martin, Fabric Development, Inc.; Robert Mathis, Hoechst Celanese; Bob McIlvane; The McIlvane Co.; Mansour Mohamed, North Carolina State University; Terry Montgomery, Springs Industries, Inc.; Bill Oxenham, North Carolina State University; Richard Pinelle, Mann Industries, Inc.; Leonard Pinchuk, Corvita Corporation; Michael Ravnitzky, IFAI; Arthur Schwope, Arthur D. Little, Inc.; Robert Seiner, Cooley, Inc.; Gerhard Seyffer, LTG Technologies; John Skelton, Albany International; Bill Smith, Industrial Textile Associates; Gary Smith, North Carolina State University; Tommy Strength, Wellington Sears Utilization Plant; Jeffrey Stull, International Personnel Protection Inc.; Thomas Tassinari, U.S. Army Natick R, D & E Center; Jerry Tew, AATCC; Rick Walker, McMurray Fabrics, Inc.; William Walsh, Auburn University; Stephen Warner, IFAI; Adam Varley, Vartest Laboratories, Inc.; Kay Villa, ATMI.

In addition I would like to thank many individuals, R&D centers, associations, government agencies, institutions and companies who provided photos, pictures, data and information for the handbook. Several pages would be needed to list their names.

Many of my students have contributed to the handbook with literature search, critical review, drafting and computer work. I thank all of them for their useful work. Special thanks are extended to Wanda Barnes, Patricia Smith and Cynthia Jones for administrative work and clerical assistance.

My sincere apologies to the individuals whose names I might have failed to mention for their valuable time and contributions.

SABIT ADANUR

Introduction

This handbook is designed to serve a dual purpose: a resource for people associated with the textile industry as well as a textbook for students. The technical level of the book is attuned accordingly to accommodate both technical and non-technical professionals.

The format of the handbook follows the natural flow of textile manufacturing processes. In Chapter 1.0, definitions, classification and an overview of industrial textiles are given. At the end of the chapter, several color photographs are included to show the typical application areas of industrial textiles.

Although this is mainly an industrial textiles book, some concepts and information related to traditional textiles are also included for convenience to the reader who may be new to the field of textiles. Polymers and fibers are covered in Chapter 2.0. Chapter 3.0 includes fiber and yarn manufacturing techniques. Fabric manufacturing is included in Chapter 4.0. Chapter 5.0 describes fabric finishing and further processing such as coating and laminating.

Design, structure, properties and applications of industrial textiles for specific markets are described in Chapters 6.0 through 16.0. Industrial textiles were grouped under usage: architecture and construction, textile structural composites, filtration, geotextiles, medical textiles, military and defense, paper machine clothing, safety and protective textiles, sports and recreation, transportation and general industrial textiles. The task of grouping industrial textiles was a challenging one because quite often an industrial textile product may fall under more than one category.

Fiber properties and technology are described in Chapter 17.0. Chapter 18.0 includes yarn nomenclature, properties and technology. Fabric properties and technology are discussed in Chapter 19.0. Textile test methods are listed and discussed in Chapter 20.0. To a certain extent, these chapters are built on the 1963 edition of the *Wellington Sears Handbook of Industrial Textiles*. The author would like to acknowledge the excellent job done by Mr. Ernest R. Kaswell in fiber, yarn, fabric properties and testing more than thirty years ago. Some of that information is still valid today.

Several chapters are included to address the new and emerging technologies in industrial textiles. Textile waste management and recycling are included in Chapter 21.0. Chapter 22.0 includes an introduction and discussion of computer technology used in textiles. Standards and regulations, which have increasingly more effect on industrial textile markets, are covered in Chapter 23.0. Chapter 24.0 gives a brief description of what textiles might be in the near future.

Every effort was made to include the most recent information and technology available throughout the book. Industrial textile markets are highly competitive and more proprietary than traditional textiles, which makes writing a book in the field a real challenge. The information available in the literature on industrial textiles is limited compared to consumer or clothing textiles. There is hardly any reference in the literature that provides comprehensive coverage of industrial textiles. There may be several reasons for this, including the relative young age of industrial textiles, lack of comprehensive expertise in the field, and the highly competitive nature of the industrial textiles market. This book is intended to fill this vacuum and provide the necessary information to both industry and academia in a comprehensive package.

This handbook could not have been a reality without help. I thank God for everything that made this book possible. I would like to acknowledge the support of a few people who made a big difference. First of all, I would like to thank Mr. Allen Hinkle, President and C.O.O. of Wellington Sears Company, who had the vision to realize the need for the handbook and initiated this work. His continuous support and encouragement during the preparation of this handbook have been very comforting, motivational and are deeply appreciated. Dr. William K. Walsh, Professor and Head of the Textile Engineering Department, gave invaluable and timely advice and suggestions from start to finish. His vast experience and pleasant management style helped greatly to secure the successful completion of the handbook. I would like to thank Mr. Stephen M. Warner, President of the Industrial Fabrics Association International (IFAI), who showed a personal interest in the successful outcome of the handbook and made his company's human and technical resources freely available for it. Last but not least, this handbook would not have been possible without my wife Nebiye's continuous support and encouragement.

As can be seen throughout this handbook, technical textiles are becoming one of the most important elements of modern technology and lifestyle. There is hardly any place that textiles are not used in today's world. It is my hope that, when the reader is flipping through the pages, the wide application examples of industrial textiles will create brainstorms for new ideas, developments and application areas of textiles. Finally, I hope that this handbook will help change the minds of those who unfairly consider textiles a "sunset" industry. The last quarter of the 20th century witnessed the "dawn" of modern industrial textiles; the 21st century may very well be the "zenith" of industrial textiles.

SABIT ADANUR

OVERVIEW OF INDUSTRIAL TEXTILES

Overview of Industrial Textiles

S. ADANUR

". . . textiles go to war, textiles go to space, become roof, choke an oil spill, imitate a heart, hold you safely in your seat, hoist tea bags, support tires, diaper babies, line roads, keep you dry, wrap wounds . . ."

American Textile Manufacturers Institute (ATMI)

1. DEFINITION OF INDUSTRIAL TEXTILES

Industrial textiles are specially designed and engineered structures that are used in products, processes or services of mostly non-textile industries.

According to this definition, an industrial textile product can be used in three different ways:

(1) An industrial textile can be a component part of another product and directly contribute to the strength, performance and other properties of that product, e.g., tire cord fabric in tires.
(2) An industrial textile can be used as a tool in a process to manufacture another product, e.g., filtration textiles in food production, paper machine clothing in paper manufacturing.
(3) An industrial product can be used alone to perform one or several specific functions, e.g., coated fabrics to cover stadiums.

Another indication of the definition above is that unlike ordinary textiles which have traditionally been used by the consumer for clothing and furnishing, industrial textiles are generally used by professionals from industries of non-textile character in various high-performance or heavy duty applications.

The term "industrial textiles" is the most widely used term for non-traditional textiles. Other terms used are "technical textiles," "high performance textiles," "high tech textiles," "engineered textiles," "industrial fabrics" and "technical fabrics."

2. HISTORY

Although the beginning of industrial textiles may be as old as traditional textiles dating back to several thousand years ago, industrial textiles are considered to be a little "younger" than traditional textiles. The history of modern industrial textiles probably began with the canvas cloth used to sail ships from the old world to the new across the ocean. Later, hemp canvas was used on covered wagons to protect families and their possessions across the land. Fabrics were used in early cars as "rag-tops" to keep out the weather and as seat cushions for passenger comfort. Fabrics offered the advantage of light weight and strength for early flying crafts in the air. The wings of the earliest airplanes were made of fabrics. Industrial textiles are still used in hot air balloons and dirigibles.

The invention of man-made fibers in the first half of the 20th century changed the industrial textiles market forever. The first truly man-made fiber, Nylon, was introduced in 1939. By development of exceptionally strong high performance fibers in the 1950s and 1960s, the application areas of industrial fibers and fabrics were widened. Man-made fibers not only replaced the natural fibers in many applications, but also opened up completely new application areas for industrial textiles. Synthetic fibers offered high strength, elasticity, uniformity, chemical resistance, flame resistance and abrasion resistance

3

among other things. New fabrication techniques also contributed to the improved performance and service life of industrial textiles. Application of new chemicals help the design engineers to tailor their products for special uses.

Industrial textiles have played a critical role in space exploration. Spacesuits are made of a layered fabric system to provide protection and comfort for the astronaut. Engineered textiles provided strong and lightweight materials for the lunar landing module and for the parachutes used to return the astronauts to earth in 1969.

Military applications, especially during the global conflicts, expedited the development of technical textiles to better protect the soldiers. Today technical textiles are used extensively in military equipment and protective structures.

Technical textiles have met the various challenges created by the advancement of the society and by the ever increasing needs of mankind. Industrial textiles have been entering every aspect of human life. Thanks to advanced medical technology, today minute bundles of fibers are implanted in human bodies to replace or reinforce parts of the human body. Specially engineered textiles are used in airplanes, under highways, in transportation, and for environmental protection to name a few.

3. SIGNIFICANCE OF INDUSTRIAL TEXTILES

Industrial textiles make a vital contribution to the performance and success of products that are used in non-textile industries. For example, 75% of the strength of an automobile tire comes from the tire cord fabric used in the tire. Pure carbon fibers that are used in textile structural composite parts for aerospace, civil and mechanical engineering applications are on average four times lighter and five times stronger than steel.

Some of the modern industries simply would not be the same without industrial textiles. For example, the U.S. Department of Defense (DOD) has in its inventory some 10,000 items which are made entirely or partially from industrial textiles. The artificial kidney used in dialysis is made of 7,000 hollow fibers and has a diameter of only two inches. Heat shields on

TABLE 1.1 Percent Share of Technical Textiles Production in Total Textile Production.

	1980	1985	1990
U.S.A.	–	25	33
Japan	–	27	36
France	11	16	22
China	13	–	22
Germany	–	17	22
Western Europe	9	16	22

space vehicles are made of textile fibers that can withstand several thousand degrees Fahrenheit. These are just some of the many examples which show the significance of industrial textiles in the journey of mankind.

Table 1.1 shows the percent share of technical textiles in total textile production for some industrialized countries. Industrial textiles account for one-third of the total textile production in highly industrialized countries such as the United States and Japan. The rapid increase of industrial textiles' share within the last two decades is a good indication of their significance.

4. PRINCIPLES OF INDUSTRIAL TEXTILES

Industrial textiles are truly engineered structures. These materials offer several advantages simultaneously in the same product that no other industrial product could provide: flexibility, elasticity and strength. Another beauty of the industrial textiles is that they have so many variables at the disposal of the designer which theoretically offer an infinite number of design possibilities. An industrial textile product involves several variables.

4.1 Polymers (Chapter 2.0)

Man-made fibers are made of polymers. Polymers can be classified in different ways such as thermoset and thermoplastic. By using different polymer types and fiber manufacturing conditions, certain properties can be programmed into the fiber. Crystallinity and molecular weight are among the typical characteristics that influence fiber properties.

4.2 Fiber Type and Structure (Chapters 2.0, 3.0, and 17.0)

Natural and manufactured fibers are the two major categories of fibers. Natural fibers are derived from sources in nature such as wool from sheep and cotton from cotton plants. Cotton and other natural fibers such as hemp and jute were extensively used in early industrial fabrics. Development of manufactured fibers presented new opportunities to engineer special properties into an industrial textile product. Today, most of the industrial textile products are made of manufactured or synthetic fibers.

Synthetic fibers can be used in many shapes or forms. They can be in the long continuous form (filaments) or be chopped into shorter lengths (staple fiber). Filaments provide a smooth surface and high strength. Fibers can be made with different diameters, cross-sectional shapes and a combination of different polymeric materials (hybrid or bi-component fibers).

It is estimated that between 4 and 5 million tons of fibers are used annually in the world for technical purposes. The percentage of fibers going into technical textiles is 38% in Japan, 28% in the U.S. and 21% in Europe [1].

4.3 Yarn Type and Structure (Chapters 3.0 and 18.0)

A yarn can consist of a single fiber (monofilament yarn) or many fibers (multifilament yarn). Staple fibers (natural and manufactured) can be spun into yarns by twisting (staple yarns). Staple yarns have more surface texture and softness. A yarn can be made by blending several different fibers for unique properties (hybrid or blended yarn).

Different staple yarn manufacturing methods (ring spun, open-end, friction spun, air-jet spun) provide different yarn structures and hence properties. The twist level of a yarn is another variable that can affect texture and strength.

4.4 Fabric Type and Structure (Chapters 4.0 and 19.0)

A fabric structure can be woven, nonwoven, knit, braided, laminated, or stitched. There are

other non-traditional fabric manufacturing methods as well.

A fabric structure can consist of a single layer or multilayers. Each layer can be made of one or more types of fibers or yarns. There are endless numbers of fabric design patterns that provide different properties.

4.5 Manufacturing Parameters (Chapters 3.0, 4.0, and 5.0)

Special properties can be given to industrial textiles by different manufacturing techniques such as heatsetting, coating, and application of different finishes.

A textile engineer can select each of the above variables to produce the best product for its intended use. The raw materials for industrial textiles can be in various forms such as wires, spun yarns, pulp, fibers, monofilament, multifilament, rovings and mats. Major textile manufacturing processes include spinning, twisting, texturizing, weaving, knitting, braiding, needle felting, bonding, tufting, laying and knotting. The resulting textile products can be ribbons, felts, laid webs, woven fabrics, knitted fabrics, scrims, belts, cord fabrics, mats, nets, nonwovens, hoses, narrow fabrics, ropes, screens, tapes, cables, carpets, wadding and cords. Dyes, films, auxiliaries, impregnating agents, adhesives, plastics and other chemicals can be added to these textile products to obtain specific properties. Several chemical, physical and mechanical methods may be involved during the process such as coating, extrusion, dyeing, filling, impregnation, calendering, backing, adhesive bonding, lamination, metallization, stitching, perforation, embossing, prepregging, pressing, punching, molding, welding, stretching, vulcanization and cutting.

5. DIFFERENCES BETWEEN INDUSTRIAL AND NON-INDUSTRIAL TEXTILES

It is reasonable to state that, compared to textiles for apparel, there is a communication gap between industrial textile manufacturers and the end users, who are often in other industries. A typical example for this is the situation in geotextiles. The major geotextiles test methods and standards have been developed by civil engineers, most of the time without the involvement of textile engineers. This situation is generally true for almost all of the industrial textile products. Therefore, a direct feedback from the end user is not always readily available to the industrial textile manufacturer. The major reason for this might be the competitive nature of the markets and resulting confidentiality in other industries. For proper development of industrial textiles for a specific end use, the textile engineer should be knowledgeable enough about the application of his/her product. On the other hand, a civil or mechanical engineer should know something about design and manufacturing of industrial textiles so that he/she could give better insight to what needs to be developed.

There are several differences between industrial and traditional textiles that make industrial textiles unique:

(1) Application areas and end users of industrial textiles—Industrial textiles are usually used in non-textile industries. Table 1.2 gives a list of the end users of industrial textiles. Traditional textiles are used mainly for clothing and home furnishing. A traditional textile product such as a garment is purchased and used by the consumer. On the other hand, most of the time the purchaser of an industrial fabric product does not use it for himself directly. Therefore, the direct user of industrial textiles is usually not the individual consumer.

As the table indicates, there is hardly any industry or human activity that does not involve the use of industrial textiles in one way or another. Almost every modern non-textile industry uses industrial textiles. Five billion yards of industrial fabrics are used in the United States every year.

(2) Performance requirements—Depending on the application areas, industrial textiles are designed to perform for heavy duty and demanding applications. Failure of an apparel textile during use may cause some embarrassment for the user at worst. However, the consequences of failure of an industrial

TABLE 1.2 Major Industries That Use Textiles.

• Advertising	• Leather
• Agriculture	• Mechanical engineering
• Automotive	• Medical
• Aviation	• Mining
• Building	• Oil industry
• Ceramic	• Packaging
• Chemical	• Paper
• Computer	• Pharamceutical
• Electrical	• Plastics
• Environmental	• Printing
protection	• Recycling
• Fishing	• Rubber
• Food	• Space
• Furniture	• Textile
• Home textile	• Transportation
• Horticulture	• Wire
• Landscaping	• Wood processing

textile can be devastating. For example, failure of an air bag in a car accident or an astronaut's suit during a space walk may be fatal.

(3) Constituent materials—Due to different application areas and performance requirements, generally higher performance fibers, yarns and chemicals are used in industrial textiles. Materials with exceptional strength and resistance to various outside effects in turn give industrial textiles the necessary strength and performance characteristics. In traditional clothing textiles, the physical performance requirements are less demanding compared to industrial textiles. However, comfort and appearance play a critical role in material selection for apparel. Functionality is the most important criteria for technical textiles; aesthetics, beauty, color, etc. are of secondary importance. For traditional textiles such as clothing, furnishing and household items, aesthetics and color are important properties along with functionality.

(4) Difference in manufacturing methods and equipment—Since industrial textiles consist of stronger constituent materials, manipulation of those materials is usually more difficult than weaker materials. Moreover, due to performance requirements, some industrial textiles may need to be built with high mass, therefore they are usually heavier than

traditional clothing textiles. As a result, the manufacturing equipment and methods of consumer textiles may not be adequate for the manufacture of industrial textiles. For example, weaving of monofilament forming fabric, which is used on a paper machine to make paper, on a regular textile loom is impossible due to the very high weaving forces involved, resulting from the high strength and mass of polyester monofilament warp and filling yarns. The width of the forming fabric, which is determined by the width of the paper machine, can be up to 500 inches. For these reasons, specially built, very heavy and wide looms (up to 30 yards) are used in paper machine clothing manufacturing.

(5) Testing — Testing of industrial fabrics also presents another challenge. Quite often, once placed in an operation, industrial textiles cannot be changed or replaced easily. For example, geotextiles that are used under roads for reinforcement and stabilization cannot be replaced without complete destruction of the road. On the other hand, one cannot use a geotextile under a ''test road'' first because it may take many years to get the results. Failure of a geotextile in construction of a dam can cause devastating results.

Many times it is impossible to simulate the actual field conditions of industrial textiles in the laboratory. Therefore, empirical methods are either difficult to achieve or not as reliable for industrial textiles. As a result of these, the design engineer of an application involving industrial textiles quite often has to rely on laboratory test results. This increases the need for accuracy and reliability of the test results for industrial textiles. Conventional textile test methods are often not suitable for industrial textiles. Therefore, new test methods and procedures have been developed for industrial textiles. Simulation and modeling of field conditions with computer aided design systems is also becoming a more common practice to be able to determine the optimum structure and performance of industrial textiles for a particular end use.

In addition to physical properties, the ''performance'' or ''quality'' of clothing textiles depends on other variables such as consumer perception and taste. Therefore, evaluation and specification of clothing textiles is rather subjective and may be difficult. For most of the tests of industrial textiles, the performance and evaluation are well defined based on the end use. This fact makes it relatively easy to develop exact measuring and testing methods for industrial textiles. As a result, development of performance data bases is also easier.

(6) Life expectancy — In general, industrial textiles are expected to last much longer than traditional consumer textiles. Unlike consumer textiles, fashion trends do not play a role in the life of traditional textiles. In large structures such as buildings, highways, stadiums or airplanes, textiles are expected to last many years.

Although longer life is desirable for industrial textiles, it is not always attainable and some industrial textiles may have a shorter useful life than traditional textiles. Moreover, in some applications, it is desirable to control or limit the life of an industrial textile. For example, some textile materials that are used inside or outside the human body during surgery are expected to degrade after they complete their task (i.e., when the natural organ or tissue grows and gains enough strength) which may take a few weeks or months. The dissolved material is carried away by the body fluids and discharged from the body.

(7) Price — As a result of all the wonders that industrial textiles offer, it is expected that they come at a higher initial cost than traditional textiles. However, considering the long life and benefits they offer and contributions they make to the national infrastructure and economy, the cost of industrial textiles should not be a concern in the overall picture. In fact, when properly constructed and applied, industrial textiles can be used in place of more expensive materials and therefore may bring substantial savings to the overall economy in the long range.

6. CLASSIFICATION OF INDUSTRIAL TEXTILES

Classification of industrial textiles is a challenging task. Industrial textiles is a very diverse market segment of the textile industry. There are some products that can be used in many different applications.

Classification of industrial textiles can be done in several ways:

- according to the raw material processed (e.g., industrial textiles made of *glass* fibers)
- according to the manufacturing sector and/or production technique (e.g., *nonwoven* industrial textiles)
- according to the basis of main industrial textiles groups (e.g., canvas, filter cloth)
- according to the end use (e.g., geotextiles, medical textiles, paper machine clothing)

Each classification has advantages and disadvantages which can be a lengthy discussion. In this book, the classification of industrial textiles is made mainly on the basis of the final application. The main reason for this selection was the convenience to the reader. Therefore, the chapter titles carry the name of the end use market. There is one exception to this which is the "Textile Structural Composites" in Chapter 7.0. Textile composites are increasingly used in so many different applications that make it easier to classify them as an industrial textiles group.

6.1 Textiles in Agriculture

- textiles for landscaping
- textile reinforced plastic and concrete parts, pipes and containers
- sacks
- insect and bird netting, crop covers
- belting
- rope wear
- hoses
- transport and lifting materials
- tarpaulins
- flexible and rigid containers

- silage protection systems
- flexible silos
- textiles for seed bed protection
- temporary agricultural buildings
- textiles for subsoil stabilization
- soil covering systems
- textile drainage and irrigation systems
- moisture-retaining soil mats
- shading textiles
- scrims for protection from hail and ground frost
- soil sealing systems
- sealing systems for liquid manure pits
- textiles for animal husbandry
- horticultural textiles
- erosion prevention textiles
- protective work clothing
- shade fabrics
- textiles in greenhouses

6.2 Textiles in Architecture and Construction

- reinforcement fibers for concrete and plastics
- reinforcing domes and covers for stadiums
- filaments, yarns, cords and small-wear for reinforcement purposes
- textile sheet products for reinforcement purposes
- textile reinforced structural parts, profiles and pipes
- textile reinforced molded articles
- textiles for reinforcing building materials, concrete and cement
- bridges
- textile reinforced containers
- textile reinforced lightweight building materials
- textiles for subsoil stabilization
- textile drainage systems
- fabric sculptures to beautify, reinforce and protect
- noise control in offices
- textiles in public buildings and convention halls
- textile shuttering materials
- textile facade packing systems

- textile roofs and roof sheeting
- textile products for building electric systems
- insulation against cold, heat and noise
- tents and tent frames
- temporary buildings, inflatable buildings for warehousing
- membranes for lightweight plane load-bearing structures
- pneumatic structures
- winter building systems
- stay ropes and systems
- textile acoustic protection systems
- awning textiles and systems
- textile heating, cooling and air conditioning systems
- textile planting and irrigation systems for terraces, roof gardens and courtyards
- textile reinforced plastic for interior decoration
- fire protection and rescue systems

6.3 Textile Structural Composites

- textile reinforced lightweight building materials
- textile reinforced structural parts, molded articles and profiles
- textiles for use in corrosive media
- textile reinforced motor and machine parts

6.4 Filtration Textiles

- textiles for cleaning and separating of gases and liquids
- textiles for recovery of products
- textiles for hot air/gas industrial filtration
- filtration in cigarettes
- textiles for filtration in food industry
- textiles for sewage filtration

6.5 Geotextiles

- textiles for earthworks and roadmaking
- textiles for bank and coastal stabilization
- textiles for hydraulic engineering
- fabrics for erosion control
- textiles to reinforce and line waste ponds and pits

- reinforcing materials for soil stabilization
- materials for landfills and the waste industry
- textiles in drainage systems
- membrane materials
- environmental protection products
- reinforcing textiles for plastics
- reinforcing textiles for concrete

6.6 Medical Textiles

- textiles of bactericidal fibers
- hygiene nonwovens
- bandage materials
- sutures
- operating and emergency room textiles
- textile products for surgery
- textile reinforced prostheses
- operating sheets
- hospital bed linen and blankets
- mattresses and protective mattress covers
- medical cushions
- dental floss
- synthetic skin
- general textiles for institutional and hospital use
- doctors' and nurses' clothing
- rescue services equipment
- textiles for medical equipment

6.7 Military and Defense Textiles

- textile armor
- parachutes for spaceships
- personnel protection
- aerospace and electronics components
- chemical suits
- covers
- helmets
- air conditioned suits
- ballistic protection
- military tents
- inflatable buildings
- fabrics for bullet-proof vests
- medical devices
- flight and tank driver garments
- marine applications
- rescue systems for air, water and land vehicles

6.8 Paper Machine Clothing

- monofilament paper forming fabrics for drainage, support and transportation
- press felts and fabrics
- dryer fabrics

6.9 Safety and Protective Textiles

- breathable waterproof textiles and barrier laminates
- protective work clothing
- textiles for impact and pressure protection
- protection from ionizing and non-ionizing radiation
- weatherproof and winter clothing
- textiles for protection from heat and fire
- chemical protection equipment
- equipment for rescue services
- space shuttle uniforms
- fire protection equipment
- survival equipment
- textiles for protection of possessions
- textile packaging
- protective covering systems
- indoor and outdoor textile acoustic protection systems
- vinyl coated life jackets
- safety flags

6.10 Textiles in Sports and Recreation

- active wear fabrics
- covers for domes and stadiums
- stadium blankets
- inflatable buildings for sports
- tennis rackets
- golf clubs
- football, tennis ball felt
- roller skates
- water and snow skis, ski ropes
- helmets
- breathable waterproof uniforms
- tennis nets
- tennis court curtains
- fabrics for hunting vests
- race car drivers' uniforms
- fabrics for hot air balloons
- fabrics for sport shoes
- fishing nets and line

- swimming pool covers and liners
- sleeping bags

6.11 Textiles in Transportation

- materials for automotive applications
- materials for aerospace industry
- materials for marine applications
- materials for railway vehicles
- materials for bicycles
- seat belts
- air bags
- tire cord
- canvas
- textile reinforced interior coverings
- textile sealing and wall coverings, sound damping
- curtain materials
- boat and car covers
- seat covering materials
- fire-resistant textiles
- industrial carpeting
- hood fabrics, headliners
- hoses and drive belts
- gaskets and brake linings
- textiles in mufflers
- filters
- seals, insulation materials
- ropes, cordage, netting
- container systems
- reinforcing fibers for plastics
- textile products for plastic reinforcement
- textile products for rubber reinforcement
- textile reinforced molded and structured articles
- textile reinforced pipes
- textile reinforced containers
- protective covering systems for aircraft, floating vessels, cars and agricultural machines
- rescue systems for aircraft, water and land vehicles

6.12 General Industrial Textiles

- reinforcing fibers for curing wraps
- textiles for protection from hot and cold
- electrically conducting textiles
- antistatic textiles
- metallized products

- surface-finished products
- textiles for electronics and data systems technology
- fiber optics
- drive systems
- hoses and textile reinforced pipes
- fabrics for timing gears
- rigid and flexible containers
- hollow pneumatic systems
- oil spill absorbing blankets
- textile reinforced rubber products
- filters
- abrasive fabrics for sandpaper
- fabrics for movie screens
- typewriter ribbon
- sorption systems
- seals and fiber reinforced sealing materials
- textile reinforced adhesives
- fabrics for luggage
- laundry textiles

Various examples of the application of industrial textiles are shown in Figures 1.1 – 1.16 (see pp. 13 – 28).

7. THE FUTURE OF INDUSTRIAL TEXTILES

Research and development work is being done continuously in the industrial textiles area. Both basic and application oriented research is conducted to find new materials, to improve the properties of the existing materials and products and to find new application areas for industrial textiles.

High-tech research is continuously being done at various levels of industrial textiles: polymers, fibers (the share of natural fibers is not expected to grow in industrial fabrics markets in the future), yarns, manufacturing methods, finishing and coating. Applied research has been proven very valuable to solve major industry problems.

The end of the cold war will affect industrial textiles more than traditional textiles. The demand for high cost, high performance industrial textiles such as textile structural composites for military, defense and aerospace applications is expected to decrease. The possible factors that will affect the industrial textiles market in the near future are [2]:

- defense spending reductions
- environmental requirements
- advancements in fiber development
- competition
- new uses for "old" products
- new products for old uses
- global requirements (impact of ISO 9000)
- trade agreements such as General Agreement on Tariffs and Trade (GATT) and North America Free Trade Agreement (NAFTA). The GATT body has overseen international trade since 1947. Effective on January 1, 1995, the GATT is being replaced by the new World Trade Organization (WTO).
- de-classification of hi-tech information
- governmental regulations

The rapid progress in industrial textiles will increase the demand for highly technically skilled people for both production and research and development. The technical textiles industry offers careers in fiber chemistry, fiber, yarn and fabric production, product design and quality management in several engineering disciplines including mechanical, civil, chemical and materials engineering.

References and Review Questions

1. REFERENCES

1. Janecke, M. "Positive Signs for Industrial Textiles," *International Textile Bulletin,* Vol. 3, 1994.
2. Lohne, W. E. and Howard, R. C. "High Performance Textiles in the 90's," *IFAI 2nd High Performance Fabrics Conference,* November 12–13, 1993, Boston, MA.

1.1 General References

The World of Industrial Fabrics, Industrial Fabrics Association International (IFAI).

The World of Industrial Textiles, Techtextil Symposium, 1994.

Svedova, J., *Industrial Textiles,* Elsevier, 1990.

The High Tech World America's Textiles, American Textile Manufacturers Institute (ATMI), Washington, D.C., 1994.

2. REVIEW QUESTIONS

1. Based on the information given in this chapter, what could the other possible definition(s) of industrial textiles be?
2. State and explain the major differences between industrial and non-industrial textiles.
3. What are the effects of raw material properties on the performance of industrial textiles?
4. How can you change the performance properties of a typical industrial textile product?
5. Classify the industrial textiles according to:
   - raw materials processed
   - the manufacturing sector and/or production technique
   - the basis of main industrial textiles group

FIGURE 1.1 Space suit: Astronaut Bruce McCandless II flies first ''solo'' in space in February 1984. He was wearing a specially designed space suit for protection from heat, cold, chemicals, micrometeoroids, pressure fluctuations and other hazards. Space suits are considered to be the ultimate protective clothing in existence today (Chapter 13.0). Photo courtesy of NASA.

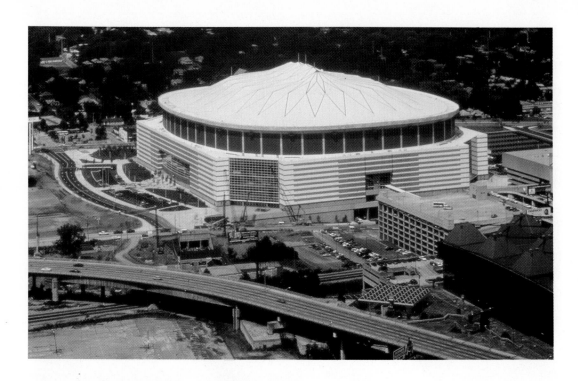

FIGURE 1.2 Textiles in architecture and construction: Georgia Dome in Atlanta, GA. The roof is made of coated textile fabric (Chapter 6.0) and the playing field is made of synthetic fibers (Chapter 14.0). Photos courtesy of Birdair, Inc.

FIGURE 1.3 Geotextiles are used in various civil engineering applications (Chapter 9.0). Top: road construction, bottom: erosion control. Photos courtesy of Hoechst Celanese.

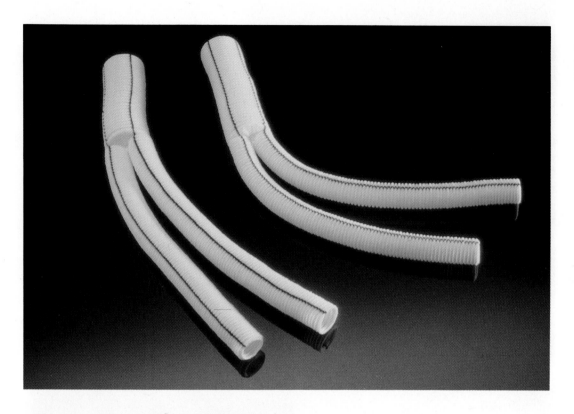

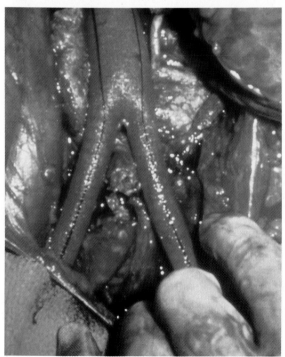

FIGURE 1.4 Two polyester vascular grafts (top) and implanted vascular graft in the aorto-bi iliac position (left). Medical textiles are used in surgery, wound healing, extracorporeal devices and in healthcare and hygiene products (Chapter 10.0). Photo courtesy of Meadox Medical, Inc.

FIGURE 1.5. An air-to-air view of a F-16C Falcon. Textile structural composites are extensively used in military and commercial airplanes (Chapter 7.0). They offer light weight, strength and stiffness. Photo courtesy of the U.S. Air Force.

17

FIGURE 1.6 A Fourdrinier paper machine. Engineered fabrics are used in forming, press and drying sections of paper machines (Chapter 12.0). In the forming section, forming fabrics perform drainage, support and transportation functions. Photo courtesy of Beloit.

FIGURE 1.7 The Personnel Armor System for Ground Troops (PASGT) vest and helmet are based on Kevlar® aramid fibers. Soldier is wearing the Individual Microclimate Cooling System (IMCS). The unit is powered by an internal combustion engine that uses JP-8 fuel. The U.S. Department of Defense (DOD) has in its inventory some 10,000 textile items (Chapter 11.0). Photo courtesy of the U.S. Army Natick Research, Development and Engineering Center.

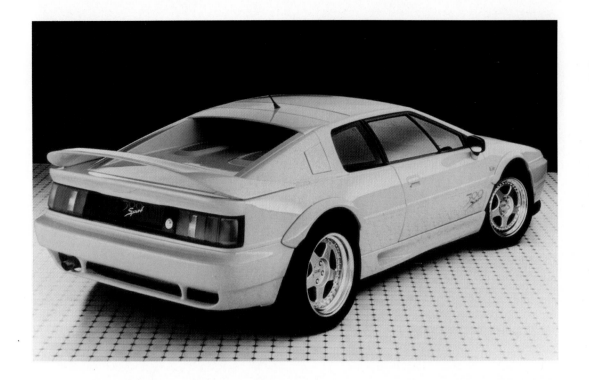

FIGURE 1.8 Textile fibers, fabrics and composites are extensively used in transportation (Chapter 15.0). Application areas include tires, interior and exterior trim, airbags, seat belts, aircraft interiors, belts and hoses. Photos courtesy of Tech Textiles USA and Industrial Fabrics Association International (IFAI).

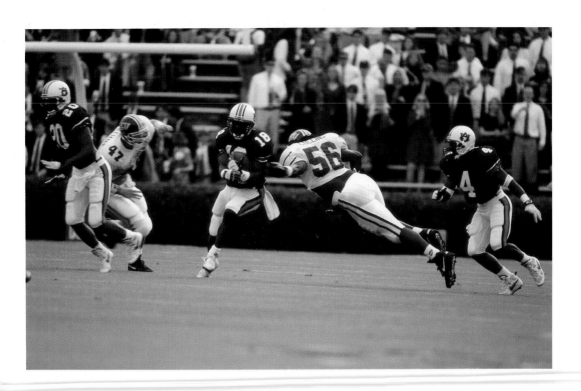

FIGURE 1.9 Almost every sport relies heavily on textiles. In football, uniforms, ball and helmets are made of highly engineered textile materials (Chapter 14.0). Sports and recreation would be quite different without textiles. Photos courtesy of Auburn University Photographic Services and Johnston Industries.

FIGURE 1.10 Tension fabric structure at Denver International Airport. The roof is made of a two-layer Teflon® coated woven fiberglass fabric. The fabric is designed to allow enough light transmission for plants. The roof is suspended by steel masts and cables, covering an area of 20,000 m² (Chapter 6.0). Photo courtesy of Birdair, Inc.

FIGURE 1.11 Industrial fabric bags are used in gas and liquid filtration (right) (Chapter 8.0). Various rubber products such as belts, are reinforced with high strength fibers and fabrics (bottom) (Chapter 16.0). Photos courtesy of Wellington Sears Company.

FIGURE 1.12 Oceandome, an all-weather indoor water park in Japan. The two roof panels in the middle of the dome retract horizontally and can open or close in 10 minutes. The translucent fabric creates a natural "outdoor" atmosphere throughout the year. Fabric is PTFE coated fiberglass (Chapter 6.0). Photos courtesy of Taiyo Kogyo Corp.

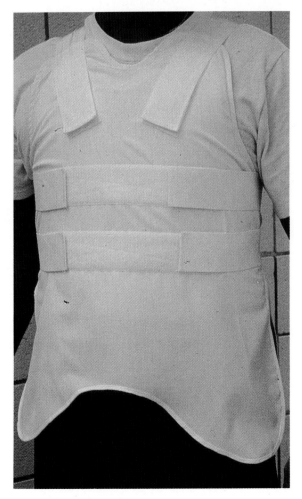

FIGURE 1.13 Industrial fabrics provide strength and flexibility for safety and protective clothing (Chapter 13.0). Shown in the pictures is fire fighter gear (top) and ballistic vest (bottom). Photos courtesy of Wacker Silicones and IFAI.

FIGURE 1.14 Computer controlled, retractable umbrellas in the Masjid of the Prophet in Medina, Saudi Arabia. The 296 m^2 screens with a 25 m diagonal are the largest in the world. The fabric is Tenara™ specially developed by W. L. Gore & Associates using high strength PTFE fibers. The construction can withstand wind speeds of up to 155 km/h (Chapter 6.0). Photos courtesy of Dr. Bodo Rasch.

FIGURE 1.15 Filament wound rocket motor cases (top) and TDRSS satellite antenna ribs (bottom) made of textile structural composites (Chapter 7.0). Photos courtesy of Hercules.

FIGURE 1.16 Various applications of industrial textiles. From top, clockwise: greenhouse, marine products, luggage and children's playground. Photos courtesy of IFAI and Wellington Sears Company.

POLYMERS AND FIBERS

2.1

Fiber Forming Polymers

R. M. BROUGHTON, JR.
P. H. BRADY

The purpose of this section is to introduce the structure of fiber forming polymers used in the production of industrial textiles.

As atoms are the smallest unit of construction for molecules, polymer molecules are the smallest unit of construction for fibers. Although not all polymer structures are fibers, regardless of whether natural or man-made, nearly all textile fibers are polymer structures. Most of the polymers have a predominantly carbon based backbone and are, therefore, organic structures.

1. BASIC DEFINITIONS

The term *polymer* means "many parts." A polymer is a large molecule containing a limited number of types of repeating units. A repeating unit is a small grouping of atoms which, if linked together in a repetitive sequence, will reproduce the polymer structure. Repeating units are derived from, and similar to, the small molecules which react together to produce the polymer. The small starting molecules are called monomers (sometimes the repeating unit is referred to as a mer). The definition of a polymer could be stated as: a large molecule made from a limited number of different kinds of monomers.

How large is a polymer? Generally if a molecule contains as many as fifty similar repeating units it would be considered as a polymer. An oligomer (meaning "few") is similar to a polymer with the exception that it will typically have no more than ten repeating units in its chain sequence. Dimer, trimer, and tetramer are sometimes used to describe blocks of two, three, or four monomer units respectively. Most of the polymers in fibers will have at least 100 repeating units, and some may have thousands.

The type of repeating unit, or mer, defines the overall polymer. Before reacting to form the polymeric structure, each repeating unit exists as a monomer. Therefore a monomer is a precursor to the polymer. If a monomer is reacted with and caused to attach to another monomer, and then the group of two reacts with a third and so on, a polymer is formed. Because the repeating unit or mer defines its resultant polymer, if the attributes of the repeating unit can be interpreted, many of the performance characteristics of its resultant polymer structure can be predicted.

2. POLYMER STRUCTURAL CLASSIFICATIONS

A polymer structure may be classified as either a homopolymer or copolymer. The homopolymer configuration indicates that all of the repeating units in the polymer are the same, or *homogeneous*, in nature. Many polymers used in textile applications are homopolymers. A copolymer configuration requires that two or more different types of repeating units are present in the structure in some sequence.

2.1 Homopolymers

If the monomer unit may be represented as A, a homopolymer structure can be depicted as

$$A-A-A-A-A-A-A-A-A- \ldots$$

where A would be repeated at least one hundred times for most fiber applications. It is important to note that the terminal or end units of a polymer should not be assumed to be the same as the main body units. Depending on the polymerization technique, or patented technology of the manufacturer, the terminal units may differ, e.g.,

$$X-A-A-A-A-A-A-A-Y$$

Although the end units may differ from the primary chain constituents, it is still considered a homopolymer because the end units exist in such a small proportion relative to the others. Since A, repeated over and over, will reproduce the entire chain (except the ends), the homopolymer depicted above may be abbreviated as

$$-[A]_n- \text{ or poly(A)}$$

The subscript n represents the number of times the unit is repeated, and is defined as the degree of polymerization (DP) for the polymer. A linear arrangement of repeating units, or atoms through the sequence of repeating units, is referred to as the polymer chain. The term backbone is frequently used to describe the chain of repeating units in the direct linear sequence. The backbone chain is defined without regard to pendant groups or side chains. Pendant groups are chemical constituents attached to the "side" of the repeating units. Side chains are also polymer structures which have been chemically attached to the main polymer chain to form a branch. Impurities or unintended comonomers are sometimes formed in polymers. Even though they may have some effect on a polymer's properties, if present in small quantities, the impurities are ignored in classifying the material as a homopolymer.

Some homopolymers are made from repeating units which are the result of a combination of two different molecules. An example of this type polymer is polyester. A polyester monomer is typically the residual of a condensation reaction between ethylene glycol and terephthalic acid. Since the repeating unit is essentially composed of two molecules, the repeating unit may be designated as AB, and the resultant polymer form is then

$$AB-AB-AB-AB-AB-AB- \ldots$$

with the abbreviated form

$$-[AB]_n- \text{ or poly(AB)}$$

This form of polymer is also considered a homopolymer, because the polymer has a single repeating unit.

2.2 Periodic Homopolymers

Envision the repetitive sequences of three or more different repeating units. Such repetitive sequences can be produced by creating and purifying large monomer units from combinations of smaller molecules. The large monomers can be reacted together to produce a polymer having a repetitive sequence as a repeating unit, and be classified as a homopolymer. The terminology for such is periodic polymers.

$$A-B-C-A-B-C-A-B-C- \ldots$$

or

$$A-A-B-B-C-C-A-A-B-B- \ldots$$

abbreviated as

$$\text{poly(A-per-B-per-C)}$$

2.3 Copolymers

Copolymers are polymer structures composed of two or more different monomer precursors. A copolymer will exist in one of three structural configurations: alternating, random, or block.

2.4 Alternating Copolymers

An alternating copolymer structure has two monomer units which alternate sequentially along the polymer chain. An example of an alternating copolymer produced from monomers A and B may be written as

$$A-B-A-B-A-B-A-B-A-B- \ldots$$

The abbreviated notation for this type of polymer is

$$[A\text{-}alt\text{-}B]_n \text{ or } poly[A\text{-}alt\text{-}B]$$

The question that may arise now may be "What is the difference between homopolymer AB and the alternating copolymer AB?" The alternating copolymer is generally less perfect in its alternation than the homopolymer. If the alternation of the repeating units were perfect, it would be called a homopolymer. In the alternating copolymer, while there is a distinct preference for the repeating sequence A−B−, there is a finite possibility that the grouping A−A or B−B could be found in the polymer sequence.

2.5 Random Copolymers

A random copolymer has two or more monomers polymerized into no particular sequence. The individual repeating units are randomly distributed along the polymer chain. A random copolymer of units A, B, and C may be shown as

$$A−C−A−B−B−C−B−A−A−A−C− \ldots$$

or

$$poly[A\text{-}ran\text{-}B\text{-}ran\text{-}C]$$

2.6 Block Copolymers

Block copolymers have their similar repeat units segregated into distinct segments. An example of this form of polymer can be represented as

$$A−A−A−A−B−B−B−B−B−A−A− \ldots$$

or for three monomers

$$A−A−A−B−B−B−B−B−C−C−C− \ldots$$

Block copolymers can be abbreviated as

$$poly A\text{-}block\text{-}poly B\text{-}block\text{-}poly C$$

If a given structure is known to be a copolymer, but the sequential arrangement is not predictable nor significant to its identification, it may be written simply as

$$poly(A\text{-}co\text{-}B)$$

2.7 Graft and Branched Copolymers

A polymer molecule that has only two ends is considered a linear polymer. Polymer molecules with three or more end groups must have other chains branching off of the primary chain. A graft copolymer has a backbone constructed of one polymer type with a branch of another type grafted onto the backbone chain. Many graft copolymers also fit the definition of a block copolymer because the grafted branch is often a block of a different type of repeating unit. As the addition of a graft requires a chemical reaction, the backbone polymer must have at least one reactive site along its chain to allow attachment of the chain to be grafted. If the primary constituent of the backbone is poly(A), the reactive site is a, and the grafted is poly(B), then the extended representation is

$$A−A−A−A−A−a−A−A−A−A−A− \ldots$$
$$\diagdown$$
$$B−B−B−B−B− \ldots$$

The abbreviated form for the graft copolymer is

$$poly A\text{-}graft\text{-}poly B$$

Again, it may be noted that when a given monomer represents a very low percentage of the total population, such as "a" in the above description, it is ignored in designating the polymer.

Branched copolymers are similar to graft copolymers in that both have side chains extending from their backbone chains. The primary difference between the two is the species of the side chain. A graft copolymer generally denotes that it is a hybridization of two species: one main component comprising the backbone, with a different component comprising the branch. A branched polymer's side chain is typically composed of the same species as the backbone chain. There is no common abbreviation for the branched polymer.

Crosslinking and Gelation

Most of the polymers used in fibers are linear, at least when the fiber is first produced. Sometimes the polymers must be crosslinked to be useful. Crosslinking is the process of producing primary chemical bonds between polymer molecules. A di-functional (or multi-functional) small molecule is caused to react with adjacent polymer molecules, tying them together. When crosslinking is extensive, the mass becomes a single, large molecular network. These polymer networks restrict molecular mobility and cannot be melted or dissolved. Extensive polymer branching may result in a type of structure similar to that obtained by crosslinking. The term applied to a mass of highly branched polymer is gel. When, in the course of polymer manufacture, branching becomes extensive enough that a large fraction of the polymer mass is tied into infusible, insoluble material, the polymer reaches the gel point. Figure 2.1 shows the schematic representation of branching and crosslinking.

3. DIMENSIONAL ORDERING

The arrangement of a polymer's elements of construction along its chain length is referred to as its configuration. The spatial or three-dimensional orientation of the chain segments along the length of the polymer is designated as the conformation. The configuration is generally deter-

minable through the description of the repeating unit. However, the conformation may not be so obvious, especially in those polymers composed primarily of singly bonded atoms along the backbone chain. Single bonds occurring between atoms allow for 360° bond-axis rotation of the atoms relative to each other. Carbon atoms in a chain of carbon-carbon single bonds assume an angular displacement of 109° from each other. When drawn on paper, the conformation is typically shown as a planar arrangement, but in reality the conformation is three-dimensional. The importance of the spatial arrangement of the chain segments is in defining the chain segments' ability to assume a position in a regular or crystalline array.

When a polymer repeating unit contains pendant or side groups, the orientation of the elements becomes much more critical in the determination of its ability to assume a position in a crystalline array. This is particularly true of a polymer whose backbone chain is composed predominantly of methyl groups (CH_2). The orientation of the pendant groups along the polymer chain is called the tacticity of the structure. In order to simplify the concept of tacticity, it is considered in a two-dimensional plane only, where the pendant groups are arranged either up or down relative to the main polymer chain. The atoms to which the pendant groups are attached are called chiral centers, and therefore tacticity is related to chirality.

Complex chemistry is required to control the chirality and therefore the final molecular arrangement of a polymer. If the chiral centers formed during polymerization are all identical, the conformation of its pendant groups are the same, and the polymer formed is isotactic. If the chiral centers randomly alternate, the resultant poly-

branching

crosslinking

FIGURE 2.1 Branching and crosslinking.

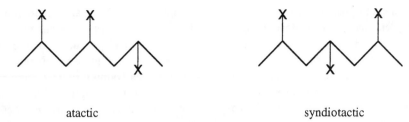

atactic   syndiotactic

isotactic

FIGURE 2.2 Tacticity in polymers.

mer is atactic. If the two possible arrangements coherently alternate, a syndiotactic structure is formed.

The effect of chirality on tacticity can be seen in the polymerization of a substituted ethylene (Figure 2.2). If all of the pendant methyl groups appear on the same side of the polymer chain, the arrangement is isotactic. If the groups alternate from side to side, it is syndiotactic, and if they are random, it is atactic.

4. POLYMER NOMENCLATURE

As a general rule, any polymer can be named according to the chemical components comprising the chain. This type of polymer naming is referred to as IUPAC (International Union of Pure and Applied Chemistry) nomenclature, and is an internationally accepted method of naming chemical structures. Many polymers have become so commonly recognized that their generic or trade names have been accepted as an adequate description. The decision of whether to use the IUPAC or generic name generally rests upon how accurate the description must be. Polyester is a good example of a commonly recognized generic name, usually implying poly(ethylene terephthalate), but in fact there are other chemical forms of polyester.

Homopolymers (those having only one kind of repeating unit) are usually named according to their monomers' chemical designation. The most familiar of this type are probably the polymer

TABLE 2.1 A List of Common Polymer Names.

Molecular Formula	IUPAC Name	Common Name
$-[CH_2CH_2]-$	Poly(ethene)	Polyethylene
$-[CH_2CH(CH_3)]-$	Poly(propene)	Polypropylene
$-[CH_2CH(Cl)]-$	Poly(chloroethene)	Polyvinylchloride
$-[CH_2CH_2OCOC_6H_4COO]-$	Poly(ethylene terephthalate)	Polyester
$-[NH(CH_2)_6NHCO(CH_2)_4CO]-$	Poly(hexamethylene adipamide)	Nylon 6,6
$-[NH(CH_2)_5CO]-$	Poly(hexanolactam)	Nylon 6
$-[CH_2CH(CN)]-$	Poly(cyanoethene)	Polyacrylonitrile

forms of ethylene and propylene, which are of course polyethylene and polypropylene. In these simplest cases, it is obvious that the polymer names are derived by adding the prefix poly to the common chemical name of the monomer. Therefore, polymerized vinyl chloride becomes poly(vinyl chloride) (PVC), and tetrafluoro-ethylene becomes poly(tetrafluoroethylene) (PTFE or Teflon®). Though not always common practice, it is more correct to include the name of the monomer in parentheses; poly(propylene) rather than polypropylene.

As would be expected, the more complex the monomer's precursors, the more complex the name of the polymer. It is usually these more complex names that are generically renamed. The polymerization of hexamethylene diamine with adipic acid produces poly(hexamethylene adipamide), but is more readily recognized when described as Nylon 6,6. The complexity of the name is further complicated by copolymeriza-tion. For example, a copolymer of ethylene and ethylene oxide would be written as poly(eth-ylene-*co*-ethylene oxide). In Table 2.1, a list of some common polymers is shown by molecular makeup, proper chemical nomenclature, and an accepted common name.

Polymer chemists and textile scientists usually use the common name rather than the name approved by IUPAC.

Natural and Man-Made Fibers

D. M. HALL
S. ADANUR
R. M. BROUGHTON, JR.
P. H. BRADY

1. INTRODUCTION

Mankind has been fortunate that over the eons, natural materials could be found with suitable fiber properties to fabricate clothing for warmth, comfort and style depending upon the needs at the time. The most important fibers in the world until about 1910 were the protein fibers wool and silk, and the cellulosic fibers cotton and linen. All of these products are produced from agriculture. The production of wool, silk and linen has remained static for the last decade owing to the fact that land and facilities for cultivating these fibers are fixed (and for some they are declining). With the world population growing rapidly and the supply of these fibers either diminishing or fixed, the cost of using these fibers for ordinary textiles is prohibitive. Only cotton is grown in quantities that enable it to continue to be an economical source for textile manufacturing.

About 1910, discoveries were made which allowed fibers to be spun from special solutions of cellulose as continuous filaments. Unlike the natural products, these fibers could be spun into varying diameters that were essentially the same for every fiber in the bale. Further, the filaments could be cut into any length uniformly unlike that of cotton and linen. Later it was found that they could be made to have cotton-like properties of softness and hand, have ignition, sunlight and rot resistance among other properties that could be built into the fibers as they were being spun. Thus, it was found that the fibers could be engineered to have specific properties depending upon the desired end uses of the product. Today, the number of different genera of fibers that have been produced for textile purposes is quite high.

1.1 Generic Names of Fibers

In addition to the chemical nomenclature (Section 2.1), the Federal Trade Commission (FTC) has defined generic categories of fibers based on chemistry and properties. Each category has been assigned a generic name. Basically, any consumer textile item sold in retail commerce must carry a label declaring the generic name and the percent by weight of all the component fibers in the composition (which exceed 5% of the total). Certain items like luggage, carpet backing, hats, wiping rags, furniture stuffing, and tarps are exempted, as are items intended for industrial applications. A manufacturer may label the fibers with trade names and other information specific to the manufacturer, but the generic names of fibers and the percent composition must always be present. The purpose of a generic name is to provide the consumer with a recognized name with which to associate a set of expected fiber properties. Some fibers find little or no use in consumer textiles, and may not have an established generic name.

Generic names for manmade fibers are defined by the chemistry of the fiber and its physical properties. New names may be added and older ones modified in definition as demanded by technology and developments. When believed required, a manufacturer can petition the FTC for the establishment of a new generic name. In response, the FTC will examine the chemistry of the new fiber as well as its properties, and then decide whether a new name is warranted. Natural fibers are labeled according to their origin. Thus cotton, linen, silk, wool, etc., are legitimate generic names.

1.2 Trade Names and Trademarks

The terms *trade name* and *trademark* may be defined as follows [1].

trade name: 1 a: the name used for an article among traders, b: an arbitrarily adopted name that is given by a manufacturer or merchant to an article or service to distinguish it as produced or sold by him and that may be used and protected as a trademark, **2:** the name or style under which a concern does business.

trademark: 1: a device (as a word) pointing distinctly to the origin or ownership of merchandise to which it is applied and legally reserved to the exclusive use of the owner as maker or seller.

Trademark is registered with the U.S. Department of Commerce, Patent and Trademark Office (Washington, D.C. 20231) as a trademark of that company. A product with a trademark is indicated with the sign ® on the right-hand corner of the trademark. A trademark owned by a company cannot be used by other companies for the same specific type of product.

The Patent and Trademark Office publishes the *Trademark Official Gazette* regularly which contains an illustration of each trademark published for opposition, a list of trademarks registered, and a classified list of registered trademarks and patent office notices.

There may be several trade names for the same generic fiber manufactured by different companies. The properties of these fibers may or may not be the same, but they all fit the generic definition of the fiber.

2. NATURAL FIBERS

As mentioned previously, the FTC generic name for natural fibers is generally the same as the common name. Cotton, wool, silk, linen, jute, etc., are all acceptable generic names.

2.1 Cotton

Cotton is the name given to a seed hair fiber from the plant genus *Gossypium*. It is the most widely used textile fiber. The structure is more than 90% cellulose, with the remainder being waxes, monomeric and polymeric sugars, residual protoplasm, and minerals. The structure of cellulose is shown below.

Cotton fiber has been around for thousands of years, hence virtually every type of fabric that may have industrial applications has been manufactured at one time or another using this agricultural based cellulose containing fiber. The fact that it has an agricultural base is the reason why the fiber varies so much in diameter (fineness), maturity level, fiber length and uniformity among others (Figure 2.3). Textile mills are quite adept at selecting the proper mix of fibers from the billions of pounds of the various varieties (American Upland, American Egyptian, Sea Island and Pima etc.), grown in the United States alone, to obtain those properties required in the final product. Because of the technical base and know-how for the processing of this fiber, the industrial fabric can usually be made less expensively than using other fiber types. Further, it is suitable for blending with many of the other fiber types in order to achieve a unique combination of properties not available in the cotton alone.

The cotton fabrics can be mercerized to enhance their strength, luster, dyeability and resistance to shrinkage. They can be colored by a wide variety of dye classes to give the desired resistance to fading or color changes due to sunlight or atmospheric contaminants. They are resistant to strong alkali, organic solvents, bleaches and heat (such as during process treatments and ironing). The fibers can be formed into woven, braided, nonwoven or knitted fabric types.

Cotton fibers' disadvantages are that they lack elasticity and resiliency, are weakened by strong mineral acids, are affected by mildew and are attacked by silverfish-type insects, have poor sunlight and atmospheric resistance, i.e., do not weather well, and are not ignition resistant, although chemical treatments are available that will mitigate most of these problems.

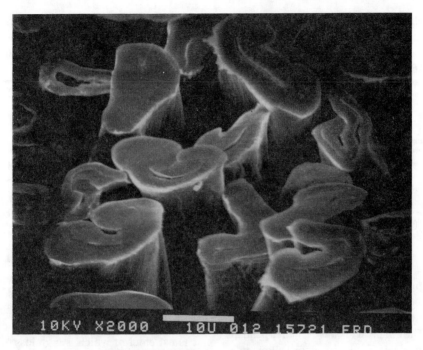

FIGURE 2.3 Photomicrograph of cotton fiber.

2.2 Bast Fibers

These fibers are linen, jute, hemp, ramie, etc., that come from either the stems or leaves of plants. These fibers also have a cellulose base in addition to containing varying amounts of lignin. These fibers provide some structural support to the plant. Fibers which provide support to plant structures are usually a mixture of cellulose and lignin, forming a fiber reinforced composite. Lignin is mainly a highly branched, substituted phenol, somewhat similar to the crosslinked structure of novoloid. It is a stiff molecule which acts as a resin or glue in these natural composites. The amount of lignin present ranges from 10% in small, short plants like ramie and flax (linen) to more than 50% in trees.

Since bast fibers have cellulose content, many of the chemical properties of them will be the same as for cotton. With the exception of linen, these fibers are coarser and stiffer than the cotton fibers. They are primarily employed in industrial applications because they are stronger fibers than cotton; however their availability may be the controlling factor in their use since they must be imported, generally as coarse fabrics in the form of canvas, carpet backing fabrics and bagging, or as twine, cordage and the like.

2.3 Protein Fibers

Wool

Wool is a hair fiber taken from sheep during shearing. Before cleaning, the collected shearing is about 50% by weight in dirt, grease and vegetable matter. The molecular structure of a hair fiber is complex. A single hair fiber is a multi-celled array wrapped with a surface layer of small scales reminiscent of fish scales (Figure 17.15, Chapter 17.0). Typically pigments will be found scattered throughout the fiber. The fiber is constructed of proteins, having 10 to 20 different kinds of amino acid monomers in its chain. The pendant chemical groups on the amino acids force the protein into a helical arrangement (alpha helix) similar to a spring. Cystine crosslinks between adjacent protein molecules keep the helices tied together. Like most proteins, wool is sensitive to heat and a variety of chemicals, and does not have extensive applications in industrial fabrics. Other hair fibers will differ in size, pigmentation, scale frequency, and harvesting methods. Hair fibers are most often used for apparel and carpeting. A generalized structure of a protein is shown below.

R = | Alanine | Glycine | Phenylalanine | Tryptophan |
Arginine	Histidine	Proline	Tyrosine
Aspartic Acid	Leucine	Serine	Valine
Cystine	Lysine	Threonine	Isoleucine
Glutamic Acid	Methionine		

Silk

Silk is a protein fiber like wool but with a much simpler structure. It is an extruded fiber which is chemically and structurally uniform across its diameter. The amino acids have smaller pendant groups than those found in wool, allowing a pleated-sheet structure rather than helical to occur.

R = Alanine, Glycine, Serine, or Tyrosine

Because no cystine is present, no crosslinking occurs. Silk has no cellular structure, no scales, and no pigment. The silkworm uses the silk to construct a home in which to undergo metamorphosis. A silkworm's cocoon is unwound to produce a natural, continuous filament fiber. Observation of the silkworm at work gave man the idea for extrusion of fibers. Although wool and silk are generally weak, other insects, such as spiders, can produce a very strong protein fiber. Indeed it seems that with a greater understanding of protein structures, a fiber having an optimal choice of physical properties may be designed.

3. MAN-MADE FIBERS

Man-made fibers are those produced by human endeavor. They may be further categorized into regenerated and synthetic classes. Regenerated fibers (such as rayon) are those created from polymeric materials produced in nature. Synthetic fibers have no natural polymeric precursors. Thermoplastic polymers such as nylon or polyester are examples of this category.

Table 2.2 lists common man-made fibers. FTC definition, chemical structure and important properties of the fibers are given below.

3.1 Acetate

"A manufactured fiber in which the fiber-forming substance is cellulose acetate. Where not less than 92% of the hydroxyl groups are acetylated, the term triacetate may be used as a generic description of the fiber." Two varieties of acetate are produced which are secondary cellulose acetate and cellulose triacetate. As the name implies, they describe the approximate number of hydroxyls of the cellulose which are replaced by acetyl groups. These acetyl groups confer new properties to the cellulosic polymeric chains, chief of which is high loss of water absorbency, i.e., the fibers are hydrophobic (water hating) in nature. They are dry spun by spinning solutions of the cellulose derivatives dissolved in low boiling volatile solvents into hot air whereupon the solvent is evaporated resulting in a solid continuous thread of the cellulose acetate polymer.

During the fiber extrusion process, either pigments can be added to the spinning dope to color the threads or delustering agents in order to modify (dull) the luster characteristics of the fibers. In addition, other agents can be added to the spin solutions to impart sunlight and ignition resistance to the fibers. The fibers of the yarns can be spun into any denier, cut into any staple length. Further, they are resistant to mildew and insects that damage cellulosic fibers. The fibers are lighter in weight (less dense) than cellulosic fibers (density is 1.32 compared to 1.52 for cotton and rayon).

TABLE 2.2 Generic and Trade Names of Common Man-Made Textile Fibers.

Fiber Type	Name	Manufacturer
Acetate	Acetate	
Acrylic	Acrilan®	Monsanto
	Creslan®	Cytec
	MicroSupreme®	Cytec
Aramid	Kevlar®	DuPont
	Nomex®	DuPont
Carbon	Thornel®	Amoco
Fluorocarbon	Gore-Tex®	W. L. Gore
	Teflon®	DuPont
Glass	Glass	
Lyocell	Tencel®	Courtaulds
Modacrylic	SEF®	Monsanto
Nylon	Nylon 6	
	Nylon 6,6	
Olefin	Polyethylene	Hercules
	Herculon®	Hercules
	Marvess®	Amoco
	Alpha®	Amoco
	Essera®	Amoco
	Marquesa Lana®	Amoco
	Patlon III®	Amoco
	Spectra 900®	Allied
	Spectra 1000®	Allied
	Fibrilon®	Synthetic Industries
Polybenzimidazole	PBI	Hoechst Celanese
Polyester	ACE®	Allied
	Compet®	Allied
	Dacron®	DuPont
	Fortrel®	Wellman
	4DG	Eastman
	Tairilin®	Nan Ya Plastics Corp., America
	Trevira®	Hoechst Celanese
Rayon	Fibro®	Courtaulds
	Cuprammonium	
	Rayon	North American Rayon Corp.
Saran	Saran®	Pittsfield Weaving
Spandex	Glospan®/Cleerspan, S-85	Globe
	Lycra®	DuPont
Sulfar	Ryton®	Amoco (licensed to Phillips)

Their major disadvantages are that they are quite weak compared to cotton or rayon, as well as being more flammable (unless treated). They are considered thermoplastic fibers hence are more affected by heat (secondary acetate should not be heated for any appreciable time over 275°F, triacetate over 400°F), have higher static propensity, and are dissolved or badly affected by common solvents such as acetone (found in nail polish removers), acetic acid, some alcohols and other common ketones and chlorinated hydrocarbons.

Numerous non-textile uses for these materials include plastics and films, cigarette filter fibers, coatings (such as fingernail polish), finishes, and the like.

3.2 Acrylic

"A manufactured fiber in which the fiber-forming substance is any long-chain synthetic polymer composed of at least 85% by weight of acrylonitrile units." The commercially produced poly(acrylonitrile) (PAN) polymers are typically formed through free radical polymerization. The polymer produced by this method is considered mostly atactic, exhibiting little conventional crystallinity. Typically the homopolymer is not used for the production of fibers. Fibers usually contain up to 15% of a comonomer (vinyl acetate, vinyl chloride or an amine functional group) to improve the dyeability or the flexibility of the fiber.

Acrylic fibers have a wool-like hand and in consumer use find use as a wool substitute because of their soft hand, high resiliency, good strength properties and durability particularly to sun and weathering, as well as being resistant to rot, mildew and insects. As a class, they have good elasticity and high bulking power and are capable of being heat set for greater dimensional stability. Some of these factors are important in industrial applications due to its relative light weight and its ease of blending with other fiber types. They dry quickly when wet. Chemically, they are resistant to laundering, bleaching and dry cleaning solvents and to all except the strongest acids and bases.

Their primary disadvantages are that some acrylics pill readily, have a tendency to hold oily stains and accumulate static charges and are relatively heat sensitive, a property that often results in excessive stretching during use. They are considered flammable fibers and burn with a yellow flame that leaves a hot residue.

3.3 Anidex

"A manufactured fiber in which the fiber forming substance is any long-chain polymer composed of at least 50% by weight of one or more esters of a monohydric alcohol and acrylic acid ($CH_2=CH-COOH$)." Anidex is not currently produced in the United States.

3.4 Aramid

"A manufactured fiber in which the fiber forming substance is a long-chain synthetic polyamide in which at least 85% of the amide linkages are attached directly to two aromatic rings."

Commercially available aramid fibers are Nomex® and Kevlar® produced by DuPont and Twaron® produced by Akzo.

Meta-Aramid (Nomex®)

Para-Aramid (Kevlar®)

Aramid fibers are chemically similar to nylons in that they contain amide groups. They differ in that the amides are separated by aromatic rings instead of aliphatic methylene units as found in ordinary nylons. These structures impart significant tensile strength (only glass, graphite and PBI-type fibers are stronger) and temperature resistance to the fibers such that they are considered to be inherently ignition resistant as well as have high resistance to dry heat and good toughness characteristics. They have a higher density than nylon but less than that for cotton. They were initially developed for space applications but have achieved widespread use in consumer and industrial applications. As a class, they have better resistance to acids than the nylons (but not as good as acrylics and polyester), have good resistance to strong bases, organic solvents and bleaches, and are resistant to moths, insects and mildew. Figure 2.4 shows a microphotograph of Kevlar® 49 fiber.

Their major disadvantages are their lack of resistance to sunlight or other sources of ultraviolet light. They are not dyeable by the usual dyeing methods (they have some affinity for cationic type dyes). Table 2.3 shows types and industrial application areas of Nomex® and Kevlar® fibers.

3.5 Azlon

"A manufactured fiber in which the fiber-forming substance is composed of any regenerated naturally occurring proteins." Wool and silk are naturally occurring protein fibers

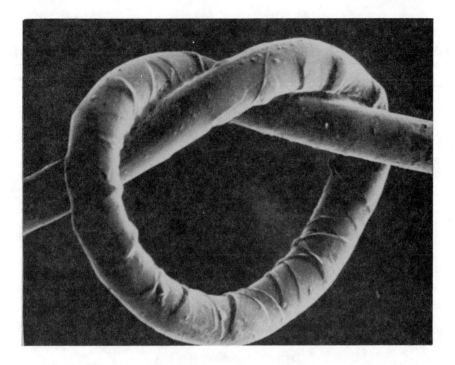

FIGURE 2.4 Kevlar® 49 fiber.

which have been used mostly for clothing and have had limited use in industrial textiles. Originally it was hoped that a man-made protein fiber might have properties as good as these natural fibers, but the results have been disappointing. Research has continued sporadically due to the potential that a fiber might be developed with the strength or elasticity of that produced by spiders and certain caterpillars. Although man-made protein fibers have been produced in the past from bean, milk and corn protein, none is currently being produced in the United States.

$$\begin{array}{ccccc} & O & & O & & O \\ & \parallel & & \parallel & & \parallel \\ +N-C-C\!\!+\!co\!\!+\!N-C-C\!\!+\!co\!\!+\!N-C-C\!\!+ \\ & | & & | & & | \\ & R_1 & & R_2 & & R_{n+1} \end{array}$$

3.6 Glass

"A manufactured fiber in which the fiber-forming substance is glass." Glass structures are composed of silicon dioxide (sand) and a variety of metallic oxides. Its exact composition is dependent on the desired end properties (heat resistance, chemical resistance, etc.). The fiber is produced by a number of manufacturers and is one of the most commonly used industrial fibers.

Glass fibers find widespread use as an industrial fiber because of its excellent ignition resistance, high strength (being stronger than steel of the same diameter), sunlight weather, insect, mildew, bleach and other chemical resistance. It is not affected by water, has virtually no shrinkage and is considered fireproof.

Its major disadvantage is its low abrasion resistance and easy damage by mechanical action. Nonetheless it can be made into woven and nonwoven fabric forms. Glass fibers can cause skin irritation when the broken fibers penetrate the skin pores.

3.7 Metallic

"A manufactured fiber composed of metal, plastic coated metal, metal coated plastic or a cord completely covered by metal." Solid metal fibers are difficult to produce by normal extrusion techniques due to the high melting temperature of the material and its high surface tension when molten. A number of unique production techniques (drawing of large bars, or

TABLE 2.3 *Types and Application Areas of Nomex® and Kevlar® Fibers from DuPont [2].*

	Type	Application
Nomex®		
	430	Standard, bright yarn for high temperature uses
	432	Same as Type 430 but producer-colored "Olive Green"
	433	Same as Type 430 but producer-colored "Sage Green"
Kevlar®		
Kevlar	950	Tire belt reinforcement and radial-tire carcasses
	956	High performance mechanical rubber goods, such as hoses, V-belts, precision drive belts, and conveyor belts
Kevlar 29	959	Fish nets
	960	Ropes and cables, with special overlay finish for improved abrasion resistance
	961	Ropes and cables, with standard finish
	962	Ropes and cables, with no secondary finish
	963	Non-apparel ballistic armor
	964	Soft ballistic armor, coated fabrics and other weaving and knitting applications
	971	Ignition cables
	972	Friction product reinforcement
	973	Pultrusion and chopping
	974	Stretch-break spinning applications
	977	Pump packing applications
	980	Industrial filament winding applications
	984	Certified product for aerospace industry
	985	Tapes and strapping
Kevlar 49	965	Woven reinforcement in aerospace composites, hard ballistic armor, and printed circuit boards
	967	Pultrusion and chopping
	968	Marine composites, fiber optics, ropes, and filament-wound composites; no secondary finish
	969	Certified product for filament-wound composites in special aerospace uses; no secondary finish
	978	Ropes and cables, with special overlay finish for improved abrasion resistance
	989	Tailored length packages for fiber optic cable reinforcement
Kevlar 68	956B	High performance mechanical rubber goods, such as hoses, V-belts, precision drive belts and conveyor belts
	965B	Woven reinforcement in aerospace composites, hard ballistic armor and printed circuit boards
	968B	Marine composites, ropes, filament-wound composites; no secondary finish
	975B	Fabrics used in pipe liner reinforcement
	989B	Fiber optic cable reinforcement
Kevlar 149	965A	Woven reinforcement in aerospace composites, hard ballistic armor, and printed circuit boards
	968A	Marine composites, fiber optics, ropes, filament-wound composites; no secondary finish
Kevlar HT	964C	Highest tenacity for advanced needs in ballistic protection
	965C	Similar to 964C

vacuum evaporation to coat a plastic film, etc.) are employed to produce metallic fibers. Since they are a non-traditional textile product, metallic fibers are usually produced by companies not normally associated as a textile operation.

3.8 Modacrylic

"A manufactured fiber in which, when not qualified as rubber or anidex, the fiber-forming substance is any long-chain synthetic polymer composed of less than 85%, but at least 35% by weight of acrylonitrile units." Most often, the comonomer used with acrylonitrile is vinyl chloride. The main effect achieved through the inclusion of this comonomer is a reduction of acrylic's flammability. It tends to shrink when heated but can be blended with other fibers to improve ignition resistance.

3.9 Novoloid

"A manufactured fiber containing at least 85% by weight of a crosslinked novolac." A novolac is a low molecular weight resin produced by the reaction between phenol and formaldehyde. The phenol-formaldehyde resin produces a branched polymer and eventually a crosslinked product if the molecular weight is increased. It is spun as a low molecular weight product and crosslinked in fiber form to produce the final product. The formaldehyde can react at the 2, 4 and 6 carbons on the phenol ring, and each ring is, therefore, a potential branching site.

3.10 Nylon (Polyamide)

"A manufactured fiber in which the fiber-forming substance is a long-chain synthetic polyamide in which less than 85% of the amide

(−CO−NH−) linkages are attached directly to two aromatic rings." Nylon is the first truly man-made synthetic fiber. It is produced by either the reaction of an amino acid with itself or between diamines and diacids. They are combined under the influence of high temperature and pressure via amide linkages to produce the fiber forming polymer. The resulting nylon is melt spun into fibers which are oriented by stretching to give variants having a wide range of fiber diameters and cross-sectional shapes, strengths, weight, elasticity, and luster among other properties that give high versatility to these fibers (Figure 2.5).

A large number of distinctly different structures is possible and has been commercially developed. These structures are generally represented by the word nylon followed by a number designation. If the nylon is formed solely from an amino acid, a single number suffix represents the number of carbons in the acid. A two number suffix is used if the nylon is made from a diamine and a diacid. The first number represents the number of carbons in the diamine, and the second number represents those in the diacid. If the diacid is terephthalic acid, the letter "T" is used instead of a number. Nylon 6 and nylon 6,6 comprise over 90% of commercial nylon fiber production.

Poly(hexanolactam) (Nylon 6)

Poly(hexamethylene adipamide) (Nylon 66)

Nylon can be heatset to impart both crimp (texturizing and bulking) and high dimensional stability to either the yarns or fabrics. Industrial applications include air springs, belting, nets and twine, hose fabrics, filter fabrics, nonwoven felts and reinforcing fabrics which can be either coated or used alone in such applications as geotextile uses, artificial grass on athletic playing fields, tentage and in many types of architectural and construction applications. These few illustrations of the present day industrial uses for nylon containing fabrics serve only to illustrate its wide potential as an industrial textile material.

ANTRON® III TRILOBAL/ROUND FIBER CONTAINS POLYMERIC CONDUCTIVE MATERIAL

FIGURE 2.5 Various cross-sectional shapes of nylon fibers.

Nylon has exceptional abrasion resistance and is often blended with other fibers to decrease weight (density is about 1.14) or impart strength as well as high flexibility in addition to the abrasion resistance to the final textile product. Other major benefits include its resistance to most strong bases, dilute acids and solvents, insects and most micro-organisms. They have a low moisture regain and dry quickly when wet.

Nylon's major disadvantages are its solubility in strong acids such as hydrochloric, formic and sulfuric acid. At low moisture levels it develops and holds a static charge; however, this can be mitigated by addition of static dissipating fibers to the yarns. It is more resistant to sunlight than the natural fibers but must be treated to obtain resistance in the 3,200−4,000 Angstrom range. Nylon 6 melts at about 212−215°C while nylon 6,6 has a higher melting point (250−265°C). Both nylon types melt and drip away from the burning site unless held within the textile fabric by the presence of other fibers. Spun yarns of nylon have a tendency to undergo pilling.

3.11 Nytril

"A manufactured fiber in which at least 85% of the polymer constituents are vinylidine dinitrile, or where the vinylidine dinitrile occurs no less than every other unit in the polymer chain." Nytril fibers are not currently produced in the United States.

3.12 Olefin

"A manufactured fiber in which the fiber-forming substance is any long-chain synthetic polymer composed of at least 85% by weight of ethylene, propylene or other olefin units." Ethylene and propylene are produced in abundance during the cracking of oil to produce petroleum. Originally, propylene was considered a waste product of little or no value. Fibers made from by-products have been among the least expensive available. The low cost and ease of production has led to the use of olefin fibers in a variety of industrial applications.

poly(ethylene)   poly(propylene)

Olefin fibers have virtually no moisture sorption and are manufactured into strong melt spun fibers having a wide range of deniers and fiber cross-sectional shapes (the most predominant is still round). They are extremely resistant to acids, alkalis, bleaches, solvents, mildew and insects and have high strength, resiliency and elastic properties. They have a low specific gravity that gives them excellent weight advantage. Although flammable, they are slow burning and also have excellent electrical resistance. These fibers find important use in non-woven applications. Further, they can be spun into microdenier variants that have very high surface-to-weight ratios.

Their major disadvantage is their low temperature resistance. The fiber melts at 330°F and undergoes considerable shrinkage above 140°F. The fiber is also quite sensitive to oxidation and ultraviolet light (which can be corrected with the use of inhibitors). These drawbacks require that considerable care should be exercised in selecting the proper end use conditions for the fibers.

Recent developments in the polymerization techniques of polyethylene have provided for the production of the highest strength and modulus fibers ever manufactured, i.e., Spectra® by Allied. The process involves the use of very high molecular weight polymers extruded from a highly viscous solution. The process is referred to as gel spinning because of the high viscosity of the spinning solution. With the advent of this process, the conditions necessary for extremely high strength and modulus fibers have become clearer. A highly linear, crystallizable polymer is required, and it must be "stretched" in a semi-solid state to produce the high strength and modulus. The semi-solid state can be either a high viscosity gel or liquid crystalline form.

3.13 Polybenzimidazole (PBI)

"A manufactured fiber in which the fiber forming substance is a long chain aromatic polymer having recurring imidazole groups as an integral part of the polymer chain." Fairly recent in development, this fiber has found use in a variety of high temperature applications, or where non-flammability is desirable. The most familiar usage of PBI has been in protective suiting for astronauts.

3.14 Polyester

"A manufactured fiber in which the fiber-forming substance is any long-chain synthetic polymer composed of at least 85% by weight of an ester of a substituted aromatic carboxylic acid, including, but not restricted to, substituted terephthalate units $(-R-O-CO-C_6H_4-CO-O-)$, and para-substituted hydroxybenzoate units $(-O-R-C_6H_4-CO-)$." Unlike the aliphatic nylons which have a melting temperature sufficiently high enough to be useful for textile fibers, polyesters require the inclusion of an aromatic ring to produce a fiber with a useful melting temperature. Because of its versatility, polyester has the highest volume consumer usage of all man-made fibers, and finds application in many industrial textile products.

These fibers have high wet and dry tensile strengths as well as good chemical, insect and microbial resistant properties. They can be heat-set to give exceptional dimensional stability to the yarn or fabric. Being melt spun, they can be manufactured into a wide range of deniers and a variety of cross-sectional shapes that allow polyester to blend well with other fiber genera. They are essentially insensitive to water, have good weather and sunlight resistance (compared to nylon) and are not affected by most strong acids, alkalis, and common industrial and household solvents. Polyester fibers have good abrasion resistance, have good electrical insulation properties, low creep, good wearability and good resistance to exposure to elevated tempera-

tures. Thus, they are ideal candidates for many industrial uses. Typical applications include floor coverings, conveyer belts, hoses, ropes, tire cord, papermaking fabrics, sewing thread and the like.

Their major disadvantage is their propensity to absorb oily liquids and soils which are difficult to remove under typical laundry conditions. They are considered flammable having a low softening temperature and will melt and drip (similar to the nylons) before ignition occurs. Their low moisture absorption can result in high static accumulations but this can be mitigated by employing graphite (or other conductive carbon) core fibers such as those employed to reduce static with the nylons. Their pilling tendency is quite similar to that for nylon.

3.15 Rayon

"A manufactured fiber composed of regenerated cellulose, as well as manufactured fibers composed of regenerated cellulose in which substituents have replaced not more than 15% of the hydrogens of the hydroxyl groups."

Rayon is a regenerated cellulose polymeric fiber that took some time to develop into a seasoned alternative to cotton. It is presently manufactured by one of two methods neither of which are environmentally friendly. These are the Cuprammonium process which employs a copper-ammonium complex to dissolve the cellulose where it can be wet spun into a water bath to precipitate the cellulose as a continuous filament. The Viscose process employs carbon disulfide to convert the cellulose into an alkali soluble "Xanthate" derivative which can be spun into an acid/salt bath to regenerate the cellulose in fiber form. A new manufacturing method recently introduced by Courtaulds should produce a more environmentally safe spinning alternative. These fibers, because of their cellulosic nature, have many of the same properties (density, flammability, mildew propensity, moisture absorption) as cotton; in addition, it competes well with cotton in cost. Figure 2.6 shows a photomicrograph of rayon fibers that are used in tire cords.

Rayon fibers can be manufactured into any desired denier either as a continuous filament or cut into staple of any length and are not subjected to bysinosis problems. They can be obtained in varying luster, have either wet or dry modulus, can be made to have fair to excellent strength, can be manufactured with self crimping variants, can be engineered to have built-in ignition resistance, cover factor, surface area (important in absorption) among other properties. It is an ideal replacement for cotton in a wide range of industrial (especially nonwoven) applications. It certainly should be considered for blending with cotton to improve the yarn uniformity, surface sorption properties, absorbency, bulk and elongation properties of the blend.

Rayon's relatively low strength and elongation have limited its use in most industrial applications. Recent developments in solvent-spun cellulose may provide more opportunities in industrial textiles. In addition to rayon, cellulose is the basis for almost all of the fibers produced in nature by plants.

3.16 Rubber

"A manufactured fiber in which the fiber forming substance is comprised of natural or synthetic rubber." The definition may be further categorized as follows:

$$\left[CH_2-C(CH_3)=CH-CH_2\right]_n$$

- a manufactured fiber in which the fiber forming substance is a hydrocarbon such as natural rubber, polyisoprene, polybutadiene, copolymers of dienes and hydrocarbons or amorphous (non-crystalline) polyolefins
- a manufactured fiber in which the fiber forming substance is a copolymer of acrylonitrile and a diene (such as butadiene) composed of not more than 50% but at least 10% by weight of

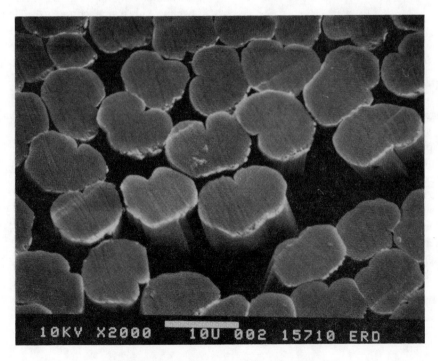

FIGURE 2.6 Cross-section of rayon tire fibers.

acrylonitrile units – The term "lastrile" may be used as a generic description for this form of rubber.

$$\left[CH_2-\underset{\underset{C\equiv N}{|}}{CH}\right]_n co \left[CH_2-CH=CH-CH_2\right]_n$$

- a manufactured fiber in which the fiber forming substance is a polychloroprene or a copolymer of chloroprene in which at least 35% by weight of the fiber forming substance is composed of chloroprene units $(-CH_2-(Cl)C=CH-CH_2-)$.

$$\left[CH_2-\underset{\underset{Cl}{|}}{C}=CH-CH_2\right]_n$$

3.17 Saran

"A manufactured fiber in which the fiber forming substance is any long chain synthetic polymer composed of at least 80% by weight of vinylidene chloride units." Saran is usually produced in a specialized monofilament form for flame retardant and chemical resistance applications. The most familiar consumer usage is in film or coating form.

$$\left[\underset{\underset{H}{|}}{\overset{\overset{H}{|}}{C}}-\underset{\underset{Cl}{|}}{\overset{\overset{Cl}{|}}{C}}\right]_n$$

3.18 Spandex

"A manufactured fiber in which the fiber-forming substance is a segmented polyurethane." Spandex is an elastomeric fiber used for clothing materials, and finds application in industrial textiles in such places as elastic bands in protective apparel. It is a block copolymer composed of stiff rigid blocks interspersed with soft flexible blocks. In general, the soft blocks have been either aliphatic polyesters or polyethers. A single chemical structure cannot be drawn for Spandex, but a generalized structure is shown below.

$$\left[\underset{\underset{O}{\|}}{C}-\underset{\underset{H}{|}}{N}-R_1-\underset{\underset{H}{|}}{N}-\underset{\underset{O}{\|}}{C}-O-R_2-O\right]_n$$

3.19 Sulfar

"A manufactured fiber in which the fiber-forming material is a long synthetic polysulfide in which at least 85% of the sulfide linkages are attached directly to two aromatic rings." As with other aromatic based fibers, sulfar finds utility from its special stability to environmental challenges. Sulfar is particularly chemically resistant, and finds most of its applications in industrial fabrics.

$$\left[\bigcirc-S\right]_n$$

3.20 Vinal

"A manufactured fiber in which the fiber forming substance is any long chain synthetic polymer composed of at least 50% by weight of vinyl alcohol units $(-CH_2-CHOH-)$, and in which the total of the vinyl alcohol units and any one or more of the various acetal units is at least 85% by weight of the fiber." Although the polymer is manufactured in large qualities for a variety of glue and coating applications, there is no vinal fiber currently manufactured in the United States. The water solubility of the vinal fiber must be overcome (usually by formaldehyde crosslinking) in order to produce a durable product.

$$\left[\underset{\underset{H}{|}}{\overset{\overset{H}{|}}{C}}-\underset{\underset{OH}{|}}{\overset{\overset{H}{|}}{C}}\right]_n$$

3.21 Vinyon

"A manufactured fiber in which the fiber forming substance is any long chain synthetic polymer composed of at least 85% by weight of vinyl chloride units." Vinyl (PVC) is almost the universal plastic film, finding application in almost everything, e.g., notebooks, electrical cords, outdoor furniture, and automobile trim.

$$\left[\underset{\underset{H}{|}}{\overset{\overset{H}{|}}{C}}-\underset{\underset{Cl}{|}}{\overset{\overset{H}{|}}{C}}\right]_n$$

3.22 Fluorocarbon

"A fiber formed of long-chain carbon molecules, having all available bonds saturated with fluorine."

$$\left[\begin{array}{cc} \overset{F}{\underset{F}{C}} - \overset{F}{\underset{F}{C}} \end{array}\right]_{n}$$

The fluorocarbon, poly(tetrafluoro ethylene) (PTFE), also known by the trade name Teflon®, can be made into a fiber and finds application in chemically and environmentally resistant products. Very little fluorocarbon is made in fiber form for apparel applications. It finds a niche in woven, knitted, braided and felted forms for specialty applications. Its widest uses are as coatings for antistick applications such as on cookwear and the drying cylinders in the textile mill for sizing and in dyeing/finishing applications. It can also be made as an expanded film for water and chemical barrier applications under the trade name Gore-Tex®. These films have the advantage of being breathable, allowing water vapor to transpire through the films but not water droplets. Other non-fiber uses include finishes for antisoil and waterproofing applications, for example in raincoats, soil resistant carpet and upholstery fibers and the like. The material also has a very low coefficient of friction and can be used to give nearly friction free surfaces. It also has high ignition resistance and cannot be ignited even in a 100% oxygen atmosphere. When ignited or decomposed at temperatures above 400°C, the fibers will generate dangerous gases, hence the materials are not used in extreme heat applications.

3.23 Fibers without Generic Names

A number of fibers are sold only into industrial and high technology applications and therefore, do not require generic names.

Carbon

Carbon appears to be unique among elements in its ability to form different structures. The range of structures includes amorphous carbon, activated carbon, carbon black, graphite, diamond and a variety of basically hollow three-dimensional structures called Fullerenes.

Typical carbon fiber manufacture involves the heating of a precursor fiber in an inert atmosphere. The precursor typically is a ring-type structure which can, under the appropriate conditions, be converted to graphite. Precursors include rayon, PAN, and mesophase (liquid crystalline) pitch. The characteristics of the carbon fiber depend on the precursor as well as the heating conditions. The structure of graphite is a planar array of benzene rings. Highly graphitic carbon has a very high tensile modulus.

Carbon fibers have super high strength properties and are considered ignition resistant as well as high temperature resistant. Their major disadvantages are their lack of elasticity and abrasion resistance. Some typical uses which illustrate their potential performance for industrial applications include reinforced resin (and metals), golf club shafts, crossbows, rotor blades and other parts for helicopters, skis, jet engine parts, ship keels among others. The fiber can be woven and fabricated as nonwoven textiles. Their electrical conductivity can be a problem in some applications.

Poly Ether-Ether-Ketone (PEEK)

$$\left[\begin{array}{c} O - \bigcirc - \overset{O}{\underset{\parallel}{C}} - \bigcirc - O - \bigcirc \end{array}\right]_{n}$$

PEEK is a high performance thermoplastic fiber used frequently in fiber reinforced composites and in dryer fabrics for papermaking.

Polystyrene

When made into fibers, polystyrene is usually stiff and brittle and therefore unsuitable for conventional textiles. It can be made into isotactic as well as atactic forms. The typical commercial product is atactic. There is interest in the fibers for use in optical devices.

3.24 Liquid Crystalline Aromatic Polyesters

They have sufficiently high concentration of the proper kind of aromatic units in the backbone to permit the formation of a liquid crystalline melt. Similar to the liquid crystalline para-aramid Kevlar®, fibers spun from liquid crystalline polyesters have high strength and modulus characteristics. Its producers also cite ad-

vantages in flexibility and fatigue, and improved chemical resistance. The structure for Vectran® is shown below.

Polyimides

They are a high temperature fiber developed through military sponsored research and produced mainly for defense applications.

2.3

References and Review Questions

1. REFERENCES

1. *Webster's Ninth New Collegiate Dictionary,* Merriam-Webster Inc., 1988.
2. Properties of DuPont Industrial Filament Yarns, *Multifiber Bulletin,* X-272, July 1988.

1.1 General References

Deanin, R., *Polymer Structure Properties and Applications,* Cahners Books, Boston, 1972.

Dictionary of Fiber & Textile Technology, Hoechst Celanese Corporation, Charlotte, NC, 1990.

Handbook of Chemistry and Physics, 65th Ed., CRC Press, Inc., Boca Raton, FL, 1984.

Isaacs, M. III, Textile World Manmade Fiber Chart 1992, Maclean Hunter Publishing Co., Chicago, 1990.

Seymour, R. B. and Carraher, C. E., Jr., *Polymer Chemistry, an Introduction, 3rd Ed.,* Marcel Dekker, Inc., New York, 1992.

Sperling, L. H., *Introduction to Polymer Science, 2nd Ed.,* John Wiley & Sons, New York, 1992.

Chapman, C. B., *Fibers,* Textile Book Services, Plainfield, NJ, 1974.

2. REVIEW QUESTIONS

1. Define the following terms:
   - polymer
   - monomer
   - degree of polymerization
   - atactic structure
   - syndiotactic structure
   - homopolymer
   - block copolymer
   - random copolymer
2. What is the effect of crosslinking on polymer behavior? Explain.
3. What is the difference between a fiber "generic name," "trademark," and a "trade name"?
4. What is the major effect of acetyl groups on acetate properties? Explain.
5. What are the differences between polyamide (nylon) and aramid fiber structures? What are the effects of these differences on aramid fiber properties?
6. How many different ways are there to make carbon fibers? Is there any property difference among carbon fibers made with different methods?

FIBER AND YARN MANUFACTURING

Manufacture of Man-Made Fibers

R. M. BROUGHTON, JR.
P. H. BRADY

1. INTRODUCTION

Textile fibers are solids with distinct shapes. The primary task in fiber manufacture is to transform solid materials into a fiber configuration. The liquid state is the only condensed state that can easily deform or have its shape changed. In its simplest form, fiber manufacturing processes consist of liquefying a solid polymer, transforming it into the shape of a fiber, and then resolidifying the liquid.

In the 1600s, Robert Hooke and other scientists noted that spiders and silkworms had developed the process of converting solids to liquids, and back to solids, but reshaped as filaments. These early scientists speculated that man would one day be capable of duplicating this natural extrusion process. Early researchers were able to extrude fibers by drawing or pulling liquid threads from solutions of natural gums and resins. An industrialist in Manchester, England, is credited with the design and construction of the first fiber producing machine. However, the materials used in this instrument were not suitable fiber forming polymers. Audemars was awarded a patent for the production of cellulose nitrate fibers in 1855, but the first commercially successful concept of production was not developed until the 1880s. In 1891, Chardonnet's facility began production of regenerated cellulose fibers. This success was followed by the development of the cuprammonium process by Despeisses, which was commercialized in 1897. Cross and Bevan invented the viscose process about 1892, which was further refined to a more practical process by Stearn and Topham between 1895 and 1900. These early pioneers not only had to develop the chemistry necessary to dissolve and resolidify cellulose without severe degradation, they also had to invent the machinery necessary for fiber production. Table 3.1 shows the important events in the history of man-made fibers.

Energy is required to convert a solid polymer to its liquid state. That energy can be developed either from heat, chemical solvents, or a combination of the two. If heat supplies the energy, the polymer is solidified in fiber shape simply by cooling. If chemical solvents are used, there are two ways of resolidifying the polymer into a fiber shape. The solvent can be evaporated, or the polymer solution can be precipitated by immersion in a non-solvent. The three ways of resolidification are used as a basis to classify fiber extrusion processes (Table 3.2). This rather simple classification scheme does not reveal the true complexity of the processes. Some of the more recently developed polymers require very sophisticated processes. For example, high strength aramid fibers are produced through a wet spinning process, but the polymer solution exists in a liquid crystalline state. High strength polyethylene fibers are created through the extrusion of an ultra high molecular weight melt or gel state. This melt/gel is highly viscous, having some of the characteristics of a solid. Even the viscose process has chemical reactions proceeding during the solidification, so it is not entirely a precipitation.

2. MELT SPINNING (EXTRUSION)

A diagram of the basic melt spinning (extrusion) process is shown in Figure 3.1. Some operations which proceed directly from polymer manufacture to extrusion may use a pump instead of a screw type extruder. All fiber spinning sys-

TABLE 3.1 *Significant Events in the History of Man-Made Fibers.*

1664	Robert Hooke proposed invention of artificial fibers better than silk.
1845	Schoenbien invents nitrocellulose.
1855	Audemars invents process for making nitrocellulose fiber.
1868	Hyatt develops plasticized cellulose nitrate (celluloid).
1884	Chardonnet produces first regenerated cellulose.
1889	Chardonnet displays first textiles from regenerated cellulose.
1890	Despeisses invents the cuprammonium process.
1892	Cross, Bevan, and Beadle patent the viscose process chemistry.
1893	Libby produces glass fiber.
1898	Stearn and Topham make the viscose process work to produce fibers.
1907	Production of cellulose acetate "dope"
1919	Henri and Camille Dreyfus begin production of acetate fiber.
1929	Carothers produces high molecular weight polymers for DuPont.
1930	Polystyrene commercialized by I. G. Farben/Dow.
1936	Poly(methyl methacrylate) (PMMA) commercialized by Rohm & Haas.
1937	Vinyon (PVC) fibers produced by American Viscose Corporation.
1938	Carothers develops Nylon 66.
1940	Saran invented by Dow Chemical.
1941	Low density poly(ethylene) developed.
1946	Polyester fiber first produced by DuPont and ICI (Polyester is invented by Wheetfield and Pickinson).
1948	Acrylic fiber commercialized by DuPont.
1954	Isotactic poly(propylene) process invented.
1955	High density poly(ethylene) process invented.
1966	Poly(p-benzamide) developed.
1968	Poly(p-phenylene terephthalamide) (Kevlar®) commercialized.
1979	Stamylan UH® patented.

TABLE 3.2 *Major Fiber Manufacturing Processes.*

Melt spinning	Heat to melt, cool to solidify
Solution spinning	
Dry spinning (evaporative spinning)	Solvent dissolves polymer. Solvent evaporated to solidify.
Wet spinning	Solvent dissolves polymer. Polymer precipitates as solvent diffuses into non-solvent.

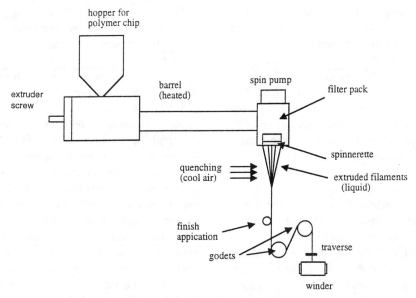

FIGURE 3.1 Schematic of melt spinning process.

tems filter the polymer before it passes through the fine holes of the spinnerette (Figure 3.2). If not filtered out, particulate or gelatinous impurities in the polymer can produce two undesirable effects. The impurities may be large enough to block or restrict flow through the spinnerette holes. The second, and equally disastrous effect would involve the passing of the impurity through the spinnerette, allowing it to be incorporated into the fiber structure. Inclusion of an impurity creates a discontinuity in the structure which may promote breakage of the filaments during the subsequent manufacturing processes. This is a particular problem with spinnerettes for non-circular fibers. Added benefits of the filter or screen before the spinnerette in plastic fiber spinning, are that the mixing action helps to reduce non-uniformities (temperature and viscosity) in the melt as well as removing what is referred to as screw memory. Screw memory is caused by the long chain polymers being twist oriented by the screw rotation. Most extrusion systems also use static mixers in-line before the spinnerette to ensure that the polymer melt is uniform in temperature and viscosity.

The spin pump is a positive displacement gear pump and provides a constant flow rate of polymer through the spinnerette. Typically one pump is paired with only one spinnerette. Any flow rate variations will show up as variations in linear density along the filaments and subsequently along the yarn or tow.

As filaments emerge from the spinnerette, they are accelerated away from the outlet, stretching the polymer before it cools. The stretched, semi-molten filaments are then cooled by a transverse air stream. The speed and temperature of the air stream are controlled to help ensure uniformity along the filament. Thermoplastic filaments may be stretched up to 100 (typically at least 25) times their original length before solidification. One to ten meters below the spinnerette (after solidification), the filaments

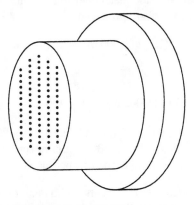

FIGURE 3.2 An extrusion spinnerette.

are brought together, passed around a series of godets, and then wound onto a yarn package. As the filaments are brought together before the godets, a lubricating spin finish is applied. The finish is applied by passing the filaments over a hard surface coated with the desired fluid. The yarn collects a small amount of liquid as it passes through the liquid film on the coated surface. The surface may either be a kiss roll, which rotates through the finish and just touches the yarn, or a thread guide with the finish pumped through a hole in the guide.

The godets positioned just below the finish applicator set the speed at which the fiber is produced. The take-up speed of the godets, in concert with the extrusion pump, controls the diameter of the filaments. If a series of heated godets operating at slightly differing speeds are used, then drawing of the filaments takes place which increases polymer orientation, and thereby improves filament strength. Uniform speed of the godets is critical to the filament uniformity. Package winders which operate in concert with the godet system, run at or just above the speed of the godet. The speed of the winder determines the tension in the thread line between the godet and winder. Slight tension is required for uniform yarn package formation. The control systems of some winders regulate their speed with such accuracy that the need for godets may be eliminated.

The ratio between the weight of polymer pumped through the spinnerette and the godet (or winder) speed determines the linear density of the extruded filament. If a pump delivers 15 grams of polymer per minute, and the filament is taken up at 1,000 meters per minute, the linear density of the yarn produced is

$$(15 \text{ g})/(1,000 \text{ m}) = 15 \text{ tex} \qquad (3.1)$$

or

$$(15 \text{ g})/(1,000 \text{ m}) \times 9 = 135 \text{ denier} \quad (3.2)$$

The definitions of tex and denier are given in Chapter 18.0. If there are 30 holes in the spinnerette, then the average linear density per filament is

$$135 \text{ denier}/30 \text{ filaments} = 4.5 \text{ denier per}$$
$$\text{filament (dpf)} \qquad (3.3)$$

The design of spinnerettes and spinnerette packs is critical, because the process depends upon a hydraulic split to ensure delivery of the same volume of polymer to each spinnerette hole. Subtle differences in flow paths can create major differences between filaments.

Most melt extruded fibers are further processed by drawing (sometimes referred to as cold drawing). Drawing of fibers is a stretching process which increases the strength of the fiber by increasing the orientation of the polymer molecules to the parallel axis of the fiber. Drawing in the solid state is much more effective at producing orientation of molecules than the stretching occurring in the molten state during extrusion. Drawing of a yarn must be done at a temperature above the glass transition temperature (T_g), and below the melting temperature (T_m). If heat is needed to allow stretch orientation, it may be supplied by heated godets, surface heaters, steam, or in some cases, by the stretching process itself. The process of drawing can reduce elongation from several hundred percent of the undrawn original length, to as little as 10 to 50%. Drawing is physically accomplished by the strain created by two godets (or pairs of godets) operating at different preset speeds. The continuous filament yarn from the extruder is wrapped around the first pair of godet rollers running at the lower speed, and is then wrapped around the faster, second godet pair as shown in Figure 3.3. Figure 3.4 depicts one model of the molecular processes occurring during drawing. Figures 3.5 and 3.6 show the processes as they occur in filaments and films.

Recent technological advances in extrusion allow for the combination of the spinning and drawing processes. This is accomplished by placing a second set of godet rolls between the spinnerette and the winder. For the purpose of speed synchronization, the spinning must usually be slowed, and the winding speed increased to greater than 5,000 meters/minute.

Often, a normally flat yarn is crimped or textured in order to produce a more interesting woven or knitted fabric texture. The basic premise of the fiber texturing process is the deformation and heating of the fiber above its glass transition temperature, and then allowing it to cool in the deformed state. Depending on the

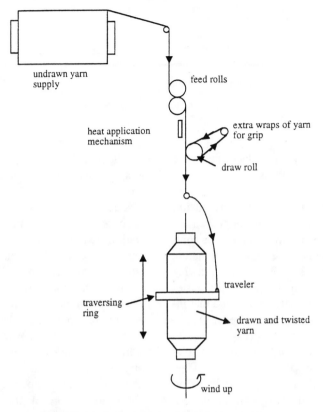

undrawn yarn
supply

feed rolls

heat application
mechanism

extra wraps of yarn
for grip

draw roll

traveler

traversing
ring

drawn and twisted
yarn

wind up

FIGURE 3.3 Schematic of a draw twisting machine.

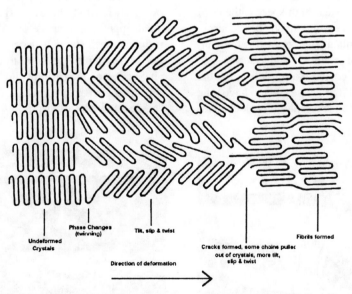

Undeformed
Crystals

Phase Changes
(twinning)

Tilt, slip & twist

Cracks formed, some chains pulled
out of crystals, more tilt,
slip & twist

Fibrils formed

Direction of deformation

FIGURE 3.4 Schematic model for drawing processes in a crystalline fiber (Peterlin).

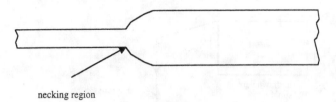

necking region

FIGURE 3.5 Necking during drawing.

texturizing process, the yarn produced may have a high degree of stretch and recovery, or it may just simply appear more like a staple yarn. Figure 3.7 depicts both flat and textured yarns. Again, the technology has changed as improved efficiency is sought. Texturing is sometimes combined with drawing to create a combination draw-texturing process. With the drawing process uncoupled from the potentially faster spinning process, producers have found that increased spinning speed equates to better performance in the draw-texturing process.

A partially oriented yarn (POY) produced by spinning with a high take-up speed (3,000–5,000 meters/minute), may require only 100 to 200% additional stretch during drawing as compared to 300 to 500% stretch for normal undrawn yarns. Researchers are experimenting with winding speeds in excess of 10,000 meters/minute during spinning. At such high speeds, the need for secondary drawing may be entirely eliminated. The molten polymer exiting the spinnerette is deformed so quickly and to such a high degree and then cooled so rapidly that the molecular relaxation processes do not have time to take effect before the orientation is locked in.

The preceding descriptions of fiber melt spinning processes have concentrated on the manufacture of filaments. Conversion of continuous filaments into yarn is covered in Section 3.2. Continuous filament yarns may be twisted, plied and/or textured before being used in fabric formation processes. Sometimes, the filament yarns are cut into staple fiber form, and then reassembled back into a yarn form. In the manufacture of staple fibers, there is no need to maintain a yarn-sized bundle of filaments. The combination of filament bundles into a large tow allows certain economies to be realized in manufacturing. Spinnerettes with more holes can be used, thus requiring fewer pumps, packs and filters for the same volume of production. A large tow containing many ''yarns'' can be drawn on one godet frame rather than a separate set for each yarn, and no winders are required. The tow is cut into short lengths after it is crimped (textured), and dropped into a press for baling. The length of the chopped staple is dependent upon its end use. If the staple fibers are to be blended with cotton or wool, they are cut into lengths about the same as the fibers with which it will be mixed. The development of crimp in the tow, which is

FIGURE 3.6 Drawing/elongation of a plastic film.

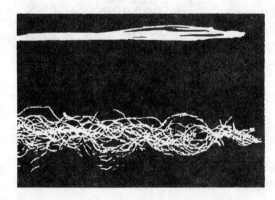

FIGURE 3.7 Flat (top) and texturized yarns.

the equivalent of texturizing in filament yarns, is necessary to give the fibers enough cohesiveness to be useful in yarn spinning.

3. SOLUTION SPINNING

Spinning from solution is usually done only for polymers which are unstable in the molten state. The instability may be manifest when the polymers melt or by the fact that they decompose without melting. There are two forms of solution spinning: dry and wet. The solidification in wet spinning occurs by coagulating the polymer out of the solution by introducing the solution into a liquid in which the solvent is soluble, but the polymer is not. In dry spinning, the solvent is removed from the fiber by evaporation in a chamber of recirculating hot air.

3.1 Dry Spinning

The dry, or evaporative spinning process is similar to the melt spinning process, except that the polymer is liquefied by dissolution in a volatile solvent, and the liquid filament is solidified by evaporation of the solvent rather than cooling. This type of spinning process was used by Chardonnet in the 1880s to manufacture his cellulose nitrate fibers. A schematic diagram of the process is shown in Figure 3.8.

The concentration of polymer in the solvent depends on the solution viscosity characteristics of the polymer-solvent system. In most cases, the polymer concentration does not exceed 25%. Viscosity is important to uniform flow of the polymer through the extrusion system, particularly in the zone immediately below the spinnerette. There is little volume change when a molten fiber cools and solidifies, but when the solvent evaporates from a polymer solution, the volume decreases usually by at least 75%. In the shrinkage process, control over fiber cross-sectional shape is seriously diminished. For economical, environmental, and safety factors, recovery of the large volume of evaporated solvent is necessary. Drawing of a dry solution spun fiber is possible only if the fiber has a structure (low T_g) which will allow the chain dislocation required in drawing.

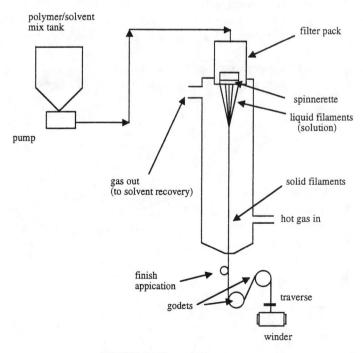

FIGURE 3.8 Dry spinning process.

3.2 Wet Spinning

Wet spinning is required when the polymer's solvents are not sufficiently volatile to permit rapid evaporation. To wet spin a polymer, the polymer is first dissolved by a solvent into solution form. The solution is extruded into a second solvent (the spin bath). The first solvent is soluble in the spin bath, but the polymer is not. When the solvent is solubilized out, or neutralized, the polymer structure precipitates or coagulates into solid form. A schematic of a wet spinning process is shown in Figure 3.9.

The type of solvent used in the spin bath is dependent upon the polarity of the polymer solvent. Water based solutions are preferred over organics, and are often used in the spinning bath when the polymer solvent is miscible with water. Volatile solution solvents may also be removed by wet spinning methods even though they could instead be evaporated. Polymers having high melting points also tend to be difficult to solubilize. These polymers are likely to require highly polar solvents such as mineral acids, or alkalimetal basic solutions. The polar solvents are typically extracted from the polymer solution either by dilution or neutralization in water. As with dry spinning, avoiding environmental problems in wet spinning is critical. The used spin-bath liquid must be recovered or at least treated before release to the environment.

Similar to the fiber formation in dry spinning, the precipitated solid occupies only a small fraction of the original solution volume. The outside layer of the fiber liquid precipitates first, and as the interior volume decreases with precipitation, a wrinkled or crenelated surface may be caused. The complications of the involved chemical reactions and shrinking filaments present both problems and opportunities. The process is much more difficult to control than melt spinning; however, the complicated structures which are possible offer great opportunities for the variation of fiber properties. The makers of viscose rayon have exploited these opportunities to produce High Wet Modulus (HWM) variants as well as fibers with an inherent crimp.

Drawing of fiber to obtain molecular orientation and strength is also desirable for wet spun fibers. Unfortunately, many of the polymers which require wet spinning procedures, also lack the mobility in the solid state which would allow normal drawing to occur. For these materials, any orientation which is obtained, must occur during the extrusion and coagulation processes.

Most wet spinning procedures have the spinnerette submerged below the surface of the coagulating bath. The rapid coagulation may restrict the amount of stretch possible in this type of spinning process. Certain products are best formed by placing the spinnerette some distance above the coagulating bath and attenuating or stretching the fiber solution as it falls through the air before the coagulation process begins. Such a process has been named dry-jet wet spinning.

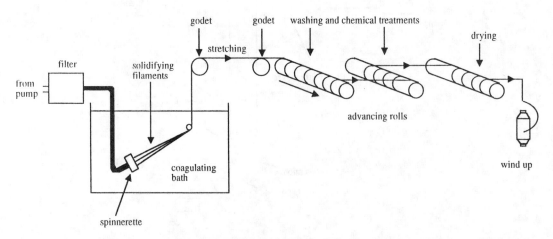

FIGURE 3.9 Wet spinning process.

4. OTHER SPINNING PROCESSES

The use of a liquid to help dissolve a material does not always produce what is normally thought of as a solution. Some polymer-solvent systems require high pressure and high temperature before dissolution will occur. Such processes might be considered as an intermediate between melt and solution spinning processes. Many other processes form solutions with unusual properties. Certain polymers can be induced to form liquid-crystalline solutions. These have been shown to aid in obtaining a high degree of fiber-parallel molecular orientation, resulting in very high strength structures. Solutions of ultra-high molecular weight polymers tend to form viscous gels rather than liquid solutions. Extruding from the gel state seems to contribute to both molecular orientation and crystallinity within a fiber structure. Ultra-high molecular weight polyethylene, marketed as Dyneema® or Spectra® is an example of a gel-extruded fiber.

Molten rock can be fiberized in a high temperature version of a cotton candy machine to produce a rock fiber suitable for a variety of high temperature insulation materials.

Solutions or vapors of some materials tend to spontaneously condense or precipitate into fibrous form. The formation of asbestos is an example of such a process occurring in nature. Indeed, almost any solid material which can be melted or dissolved can be put into a fiber form, and subsequently solidified.

Many materials consisting of small particles, which have a high melting point and are insoluble in common solvents, can be formed into fibers by extruding a filled carrier material (grease or polymer) and subsequently burning out the grease and sintering the particles together into a fiber shape.

A chemical process similar to the natural process of producing petrified materials may also be used to produce fibers. If a normal fiber is chemically reacted to a high extent with another material, the fiber may take on the characteristics of the reacting material.

These specialized approaches and others have been used to produce whiskers or longer fibers out of materials which cannot be processed in the normal melt or solution extrusion processes. Materials available from such processes include metal oxides, carbides, silicates, pure elements or metallic alloys. Such materials are usually not available from typical fiber companies, but rather from sources specializing in ceramics, metals and inorganic chemistry.

Manufacture of Continuous Filament Yarns

S. ADANUR

The manufacture of continuous filament yarn is a relatively simple matter of collecting the number of individual filaments necessary to produce the desired yarn size. As manufactured by the fiber producing companies, they are called "producer's yarns." They contain minimum twist, ranging from about zero to 2.5 turns per inch, which is just sufficient to maintain the yarn's integrity.

Most producer's yarn is delivered with a thin resinous finish or size which protects the filaments from damage due to abrasion and snagging. The finish, amounting usually to less than one percent by weight, may or may not be water soluble. Such finishes should not be confused with water soluble sizes such as starch, gelatin, or synthetic resins, which are applied to warp yarns at the mill to give additional protection during weaving. Sometimes a light lubricant is also applied to the yarn by the producer or the mill. This improves running quality by reducing static and friction, and reduces abrasion of the yarn and wear on the textile machinery guides, rollers, etc.

Because of filament uniformity and the complete absence of protruding fiber ends, continuous filament yarns are particularly smooth and lustrous. Such properties are advantageous in the manufacture of many fabrics, but a high degree of filament and yarn uniformity is necessary. Even minor irregularities will be observed as fabric defects due to changes in luster, dye pickup, irregular yarn twist or yarn spacing. Producers must always be on the alert to insure yarn uniformity, both within a package and among packages. Any differences in the amount that the yarn is drawn during manufacture will be manifested as differences in optical and physical properties, for example, dye absorption and residual rupture elongation. Excessive elongation at the beginning or end of the yarn package can result in fabrics with visually obvious defects. Staple yarns, being less uniform, can afford more irregularities, without the danger of the resulting fabrics being considered "defective."

1. THROWING AND TWISTING

There are so many different filament yarn constructions required by textile mills that it is quite impossible for the man-made fiber manufacturer to have all of them available, or make them on order. Instead the fiber producer sells several popular sizes, packaged usually on a standard spool. The textile mill must then arrange to have the producer's yarn converted into the desired yarn of proper weight, twist and ply, properly sized, lubricated, and packaged for subsequent mill operations. These procedures are collectively called "throwing." Throwing may be carried out by the mill which will ultimately weave or otherwise use the yarn, or by a commission "throwster." The term usually applies to the preparation of relatively lightweight yarns, in contrast to "twisting" which pertains to the preparation of heavier yarn constructions. More recently, the term also applies to a company that specializes in texturing yarns.

It is usually impractical to make a heavy yarn by twisting many units or ends of producer's yarn together in the same direction. Such a yarn would be soft, bulky, unstable, and might have low strength. Instead, plied yarns are constructed. Several turns of twist are inserted in one direction into the singles producer's yarn, and then several of these are twisted together, usually in the opposite direction, to make the plied yarn. Several plied yarns can then be twisted together to form a cord. By properly designing the twist relationships among singles, ply, and cord, large size stable and balanced yarns, cords and ropes can be manufactured. The twist relationships are discussed in Chapter 18.0. The two common methods for twisting are called "uptwisting" and "downtwisting." Downtwisters operate on the ring traveler basis, the name being derived from the fact that the input yarns are placed above the twisting unit. They feed down to the twister bobbin where, via the ring and traveler, they are twisted and wound up. The number of turns per inch inserted is a function of the traveler velocity and the feed roll speed. The uptwister utilizes a vertical bobbin on a spindle which rotates at high speed. The yarn is drawn up over the end of the bobbin at a constant rate, and is wound upon a take-up bobbin. The ratio of spindle speed to windup speed determines the number of turns per inch inserted.

2. MULTIPLE PLIED YARN, CORDS, CABLE, AND HAWSERS

Several plied yarns may be further plied and twisted to form increasingly heavier yarns called cords. Depending upon the direction of twist in the singles and in the plies, the resulting cords are called cables or hawsers. For cables the twist direction is alternated between singles, plies, and cord, for example:

- singles – Z twist
- plies – S twist
- cables – Z twist

The definitions of S and Z twist are given in Chapter 18.0. In the hawser, the primary and secondary twists are always in the same direction, while the cord twist is always opposite. Chapter 18.0 discusses the relationships among singles, ply, and cord twists.

3. TEXTURED OR BULK FILAMENT YARNS

Fabrics made from continuous filament yarns are normally smooth, dense, lustrous, and look and feel cold and "metallic," making them unacceptable for some textile purposes. One way of making filament yarns more spun-like in character is to insert into them a multiplicity of random permanent bends, loops, or crimps. Texturing produces yarns (and fabrics) which are less smooth and uniform than conventional filaments, but which are soft, lofty, dull, stretchy, warmer, and generally approach the look and feel of spun fabrics.

There are several methods of texturing yarns including air-jet, edge-crimping, false-twist, gear crimping, knit-de-knit and stuffer box methods.

Most bulk or textured yarns can be extended or stretched because of the incorporation of so much geometric elongation, recovery occurring when the extending force is removed. This geometric elongation is independent of any intrinsic elongation resulting from the stretching of the fiber substance per se. Except for isolated cases, textured yarns are used entirely for apparel rather than industrial purposes.

4. CORE, WRAPPED, AND COMBINATION SPUN-FILAMENT YARNS

In cases where a lofty or soft yarn of high breaking strength is required, a cored or wrapped yarn may be used. This usually consists of a high strength continuous filament core, about which is wrapped a soft low strength cover yarn. The cover may protect the core from abrasion, or give the yarn necessary bulk. Cotton wrapped glass or nylon core yarns are examples. Often wrapped yarns are made with a low cost core yarn, which does not show, and an expensive cover yarn. Another type is elastic yarn, where a cotton wrapper surrounds a rubber core. The helix of the cotton wrapper yarn permits the rubber to elongate under stress to a predetermined maximum. The rubber causes the combined structure to retract when the stress is removed. Plied yarns of spun and filament singles are also prepared for special applications.

5. HOT STRETCHED AND HEAT SET YARNS

Synthetic yarns are usually thermally sensitive, in that they may shrink or become more plastic as temperature is increased. The fiber producer normally heat treats and mechanically draws the fiber to convert it from a weak, heavy, amorphous filament into a strong, tough, textile filament. A heat treating and drawing procedure is selected which will permit the subsequent yarn or fabric manufacturer further to heat treat and tension his product to attain thermal stability. In the manufacture of a 3-ply nylon thread, for example, after plying, the thread is hot stretched at a temperature of about 300°F. This sets and balances the twist in the plied thread structure, makes it heat stable, and in addition produces greater strength per weight and lower rupture elongation—desirable properties in most sewing thread. By proper manipulation of temperature, tension, strain (elongation), and relaxation, the manufacturer can, with certain limits, incorporate desired strength, elongation, and thermal shrinkage properties in the hot stretched yarn.

3.3

Manufacture of Staple Yarns

Y. E. EL-MOGAHZY

1. INTRODUCTION

In the traditional textile process, natural or man-made staple fibers are converted into yarn through a series of processes which, in principle, may be divided into two stages: (1) spinning preparation, and (2) spinning. In the first stage, a tightly packed mass of staple fibers is gradually opened, cleaned, mixed, and finally attenuated to form a fiber strand that is ready to be spun. In the second stage, the fiber strand is further attenuated to obtain the desirable yarn thickness, and some form of fiber-to-fiber cohesion is introduced to provide the necessary strength to the yarn.

The purpose of this section is mainly to review the objectives and working principles of the main operations involved in staple yarn manufacturing.

2. SPINNING PREPARATION

In order to convert the bulk of fibers represented by a fiber bale (about 50 billion fibers) into a long thin yarn (typically, 100 fibers/cross section), a number of preparatory processes are used. These processes are listed in Table 3.3. Some of these processes are mandatory in all yarn making operations and others may be needed depending on the type of fibers to be spun, yarn count, and the spinning system utilized.

2.1 Blending, Opening, and Cleaning

The initial process of spinning preparation involves three main operations:

(1) Blending: to assemble and combine together predetermined proportions of different fiber components so that the fiber strand exhibits a high level of uniformity
(2) Opening: to divide the tightly packed fiber assembly into small fiber tufts or clusters of very few fibers
(3) Cleaning: to remove trash particles and foreign matter such as leaves, seed fragments, etc.

Prior to the blending process, fiber bales should be selected carefully so that their average fiber characteristics are uniform on mix-to-mix basis. The purpose of fiber selection is twofold: (1) to ensure a consistent quality of the input fibers on mix-to-mix basis, and (2) to control the output yarn quality by utilizing fiber-to-yarn quality prediction models. Software programs are now available to perform this critical task.

2.2 Blending

Blending of fibers is done for technological and/or economical reasons. The main technological reason is to produce a yarn of specific characteristics. In some cases, two fiber types are blended with one of them placed on the yarn surface while the other is largely presented in the yarn core. The advantage of this arrangement is to produce a yarn of different surface and core characteristics. For example, when cotton fibers are on the surface while polyester fibers are in the core, a yarn of cotton-like surface hand and polyester-like strength can be produced. Economical objectives of blending include mixing of fiber components of different prices and different values of quality characteristics to produce

TABLE 3.3 *Spinning Preparatory Processes.*

Process	Function	Application
Blending, opening, and cleaning	Opening and cleaning of fibers Mixing and blending of different fibers	Mandatory
Carding	Removal of trash Additional opening Removal of neps Sliver formation	Mandatory for some fiber types
Drawing	Straightening and orientation of fibers Reduction of the thickness of fiber strand Blending by doubling	Mandatory
Combing	Further straightening and orientation of fibers Removal of trash and short fiber content Removal of neps	Essentially used for high quality fine ring spun yarns It may be used for medium to fine open-end spun yarns It may be used for air-jet spun yarns
Roving	Further reduction in fiber strand thickness	Used primarily for ring spinning

a yarn of specified price with respect to its expected quality.

The blending process should produce a uniform fiber flow throughout the spinning line. Perfect blending may be defined as a condition in which every fiber characteristic is found in the same proportion within every unit cross section of the yarn. Several problems occurring during manufacturing are results of improper blending of the fed fibers. Blending is, therefore, considered to be the most critical stage of processing in staple yarn manufacturing.

In practice, there are two types of blending: (1) blending of nominally similar or dissimilar components of the same type of fibers, and (2) blending of different types of fibers (e.g., cotton and polyester). The first type of blending is normally achieved at the initial stage of processing where bales of the same type of fiber (e.g., cotton) but of dissimilar fiber properties are arranged in a bale laydown and more or less equal proportions of fibers are taken off from each bale and transported into a blending reservoir. Blending of different types of fibers may be achieved in the initial stage of processing to provide intimate blending or later in the drawing process (draw blending) where slivers of the two types of fibers are drawn together.

The advantage of blending in the early stage of processing (at the opening line) is that the blend produced is more intimate and more homogenous. When dissimilar fibers are blended (e.g., cotton and polyester), the opening and cleaning equipment for each type of fiber should be different in type and/or settings. While natural fibers require a great deal of cleaning, man-made fibers require no cleaning. Since blending equipment involves opening and cleaning elements, settings and speeds of these elements must be selected properly to produce an intimate blend of the two fiber types while preserving the unique characteristics of each type.

The advantage of draw blending is better control of the proportions of different fiber types in the blend determined by the sliver weight and the number of slivers of each fiber type. However, the parallel alignment of the fiber strands of each fiber type throughout the drawing process produces a less intimate blend than that of the initial stage.

Some of the effects that should be taken into consideration during blending are as follows:

- Finer yarns appear less intimate or more streaky than coarser yarns with the same fiber content and blend ratio. This is due to the fact that in finer yarns, fewer number of fibers per cross section are used which exaggerate the non-uniformity or irregularity.

- For a given yarn count, coarse fibers lead to more streaky yarns than fine fibers, again, due to the fact that with fine fibers, more fibers per cross section can be utilized leading to a better doubling effect.
- At similar conditions and for the same types of fibers, an unbalanced blend (e.g., 70/30) will appear to be more streaky than a balanced blend (e.g., 50/50).
- Blending of equal proportions of fibers of different color does not result in an intermediate color but rather a color which is more biased to the darker component in the blend.
- Optimum or intimate blending occurs when the number of doubling after blending exceeds the number of fibers per yarn cross section.
- Fabrics made from blends of different fiber types are unlikely to be superior in any one property to fabrics produced from an individual fiber type, but fiber blending does enable fabrics to be produced with combinations of properties which would be unattainable by using only one fiber type.

2.3 Opening

The objective of opening is to divide the tightly packed fiber assembly into small fiber tufts. Opening is normally achieved using mechanical means in which conveyor lattice or rollers covered by metallic wires act against the fiber mass for separation into small tufts.

Opening equipment may range from mild to fine openers depending on the type of fibers used and the density of the input fiber mass (the tuft size). In general, as the number of fibers of a given mass increases, the number of opening points (the wire density) should increase. The effectiveness of an opening unit should be measured by the tuft size produced from this unit. In principle, the opening process should provide a gradual reduction in tuft size. Theoretically, the tuft size (g/tuft) is a function of the throughput,

the circumference speed of the opening roll, and the number of different working elements (teeth or pins) per unit area.

Details on effects of these factors vary depending on the type of machinery utilized. Nevertheless, each yarn manufacturer in conjunction with machinery makers should evaluate opening performance through developing relationships between tuft size and the factors mentioned above.

2.4 Cleaning

Cleaning is mainly required for natural fibers. Cleaning may be achieved using mechanical or chemical means depending on the type of natural fibers. When natural fibers are to be blended with synthetic fibers, their cleaning should be performed prior to the blending process with those fibers.

Cotton fibers brought to the mill may contain several types of foreign matter. These include: (1) sand, soil and dust which are easy to remove because of their density and shape, (2) seed coat fragments, bark, and grass fragments; these are often difficult to remove during textile processing. In general, large particles or those of high density can be effectively removed in the opening and cleaning process. On the other hand, small particles are extremely difficult to remove and they are often observed in the final spinning process. The International Textile Manufacturers Federation (ITMF) has defined cotton trash and dust in terms of the particle size as follows:

Trash	$> 500\ \mu m$ (2/100th inch)
Dust	$< 500\ \mu m$
Fine Dust	$< 50\ \mu m$ (2/1,000th inch)
Microdust	$< 15\ \mu m$ (6/10,000th inch)

Cotton trash particles are typically removed using mechanical means. The cleaning process begins in the ginning process where fibers are separated from the seeds. Normally, two passes of lint cleaning are used after ginning to remove trash. Cleaning is then continued in the textile process in the opening and cleaning line. The final stage of cleaning is the carding process.

Raw wool fibers also contain a great deal of impurities. These include:

(1) Natural impurities, produced by the sheep itself; these include wool grease, dried sweat and the excretion stains caused by dung and urine in the breech area

(2) Acquired impurities such as those picked up by the sheep from its habitat; these include minerals such as sand, dust, soil, and lime; vegetable matter such as burrs, and grass; and animal impurities such as insect parasites, and pests

Wool fibers are normally cleaned using both chemical and mechanical means. Wool grease may be removed by emulsion scouring or by organic solvents. Most of the mineral impurity is attached to the fibers by wool grease and so is released when the grease is removed. Mechanical cleaning is used to remove vegetable impurities. Typically, they are removed during carding or combing (in the worsted system). Vegetable impurities may also be removed by carbonizing. This process is commonly used in woolen processing, and occasionally in worsted processing.

2.5 Carding

The objective of the carding process is to further open and clean the fibers delivered from the opening and cleaning line. Fibers undergo a series of dividing and redividing actions using a large number of pins in an attempt to separate each fiber from its neighbors and form a fluffy but coherent mass (fiber web). The idea is to facilitate further fiber manipulation and ease removal of impurities. The critical importance of carding lies in the fact that it may represent the last process of cleaning in the spinning line (additional cleaning is achieved if combing is used). After this process, unremoved trash or foreign matter is likely to adhere to the fibers down the chain of processing.

The carding process removes a great deal of trash, short fibers, and neps. Accordingly, it is evaluated using three criteria: (1) the extent of fiber damage, (2) the cleaning efficiency, and (3) the amount of neps in output fibers.

The output fiber strand from the carding process is called the card sliver. Quality-related parameters associated with card slivers are the sliver weight (grains per yard) and the sliver irregularity. The sliver weight is determined by the amount of mechanical draft (the ratio between the delivery speed and the feeding speed). Variation in sliver weight may result from improper machine settings, improper autolevelling (fiber control), or inherent variability in fiber mass. Irregularity in card sliver can be measured by using on-line or off-line evenness monitoring equipment.

2.6 Drawing

The main objective of the drawing process is to improve fiber orientation by aligning the fibers in one direction. Drawing may also be used for blending slivers of cottons and other synthetic fiber slivers (draw blending). The working principle of this process is based on sliding fibers over one another without elongating or stretching them. The drawing (or drafting as it is commonly called) is gradually achieved by feeding several card slivers (six or eight slivers) to a series of drafting rollers. The ratio between the delivery roller speed and the feed roller speed determines the amount of draft or the draft ratio, consequently, the size of the output sliver.

The purpose of feeding several slivers to the drawing process is to improve the uniformity of the output sliver (by applying a "doubling" effect) and to enhance fiber blending. Doubling is defined by the number of slivers fed together to the drafting system. For example, if six slivers are fed to the drafting system, the doubling value will be 6. If another drawing pass in which six of the slivers produced from the first pass is utilized, the total doubling value will be $6 \times 6 = 36$, and so on. The drawing process reduces the size of the fed slivers (collectively) down to the size of a single sliver by the amount of draft utilized.

Fiber quality has a great impact on the drawing process and the drawn sliver. Fiber characteristics related to drafting include fiber length, fiber fineness, and fiber cohesion. Brief discussions on the effects of these characteristics are given below.

The draft settings or the distance between different pairs of drafting rollers is determined by the staple length. The rule of thumb is that these settings should not be too wide (longer than the staple length) to avoid floating fibers or fibers flowing at undetermined speeds. The presence of floating fibers is one of the primary causes of sliver irregularity. On the other hand, draft settings should not be too tight (shorter than the staple length) to avoid stretching of fibers and fiber breakage.

The effect of fiber fineness on the quality of drawn sliver lies in the fact that high variability in fiber fineness is likely to result in a high weight variation of the sliver. High variation in fiber fineness is often attributed to improper mixing of fibers. When a mix of very fine and very coarse fibers is presented to the drafting process, potential problems resulting from uncontrolled clinging forces would be expected.

Since drafting is based on sliding fibers over one another, it is desirable that the fiber strand has low resistance to the drafting force. Very fine or very long fibers are likely to increase fiber cohesion; gradual drafting usually assists in overcoming the high cohesion resulting from the presence of fibers of such extreme characteristics. Different fibers may also have different surface characteristics which result in variation in fiber cohesion. This problem is often encountered in cotton/synthetic blending.

2.7 Combing

The combing process is used to produce a smoother, finer, stronger, and more uniform yarn than otherwise would be possible. Combing is, therefore, confined to high grade, long staple natural fibers. Basic objectives of combing are:

- reduction in short fiber content
- removal of trash and neps
- production of more straight, parallel, and uniform sliver

In general, combing is used for high quality fine ring spun yarns. The high quality of combed yarn is the primary justification for the addition of this process to the spinning line and its associated cost. Combing may be required for cotton when it is draw blended with synthetic fibers. This is particularly true for air-jet spinning. In recent years, combed yarns have been produced on open-end (rotor) spinning. This new development has been a result of the continuous strive to produce fine yarns on open-end spinning. Both trash content and short fiber content represent obstacles against the production of fine yarns on rotor spinning. Removal of these problems by the combing process was found to improve both the spinning limit and the quality of rotor yarns.

The combing process is usually preceded by a lap forming process in which the drawn sliver is converted into a large roll of fiber lap which is a flat sheet of fibers. A series of laps is then fed to the comber in an intermittent fashion. The laps are advanced and then gripped so that fringe of fibers is presented to the combing cylinder (toothed half roll) which combs the fibers and removes short fibers. The output product of the combing process is called combed sliver.

2.8 Roving

The roving process prepares the drawn sliver for the ring spinning process by converting it into an intermediate strand called roving. It involves drafting down to about one cotton count through a set of drafting rolls. The produced strand is then slightly twisted to permit winding it on the roving bobbin.

3. SPINNING

The spinning process is the final stage of staple yarn manufacturing. It is the final opportunity for a fiber to interact with the machine. In the spinning process, the goodness of fiber preparation through the different preceding processes can easily be evaluated. A failure in spinning is often a result of a default in the preparatory process.

There are different spinning techniques available in today's technology. Each technique is unique in its principle and in its requirements of fiber quality. New spinning techniques involve high drafting where the input fiber strand (the drawn sliver) is separated, partially or fully, into approximately single fibers, flown in an air stream, and reconsolidated to form the yarn. These techniques are, therefore, highly sensitive to fiber quality, and to the presence of fine trash and dust.

3.1 The Principles of Spinning

In any spinning method, three main mechanisms should be used to convert fibers into a yarn (Figure 3.10). These mechanisms are

- drafting mechanism
- fiber coherence mechanism
- winding mechanism

The drafting mechanism works on the same principle as the drawing process: sliding fibers over one another without elongating or stretching them. The objective of drafting in the spinning process is to reduce the size of the fiber strand down to the desirable size of yarn.

The fiber coherence mechanism produces cohesive forces to hold the fibers together in the yarn by introducing interfiber three-dimensional crosslinking. In any spinning system, the coherence mechanism is responsible for providing yarn strength.

The winding mechanism involves winding of

yarn on a package (bobbin or a cone), and building the yarn along the length of the package. Proper yarn winding is extremely important particularly in weaving preparation. Yarn tension should be uniform and the appearance of the package is a critical factor.

3.2 Classification of Spinning Techniques

There are four major spinning technologies:

- ring spinning
- rotor spinning (open-end spinning)
- air-jet spinning
- friction spinning

These spinning techniques may be classified in many different ways. In principle, spinning techniques may be divided into two main categories (Figure 3.11):

(1) Continuous spinning

(2) Interrupted spinning

In continuous spinning, the fiber flow is continuous from the feeding point to the delivery

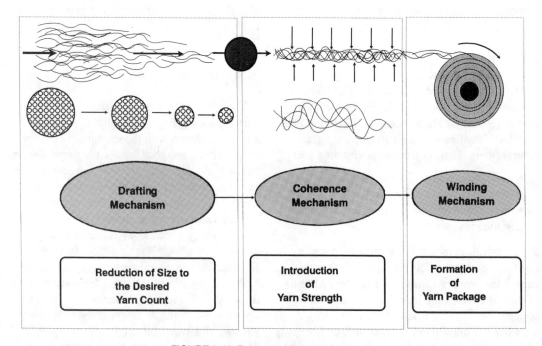

FIGURE 3.10 Basic mechanisms of spinning.

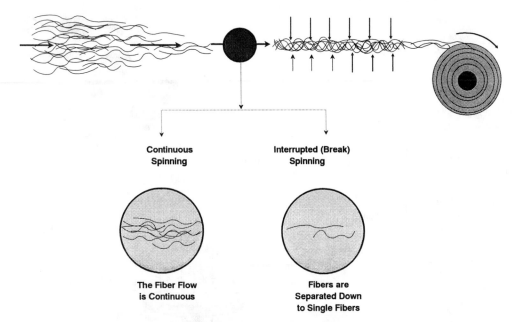

Continuous Spinning

Interrupted (Break) Spinning

The Fiber Flow is Continuous

Fibers are Separated Down to Single Fibers

FIGURE 3.11 Classification of spinning techniques.

point; and fibers are under a full mechanical control. The conventional ring spinning is a continuous spinning process. In interrupted spinning, fibers undergo a complete or partial separation (rotor, and air-jet spinning, respectively) before they are reconsolidated into a yarn. The primary reason for the interruption in fiber flow is to allow separation of the fiber coherence mechanism and the winding mechanism. This separation results in producing larger yarn packages, increasing production, and introducing strength at minimum energy consumption. Table 3.4 shows comparison of the spinning methods. Figure 4.21, Chapter 4.0, shows structures of yarns, with equal count, produced with different yarn spinning methods.

3.3 Ring Spinning

Ring spinning is characterized by the continuity in the fiber strand from roving (the input strand) to yarn. The input fiber strand is drafted using roller-drafting to reduce its size down to the desirable yarn count. The fibers being delivered at the nip of the front roller form a triangle called the spinning triangle (Figure 3.12). The bottom end of this triangle represents the twisting point or the point at which twisting begins. In the spinning triangle, different fibers have different tensions depending on their position in the triangle. Fibers in the center of the triangle are usually slack and those in the outer layers are under maximum tension. When fibers

TABLE 3.4 Comparison of Different Spinning Techniques.

Type of Spinning	Input Strand	Type of Rotating Element	Speed of Rotating Element	Yarn Delivery Speed (m/min)	Yarn Count (Ne)	Yarn Structure
Ring	Roving	Bobbin	Up to 25,000 rpm	up to 40	3 – 200	True twist
Rotor	Sliver	Rotor	Up to 130,000 rpm	up to 179	5 – 30	True/partial/belts
Air-jet	Sliver	Air	3 million rpm*	up to 250	15 – 80	Parallel core/wrapping fibers
Friction	Sliver	Yarn	250,000 rpm	up to 300	10 – 30	True twist/loops

*The yarn rotational rate in the nozzle is 2 – 300,000 rpm.

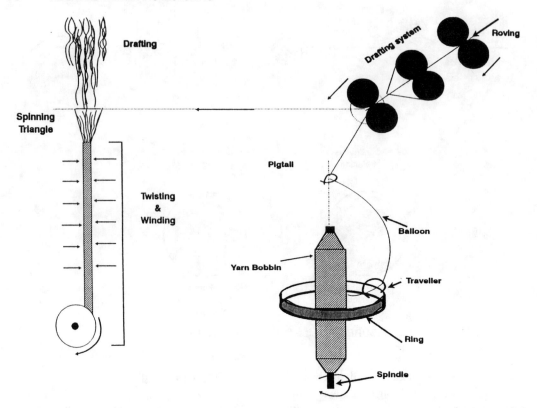

FIGURE 3.12 Schematic of ring spinning.

are released from the nip of the front roller, those exhibiting high tension tend to move toward the center displacing the initially central fibers to the outer layers. This phenomenon is called fiber migration. The effect of fiber migration is to enhance fiber crosslinking, and consequently yarn strength.

The coherence mechanism in ring spinning is twisting. Twist is inserted to the fibers by a traveler rotating around a ring. Winding is made simultaneously with twisting. For each full turn of twist, the yarn package must make a full turn around its axis. Thus, yarn strength is a result of twisting, and fiber migration.

Strength-Twist Relationship

The effect of twist on yarn strength is illustrated in Figure 3.13. Initially, as the twist level (number of turns per unit length) increases, yarn strength will also increase. The reason for this is the increase in the cohesion of fibers as the twist is increased. This effect holds only up to a certain point beyond which further increase in twist causes the yarn to become weaker. This results from a decrease in the effective contributions to the axial loading of the yarn due to fiber obliquity. Thus, there is a point of twist at which yarn strength is at its maximum value. This point is known as the optimum twist.

The amount of twist is determined by the ratio between the traveler rotational speed (or the rotational speed of the spindle carrying the bobbin) and the delivery speed. Since spindle speed is normally constant, an increase in twist would mean a reduction in delivery speed, consequently, a reduction in production. This point emphasizes the importance of selecting fiber characteristics that permit lower optimum twist (twist at maximum strength). Generally, the use of longer, finer and stronger fibers permits the use of lower optimum twist (Figure 3.14).

The main limitation in ring spinning is the low

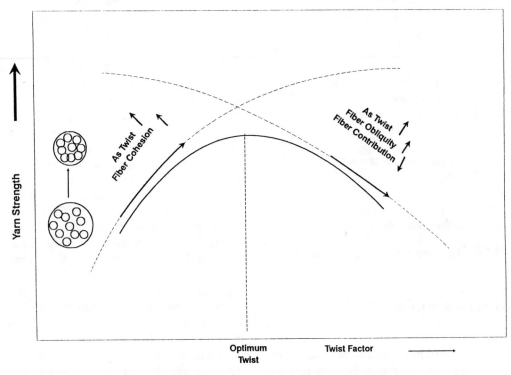

FIGURE 3.13 Yarn strength-twist relationship.

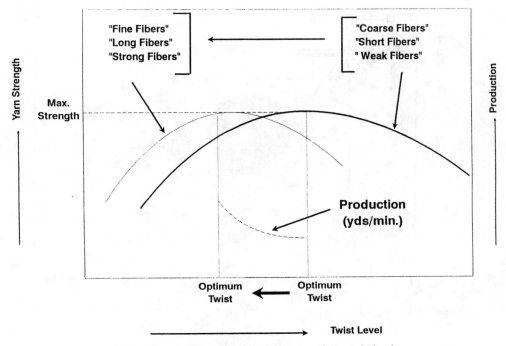

FIGURE 3.14 Effect of fiber properties on optimum twist level.

production rate resulting from having to rotate the yarn package to simultaneously introduce twist and wind the yarn. The yarn bobbin is of a small size because of the geometrical limitation of ring spinning (the ring size). One solution to these limitations is to separate the coherence mechanism from the winding mechanism. This approach represents the underlying concept of the new spinning techniques.

3.4 Rotor Spinning (Open-End, OE, Spinning)

In rotor spinning, the drafting mechanism consists of three different zones:

(1) Mechanical drafting using an opening roll
(2) Air drafting using an air stream and transporting duct
(3) Condensation mechanism

The use of a fiber sliver requires a large amount of draft to reduce its size down to that of the yarn size. A sliver may have more than 20,000 fibers in its cross section. Before this number of fibers is reduced to a value of about 100 fibers per cross section in the yarn, it must first be reduced to as low as two fibers (draft ratio of 10,000:1) to initiate the interruption or the discontinuity in fiber flux. This immense reduction requires high speed mechanical drafting supported by air drafting (Figure 3.15). The mechanical drafting is achieved using a toothed opening roll. This opening roll also allows removal of trash in the input sliver. Upon drafting the strand down to about two fibers, condensation of fibers back to the desired number of fibers per yarn cross section is achieved on the inside wall of the rotating rotor.

The coherence mechanism in rotor spinning is achieved by twist insertion resulting from the high speed rotation of the rotor. The amount of twist is determined by the ratio between the rotor speed and the yarn delivery speed. The lack of tension differential similar to that in ring spinning yields no significant fiber migration. In addition, the lack of full mechanical control, resulting from the creation of an open end, leads

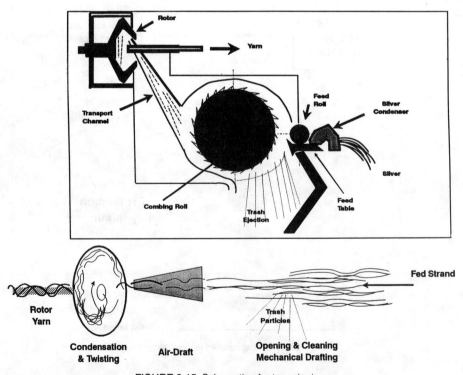

FIGURE 3.15 Schematic of rotor spinning.

to less fiber orientation and a yarn structure consisting of a core that is fully twisted (similar to ring spun yarns), and an outer layer that is partially twisted (Figure 4.21, Chapter 4.0).

Another disturbing structural feature of rotor yarns is the presence of fiber belly bands (bridging or belt fibers). These fibers result from the interfacing between the processes of laying fibers on the rotor collecting surface and the peeling of the yarn from the collecting surface. This interface occurs once per rotor revolution. Fiber belly bands are laid at these times. They do not contribute significantly to the coherence of rotor yarns, and they result in clear departure of rotor yarn structure from that of the conventional ring yarn structure.

The condensation action, on the other hand, resembles a doubling effect of several fibers which enhances the uniformity of rotor spun yarns. This is one of the reasons that rotor spun yarn exhibits better uniformity than comparable ring spun yarns.

The winding mechanism in rotor-spinning is separated from the other two mechanisms. The yarn package is no longer the rotating twisting element as in ring spinning. The rotating twisting element is the rotor which is of much smaller size than the yarn bobbin. This allows much higher speed in rotor spinning than in ring spinning (Table 3.4), and consequently higher production. Furthermore, there is no limitation on the size of yarn package; this eliminates further winding of yarn in the weaving preparation. Figure 3.16 shows a rotor spinning machine.

3.5 Air-Jet Spinning

Air-jet spinning utilizes high roller-drafting to reduce the size of the input sliver down to the desired yarn size (Figure 3.17). The coherence mechanism in air-jet spinning is quite unique. It is achieved by blowing out compressed air through air nozzle holes of about 0.4 mm diameter to form an air vortex. The air revolves at high speed (more than 3 million rpm).

To simplify the principle of the coherence mechanism, let us examine the case of only one air nozzle (nozzle b). When air rotates in the clockwise direction, it twists the fibers fed to the nozzle. When the yarn leaves the nozzle, un-

twisting takes place. Thus, with one air nozzle, a case of false twisting would be achieved.

A second nozzle, a, is positioned between the nip of the front roller and nozzle b, with air rotating in a counterclockwise direction. Now, we have two nozzles applying air rotation in two opposite directions. Nozzle b normally has a higher rotational speed than nozzle a to avoid complete false twist. Thus, the opposite rotation of air in nozzle a assists in detaching some fibers from the input strand. In other words, fibers are partially separated at this point.

The coherence mechanism involves two main actions: (1) false twist action, and (2) end opening action. The idea is to transmit the twist inserted by air rotation in nozzle b to the fibers at the nip of the front roller, and to detach some fibers from the twisted strand by the rotation of air in nozzle a in the opposite direction. This end opening action takes place at the moment nozzle b twist is imparted. As the strand passes through nozzle b, it will consist of detached fibers (outer layer) and truly twisted fibers (the core). When the fiber strand exits nozzle b, the twisted core will immediately tend to untwist and the detached fibers will wrap around the core fibers. The net coherence in air-jet spun yarn is introduced by the wrapping fibers resulting from the end opening action. Accordingly, yarn strength is mainly a result of a wrapping effect rather than a twist effect.

Since only a portion of fibers is detached from the fiber strand, interruption in fiber flow in air-jet spinning is of a different nature than that of rotor spinning. Both systems are similar in the principle of winding mechanism; i.e., complete separation from the coherence mechanism which allows large yarn package. The rotating part in air-jet spinning is air. This allows much higher rotational speed than that of the rotor. Consequently, the production rate of air-jet spinning is higher than that of rotor spinning.

3.6 Friction Spinning

As in the case of rotor spinning, friction spinning utilizes an opening roll for drafting and separating fibers of the input fiber strand (sliver). After passing the opening roll, fibers are carried to a collecting point between two friction drums

FIGURE 3.16 Rotor (open-end) spinning machine (courtesy of Wellington Sears Company).

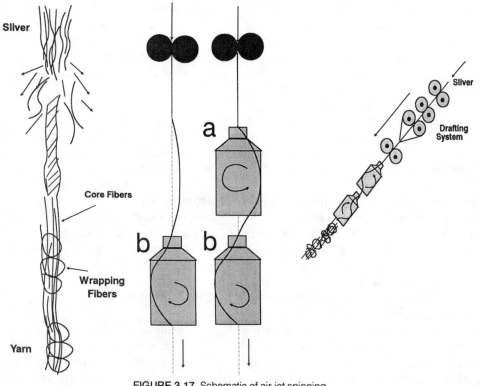

FIGURE 3.17 Schematic of air-jet spinning.

TABLE 3.5 *Ranking of Fiber Properties According to Their Contribution to the Yarn Quality for Different Spinning Systems.*

Ring Spinning	Rotor Spinning	Air-Jet Spinning	Friction Spinning
Fiber strength	Fiber strength	Fiber length	Fiber strength
Fiber length	Fiber fineness	Fiber fineness	Fiber fineness
Fiber fineness	Fiber length	Fiber strength	Fiber stiffness
	Small foreign particles	Small foreign particles	Fiber friction
		Fiber friction	

by means of air current (Figure 3.18). At this point, the yarn is formed by twisting imparted by the relative rotation between the surface of the drums and the yarn end. Thus, the coherence mechanism is friction between fibers and the drum. The rotating element in this case is the yarn itself. At each turn of the yarn end, the yarn is given one turn of twist. The amount of twist in friction spinning is determined by the ratio between the yarn diameter and the drum diameter. Consequently, a very high degree of twist may be imparted.

Since friction spinning requires interruption of

fiber flow by collecting fibers on friction drums prior to making the yarn, a loss of fiber orientation may be expected. Another factor of disorientation results from the landing of fibers at high speed on slow moving friction drums. This mode of landing immediately prior to twisting makes it difficult to adjust for fiber disorientation. The impact nature of this landing compresses the fibers against the friction drum which results in a looped yarn structure. Since these loops do not effectively contribute to the strength of the yarn, friction-spun yarns have less strength than comparable rotor spun yarns.

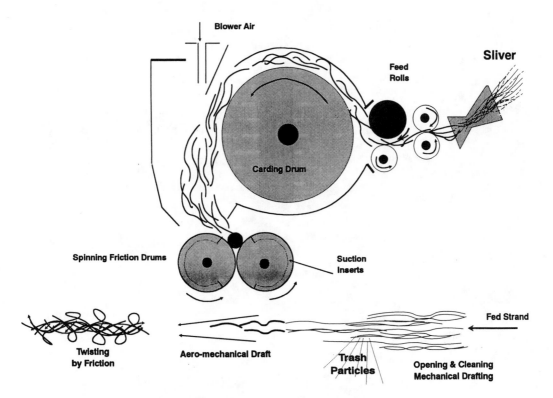

FIGURE 3.18 Schematic of friction spinning.

3.7 Required Fiber Characteristics for Different Spinning Systems

There have been many attempts to establish a ranking of fiber characteristics according to their contributions to the quality of yarns produced using a particular spinning system. These attempts have mainly been based on long experience and understanding of the principle involved in converting fibers into yarns in a particular spinning system. Table 3.5 provides a suggested general ranking of fiber properties according to their contributions in different spinning systems.

References and Review Questions

1. REFERENCES

1.1 General References

O'Sullivan, D., "DSM High-Performance Fiber Attracts Growing Interest," *Chemical & Engineering News,* March 25, 1991.

Seymour, R. B. and Carraher, C. E., Jr., *Polymer Chemistry, an Introduction, 3rd Ed.,* Marcel Dekker, Inc., New York, 1992.

Sperling, L. H., *Introduction to Physical Polymer Science,* John Wiley & Sons, New York, 1992.

Bischofberger, J., "Rotor Yarn-Combed System," *International Textile Bulletin,* 2, 1990.

Ferdinand, L., "Modern, Sophisticated Preparation Plants for Ring Spinning," *Textil Praxis,* 2, 1992.

Hunter, L., "The Production and Properties of Staple-Fiber Yarns Made by Recently Developed Techniques," *Textile Progress,* The Textile Institute, Vol. 10, 1/2, 1978.

El-Mogahzy, Y., "Optimization of Cotton Value Using EFS," *3rd Research Forum Proceedings,* Cotton Incorporated, 1990.

2. REVIEW QUESTIONS

1. What is the difference between fiber spinning and yarn spinning? Explain.
2. What are the major technologies for fiber spinning? Explain.
3. Discuss the comparative advantages and disadvantages of fiber spinning techniques.
4. What are the main factors to consider when selecting a fiber spinning system for a particular polymer?
5. Explain the following terms:
   - screw memory
   - spin finish
   - glass transition temperature
   - draw texturing
   - dry-jet wet spinning
6. Discuss some possible reasons why textured yarns are not suitable for industrial fabrics.
7. What are the main differences between a plied yarn, cord and a cable?
8. What are the principal procedures in making staple fiber yarns?
9. Compare the advantages and disadvantages of intimate and draw blending.
10. What are the factors affecting the cleaning efficiency?
11. Discuss the importance of fiber damage in the opening process. Suggest ways to measure the extent of this damage.
12. Discuss the reasons for the superiority of combed yarns over that of carded yarns.
13. What are the basic mechanisms of spinning? In view of these mechanisms, discuss the differences between continuous and interrupted spinning techniques.
14. Explain the strength-twist relationship. What are the effects of fiber fineness, fiber strength, fiber length, and fiber friction on optimum twist?
15. Explain the principles of the coherence mechanisms in ring and rotor spinning.
16. In view of the principle of each spinning technique, explain the reasons for the ranking of fiber properties listed in Table 3.5.

FABRIC MANUFACTURING

4.1

Classification of Fabrics

R. P. WALKER
S. ADANUR

A fabric may be defined as a planar assembly of fibers, yarns or combinations of these. There are many different methods of fabric manufacturing, each capable of producing a great variety of structures dependent upon the raw materials used and the setup of control elements within the processes involved. The particular fabric selected for a given application depends on the performance requirements imposed by the end use and/or the desired aesthetic characteristics of the end user with consideration for cost and price. Fabrics are used for many applications such as apparel, home furnishings and industrial. The most commonly used fabric forming methods are interlacing, interlooping, bonding and tufting.

1. INTERLACING (WEAVING AND BRAIDING)

Weaving—interlacing of a lengthwise yarn system (warp) and a widthwise yarn system (filling) at 90 degrees to one another with fabric flowing from the machine in the warp direction [Figure 4.1(a)].

Braiding—interlacing of two yarn systems such that the paths of the yarns are diagonal to the fabric delivery direction forming either a flat or tubular structure [Figure 4.1(b)].

2. INTERLOOPING (WEFT AND WARP KNITTING)

Knitting—interlooping of one yarn system into vertical columns and horizontal rows of loops called wales and courses respectively with fabric coming out of the machine in the wales direction [Figure 4.1(c) and Figure 4.1(d)].

3. TUFTING

"Sewing" a surface yarn system of loops through a primary backing fabric into vertical columns (rows) and horizontal lines (stitches) forming cut and/or uncut loops (piles) with the fabric coming out of the machine in the rows direction as shown in Figure 4.1(e). Fabric must be back-coated in a later process to secure tufted loops.

4. BONDING (NONWOVENS)

Nonwovens—using either textile, paper, extrusion or some combination of these technologies to form and bond polymers, fibers, filaments, yarns or combination sheet into a flexible, porous structure [Figure 4.1(f)].

(c) weft knit

(f) nonwoven

(b) braided

(e) tufted

(a) woven

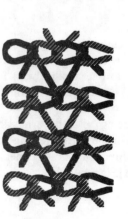

(d) warp knit

FIGURE 4.1 Types of fabrics.

Yarn Preparation

R. P. WALKER
S. ADANUR

Before forming fabrics that involve the use of yarns, the yarns must undergo various stages of preparation. Yarns are produced on various types of packages (bobbins, pirns, cones, tubes, etc.) and in many cases the yarn and/or the package is not suitable for further processing. Changing the yarn itself or altering the package is often required before fabric can be produced. The specific preparatory sequence required depends on the type of fabric being produced. Yarn preparation processes for selected fabric types are as follows:

- weaving: winding, warping, slashing, drawing-in
- knitting: winding, waxing, steaming, warping
- tufting: plying, heat setting, winding or beaming

The processes used to prepare yarns for fabric forming vary not only for fabric types, but also for yarn types. The more common yarn preparation processes are described below:

- winding – transferring a yarn from one type package to another to combine smaller packages and clear defects such as thick and thin places from the yarn (also called spooling); used in one way or another for all fabric forming types (Figure 4.2)
- quilling – transferring a yarn from a larger package to small quills which are filling bobbins in a shuttle; used in shuttle weaving
- warping – transferring many yarns from a creel of single-end packages forming a parallel sheet of yarns wound onto a beam or a section beam; used in weaving and warp knitting (Figure 4.3)
- section warping – winding parallel sections of a yarn sheet side by side into a wider full sheet; used in weaving
- ball warping – combining yarns from a creel into a condensed rope which is wound as a single strand forming a log-type package called a ball, several of which are then used in a rope dyeing process; used in weaving typically with the indigo dyeing process in the production of denim fabric
- heat setting – subjecting many single yarns to controlled heat conditions to achieve uniform dyeability; used in weaving and knitting of thermoplastic yarns and particularly for tufted fabrics
- draw warping – combining the drawing of filament yarns with heat setting and warping processes to achieve uniform stretching and heating for improved dye uniformity, end to end; used for weaving and knitting of thermoplastic yarns
- plying – twisting together two or more single yarns; used for any system that requires plied yarns, most often for industrial woven fabrics and tufted fabrics
- waxing – application of lubricant to single yarns during winding to reduce frictional forces during further processing; used for weft knitting yarns (Figure 4.2)
- conditioning – subjecting yarn to moisture and heat to offset the liveliness created by twist during yarn forming; used for weft knitting yarns

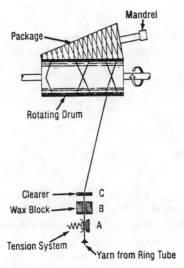

FIGURE 4.2 Winding (courtesy of Hoechst Celanese).

- beaming—combining yarns either from single-end packages or section beams (multi-end sheets) forming a final sheet of yarns for fabric forming; used for weaving, warp knitting and tufting yarns
- slashing—application of sizing chemicals to yarns to improve yarn properties and combining of section beams, sheet to sheet, forming a final beam for fabric forming; used for warp yarns in weaving

- drawing-in—entering of yarns from a new warp into the weaving elements of a weaving machine (drop wires, heddles and reed) when starting up a new fabric style (Figure 4.4). Tying in the new warp ends to the depleted warp is done when a new pattern is not required.

All the above processes may be summarized as necessary steps in fabric forming to render yarns more suitable for further processing. The following fundamentals must be considered before fabric forming is attempted:

- Does the yarn length need to be changed (longer or shorter)?
- Does the fabric process require a single-end or multi-end package (cones, cheeses, tubes, spools, section beams, loom beams, etc.)?
- Are the yarns to be dyed before fabric forming (single-end package, multi-end rope or beam dyeing)?
- Must dye uniformity be enhanced before fabric forming (filament or thermoplastic fiber yarns)?
- Should yarn defects be removed (splicing or knotting)?
- What yarn properties need to be altered (kinkiness, hairiness, lubricity, strength,

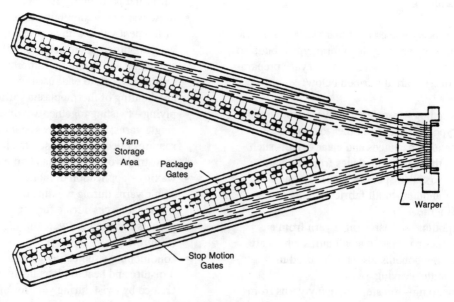

FIGURE 4.3 Warping.

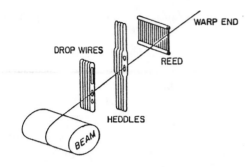

FIGURE 4.4 Schematic diagram for drawing-in.

elongation, abrasion resistance,
flexibility)?

Regardless of the processes employed, a second concept of quality has to be embraced. Not only must the quality of the yarn itself be maintained and enhanced, but also the quality of yarn packages is extremely important to further processing. Many, if not most, of the quality problems encountered during fabric forming are directly related to mistakes made during yarn preparation.

Since most fabric in use is woven and the weaving process is particularly abusive to lengthwise yarns in a woven fabric, the technology surrounding the preparation of yarn for weaving must be given special attention.

1. YARN PREPARATION FOR WEAVING

As stated earlier, a woven fabric is composed of two yarn systems, the filling or widthwise yarns and the warp or lengthwise yarns, which are interlaced at near 90 degrees to one another in the weaving process. The filling yarns are not subjected to the same type stresses as are the warp yarns and thus are easily prepared for the weaving process. Most often the filling yarns are not prepared at all, but rather are taken straight off the spinning process and transported to the weaving process. This is the case with open-end (rotor) and air-jet spinning systems which provide a large single-end package suitable for inserting during weaving. With ring spun filling yarns small bobbins are combined by splicing or knotting several shorter lengths of yarn onto a

single package to provide an opportunity to clear the yarn of defects (thick and thin places) and to provide a longer lasting supply package for weaving. In the case of shuttle weaving the filling yarns are either spun directly onto quills or quills are produced from larger wound packages of cleared filling yarn.

The preparation of warp yarn is another matter entirely. A very large body of knowledge has evolved concerning the preparation of warp yarns for weaving involving the interaction among machines, yarn, chemistry and practical experience. An entire segment of the textile industry has developed around this one aspect of woven fabric production creating an intensely competitive atmosphere for the many suppliers of weaving support equipment, instrumentation, chemicals and services. Warp yarn preparation, called slashing, is explained in Section 4.3.

Successful warp preparation depends on a fundamental understanding of the prior influences of yarn forming and a sound comprehension of the stresses of weaving. Experience offers the following observations for consideration:

(1) The increasing use of newer spinning technologies creates a situation where the old concept of yarn clearing and package quality now has become a part of the spinning process rather than part of a separate winding process which provides an opportunity for neglect of important quality standards. Properly formed packages of defect-free spun yarn are an even more critical factor. Package considerations include condition of the package core, the proper provision of yarn tails, properly formed splices or knots, elimination of internal defects (such as sloughs, tangles, wild yarn, scuffs and ribbon wind), elimination of external defects (such as over-end winding, cobwebs, abrasion scuffs, and poor package shape or build), proper density (hardness) and unwindability.

(2) Spun yarn quality characteristics which are most important for good weaving performance include short- and long-term weight uniformity, imperfections, weight variation, tensile properties and hairiness. It

should be noted that variation in a property is almost always more important than the average value of that property.

(3) Spun yarn properties positively affected by slashing are strength (increased though not critically needed), elongation (reduced but controlled), flexibility (reduced but reasonably maintained), lubricity (improved), hairiness (reduced), and most importantly, abrasion resistance (greatly improved).

(4) Modern weaving machines have placed increased demands on warp preparation due to faster weaving speeds and the use of insertion devices other than the shuttle.

(5) The cost to repair a yarn failure is much less if made to occur prior to the weaving process where a yarn failure during production also provides an increased opportunity for off-quality.

(6) Often, 80% or more of yarn failures in weaving will be caused by 20% or less of the yarns in a warp, called repeater ends.

(7) Running the slasher at creep speed which is sometimes necessary creates a very undesirable condition for proper sizing and should be minimized in every way possible.

(8) Stretch of warp yarns during slashing should be controlled accurately to maintain residual elongation in the yarn which is needed for good weaving.

(9) Process studies to determine causes for inefficiency should be conducted with strict cause analysis techniques by an experienced practitioner and not as part of a typical stop frequency study for job assignments. Such studies when combined lead to inaccurate data on the causes for stops.

(10) Weaving tensions should be maintained at minimum levels for best weaving performance.

(11) The slashing process deals with enhancing individual warp yarn properties not with improving the characteristics of the warp sheet. If done improperly, slashing can worsen yarn sheet characteristics.

(12) Yarn spacing at the slasher size box and on the drying cylinders is very important.

(13) Improper splicing and/or knotting can become critical to good weaving performance.

(14) A practical understanding of the importance of size penetration, size encapsulation, yarn hairiness, residual yarn elongation, and yarn abrasion resistance is essential to good slashing practice.

(15) Factors influencing yarn hairiness include hairiness created by the winding process, spinning tensions, location of the yarn on the spinning package, yarn balloon shape, yarn twist, spindle speed, the yarn size (count), % synthetics in a blend, end spacing at slashing, size add-on, slasher creep speed and bottom squeeze roll cover.

(16) Each spot in a warp yarn must undergo several thousand cycles of the various stresses applied by the weaving machine. Weaving stresses include dynamic extension/contraction, rotation (twist/untwist), and clinging of hairs. Additionally, there are metal-to-yarn and yarn-to-yarn flexing and metal-to-yarn and yarn-to-yarn abrasion stresses.

(17) Warp yarns closest to the selvages of the fabric undergo more stress due to widthwise contraction of the fabric toward the center causing linear angular displacement of these outermost yarns. Control of this contraction by the ''temples'' of the loom is another critical aspect of good weaving performance (Figure 4.5).

(18) Warp yarns controlled by lifting devices closest to the back of the weaving machine receive more stress since these yarns have to be raised and lowered a greater distance to form openings for filling yarn insertion (Figure 4.6).

(19) Spun warp yarns are sometimes stressed more when they are lowered than when they are raised during weaving because this helps to remove ''reed'' marks giving the fabric a better appearance (see Figure 4.7).

(20) Spun warps are also timed to interlace the filling yarn earlier increasing the force of beat-up and increasing the stress on warp yarns to enhance fabric appearance (Figure 4.8). These extra stresses generally are not

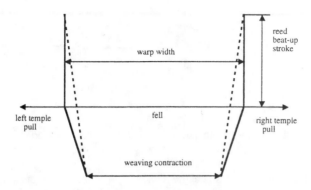

FIGURE 4.5 Temple function in weaving.

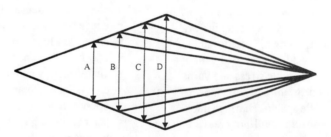

FIGURE 4.6 Shed geometry showing A < B < C < D (A: front harness, D: back).

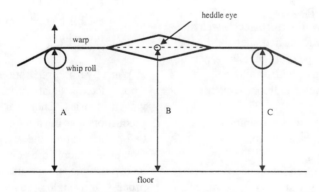

FIGURE 4.7 Raising whip roll makes bottom shed tighter (figure shows ''perfect'' shed line, A = B = C).

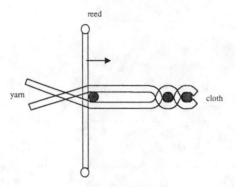

FIGURE 4.8 Shed timing, beat-up of interlaced pick (early shed timing).

applied to filament warp yarns since the uniformity of yarn dyeability would be detrimentally affected.

(21) The straightness of individual warp yarns and their freedom to act independently as they pass through a weaving machine are important to good weaving performance. Yarns that are crossed and tangled cannot proceed without excessive stress and yarns that are restricted or influenced by dropwire activity, heddle spacing, harness interference or reed spacing will not weave at top performance.

(22) Attention to details in preparation processes and in the weaving process should yield good results. An old rule is that if some "perfect" yarn packages can be produced and if some "good performance" weaving machines can be identified, then these good practices should be reproducible. A partial yarn preparation checklist follows:
- check yarn path conditions (rough surfaces, etc.)
- check package holder alignment
- check package spacing (balloon interference)
- check stop motion function
- check creel to head-end configuration
- check beam to head-end alignment
- check braking function (eliminate lost ends)
- check yarn accumulator function
- check comb alignment (especially selvages)
- check yarn tension controls
- check static eliminator function
- study speed level versus performance level
- check beam density (press roll)
- check defect scanners and/or counters
- check run-length measurement devices
- check leasing and taping procedures
- check doffing and handling procedures
- check housekeeping (especially fan operation)
- use monitoring system information properly
- check maintenance schedules and procedures
- conduct meaningful cause analysis studies
- establish good operator training procedures

Warp yarn preparation is considered a "bottleneck" situation in the production of woven fabric. There are many individual production units in yarn forming and in weaving, before and after the preparation processes. However, there are only a few machines involved in preparing warp yarns for the weaving process. If a single production unit is producing bad work in spinning or weaving, the consequence is relatively small. A similar situation in warp preparation can be disastrous. The key word is "consistency," having every warp yarn the same as all others. The "chemistry" of slashing is discussed in the following section.

4.3

Slashing

D. M. HALL

1. INTRODUCTION

The purposes of the slashing process in textiles are to significantly reduce the yarn hairiness that would interfere with the weaving process, to protect the yarn from various yarn-to-yarn and yarn-to-loom abrasion and to increase the strength of the yarns so that the yarn will be capable of negotiating the tortuous path it must take through the loom in order to be woven into finished cloth. These various tension and abrasion zones of the loom are summarized in Figure 4.9. It is important for today's high speed looms to have size products that can withstand the loom abrasion. Thus the quality of the yarn coming into the slashing operation must be quite good if an effective job of slashing is to be accomplished.

To accomplish these goals, a protective coating of a polymeric film forming agent is applied to the warp yarns prior to their being placed on the loom. Hence only about half of the yarns in the cloth (depending upon the ends/inch in each direction) will have the protective coating that must later be removed once their job of protecting the warp during weaving has been accomplished. Thus the filling yarns will be free of size, and special finishing considerations will not need to be addressed to these yarns in the cloth. An illustration of how the film former must lay down the hairiness in yarns while improving the strength and abrasion resistance of the yarns is given in Figure 4.10. If the sizing is improperly done, the long hair fibers protruding from one yarn will be glued to adjacent yarns. When the yarn sheets are separated back into individual fibers at the bust rods on the slasher, the size film will be damaged and the yarns may not negotiate the loom prior to causing a yarn break. Ideally,

the hairiness will be laid along the body of the yarn bundle where they do not interfere with the weaving process.

Figure 4.11 shows actual longitudinal and cross-sectional photomicrographs of properly sized yarns illustrating good fiber lay along with the proper level of size penetration and film coverage. For proper sizing, the size film must coat the yarn surface without excessive penetration into the body of the yarn bundle. Only enough penetration should occur to achieve anchoring of the size film to prevent removal during weaving. Proper sizing requires that the yarn hairiness be properly laid along the yarn bundle. It can be noted that a single hair fiber has the potential to become entangled in a neighboring yarn.

2. POLYMERS USED IN SIZING

The film forming polymers used in the sizing process will be either natural products such as starch and derivatives of cellulose, or they will be synthetic polymers derived from petroleum such as polyvinyl alcohol, other vinyl polymers such as acrylates and acrylamides or addition polymers such as polyester resins. In each case these film forming agents will be applied to fibers to which they will have the greatest adhesion. There is an adage or saying that "like likes like." Chemicals which are similar in their constitution will mix with or dissolve into each other and have a strong attractions for one another. Thus a film former that is similar in its chemical makeup to say cotton, a cellulose containing natural polymeric fiber, will have good adhesion to that fiber. An example of such a polymer is starch which has an identical chemical structure to cellulose. After the cloth is woven these materials

95

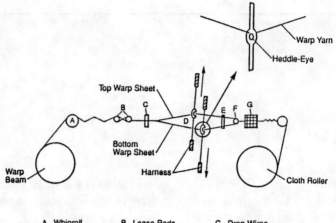

A. Whiproll
B. Lease Rods
C. Drop Wires
D. Heddles
E. Reed
F. Shuttle or Air Space
G. Cloth

FIGURE 4.9 Tension and abrasion zones in a typical loom.

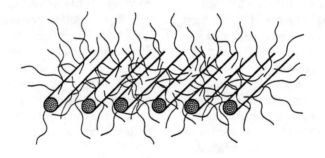

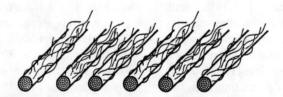

FIGURE 4.10 Control of yarn hairiness with sizing; top: unsized; middle: improperly sized; bottom: properly sized.

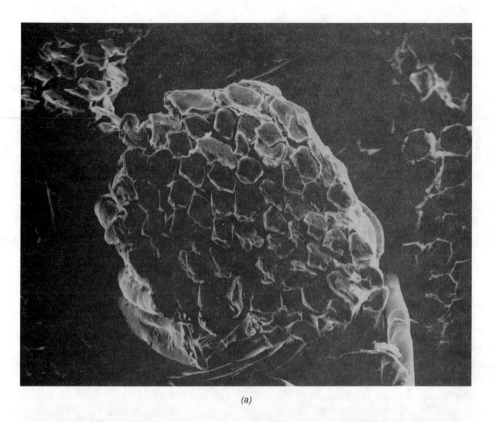

(a)

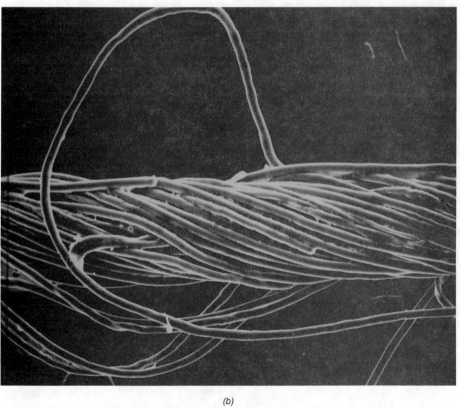

(b)

FIGURE 4.11 Longitudinal and cross-sectional photomicrographs of properly sized yarns.

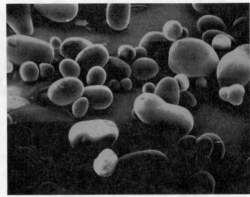

FIGURE 4.12 Cornstarch granules (left) and potato starch granules (right).

must be removed from the fabric during the finishing operation.

Along with the film formers, other products are included as part of the size mixture. One important ingredient that is generally always added to the size mix is a lubricant in order to assist in reducing abrasion; specifically that resulting from yarn to loom interactions. Some lubricant products can be especially troublesome in later operations such as heat setting or dyeing if they are not completely removed during finishing. Other ingredients will be added to the size mixture depending upon the particular loom requirements (air-jet, projectile, etc.) or if a particular finishing requirement is required once the cloth is woven. These other size additives include humectants, antistatic and antifoam agents, removable tints (for warp or style identification), binders, preservatives (if the warps or the cloth is to be stored for long periods of time), penetrating agents (to allow the size to penetrate into tightly constructed styles), viscosity modifiers, weighting agents (to make the cloth heavier, usually used on loom finished goods), among others. Thus the size mixture can be quite complex. The general rule of thumb is to keep the sizing mixture as simple as possible and only add other ingredients when there is a definite need for them.

2.1 Starch

Starch was the very first film forming material which was used as a sizing agent for cellulosic fibers. In the United States the starch of choice is cornstarch derived from yellow dent corn, while in Europe, potato starch is employed (Figure 4.12).

The starch granules are shipped to the plant in bulk or in bags. In order to be useful for sizing, the starch must first be cooked to release the starch molecules within the granules to form a water dispersion. The yarn is passed through the hot size solution followed by drying to form a protective film or coating around the yarn bundle. This coating will glue down the protruding fibers and act to protect the yarn bundle from the loom's abrasive forces.

The starch granules contain two types of chain structures. About 25% of the chains are linear assemblies of glucose molecules (called amylose) and the remaining chains are highly branched chains (called amylopectin) (Figure 4.13). The starch granule is not water soluble. Thus, the granules must be cooked at temperatures at or near the boil in order to release the compressed chains within the granule into the size solution.

The monomer or building block of the chain structure of starch is identical to that of cellulose; thus, the adhesion of starch films to cotton, rayon, linen or other cellulosic fibers or blends containing high levels of these fibers is quite good. Figure 4.14 compares the polymer chain structure of cellulose and starch. These two materials differ only in the manner in which the glucose units of the polymers are attached to each other. Yet, the manner of attachment provides for considerable differences in their properties.

After the cloth has been woven all of the starch

based size must be completely removed by a finishing process called desizing. Fabrics containing starch sizes are treated with chemicals that will preferentially break down both the linear and branched chains into shorter fragments which then become soluble in the hot washing step that follows and thus are removed from the fabric. These chemicals employed will have a minimal effect on the other fibers of the fabric.

Starch is a food material (carbohydrate). This allows the starch to be broken down in the same manner as the starch eaten by animals and bacteria. Thus weak acids (similar in concentration to that found in the stomach) and enzymes (sim-

ilar to the ones in the mouth) can be used to break down the starch chain structure without damaging the cotton cellulose (Figure 4.15). That is, the glucosidic bonds in the cellulose molecule are stable to these reagents but those in the starch polymer are not. Thus the insoluble chains are broken down into smaller water soluble fragments that can be washed out of the cloth. Both amylose and amylopectin are broken down in a similar fashion with the exception that the amylopectin residues may still have some branching. Use of these reagents allows the desizing process to be accomplished at near room temperature although the desizing reaction can be accomplished at a faster rate if these processes

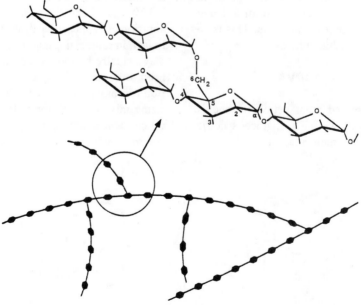

Linear Starch—Amylose

Branched Starch—Amylopectin

FIGURE 4.13 The starch granule contains about 25% linear chains (amylose) and about 75% branched chain (amylopectin).

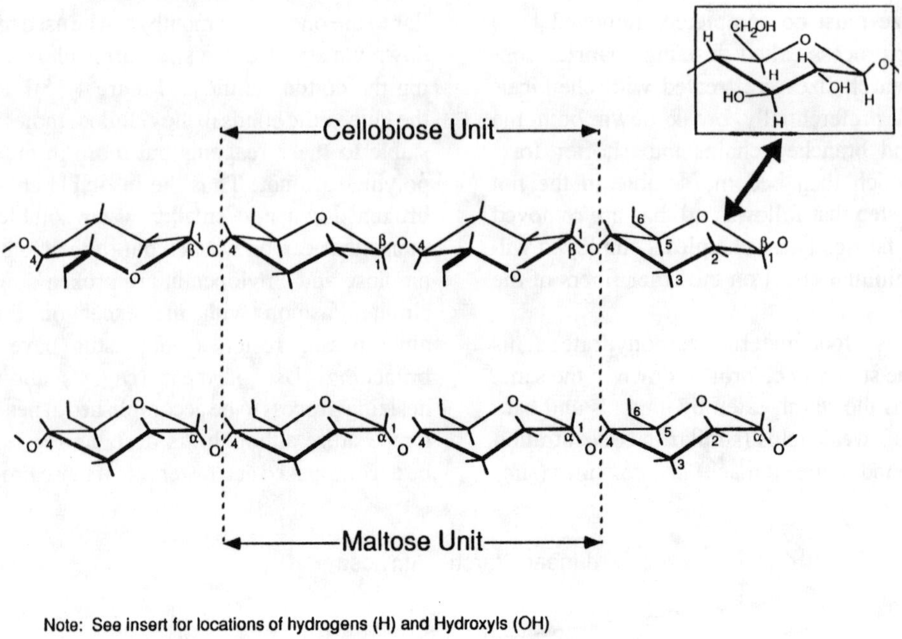

←————Cellobiose Unit————→

Note: See insert for locations of hydrogens (H) and Hydroxyls (OH)

←————Maltose Unit————→

FIGURE 4.14 Comparison of the glucosidic chain structures of cellulose (β) and starch (α).

are done at elevated temperatures. This is the prerogative of the finishing plant.

Starch consists of polymerized sugar (glucose). Thus, the wastewater contains nutrients having a high Biological Oxygen Demand (BOD) which can affect the ecological balance of rivers; hence, the water must be treated to remove (destroy) these materials prior to being released into the nations waterways. This is an added cost of the slashing process.

2.2 Polyvinyl Alcohol (PVA)

PVA is synthesized from petroleum and is man-made. This allows one to engineer specific properties into the molecule. Starch being a natural product has limitations that must be accepted if the product is to be employed for sizing. For example, the size films are very sensitive to weave room moisture levels and require high weave room humidities in order not to dry out and lose their flexibility. They are also sensitive to mildew and have other shortcomings.

PVA gives size films that are very strong, abrasion resistant and easily desized in hot water. In fact, in some sizing applications the PVA films may actually be too strong and the films may need to be modified by the addition of weaker film forming polymers (including starch) to the sizing mixture. PVA forms true solutions with water, hence, desizing requires only that the polymer redissolves in the hot desize water.

FIGURE 4.15 Desizing of starch molecules is accomplished by treatment of the cloth with acids, enzymes, or oxidizing agents.

During desizing it is not necessary to degrade the PVA chains in order to remove the size film, thus it is possible to recover the size by one of several recovery processes for reuse. This is probably the single most important property of PVA. About 10−12 U.S. textile companies are routinely recovering and reusing PVA sizes.

PVA can be obtained in one of several modifications or "grades." It is synthesized from polyvinyl acetate. If all of the acetate groups are removed from the vinyl chains after the polymerization, a PVA grade called "Fully Hydrolyzed" (FH) is obtained. Some acetate groups can be left on the molecule resulting in a family of "Partially Hydrolyzed" (PH) grades. The PVA can also be co-polymerized with methyl methacrylate among other comonomer groups to give grades of PVA having unique sizing properties and recovery stability. The various reactions are summarized in Figure 4.16.

2.3 Other Film Formers

Starch and PVA constitute the bulk of the size materials used in the United States. Other sizes

have been developed for specific uses. One size that was utilized in high volume years ago is carboxymethyl cellulose or CMC (Figure 4.17). This size had high water solubility and resolubility, hence, was easily desized as well as having low BOD-type pollution. It has been largely supplanted as a size by PVA. CMC has since found many other uses including uses as food additives and in laundry detergents such that the textile sizing market could not compete because of price considerations except where specific requirements dictate its use.

Another family of sizes based upon poly-acrylic acid (polyacrylates and polyacryl-amides), find use in sizing the hydrophobic fibers and their blends such as nylon, acrylics, acetate, polyester and the like. Their advantage, in addition to having good binding to these fibers is that they can be made water insoluble during the sizing operation which allows them to be used for water jet weaving applications. During the desize operation the size is resolubilized using an alkaline desize. If acetate yarns are being sized, a solvent desize may be necessary. The structures of the various acrylic-type sizes are given in

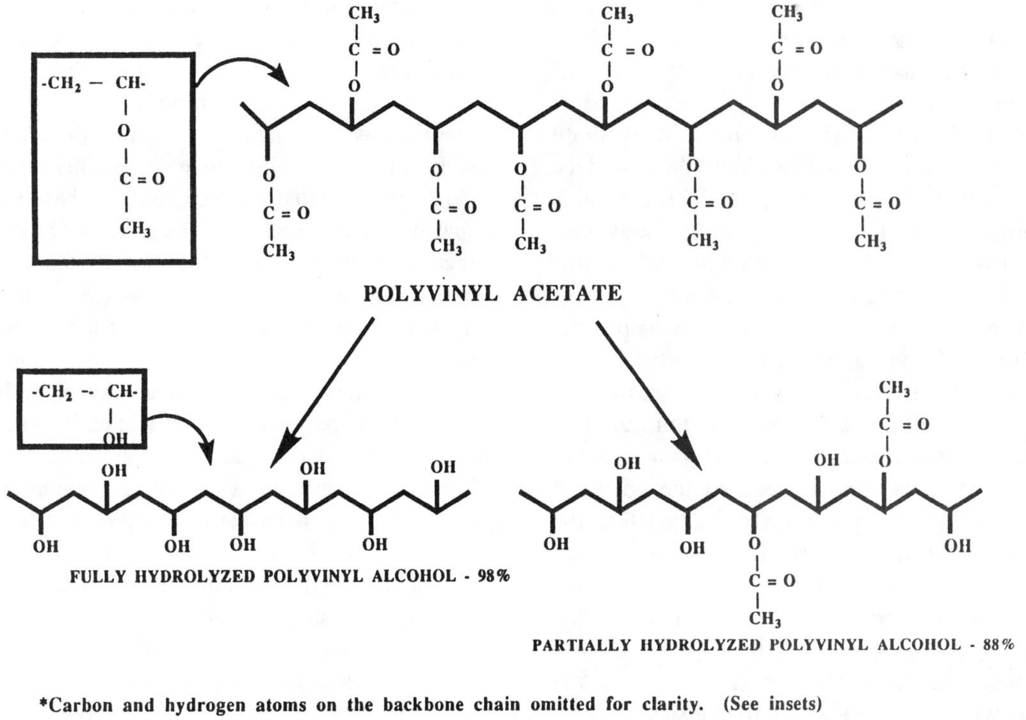

POLYVINYL ACETATE

FULLY HYDROLYZED POLYVINYL ALCOHOL - 98%

PARTIALLY HYDROLYZED POLYVINYL ALCOHOL - 88%

*Carbon and hydrogen atoms on the backbone chain omitted for clarity. (See insets)

FIGURE 4.16 The chemistry of polyvinyl alcohol.

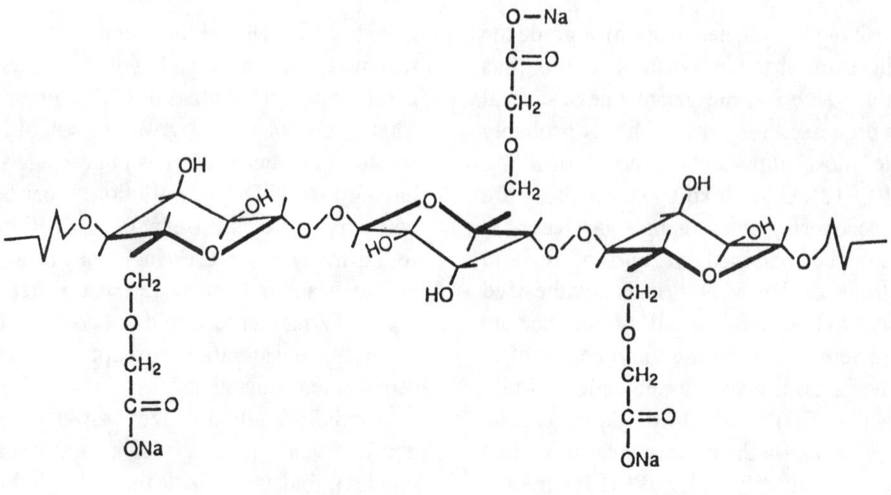

FIGURE 4.17 The structure of carboxymethyl cellulose (courtesy of CMC).

Figure 4.18. Other vinyl polymers and copolymers are provided in Figure 4.19 while Figure 4.20 shows the synthesis and structure for polyester type binders. Table 4.1 compares the properties of various size materials while Table 4.2 compares their BOD contents.

2.4 Other Factors Affecting Sizing

The size must coat the yarns and protect them during weaving. Yarns can be made by one of several spinning methods (air-jet, open-end or ring spinning). The yarns produced by these methods differ considerably in their surface structure and porosity. Figure 4.21 compares yarns spun by the three spinning methods. Obviously, the sizing considerations will be different for each of these yarns even though they may have the same yarn count. For example, the more open and porous surface structure of the open-end spun yarn will require the size mixture to have a higher viscosity in order to insure that the size penetration into the yarn bundle is not excessive. Also the diameter of the open-end yarn will generally be $10-15\%$ larger than the equivalent ring spun yarn. This affects the number of yarns that can be sized in a single size box without undue crowding, the amount of yarn that can be placed on a section beam or loom beam among other considerations. All of the yarn factors will need to be known by the slashing personnel if they are to do an effective job of sizing the yarns.

For a ring spun yarn the amount of twist (twist per inch, tpi) will affect the ability of the size to penetrate into the yarn for proper anchoring of the size film. Tightly twisted yarns may require less viscous size solutions or the use of a penetrating agent to achieve the proper penetration of the size to effectively anchor it onto the yarn surfaces. Thus the slashing personnel should be notified when even small changes in the yarn specifications are to be made since they will invariably affect the sizing process.

Numerous other factors, including the blend levels of fibers, will influence sizing. Obviously a 50/50 polyester/cotton blend will not have the same sizing characteristics as a 65/35 blend. Since the polyester has lighter density than the cotton (polyester = 1.38, cotton = 1.52), there will actually be more polyester fibers in the yarn bundle in the 50/50 blend and not the same number of each fiber as the blend level may imply. In fact the 65/35 polyester/cotton is actually closer to an 80/20 blend based upon the number of fibers actually present. Thus the yarn should be sized with a size material that has good adhesion to the polyester fibers in order to have the size film adhere adequately to the surface of the yarn bundle and not shed off during weaving.

Even the shape and diameter of the fibers can cause changes in the sizing characteristics of the yarn. Smaller fibers actually have more surface

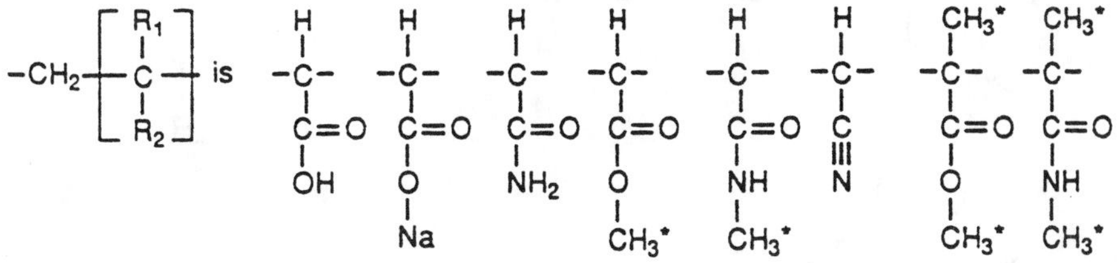

$-CH_2-\begin{bmatrix} R_1 \\ | \\ C \\ | \\ R_2 \end{bmatrix}$ is

$\begin{matrix} H \\ | \\ -C- \\ | \\ C=O \\ | \\ OH \end{matrix}$ $\quad$ $\begin{matrix} H \\ | \\ -C- \\ | \\ C=O \\ | \\ O \\ | \\ Na \end{matrix}$ $\quad$ $\begin{matrix} H \\ | \\ -C- \\ | \\ C=O \\ | \\ NH_2 \end{matrix}$ $\quad$ $\begin{matrix} H \\ | \\ -C- \\ | \\ C=O \\ | \\ O \\ | \\ CH_3{}^* \end{matrix}$ $\quad$ $\begin{matrix} H \\ | \\ -C- \\ | \\ C=O \\ | \\ NH \\ | \\ CH_3{}^* \end{matrix}$ $\quad$ $\begin{matrix} H \\ | \\ -C- \\ | \\ C \\ ||| \\ N \end{matrix}$ $\quad$ $\begin{matrix} CH_3{}^* \\ | \\ -C- \\ | \\ C=O \\ | \\ O \\ | \\ CH_3{}^* \end{matrix}$ $\quad$ $\begin{matrix} CH_3{}^* \\ | \\ -C- \\ | \\ C=O \\ | \\ NH \\ | \\ CH_3{}^* \end{matrix}$

(A) $\quad$ (B) $\quad$ (C) $\quad$ (D) $\quad$ (E) $\quad$ (F) $\quad$ (G) $\quad$ (H)

(A) Poly Acrylic Acid (PAA)

(B) PAA Sodium Salt (or Ammonium Salt $\begin{matrix} O \\ || \\ -C-O-NH_4 \end{matrix}$)

(C) Acrylamide

(D) Methyl Ester (Methyl Acrylate)

(E) Methyl Amide (Methyl Acrylamide)

(F) Nitrile (Acrylonitrile)

(G) C2 Substituted Ester (N-Methyl Methacrylate)

(H) C2 Substituted Amide (N-Methyl Methacrylamide)

Water Soluble (A), (B), & (C)

Water Insoluble (D), (E), (F), (G), (H)

*Derivatives other than methyl include ethyl, propyl, n-butyl, isobutyl, and 2-ethylhexyl among others

FIGURE 4.18 Some of the acrylic family of copolymers useful in sizing.

103

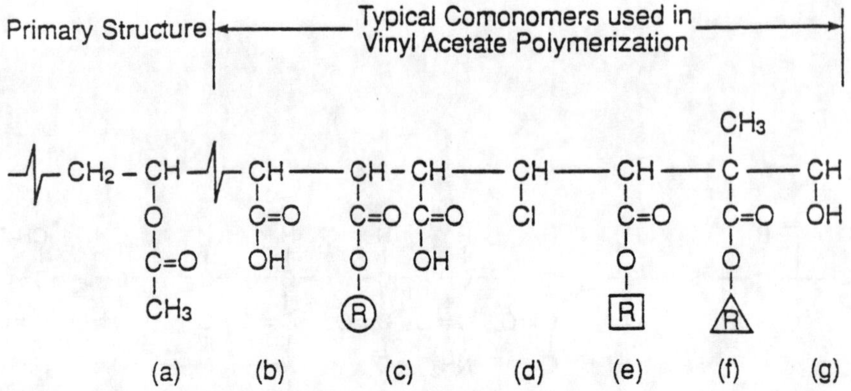

Primary Structure | Typical Comonomers used in Vinyl Acetate Polymerization

(a)　(b)　(c)　(d)　(e)　(f)　(g)

R = Methyl, Ethyl, etc.
a) Basic Vinyl Acetate Monomer (Homopolymer)
b) Acrylic Acid (Ammonia or Sodium Salt)
c) Monoester of a Dicarboxylic Acid
d) Vinyl Chloride
e) Acrylate Ester
f) Methyl Acrylate Ester
g) Vinyl Alcohol

FIGURE 4.19 Polyvinyl acetate (PVAc) copolymers.

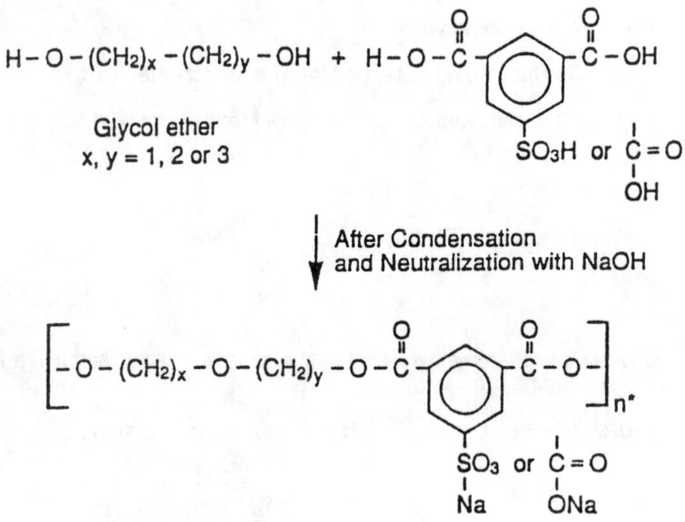

Glycol ether
x, y = 1, 2 or 3

After Condensation and Neutralization with NaOH

*n = 3–5.5 (oligomer types)
n = 18–26 (medium MW type)

FIGURE 4.20 Synthesis of a typical polyester resin.

104

TABLE 4.1 *Comparison of the Properties of Some Film Forming Sizes.*

Size	Tensile Strength (PSI)	Elongation (%)	Moisture Content at		
			50% RH	65% RH	80% RH
PVA*	7,000 to 15,000	100 to 150		8−9**	16−17**
Starch	600 to 900	8 to 12		15−20	19
CMC	2,000 to 4,000	10 to 15	14	15−20	30.5
Acrylic	1,000 to 2,000	100 to 600			17−21

*At 70°F and 65% RH.
**Depending upon the degree of hydrolysis.

TABLE 4.2 *Waste Treatment Requirements for Various Sizes [1].*

Size Material	BOD$_5$ (ppm)*
Starch	
Pearl corn	500,000
B2 gun (starch dextrins)	610,000
Keofilm No. 40	550,000
Penford Gum 300 (starch ether)	360,000
wheat starch	550,000
Polyvinyl alcohol (PVA)	10,000 to 16,000
Carboxymethyl cellulose (CMC)	30,000
Polyvinyl acetate (PVAc)	10,000
Hydroxyethyl cellulose (HEC)	30,000
Sodium alginate	360,000
Acrylic	205,800

*5 days, parts per million.

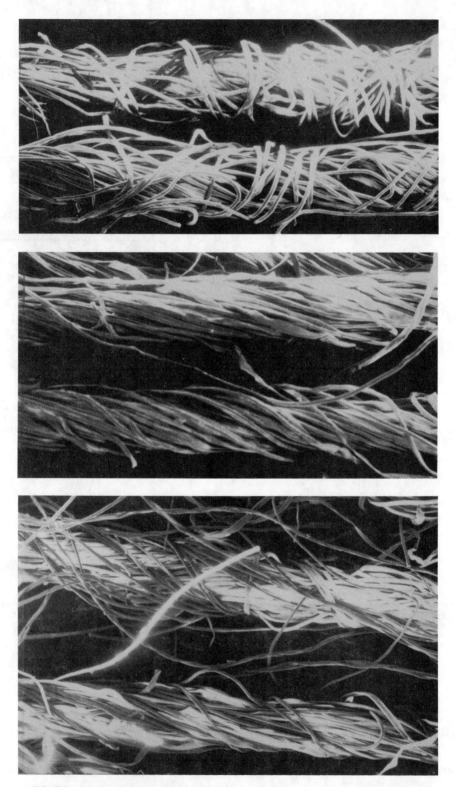

FIGURE 4.21 Comparisons of open-end (top), ring spun (middle) and air-jet spun yarns.

area/fiber than do larger diameter fibers. Trilobal fibers have more surface area than round fibers. Surface roughness due to high delusterant content can also affect the amount of size picked up by the yarn. Thus changes in the shape, denier, and delusterant level of the fibers used to manufacture the yarn can cause a change in the size pickup of the yarn.

One final factor that should be addressed is yarn quality. Excessive hairiness, thick and thin places and other non-uniformities in the yarn will result in a poorer job of sizing. A well sized quality warp on the loom is said to be half woven.

3. FILAMENT SIZING

Multifilament yarns have essentially a continuous thread line that is as free as possible from individual broken filaments. The yarns are generally smooth and have a producer applied finish that protects the yarns against abrasion and static during processing. Because of their smooth and lubricated surfaces, high twist filaments may not require sizing. The norm however, is low twist, thus should a single filament become broken it can peel back to initiate a fuzz ball, float or skip that will ultimately cause a loom stop.

If the multifilaments were spot welded (glued together at multiple points along the yarn length, a single broken filament could peel or strip back only to the spot weld and stop and not continue to form a loom stoppage.

At the lease rods when the yarn sheet is separated in the "dry" state, the spot welds should be weak and brittle for easy separation without causing any monofilament breaks. But, after regaining moisture on the loom beam, the welds should now be strong and flexible and possess good multifilament bundle cohesion so as to prevent any broken single filament separation on the loom. Size requirements for filament yarns are as follows:

- The size solution must be wet out and penetrate the filament bundle. This may not be an inherent property of the size since this may be achieved through the use of binders and additives such as emulsifiers, wetting agents, etc.
- The viscosity of the size solution must be low enough to allow for good penetration into the filament bundle.
- The size must have good adhesion to the particular filaments being sized.
- The sizing agent must be quick drying without delayed set or producing a tacky surface.
- The size should produce elastic and flexible films that match the elasticity and flexibility to which the yarns are subjected during weaving.
- The size should be antistatic or at least not contribute toward additional static buildup.
- Size shedding on the loom resulting in buildup on the heddles, reeds or other loom parts should be avoided, or held to a minimum.
- Size film should not be subject to drastic changes in properties due to the extremes of humidity changes. Ideally, the size should be brittle when bone dry on the slasher in order to achieve a soft break at the bust rods and when equilibrated to the weave room conditions the size should be tough and flexible.
- The size should be easily removable during desizing.
- The size should not be detrimental to the yarn, processing equipment or human health.
- The size should be easily prepared (mixed) in a short time without the need for special equipment or costly preparation controls.
- The size properties should not be changed by the producer spin finish oils.

Table 4.3 shows size considerations for filament yarns.

The concentration of the size, which controls the number of spot welds, depends on [2]:

- yarn denier—Lower deniers have more fiber surface area and require more size.
- number of ends/inch—A higher sley requires more size.
- type of weave—Plain weaves are more difficult to weave than those with long floats, e.g., satins.
- the loom type and condition—Some

TABLE 4.3 Size Considerations for Filament Yarns.

Size	Fiber	Type of Loom		
		Shuttle	Air-Jet	Water-Jet
Dispersible polyester	Polyester	S	S	S
Polyacrylates (esters)				
Sodium	Polyester	S	S	N
Ammonium	Polyester, nylon	S	S	S
Polyacrylic acid	Nylon	S	S	S
Polyvinyl alcohol				
FH	Viscose rayon	S	S	N
PH	Nylon*, acetate, polyester	S	S	N
Styrene/maleic anhydride				
Sodium	Acetate	S	S	N
Ammonium	Acetate	S	S	S
Polyvinyl acetate	Acetate, polyester	S	S	S

*Fabrics requiring a neutral pH.
S: suitable, N: not suitable.

shuttleless looms are more abrasive to the warp.
- type of slasher and drying arrangement employed—Each type (conventional, predryer-type, single end beamer) has different constraints.

4. THE SLASHER

The size material is applied to the yarns on a machine called the slasher. The major parts of the slasher are the creel, where the supply of yarns on section beams is passed through the size box containing the hot water solution or mixtures of sizing agents. The yarn picks up the requisite quantity of size solution, any excess being squeezed off as the yarns pass through squeeze rolls. The yarns are then passed into a drying section. The wet yarns are preferably predried by using hot air, infrared radiation or the like and the drying finished on steam heated hot rolls (called drying cylinders). Exiting the drying section, the sheets of yarn are stuck together and must be separated back into their individual ends. This is accomplished using bust rods to first separate the individual sheets of yarns from each section beam followed by pins to separate the yarns within each sheet. The now sized, dried and separate ends enter the beaming section of the slasher where they are wound onto a loom beam ready for weaving.

A slasher machine and schematic of a typical slasher configuration are shown in Figure 4.22. The configuration of the slasher can be changed such that each section will be designed to provide optimum results for whatever the mill may be running. For example, several types of creel designs (over and under, stationary, equitension) are available. The mill can employ multiple size boxes. The slasher can have a wide variety of dry can designs for more efficient drying of the yarns. In addition, the slasher will be designed to reduce yarn tensions and have electronic readout indicators of speeds, tensions, temperatures etc.

The slasher is a fairly complicated machine where numerous variables must be controlled for effective and efficient slashing. For example, excessive tensions in any of the slashing zones, but more specifically when the yarn is wet can cause yarn stretch which will reduce the ability of the yarns to elongate on the slasher. Temperature control in the size box is important for proper size pickup. Excessive dry can temperatures can cause the size to become overdried and brittle. Excessive size pickup or penetration into the yarn bundle will cause the energy of break at the bust rods to be high and can lead to excessive hairiness and even yarn breakage. Slasher personnel who are skilled in their job are perhaps one of the most important assets of a mill. A poorly slashed warp that does not weave well, resulting in numerous loom stops will increase the costs of weaving and could cause an increase in the seconds produced.

5. FINISHING CONSIDERATIONS

After the size has done its job, it must be removed in the finishing process unless it is a loom finished material such as denim. Removability is perhaps one of the most important properties a size material must have. If the size is not to be recovered then the effluent from the finishing plant will contain the size and will require treatment before it can be discharged. Such factors as ease of removal and the associated costs of desizing (chemicals to break down the size into water soluble fractions, the hot water and steam needed to dissolve the size and remove it from the cloth, the wastewater treatment process to allow it to be discharged) will differ for each size material. Also to be considered are the other ingredients in the size mix-

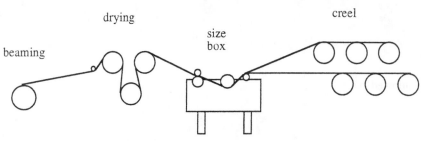

FIGURE 4.22 A slasher machine (courtesy of West Point Foundry) and schematic of a simple slasher configuration.

ture. For example, the type of lubricant employed will affect the finishing process since these materials should be completely removed in order not to have them interfere with other finishing and dyeing processes.

Some sizes are affected by the heat setting treatment given to the hydrophobic fibers such as nylon and polyester. This process is usually done prior to the desizing step in the finishing plant. If such sizes are employed, a change in the usual finishing procedures may be required. It is important that the finishing plant be made aware of any changes in the sizing process that could result in incomplete or a more expensive finishing process. In this case it is quite necessary that the right hand knows what the left hand is doing in order to obtain the best possible finishing process for the woven cloth. It does not make sense to use materials in the slashing process because they seem to save money, then to lose the advantage because of increased finishing costs.

4.4

Weaving

R. P. WALKER
S. ADANUR

1. INTRODUCTION

The weaving machine provides the means to interlace warp and filling yarns to form woven fabric. The warp yarn is that yarn system in woven fabric lying parallel to the lengthwise direction of the cloth or parallel to the selvages (side edges). Warp yarns are often referred to simply as ends. The filling is the yarn system in a woven fabric lying across the width of the cloth perpendicular to the selvages. The filling is also referred to as weft. Individual filling yarns are called picks because in early hand weaving a filling yarn was inserted by tying it to a pointed device which was used to pick the filling yarn through the warp yarns. As depicted in Figure 4.23, the weaving machine (loom) provides mechanisms required to deliver and manipulate warp yarns for interlacing with filling yarns which are inserted into the warp. Warp yarns are delivered to the loom in sheet form on a warp beam while filling yarns are delivered (usually one pick at a time) from smaller single-end packages.

2. FABRIC CONTROL AND DESIGN

The loom has other mechanisms that perform important functions related to efficiency of the process or quality of fabric produced, but only five basic motions are necessary for weaving to occur. The five basic mechanisms are

(1) Warp let-off
(2) Warp shedding
(3) Filling insertion
(4) Filling beat-up
(5) Fabric take-up

Warp let-off and fabric take-up provide continuous weaving.

2.1 The Let-off

The let-off motion delivers warp to the loom while maintaining a desired level of tension on warp yarns. This control of tension produces a ratio of crimp (bending of yarn during interlacing) between warp and filling yarns in the fabric.

Increasing tension on the warp causes it to lie straighter in the fabric transferring the crimp to the filling. Fabric thickness is affected by the crimp ratio of the two yarn systems. When warp and filling crimp are equal, the fabric thickness will be at a minimum. Any change that causes crimp of warp and filling to become unequal makes the fabric thicker, assuming yarn diameters stay the same.

2.2 Warp Shedding

The opening formed by the separation of the warp yarns during weaving is called shed. The warp shedding mechanism manipulates warp yarns up and down to dictate the warp to filling interlacing pattern called the weave. Referring to Figure 4.24, every loom requires a control device for each warp yarn. This control device is called a heddle and consists of a thin metal strip which has been fashioned with a small eye through which a yarn may be drawn. Heddles controlling warp yarns that always follow the same interlacing pattern can be grouped together into a common frame called a harness.

There must be a different harness provided for each group of warp yarns or heddles that follows

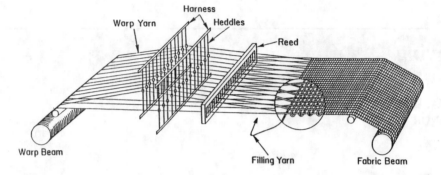

FIGURE 4.23 Loom elements dealing with warp delivery, filling insertion and fabric removal.

a different weaving pattern. In the case where every end weaves a different pattern a harness cord is provided for each heddle. There are three common systems used to provide manipulation to the warp yarns; harnesses controlled by cams, harnesses controlled by a dobby and harness cords controlled by a jacquard. Cams with weave pattern profiles rotate to deliver lifting and lowering instructions to harnesses in a cam system which typically can handle weave patterns with no more than fourteen different harnesses and patterns that repeat in no more than ten picks. Recent, more modern dobbies and jacquards utilize electronic systems for input of the harness lifting and lowering patterns. Dobbies can weave patterns requiring up to about thirty harnesses

and repeating on as many as 5,000 picks. Jacquard heads generally are equipped to handle over 1,200 harness cords with patterns repeating on about 5,000 picks and multiple heads can be employed over a single weaving machine to increase the weave pattern capability.

To control the weave of a fabric, two aspects of the shedding mechanism must be specified. First, the order of entering in which warp yarns are threaded or drawn into heddles (or harnesses) and second, the order of lifting by which the harnesses are raised and lowered. For the illustration in Figure 4.24, the most common weave pattern, a plain weave, was chosen. The warp yarns are drawn into the two harnesses in a "harness 1–harness 2" order since the plain

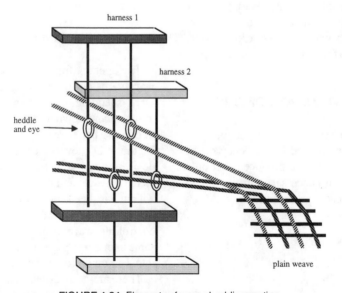

FIGURE 4.24 Elements of warp shedding motion.

weave contains only two different warp yarn weaving patterns; those that weave above and those that weave below a given filling yarn. These two groups of warp yarns swap positions on each successive pick insertion as seen in the fabric of Figure 4.24. Two warp yarns and two filling yarn insertions are required to repeat the complete interlacing pattern order. One unit of an interlacing pattern is called a repeat and is referred to as the order of interlacing, a definition for weave. The minimum number of harnesses required for a weave is determined by this repeat size, e.g., two harnesses are required for the plain weave.

As can be seen in Figure 4.25, drawing the warp yarns in an order different from the order shown in Figure 4.24 results in a different weave, in this case a filling rib or oxford weave.

In Figure 4.25 ends are shown weaving in pairs, first up then down or vice versa. This new weave was accomplished simply by redrawing warp yarns from the "harness 1−harness 2" order of Figure 4.24 to the "harness 1, harness 1−harness 2, harness 2" order of Figure 4.25. Notice that the two weaves were accomplished with the same order of lifting.

The order of entering shown in Figure 4.24 is used again in Figure 4.26, but with a different lifting sequence of two-up, two-down or vice versa. This change in lifting order produces yet another weave, a warp rib, different from either of the other two.

By combining the draw for the filling rib with the lifting pattern of the warp rib, a basket weave is produced (Figure 4.27).

In designing of woven fabrics, a symbolic notation is used to illustrate weave patterns. Vertical columns of squares on a grid or point paper represent warp ends while horizontal rows of squares represent picks. A marked square on the point paper symbolizes that the warp end represented by that particular vertical column is raised above the pick represented by that particular horizontal row. Leaving a square unmarked or blank means the warp end is lowered under the pick during weaving. The same four weaves illustrated graphically in Figures 4.24 through 4.27 are shown symbolically in Figure 4.28.

The number of weave patterns is endless. Other basic or common weaves include the twill and satin weaves which are illustrated in Figure 4.29. At least three harnesses are required to weave a twill and at least five harnesses are needed for a satin. Satin weave is a warp face weave and sateen is a filling face weave.

2.3 Filling Insertion (Picking)

The picking motion places filling yarns into the warp sheet between those warp yarns that

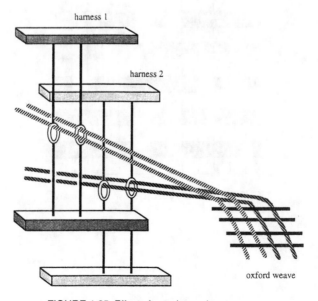

FIGURE 4.25 Effect of entering order on weave.

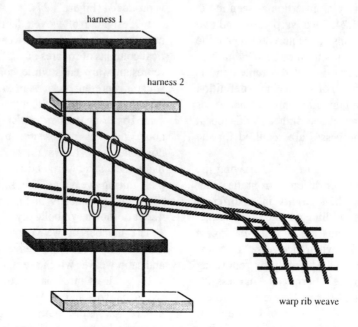

FIGURE 4.26 Effect of lifting order on weave.

FIGURE 4.27 A two-up, two-down basket weave.

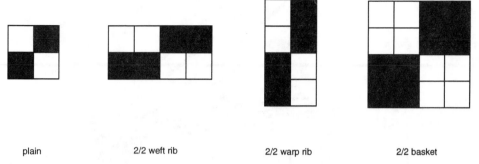

plain   2/2 weft rib   2/2 warp rib   2/2 basket

FIGURE 4.28 Symbolic notation for plain, rib, and basket weaves.

have been raised and those warp yarns that have been lowered as shown in Figure 4.30. For fabrics requiring different types or colors of filling yarns, a separate filling selection mechanism assists in presenting the proper filling yarn in the correct order to the insertion device.

There are several types of filling insertion, which is the main factor in classification of weaving machinery:

- shuttle looms
- air-jet
- rapier
- projectile
- water-jet

- needle looms (for narrow fabric weaving)
- multi-phase looms
- circular looms
- others

Weaving machines for many years of textile history depended on the shuttle as the primary insertion device. Although the shuttle (Figure 4.31) is still used in existing shuttle looms, new shuttle looms are no longer manufactured for consumer textiles by the major weaving machine producers in the world. However, manufacture of some industrial textiles such as paper machine clothing still depends on shuttle technology. The modern technology of weaving machines has

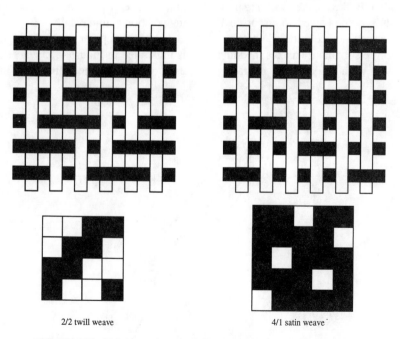

2/2 twill weave   4/1 satin weave

FIGURE 4.29 Graphic and symbolic illustrations for twill and satin weaves.

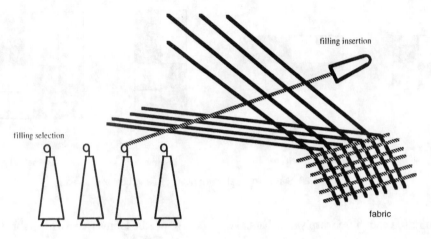

FIGURE 4.30 Filling selection and insertion.

turned to three other prominent devices for filling insertion as shown in Figure 4.32: projectile, rapier and air-jet.

A gripper projectile transports a single filling yarn into each shed as shown in Figure 4.33. The gripper projectile draws the filling yarn into the shed. Energy required for picking is built up by twisting a torsion rod. On release, the rod immediately returns to its initial position, smoothly accelerating the projectile through a picking lever. The projectile glides through the shed in a rake-shaped guide. Braked in the receiving unit, the projectile is then conveyed to its original position by a transport device installed under the shed. The projectile's small size makes shedding motions shorter which increases operating speeds over wide widths of fabric, often weaving more than one panel of fabric with one insertion mechanism.

Figure 4.34 illustrates filling insertion by two flexible rapiers with filling carriers, a giver and a taker. The filling is inserted halfway into the shed by one carrier and taken over in the center by the other carrier and drawn out to the opposite side of the fabric. A spatial crank gear drives the oscillating tape wheels to which the rapier tapes are attached. In the shed, the tapes move without guides. The grippers assume the correct clamping position automatically.

The most popular method of filling insertion is illustrated in Figure 4.35 where a jet of air is used to "blow" the filling yarn into the shed.

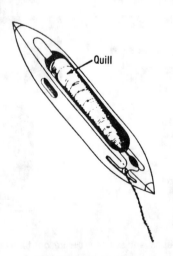

FIGURE 4.31 Shuttle.

FIGURE 4.32 Filling insertion devices: projectile (left), rapier (center), and air-jet (courtesy of Sulzer Ruti).

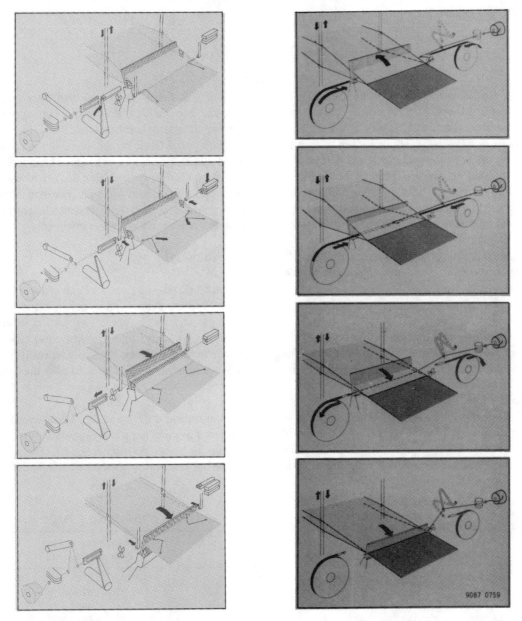

FIGURE 4.33 Schematic of filling insertion with projectile (courtesy of Sulzer Ruti).

FIGURE 4.34 Schematic of filling insertion with two flexible rapiers (courtesy of Sulzer Ruti).

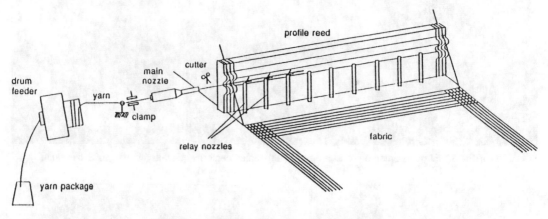

FIGURE 4.35 Schematic of air-jet filling insertion [3].

This small mass of insertion fluid enables the mechanism to operate at extremely high insertion rates. The picks are continuously measured and drawn from a supply package, given their initial acceleration by the main air nozzle and boosted or assisted across the fabric width by timed groups of relay air nozzles. Other fluid systems use water as the insertion medium, but the use of a water jet is generally limited to hydrophobic yarns such as nylon or polyester filament.

A comparison of approximate insertion rates for the various picking methods is given in Table 4.4.

2.4 Beat-up

The filling yarn beat-up motion pushes with a reed each filling yarn through the warp sheet and into the edge of the fabric called the fell (Figure 4.36). The reed is also used to control warp yarn density (closeness) in the fabric. Warp density is expressed as either ends per inch (epi) or ends per centimeter (epc). The reed is a comb with metal wire blades spaced equidistantly apart identified by a reed number corresponding to the number of spaces or "dents" per inch. Specifying the number of ends per dent with a certain reed number dictates the construction (density) of ends per inch in the fabric on the loom. It should be noted that interlacing causes a natural contraction of yarns in the fabric such that density of warp ends (epi) off the loom will be higher than in the reed, generally about 5% higher depending on the weave, tensions and yarn sizes involved.

2.5 Take-up

The fabric take-up removes cloth at a rate that controls filling density, picks per inch (ppi) or picks per centimeter (ppc). Two factors determine filling density: filling insertion rate (loom speed) and rate of fabric take-up. Generally, the pick insertion rate of a loom is fixed at the time of purchase based on the range of fabrics it is intended to produce, the type of insertion mechanism and the loom width. Loom speed is expressed as picks per minute (ppm) and rate of take-up as inches per minute (ipm). The ppm divided by the ipm equals the ppi. Warp density

TABLE 4.4 Approximate Rates for Different Filling Insertion Methods (Fabric Width = 60 inches).

Insertion Method	Machine Speed (picks/min)	Filling Insertion Rate (yds. of filling/min)
Shuttle	250	417
Rapier	550	916
Projectile (two panels)	425	1,416
Fluid (air, water)	1,000	1,667

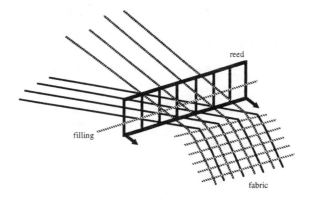

FIGURE 4.36 Beat-up of filling.

and filling density together are referred to as the "construction" of the fabric, expressed as epi × ppi as illustrated in Figure 4.37.

Again, it should be emphasized that both the ends and picks contract because of interlacing causing construction in the loom and off the loom to be different. Subsequent fabric finishing steps also introduce changes in the fabric construction which must be considered in setting up loom specifications.

2.6 Auxiliary Functions

In addition to the five basic motions of a loom, there are many other mechanisms on typical weaving machines to accomplish other functions. These include

- a drop wire assembly, one wire for each warp yarn, to stop the loom when a warp end is slack or broken
- a tension sensing and compensating whip roll assembly to maintain tension in the warp sheet
- a mechanism to stop the loom when a filling yarn breaks
- filling feeders to control tension on each pick
- pick mixers to blend alternate picks from two or more packages (up to eight)
- filling selection mechanism for feeding multi-type filling patterns
- filling selvage devices such as trimmers, tuckers, holders and special weave harnesses for selvage warp ends

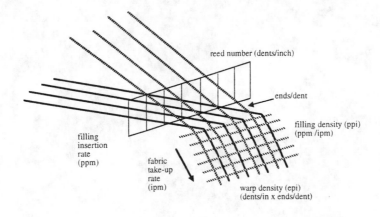

construction = epi x ppi

FIGURE 4.37 Control of fabric construction.

- filling replenishment system to provide uninterrupted filling insertion by switching from a depleted to a full package
- a temple assembly on each selvage to keep fabric width at the beat-up as near the width of the warp in the reed as possible
- sensors to stop the machine in the event of mechanical failure
- a centralized lubrication control and dispensing system
- a reversing mechanism to avoid bad start-ups after a machine stop
- a color coded light signal device to indicate the type of machine stop from a distance
- a production recording system

3. MODERN WEAVING MACHINERY AND TECHNOLOGY

Some of the more modern weaving machines incorporate other systems that move the weaving process toward total automation. These systems include

- an automatic broken pick repair unit
- an automatic broken warp end repair unit
- a partially automated warp replenishment system
- a computer monitoring system for production efficiency measurement and control
- a computerized pattern control system (for dobby and jacquard looms)
- a computerized maintenance control interface
- an automated cloth roll doffing unit
- an automated cloth inspection system
- quick warp change (style change) system. With the quick warp change system, the entire assembly of "warp beam, harnesses and reed" is removed from the machine. A new "warp beam, harnesses and reed assembly" with a different style is mounted on the machine. The whole process can be done by a single person in less than half an hour. Figure 4.38 shows quick warp change system on an air-jet weaving machine. The advantages of quick warp change are savings in machine

FIGURE 4.38 Quick warp change system (courtesy of Picanol).

FIGURE 4.39 Projectile weaving machine (courtesy of Sulzer Ruti).

downtime, fewer operators in the weave room and therefore lower weaving costs.

Typical characteristics of modern weaving machines are given below. These characteristics are not specific to any particular manufacturer's machine except the projectile machine which is manufactured only by Sulzer Ruti. The information below is given to demonstrate the state-of-the-art technology in weaving.

3.1 Projectile Weaving Machines

Figure 4.39 shows a P 7200 projectile weaving machine by Sulzer Ruti. P 7200 is designed to be functional, modern with central microprocessor control and automated subunits. The microprocessor terminal informs about all logged data and programmed parameters, and receives commands for the modification of settings. The monitor display shows the operational state of the machine, and, in case of stoppage, its cause.

One of the important new developments is the projectile made of composite material. This reduced the weight of the projectile and increased its speed. Composite projectile is especially suited to economically produce very delicate fabrics. Figure 4.40 shows steel and composite projectiles.

The automatic filling break repair motion enables repair of the faulty pick within a short time without damaging the warp threads. The electronically controlled filling thread brake enables the braking force and braking duration to be adapted to the requirements of the individual filling yarn. High fabric quality is assured by the reversible, electronically controlled warp let-off and the programmable automatic shed leveling motion, which, by closing the shed at stoppages, relieves the warp threads.

For shed formation, there is a range of mechanical or electronically controlled shedding motions available:

- tappet motions for up to fourteen shafts, with weave repeats of up to ten picks dobbies, electronically controlled, for up to ten shafts
- name-selvage jacquard machines, electronically controlled
- jacquard machines, electronically controlled, with all standard numbers of hooks

Warp stop motion pinpoints the broken warp end on the digital display. The electronic filling stop motion can be adapted to any type of filling yarn. Projectile weaving machines are very versatile and adaptable to various fabric styles from

FIGURE 4.40 Composite (top) and steel projectiles (courtesy of Sulzer Ruti).

finest silk shirting to heavy denim styles and high performance industrial fabrics.

3.2 Rapier Weaving Machines

Although there have been different variations of rapier insertion mechanisms such as single rigid rapier and double rigid rapier, double flexible rapier machines are the most common.

Rapier machines are very versatile weaving machines. The central microprocessor system regulates and monitors all electronically controlled functional units, logs the machine speed, the number of picks and calculates the machine's efficiency. Permanently stored and frequently used weaves and color sequences as well as various auxiliary functions necessary for starting to weave, can be readily activated via the microprocessor terminal. If the filling yarn breaks before the feeder, the microprocessor simply switches over to another feeder with the same and intact filling yarn, without stopping the loom. Electronically controlled cloth take-up with freely programmable filling densities optimizes the fabric quality. Any filling density from 6 to 200 picks/cm is infinitely adjustable at the microprocessor terminal. Cloth take-up and warp let-off are synchronized. The warp tension to be set, is calculated and displayed by the microprocessor according to the style data. Optimum compensation of the warp tension is maintained by the lightweight warp tensioner. Together with the electronically controlled warp let-off, the lightweight warp tensioner guarantees optimum warp tension. The electronically controlled filling thread brake reduces yarn strain measurably.

Rapier weaving machines can weave a wide range of demanding fabrics with natural and manmade yarns and their blends. Some machines can accommodate dobbies of up to sixteen, alternatively twenty-eight shafts as well as for combination with jacquard machines and positive tappet motion. Selvages can be formed either by means of leno motion, sealing or tucking units. Figure 4.41 shows a rapier weaving machine.

3.3 Air-Jet Weaving Machines

Figure 4.42 shows a microprocessor controlled air-jet weaving machine. An extremely flexible program of styles, short style changing

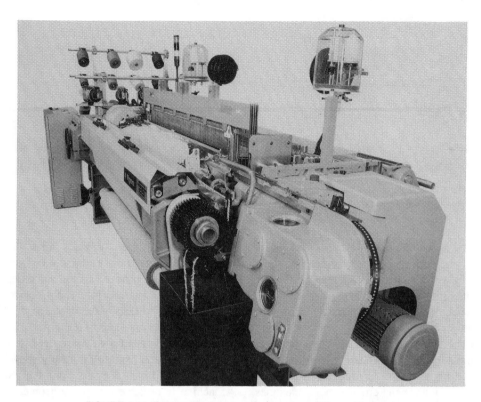

FIGURE 4.41 Rapier weaving machine (courtesy of Nuovo Pignone).

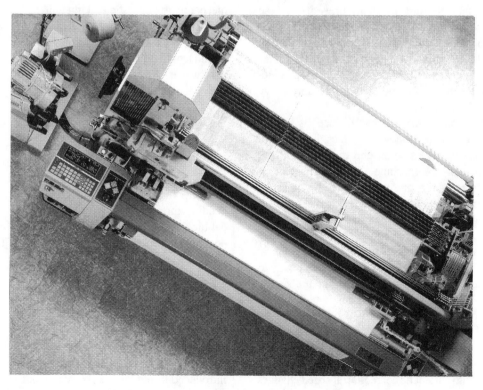

FIGURE 4.42 Air-jet weaving machine (courtesy of Tsudakoma).

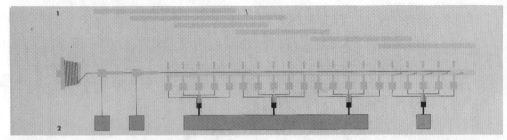

FIGURE 4.43 Programmable air blowing times (1) with automatic correction at speed changes and automatic pressure optimization due to individual air supply (2) (courtesy of Sulzer Ruti).

times and a minimum of attendance and maintenance requirements guarantee high economical efficiency with air-jet looms. Gentle handling of warp and filling yarns permits highest speeds. In these machines, due to optimized air consumption, power requirement is modest and filling yarn waste is surprisingly little. Compressed air control and warp let-off are processor controlled. The automatic filling fault removal reduces downtimes and enables more machines to be assigned to a weaver.

The robust and compact construction allows for high speeds as well as the weaving of the most heavy fabrics. Start marks are prevented by the swift acceleration of the main motor of the weaving machine to full speed. The warp stop motion with LED display for each contact bar facilitates locating broken filling threads. The optoelectronic filling stop motion detects reliably every filling yarn breakage.

Integrated artificial intelligence is incorporated in some of these machines. The control terminal serves as communication element between man and machine. Setting data can be transferred from one machine to the other by means of the memory card. If no setting data is available for a style which had not been woven previously, the microprocessor can calculate standard parameters from the entered fabric specification and transfers them automatically to the electronically controlled function units.

Profile reed and relay nozzles, which are the most common, assure safe filling insertion even with larger weaving widths. The air pressure in the nozzles adjusts itself automatically to the insertion conditions. Programmable blowing times with automatic correction at speed changes and automatic pressure optimization due to in-

dividual air supply provides air-saving and gentle filling insertion (Figure 4.43). During the filling insertion process, thread velocities of 50 to 80 m/s are attained [4]. The processor-controlled filling thread brake dampens the filling tension peak by approximately 50%. The electronically controlled warp let-off assures uniform tension from full to empty warp beam. Changing filling densities on the running machine is possible due to electronically controlled cloth take-up. High quality of selvages is assured by the various selvage forming units such as full and half leno selvage devices and mechanical-pneumatic tucking unit.

The electronically controlled filling feeder guarantees exact metering of the required filling thread length, thus ensuring a minimum of filling yarn waste.

3.4 Water-Jet Machine

Water-jet machines offer very high speed which can be around 1,500 ppm. Features like filling insertion with a minimum influence over the warp yarn, shedding devices with much less heddle vibration, fully super high speed compatible let-off and increased durability in the beating motion, improve fabric grade. Figure 4.44 shows a water-jet machine. Figure 4.45 shows a modern weaving mill.

3.5 Multi-Phase and Circular Weaving Machines

Innovative approaches to weaving have been introduced through several design modifications of shedding and filling insertion components of

FIGURE 4.44 Water-jet machine (courtesy of Nissan).

FIGURE 4.45 Modern weaving mill (courtesy of Wellington Sears Company).

the traditional weaving machine. These developments include the circular loom and several different multi-phase weaving systems such as warp direction shed waves and weft direction shed waves. These multi-phase concepts offer an advantage in production rate, gained by providing segmented sheds along either the warp or the filling direction of the structure coupled with progressive, simultaneous filling insertions. Although some of these new multi-phase systems and circular looms are being used in limited fabric markets, none of them has challenged traditional weaving machines in major fabric markets so far [5].

4.5

Knitting

R. P. WALKER
S. ADANUR

1. INTRODUCTION

As defined earlier, knitting involves the inter-looping of one yarn system into continuously connecting vertical columns (wales) and horizontal rows (courses) of loops to form a knitted fabric structure. There are two basic types of knit structures as shown in Figure 4.1: weft knit and warp knit. Figure 4.46 shows a circular weft knitting machine.

In weft knitting, the yarn loops are formed across the fabric width, i.e., in the course or weft direction of the fabric. In warp knitting, the loops are formed along the fabric length, i.e., in the wale or warp direction of the cloth. In both knitting systems the fabric is delivered in the wale direction. Special needles are used to form the yarn loops as shown in Figure 4.47. The latch needle is the most common type in use for weft knitted fabrics and the compound needle is used mostly in warp knitting. Spring beard needles are becoming obsolete.

The basis of knit fabric construction being the continuing intersecting of loops, any failure of a loop yarn will cause a progressive destruction of the loop sequence and a run occurs. Thus, knitting yarns must be of good quality in order that yarn failures be kept at a minimum. Other important geometrical definitions relating the knit structures are as follows:

- count: total number of wales and courses per unit area of the fabric
- gauge: the number of needles per unit width (the fineness or coarseness of the fabric)
- stitch: the loop formed at each needle

(the basic repeating unit of knit fabric structure)
- technical face: the side of the fabric where the loops are pulled toward the viewer
- technical back: the side of the fabric where the loops are pulled away from the viewer

Industrial application areas of knit structures include medical products such as artificial arteries, bandages, casts, and surgical gauze and flexible composites. Knit fabrics are used as reinforcing base for resins used in cars, boats, and motorcycle helmets.

2. WEFT KNITTING

Weft knit goods are made by feeding a multiple number of ends into the machine. Each loop is progressively made by the needle or needles. Figure 4.48 shows the loop forming process with a latch needle. The previously formed yarn loop actually becomes an element of the knitting process with the latch needle. This is why the latch needle is referred to as the "self-acting" needle. As the needle is caused to slide through the previous yarn loop, the loop causes the swiveled latch to open, exposing the open hook (head) of the needle. The newly selected yarn can now be guided and fed to the needle. If a simple knitted loop is to be formed, the previous loop (the one which opened the latch) must slide to a point on the needle stem allowing it to clear the latch. Having the needle reach this clearing position allows a reversal of the sliding action which

FIGURE 4.46 Circular weft knitting machine.

in turn pulls down on the new yarn and uses the previous yarn loop to close the latch trapping the new yarn inside the hook. The previous loop is now in a position to ride over the outside of the latch and be cast off the needle head, thus becoming a part of the fabric while the new yarn loop is pulled through the previous loop.

Depending on the structure in weft knitting, several types of knitting stitches are used including plain [Figure 4.1(c)], tuck, purl (reverse), and float (miss) stitch which are shown in Figure 4.49. The plain stitch fabric has all of its loops drawn through to the same side of the fabric. The plain fabric has a very smooth face and a rough back. Other stitches produce different effects depending on the arrangement of the loops. Special stitches are also available to prevent runs.

Weft knitting machines may be either flat or

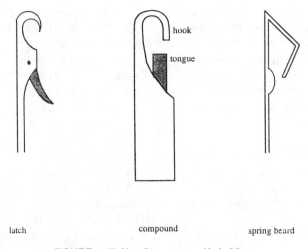

FIGURE 4.47 Needle types used in knitting.

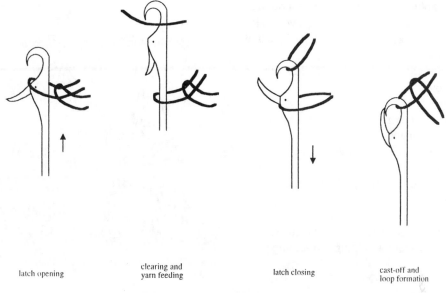

| latch opening | clearing and yarn feeding | latch closing | cast-off and loop formation |

FIGURE 4.48 Loop forming on a latch needle.

circular, the former knitting a flat single layer of fabric, the latter knitting a continuous tube. No matter which machine configuration is used, weft knit manufacturing involves the same fundamental functions:

- yarn selection and feeding
- needle knitting action
- fabric control during knitting
- needle selection
- fabric take-up and collection

There are several devices added to weft knitting machinery for improving quality of the product and/or operation of the process. A partial list of these added features includes:

- yarn break sensors
- fabric hole detectors
- needle "closed latch" sensors
- air blowing systems to keep needles clear of lint
- centralized lubrication dispensing unit
- computer interfaces for production monitoring
- computer interfaces for pattern entry
- computer aided design systems

Knit fabrics can be classified as single knits and double knits. Single weft knits have one layer of loops formed with one yarn system. Three major types of single weft knits are jersey, rib and purl structures. Double knits have two insepa-

| purl (reverse) stitch | float (miss) stitch | tuck stitch |

FIGURE 4.49 Types of stitching in weft knitting.

rable layers of loops. Each yarn forms loops that appear on both faces of the fabric. Two major types of double knits are rib double knit and interlock double knit.

The major characteristics of weft knit fabrics are as follows:

- can be either manufactured as net-shape or cut to shape and sewn
- form a run the wale direction if a yarn breaks
- have good stretch especially in the course direction
- do not ravel
- do not wrinkle easily and have good recovery from wrinkling and folding

3. WARP KNITTING

Warp knit fabrics are manufactured by preparing the equivalent of a warp beam containing several hundred ends. Each end passes through its own needle and is formed into loops which intersect with adjacent loops. Thus, a flat looped fabric is knitted using only "warp" yarns without the necessity of "filling" yarns being interwoven.

The two major types of warp knits are tricot and Raschel. Based on the number of yarns and guide bars used, tricot knits are identified as single [Figure 4.1(d)], two (Figure 4.50), three and four (or more) bar tricots. Raschel knitting is suitable for making highly patterned, lacy, crocheted or specialty knits (Figure 4.51). In general, Raschel machines are used for the production of knit structures for industrial applications. For increased structural support in the filling direction, additional filling yarns can be inserted as shown in Figure 4.52.

Figure 4.53 shows a schematic of a warp knitting machine. The knitting elements required for a warp knitting machine include:

- needles arranged in one or more solid bar to function as a unit (called a needle bar)
- yarn guides, one for each warp yarn, arranged in solid bars, one for each different warp, to function as a unit (called guide bars)

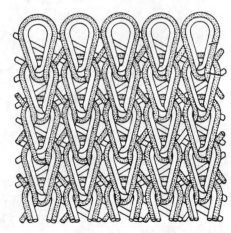

FIGURE 4.50 A two-bar tricot (courtesy of Noyes Publications).

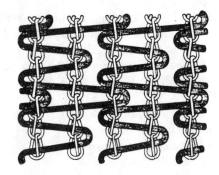

FIGURE 4.51 Simple Raschel crochet knit (courtesy of Noyes Publications).

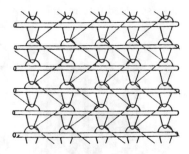

FIGURE 4.52 Weft inserted warp knit structure (courtesy of Karl Mayer).

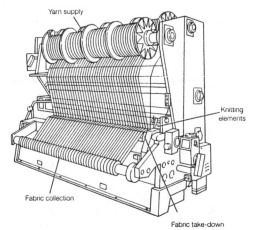

FIGURE 4.53 Schematic of a warp knitting machine (courtesy of Noyes Publications).

- sinkers arranged in a solid bar to function as a unit (called a sinker bar)
- presser bar functioning on spring beard needle bars only
- pattern chains or cams to control the side-to-side motion of guide bars
- in Raschel knitting, over the self-acting latch needle, no pressure bar is needed; in its place, a latch guard and a face

plate, over which the old loops are "knocked-over," are used

Stitch formation on warp knit machines differs from weft knitting in that a complete course of loops is formed by one cycle of the needle bar(s) rather than individually acting needles forming loops within a course.

It is estimated that 85% of the knit structures used directly or as composites in the field of industrial textiles are warp knits.

Open and closed warp knit structures are used in protection from insects, sun protection, harvesting nets and heavy fishing nets.

Three dimensional warp knit structures such as sandwich structures are made on machines with two separate needle bars working independently of each other and forming a separate fabric on each side. Another fully threaded guide bar fills the sandwich by overlapping both needle bars. Knit fabrics can also be coated and molded into three-dimensional honeycomb structures.

Mono- and multi-axial knit structures are produced by inserting layers of straight reinforcement yarns into the knit construction. These structures are also called directionally oriented

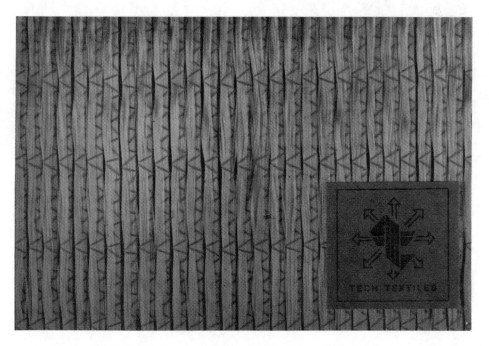

FIGURE 4.54 No-crimp fabric structure (courtesy of Tech Textiles).

structures (DOS) or no-crimp fabrics. The reinforcement fibers can be oriented at various directions including 0°, 90°, 45°, etc., as shown in Figure 4.54. Some of these structures are used for composite reinforcement (Chapter 7.0).

4. KNIT FABRIC CHARACTERISTICS

The basic knit structures have the following general characteristics:

- High extendibility is seen.
- elastic recovery: Knit fabrics recover well from deformation since the loops attempt to return to their original position.
- shape retention: When the loops do not recover from deformation, the fabric stretches out of shape or it bags.
- Knits generally crease or wrinkle less than woven fabrics since the loops act as hinges (which is a problem when crease is needed).
- Relaxation shrinkage of knit fabrics can be high, especially if the fibers are hydrophilic.
- Pliability and form-fitting due to the loop structure and to the low twist in some weft knit yarns.
- Pilling and snagging due to ease of catching a loop and pulling it to the fabric surface.
- insulative ability and air permeability: Knit fabrics, which are open structures, are ideally suited for providing thermal insulation in still air conditions. Knit structure contributes a desirable cooling effect in summer and a chilling effect in winter due to wind.
- running or laddering: Knit fabrics may run or ladder if a yarn in the structure is broken.

Braiding and Narrow Fabrics

S. ADANUR

1. BRAIDING

Braiding is the simplest form of fabric formation and probably it is older than weaving. A braid structure is formed by the diagonal intersection of yarns. There are no warp and filling yarns in the sense of a woven fabric. Braiding does not require beat-up and shedding; the yarns do not have to go through heddles and reed.

Braiding is more significant for industrial fabrics than consumer textiles. Although the volume of braided structures is small compared to weaving and knitting in consumer textiles, braiding is one of the major fabrication methods for composite reinforcement structures. With increasing applications of industrial textiles, the use of braided fabrics is also increasing. Traditional examples of the braid structures for industrial applications are electrical wires and cables, harnesses, hoses, industrial belts and surgical sutures. Examples of the relatively new application areas for braiding include reinforcement structures of sporting goods (baseball bats, golf clubs, water skis, snow skis), aerospace and automotive parts.

The geometry of the braided fabric is directly related to the machinery which forms the fabric. By understanding the relationship between the machinery and the yarns, it is possible to construct diagrams which describe the structure of the braid. Braiding can be classified as two- and three-dimensional braiding. Two-dimensional braid structures can be circular or flat braids. Three-dimensional braiding is relatively new and was developed mainly for composite structures. Although circular and flat braids have thickness, it is small compared to the other two dimensions; therefore they are considered as two-dimensional.

2. TWO-DIMENSIONAL BRAIDING

A two-dimensional circular or flat braid is formed by crossing a number of yarns diagonally so that each yarn passes alternately over and under one or more of the others. The most common designs in two-dimensional braids are as follows (Figure 4.55):

- diamond braid: 1/1 intersection repeat
- regular braid: 2/2 intersection repeat
- Hercules braid: 3/3 intersection repeat

Referring to Figure 4.56, the width L is called a line and the length S is called a stitch or pick. W is the width of yarn from which the braid is made. Braid structures are specified by the line and stitch numbers. α is the braid angle which specifies the angle by which the yarns lay from the direction of machine axis. Braid angle is determined from the relation between the take-up speed and carrier speed.

In a special type of braid for composites, called triaxial braiding, axial yarns are introduced to the structure as shown in Figure 4.57. The axial yarns do not interlace or intertwine with other yarns and are trapped between the two sets of yarns in the structure.

2.1 Circular Braiding

Circular (tubular or round) braids are formed hollow or around a center core. A circular braid-

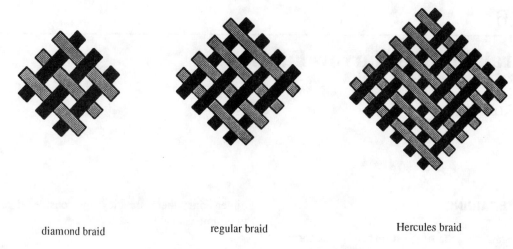

diamond braid   regular braid   Hercules braid

FIGURE 4.55 Common types of braiding patterns.

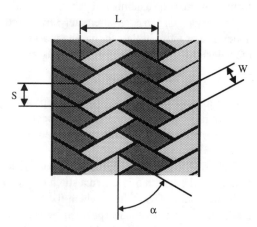

FIGURE 4.56 Braiding fabric parameters.

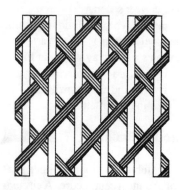

FIGURE 4.57 Triaxial braid.

ing machine consists of two sets of an even number of spools containing the braiding yarns. One set runs clockwise around the center of the machine and the other set turn in counterclockwise direction as shown in Figure 4.58. While revolving in opposite directions, the carriers are diverted to pass alternately inside and outside (under and over) one another. The clockwise and counterclockwise paths cause the two sets of yarns to intersect, thus producing a tubular braid. The yarns from the bobbins are collected above the hub of the circular track in which the bobbins travel. Since the speed of the yarn carriers is constant, the openness of the fabric is changed by changing the take-up speed of the fabric. Circular braiders are also called ''maypole'' braiders since their motion is similar to the maypole dance.

The tube can be braided about a central core of yarns, thus producing a solid braid composed of a core and ''sleeve.'' The core, in fact, can be any shape and material. This is the reason why circular braiding is widely used in composite preform manufacturing: unlike weaving or knitting, the braiding structure conforms to the core very well thus making it easy to develop braided structures of complex shapes (Figure 4.59). Depending upon the number of spools (carriers), the nature of the yarn, the path of the spools and the core, endless numbers of braided structures can be produced.

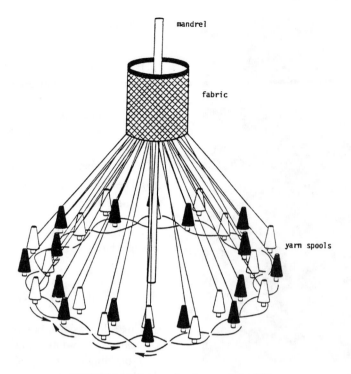

FIGURE 4.58 Schematic of maypole braiding.

Figure 4.60 shows a maypole braiding machine. The mechanical components of a maypole braiding machine can be grouped under four categories:

- track plate
- horn gears
- spool carriers
- fabric take-up mechanism

Another type of braiding machine to produce two-dimensional braids is the rotary machine. The rotary braiding machine is faster than the maypole braider. However, rotary machines are less versatile in terms of making different shapes and they have less number of carriers.

2.2 Flat Braiding

Flat braids are made in the form of a flat strip or tape. In flat braiding, instead of following two continuous paths, the carriers turn around or reverse direction at two given points called "terminals" and then continue on the opposite track, i.e., the track does not complete a circle as shown in Figure 4.61.

FIGURE 4.59 Braiding complex structures (courtesy of Fiber Innovations, Inc.).

FIGURE 4.60 Maypole braiding machine (courtesy of North Carolina State University).

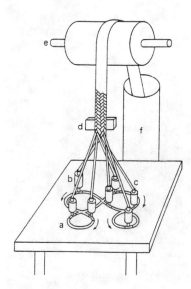

a track plate
b spool carrier
c braiding yarn
d braiding point and former
e take off roll with change gears
f delivery can

FIGURE 4.61 Flat braiding (courtesy of Ciba Geigy).

The size of a braid is governed by the following factors:

(1) The number of carriers: Tubular braiders have an even number of carriers, and flat braiders usually have an odd number of carriers. The minimum number of carriers is three which gives a basic diamond braid similar to a girl's plaited hair.

(2) The diameter of the yarns

(3) The number of yarn ends per carrier

(4) The number of yarns per unit length

Figure 4.62 shows various two-dimensional braided fabric structures.

3. THREE-DIMENSIONAL BRAIDING

Three-dimensional braiding is relatively new compared to two-dimensional braiding. The first 3-D braiding machine was developed in the 1960s. The three-dimensional braiding concept has been developed mainly for textile structural composites. There is no three-dimensional braiding machine that is commercially available. The main reason for this is that every different three-dimensional braided structure requires a different machine with specific characteristics and dimensions. Therefore, companies and research institutions custom-build their 3-D braiding machines. Three-dimensional braiding is explained in Chapter 7.0, Textile Structural Composites.

4. NARROW FABRIC WEAVING

Ribbons, tapes and webbings are all considered as woven narrow fabrics if they contain woven selvages, and are less than 12 inches. They are woven on special narrow fabric looms, using the basic principle of warp and filling interlacing. Several sets of warp yarns may be beamed to make several narrow fabrics, side by side, on the same loom. Some tapes and ribbons are prepared by cutting full width fabrics into strips, and sealing the edges. Thermoplastic fiber ribbons can be made this way. Elastic webbing or tape is made by using bare or wrapped rubber

FIGURE 4.62 Two-dimensional braided fabric structures (courtesy of Wardwell Braiding Machines Co.).

FIGURE 4.63 Narrow needle loom (courtesy of Jakob Muller Ltd.).

warp yarn. Heavy webbings are often made with *stuffer* warps embedded between the webbing face and back. In essence, such webbings are double or tubular fabrics with reinforcing warp yarns lying between the two layers, and intersecting both sufficiently to bind the composite structure together, forming an integral unit.

Although some shuttle looms are still used, narrow woven fabrics are generally manufactured on needle looms. Because of the importance of a good selvage, shuttleless methods are generally not used. Figure 4.63 shows a needle loom whose functions are very similar to a regular loom. In narrow fabric weaving, a single warp beam or multiple warp beams can be used. The warp yarn can also be taken directly from the creel for simple designs with small number of ends.

In narrow fabric weaving, a hairpin loop of filling is inserted into the shed at very high speed by the needle. Then the filling yarn is locked in at the opposite selvage by a knitting latch needle. Due to the loop of the filling yarn, the fabric always has double picks. The selvage on the filling insertion side is a woven selvage and the selvage on the latch needle side is a knit selvage.

4.7

Tufting

R. P. WALKER
S. ADANUR

A tufted fabric involves a system of yarns that are sewn into a fabric (primary backing) leaving yarn loops below the primary backing (Figure 4.64). Sewing is accomplished by a solid bar of needles extending the full width of the primary backing fabric with each needle threaded with a yarn. Because the needles are spaced a certain distance apart (called the needle gauge), the yarns will be aligned in the structure in precise vertical rows across the width (or horizontal direction) of the fabric. The rate of reciprocal up and down motions by the needle bar is referred to as "stitches per minute" or the speed of the needle bar in revolutions per minute (rpm). Spacing of these needle bar stitches is controlled by the feeding rate of the primary backing fabric resulting in a specified number of "stitches per inch" along the machine (or vertical) direction of the fabric. The primary backing fabric is supported by a bed plate which may be lowered or raised to control the depth of needle penetration through the fabric carrying the yarn loops to a desired distance below the primary fabric (the fabric is made upside down). The distance of the loops below the primary fabric constitutes the "pile height" of yarn tufts. These surface tufts represent the face or use side of the fabric after the fabric is removed and finished.

Different type tufting elements are used to produce various surface effects. Fabrics may have surfaces of level loops, multi-level loops, level cut loops, multi-level cut and uncut loops or level cut and uncut loops. These variations are made possible by the types of loopers, knives and yarn feed controls used. A simple uncut loop surface is produced as shown in Figure 4.64(a) when the loopers take yarns from the needles and hold the loops formed while the needles retract.

Notice that the looper is facing in the direction

of fabric motion to allow the formed loops to move off the looper after forming. Cut pile surface is produced by changing the type of looper with corresponding knives. As shown in Figure 4.64(b), the loops are formed on the looper and must progress through the knives and be cut free of the looper which points opposite the direction of fabric flow. In all cases of level loop surfaces, the feed rate of yarns must be adjusted to accomplish the desired pile height (in conjunction with the bed plate setting) and to obtain a secure (tight) backstitch between surface stitches.

Multi-level textured designs can be created on the fabric surface by introducing variable yarn feed to selected needles. In these patterned surfaces a special looper and knife assembly is used which allows yarns with unrestricted feed rates to process as normal length loops, either cut or uncut. Restriction of the yarn feed causes loops to be drawn back off the looper allowing them to pass over the looper as lower level uncut loops. In all cases a low level loop is produced by a process called "stitch robbing" where a needle with restricted yarn feed steals yarn from the previous stitch causing the previous stitch to be low. Two yarn feed systems are commonly used to produce these high-low stitch patterns.

Another feed system called a "scroll" uses photoelectric cells to read an acetate film pattern which in turn controls two drive rolls for fast or slow yarn feed to produce high or low loops. The slat system is more suited to long production runs of the same pattern while the scroll system works better for short runs of often changing styles. Both systems must operate with an individual package creel rather than a beam creel since each yarn is supplied to the machine at a variable rate. Still other special looper assemblies can be

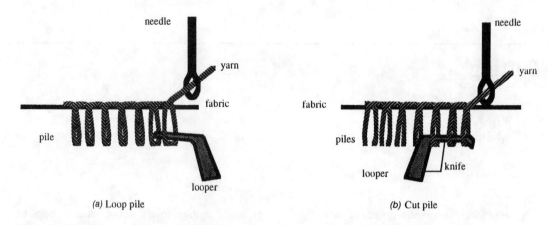

(a) Loop pile   (b) Cut pile

FIGURE 4.64 Tufting elements (a) loop pile and (b) cut pile.

employed to create cut and uncut loops at the same level.

Since the fabric is tufted upside down, a system of rolls is used to guide the fabric up and over an inspection zone so that the surface of the fabric can be observed prior to final roll up. This provides not only an opportunity for inspection but also an opportunity to repair lines in the surface which may have been too tight, too loose or missing entirely. This surface mending is done with a single needle, hand-held gun and scissors (for cut pile).

After the tufting process, the unfinished rolls of fabric are stored to await orders for coloration before final back-coating.

Color may be added prior to fiber manufacturing, to the fiber or to the yarn before tufting. Usually, coloration is done after tufting. However the color is applied, the fabric is processed through a final back-coating where latex and a secondary backing fabric is added to secure the entire surface yarn, primary backing and secondary backing assembly. In some cases only the latex is added and in others a foam rubber coating is applied. Most tufted fabrics are composed of a woven primary backing fabric (split film polypropylene yarns) and a woven secondary backing fabric (either spun jute or fibrillated polypropylene yarns).

4.8

Nonwoven Fabrics

R. M. BROUGHTON, JR.
P. H. BRADY

1. INTRODUCTION

A fabric is a unit of solid matter, generally composed of fibers, and is essentially two-dimensional in shape (i.e., large area but small thickness). Because of its small thickness and fibrous nature, a fabric is usually flexible. The word nonwoven is not very descriptive in that, at best, it describes what fabric is not. Nonwoven usually implies that the fabric has been formed by a procedure other than weaving, knitting, braiding or tufting. Weaving, knitting, braiding and tufting are considered conventional methods for the production of fabrics, so the term non-conventional fabric is sometimes used for those formed by other methods. A reasonable definition of a nonwoven fabric is a fabric made directly from a web of fiber or film, without the intermediate step of yarn manufacture (necessary for weaving, braiding, knitting or tufting). Although paper also fits this definition, it is not traditionally considered to be a nonwoven fabric. Frequently, woolen felts are also excluded from the nonwoven definition.

Often, the word fabric is omitted in the description of nonwoven materials, and they are simply referred to as nonwovens. A schematic definition of nonwoven processes is shown in Figure 4.65.

A composite fabric (several fabric components combined into a single fabric) is usually considered to be a nonwoven if one of the components qualifies by definition. Because of manufacturing and product similarities between nonwovens and paper, a nonwoven fabric is often considered as an intermediate form between paper and conventional textile fabrics. Table 4.5 shows a description of textile fabric properties and paper properties. As may be expected, non-woven fabric properties are frequently intermediate between that of paper and conventional textiles.

1.1 Why Nonwovens?

In general, nonwovens compete with textiles or paper in already established markets. One might wonder why companies would undertake to produce nonwovens. Cost advantages have been one of the many reasons. The production of a nonwoven blanket, for example, requires less fiber, labor, equipment, and time than the production of a woven blanket. To illustrate the time savings, Table 4.6 shows the production speeds of various textile fabric manufacturing operations. Of course, the nonwoven blanket is substantially less durable, but with proper care in washing, the nonwoven can last five to ten years, and that is sufficiently durable to satisfy most customers. Other textile companies have envisioned nonwovens to be a profitable way of disposing of waste fiber from conventional textile manufacturing operations. Even yarns and fabrics can be chopped and pulled apart to make suitable raw material for some nonwoven production.

Paper companies have a different motivation for moving toward nonwovens. Paper is a relatively low-priced commodity. Any change which will allow for the production of a more fabric-like material generally increases the value of the commodity, and thereby its profitability.

In most cases, the cost of a nonwoven article is less than a commensurate, conventional textile. However, if the nonwoven product has been designed to be disposable, the cost per end-use may actually be higher. Obviously, considera-

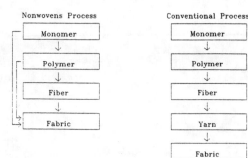

FIGURE 4.65 Major steps in nonwoven and conventional fabric manufacturing.

tions other than cost are important in justifying the selection of a nonwoven product. The advantages offered by a nonwoven product depend on the end-use and the competing product as shown in Table 4.7.

1.2 Markets for Nonwovens

Table 4.8 presents a list of various end uses for nonwoven fabrics and their production methods. The total of nonwoven products consumed in the U.S. in 1986 was 1.15 billion pounds having a total value of about 1.8 billion dollars. This is almost double the volume consumed ten years earlier. This represents an increase of about 7% per year. A 1994 listing of the top 40 nonwovens companies worldwide allows an estimate of the 1993 U.S. sales of nonwoven roll goods at about $2.2 billion, with worldwide sales over $6 billion [6]. This represents mill value of the nonwoven fabric rather than the retail value of the products made from these nonwoven materials. Also omitted from these 1993 statistics are the producers of glass mat (roofing and insulation), and those manufacturers who use all of their nonwovens production internally in the manufacture of other products.

Deciding which of these products should be called industrial fabrics is not a simple process. Fabrics used in automobiles are sometimes considered as industrial fabrics even though some of them actually qualify as upholstery. Carpet components, particularly backing fabrics, are included in some lists of industrial fabrics, but not in others. A coated fabric used for rainwear

TABLE 4.5 Comparison of Paper and Textiles.

Paper	Textiles
Cheap	More expensive
High production rate	Low production rate
Short fiber	Long fiber
Simple structure	Complex structure
Smooth surface	Textured surface
Dense	Bulky
Inextensible	Some extensibility
Able to hold sharp folds	Non-creasing
Stiff, little drape	Flexible, drapes easily
Relatively non-porous	Porous
Low wet strength	Wet strength is essential
Low tear strength	High tear strength

TABLE 4.6 Productivity of Fabric Forming Methods.

Process	Linear Speed (m/hour)
Weaving	5–40
Knitting	100
Mechanical web formation	3,000
Spunbond	4,500
Papermaking	50,000

TABLE 4.7 Advantages of Nonwovens.

End Use	Competes with	Nonwoven Advantage
Drapes	Woven fabric	Cost
Blinds, shades	Woven fabric, aluminum, wood	Cost
Carpet backing	Woven jute	Cost, supply
Wall covering	Paper, paint	Strength, wear
Packaging	Paper	Strength
Linings	Woven fabric	Cost
Fusible lining	Woven fabric	Application ease
Filter media	Woven fabric	Performance
Bandages	Woven fabric	Cost, performance
Surgical wear	Woven fabric	Reduced infection
Wipes	Paper	Durability, less lint
Wiper	Woven fabric	Disposable, cost

TABLE 4.8 *End Uses and Production Methods of Selected Nonwovens [7].*

Fabric End Use	Production Method
Cover stock	Card, air lay, wet lay, spunbond
Interlinings/interfacings	Card, spunbond, spun laced
Filtration	All types
Surgical fabrics	Composite, spunbond, melt-blown, spun laced
Wipes/towels	All types
Coating/laminating	Card, air lay (needled), spunbond, spun laced
Bedding/home furnishings	All except wet lay
Durable papers	Spunbond
Geotextile/construction	Wet lay (roofing), spunbond/needled (drainage)
Other	All types

would be a consumer product, while the same fabric used for a tent or tarp would be an industrial fabric. There does seem to be general agreement that filtration products, medical textiles, coating and laminating substrates, and geotextiles all qualify as industrial textiles.

2. CLASSIFICATION OF NONWOVENS

Nonwovens can be broken into groups based on the fiber type, method and type of reinforcement, web preparation, and method of bonding between fibers. The presence of reinforcement implies a composite structure. Table 4.9 is a classification scheme for nonwoven fabrics. Other schemes have been proposed, but, this one

TABLE 4.9 *Classification of Nonwovens.*

By . . .	
Fiber	
Chemical type	Rayon, polyester, etc.
Physical form	Filament, staple, file, foam
Web formation	Mechanical
	Fluid: air, water
	Electrostatic
	Film extrusion
	Monomer casting
Bonding of fibers	Mechanical: needle, fluid
	Chemical: adhesive, self
Reinforcements	None
	Filaments
	Fabric
	Foam

is more favorable because a nonwoven fabric usually finds one place where it definitely fits within the system.

Many of the properties of a nonwoven fabric are derived directly from a similar property in its fibers. The translation of fiber properties into fabric properties is dependent upon the geometrical arrangement of the fibers in the fabric and the properties of any adhesive or other bonding agent present. The methods of producing a nonwoven fabric depends upon the form of the raw material. Nonwoven textiles are usually made from fibrous materials, and therefore, most of this section will center on the production of fibrous webs. In order to be considered a nonwoven, a film must either be treated to make it fiber-like, or the film must be combined with fibrous material. Likewise, a sheet of foam is not considered a nonwoven unless combined with fibrous materials.

3. PRODUCTION OF NONWOVENS

There are two major steps in the manufacture of nonwovens: web formation and bonding of fibers in the web.

3.1 Web Formation

Figures 4.66 through 4.69 show some of the mechanical methods of nonwoven fabric formation. All of these methods involve the use of rotating, wire covered cylinders to pick up individual (staple) fibers, and transport and deposit

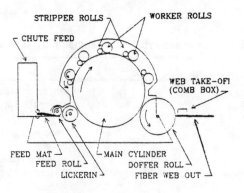

FIGURE 4.66 Roller-top card.

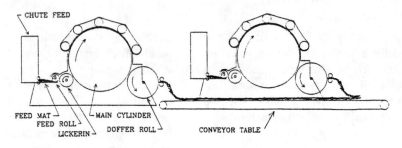

FIGURE 4.67 Flat-top cards in tandem.

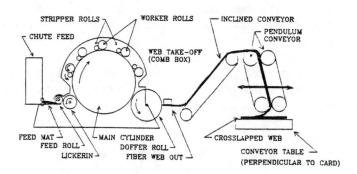

FIGURE 4.68 Carding and crosslapping.

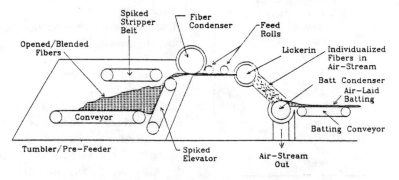

FIGURE 4.69 Air-lay process (courtesy of Rando-Webber®).

144

them in a sheet or batt form. A typical roller-top or flat-top card produces a web too thin to be used in a single layer form for most applications. Consequently, card webs from multiple cards are often stacked or the web from a single card is folded in a crosslapper form to create a multi-layered batting.

A card usually operates to align fibers in the direction of web travel (machine direction or MD). Card webs have a higher strength in the machine direction than in the cross-machine direction (CD). The ratio of MD/CD strength may be as high as 10/1. If the web is crosslapped, the orientation is changed from the direction of web travel to the cross direction. There is some dispersion of fiber orientation as the web conveyance continues while the web is being folded (Figure 4.68). Still there is a high ratio of CD/MD strength in crosslapped webs. Through the addition of extra cylinders or rollers, some cards allow for the manipulation of roll speeds in an attempt to minimize the fiber alignment. A garnet is a card-like machine, but having a heavier wire than a typical card. It is designed to tear yarn and small pieces of fabric into individual fibers and form a web or batt.

The most effective way of minimizing fiber alignment is to sweep the opened fibers from a wire wound roll (either card or opening system) into a stream of air, and then condense the fiber on a slowly moving screen or perforated drum. This type of product is referred to as an air-laid web. Normal cards with add-on air handling and condensing attachments were often constructed in-house in yarn manufacturing operations to allow the production of a marketable product from manufacturing waste. As the products became more sophisticated and profitable, machinery specifically for air laying was designed. Air-lay machinery is now offered by a variety of companies including Rando Machine Co., D.O.A., and Fehrer. Figure 4.69 shows the internal operation of a typical air-lay machine.

Air can also be used as a dispersing medium for continuous filament fibers. This type of process, producing a spunbonded material, links the fiber manufacturing and web formation processes into a single operation. A schematic of a spunbond process appears in Figure 4.70. If continuous filament fibers are allowed, or forced

to fall through the air and land on a moving conveyor, a nonwoven batt will be produced. A large number of filaments (perhaps from multiple spinnerettes) are required to produce a web of reasonable width, and care must be taken to promote random dispersion of the filaments. The use of electrostatics may be used to keep the filaments separated and dispersed in the web. Bonding of the fibers (usually thermal bonding or needlepunching) is done in line before take-up of the batt.

Melt blowing is similar to spunbonding, except that a high velocity stream of air is used to force the filaments away from the spinnerette face (Figure 4.71). The air stream can attenuate the fibers to a degree of fineness smaller than typically extruded textile fibers. The fibers (normally less than one denier) are deposited onto a moving condenser/conveyer belt and bonded before take-up. This type of process allows for the creation of a web having a very fine pore structure and large surface area. Such fabrics are generally suitable for absorption and filtration applications.

Air is not the only fluid which can be used to deposit a fibrous web onto a condenser screen. Paper has been produced for centuries by suspending fibers in water and then draining the suspension through a condensing screen (Chapter 12.0). The high density, viscosity and surface tension of water (compared to air) causes wetlaid nonwovens to be very dense in structure compared to those which are air laid. Wet-laid fabrics have some directional character, being stronger in the machine direction than the cross direction. The directionality (usually less than carded systems) is introduced when the fiber suspension is accelerated by contact with the

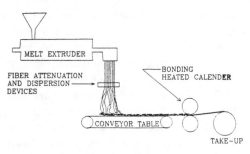

FIGURE 4.70 Spunbond process.

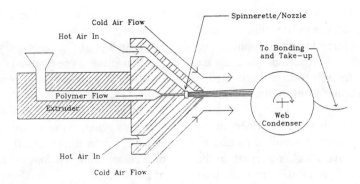

FIGURE 4.71 Melt blowing process.

moving screen. The greater the acceleration to machine speed, the greater is the directionality in the fabric.

A second major difference between wet-lay and air-lay systems is that in the wet-lay system, water must be removed in a drying zone. Several geometries of the condensing zone are possible. Fourdrinier, inclined wire, and cylinder machines are typical of the various types of machinery in use. Figures 12.2 and 12.3 (Chapter 12.0) show schematics of Fourdrinier machine. After formation, water is removed from the web by gravity, vacuum extraction, pressing and finally by heating.

Paper is typically made from fibers which are less than 0.25 inch in length. Textiles normally use fibers which are an inch or more in length. When flexible fibers of longer than an inch in length are suspended and stirred in water, they tend to become entangled. Only through very high dilution ratios can long fibers (0.5 to 1 inch) be processed on a paper machine. The processing of fibers longer than one inch is impractical on ordinary paper machines. A variety of saturation substrates, containing glass fiber or mixtures of wood pulp and other fibers are made on paper machines. These mixtures of wood and other longer fibers are often used to create absorbent wiping materials. Glass fiber mats produced on wet-lay machinery are used for roofing products.

Some production processes bypass the fiber manufacturing step and produce a flat sheet (film) directly. Since some porosity is usually required of a fabric, a flat, impervious film would not likely be considered to be a fabric. However, if there are a sufficient number of holes in the film, making it porous and flexible, it may qualify as fabric. There are two manufacturing processes which can produce a film having sufficient flexibility and porosity. These are the fibrillated film process and a process originally developed by Smith and Nephew Ltd., which embosses and then stretches the film until it ruptures in its thin spots. The embossing may, in the extreme, punch holes in the film. A diagram of the fibrillated film extrusion process appears in Figure 4.72.

If a film of polyolefin is stretched to a high degree in the machine direction, it develops strength in that direction, and weakness in the cross-machine direction. If the film is then mechanically worked, splits may appear along the machine direction, causing the film to become more like a web of fiber. The splitting (or fibrillation) process can be accelerated by either abrading the film along the machine direction, passing the oriented film over a pinned or bladed roller, or extruding the film onto a profiled roller, which has grooves parallel to the machine direction. A highly embossed film, or a film with an array of holes punched in it tends to form a porous network when stretched biaxially. This may produce a fine netting, or even a tough plastic netting suitable for fences and sieves.

In the extrusion of film, the polymer is processed through an extruder equipped with a thin, flat die. It is possible to start the production of a film from a monomer spread on the surface of a solid or liquid. In this case, the monomer is polymerized directly into the film shape. This technique is particularly useful when producing microporous films or film coatings on substrates.

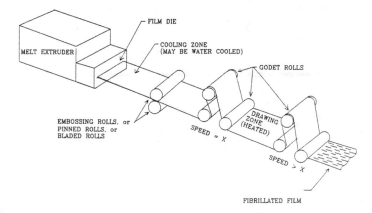

FIGURE 4.72 Fibrillated film process.

Such processes are sometimes referred to as monomer casting.

Flocking operations involve the perpendicular arrangement of short fibers on a substrate sheet (Figure 4.73). Velvet is an example of a woven fabric with short fibers perpendicular to the fabric surface. Flocking attempts to reproduce a velvet appearance by gluing short fibers on end to a fabric substrate. The fibers can be made to stand on end either by vibration of the sheet, or by electrostatics which cause the fibers to fly through the air, impinging end first into the glue on the fabric surface. Patterning possibilities are introduced when the glue is applied in the desired pattern on the fabric surface.

A particularly ingenious, nonwoven production method involves the direct extrusion of a fabric. Two concentric circles of spinnerette holes which rotate in opposite directions can produce a web that appears similar to a loosely woven, bagging material. The cylindrical material is frequently used for food packaging. Larger versions, when split, can be used for netting, fencing and geotextile grids. This method of production is referred to as the Netlon process. A schematic diagram of the process is shown in Figure 4.74.

3.2 Fiber Bonding

When fibers are laid down in a batt without the interlacing provided by knitting or weaving, the batt has little strength. The entanglement of fibers, or the addition of a bonding agent is

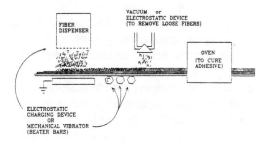

FIGURE 4.73 Flocking.

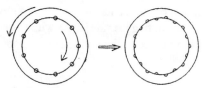

Rotating concentric spinnerettes

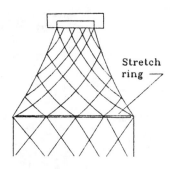

Stretch ring

FIGURE 4.74 Netlon process.

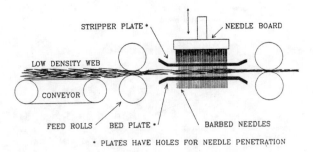

FIGURE 4.75 Needlepunching.

necessary if strength is to be developed. Fiber entanglement can be provided by either needlepunching (Figure 4.75) or by hydroentanglement (Figure 4.76).

In needlepunching, barbed needles are pushed through the web. The barbs catch fibers on the surface of the batt, and push them into the center, densifying the structure and producing strength through entanglement. The machine requires a bed plate to support the web as the needles penetrate, and a stripper plate to strip the fabric off of the needles as they are withdrawn. Needling can be done from both sides of the fabric, and penetration of needles does not have to be perpendicular to the sheet structure.

Hydroentanglement employs fine jets of water to push fibers from the surface toward the interior of the fabric. A more powerful jet may produce an apertured fabric, while less forceful jets tend to produce entanglement without completely penetrating the fabric. As with the needlepunching process, the batt must be supported during the impingement of the water stream. The size, spacing, and height of the knuckles on the conveyor or plate surface have a major influence on ability of the fibers to entangle, and the appearance of the fabric. In general, a stream of water does not exert as great a force on fibers as that produced in needlepunching. Therefore, hydroentangled structures are usually less dense and more flexible than those produced by needlepunching. Hydroentangled nonwovens are often referred to as spun laced fabrics. Hydrotanglement has an advantage over needlepunching in that it can be used to bond relatively thin fabrics.

An alternative to mechanical entanglement to provide structural integrity to a nonwoven material is the addition of an adhesive to bind the fibers together. The adhesive may be included as either a secondary treatment, or in the form of adhesive-containing fibers which are manufactured into the structure. Adhesives in the finished fabric are solid like the fibers they bind together,

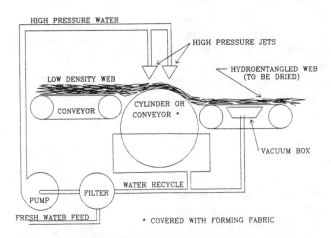

FIGURE 4.76 Hydroentanglement.

and are frequently applied to the web in solid form (powders or fibers). Because only liquids can stick to a solid surface (like a fiber), for the batt's fibers to be bound together by the adhesive, it must exist as a liquid at some predetermined moment while in the batt. A liquid state can exist in several forms: melts, solutions, emulsions, suspensions, pastes and foams. The form of the liquid is directly related to the process used for application and bonding of the fibers.

The application of liquid adhesives (subsequently solidified in the batt) may involve spraying, dipping and squeezing, printing, "kiss roll" or foam application. Figures 4.77–4.81 show schematics of representative application techniques (many system geometries are possible). An unbonded fiber batt is too fragile to survive most dipping and printing processes, so open-mesh screens may be used to support the batt through the bonding machinery. Alternatively, light needle tacking may be used to provide some mechanical integrity. Liquid adhesives applied in molten form (hot melts), may be sprayed or printed from heated rolls, and require only cooling to solidify and bind the fibers together. For adhesive solutions and dispersions, a solvent must be removed, usually requiring an oven for evaporation (or curing in the case of thermosetting adhesives).

The use of solid adhesives usually avoids the problems of solvent removal incurred by the use of adhesive solutions, emulsions, etc. Activation of the solid adhesive usually involves heating a thermoplastic material above its melting point; although softening by solvents or solvent vapors is occasionally used. Figures 4.82 and 4.83 show

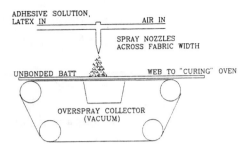

FIGURE 4.77 Spray application of adhesives.

two application methods for powdered solid adhesives. Adhesive powders tend to migrate or even fall out of the web as it moves along a conveyor until after the activation step. Small particles, fluidized bed application, and electrostatics may help in avoiding migration.

Solid, thermoplastic fibers may act as an adhesive in batts of higher melting fibers. The adhesive fibers are mixed with other fibers during batt formation and require only heating to produce bonding. A fiber adhesive tends to remain dispersed in the batting, thus avoiding the migration problems associated with powders. A particularly elegant example of adhesive fiber application is the bicomponent fiber, in which a low melting glue is combined with, or coats a much higher melting point material. Figure 4.84 shows a sheath-core bicomponent fiber alone and in a web.

At least one commercial manufacturing process uses solvent vapor activation of the fiber to produce a "glue" layer on the surface of each fiber in a batting, causing them to stick together. This process, known as the Cerex process is very

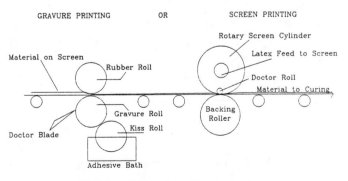

FIGURE 4.78 Printing of liquid adhesives.

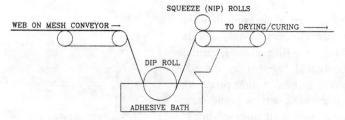

FIGURE 4.79 Dip and squeeze application of adhesives.

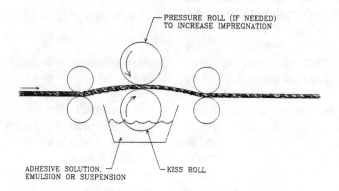

FIGURE 4.80 Kiss roll application of adhesives.

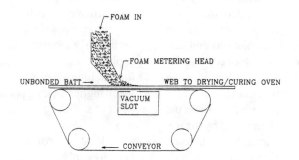

FIGURE 4.81 Foam application of adhesives.

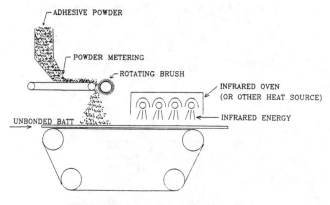

FIGURE 4.82 Powder adhesive sprinkling.

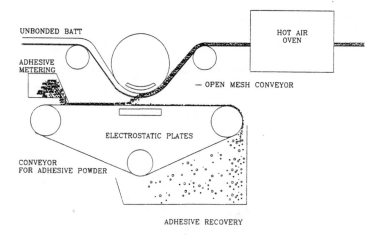

FIGURE 4.83 Electrostatically assisted powder adhesive application.

specific for a particular solvent and fiber type (HCl vapor and nylon).

Except for vapor-activated adhesives, all solid application processes require heating at some point during the process. Thermoplastic adhesives must be heated to cause flow and sticking. Adhesive/solvent systems require heating to remove the solvent after adhesion. Heat may be applied from hot, circulating fluids (usually air, shown in Figure 4.85), heated and rotating cylinders (Figure 4.86), or radiant (infrared) sources (Figure 4.82). Microwave/RF heating has also been investigated for use in bonding.

Thermoplastic adhesives may be activated by either convective, conductive, or radiant heating methods. Passing the web over a heated cylinder (conduction) can provide the required heat, but also compresses the fabric (pressure is an essential ingredient in this type of bonding). A calender with engraved roll provides point bonding, while a smooth roll tends to bond the entire area (Figure 4.86). Hot air can be forced through the fabric (convection) to melt the adhesive without producing excessive compression. Infrared heating (radiant) also minimizes compression, but works best on relatively thin or low-density materials. Ultrasonic energy can also be used to fuse materials, usually in a point bonded pattern.

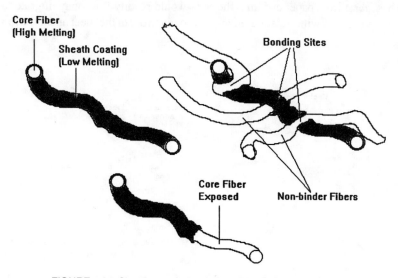

FIGURE 4.84 Sheath-core bicomponent fiber alone and in batting.

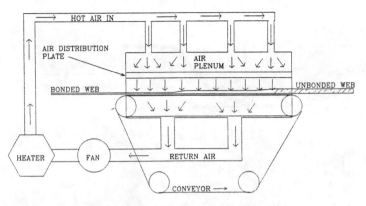

FIGURE 4.85 Through-air thermal bonding.

Several factors are important in determining the effectiveness of an adhesive in a given fabric:

- distribution of adhesive in web
- viscosity of adhesive while in the liquid state
- specific adhesion to fiber
- physical and chemical properties of adhesive (particularly T_g, solvents)
- adhesive stability to environment

The distribution of adhesives can range between the extremes of complete saturation of the structure, to bonding only at inter-fiber intersections (Figure 4.84). Area and point bonding are terms sometimes used to describe the distribution of the bonding sites. One of the characteristics that distinguishes fabric from paper or film is the ability of fibers to move with relative independence in the structure. Application of any adhesive interferes with fiber movement. Small islands of bonding separated by unbound areas provide for fabric-like properties. Bonding only at fiber intersections (as with the sheath/core bicomponent fibers) may also produce textile-like properties, provided that the intersections are not too numerous. As an adhesive begins to fill the spaces between fibers, the structure begins to behave more like a fiber reinforced plastic than a fabric.

The viscosity of the adhesive while liquid within the web is significant. As viscosity increases, the ability of a given material to flow is reduced. Absent proper flow, the adhesive will not sufficiently wet the surfaces that are desired to be bonded. Low viscosity materials, which would readily flow, may migrate to the bottom, or even out of the sheet under the force of gravity.

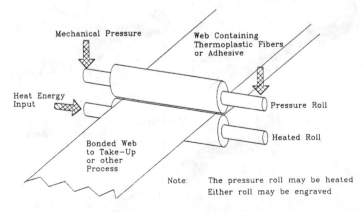

FIGURE 4.86 Calender bonding.

Some materials adhere well while others do not. Obviously, a desirable adhesive property is adherence to the batting fibers. Similarities in chemical polarity, or the matching of solubility parameters usually permits adequate adhesion between two materials.

Most adhesives are amorphous solids (except when a semicrystalline fiber is itself an adhesive). The strength of the adhesive in its solid form is obviously important in determining the strength of the resulting bonded fabric. Adhesive strength may be usually increased through an increase in molecular weight of the adhesive. Aside from molecular weight, the property which dominates the behavior of amorphous polymers is the glass transition temperature (T_g). An amorphous material with a T_g well below its end-use temperature will behave as a flexible, stretchable elastomer. If the T_g is above the end-use temperature, the material will be stiff and rigid. The T_g for most typical adhesives ranges from about $-100°C$ to $+100°C$.

Other important considerations in the selection of adhesives are color, odor, ease of application, stability in the environment, and the physical and environmental safety aspects of any required solvent. The ideal stability for most materials is complete stability during use, and instability and degradation as soon as it is discarded. A fabric must be stable to its use environment, but after use, biodegradability is usually considered helpful in minimizing the environmental impact of solid waste. The recovery and reuse of materials is increasingly important, but is usually difficult for intimately mixed materials like fiber and adhesive blends. Calender bonded thermoplastics of a single fiber type are ideal recycling candidates.

3.3 Reinforcement of Nonwovens

If a desirable, nonwoven fabric is too weak to be used in a particular application, it may be strengthened by the incorporation of yarn, film, foam, or other reinforcements. In some cases, the reinforcement actually serves as the binding agent holding the nonwoven together, and the fabric could not be made without the reinforcement. For example, a flocked fabric cannot be produced without the substrate, so the substrate/reinforcement is an integral component. Likewise, if an unbonded batt of loose fibers is passed through a sewing or knitting machine, the stitching will hold the batt together in a sheet known as a stitch-bonded fabric. The stitching yarn literally holds the fabric together, but reinforcement is not an adequate description of its purpose. The fabric is an open mesh interlocked yarn structure in which the interstices have been filled with nonwoven fiber batt. The absence of any adhesive bond between fibers gives these structures textile aesthetics missing from many other types of nonwoven fabrics. A wide variety of fabrics can be produced by stitch bonding technology. The process names for this technology include Mali® and Arachne®.

4. PROCESS SELECTION

Selection of the appropriate process technology for production of a nonwoven product may range from the simplest to the most complex ventures. For the company with waste fiber to dispose of, and surplus carding capacity, the addition of crosslapping and bonding may allow for the production of a material suitable for simple applications like sound insulation or upholstery padding. Conversely, the production of a clean room air filter or disposable apparel for protecting military personnel from chemical and biological threats, may require a process engineered specifically for the production of that product.

The raw materials required, or available, may dictate which manufacturing process options are feasible. Fibers of only 0.5 inch length are not likely candidates for carding or needlepunching. On the other hand, fibers over one inch in length would not be very suitable for a wet-lay process. The presence of thermoplastic fibers may allow the use of calender or ultrasonic bonding.

A filter for removing fine particles is likely to require very fine fibers for its construction. The incorporation of very fine fibers may then dictate the use of the melt blowing process. Filtration of stack gasses from coal fired boilers requires fibers and binders (if present) that resist both heat and acidic vapors. These requirements might dictate a needlepunched sulfar batt, or a thermosetting, heat resistant binder.

TABLE 4.10 *Industrial Applications of Nonwovens.*

Abrasives	Lofty or dense construction, large denier fibers, binder holds both fibers and abrasive
Acoustic materials	Dense to moderately lofty, often contain an impermeable film layer
Agricultural covers	Lightweight, thin, controlled porosity depending on temperature
Athletic mats	Lofty, resilient, relatively thick
Protective apparel	Limited porosity, restrict the passage of particles, liquids, gases
Automotive	Liners, carpet, mostly needlepunched, colored
Bandages	White, sterilizable, frequently hydroentangled
Brake linings	Resin impregnated, heat resistant
Thermal insulation	Glass, mineral wool, specialty fibers, moderate loft, mostly air laid, resin bonded
Envelopes	Typically spunbonded polypropylene, better tear resistance than paper
Ceiling tiles	Typically resin bonded fiberglass, non-flammable
Erosion control	Temporary—biodegradable, inexpensive; permanent—environmentally resistant
Drainage (geotextile)	Acts as a filter to keep soil out of drainage channel, bulky, environmentally resistant
Roofing shingles	Wet-laid fiberglass, some resin bonding, saturated with asphalt
Roofing felt	Typically polyester spunbonded or needled felt, moderate loft, saturated in situ
Filters	For oil, gas, food, air, etc. Loft and porosity depend on the materials to be filtered. Highly engineered product in terms of porosity, pore size and material chemistry
Papermaker's felt	Smooth needled layer on top of woven conveyor belt used in paper manufacture
Pond liners	Provide cushion to prevent puncture of the plastic liner
Road bed "support"	Needled heavyweight felt; provides permanent separation between soil and ballast
Flooring	Base material for asphalt or vinyl saturant in flooring for kitchens, etc.
Tents	Strong, water repellent, frequently layers of different fabric types
Paint applicators	Flocked or specially needled materials
Upholstery padding	High resilience, moderate to low loft, sufficient to support sitting or reclining person
Upholstery separators	Thin spunbonded or carded and resin bonded, separates layers of materials
Instrument wrap	Keeps surgical instruments sterile, similar to barrier fabrics, very water repellent
Computer disks	Extreme product uniformity required
Reinforced plastics	Glass or specialty fibers saturated with plastic resin (frequently thermosets)
Synthetic leather	Relatively dense, frequently needled, saturating substrate for vinyl or polyurethane
Home, industrial wipes	Bonded, reinforced air-laid pulp, or melt blown calender bonded absorbent sheets

While many of these decisions are primarily based in common sense, the selection of the best materials and manufacturing processes to use in a particular situation require extraordinary attention to detail.

5. INDUSTRIAL APPLICATIONS OF NONWOVENS

As mentioned earlier, the classification of products into industrial or consumer products is quite subjective. The authors have surveyed a list of nonwoven products compiled by Holliday and have selected a good cross section of industrial nonwoven fabrics [8]. The list appears in Table 4.10 with some comments about these products. Obviously Table 4.10 omits many of the industrial nonwoven products, but it should give the reader an idea of the diverse industrial applications of nonwoven products.

4.9

References and Review Questions

1. REFERENCES

1. McAllister, I., "How to Make Those Sizing Calculations," *Textile World,* Vol. 130, No. 8.
2. Mullins, S. M., "Filament Yarn Sizing," *Textile Slashing Short Course Proceedings,* Auburn University, 1984.
3. Adanur, S., "Dynamic Analysis of Single Nozzle Air-Jet Filling Insertion," Ph.D. Thesis, North Carolina State University, May 1989.
4. Adanur, S., "Air-Jet Filling Insertion: Velocity Measurement and Influence of Yarn Structure," Masters Thesis, North Carolina State University, 1985.
5. Adanur, S., Walker, R., Broughton, R., and Beale, D., "Weaving Technology—What Next?" *Melliand Textilberichte,* Vol. 75, No. 4, April 1994.
6. Noonan, E. and Sullivan, S., "The International Nonwoven Roll Goods Companies, The Top 40," *Nonwovens Industry,* 25:9, September 1994.
7. *The Nonwovens Handbook,* INDA, Association of the Nonwoven Fabrics Industry, New York, 1988.
8. Holliday, T., "End Uses for Nonwovens," *Nonwovens Industry,* Vol. 23 December 1993.

1.1 General References

Spencer, D. J., *Knitting Technology,* Pergamon Press, Elmsford, NY, 1983.

Marschner, M., "Warp Knitted Technical Textiles with Expected High Increase in Volumes," in *Industrial and Technical Textiles: Fibers, Fabric, Machinery,* July 1993.

Brunnschweiller, D., "Braids and Braiding," *Journal of the Textile Institute,* Vol. 44, 1953.

Brunnschweiller, D., "The Structure and Tensile Properties of Braids," *Journal of the Textile Institute,* Vol. 45, 1954.

Ko, F. K. et al., *Handbook of Industrial Braiding,* Atkins & Peirce, 1989.

Kipp, H. W., *Narrow Fabric Weaving,* Sauerlander, 1989.

Morton-Jones, D. H., *Polymer Processing,* Chapman and Hall, New York, 1989.

Dictionary of Fiber & Textile Technology, Hoechst Celanese Corporation, Charlotte, NC, 1990.

Turbak, A. F., *Nonwovens: Theory, Process, Performance and Testing,* TAPPI Press, Atlanta, 1993.

2. REVIEW QUESTIONS

1. Reproduce Figure 4.1 on a computer using a computer aided design (CAD) or any drawing software program.
2. Discuss the major differences between woven, knit, braided, tufted and nonwoven structures.
3. What type of fabric structure (i.e., woven, knit, nonwoven, braided, tufted) has better resistance to each of the following:
   - abrasion
   - tension
   - compression
   - shear
   - bending
   - impact

4. Match the proper terms from the list with the statements that follow.

Terms
1. Winding (spooling)
2. Warping
3. Slashing
4. Drawing-in
5. Repeater End
6. Cause Analysis
7. Abrasion Resistance
8. Creep Speed
9. Shed Timing
10. Shed Geometry
11. Woven Fabric
12. Package Quality
13. Variability
14. Section Beam
15. Size Box
16. Encapsulation
17. Braided Fabric
18. Draw Warping
19. Section Warp
20. Cloth Fell
21. Reed Plan
22. Leasing
23. Lost End
24. Tying-in
25. Creel

__ Product of the warping process

__ Process used to replenish a warp of the same style

__ The interlacing of a pick relative to beat-up

__ A broken warp yarn embedded in a beam

__ Term used to describe the beginning of the cloth

__ Two yarn systems interlaced at 90 degrees with fabric produced parallel to the warp direction

__ Ends per dent and dents per inch

__ Technique used to determine reason for yarn failure

__ Filament yarn process to provide better end-to-end dye uniformity

__ The configuration of warp yarns resulting from their manipulation by multiple harnesses

__ Most important warp yarn property enhancement

__ A loom beam made up in sections arranged side by side

__ Measurement of degree that size encircles warp yarn

__ Entering of ends through weaving elements

__ Process where warp stripe patterns are developed from different solid colored sheets of yarn

__ Passing of cord through a yarn sheet in order to separate sheet from sheet or end from end

__ Usually a more important statistical consideration than the ''average value'' of a yarn property

__ Phrase used to describe slowing of slasher

__ Combining yarns from single-end packages onto a beam

__ As or more important to good weaving as yarn quality

__ Increases the length of yarn on a single-end package by combining smaller packages, spliced end to end

__ Warp yarn that breaks over and over during weaving

__ Section of the slasher where chemicals are added

__ Two yarn systems interlaced at 90 degrees with fabric produced 45 degrees to either yarn direction

__ Section of the slasher where section beams are placed

5. Why is yarn preparation and slashing necessary for weaving? Explain.

6. What are the main factors to consider when selecting a sizing agent for a particular yarn?

7. How does the yarn structure (i.e., ring spun, open-end, air-jet spun, friction spun) affect size pickup?

8. Match the proper terms from the list with the statements that follow:

Terms
1. Tucked Selvage
2. Cotton Count
3. Construction
4. Weft Density
5. The Fabric Weave
6. Weft Yarn
7. Drop Wire

8. Shedding Motion
9. Heddle
10. Harness
11. Reed Number
12. Take-up Motion
13. Multi-Phase
14. Temples
15. Whip Roll
16. Let-off Motion
17. Reed
18. Beat-up Motion
19. Weft Selector
20. Picking Motion

___ Component that dictates color stripes in the filling direction
___ Component of the beat-up that controls ends per inch (epi)
___ Component of the let-off that maintains desired tension on warp yarns by removing slack when necessary
___ Provides warp yarn as needed for weaving to continue
___ Houses heddles that control warp yarns that weave alike on a cam or dobby loom
___ Yarn system perpendicular to selvages
___ Produces double weft density at the fabric edge
___ Loom motion that controls weft density
___ Causes loom to stop if warp yarn breaks
___ The order of interlacing between ends and picks
___ Inserts weft yarn into shed
___ Holds fabric to same width as warp in the reed
___ Picks per inch
___ Causes the reed to reach the fell of the cloth
___ Motion that controls the weave pattern
___ Weaving machine with simultaneous weft insertions
___ The number of hanks per pound
___ Dents per inch
___ ends per inch (epi) × picks per inch (ppi)
___ Weaving element that controls an individual warp end

9. Compare the warp shedding control mechanisms for design flexibility and loom speed.
10. Explain the effects of entering order and lifting order on fabric structure.
11. What are the advantages of shuttleless looms over shuttle looms?
12. What are the major parameters to control woven fabric construction? Explain the effect of each.
13. Can you weave a monofilament filling yarn with air jet loom? Explain why.
14. Match the proper terms from the list with the statements that follow.

Terms
Types of Knit Fabrics
1. Weft
2. Single
3. Jersey
4. Double
5. Rib
6. Interlock
7. Purl
8. Tricot
9. Raschel
10. Warp

Descriptions, Definitions or Functions
11. Clearing
12. Cut
13. Course
14. Dial
15. Float
16. Stitch Cam
17. Guide Bar
18. Pattern Chain
19. Land
20. Overlap/Underlap

___ Weft knit with alternating courses of face and back loops
___ If a needle doesn't rise to this position, it makes a tuck
___ Weft knit made on two sets of needles, any gaiting
___ Component in warp knit machine that puts yarn on needles
___ Gaiting with all needles of both sets unable to knit one yarn
___ Moving warp yarns across the needles
___ Warp knit made on spring beard needles

__ Type stitch formed if a needle is not activated

__ Yarns interlooped in the fabric length direction

__ One horizontal row of loops in a knit fabric

__ Setting of this determines courses/inch in weft knits

__ Yarn system interlooped in the fabric width direction

__ The system of weft knit needles that forms back loops

__ Machine direction yarns interlooped with latch needles

__ Placing a previously formed loop over closed beard

__ Weft knit on one set of needles with tucks or floats

__ Causes yarns to be moved across needles

__ Weft knit with alternating wales of face and back loops

__ Another word for gauge or how wales/inch is controlled

__ Single knit with only knitted loops

15. Discuss the application areas of knit fabrics in industrial textile markets.

16. Explain the main differences between weft and warp knitting.

17. Compare the woven and knit fabric structures for
- elastic recovery
- extendibility
- shape retention
- drapeability
- modulus

18. What are the differences between 2-D and 3-D braided structures?

19. Match the proper terms from the list with the statements that follow:

Terms
1. Primary Back
2. Surface Pile
3. Bed Plate
4. Row
5. Pile Height
6. Latex
7. Looper
8. Slats
9. Secondary Back
10. Loop Pile
11. Cut/Loop Pile
12. Stitch
13. Stitch Robbing
14. Needle Bar
15. Knife
16. Scroll

__ A vertical line of tufts made by a single needle

__ Type pile made with knives and slats

__ How a low loop is made in carpet pile

__ Electronic high/low pattern device

__ Takes yarn from needle to form surface pile

__ Length of yarn above the primary backing

__ A single line of tufts in the cross-machine direction

__ Part of the carpet on the "down" or "under" side as it is produced at the tufting machine

__ Works against side of looper to form scissors action

__ Gauge of this determines rows per inch

__ Rate of take-up on this determines stitches per inch

__ Secures the secondary backing and locks in tufts

__ Pile height is changed by adjusting position of this

__ Final carpet component added at back-coating

__ Yarn feeding device that provides yarn displacement for individual needles at each stitch

__ Type carpet made with no knives

20. Explain the reasons for the differences between paper and textiles listed in Table 4.5.

21. Describe the following processes:
- air lay
- spunbonding
- melt blown
- hydroentanglement

22. What are the factors to consider when selecting a production process for a particular nonwoven fabric. Explain.

FABRIC FINISHING AND COATING

Dyeing, Printing and Finishing

W. PERKINS

Greige fabric as it comes from the manufacturing machine may or may not be ready for its end-use function. If fabrics are to be dyed, coated, impregnated, preshrunk or otherwise finished, it is usually necessary to remove the warp size and other impurities which may interfere with dyeing or prevent proper adhesion of a coating or finish.

1. PREPARING FABRICS FOR DYEING AND FINISHING

Most textile materials and fabrics require pretreatments before they can be dyed and finished. The required preparatory treatments depend on the type of fiber in the material and particular dyeing and finishing treatments that are to be done. Generally, fibers containing the most types and the greatest amount of impurities require the greatest amount of preparation for dyeing and finishing.

Most preparatory processes for dyeing and finishing involve heating the fabric or treating it with chemicals. Therefore, the potential is present for thermal and chemical damage to the fibrous polymer comprising the fabric. Fabrics can also be damaged mechanically in most preparatory processes. High temperature thermal treatments are often beneficial to fabrics containing thermoplastic fibers while these treatments are not beneficial or desirable on fabrics containing only non-thermoplastic fibers.

The following discussion of preparation for dyeing and finishing is general and reference is made to many types of fibers and textile materials.

Typical processes for preparation of materials for dyeing and finishing are as follows:

- heat setting
- singeing
- desizing
- scouring
- bleaching
- mercerizing

The sequence shown is common but many variations may be used. Virtually all materials go through some of these processes prior to dyeing and some materials in fabric form are subjected to all of them.

1.1 Heat Setting

The dimensional stability, dyeability, and other properties of thermoplastic fibers are affected by repeated heating and cooling, or the heat history, of the material. The main purposes of heat setting are as follows:

- to stabilize the material to shrinkage, distortion, and creasing
- to crease, pleat, or emboss fabrics
- to improve the dyeability of fabrics

Heat relieves stresses in the amorphous regions of thermoplastic fibers. When the fiber is heated above its glass transition temperature, the molecules in the amorphous regions can move, and the material can be formed into a new shape. When the temperature is decreased, the material stays in its new shape. Thus, creases that have developed in the fabric can be pulled out, and the width of the fabric can be changed somewhat in the heat setting process. Creases can be permanently set in the fabric by heat setting if desired. Heat setting is used to permanently set twist and crimp in yarns.

Problems or defects that may be caused by heat setting or improper control of the heat setting process include the following:

- permanent set wrinkles
- strength loss
- improper hand or feel in the fabric
- permanent set stains
- non-uniform dyeing
- improper fabric width or weight

1.2 Singeing

In singeing, the fibers that protrude from the fabric are burned away to give the fabric a smoother surface. Singeing is usually done by passing the fabric through a burning gas flame at high speed followed by quenching in water or the desizing bath to extinguish the smoldering fibers. Alternatively, the fabric may be passed close to a very hot plate to ignite the protruding fibers. Singeing is sometimes done after scouring since heating of the fabric in the singer can increase the difficulty of removing size and soil from the fabric.

Extreme care must be taken when singeing fabrics containing thermoplastic fibers. Thermoplastic fibers such as polyester melt when singed, and the fiber ends form beads on the fabric surface.

Yarns, sewing threads, felts, and carpet backing can also be singed. Yarn singeing is usually called gassing.

1.3 Desizing

Desizing is the process of removing the size material from the warp yarns in woven fabrics. Most of the size must be removed before the fabric can be dyed satisfactorily. Residual size prevents the yarns and fibers from wetting quickly and can affect dye absorption in either batch or continuous dyeing. Most synthetic sizes are water soluble by design so that they can be easily washed from the fabric. Typical synthetic sizes are polyvinyl alcohol, acrylic copolymers, and carboxymethyl cellulose. Starch is also com-

mon in size formulations. Starch is not very soluble in water and must be chemically degraded in order to remove it from the fabric. Starch used in sizing is often modified to improve its properties and removability in desizing. Lubricants added to size formulations to enhance the fabric manufacturing process may be more difficult to remove than the size itself. Virtually all size formulations contain lubricants derived from natural fats and waxes. Virtually complete removal of these lubricants is required before the fabric can be dyed. The desizing step removes mostly size and not much lubricant. Most of the lubricant is removed in the scouring process.

Fabrics containing only water soluble sizes can be desized using hot water perhaps containing wetting agents and a mild alkali. Fabrics that contain starch are usually desized with enzymes. Enzymes are complex organic substances that catalyze chemical reactions in biological processes. They are formed in the living cells of plants and animals. A particular enzyme catalyzes a very specific reaction. An enzyme is usually named by the kind of substance degraded in the reaction it catalyzes. Thus, enzymes that hydrolyze and reduce the molecular weight of starch are called amylases because they hydrolyze the amylose and amylopectin molecules in starch. The hydrolysis of starch catalyzed by amylase enzymes produces starch fragments that are soluble enough to be washed from the fabric. Mineral acids and oxidizing agents can also be used to degrade starch so that it can be removed in desizing.

1.4 Scouring

Scouring of textile materials refers to removal of impurities by wet treatments so that the impurities do not interfere with dyeing and finish applications. The amounts and types of impurities present depend on the type of fiber in the material. Materials containing only synthetic fibers usually contain only the lubricants that have been added to aid in manufacturing of the material and soil deposited on the material in manufacturing processes and handling. Up to 2/3 of raw wool is impurities such as suint (dried perspiration), dirt, and fat or grease. Most of these impurities must be removed; so preparation of wool is quite different from and more extensive than preparation of cotton and synthetic fibers.

Cotton contains natural impurities which must be removed in scouring and bleaching. The exact composition of cotton fiber depends on its source, maturity, and other factors. Typical composition of dry cotton fiber is as follows [1]:

	Composition of Cotton (%)
Cellulose	94
Protein	1.3
Pectic matter	0.9
Ash	1.2
Wax	0.6
Organic acids	0.8
Sugars	0.3
Other materials	0.9

The waxy substances on cotton are esters of fatty alcohols or glycerol and have relatively high melting points of 80−85°C. Therefore, high temperature treatments assist greatly in removal of natural waxes from cotton. The exact nature of the protein material in cotton fibers is unknown but these materials are removed by hot alkaline treatments. The pectins are high molecular weight carbohydrates and appear to exist as calcium and magnesium salts of pectic acid or its derivatives. Because of the presence of acidic carboxyl groups, alkaline treatments are effective in removing pectic substances.

Cotton can be scoured in either batch or continuous processes. The chemical formulation usually contains caustic soda (sodium hydroxide) and surfactants. Other ingredients such as organic solvents and builders may also be added. The caustic soda solubilizes many of the impurities in cotton, making them removable in aqueous medium. The process is done at elevated temperature for an extended period of time. The chemical concentrations, temperature, and time required vary with the particular process being used. Since polyester is subject to alkaline degradation, the treatment conditions are sometimes milder when processing blends of cotton with polyester. Lower temperatures may be used and soda ash (sodium carbonate) may be substituted for caustic soda.

Cleaning of textile materials requires both removal of the soil and suspension of the soil in the bath. Suspension of soil in the bath to prevent its redeposition on the substrate is just as important as removal of the soil in most cases.

1.5 Bleaching

Cellulose and most other fiber forming polymers are white in their natural state. However, impurities in fibers may absorb light causing the fibers to have a creamy, yellowish, or dull appearance. Cotton fibers usually require bleaching unless the material will be dyed very dark or dull shades. Synthetic fibers are often very white as supplied by the fiber producer but may require bleaching in some cases. The goal of bleaching is to decolorize the impurities that mask the natural whiteness of fibers. Oxidizing agents are used to bleach fibers. Bleaching processes must be closely controlled so that the color in the fibers is destroyed while damage due to oxidation of the fibrous material is minimized.

Cellulose may be damaged by oxidation during bleaching. Oxidation of cellulose produces carbonyl and carboxyl groups that can affect the ability of the fiber to accept dyes. Oxidized cellulose is also subject to depolymerization under alkaline conditions so oxidative damage can decrease strength and change other physical properties of the fibers.

The most widely used agents for bleaching are hydrogen peroxide, sodium hypochlorite and sodium chlorite.

Industrial bleaching of cotton is most commonly done using hydrogen peroxide (peroxide). Commercially available peroxide is usually either a 35 or 50% solution in water. Hydrogen peroxide is stable under acidic conditions and unstable in alkaline medium. The activity of peroxide in bleach baths is regulated by controlling pH and alkalinity of the bath. The time, temperature, and concentration of peroxide used vary depending on the process used in peroxide bleaching. Continuous, batch, and pad/batch processes can all be used in peroxide bleaching. High temperature greatly accelerates peroxide bleaching.

Sodium hypochlorite is available commercially as a solution in water. Lowering the pH of a solution of sodium hypochlorite causes the formation of hypochlorous acid which is believed to be the active species in hypochlorite bleaching. The pH chosen for hypochlorite bleaching is usually 9.5 to 10.0, but the bleaching action is greatly accelerated by lowering the pH slightly. Hypochlorite bleaching is most often done at room temperature, but slight heating greatly accelerates the bleaching rate and reduces the concentration of hypochlorite required. After hypochlorite bleaching, residual chlorine must be removed to prevent damage to the fibers. This treatment is called the antichlor and is done with a mild reducing agent such as sodium bisulfite.

Sodium chlorite is an effective bleach for both natural and synthetic fibers. Sodium chlorite is a white powder which when mixed with an alkali such as sodium carbonate can be stored indefinitely. A solution of sodium chlorite in water decomposes when acidified and heated forming chlorine dioxide, the active bleaching species. Cotton may be bleached with sodium chlorite at pH of 4.0 to 4.5. Synthetic fibers usually require lower pH of 2.0 to 4.0 and higher concentration of sodium chlorite than does cotton. Temperature near the boiling point of water is generally used in chlorite bleaching. Since chlorine dioxide is very corrosive to metals (including stainless steel), fiber glass equipment is preferred for bleaching with sodium chlorite. Sodium chlorite is used mostly on synthetic fibers such as acrylic, polyester, and nylon which are difficult or impossible to bleach with hydrogen peroxide. Acrylic fibers may yellow if heat set after bleaching with sodium chlorite.

Various peroxygen compounds can also be used as bleaches. Peracetic acid may be used to bleach polyamides. Accelerated and combination desize/scour/bleach processes based on persulfate and perphosphate chemistry have been devised for cotton and polyester/cotton blends [2].

1.6 Use of Fluorescent Whitening Agents

The creamy, yellowish color of textile materials can be partially, but not completely, destroyed by bleaching making the material appear whiter. Fluorescent whitening agents can be used to further improve the whiteness of textile materials. Fluorescent whitening agents, which are sometimes called optical brighteners, absorb ultraviolet (UV) light. Some of the absorbed energy is emitted at longer wavelengths. If the wavelengths at which the energy is emitted are in the visible region, the brightness of the textile substrate is increased. If the emitted energy is in the blue region, its addition to the light reflected by the yellowish substrate makes the substrate appear whiter.

Fluorescent whitening agents are like dyes in that they have affinity for the fibers to which they are applied. Therefore, the chemical nature of the fiber must be considered in selection of a fluorescent whitening agent.

Fluorescent whitening agents may be added to the bleach bath or applied in a separate step. Fluorescent whitening agents are often added during the manufacture of synthetic fibers and are a common ingredient in laundry detergents.

1.7 Tests for Scoured and Bleached Fabrics

Many different analytical methods may be useful in evaluating quality of preparation, controlling preparation processes, and detecting damage caused by preparatory treatments [3]. Following are some tests commonly done for quality control and detection of problems in preparation processes.

Whiteness

Reflectance of visible light from the fabric is compared to that from a white standard such as magnesium oxide. Standard methods for measurement and reporting of whiteness are available [4].

pH

Fabric is extracted with water, and the pH of the extract is measured. A standard method for pH measurement is the AATCC Method 81, pH of Water-Extract from Bleached Textiles.

Absorbency

Absorbency is affected by many variables. AATCC Method 79, Absorbency of Bleached

Textiles uses the length of time required for the fabric to absorb a droplet of water carefully placed on the surface of the fabric as a measure of absorbency.

Fluidity

Fluidity is inversely related to viscosity. The viscosity of a polymer solution is a function of the molecular weight (and degree of polymerization) of the polymer. Viscosity of a solution of cellulose in cuprammonium hydroxide (CUAM) or cupriethylene diamine (CUEN) can be used to detect depolymerization of cellulose. The results are usually expressed as fluidity values where higher fluidity means that the molecular weight of the polymer is lower, indicating that damage has occurred. Fluidity measurement is difficult and time consuming and is not done routinely in most plants. A standard test method is AATCC 82, Fluidity of Dispersions of Cellulose from Bleached Cotton Cloth.

Presence of Aldehyde Groups

Chemical damage from overoxidation or hydrolysis of cellulose produces aldehyde groups. Fehling's test may be used to detect aldehyde groups in cellulose. Aldehyde groups in oxidized cellulose reduce copper ions to copper which is deposited on the fibers as a reddish-brown solid. The test can be used as a qualitative spot test or can be done quantitatively. When done quantitatively, the result is called the copper number.

Presence of Carboxyl Groups

Severe overoxidation of cellulose produces carboxyl groups. Several methods have been devised to detect carboxyl groups in cellulose. In one of these methods, the fibers are dyed with methylene blue, a basic dye. Since the carboxyl groups are dye sites for methylene blue, cellulose containing carboxyl groups is dyed. The depth to which it dyes with methylene blue is a measure of the carboxyl group content of a cellulose sample.

1.8 Continuous Preparation Range

The preparatory processes described in the preceding sections may be done in a continuous manner by placing the appropriate stages in tandem so that the material proceeds from one stage to the next without interruption. Figure 5.1 shows a schematic of the process. The wet steps of the process are usually done in three separate stages – desizing, scouring, and bleaching. Knit fabrics that normally do not contain sized yarn would require only the scouring and bleaching stages.

Continuous preparation processes provide the time required for diffusion of chemicals and chemical reactions by using J-boxes or bins to accumulate and hold fabric for extended periods of time in a stage. J-boxes are often heated to accelerate chemical reactions and lower the dwell time required. Open width steamers may also be used to heat and hold fabrics for the time required for chemicals to accomplish their function.

Washing of the fabric between stages is important so that carryover of decomposition products and chemicals from one stage to the next is minimized.

Sometimes two, or even all three, stages in a preparation range are combined to save energy and processing time. Combination of stages usually results in a compromise between better quality of preparation with the three stage pro-

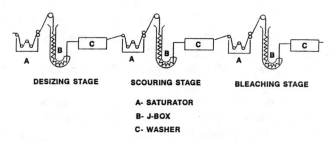

DESIZING STAGE   SCOURING STAGE   BLEACHING STAGE

A- SATURATOR
B- J-BOX
C- WASHER

FIGURE 5.1 Schematic diagram of continuous preparation range.

cess and savings of time and energy that may be achieved in a combined stage process.

1.9 Mercerization

The process of treating cotton with a concentrated solution of sodium hydroxide is called mercerization because it was discovered by John Mercer in 1850. Treatment of cotton with alkali has many beneficial effects including:

- increased tensile strength
- increased softness
- increased luster (if done under tension)
- improved affinity for dyes
- improved dyeability of immature fibers
- higher water sorption

The chemical effect of concentrated caustic soda solutions is unusual because the crystalline as well as the amorphous regions of the fiber are affected while most beneficial treatments of cotton affect only the amorphous regions of the fiber. Treatment of cellulose with caustic soda forms soda cellulose.

$$Cell-OH + NaOH \rightarrow Cell-O^-Na^+ + H_2O$$

$$(5.1)$$

Subsequent neutralization of the soda cellulose regenerates cellulose forming a hydrated cellulose which when dried has a slightly different arrangement of the glucose units in the crystal structure than does unmercerized cellulose. This change in the crystal structure is responsible for the difference in properties of mercerized and unmercerized cotton. Mercerization of fabrics is usually done in a continuous process.

2. DYEING

2.1 Dyeing Principles

Colorfastness properties of colored products have a direct impact in the everyday use of textile products. The color must be durable for the life of the product in many cases or in other cases must fade in a predictable and pleasing manner with use. Colorfastness of the dyed material to washing, laundering, dry cleaning, light, perspiration, rubbing (crocking), atmospheric contaminants, weathering, and other conditions are all important. The end use of the product determines the level of importance and what particular type of fastness properties a textile product must have. For example, colorfastness to dry cleaning may not be very important for fabrics going into automotive seat covers, but lightfastness requirements might be very high for this application.

Uniformity and colorfastness requirements are major determining factors in selection of dyes and dyeing methods for textile materials. The task that must be accomplished in dyeing is to transfer dye from the dyebath to the fiber. The term exhaustion is used to express the degree of dye transfer from dyebath to fiber. Exhaustion is usually expressed as a percentage of the amount of dye originally placed in the dyebath. For example, if 3/4 of the dye originally added to the dyebath transfers to the fiber, the exhaustion is 75 %. For economic and environmental reasons, a high degree of exhaustion is desirable. Dyes are expensive, and dye which is left in the bath is wasted. Furthermore, dye left in the dyebath is a pollutant which must be controlled and disposed of along with the wastewater from the plant. Often auxiliary chemicals are added to the dyebath to improve exhaustion.

Dye distributes between the dyebath and the fiber because it has an inherent attraction for both of these phases. The attraction of dye to the fiber is often referred to as the affinity of the dye. Actually, the dye has affinity for both the fiber and the dyebath. The driving force for transfer of a dye molecule from one phase to the other (dyebath to fiber or fiber to dyebath) is the concentration of dye in the two phases. Since dye adsorption is concentration dependent, the relative amounts of dyebath and fiber used influences the exhaustion.

The term liquor-to-fiber ratio or just liquor ratio is used to express the relative amounts of dyebath and fiber. The liquor ratio is the mass of dyebath used per unit mass of material being dyed. If one kilogram of dyebath is used per 0.1 kilogram of material being dyed, the liquor ratio is 10 to 1. Liquor ratio varies over a wide range depending on the type of dyeing process and equipment used. Typical values range from about $50-1$ for some batch processes to as low as $0.3-1$ for some continuous processes. Low liquor ratio, smaller amount of bath relative to the amount of fiber being dyed, gives higher dyebath exhaustion other factors being equal. Therefore, utilization of dye is usually better under lower liquor ratio dyeing conditions. Many factors determine the optimum liquor ratio for a given dyeing process.

2.2 Dyes in Water

The two general types of dyes with regard to their behavior in water are ionic dyes and non-ionic dyes. Ionic dyes may be either anionic or cationic. In anionic dyes the part of the molecule primarily responsible for color has a negative charge, Dye^-Na^+. In cationic dyes, the colored part of the dye molecule is positively charged, Dye^+Cl^-.

Non-ionic dyes do not interact strongly with water and are usually used as dispersions in water. Although non-ionic dyes are usually thought of as being insoluble in water, they are often very slightly soluble in water. This solubility, although small, may be vital to adsorption and diffusion of dye in the fiber. Non-ionic dyes are manufactured to have very small particle size and are formulated with surfactants so that they are easily dispersible in water.

2.3 Fibers in Water

Water is usually the dyeing medium. Hydrophilic fibers like cotton, rayon, and wool attract water. Water molecules diffuse into the amorphous regions of the fiber and break internal hydrogen bonds. Swelling of the fiber results. Swelling of the fiber by water may also be important in dyeing of some of the more hydrophilic synthetic fibers such as nylon. Swelling increases the size of openings and increases the mobility of the polymer molecules in the amorphous regions of the fiber making possible the diffusion of dye into the fibers. Swelling of the fiber in water is enhanced by higher temperature. The effect of temperature on swelling of fibers in water is probably a major reason why increasing dyeing temperature increases dyeing rate.

Hydrophobic fibers like polyester and some polyamides do not swell much in water. Water plays a less active role in the dyeing of these fibers. Water may still be needed as the medium to dissolve the dye so that the particles or molecules of dye are small enough to diffuse into the fibrous polymer. Water may also serve as the heat transfer medium in dyeing of polyester which is often dyed at temperatures of about 130°C. In continuous dyeing of polyester, water is simply the medium through which dye is deposited on the surface of the fibers and serves no active role in transporting dye into the fibers.

2.4 Dyes in the Fiber

Dyes and the fibers to which they are applied usually have an inherent attraction for one another. This natural attraction promotes transfer of dye from the dyebath to the fiber and is sometimes important in holding the dye in the fiber. Dyes are attracted to fibers because of chemical interactions between the dye molecule and the fiber. The bonding between the dye and fibers may result from weak secondary interactions such as hydrogen bonding or van der Waals forces.

2.5 Variables in Dyeing

Because of the complex interactions of the components in a dyeing system, many variables must be controlled in order to produce high quality dyeing. Variables in dyeing may be of several types including substrate variations, variations in chemicals (including water), variations in preparation of the substrate for dyeing, and procedural variations.

Several types of substrate variations may cause variations in dyeability of textiles. Cotton maturity is an excellent example of a variation in raw material that can have a major effect on dyeability. The secondary wall in the immature cotton fiber is not completely developed. Since dyes adsorb on the cellulose molecules in the fiber, immature fibers do not dye the same as mature fibers. Different dyes are adsorbed to different degrees on very immature cotton fibers.

Although the synthetic fiber manufacturer may have good control over the manufacturing process, synthetic fibers are subject to some variations which affect dyeability. For example, the manufacturer of synthetic fibers such as nylon and acrylics builds in a specific number of dyesites during polymer synthesis. A small variation in the number of dyesites in these fibers can affect the dyeability of the fibers with certain types of dyes. Variation in dyeability resulting from differences in number of dyesites in the fiber come from differences in affinity of the fiber for the dye. Such differences cannot be corrected by dyeing for longer times or at higher temperatures. By judicious dye selection, the dyer can sometimes compensate for these fiber differences that are related to affinity for dyes.

Variation in yarn and fabric structure can also cause dye defects. For example, wetting and dye uptake in highly twisted yarns will usually be slower than that in soft twist yarns. In continuous dyeing or non-equilibrium batch dyeing, these differences can cause streaks or other dye defects. Incorrect blend level in a blended yarn can cause dyeing differences. Contamination by either fibrous or non-fibrous substances can also cause dyeing defects.

2.6 Adsorption of Dye from the Dyebath

Several distinct identifiable events take place in the dyeing of a textile material. The events are as follows:

(1) Diffusion in solution—Dye must move or diffuse through the dyebath in order to establish contact with the textile material being dyed.
(2) Adsorption on fiber surface—Dye molecules are attracted to the fiber and are initially deposited on the fiber surface.
(3) Diffusion in the fiber—Dye deposited on the surface creates a concentration gradient which is the driving force for movement of dye from the surface toward the interior of the fiber. During diffusion, dye molecules migrate from place to place on the fiber. This migration tends to have a leveling effect on the dye application. Dyes that migrate readily are easy to apply uniformly. However,

dyes that migrate and level easily also tend to have poorer washfastness than dyes that do not migrate and level well.

A hypothetical dye cycle is shown schematically in Figure 5.2. Actual dye cycles vary greatly depending on the type of dye and fiber as well as the type of equipment used. In a typical dyeing procedure, the dyebath and the material to be dyed are placed in contact. The system is then gradually raised to some predetermined temperature over a predetermined period of time.

The system is held at this elevated temperature for some period of time after which the dyebath is drained or washed away. The system may be cooled some before the dyebath is drained. Required aftertreatments are then done to the dyed material.

2.7 Dyes

Colorants for textile materials may be classified as either dyes or pigments. The terms dye and pigment, while almost interchangeable in common use, have distinctly different meanings in coloration of textiles. A dye is a substance which at least during some stage of its application has inherent affinity for the textile material. Dyes are soluble in the dyeing medium during or at least in some stage of dyeing process. A pigment is simply a substance used to impart color and which does not have inherent affinity for the

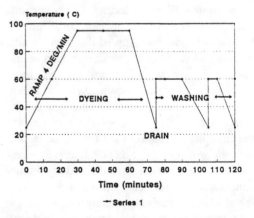

FIGURE 5.2 Hypothetical dyeing profile (dyeing time/temperature profile).

textile material. Both dyes and pigments can be used to color textile materials. Dyes can diffuse into fibers and interact with the polymer structure of the fiber. Pigments are simply bonded to the surface of the fiber, fabric, or yarn by other chemical agents. Pigments can be either organic or inorganic substances. All textile dyes are organic chemicals.

Dyes may be classified according to chemical structure or according to their method of application. Classification of dyes according to chemical structure is most useful to the dye chemist who may be interested dye synthesis and the relationship between chemical structure and properties of the dye. Classification according to method of application is most useful to the technologist concerned with coloration of textile products.

Eight major dye classes according to method of application are commonly used in textiles. The five classes used mainly on cellulose fibers are direct dyes, sulfur dyes, azoic dyes, reactive dyes, and vat dyes. The three classes used mainly for protein and synthetic fibers are acid dyes, basic dyes, and disperse dyes. All of these dyes are synthetic organic compounds that have been discovered since the synthesis of the first synthetic dye by W. H. Perkin in 1856. Several important synthetic dyes produced today have natural counterparts, but synthetic manufacture of the products is more economical than collecting the naturally occurring dyes. Following is a brief introduction to these eight major classes of dyes.

Direct Dyes

Direct dyes are so named because they have natural affinity for cellulose and can be applied without using any auxiliary chemicals. Direct dyes have been used to dye cellulose for over 100 years. Because of the simplicity of application and the great choice of products available, direct dyes are one of the dye classes most used today on cotton and rayon. Direct dyes range from moderate to poor in washfastness. Lightfastness varies from poor to excellent depending on the particular dye. The greatest limitation of direct dyes is that their fastness to washing is not good enough for some purposes.

Sulfur Dyes

Sulfur dyes are complex organic compounds synthesized by heating simple amines or phenolic compounds in the presence of sulfur. Sulfur dyes exist as a pigment form which does not have affinity for cellulose. They are converted to a water soluble form having affinity for cellulose by treatment with a reducing agent under alkaline conditions. The greatest advantage of sulfur dyes is low relative cost. The biggest limitation of sulfur dyes is that they are not bright enough in color for many uses. Washfastness and lightfastness are moderate. Fastness of sulfur dyes to chlorine is usually poor.

Azoic Dyes

Azoic dyes are pigments that are synthesized inside the fiber by coupling of two components neither of which is a dye itself. The two components are an aromatic diazonium salt and an aromatic hydroxyl compound, often a naphthol. Because of the use of naphthol as a component in the reaction, this class is sometimes called the naphthol dyes. Although the colored material produced in azoic dyeing is a pigment, the azoics are classified as dyes because the naphthol component has affinity for cellulose and is applied like a dye before the diazo component is added. The greatest advantage of azoic dyes is that they provide an economical way to obtain certain shades, especially red. The greatest limitations of azoic dyes is that they sometimes possess poor fastness to crocking and the possible color range is limited.

Reactive Dyes

Reactive dyes are relatively new, having been developed in the 1950s. They are sometimes called fiber reactive dyes. As the name implies the reactive dyes chemically react with the fiber forming covalent bonds. Since the covalent bonds between the dye and the fiber are strong, reactive dyes have excellent washfastness. Outstanding washfastness is the greatest, but by no means the only advantage of reactive dyes. Limitations of reactive dyes include higher cost than some other classes and poor fastness of some particular reactive dyes to chlorine.

Vat Dyes

Vat dyes are like sulfur dyes in that they are pigments which must be reduced and oxidized during application. However, the similarity between sulfur and vat dyes ends there. Vat dyes have outstanding washfastness and lightfastness as a class. The biggest disadvantage of vat dyes is their relatively high cost.

Acid Dyes

Acid dyes contain acidic groups, usually $-SO_3H$, and are used on fibers containing basic groups that can interact with these acidic groups. The fibers that are dyeable with acid dyes are polyamides containing some free amino, $-NH_2$, groups. Nylon is the most important synthetic fiber having this characteristic. Wool, silk, and other protein-based natural fibers also have amino groups that can bond with acid dyes. The bonds formed between the dye and fiber are salt linkages.

$$Fiber-NH_2 + HSO_3-Dye \rightarrow Fiber-NH_3^+ - SO_3-Dye$$

| Fiber with basic amino group | Dye with acidic sulfonic acid group | Dyed fiber with salt linkage between dye and fiber |

(5.2)

In nylon the dye sites are mainly the amino end groups on the polyamide molecule. Wool and other protein fibers have amino groups in the side chains of amino acid units which make up the polypeptide molecules.

Nylon is a thermoplastic fiber. The rate of dyeing of thermoplastic fibers is very dependent on dyeing temperature. Acceptable rate of dyeing is achieved only at temperatures well above the second order transition temperature of the polymer. In the case of nylon, the absorption of dye by the fiber begins as low as 40°C and is rapid at temperatures below the boiling point of water. Therefore, dyeing can be done at temperatures below 100°C and pressurized dyeing equipment is not required. Although nylon is readily dyeable at temperatures below 100°C, higher dyeing temperature increases the migra-

tion and leveling tendency of the dye. High dyeing temperature and prolonged dyeing time help to produce uniform dyeing of nylon with physical irregularities because these dyeing differences are caused by differences in the rate of dyeing. However, dye defects caused by variations in amine end group content in the nylon polymer are not corrected by extending the dyeing time or using higher dyebath temperature.

Basic Dyes

Basic dyes are sometimes called cationic dyes because the chromophore in basic dye molecules contains a positive charge. These basic or cationic groups react with acidic groups in acrylic, cationic dyeable polyester, cationic dyeable nylon, or occasionally protein fibers. A limitation of basic dyes is that their fastness to light is sometimes not satisfactory, especially on protein fibers.

Although they can be applied to other fibers, the major use of basic dyes is on acrylic fibers which have sulfonic acid groups in their structures. Since the number of dye sites is limited, the amount of dye that can be absorbed by acrylic fibers is limited. Basic dyes in mixtures must be carefully selected to avoid incompatibility which may cause non-uniform or otherwise unsatisfactory dyeing.

Disperse Dyes

Disperse dyes are used mostly for polyester, nylon, and cellulose acetate although they will dye some other fibers. The name disperse dye comes from the fact that these dyes are almost insoluble in water and have to be dispersed in water to make the dyebath. Disperse dyes were developed when cellulose acetate was first marketed. Disperse dyes are the only acceptable dye class for acetate and unmodified polyester fibers.

The most widely accepted mechanism for disperse dyeing of hydrophobic fibers contends that the fiber absorbs dye molecules that are dissolved in the dyebath. When a dissolved dye molecule is absorbed from the dyebath, it is replenished by a molecule of dye from the dispersion of dye particles in the dyebath.

The temperature required to dye synthetic fibers at an appreciable rate is always greater than the second order transition temperature. Since the second order transition temperature of nylon is lower than that of polyester, nylon can be dyed at lower temperatures than polyester. Actually, the temperature required to achieve an acceptable dyeing rate is usually much above the second order transition temperature of the fiber. The term dyeing transition temperature is sometimes used to refer to the temperature at which the dyeing rate of a synthetic fiber becomes rapid. The following dyebath temperatures are typical in application of disperse dyes:

Fiber	Typical Dyeing Temperature ($°C$)
Polyester	$100-140$
Nylon	$80-120$
Cellulose acetate ($2°$)	$85-90$
Cellulose triacetate	115
Acrylics	$95-110$

The temperature required to dye with disperse dyes depends on the thermal characteristics of the fiber being dyed and the particular disperse dyes being used.

Polyester dyes slowly at temperatures of $100°C$ or lower. The preferred dyeing temperature is about $130°C$ which requires a pressurized dyeing vessel. The dyeing rate of polyester can be increased by the addition of a carrier to the dyebath. Carriers are relatively small organic substances that usually have affinity for polyester. Chlorobenzenes, orthophenyl phenol, biphenyl, chlorinated benzene, aromatic esters, chlorinated hydrocarbons, and many other substances increase the rate of dyeing of polyester and have been used as carriers. The greatest disadvantage in use of carriers is the potential for pollution of air.

Certain non-ionic surfactants are beneficial in high temperature batch dyeing of polyester. These non-volatile, ethoxylated materials improve the compatibility of mixtures of disperse dyes and have been promoted as substitutes for conventional carriers in high temperature dyeing of polyester.

2.8 Dyeing Processes

Textile materials may be dyed using batch, continuous, or semicontinuous processes. The type of process used depends on several things including type of material (fiber, yarn, fabric, fabric construction, garment), generic type of fiber, size of dye lots, and quality requirements in the dyed fabric.

Machinery for dyeing must be resistant to attack by acids, bases, other auxiliary chemicals, and dyes. Stainless steel is usually used as the construction material for all parts of dyeing machines that will come in contact with dye formulations.

Batch Dyeing Processes

Batch processes are the most common method to dye textile materials. Batch dyeing is sometimes called exhaust dyeing because the dye is gradually transferred from a relatively large volume dyebath to the material being dyed over a relatively long period of time. The dye is said to exhaust from the dye bath to the substrate. Textile substrates can be dyed in batch processes in almost any stage of their assembly into a textile product including fiber, yarn, fabric, or garment. Generally, flexibility in color selection is better and cost of dyeing is lower the closer dye application is to the end of the manufacturing process for a textile product.

Some batch dyeing machines operate at temperatures only up to $100°C$. Enclosing the dye machine so that it can be pressurized provides the capability to dye at temperatures higher than $100°C$. Cotton, rayon, nylon, wool and some other fibers dye well at temperatures of $100°C$ or lower. Polyester and some other synthetic fibers dye more easily at temperatures higher than $100°C$.

The three general types of batch dyeing machines are those in which the fabric is circulated, those in which the dyebath is circulated while the material being dyed is stationary, and those in which both the bath and material are circulated. Fabrics and garments are commonly dyed in machines in which the fabric is circulated. The formulation is in turn agitated by movement of the material being dyed. Fiber, yarn, and fabric can all be dyed in machines which hold the material stationary and circulate the dyebath. Jet dyeing is the best example of a machine that circulates both the fabric and the dyebath. Other examples of batch dyeing machines are becks, jigs, package dye machines, skein dye machines, beam dye machines, paddle machines, rotary drums and tumblers.

Continuous Dyeing Processes

Continuous dyeing is most suitable for woven fabrics. Most continuous dye ranges are designed for dyeing of blends of polyester and cotton. Nylon carpets are sometimes dyed in continuous processes but the design of the range for continuous dyeing of carpet is much different than that for flat fabrics. Warps can also be dyed in continuous processes. Examples of this are slasher dyeing and long chain warp dyeing usually using indigo.

Continuous Dyeing of Polyester-Cellulose Blend Fabrics

A continuous dye range is efficient and economical for dyeing long runs of a particular shade. Tolerances for color variation must be greater for continuous dyeing than batch dyeing because of the speed of the process and the large number of process variables that can affect the dye application. The process as shown in Figure 5.3 is often designed for dyeing both the polyester and cotton in a blend fabric in one pass through the range. The polyester fibers are dyed in the first stages of the range by a pad-dry-thermofix process. The cellulosic fibers are dyed in

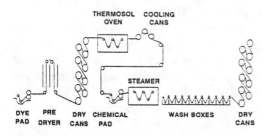

FIGURE 5.3 Schematic diagram of a continuous dye range.

the latter stages of the range using a pad-steam process.

Fabric which was previously prepared for dyeing enters the dye range from rolls. A scray is used to accumulate fabric entering the range so that the range can continue to run while a new roll of fabric is sewn to the end of the strand being run. Uniformity of application of dye requires that continuous dyeing be done in open width. Typical line speed in a continuous dyeing is 50 to 150 meters per minute.

Pad-Batch Dyeing

Pad-batch dyeing is a semicontinuous process used mainly for dyeing of cotton fabrics with reactive dyes. Both woven and knitted fabrics can be dyed using this method. Fabric is padded with a formulation containing dye, alkali, and other auxiliary chemicals. The padded fabric is accumulated on a roll or in some other appropriate container and stored for a few hours to give the dye time to react with the fiber. Time, temperature, alkalinity, and reactivity of the reactive dye all influence the process. Pad-batch dyeing is usually done at ambient temperature, but heating of the fabric during batching decreases the time required in the batching stage. Higher alkalinity and selection of more reactive types of reactive dyes also shortens the time required to complete the reaction. Typical batching times range from 4 hours to 24 hours.

2.9 Long Chain Dye Range

Warp yarns are often dyed with indigo and sulfur dyes using a long chain dye range. The process is used where the warp will be one color and the filling another color or white. Ball warps (sometimes called logs because of their cylindrical shape) are prepared as supply packages for the long chain dye range. A ball warp is a warp in which several hundred warp yarns are condensed into a rope and wound up as a single strand into a ball (log). The yarn from each ball warp constitutes a continuous rope (chain). A long chain dye range accommodates multiple ropes or chains side-by-side so that thousands of yarns are being dyed simultaneously. After exiting the long chain dye range, each rope is taken up in a separate container. After dyeing, each individual warp is back wound onto a warper beam (section beam) and becomes a supply package for the slasher.

3. PRINTING

Printing produces localized coloration of textile materials. Each color applied in a printing process must be applied in a separate step or position in the printing machine. Printing methods may be classified as direct, discharge, or resist.

(1) Direct printing–Dye in a thickened formulation is applied to selected areas of the fabric producing a colored pattern.
(2) Discharge printing–A discharging agent destroys dye on selected areas on a fabric which was previously dyed a solid shade. A white pattern remains where the dye was discharged. Alternatively, a discharge formulation containing dyes that are resistant to discharging produces a second color where the discharge is applied to the previously dyed fabric.
(3) Resist printing–Dye is applied to a fabric but not fixed. A resist formulation is printed on selected areas of the fabric. The resist agent prevents fixation of the dye in subsequent processing. The unfixed dye is washed away leaving a white pattern. If the resist agent is applied before the dye, the method is called a preprint process. If the dye is applied first followed by the resist formulation, the method is called an overprint process.

Printing methods may also be classified according to the process used to produce the pattern. Screen printing and roller (gravure) printing are the two most common printing methods used in textiles. Ink jet printing, heat transfer printing, and methods based on reliefs are also practiced in textiles.

Screen printing uses the principle of stencils. It is a resist method in the sense that the screen is made to resist penetration by the print formulation in areas where printing is not desired. Dye or pigment in a thickened formulation is forced by a blade or roller called a squeegee through a permeable screen onto a fabric underneath the screen. Screen printing may be done with either flat or cylindrical screens.

The printing mechanism in roller printing (gravure printing) is intaglio. A print roller is engraved with a pattern. Color deposited in the engravings is transferred to the fabric in the printing process.

4. FINISHING

Finishing is a general term which usually refers to treatments on textile fabrics after dyeing or printing but before the fabrics are cut and sewn into final products. However, many of the finishing principles covered in the following sections apply to treatment of yarns and garments as well.

Finishes have a wide variety of functions all of which are intended to make the fabric more suitable for its intended use. Functions of finishes include the following:

- Accentuate or inhibit some natural characteristic of the fabric. Examples are softening, stiffening (firming), delustering, brightening, and changing surface characteristics.
- Impart new characteristics or properties to the fabric. Durable press finishes, flame retardant finishes, and many other chemical treatments are examples of finishes which impart new characteristics.
- Increase life and durability of the fabric.
- Set the fabric so it maintains its shape and structure.
- Set dyes.

Finishes may be categorized as mechanical or chemical. Often, mechanical finishing is thought of as modification of dry fabric by a machine and chemical finishing as treatment of fabric with aqueous solutions of chemicals. However, this distinction sometimes fails because water and chemical formulations are often used in treatments that are best classified as mechanical finishes. Furthermore, most chemical finishes involve the use of machines which subject the fabric being finished to various degrees of mechanical action. Perhaps finishes can best be classified as mechanical or chemical depending on whether the mechanical treatment or the chemicals added are most responsible for imparting the desired characteristics to the fabric. Some overlap in classification of finishes as mechanical or chemical may be inevitable.

4.1 Mechanical Finishes

Calendering

Calendering consists of passing the fabric between rolls under pressure. Luster of the fabric is increased because the yarns are flattened making the fabric surface smoother and better able to reflect light. Adding moisture to the fabric and heating the calender rolls accentuate the effects of calendering.

Calender rolls may be steel, paper covered, rubber covered, polished or engraved, cold or heated depending on the surface characteristics desired in the fabric. The rolls on a calendering machine are sometimes called bowls. A simple calender usually has three bowls while a more complex, high production calender may have several rolls allowing multiple nips of the fabric. One of the rolls at each nip point in a calender must be a compressible material such as rubber or paper so that thick places in the fabric are not damaged.

In simple calendering the rolls run at the same speed and simply press the fabric. In friction calendering the surface speed of the rolls is different. Friction calenders polish the surface of the fabric producing a shiny effect.

Mechanical Stabilization

Mechanical stabilization is most often done to prevent excessive shrinkage of fabrics when they are laundered or exposed to heat. Manufacturing processes create stresses and tensions in fibers, yarns, and fabrics. Water and heat provide the opportunity for these stresses to relax. Relaxation of stresses in the fabric and swelling of fibers in water usually cause contraction and shrinkage of the fabric. Heat setting of fabrics containing mostly thermoplastic fibers usually stabilizes these fabrics sufficiently.

Woven fabrics containing non-thermoplastic fibers such as cotton can be mechanically stabilized by preshrinking the fabric. Since shrinkage

in woven fabrics is mostly lengthwise (warp direction) rather than width wise, compressing the filling yarns closer together stabilizes the fabric against further shrinkage. The term compressive shrinking is used because the process mechanically pushes the yarns in the fabric closer to one another. Sanforizing is the best-known compressive shrinkage process. Figure 5.4 shows the principle of compressive shrinkage.

The thick rubber belt passes around a small guide roller which causes the rubber to stretch extending the belt on its outer radius. The heated shoe presses the previously wetted fabric against the stretched surface of the rubber blanket as the fabric enters the compressive shrinkage machine. When the belt straightens as it emerges from the compression zone, it relaxes and contracts. Since the fabric is held firmly against the belt, the filling yarns move closer together so the fabric conforms to the shorter length of the relaxed belt. The thickness of the blanket determines the degree of compression achieved. Fabric is dried on a cylinder before leaving the compressive shrinkage machine. The compressive process must be controlled so that the residual shrinkage is slightly less than 1%. If compression of the fabric is excessive, the preshrunk fabric will stretch when laundered.

Raising

Raising refers to the lifting of fibers from yarns near the surface of the fabric to produce a hairy or fuzzy surface. Teaseling, napping, sueding, and shearing are names commonly used for

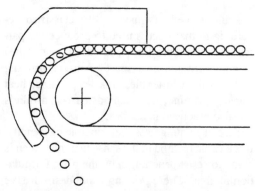

FIGURE 5.4 Principle of compressive shrinking of woven fabric.

specific raising processes. Napping uses small wires to pluck fibers from the yarns near the fabric surface.

Either one or both sides of a fabric may be raised in napping, and the fabric may be processed through the napper more than once to achieve the desired effect. Napping must be controlled well to prevent excessive strength loss in the fabric. Napped fabrics are very soft. The air pockets between the raised fibers in a napped fabric produce a good insulating effect making napped fabrics warm to wear.

Sueding subjects the fabric to sandpaper covered rolls producing a surface resembling suede leather. Suede cloth may be raised on either one side or both sides.

Shearing is used to cut away a portion of the fibers on a fabric having a raised surface to produce a level fabric surface. Cut-pile carpet is commonly sheared to level the tuft length.

Mechanical Softening

Fabrics are stiff if the fibers and yarns are not flexible and able to move freely in the fabric structure. Drying of fabrics under tension tends to set fibers making them stiffer and chemicals, agents such as size materials, printing thickeners, film forming finishes, and durable press finishes bind fibers and yarns together making the fabric stiff.

Mechanical softening simply consists of bending, flexing, or pounding the fabric to cause adhesions between the fibers and yarns to break. A variety of machines can be used for mechanical softening.

Flocking

Flocked fabric is made by gluing short fibers to the surface of a base fabric. Flock can be applied to the entire surface of the fabric to produce a raised surface or to only certain areas to produce a pattern (Section 4.8).

4.2 Chemical Finishes

Easy-Care Finishes

A severe limitation of fabrics made of all or mostly cellulosic fibers is their tendency to wrinkle. Wrinkling occurs when the fiber is severely bent. Hydrogen bonds between the molecular chains in the amorphous regions of the fibers break allowing the chains to slip past one another. The bonds reform in new places and hold creases in the fiber and fabric.

Resistance to creasing is imparted to cellulosic fibers by restricting the slippage of molecular chains. Chain slippage can be restricted by adding chemical crosslinks between the molecular chains in the amorphous regions of the fiber or by deposition of a polymeric substance in the amorphous regions. Permanent creases may be placed in a fabric by creasing the fabric at the desired location before the crosslinks are formed.

Durable press (DP), wash-and-wear, crease resistant, wrinkle free and other terms are used for easy-care finishes. DP finished fabrics may still form creases while the product is being used but are designed to return to a smooth and crease-free configuration after washing and drying.

Since crosslinking affects the mobility of the molecular chains in the amorphous regions of the fiber, many properties of the fiber are affected. Furthermore, durable press resins are reactive substances and may themselves undergo changes as treated fabrics age. Some effects of application of durable press finishes are as follows:

- Flexibility, tensile strength, tear strength, and abrasion resistance of the fabric are usually decreased.
- Swelling in water is inhibited, and the fibers are less absorbent.
- Since dyes and other chemicals cannot penetrate crosslinked fibers easily, dyeing must usually be done before the crosslinking finish is applied.
- On the positive side, crosslinking helps to trap dye molecules in the fiber and can thereby enhance the washfastness of the dye.
- Durable press finishing may accentuate soil retention by the fabric.

Use of a softener in the durable press formulation partially overcomes the loss of strength and resistance to abrasion due to crosslinking. Strength problems are less severe in cellulosic/polyester blends than in 100% cellulosic fibers since polyester is unaffected by the finishing agents. Problems with odor, chlorine retention, lightfastness of dyes, and dye shade changes are usually minimized by judicious selection of the finish and dyes.

Flame Retardant Finishes

A limitation of many textile fibers, especially cellulosics, is their tendency to burn readily. Flammability of textile products is regulated by federal and state governments. The federal Flammable Fabrics Act of 1953 requires virtually all textile products to meet minimum standards of flame retardancy. A series of amendments to the Flammable Fabrics Act of 1953, require children's sleepwear, carpets, and mattresses to meet stringent standards for resistance to burning.

Textile fibers vary greatly in flammability. All organic fibers burn if the conditions are severe enough. Cellulosic fibers ignite at relatively low temperature and burn rapidly. Polyamide and polyester fibers do not ignite easily because they melt and recede from a flame. However, in a blend with cellulosic fibers which prevent them from receding from the flame they burn vigorously. Often a blend of cellulose with a synthetic fiber will burn more readily than a fabric made from either of these individual fiber types. Acrylic fibers are difficult to ignite but burn rapidly with much evolution of heat after ignition takes place. Wool has a high ignition temperature and low heat of combustion making it naturally flame retardant. Aramid and modacrylic fibers are very flame retardant, and most of their uses are the result of their outstanding resistance to thermal decomposition.

Three conditions must exist for textile materials to support combustion. There must be a source of ignition, oxygen, and a fuel source (the textile material). Generally, not much can be done about sources of ignition and elimination of oxygen in the vicinity of textile materials so the propensity of the textile fiber to burn must be altered in order to make the material flame retardant.

When the temperature of cellulose is raised to about 350°C by the ignition source, cellulose begins to decompose producing solid char, liquid products called tar, and a mixture of combustible

gases. If the amount of heat produced by the burning gases is sufficient, further decomposition of cellulose takes place, and combustion is self-supporting. One of the tarry components of the decomposition of cellulose is levoglucosan. Upon further heating, levoglucosan decomposes to low molecular weight flammable gases.

Treatment of cellulose (and other fibers) with certain chemicals alters the mechanism of thermal decomposition so that combustion is not self-supporting. Therefore, when the ignition source is removed the flame extinguishes.

Generally, flammability standards require that textile products be self-extinguishing under specified conditions. This means that the material must not support combustion for very long after the ignition source is removed even though it may continue to burn as long as an external ignition source is present.

Several chemical treatments have been used successfully as flame retardants on textile materials. However, inherently flame retardant fibers are used in lieu of chemical treatments in many applications requiring flame retardancy because of questions regarding threats of chemical flame retardant treatments to human health.

Flame retardant chemical treatments may be non-durable or durable. Non-durable treatments may be used where the treated material will not be exposed to water or where only temporary flame retardance is needed. Durable treatments are used where the fabric must be washed. Flammability standards for apparel and household textiles require that the treatments be durable to laundering.

Simple water soluble acidic ammonium phosphates, ammonium chloride, borax/boric acid mixtures and other inorganic salts function as non-durable flame retardants. These chemicals are believed to inhibit the formation of levoglucosan when cellulose decomposes. Therefore, less volatile gases and more solid char are formed, and burning is impeded. Additionally, the borax/boric acid treatment may form a thin glassy coating on the fibers excluding the oxygen required for combustion of the fibers.

Durable flame retardant finishes based on the presence of both nitrogen (N) and phosphorus (P) are effective on cotton. Like the non-durable flame retardants, the durable N/P flame retardants are believed to inhibit the formation of volatile decomposition products when cellulose is pyrolyzed by preventing the formation of levoglucosan. The most common N/P flame retardant system uses tetrakis (hydroxymethyl) phosphonium chloride as the source of phosphorus.

The sulfate may be used in lieu of the chloride. In this case the material is referred to as THPS. THPC and THPS react with cellulose and are thus durable to washing. These substances also react with amino groups to form polymers within the fiber. Amino compounds commonly used as the nitrogen source in N/P flame retardant systems are methylolated urea and melamine. Nitrogen/phosphorus flame retardant systems containing the nitrogen and phosphorus atoms in the same molecule have also been successful.

Antimony/halogen treatments also produce durable flame retardants. The fabric is treated with a mixture of antimony oxide, a chlorinated wax, and a binder chemical. Antimony/halogen flame retardants are believed to function either by scavenging free radicals which propagate combustion or by a dehydration mechanism which promotes the formation of solid char during the thermal decomposition of cellulose.

Application of Flame Retardants

Flame retardant finishes are usually applied by a pad-dry-cure process. Some flame retardants for synthetic fibers have affinity for the fibers and can be applied by exhaust methods. These types of finishes may be applied in a batch process immediately after exhaust dyeing.

Water and Soil Repellent Finishes

Repellency is attained by limiting the wettability of fabrics. Coating of fabrics with a low surface energy film former such as polyvinyl chloride (PVC) or rubber makes the material completely waterproof. These impermeable types of coating are sometimes used to make rainwear or other protective fabrics. Since these coated fabrics are also impervious to air, they may not be comfortable as wearing apparel.

Repellent finishes, in contrast to waterproof finishes, only modify the surface of fibers and do not block the interstices. Therefore, the fabric remains porous to air and water vapor.

Some important types of water repellent finishes are wax emulsions, pyridinium compounds, N-methylol compounds, silicones, and fluorochemicals. These finishes may be used to impart water repellency to various natural and synthetic fibers.

Wax emulsion formulations usually contain a zirconium salt such as zirconium acetate in addition to the wax and emulsifying agents. Addition of film forming polymers or crosslinking reactants to wax emulsion formulations improves the durability of the wax emulsion water repellent finish.

Pyridinium-type water repellents consist of a long chain hydrocarbon group attached to a pyridinium ring. The pyridinium compound reacts with cellulose leaving its hydrocarbon group attached to the fiber. The resulting fiber surface is more hydrophobic and water repellent. N-methylol groups can also be used to chemically attach a long hydrocarbon group to fibers making the fibers water repellent.

Polymerized siloxanes are good water repellents. Dimethyl polysiloxane, methyl hydrogen polysiloxane, or a mixture of these two substances is often used.

Fluorochemical finishes impart both water repellency and oil repellency to fabrics while the silicone and repellents based on hydrocarbon hydrophobes provide only water repellency. The performance of fluorocarbon repellents results from the very low surface energy surface that they produce. The surface of the fibers must be covered by the fluorocarbon groups. Fabrics containing fluorocarbon finishes are often referred to as being stain repellent or stain resistant.

Soil Release Finishes

Soiling is mainly the result of adhesion and physical entrapment of particulate matter on fabrics. As textile products are laundered, physically entrapped particles are usually loosened and removed. However, particulate matter mixed with oily soil adheres chemically to hydrophobic fibers like polyester making its removal more difficult. Soil release finishes make the surface of the fiber more hydrophilic so that adhesion between the fiber and hydrophobic soil is lowered.

Several approaches have been used to increase the hydrophilic nature of cellulose fibers. The surface of polyester can be modified by grafting on a substance having hydrophilic tails. Partial hydrolysis of the polyester surface forms hydrophilic carboxyl groups. One approach to improving the hydrophilicity and soil release capabilities of polyesters is the use of amphiphilic substances. These substances are either non-ionic surfactants where the hydrophile-lipophile balance has been judiciously selected or block copolymers containing both hydrophobic and hydrophilic blocks. The hydrophobic part of the finish molecule has affinity for polyester and is sorbed into the fiber much like a disperse dye. The hydrophilic part does not have affinity for polyester and extends outward from the fiber surface making the surface more hydrophilic. High application temperature is required to achieve good penetration and bonding of these agents to polyester.

Soil release agents in durable press finishes are often anionic polymers, fluorocarbon polymers, or mixtures of the two. Copolymers of acrylic and methacrylic acid are effective as soil release agents when co-applied with durable press finishes or top-applied to durable press-finished fabrics. Soil release finishes for durable press finishes are often not entirely satisfactory.

Softeners, Antistatic Agents, and Hand Builders

The way a fabric or garment feels to the touch is referred to as its hand. Hand of fabrics is subjective to some extent and is often difficult to describe meaningfully with words. Nevertheless, hand is important, and hand modification is one of the most common goals of finishing of textiles.

Treatments that make the fabric more flexible and pliable impart the impression of softness. Softness generally comes from making the fibers themselves more flexible and from decreasing interfiber friction. Therefore, agents that plasticize fibers and lubricate the surface of fibers produce softness. Since softeners lubricate fibers, they may decrease yarn and fabric tensile

strength by decreasing fiber cohesion. On the other hand, softeners usually improve abrasion resistance and tear strength of fabrics. Abrasion resistance is improved because of lubricity of the fabric surface or improved mobility of fibers in the softener-treated fabric. Tear strength is improved because of better mobility of fibers so that the load imparted by tearing action is better distributed over the individual fibers in the fabric.

Synthetic fibers such as nylon and polyester tend to accumulate static charge because they absorb little water. On the other hand, cellulosic fibers having higher moisture content tend to dissipate static charge. Surface active agents help to spread the small amount of moisture on the surface of fibers into a continuous film which can dissipate charge. Therefore, surfactants used as softeners are also effective as antistatic treatments. Cationic, anionic, non-ionic surfactants all exhibit some antistatic behavior. The durability of antistatic treatments is improved by using them along with N-methylol compounds which chemically bond them to the fibers.

Hand building is the opposite effect of softening. Hand builders add stiffness to the fibers and fabric. Agents that stiffen fibers, bind fibers together, or increase interfiber friction all produce stiffness. Hand builders usually increase tensile strength and abrasion resistance of fabrics but may decrease tear strength because of increased stiffness in the fabric.

Antimicrobial Finishes

Growth of bacteria and fungi in textile fabrics is usually undesirable. Laundering with hot water and use of disinfectants such as chlorine or peroxygen bleaches destroys many microorganisms in textile products. Chemicals that prevent growth of bacteria and fungi are referred to as antimicrobial agents. Many chemicals are effective antimicrobial agents for fabrics used in environments where growth of microorganisms must be prevented. Quaternary ammonium surfactants and antibiotics function as antimicrobial agents. Many chlorinated organic compounds and organometallic compounds containing copper, silver, iron, manganese, or zinc also make textile materials resist growth of microorganisms. However, many of these agents are also toxic to higher organisms and may resist degradation in the environment for long periods of time. Chlorinated organic compounds are also effective in inhibiting attack of insects on natural fibers composed of protein and cellulose. AATCC Method 100, Assessment of Antibacterial Finishes on Textile Materials is a standard method to evaluate the above products.

5. TEXTILE CHEMICAL PROCESSES

Chemical treatment of textile materials is done in batch, continuous, and semicontinuous process. The principles discussed below can be applied to various processes including slashing, preparation of materials for dyeing, dyeing, and chemical finishing.

A batch process treats a specific amount of material with chemical formulations for a specific length of time. Several different treatments may be done sequentially by draining and refilling the bath between treatments. The complete process may all be done in one machine or the batch may be transferred from one machine to another for various steps in the process. Batch wet processing machines for fiber, yarn, fabric, and garments are discussed earlier in Dyeing Processes.

Continuous processes treat materials by passing the material in a continuous strand through one or more processing steps arranged in tandem. The amount of chemicals applied to the fabric (add-on) at the padder depends on the concentration of chemicals in the formulation and the wet pickup. Wet pickup is the amount of formulation picked up by the fabric and is usually expressed as a percentage on weight of the dry fabric

Wet pickup (%) =

$$\frac{\text{Weight of formulation picked up}}{\text{Weight of dry fabric}} \times 100$$

$$(5.3)$$

Add-on (%) =

Concentration of formulation (%)

$$\times \text{ Wet pickup (\%)} \qquad (5.4)$$

Therefore, the wet pickup must be controlled across the width and along the length of fabric in order to achieve uniform add-on of chemicals.

Applicators in textile chemical processes include padder, kiss rolls, foam applicators, wired blade applicators and vacuum extraction applicators.

5.1 Dryers

Textile materials are usually wetted and dried several times in preparation, dyeing, and finishing processes. Moisture can be removed from textile materials to some extent by mechanical methods such as squeezing, centrifugal extraction, or vacuum extraction. These mechanical methods remove only the moisture which is very loosely bound to the textile material such as that which is located in the interstices between the yarns comprising the fabric. Water droplets trapped within the fibers and yarns and water molecules bound to the fibers by secondary forces such as hydrogen bonding are not removable by mechanical means and must be vaporized in order to be removed from the material during the drying process.

Vaporization of water requires a large amount of energy so good removal of loosely bound water using mechanical means usually is desirable to minimize cost of drying. Most drying processes for textile materials use thermal energy. The thermal energy heats water in the textile material. As the temperature of the water rises, water molecules begin to vaporize and escape into the atmosphere around the material. Mechanisms for heat transfer to textile materials are classified as follows:

Indirect:	Convection
	Conduction
	Irradiation
Direct:	Dielectric
	Microwave
	Radio Frequency

The indirect methods all rely on a heat source external to the material to be dried. In convection heating, hot air is circulated around the material to be dried. Super heated steam or other hot gases can be substituted for air in convection drying.

In conduction heating, heat is transferred from a hot surface to the material to be dried by contact between the hot surface and the material being dried. Both convection drying and conduction drying often use steam to deliver the energy to the point of drying. Radiant drying, or drying by irradiation, uses a source with a high content of infrared waves. Water molecules absorb infrared energy causing the water to heat and vaporize. Dielectric heating is classified as direct heating because it is electric energy which vibrates the entire water molecule causing molecular friction and generating heat inside the wet textile material.

5.2 Steamers

Steamers are commonly used in continuous wet processes to rapidly heat fabric to a temperature of about 100°C and maintain this temperature for the required processing time. Steam is admitted from a boiler. The pressure in the steamer should be slightly higher than the pressure of the atmosphere outside the steamer to prevent entry of air into the steamer. Boiling water in the bottom of the steamer keeps the steam saturated with moisture. When cold wet fabric contacts steam, the steam condenses on the fabric transferring its heat to the fabric. If the steam is superheated (not saturated), condensation will not occur and the steam may actually dry the fabric. Drying of the fabric in the steamer usually adversely affects quality of the material being processed. Air must be excluded from a steamer since it decreases the transfer of heat to the wet fabric. Air is excluded by maintaining a slight positive pressure in the steamer and by using caution with exhaust fans near the steamer.

Coating and Laminating

C. G. NAMBOODRI
S. ADANUR

A coated industrial fabric consists of a closely woven or knit base fabric that is coated on one or both sides with a man-made or natural elastomer. For some applications, such as life rafts and diving suits, an interply of polymer may be sandwiched between two fabric layers which may be called a laminated product by coating (Figure 5.5). A simple laminated industrial fabric is a sandwich construction made of a loosely woven or knit fabric (scrim) that is locked between two sheets of thermoplastic or rubber elastomer. The open scrim fabric allows the outer sheets to bond to each other at the interstices (Figure 5.6). With the lamination process, many layers of fabric and other substrates can be combined. Coated and laminated fabrics are used in many industrial applications including architecture and construction, transportation, safety and protective systems.

1. COATING

A coated fabric is a composite textile material where the strength characteristics and other properties are improved by applying a suitably formulated polymer composition. The selection of fiber and fabric for coating depends on the type of coating and the end-use performance requirements. The performance standards are usually set up based on biaxial strengths both tensile and tear, abrasion resistance, stiffness, dimensional stability, thermal stability, chemical resistance, water repellency and air permeability requirements. To meet the above requirements proper choice of fiber, fabric construction and the polymeric coating compounds are needed.

1.1 Fibers and Fabrics for Coating

Woven, knit, tufted and nonwoven fabrics are used in coating. A woven fabric can be plain, twill or sateen construction (Section 4.3). The knitted fabrics can be warp knits or tubular knits finished and slitted. Cut edges are treated with resin to prevent stitch running. The nonwoven fabrics could be stitch bonded, spunbonded or needlepunched constructions. Tufted fabrics are also back-coated to prevent unraveling of tufts and to impart dimensional stability.

The choice of fiber, size of the yarn and type of weave for coated fabrics determine the fabric properties. The yarn and fabric construction influences the durability of coating. The fiber selection is critical as to the bonding of polymer compound since mechanical adhesion and chemical bonding vary from fiber to fiber. If the fiber is synthetic, spun yarn construction compared to filament yarn gives different mechanical properties to the coated fabric.

Cotton

Fabrics made of cotton were the traditional base material used in coating. This trend has declined due to the availability of synthetics which give improved strength characteristics as compared to cotton. The advantages of cotton fabrics include lower cost, availability and the excellent coating adhesion imparting dimensional stability to fabric. The disadvantages of cotton include low extensibility, poor rot, acid and weather resistance, flex cracking and limited shock absorption properties. Cotton has medium abrasion resistance. Extensibility of the fabric can be increased by tubular knit construction. Size material is normally removed from the cotton fabric prior to coating. Cotton is also blended with polyester for coated lighter weight fabrics having improved wear characteristics.

coating on one surface

coating on both surfaces

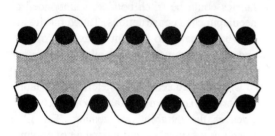

two fabrics with an interlayer polymer

FIGURE 5.5 Structure of coated fabrics.

Rayon

Although rayon gives higher tearing strength than cotton, its use for coated fabrics is declining and only a smaller share of coated rayon fabric reaches today's market. The main disadvantage of rayon fabric, including the high tenacity fabrics, is that it has poor wet shrinkage and low weather resistance.

Nylon

Nylon, due to its higher strength, has many advantages over cotton and rayon fabrics. Fabrics made of nylon have high abrasion resistance, tear and tensile strength. Nylon is widely used for lighter weight coated fabrics as it imparts higher strength, toughness, flexibility, waterproofness and durability. Coated nylon fabrics are used for survival rafts, inflatable medical products, personal floating devices, truck tarpaulins, machinery covers, tents, swimming pool covers, liners and irrigation tubing. Since lighter weight fabrics can be coated, it is easier to transport. Inflatable air structures and inflatable rubberized radomes have been widely used by the military. Filament nylon fabrics can be coated as gray fabric. A pretreatment called tie coat which is usually solvent based, is applied to fabric prior to coating for improving adhesion of coating compound. Due to environmental reasons the solvent based pretreatment is replaced with a water based system. Spun nylon/cotton blends of 6−8 oz/sq. yd, camouflage printed, fire retardant treated and waterproof coated fabrics are used as military tents.

Polyester

Polyester has high modulus, low extensibility and low rate of stress decay. Use of polyester in coating applications is increasing. Although the abrasion resistance is somewhat lower than nylon, it gives flexible tough coatings similar to nylon and has high tensile and tear strength. The improved priming methods and improved yarn manufacturing give better adhesion of coating compound to polyester and improved dimen-

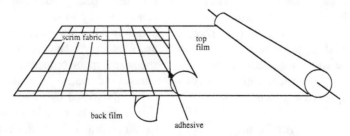

FIGURE 5.6 Schematic of a laminate.

sional stability. As in the case of nylon the polyester filament yarns can be used for coating in the gray stage. For coated lighter weight sheetings, polyester is often blended with cotton. Nonwoven polyester coated for stability and waterproof treated is extensively used in the outdoor furniture market.

Acrylics

Only small usage of acrylic fabrics is found in the coating market. Acrylic fabrics are used where outstanding weather resistance properties are required. The mechanical properties of the modified acrylics limit its application in the coating field. Acrylics have low thermal stability and low abrasion resistance.

Polypropylene

Polypropylene has excellent mechanical properties under normal conditions but has poor thermal stability. It is used for laminating with olefin fibers or for olefin melt coating. Polypropylene has high stress decay and poor coating adhesion which limits its use. Polypropylene fabric is widely used in the carpet industry replacing jute backing. Polypropylene fabric tufted with nylon or polyester yarns producing carpets, is normally back-coated for dimensional stability.

Other Fibers

Glass, Nomex®, Kevlar® and graphite fibers are used in coating where high performance composites are needed. These fibers possess inherent flame retardancy and have merit in certain applications. Polyester or nylon coated with fire retardant compounds limits the use of costly high performance fibers in coating.

Nonwoven fabrics of stitch bonded, spunbonded or needlepunched constructions are used as the base fabric for coating and film laminating. Various fiber combinations and binders are selected, depending upon end-use requirements. Certain nonwoven fabrics have a random fiber

distribution, assuring balanced strength and elongation. These base materials provide good tear strength, flexibility, and smooth surfaces for embossing, with no ''show through'' of weave pattern. They also provide high thickness-to-weight ratios at reasonable cost, and can be calendered, laminated, or electronically heat sealed to vinyl film. Polyester nonwoven fabrics, back-coated for stability and coated on face with water and stain repellents, are used for the outdoor furniture market.

1.2 Polymers and Additives in Coating

The coating polymer or blends of polymers are selected based on the end-use application. The coating compound contains different ingredients apart from the base polymer to impart the desired properties and performance characteristics. As the need arises, the coating compounds are often modified with addition of newer chemicals to meet the end-use performances of coated fabric.

The base polymer for coating can be natural or synthetic rubber or rubber-like polymers. High polymeric materials such as cellulose ester and ethers, polyamides, polyesters, acrylics, vinyl chloride, vinylidene chloride, polyurethanes and natural and synthetic elastomers and rubbers are used in compounding [5].

The mixing operation of various chemicals to produce the coating compound is commonly known as the compounding operation. The compounding can be within a coating plant or can be bought from a supplier as a compounded product. The choice of compounding chemicals and fillers are governed by the end product requirements. Besides polymers, resins and fillers, other additives such as fire retardants, thickeners and coloring agents may be included in coating formulations.

Natural Rubber

While there are many species of plants which produce rubber-like fluids, natural rubber is normally accepted as being derived from the latex of the cultivated tree, *Hevea brasiliens*. Chemically, natural rubber is composed of poly-isoprene, a stereospecific polymer. Other forms of natural rubber and rubber-like substances are guayule, chicle, gutta percha, and balata [5,6].

Balata Gum

It is a rubber-like natural gum obtained from the latex of mimusops-balata which grows wild in Central and South America. Balata is tough and water resistant, but is thermally sensitive at about 200°F and cannot be vulcanized.

Acrylic Polymers

They are commonly known as acrylics or acrylic latex and are composed of polymers of acrylic acid esters of alcohols with moderate chain length, i.e., methyl to octyl [7,8]. These products may also be copolymerized with acrylonitriles. Popular trade names are Acrylon (Borden Chemical), Hycar (Goodrich), Rhoplex (Rohm & Haas) and Flexbond and Valbond (Air Products). In recent years acrylic polymer emulsions have been developed for low temperature curing applications. These acrylic polymers are commonly based on methyl, ethyl and butyl acrylate copolymers having lower glass transition temperatures. Low temperature cure polymers have been developed for coating that give excellent solvent resistance. The same type of polymers have been developed for pressure sensitive adhesives for various end uses.

Polybutadiene

This polymer is prepared via the polymerization of butadiene. Two types of polymers may be formed, depending upon the special arrangement of the atoms in the molecule: *cis* and *trans* forms. Polybutadiene is usually not used alone as a rubber, but is copolymerized with other materials in the manufacture of commercial elastomers.

Styrene-Butadiene Rubber (SBR) and Copolymers

When butadiene and styrene are copolymerized in a specified ratio of about 72 to 28, the resulting rubber is called GR-S, the term stands for "Government Rubber—Styrene" type. The styrene-butadiene rubber is commonly known as SBR polymer in the trade. Typical trade names are Ameripol (Goodrich), FR-S (Firestone), Hycar (Goodrich), Plioflex and Pliolite (Goodyear), and Polyco (Borden Chemical).

Butadiene-Acrylonitrile Copolymers (Nitrile Rubber)

A series of copolymers with different properties, depending upon the ratio of butadiene to acrylonitrile, are known for their exceptional resistance to oils and aromatic fuels. Popular names include the German Buna-N and Perbunan, with an acrylonitrile content of 25%. American trade names are Butaprene (Firestone), Chemigum and Chemivic (Goodyear), and Hycar (Goodrich).

Polyisobutylene

This polymer has no crosslinks and is not vulcanizable. It is subject to cold flow and so cannot be used where a continuing stress is applied. It is used primarily in adhesives, particularly those that are pressure sensitive.

Butyl Rubber

Isobutylene is copolymerized with isoprene to make this rubber, which is vulcanizable. Butyl rubber is superior to GR-S with respect to heat and weather aging resistance. Formerly manufactured in U.S. government owned plants, it was then known under the name GR-I.

Polysulfide Rubber

This was one of the earlier synthetic rubbers developed. A probable formula is $(CH_2CH_2S_n)_x$, prepared from dichloro ethylene and sodium polysulfide. In the United States this rubber is manufactured by Thiokol Chemical Corporation under the trade name Thiokol, and has outstanding resistance to oils, fuels, and solvents.

Chloroprene Rubber (Neoprene)

It is the polymer of 2 chloro 1,3-butadiene. It was developed by Carrothers and his associates at DuPont, and was marketed under the trade name Neoprene. An entire series of neoprenes (now a generic term) of specific properties is manufactured. Because of their higher cost than other natural and synthetic rubbers, their uses are limited to special applications. Multi-filament nylon and polyester fabrics are neoprene coated for outdoor applications where resistances to weathering, abrasion, and flexing over extended periods of time are important factors. Colors are limited, with black and silver commonly used. Neoprene also possesses good resistance to oils, greases, gasoline, acids, and alkalies. Among the uses for neoprene coated fabrics are truck tarpaulins, mine ventilation tubing, collapsible containers, portable irrigation dams, protective clothing, air bags, and liferafts for boats and airplanes.

Chlorosulfonated Polyethylene (Hypalon®)

This synthetic rubber, which was developed by DuPont under the trade name Hypalon, was made by adding chlorosulfonyl groups (-SO_2Cl) to polyethylene. Available in a wide range of colors, including translucent forms, Hypalon coated fabrics have outstanding abrasion and outdoor weathering resistance properties. They also resist soiling and can be readily cleaned for maintenance of a bright, attractive appearance. Fabrics coated with Hypalon are used for radomes, protective clothing, truck tarpaulins, fumigation tents, inflatable warehouses, and convertible automobile and boat tops.

Silicone Rubbers

The basic structure of silicone polymers is composed of silicon-oxygen linkages, rather than carbon-carbon linkages. Organic carbon radicals or side groups are also present. Dimethyl dichlorosilane is the monomer from which most silicone rubbers are prepared. Outstanding properties include high and low temperature stability as well as heat resistance. General Electric, Dow Corning Corporation, and Union Carbide are manufacturers of silicone rubbers and other silicone polymers.

Fluorinated Polymers

Various fluorine based polymers have been developed and used where special properties are required. Some of these products are:

- polytetrafluoroethylene; $CF_2=CF_2$ (Teflon® from DuPont)
- polytrifluoro-monochloroethylene; $CClF=CF_2$ (Kel F from 3M Co.)
- trifluorochloroethylene and vinylidene chloride copolymer (Kel F 3700 and 5500 from 3M Co.)
- polyfluoropropylene and vinylidene fluoride copolymer (Viton A from DuPont and Kel F 2140 from 3M Co.)

The Scotchgard® brand of polymers from 3M and Teflon® from DuPont are both based on fluorinated rubbers used for carpet and fabric protective finishing. Viton A is used for coatings and has exceptionally good heat resistance, capable of withstanding temperatures as high as 600°F. It has outstanding chemical, solvent, fuel, and lubricant resistance at elevated temperatures.

Polyurethane Foam

For manufacturing polyurethane foam, the most popular diisocyanate used is TDI or toluene diisocyanate. This monomer has two isocyanate ($-N=C=O$) reactive groups and reacts extremely rapidly with polyhydroxy compounds or polyols commonly known in the trade as polyesters or polyethers to form polyurethanes [9–12]. In the presence of water or a blowing agent, a foam is formed. The polyesters usually produce flexible foams, and the polyethers produce rigid foams. The largest market at present for foamed polyurethanes is their use as thermal insulators for outerwear. They are bonded by heat fusion or flame lamination or by using adhesives to fabrics, and have particular success when bonded to soft knit goods.

Urethane Coating Resins

Both polyester-type and polyether-type polyurethane resins have been developed for coating, varnishes and paints, where foaming is not involved. Aqueous dispersions of polyurethane resins are formulated for coating applications. Large quantities of resins are coated from solvents. Polyurethane based coatings impart solvent resistance, improved strength, improved abrasion resistance, outdoor weathering resistance and impart flexibility and softness to the coated fabric. They can be formulated with fire retardant additives to produce flame retardant coated fabrics.

Vinyl Resins

These are synthetic resins formed by the polymerization of chemical monomers containing the group $-CH=CH_2$. Polyvinyl acetate, alcohol, butyral and chloride, and mixtures or copolymers thereof, are commonly included in this category. Polyvinyl cyanide, however, is identified by the separate term "acrylonitrile." Note that for fibers the Federal Trade Commission has assigned specific generic names to the various vinyl types; i.e., vinyon for vinyl chloride; acrylic for acrylonitrile (vinyl cyanide); and vinal for polyvinyl alcohol. The vinyl resins are also known as acrylic resins in industry.

Pyroxylin

Pyroxylin is a cellulose nitrate coating material that has declined in its use due to its flammability. It can be applied to fabrics in relatively small amounts, having excellent film forming characteristics. The use of pyroxylin coated fabrics was once popular as book binders in the past. However, it is quite flammable and with the advent of acrylics and other types of resins, cellulose nitrate declined in importance.

Starch

The relatively low cost of starch is the chief advantage of this coating material. Starch has been used in very large volume in the past, but it is declining steadily in importance. The largest markets for this type of coated fabric are in the bookcloth. Modified starch such as starch ethers is used as a viscosity builder in many water based coating systems.

Cellulose

Cellulose derivatives such as the cellulose ethers and esters are often used in coating com-position to alter viscosity and other properties. Hydroxyethyl cellulose is a cellulose ether and carboxymethyl cellulose is an ester type of viscosity modifier commonly used. Viscosity modifiers become part of the coating compound and are present in the final product.

Fillers

The amount and type of fillers used in a coating compound depend on the end product. Fillers are classified as active or inactive fillers. The common fillers are inactive and their use can alter physical properties such as tear and tensile strength and abrasion resistance. The inactive fillers include, talc, carbon black, kaolin, calcium carbonate, barium sulfate, magnesium carbonate and zinc oxide. Pure silica (SiO_2) is considered an active filler whereas calcium silicate is a semiactive filler. The inorganic and organic fire retardant additives are also considered as fillers in a compound.

Pigments

Both inorganic and organic pigments are used in coating compounds if a color is needed. A commonly used white pigment is titanium dioxide. Iron oxide is a typical example of an inorganic pigment. There are environmental problems in using chromium and cadmium based inorganic pigments. Organic pigments are based on azo dyes and phthalocyanine dyes. Brilliant colors are obtained with azo pigments and they are widely used. Generally certain blues and green colors are based on phthalocyanine dyes which are metal complexed organic pigments.

1.3 Coating Methods

There are several methods and machinery presently available for applying polymeric coating compounds to textile substrates. The selection of any one procedure is governed largely by the end-use requirements of the coated fabric. The choice is also influenced by the physical and mechanical properties of coating and the base fabric, together with the associated processing economics.

Although a great deal of cloth is coated

without preliminary treatment, for some, pre-coating preparation of the fabric is often necessary. If cotton or cotton/polyester blend substrates are used, preparatory treatment includes desizing, bleaching, dyeing, heat setting and surface preparations such as a chemical tie coat, to ensure adequate adhesion of the coating to the base fabric. Filament nylon and polyester fabrics are ordinarily coated in the gray state for many outdoor protective covering applications, but in certain critical end uses it may be necessary to scour and heat set these fabrics before coating.

Once coated, the fabric is passed through an oven where the excess solvent, if present, is evaporated, and the resin is cured. Often a lower temperature is used for solvent evaporation and resin drying, followed by a higher temperature for accomplishing the cure. Several passes through the coater may be required to build up the coating to the required thickness. The final pass through the oven is carried out at a higher temperature for curing. Neoprene coated fabrics may be cured by festooning in a hot air oven, or by steaming in an autoclave. It should be noted that solvent users must have an Environmental Protection Agency (EPA) permit that can be acquired through the state regulatory group.

The following section describes the principal coating techniques for fabrics. Figure 5.7 shows a schematic diagram of a coating range and gives location of the coating area and the oven.

Calendering

In this method, the plastic is calendered directly onto the fabric or an unsupported film is formed that is subsequently laminated to the cloth in a separate laminating operation. A compound containing resin, plasticizer, stabilizer, and pigments is fed to a preblender, where it is blended to a free flowing powder. This powder is then charged to a mixer, in which it is fused into a plastic state. The hot plastic is then fed to the calender, where the film is formed.

In the case of calender coating, the hot viscous resin is deposited on the two top rolls of the calender. The resin is picked up by an intermediate roller that transfers it to the fabric at the pressure nip between the fabric feed roller and the intermediate roller. Two calender coaters are shown diagrammatically in Figure 5.8.

The calender method of coating is adaptable to heavy vinyl coatings as well as to light coatings as low as four ounces per square yard in the case of a direct fabric calendering operation. Lighter weight films can be formed by this method for subsequent laminating to cloth. Nearly all vinyl coated fabrics for automotive interiors are made in this way.

Flexible Film Laminating

Flexible films, such as urethane, polyester and vinyl (previously formed on a calender as described above), are combined with fabrics on laminating machines (Figure 5.9). The film is first heated to make its surface soft and tacky so that proper adhesion of film to fabric will result. Fabric and film are then sent through heated pressure rolls of the machine, and combining takes place. Film patterns and surface effects can be created at this point by use of proper embossing rolls. The resulting combined flexible film laminate is then passed over cooling drums to set the film. Sometimes an open mesh fabric such as a filament nylon scrim is sandwiched between

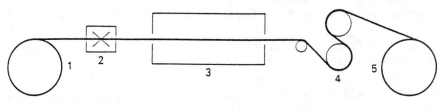

1 winding off arrangement (under tension)   4 cooling cylinders
2 coating device   5 batching arrangement
3 drying tunnel

FIGURE 5.7 Schematic diagram of coating range [13].

Inverted 'L' calender

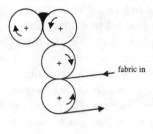

'Z' type calender

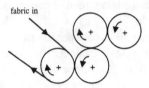

FIGURE 5.8 Schematic of calender coating.

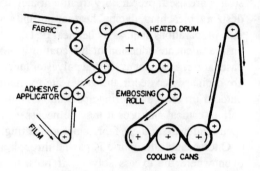

FIGURE 5.9 Flexible film laminating.

two vinyl films. These are used for outdoor protective coverings.

It is important to distinguish between these flexible film-fabric laminates and rigid laminates. The latter types are usually made by combining and curing many layers of paper, fabric, or other fibrous materials, cemented or bonded together with resins.

Knife Coating

In the knife or spread coating operation the viscous coating material is spread onto the fabric surface, which is then passed under a closely set metal edge called a coating knife. The knife serves to spread the coating uniformly across the entire width of the substrate fabric, simultaneously controlling the weight of coating material applied. Figure 5.10 shows different types of knife coating.

The knife-over-roll coater is suitable for applying thin coatings of high viscosity materials onto fairly closely woven fabric. For more delicate fabrics which should not be stretched during the coating operation, the knife-over-blanket coater should be employed. However, it, too, can only be used for fairly closely woven base fabrics due to the coating striking through.

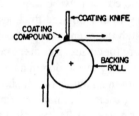

KNIFE-OVER-ROLL

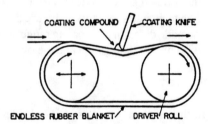

KNIFE-OVER-BLANKET

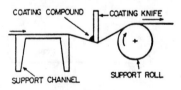

FLOATING KNIFE

FIGURE 5.10 Knife coating.

The floating knife coater is used extensively, having the advantage of being able to process open weave as well as closely woven fabrics, since strike-through of the applied coating will not adversely affect the coating operation. Unfortunately, all three knife coaters have two fundamental processing problems, one of which is streaking of the coating, and the other is the limited amount of coating which may be applied in one pass. Whereas the streaking may be eliminated by using smoothing bars or air brushes, multiple passes must be made if substantially heavy coating thicknesses are required.

In both knife coating and in floating knife, the shape of the knife has developed from various needs in coating. A broader knife edge gives greater add-on (Figure 5.11).

Both plastisol and organosol formulations may be applied by knife coating processes. Plastisols are liquid dispersions of finely divided resins in a plasticizer. Stabilizers, pigments, and fillers may also be present, but plastisols do not contain, nor are they applied from, solvents. An organosol differs from a plastisol in that it contains a non-aqueous volatile solvent.

Roller Coating

The simplest type of coating device is the roll coater where a roller, rather than a knife, applies the coating compound to the fabric (Figure 5.12). A two-roll reverse-roll coater utilizes a smaller diameter backup roll and doctor knife to exert some measure of control upon the thickness of the applied coating.

During normal operation the large diameter applicator roll picks up coating material from the trough. This picked-up coating material is then metered according to the gap that exists between the applicator and backup roll, which rotates in opposition. The metered coating is subsequently applied to the substrate fabric by the applicator roll, which runs in the reverse direction to that of the fabric. The doctor knife, running against the backup roll, serves to assist the metering operation performed by the gap between the reverse running rolls. The actual weight of coating material applied depends not only upon the reverse-roll spacing, but also on the nature of the substrate fabric, its speed, the

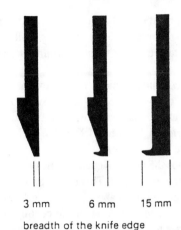

3 mm   6 mm   15 mm

breadth of the knife edge

FIGURE 5.11 Various knife shapes [13].

coating viscosity, and the relative speeds of rolls and fabric.

One of the practical limitations of this method is the necessity to maintain a proper and constant fabric tension which will, under certain circumstances, dictate low coating speeds to insure uniform coating application. Some of these disadvantages are eliminated in the pan-fed three-roll reverse-roll coater (Figure 5.12). In this machine the coating material is picked up from the trough by the applicator or transfer roll B,

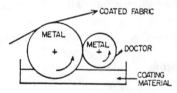

TWO-ROLL REVERSE-ROLL (KISS)

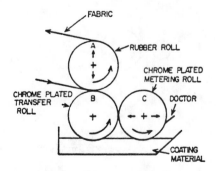

PAN-FED THREE-ROLL REVERSE-ROLL

FIGURE 5.12 Roller coating.

metered by the nip between the transfer roll B and the reverse running metering roll C, and applied to the fabric passing around the rubber roll A.

This coating technique is used extensively for the application of lacquer and vinyl coatings to paper, fabric, and nonwoven webs, but is not suitable for high speed coating due to the lack of adequate fluid control.

Perhaps the best roll coater is the pan-fed four-roll reverse-roll coater. This is similar to the three-roll coater described above, but it exerts a greater control over the metering of the coating prior to application to the fabric passing around the rubber roll. Greater control is also exerted over the coating material so that higher coating speeds may be employed in thicknesses from one to twenty mils. It cannot be used to coat open weave fabrics due to strike-through onto the rubber roll.

Nip Coating

Low viscosity coating materials can be speedily applied to fabrics using interroller pressure for both application and metering. The three-roll squeeze coater, shown in Figure 5.13, uses a pick-up or fountain roll C to apply the coating to the two pressure rolls A and B. As the fabric passes through the nip formed between rolls A and B, the coating is forced into the fabric. Surface gloss of the coating can be enhanced by using a polishing roll subsequent to coating.

Dip Coating

Dip coaters are used when complete saturation

impregnation of the base fabric is required. A typical machine is shown in Figure 5.14. Several lacing-up paths are shown. The actual one chosen for any combination of coating and fabric will depend upon the impregnation time and the number of pressure nips or reverse-roll contacts necessary to meter the coating adequately. If the coating is to be applied without the application of pressure, a reverse-roll metering action is executed at the nip between rolls A and B. Otherwise, a pressure nip between rolls B and C may be employed as an alternative to, or in conjunction with, the reverse-roll metering between rolls A and B. One advantage of this technique is that both the face and back of the fabric can be coated in one pass through the coating machine.

Cast Coating

The cast coater combines both the coating and curing operation into one process as a method of producing a high gloss coating upon a fabric web. While the curing operation may not be quite complete, it is sufficient to create the exceptionally smooth coating surface. A cast coater is shown in Figure 5.15.

The coating material specially chosen for rapid cure and viscosity is applied to the fabric web just prior to contact with the heated casting drum. During the time the coated fabric is in contact with the heated drum, the coating sets, and the coated fabric can be continuously stripped from the drum. Coating smoothness is directly related to the smoothness of the metal casting drum, the amount of coating applied, and the fabric web roughness. By using this coating technique, rela-

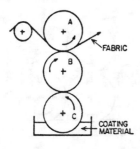

THREE-ROLL SQUEEZE

FIGURE 5.13 Three-roll squeeze nip coating.

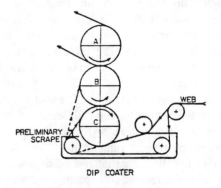

DIP COATER

FIGURE 5.14 Dip coater.

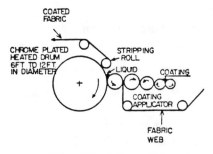

FIGURE 5.15 Cast coater.

tively open weave fabrics with a high degree of stretch may be coated satisfactorily.

Extrusion Coating

Vinyls, polyethylenes, and polystyrenes can be successfully coated to substrate fabrics by using a hot extrusion coater, illustrated in Figure 5.16. In this process, the coating material is extruded as a soft sheet into the nip formed between the chill roll and the rubber pressure roll. The fabric to be coated also passes through this nip, and the extruded sheet is pressure laminated to the substrate fabric. The resultant coating acquires the smooth surface of the chill roll and is comparable in quality to that obtained by cast coating techniques.

Spray Coating

In this coating method, the latex coating formulation—at a lower viscosity than conventional coating compounds—is sprayed through series of nozzles uniformly on traveling fabric web. The fabric is then dried and cured as in other coating methods. The method is widely used for back-coating of upholstery fabrics where the fabric weights, construction or yarns do not allow conventional knife or roller coatings. The spray coatings are also used for stabilizing nonwoven webs.

Foam Coating

The foam back-coatings are generally used to stabilize the coated base fabric, to impart opacity (prevent see-through), for bulkiness, for softness (hand) and for desired aesthetics depending on

end use. Foam coated fabrics are produced for textile applications by crushed foam coating method, uncrushed foam coating or by flocked foam coating techniques. Foam coatings for tickings, drapery, fire retardant foam coatings and certain upholstery coatings are examples where crushed foam back-coatings are used. Uncrushed foam coatings are used for carpets, mats, filtration fabrics and for upholstery fabrics. Flocked foam coatings are used for drapery, tickings and upholstery [14–17].

Typically additives in a back-coating foam contain latex, filler, foaming agent, auxiliary foaming agents, thickeners and alkali such as ammonium hydroxide to adjust required pH range.

UV-Cured Coatings

UV-curable coating systems can be used for vinyl coated fabrics. Vinyl coated wall coverings have poor soil resistance. A typical UV-curable topcoat system based on polyurethane contains the urethane oligomer with acrylate end groups, reactive acrylate monomer, and multifunctional acrylate crosslinker and photosensitizer. The UV topcoats are considered superior to conventional coated fabrics in resisting ballpoint ink stains, and other soiling materials [18].

Powder Coating

Powder coating using thermosetting epoxy resins, crosslinkable polyester resins, or acrylic resins are used for coating fabrics, wire, pipes and for bottle coatings [19]. Electrostatic application, spray method or screen methods are commonly used. Powder coating methods can also be used for applying the laminating adhesive or thermoplastic resins to fabric surface. The powder can be applied as a layer or with a

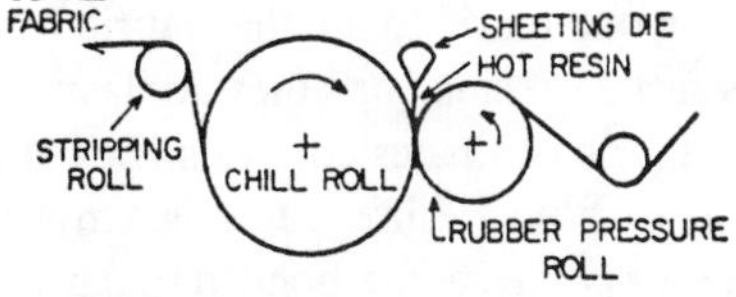

FIGURE 5.16 Hot extrusion coater.

gap or in an engraving pattern. Following powder application heat is applied for the coating of the thermoplastic material and to adhere onto the fabric.

2. FABRIC LAMINATION

Fabric lamination is the process where layers of fabric or fabric and other substrates such as foam products are combined to form a composite material. The laminated fabrics have varying end uses from upholstery to apparel fabrics. Laminates are classified as rigid laminates and flexible laminates.

2.1 Rigid Laminates

Rigid laminates are composite products made of thin layers or sheets combined by adhesives. Usually they consist of several layers of fabric, or other fibrous types of materials, cemented or bonded together with resin, to make rigid plastic-like materials of any desired thickness. Both thermoplastic and thermosetting resins are used. Thermoplastic resins soften and melt under the application of heat and harden upon cooling, the action being reversible. Resins in this category include the polystyrenes, acrylics, cellulosics, polyolefins, vinyls, nylons, polyesters, and fluorocarbons. Thermosetting resins, when subjected to heat curing, harden permanently, usually via crosslinking. The action is irreversible. Subsequent heating may soften the material, but it is not possible to restore the resin to its uncured state. Resins in this category include urea formaldehyde, melamine formaldehyde, and phenolic aldehydes, alkyds, and epoxies.

2.2 Flexible Fabric Laminates

Flexible laminating products are produced by laminating one fabric to another fabric by using thermoplastic resins or thermoplastic film. Flexible fabric laminates could also be laminates of fabric to films. Films such as vinyl, polyurethane, or polyester are bonded to fabrics with heat and pressure. Usually a primer coat on the fabric is required. This method is commonly used for flexible laminates and allows large rolls to be produced on a continuous operation.

2.3 Waterproof Breathable Laminates

Gore-Tex® is an outerwear breathable waterproof laminated fabric. This fabric is produced from microporous polymeric film of polytetrafluoroethylene having a thickness of 0.001 inch and more than 82% of the pores having size of 0.2 μm by laminating to cotton, polyester/cotton or nylon fabric. The laminated fabrics are flexible and give comfort to the wearer. The laminate is waterproof and durable to home washings.

2.4 Resins for Laminating

Phenolics

The term phenolic resins covers a variety of products formed by the chemical reaction of a phenol with an aldehyde. The most commercially significant phenolic is the one based upon phenol and formaldehyde.

Phenolic resins have been used in the manufacture of plastic laminates. However, their use is declining due to the availability of more environmentally safer acrylic coating resins. The phenolic resins are relatively low in cost, possess good mechanical and electrical properties, and have good heat and moisture resistance, as well as good chemical resistance to mild acids and alkalies.

In the production of high pressure laminates, the uncured resin is dissolved in alcohol to form a varnish. Fabric is passed through the varnish bath and out through squeeze rolls or over doctor blades. Then, passing through an oven, the excess volatile matter is driven off at a relatively low temperature, and the resin, partially polymerized, is dried on and within the fabric. To make flat sheets, the treated fabric is then cut into required lengths. Several sheets, depending upon the desired laminate thickness, are placed between polished steel plates in a hydraulic press. The laminate is cured at a temperature of about 350°F and pressures ranging from 750 to

2,000 psi. Nonwoven fibrous bases may also be molded into reinforced phenolics of various shapes.

Melamine Resins

Melamine formaldehyde resins, which are more costly than the phenolics, are used for electrical insulating laminates because of their good flame and arc resistance. These resins are essentially colorless and are predicable in a wide color range, with excellent lightfastness. Hence, they are useful in decorative applications. Melamine formaldehyde resin releases free formaldehyde into the atmosphere. Formaldehyde is considered a cancer producing agent and the use of melamine resin is declining in the laminating field.

Epoxy Resins

Epoxy resins, also more costly than the phenolics, are generally used with glass and carbon fabrics, although they may also be used with synthetic yarn or fabric reinforcements. They possess a high degree of chemical resistance, have low moisture absorption, good dimensional stability, high mechanical strength, and superior electrical properties under high humidity conditions. Helmets for football players and motorcycle riders are often made from epoxy resin laminates of glass fiber fabrics. Epoxy resins are widely used in other textile structural composites (Chapter 7.0).

Polyurethane Adhesives

Both solvent based and aqueous dispersions of polyurethanes are suitable adhesives for lamination of textile to textile materials or textiles to foam products. The laminated products are fast to washing and dry cleaning. The polyurethane based adhesives are prepared by thickening the dispersion and coating by roller or knife application. When a urethane foam is bonded to textile materials, low pressure is used for lamination.

Silicone Resins

Silicone resins are generally used in conjunction with glass fiber fabrics to obtain laminates with high temperature resistance and excellent electrical properties. Higher in cost than the phenolics, these laminates resist temperatures up to about 550°F, and have higher arc resistance and lower moisture absorption. Silicone resins are available for both high and low pressure laminating processes, but one deficiency that they exhibit is their sensitivity to organic solvents. Hydrocarbon solvents readily attack silicone resins.

Polyester Resins

Polyester resins have wide applications for low and high pressure laminates, providing good mechanical and electrical properties. These resins, with glass fiber reinforcement, are used in the fabrication of many types of products, from small electrical components to large boat hulls. Glass fiber is used in various forms, including yarn, fabric, woven roving, mat, or preforms made from chopped fibers. Laminating pressures are not critical, ranging from contact pressure to about 400 psi. Saturated polyesters having hydroxy function can be further extended with isocyanates and used in film lamination for sailcloth and other applications.

Acrylic Resins

Acrylic resins are characterized by excellent clarity, high impact strength, chemical and weather resistance, and ease of formability. Acrylic resins are widely used in many types of fabric lamination. They are used in synthetic fiber laminates for special applications, particularly in the aircraft field.

Pressure Sensitive Adhesives

Pressure sensitive adhesives and low energy cure adhesives have interesting applications. The room temperature cure function of these adhesives provides diversified end use. They are acrylic based resins and copolymers based on methyl, ethyl and butyl acrylate to obtain polymers having lower glass transition temperatures. Low temperature cure polymers have been developed in recent years having good adhesion and solvent resistance. They are used in the lamination and bonding of polyurethane foam to fabrics.

Ethylene Vinyl Acetate Copolymers

These copolymers are gaining a large market in the laminating field as they are environmentally safer than phenolics and melamine formaldehyde resins. These polymers give good adhesion for synthetic fiber lamination and are used in the manufacture of laminated furniture upholstery fabrics.

2.5 Fibers and Fabrics for Laminates

Woven, knit and nonwoven fabrics are used in producing fabric laminated products depending on end-use application and its desired properties. The fiber selection for the fabric manufacture also depends on the performance characteristics required of the end-use product. The following fibers and fabrics are used for high pressure and low pressure lamination process.

Cotton has been used for reinforced high pressure laminates requiring good mechanical properties such as high impact strength, superior bond strength, and good machining qualities. However, cotton's share has been declining due to the synthetics. The cotton fiber at relatively low cost, has the very desirable combination of good strength and relatively low extensibility, thus aiding in the production of dimensionally stable rigid laminates. Cotton blended with polyester give higher strength and is commonly used for lightweight base fabric for lamination. Cotton and cotton/polyester blended fabrics supplied in gray stage have to be prepared to remove size and often scoured and bleached followed by a thorough wash to remove salts and electrolytes prior to its use in lamination.

Glass fiber is used as a reinforcement for phenolic laminates for improved heat and electrical resistance in combination with high mechanical strength. It has broad application in the low pressure field in combination with polyester resins, and where the high pressure lamination techniques are used. Epoxy resins are used for laminating glass for higher strength and impact resistance.

Nylon fiber gives exceptional electrical resistance with high impact strength and is often used for making laminated electrical products.

Polyester has good chemical resistance, weather resistance and good abrasion resistance and its use in laminated products has increased in recent years.

Acrylics have special properties, making them suitable for many useful laminated products. Acrylics possess good resistance to outdoor exposure and have excellent acid resistance. The fabrics are often used as overlays in glass fiber reinforced laminates, produced by low pressure techniques. The acrylic overlay affords improved resistance to abrasive wear, chemical attack, and weathering.

Nonwoven fabrics are used for special laminates, particularly where smooth machining characteristics are desired in the final laminated product. These nonwovens or mats are made with fibers that are randomly distributed so that the mechanical properties are equal in all directions within the fabric plane. If a resin is used in the manufacture of nonwovens, it must be compatible to the laminating process. Nonwoven polyester mats are often laminated as backing for printed or dyed upholstery fabrics made of cotton or polyester.

2.6 Types of Laminates

Textile fabric supported plastic laminates are used in many forms. Some of the most important types are described below.

Sheet Stock

In the manufacture of sheet stock, layers of fabric impregnated with the laminating varnish are carefully assembled to form a sheet of the required thickness. In some cases, the assembly may consist of several different types of impregnated material to provide the finished product with the required mechanical or electrical properties. The assembly of impregnated sheets is then placed in the hydraulic press, where it is subjected to carefully controlled heat and pressure. The layers fuse together to form a hard, dense, homogeneous sheet.

Post Formed Laminates

Post forming is a low cost process by which relatively thin plastic laminates are formed into simple shapes. First, layers of fabric impregnated with modified phenolic resins or other laminating resins are cured to make flat laminates. These laminates are then post formed at elevated temperatures. This is accomplished by rapidly heating the sheet stock to a higher temperature than the initial curing temperature, and quickly bending or shaping it into a product of the desired contour by use of a mold. After cooling, the laminate retains the new shape. In those cases where a high degree of deformation and stretching of the fabric base is involved, loosely woven fabric constructions, as well as those with high yarn crimps, are used.

Tubes and Rods

Tubes are formed by winding resin impregnated fabric on mandrels, and then rolling them between heated pressure rolls. The tube assembly is then cured in an oven. Another method is to mold the tubing under pressure in a hydraulic press. After curing or molding, the mandrel is removed and the tube is machined and sanded to the proper size. Laminated plastic rods are formed in the same way as molding tubing. The impregnated fabric is wound on a small mandrel, which is withdrawn prior to molding. In some cases, no mandrel is used. Rods may also be formed from sheet stock by machining.

Molded Laminated Plastics

Simple shapes can be molded from resin impregnated fillers, with the resulting products possessing the strength characteristics of laminated plastics, but also having the advantages of being molded to the proper form. Gear blanks, for example, can be molded from assemblies of die-cut sections of impregnated fabric.

Honeycomb Laminates

Honeycomb laminates made from resin impregnated fabric have a high strength/weight ratio. Cotton and glass fiber fabrics are used as reinforcing materials for this type of laminate, which has many structural applications.

References and Review Questions

1. REFERENCES

1. AATCC Southeastern Section, *Textile Chemist and Colorist*, Vol. 14, No. 1, 1982.

2. Gorondy, E. J., in *Textile Printing: An Ancient Art and Yet So New*, AATCC, Research Triangle Park, NC, 1975.

3. Herard, R. S., "Equipment Developments for Napping, Shearing, Sueding, Sanding Circular Knits," *Knitting Times*, April 23, 1979, 12–17.

4. Hollen, N., J. Saddler, A. L. Langford, and S. J. Kadolph, *Textiles*, 6th Ed., Macmillan Publishing Company, New York, NY, 1988.

5. Hofmann, W., *Rubber Technology Handbook*, Hanser Publishers, Oxford Press, NY, 1989.

6. Eirich, F. R., *Science and Technology of Rubber*, Academic Press Inc., New York, 1978.

7. "Developments in Coating and Laminating," paper presented at the *Shirley Institute Conference*, Publication S41, March 27, 1981.

8. Gunnell, D. "Some Aspects of Acrylic Polymers," *Shirley Institute Conference Book of Papers*, March 1981.

9. "Coated fabrics Update," *Symposium Book of Papers*, American Association of Textile Chemists and Colorists (AATCC), March 31–April 1, 1976.

10. "Coated Fabrics Technology," *Symposium Book of Papers*, American Association of Textile Chemists and Colorists (AATCC), March 28–29, 1973.

11. Badger, J. H. and Rosin, M. L. "Poly-urethane Coatings; Chemistry and Consumer," *AATCC Coated Fabrics Conference Book of Papers*, March 1973.

12. Lerner, A., "Polyesters in Urethane Fabric Coating," *AATCC Coated Fabric Update Conference Book of Papers*, April 1976.

13. Schmuck, G., "Textile Coating Techniques," Review, Ciba-Geigy, 1974/4.

14. "Progress in Textile Coating and Laminating," paper presented at the *British Textile Technology Group (BTTG) Conference*, July 2–3, 1990.

15. Tanis, P., "Developments in Mechanical Foaming Applied to Coating," *BTTG Conference Book of Papers*, July 2–3, 1990.

16. Woodruff, F. A., *Some Developments in Coating Machinery and Processes*, Shirley Institute Publication S41, March 1981.

17. Van Mol, T. and Nijkamp, A., "Latest Developments in Coating and Finishing," *BTTG Conference Book of Papers*, July, 1990.

18. Koch, S. D and Price, J. M., "UV-Cured Coatings for Vinyl Coated Fabrics," *Coated Fabrics Technology—Volume 2*, Technomic Publishing Co., Inc., 1979.

19. Powder Coatings—Recent Developments, Noyes Data Corporation, Park Ridge, New Jersey, 1981.

1.1 General References

AATCC, *Technical Manual*, American Association of Textile Chemists and Colorists, Research Triangle Park, NC, published annually.

AATCC Publications Committee, *The Applica-*

tion of Vat Dyes, AATCC Monograph No. 2, American Association of Textile Chemists and Colorists, Research Triangle Park, NC, 1953.

Carty, P. and M. S. Byrne, *The Chemical and Mechanical Finishing of Textile Materials,* 2nd Ed., Newcastle upon Tyne, Polytechnic Products Ltd., 1987.

Datye, K. V. and A. A. Vaidya, *Chemical Processing of Synthetic Fibers and Blends,* John Wiley and Sons, 1984.

Lewin, M. and S. B. Sello, eds., *Handbook of Fiber Science and Technology: Volume II— Functional Finishes, Part A,* Marcel Dekker, Inc., 1983.

Lewin, M. and S. B. Sello, eds., *Handbook of Fiber Science and Technology: Volume II— Functional Finishes, Part B,* Marcel Dekker, Inc., 1984.

Nettles, J. E., *Handbook of Chemical Specialties,* John Wiley and Sons, 1983.

Olson, E. S., *Textile Wet Processes, Vol. 1,* Noyes Publications, 1983.

Peters, R. H., *Textile Chemistry, Volume 3,* Elsevier Scientific Publishing Company, 1975.

Trotman, E. R., *Dyeing and Chemical Technology of Textile Fibers,* 6th Ed., John Wiley and Sons, 1984.

Weaver, J. W., ed., *Analytical Methods for a Textile Laboratory,* 3rd Ed., American Association of Textile Chemists and Colorists, Research Triangle Park, NC, 1984.

Gohlke, D. J. and Tanner, J. C., "Gore-Tex® Waterproof Breathable Laminates," *AATCC Coated Fabric Update Conference Book of Papers,* April 1976.

Industrial Coatings: Conference Proceedings, ASM/ESD Conference, Chicago, November 1992.

"Developments in Coating and Laminating," *Shirley Institute Conference,* Publication S41, March 1981.

Coated Fabrics Update, *AATCC Conference Book of Papers,* April 1976.

2. REVIEW QUESTIONS

1. What are the reasons for heat setting of yarns and fabrics in industrial textile products?
2. Define the following terms:
   - copper number
   - pH
   - colorfastness
   - dyeing transition temperature
3. What kinds of chemicals are used for flame retardant finishes? Explain their performance.
4. What are the possible application areas of soil repellent fabrics?
5. Find out how each test for water repellency is conducted. What are the major differences among these tests?
6. What is the difference between coating and laminating? Explain.

ARCHITECTURAL AND
CONSTRUCTION TEXTILES

6.1

Architectural and Construction Textiles

S. ADANUR

1. INTRODUCTION

Synthetic fibers allowed development of high performance fabrics with properties such as good strength, hydrophobicity (water repellent), rot and fungi resistance which are essential properties of fabrics to be used as building and construction materials. Development of new and improved polymer coatings increased the properties and performance of these fabrics. As a result, coated fabrics are now widely used as the "envelope" of large building constructions such as airports, stadiums, sports halls, exhibitions and display halls, and storage bases for industrial and military supplies. The fabric resists extremes of sunlight, temperature, biological attack, wind, rain, and snow.

There are several advantages of using a textile structure in buildings. The major advantage of coated fabric structures is the weight. The weight of a fabric envelope can be 1/30th of the conventional envelope of bricks and mortar or concrete and steel [1]. Therefore, less structural support and reinforcement is needed which reduces the cost of the building. Another advantage is that textile structures provide large obstruction-free spans, which is highly desirable for large public gatherings, equipment and material storage, exhibition and sporting activities. The erection time of the fabric envelope is much shorter compared to conventional construction. Smaller fabric envelopes can be taken down easily and re-erected somewhere else. Since fabrics do not easily tear, mechanical damage due to wear or impact is restricted to a small area by the fabric structure. The damage can also be repaired very easily. Buildings with fabric envelopes have better resistance to major destructive forces such as earthquakes. Fabric structures also give freedom to architects to design various shapes and appearances.

Membrane structures are used in the architecture and construction industry. There are three types of membrane structures: films, meshes and fabrics. Films are transparent polymers in sheet form without coating or lamination. Examples include clear vinyl, polyester or polyethylene. They are less expensive and durable than textiles. Meshes are porous fabrics, such as woven polyester, that are lightly coated with vinyl. Knitted meshes are made of high density polyethylene, polypropylene, or acrylic yarns. Meshes are used as shelters from wind and sun, however they cannot provide adequate protection from rain. Films or meshes are not used in air structure designs.

Fabric structures are by far the most widely used membrane structures. Fabrics are typically coated or laminated with synthetic materials to improve strength and environmental resistance (Section 5.2, Coating and Laminating).

2. FABRICS FOR ARCHITECTURE AND CONSTRUCTION

Textiles for architecture and construction should be resistant to deformation and extension under tension and to wind and water. The fabric must be waterproof, impermeable to air and wind, and resistant to abrasion and mechanical damage. It should also be resistant to degradation from long-term exposure to sunlight and acid rain. Depending on the application, the fabric may need to transmit or reflect different levels of light. These requirements can be met by applying special coatings to base fabrics.

2.1 Base Fabrics

Base fabrics are usually made of synthetic fibers and form the carrier layer which provides the necessary strength to the structure. Although aramid and carbon fibers have excellent properties, they are rather expensive to use in these types of construction extensively. The most widely used materials are high-tenacity polyester, fiberglass and nylon. Use of synthetic materials allowed lower backing fabric weight compared to conventional canvas fabric made of natural fibers.

Polyester is the most widely used base fabric material for its strength, durability, cost and stretch properties. Polyester fabrics are the least expensive for longer term applications. Nylon is sometimes used for membrane structures. It is more durable than polyester, but it has higher cost and more stretch. Glass fabrics resist stretching after they are tensioned and therefore do not wrinkle or bag. Glass fabrics reflect a high percentage of the sun's heat and keep the interior of the structure cool. They do not burn or smoke. Woven fabrics made of PTFE fibers combine environment-resistant advantages with durability against flexing and folding. However, these structures are more expensive.

As in most industrial fabric applications, continuous filament yarns are preferred over staple yarns due to inherent strength and elongation resistance. Yarn twist level is selected to be low to carry higher tensile loads as well as to prevent fiber slippage which may cause early yarn rupture. Filaments with low twist levels can flatten during weaving resulting in high cover factor. Hydrophobic fiber materials are preferred.

Base fabrics can be woven, knit or nonwoven. Woven structures are usually the design of choice for fabric rigidity and dimensional stability for many applications. Generally simple weave patterns such as plain weave and low harness twills are used. Warp-knit and nonwoven stitchbonded fabrics are also used.

Special care should be taken during warping and weaving of the woven base fabrics. The tension in the warp sheet should be kept as uniform as possible since a tight threadline in warping will be visible as a defect in the coated fabric. It is recommended that the yarn tension should not exceed 9 N/tex during warping [2]. For some applications, yarns between 220 and 250 denier may need to be twisted before weaving. Twisting is usually not required for 500 or higher denier yarns. Filling yarns over 500 denier may need to be twisted when woven on projectile looms (Section 4.4). The warp yarn may require sizing or twisting if a shuttle loom is used and the fabric construction is denser than 12×12 yarns/inch. Sizing may be necessary when weaving polyester on projectile looms. Reinforcement fabrics can be woven on conventional shuttle, rapier, projectile, air-jet and water-jet machines.

With the availability of low and intermediate shrinkage yarns, heat setting is not required. However, base fabrics made of high-shrink yarns need to be heat set before coating in order to prevent excessive shrinkage during coating. Scouring base fabric to remove warp size is generally unnecessary for scrim fabrics since the laminates bond to each other and not to the fabric. For coated fabrics, scouring may not be needed if the sizing material is compatible with the coating compounds. For example, greige polyester can be satisfactorily coated with vinyls and rubbers.

Typical base fabrics for coating have a greige weight of $3.3 - 5.9$ oz/sq. yard ($112 - 198$ g/m^2). Since the tightly woven or knit structure does not allow the yarns to slip together as well as a scrim fabric under tearing load, the higher tensile strength of coated fabric does not necessarily yield a higher tear strength. The tear strength of coated fabrics can be increased up to 100% by using a rip-stop construction (a variation of plain weave having two yarns woven as one at regular intervals). However, at the doubled-yarn crossover points, the coating may be thinner and therefore more susceptible to abrasion. Typical scrim fabrics for laminating have a greige weight of $1.2 - 3.3$ oz/sq. yd ($41 - 112$ g/m^2). Weft inserted warp knit fabrics (Figure 4.52) are suitable for either coating or laminating. Table 6.1 shows typical base fabric and scrim constructions. Weights are for greige fabrics.

The width of woven fabrics can be up to 5 m. Depending on the shape and size of the construction, fabric pieces are cut and seamed. The seam area is welded by heat and pressure to attain

TABLE 6.1 *Typical Base Fabric and Scrim Constructions [3].*

Fabric	Ends × Picks (per cm)	Weight (g/m²)	Warp Yarn (denier)	Filling Yarn (denier)
Lightweight scrim	7.1 × 3.5	33	220	500
Intermediate-weight scrim	3.5 × 3.5	44	500	500
	3.5 × 3.5	85	1,000	1,000
	3.9 × 3.9	85	1,000	1,000
	3.9 × 3.9	85	1,000	1,000
Base fabric	7.1 × 4.7	149	1,000	1,000
Base fabric	7.1 × 6.3	143	840	1,000
Base fabric	8.5 × 9	197	1,000	1,000

impermeability to air, wind and rain. Special techniques are used to ensure a properly welded seam.

Coating and Laminating

Coating is made of a plastic material or synthetic rubber. Coating provides waterproofness and protects the base fabric from sunlight and weathering degradation. Common polymer coatings are natural rubber (polyisoprene), styreneutadiene rubber (SBR), nitrile rubber (acrylonitrile-butadiene copolymers), neoprene (polychloroprene), butyl rubber (isobutene-isoprene copolymers), Hypalon® (chlorosulphonated polyethylene), polyvinyl chloride (PVC), polyvinylidene chloride (PVDC), polytetrafluoroethylene (PTFE), polyurethanes (PU), silicone rubbers (polysiloxanes), polyvinyl fluoride (PVF), polyvinylidene fluoride (PVDF) and fluororubbers [4]. Mostly, polyvinyl chloride (PVC), polytetrafluoroethylene (PTFE) or silicone coatings with flame-resistance properties are used. Pigment coloration is common. To give the PVC fabric "self-cleaning" properties and added protection, a polyurethane topcoating is used.

Polyester fabrics are usually coated or laminated with PVC films. Laminates usually consist of vinyl films over woven or knitted polyester or nylon meshes (substrates). In the Precontraint process, polyester fabric is tensioned before and during the coating process which is claimed to improve dimensional stability by imparting identical characteristics to the warp and filling yarns (Figure 6.1). Fiberglass fabrics are usually coated with PTFE for

durability. PTFE coated fiberglass is the only material that meets the U.S. model building codes' definition of non-combustible materials.

Coated fabrics are usually more expensive than similar laminated fabrics but they offer higher tensile strength, higher flex resistance, higher abrasion resistance and longer life. The material and/or thickness of the coating layers on each side may or may not be the same. For applications where water and chemical resistance are the major concern, double coating may be used. Laminated fabrics are the most economical for general applications. They have

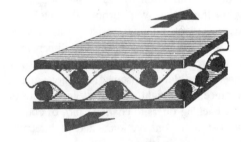

Conventional coating process

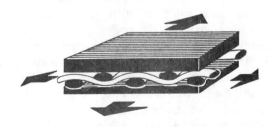

PRECONTRAINT system

FIGURE 6.1 The Precontraint coating system (courtesy of Ferrari).

TABLE 6.2 *Typical Properties of Coated and Laminated Fabrics Made with Trevira® Polyester [3].*

Property	Fabric Type					
	Light Duty	Light Duty	Medium Duty	Medium Duty	Heavy Duty	Heavy Duty
Process	laminated	laminated	laminated	laminated	coated	coated
Warp (denier)	220	220	1,000	1,000	1,000	1,000
Filling (denier)	500	220	220	1,000	1,000	1,000
Fabric weight (g/m^2)	37	44	85	85	139	200
Thickness (cm)	0.03	0.03	0.048	0.053	0.102	0.055
Yarn count (ends × picks/cm)	7.1 × 3.5	7.9 × 9	3.9 × 3.9	3.5 × 3.5	7.1 × 4.7	8.5 × 9
Grab strength	400 × 4,000	512 × 489	1,200 × 1,150	978 × 890	2,000 × 1,450	3,000 × 3000
Tongue tear strength (N)	107 × 156	107 × 120	356 × 400	400 × 445	810 × 1,290	343 × 445
Finished weight (g/m^2)	373	373	576	440	1,288	644

good tear resistance; however they may delaminate under repeated flexing or wind-whipping conditions.

2.2 Properties of Coated Fabrics for Architecture and Construction

Depending on the application area, general properties required for coated industrial fabrics are high tensile strength, adequate elongation, high melting point, waterproofness, toughness, resistance to rot and fungi, resistance to weathering effects and aging, wet and dry dimensional stability, resistance of coating to high and low temperatures, flame resistance, abrasion and tear resistance, low weight, flexibility, good adhesion of the backing fibers to coating, and, of course, low cost. For constructional and architectural coated fabrics, resistance to fatigue and time related fatigue become very important. Table 6.2 shows typical properties of coated and laminated fabrics made with Trevirafi polyester. Characteristics of commercially available air, tent and tensile fabric structures are listed in Appendix 1.

Since they are generally exposed to outside conditions, aging characteristics are critical for coated fabrics. Weathering effects cause degradation of both fibrous reinforcement and coating which, in turn, negatively affect the properties and performance characteristics. Ultraviolet radiation, particularly, degrades the properties of some synthetic fibers. The life of the coated fabric depends on how well the coating blocks the ultraviolet radiation. Therefore, it is important that the coating covers all the interlacing points in the fabric sufficiently. Figure 6.2 shows the aging performance of canvas cloth and nylon coated fabrics. Weather resistance tests are time-consuming since they usually require several years for reliable results. Therefore, rapid tests can be done on simulating machines. However, correlation of accelerated indoor testing with 15−20 year outdoor testing has not been demonstrated.

One of the most important properties of coated fabrics for buildings is the residual strength. Tensile test results of PVC coated polyester and nylon 6,6 fabrics, which were subjected to five

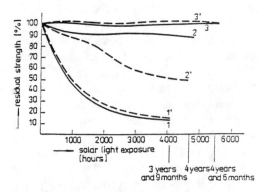

FIGURE 6.2 Aging performance of canvas cloth and nylon fabrics. Linen canvas cloth, finished, 1-warp direction, 1'-filling direction; nylon fabric (Chemlon), PVC coated, 2-warp direction, 2'-filling direction; nylon fabric (Nylon) PVC coated, 3-warp direction, 3'- filling direction (Elsevier Science [2]).

AFTER 8 YEARS
IN %

THICKNESS

BREAKING LOAD — WARP / WEFT

BREAKING ELONGATION — WARP / WEFT

TEAR STRENGTH — WARP / WEFT

0 10 20 30 40 50 60 70 80 90 100%

FIGURE 6.3 Residual properties of PVC coated polyester fabric after eight years of use in air structure (The Textile Institute [5]).

years of loading, revealed that the strength loss was around 10%. In another study on PVC coated polyester fabrics after eight years in an air structure, similar results were found which are shown in Figure 6.3. The rather low tear strength is the result of the brittleness developed in the fabric during its use.

Blumberg compared Teflon® coated fiberglass fabric with PVC coated polyester fabric. His results are shown in Figure 6.4.

Coated fabrics made of synthetic fibers have replaced conventional canvas cloth, which is generally made of cotton, almost in every application. The reason for this is the improved properties of synthetic coated fabrics. Figure 6.5 shows the stress-strain curves of canvas cloth and coated nylon and polyester fabrics.

3. APPLICATIONS OF COATED FABRICS IN BUILDING STRUCTURES

Membrane structures can be divided into four categories: tent, clear-span, air and tensile structures. Although each structure type has specific characteristics, there are some structures that may fall under more than one category. Some manufacturers categorize membrane structures by temporary or permanent usage. The common denominator in all of the membrane structures is textile fabric. The most widely used textile materials for architecture and construction are polyester coated with polyvinyl chloride (PVC), woven fiberglass coated with polytetrafluoroethylene (PTFE) and silicone-coated fiberglass. Figure 6.6 shows application examples of textiles in architecture and construction.

3.1 Tents

Tents and shelters were probably the first constructions in which textiles were used as building materials. Tents have been traditionally used for various purposes by nomadic people, traders, military, explorers, and campers. More recently,

	GLASS TEFLON-COATED	PES PVC-COATED
FABRIC APPEARANCE		→
TEAR STRENGTH		→
DIMENSIONAL STABILITY	←	
FLAMMABILITY	←	
TRANSLUCENCY		→
SEAM STRENGTH	←	
CLEANABILITY	←	
MAKING-UP PERFORMANCE		→
FLEXIBILITY		→
FLEXING LIFE		→
CHEMICAL RESISTANCE	←	
HEAT-RESISTANCE	←	
RESISTANCE TO COLD	←	

→ SLIGHTLY BETTER
⟶ DISTINCTLY BETTER

FIGURE 6.4 Comparison of Teflon® coated glass fabric and PVC coated polyester fabric (The Textile Institute [5]).

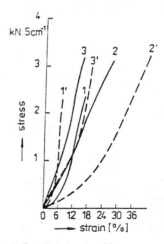

FIGURE 6.5 Tensile behavior of linen canvas cloth and synthetic coated fabrics. 1: linen canvas, 2: PVC coated nylon, 3: PVC coated polyester fabric. Prime (') indicates filling direction testing (Elsevier Science [2]).

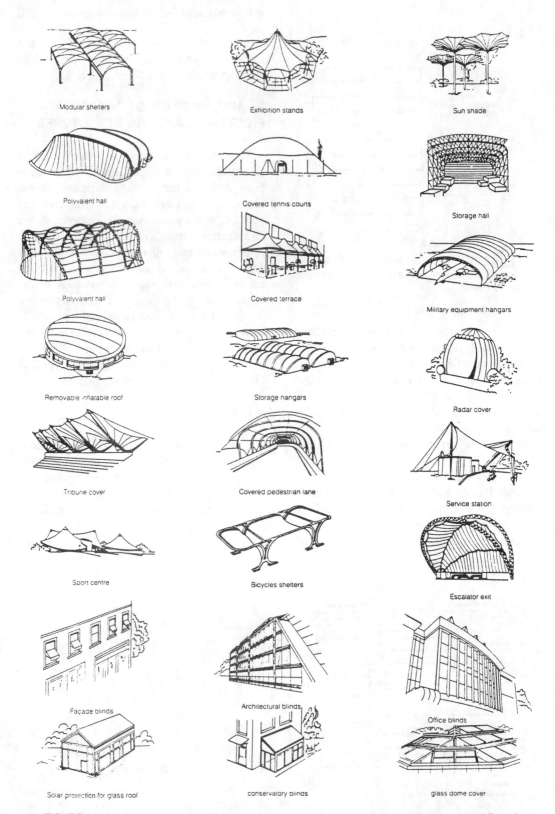

Modular shelters

Exhibition stands

Sun shade

Polyvalent hall

Covered tennis courts

Storage hall

Polyvalent hall

Covered terrace

Military equipment hangars

Removable inflatable roof

Storage hangars

Radar cover

Tribune cover

Covered pedestrian lane

Service station

Sport centre

Bicycles shelters

Escalator exit

Façade blinds

Architectural blinds

Office blinds

Solar protection for glass roof

conservatory blinds

glass dome cover

FIGURE 6.6 Application examples of industrial fabrics in architecture and construction (courtesy of Ferrari).

FIGURE 6.7 Pole tent (courtesy of Armbruster Manufacturing Co.).

new tent constructions were developed for building construction, business, exhibitions, entertainment, leisure and recreation.

In pole tents, the fabric is draped or hung rather than tensioned (Figure 6.7). Tension tents include various tensile structure characteristics and therefore fall between tents and tensile structures (Figure 6.8). Tensile tents have tensioned fabric that provides clear span and they do not require guys. Pole tents and tensile tents are mass produced.

The most widely used fabrics for tent walls are polyesters and vinyls. Table 6.3 shows typical properties of heavy duty coated tent fabrics.

3.2 Clear-Span Structures

Clear-span structures provide clear space beneath the fabric, free of poles and other supporting elements. They are also known as clear-span tents. In a clear-span, the fabric is pulled taut through channels in the frame's ribs, i.e., no poles or masts are used. Well built clear-spans are very stable and have been used in many commercial and industrial applications. They are more permanent than tents and less permanent than air or tensile structures. Clear-spans can accommodate doors, flooring, insulation, elec-

FIGURE 6.8 Tension tent (courtesy of Armbruster Manufacturing Co.).

TABLE 6.3 Typical Properties of a Heavy Duty Tent Fabric (Herculite Products, Inc.).

Weight	16 oz/sq yd
Tear strength	150 lb (warp) × 160 lb (filling)
Tensile strength	300 lb (warp) × 290 lb (filling)
Low temperature resistance	−40°F
Hydrostatic burst	450 psi
Flame retardance	0.6 sec, 3.7 inch

tricity and HVAC easier than tents. PVC coated polyester is the most widely used fabric for clear-span structures.

3.3 Tension Structures

In tension structures, metal pylons, tensioning cables, wooden or metal frameworks are used to support the fabric. A relatively minimal support system, which must be rigid, is required for these structures because the fabric carries most of the load. Tension structures are curvilinear and therefore a fabric system must have or develop curvature to resist applied loads (Figure 6.9). The fabric is highly tensioned and therefore must have good strength.

Although tension structures can be designed in various shapes, the basic building block shapes are generally hyperbolic paraboloid and the hyperboloid. The fabric is double curved, with the curvatures opposing each other from a single intersecting point. The crossed arch appears parabolic in cross section and as an "X" when viewed from above. The two main ways to attach fabric to anchorages are sleeves and clamps. Figure 6.10 shows the installation stages of a tension structure as a stadium roof. Figures 6.11−6.14 show several tension structures.

3.4 Air Structures

Air-supported structures can be built in two ways:

(1) Air pressure inside the envelope provides tensioning and maintains required configuration and stability (Figure 6.15). The main components of an air-supported system are the envelope (fabric), inflation system (fans), anchorage system (cables and foundation) and doors and access equipment. PVC coated polyester is the most widely used fabric. The collapse of some large-scale stadium domes in the early 1980s affected the acceptance of air supported structures. However, well-built air structures are quite stable and strong. Pressurized air-supported structures are modular buildings. They can be designed to form nearly any shape provided that surfaces are curved. Simple domes and "bubbles" are the most common (Figure 6.16).

(2) Air-inflated ribs (air beams) are used to support the structure (Figure 6.17). Air beams are pressurized air-inflated support tubes.

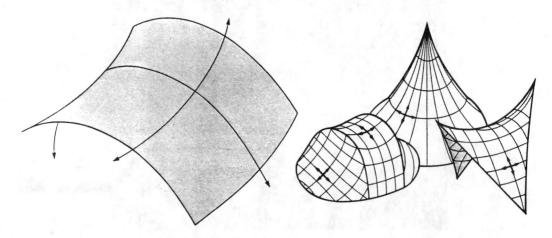

FIGURE 6.9 Schematic of tension structures (The Textile Institute [1]).

FIGURE 6.10 Installation stages of a tension structure stadium roof (courtesy of Birdair, Inc.).

FIGURE 6.11 Coated fabric is used to cover high traffic access areas at Jacksonville International Airport, Jacksonville, Florida (courtesy of Birdair, Inc.).

FIGURE 6.12 Polyester coated fabric was used to replace a worn-out stadium roof in 1993 in Stuttgart, Germany. The membrane structure consists of 35,000 m^2 membrane and 100,000 individual parts which is the largest of its kind in Europe (courtesy of Hoechst Celanese).

FIGURE 6.13 San Diego Convention Center, San Diego, California (courtesy of Birdair, Inc.).

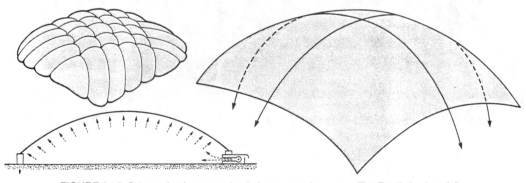

FIGURE 6.14 The roof of Orlando City Hall (Florida) was designed as a soccer ball to commemorate the 1994 World Cup Soccer finals (courtesy of IFAI).

FIGURE 6.15 Schematic of a pressurized air-supported structure (The Textile Institute [1]).

FIGURE 6.16 Pressurized air structure (courtesy of IFAI).

FIGURE 6.17 Inflatable air beam for an air-supported structure (courtesy of Albany International).

The interior of the structure itself is not pressurized. Wide diameter supports (up to 42 inch) use low air pressure (4 to 10 psi) and narrow-diameter supports (4 to 18 inch) use high air pressure (30 to 100 psi) [6]. They are currently used for lightweight, easy-to-transport military structures although civil applications may have good potential in the near future. Polyester and nylon are the materials of choice. For high pressures aramid fibers may be needed. The width/height ratio may be a limitation for a single air structure but this limitation can be overcome by conjoining several modules together.

There are approximately 150 big fabric structures in the world which were built within the last twenty years, including Georgia Dome in Atlanta (cable supported, Figure 1.2), Silver Dome in Detroit (where Teflon® coated glass fabric was used), Bullock's department store in San Mateo, California, Hoosier Dome in Indianapolis (air structure) and the Carrier Dome in Syracuse. The Carrier Dome is air supported and made of glass fiber coated with a fluoropolymer dispersion (Figure 6.18). For the large-span areas such as stadium roofs and airport terminals, tension designs are preferred. For example, over 500,000 m² of PTFE coated glass fabric was used in the construction of the Haj Terminal building for Jeddah International Airport in Saudi Arabia which was completed in 1982. It is the world's largest tensile structure [7]. The newest tension fabric structure is the terminal at the new Denver International Airport which has a coated fabric roof that consists of two layers of Teflon® coated woven fiberglass (Figure 1.10). The fabric is designed to allow enough light transmission for plants. The roof is suspended by steel masts and cables, covering an area of 20,000 m² [8].

Lighter polyester fabrics (6 to 8 ounces per square yard) are used as acoustic and insulated liners. Waterproof, flame retardant tarpaulins are used in cold weather construction. The fabric keeps heat in, so construction can continue all winter, even at 40 degrees below zero (Figure 6.19). Heavier polyester materials such as 20 to 26 ounce fabrics, are suitable for long-term ex-

terior use. Fiberglass based fabrics have been the material of choice for stadium domes (both air and cable supported) and other permanent structures especially in the United States.

3.5 Designing with Coated Fabrics

The major functional requirements for permanent architectural fabrics are listed in Table 6.4.

Mainly three types of stresses act on an architectural fabric:

(1) Stresses due to applied load to tension the fabric
(2) Weight of suspended fabric
(3) Stresses induced by natural forces due to wind, rain and snow

The calculation of the loads depends on the region where the structure will be installed. Snow loads push the fabric down while wind loads tend to pull the structure upward. Wind can affect a fabric structure in two ways: laterally across the top, or, if the structure is open-sided, up from below. Heavier loads necessitate stronger anchoring which may require additional or larger cables, stronger connecting hardware, and larger foundations or reinforced perimeter walls into which to ground the anchors. Properly

TABLE 6.4 Functional Requirements for Permanent Architectural Fabrics (Chemfab).

Tensile strength	500 lb/inch for air-supported structures
	650 lb/inch for pretensioned structures
Durability	Retain functional integrity for at least twenty years
Fire safety	Find acceptance within existing building codes
Energy efficiency	Provide adequate daylighting without artificial illumination
	Provide acceptable thermal gain/loss characteristics
Cost	Provide life cycle cost competitive to conventional construction
Aesthetics	Constitute an exciting structural alternative to traditional structural options

FIGURE 6.18 The Carrier Dome, Syracuse, New York (courtesy of John Dowling Photography, Syracuse).

FIGURE 6.19 Safety tarpaulins for cold weather construction (courtesy of Research Plastics).

designed and reinforced, fabric structures can be made to withstand hurricanes and snowstorms.

Tensioning forces on the fabric may be up to 13 kg/cm. To be able to maintain the tensioned shape of the envelope, the fibers and fabrics must have high modulus and resistance to deformation, extension and creep. Figure 6.20 shows typical uniaxial and biaxial elongation of a coated

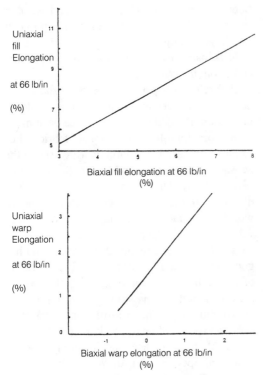

FIGURE 6.20 Typical uniaxial and biaxial elongation diagrams of an architectural fabric (courtesy of Chemfab).

fabric. They should be resistant to extreme hot and cold temperatures, water, sunlight degradation, atmospheric pollution, rotting and insect attack. These types of structures are expected to last many years.

The major challenge in construction with architectural fabrics is to achieve enough rigidity and stability that can withstand high winds, rain and snowfall. Design for multi-directional tensioning is critical, and complex engineering and computer aided design are used for this purpose. Fabric structure should have the necessary curves, sweeps and slopes to fit to the structure frame. In air-supported structures, an air-impermeable fabric assembly is anchored and sealed to the foundation or an air retaining wall. Then, air is pumped in to inflate and tension the fabric. The air pressure required for this purpose is approximately 0.3% above the ambient atmospheric pressure which can be hardly felt by the occupants. The structure is automatically reflated with an air-pumping system if the pressure drops. Entrances and exits are built with simple air-lock systems to minimize the pressure loss. Sometimes, two layer envelope fabric is used to provide stagnant air between layers for thermal insulation, to contain insulatory material, to improve acoustical properties and allow warm air circulation for snow melting. In structures supported by air-inflated ribs, semicircular ribs similar to large diameter hoses with impermeable air locks are used.

Polyester woven fabrics coated with PVC and an outer coating of polyurethane are mostly used in "semipermanent" structures. Permanent structures generally use glass woven fabrics coated with PTFE. They can last up to twenty years and are relatively more expensive than PVC coated fabrics.

Membrane structures can be constructed for short-term use (sporting events, exhibition and storage halls) or long-term use (stadiums, airports). Short-term or temporary structures can usually be built by using available and standard size coated fabrics. However, for large construction, architects and engineers (textile and others) must work together to design, produce and install large size coated fabrics. In addition to highly engineered coated textile structure, support and tensioning systems must be well calculated and established for long-term, permanent use.

Tension structure designers use Computer Aided Design (CAD) systems to calculate loads and determine their effects on the designs before building the structure itself. CAD of fabric structures reduces development time and allows designers to explore various alternatives and options in their work.

General considerations for a tension structure design are as follows [9]:

- dimensions of the structure
- permanent (15 to 20 years) versus short-term life structure
- type of use (commercial, industrial, public assembly, roof, skylight, shelter, signature, etc.)
- free standing versus anchored to a hard structure
- budget limitations and considerations
- local requirements regarding fire resistance, loads and other code-related matters

4. AWNINGS AND CANOPIES

Awnings are widely used in commercial and residential buildings for decoration, protection, or visibility. Canopies, patio covers and porch awnings are the leading products in awning business. Fabric awnings and canopies are the most common types of shading structures.

The main difference between an ''awning'' and a ''canopy'' is that a canopy requires support in addition to the anchoring that attaches it to a building. BOCA (Building Officials and Code Administrators International) defines a fabric awning as ''an architectural projection that provides weather protection, identity, and/or decoration and is wholly supported by the building to which it is attached. An awning is comprised of a lightweight, rigid, or retractable skeleton structure over which a fabric cover is attached.'' A fabric canopy is defined by BOCA as ''an architectural projection that provides weather protection, identity, and/or decoration and is ground-supported in addition to being supported by the building to which the canopy is

attached.'' Figure 6.21 shows standard awning and canopy styles. Figure 6.22 shows the main parts of an awning installation. Retractable or lateral arm awning segment is recently gaining popularity (Figure 6.23). Lateral arm awning systems, which were first introduced in Europe, provide easy installation, sun control and good looks. Retractable awning systems include a manual or electric cranking system that allows them to be folded up or retracted when not used.

Illuminated (backlit) awnings offer high visibility in commercial applications. Backlighting is usually achieved by attaching lighting to the support frame, beneath the fabric cover, as shown in Figure 6.24. Although any kind of lighting may be used for illuminating awnings, fluorescent lighting is the most widely used because of its low expense, low temperature, high durability and a more even distribution of light. Awning and canopy applications are shown in Figures 6.25 and 6.26, respectively.

Fiberoptics are also used in awnings for animation and various color possibilities. The awning industry uses a specially extruded acrylic fiber that is lightweight, durable and weather resistant. The fiber is treated with a polymer cladding. The light aimed at one end of the fiber reflects back and forth along the length and comes out the other end.

Umbrellas are primarily used as shade and decorative devices for commercial and residential patios. Recently umbrellas have been made with more high-tech frames and fabrics. For example, retractable umbrellas were used in the Mosque of the Prophet in Medina, Saudi Arabia (Figure 1.14). The 296 m^2 screens with a 25 m diagonal are the largest ever constructed in the world. The fabric is Tenara™ specially developed by W. L. Gore and Associates, Inc. It is woven from high strength PTFE fibers and has good flex life and UV immunity. The operation of the umbrellas is computer controlled at the press of a button. The shade roof also provides air conditioning. The construction is designed for wind speeds of up to 155 km/h [10,11].

ASTM performance specification D 4847 (Standard Specification for Woven Awning and Canopy Fabrics) includes test methods and specification requirements for awning and canopy fabrics.

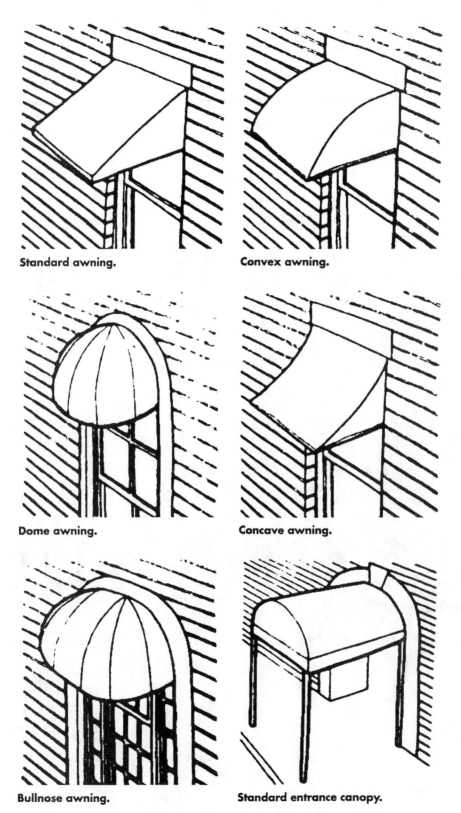

Standard awning.

Convex awning.

Dome awning.

Concave awning.

Bullnose awning.

Standard entrance canopy.

FIGURE 6.21 Standard awning and canopy styles (courtesy of IFAI).

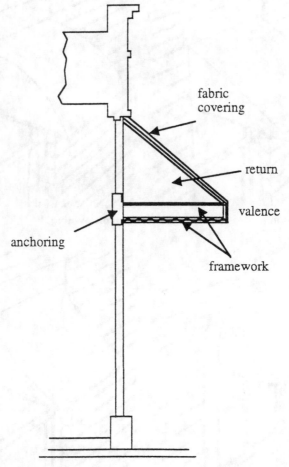

FIGURE 6.22 Main parts of an awning installation (courtesy of IFAI).

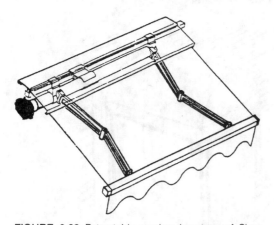

FIGURE 6.23 Retractable awning (courtesy of Simu U.S. Inc.).

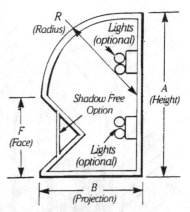

FIGURE 6.24 Illuminated (backlit) awning frame system (courtesy of Eide Industries, Inc.).

FIGURE 6.25 Awning (courtesy of IFAI).

FIGURE 6.26 Canopy (courtesy of IFAI).

TABLE 6.5 Awning Fabric Characteristics (courtesy of IFAI Awning Division).

	Painted Army Duck	Vinyl Coated Cotton	Vinyl Coated Polyester/ Cotton Blend	Vinyl Laminated Polyester	Vinyl Coated Polyester	Acrylic Coated Polyester	Solution-Dyed Acrylic	Solution-Dyed Modacrylic
Description	Acrylic-painted cotton duck fabric	Vinyl coated cotton duck fabric	Vinyl coated on each side of poly/cotton base	Tri-layer fabric; vinyl outside layers, woven polyester middle	Vinyl coated on each side of a polyester base	Acrylic coated on each side of a polyester base	Woven acrylic fibers, fluorocarbon finish	Woven, modacrylic fibers, fluorocarbon finish
Typical weight (oz/sq yd)	11	15	13	10–16	11–17	9.5–12.5	9.25	9.25
Properties	Resistant to UV light, mildew, water	Resistant to UV light, mildew, water	Resistant to UV light, mildew; water repellent	Resistant to UV light, mildew, water	Resistant to UV light, mildew, water, cleanable	Resistant to UV light, mildew; water repellent	Resistant to UV, color degradation, water, mildew	Resistant to UV, color degradation; water repellent, resistant
Colors	Stripes or solids; many colors	Stripes or solids; many colors	Solids; same color both sides	Stripes or solids; primary colors, pastels	Stripes or solids	Mostly solids, some stripes	Stripes or solids; many colors	Solid colors and tweeds
Underside	Pearl gray, green, gray with floral print	Solid pearl gray	Same as top side	Linen-like pattern, solid coordinating color or same color as top	Same as top side	Same as top side	Same as top side	Same as top side
Surface	Matte finish with linen-like visible texture	Smooth, nonglare surface with little or no texture	Textured surface	Smooth or matte, slight woven or linen-like texture	Smooth, somewhat glossy top surface	Surface is textured with cloth appearance	Woven texture	Woven texture surface

220

TABLE 6.5 (continued).

	Painted Army Duck	Vinyl Coated Cotton	Vinyl Coated Polyester/ Cotton Blend	Vinyl Laminated Polyester	Vinyl Coated Polyester	Acrylic Coated Polyester	Solution-Dyed Acrylic	Solution-Dyed Modacrylic
Transparency level	Opaque	Opaque	Opaque	Translucent, depending on color, certain styles formulated for backlighting are highly translucent	Translucent, depending on color	Translucent, depending on color	Translucent, depending on color	Translucent, depending on color
Abrasion resistance	Very good	Very good	Very good	Good; very strong base fabric	Good	Very good	Good	Good
Stretch*	Very good	Very good	Very good	Very good	Stable	Very good	Good**	Good
Mildew resistance	Good, not recommended for high humidity	Good, not recommended for high humidity	Very good	Very good, recommended for sustained high humidity	Very good	Very good	Very good	Very good
Durability†	5 – 8 years	5 – 8 years	5 – 8 years	5 – 8 years	5 – 8 years	5 – 8 years	5 – 10 years	5 – 10 years
Flame resistance (FR)	Some colors with FR treatment	Some colors with FR treatment	All colors are flame-resistant	All colors are flame-resistant	All colors are flame-resistant	All colors are flame-resistant	Non-flame-resistant	All patterns are flame-resistant

*Dimensional stability.
**Some shrinkage in cold weather; some stretch in hot weather.
†Depends on climate and proper care of fabric.

Awning Fabrics

Fabric choice depends on the application: interior or exterior, illuminated or standard, commercial or residential, etc. Fabrics used in backlit awnings and canopies have similar weight but differ greatly in translucency. To maximize light translucency, vinyl coated and vinyl laminated polyester fabrics are used.

There are several major fabric types that are laboratory and field tested for weathering, durability, flame retardancy and other performance characteristics which include vinyl laminated polyester, vinyl coated cotton, army duct, solution-dyed modacrylic, vinyl coated polyester/cotton and vinyl coated polyester. Table 6.5 shows the general properties of some awning fabric types. Appendix 2 lists the major characteristics of commercially available awning and canopy fabrics from different manufacturers.

Although traditional cotton duck or canvas is still used especially for traditional-look applications, synthetic fabrics dominate the market for both residential and commercial awnings and canopies. Synthetics have better resistance to color-fading, mildew, flame and wicking.

The most widely used synthetic is solution-dyed acrylic because of its wide variety of colors and canvas-like texture. Geographic location and climate are the major factors for selection of an awning fabric for durability. Solution-dyed acrylic and vinyl laminated/coated polyester can withstand a number of climates. Some of the other properties that influence awning selection are [12]:

- fire resistance
- mildew and UV inhibitor treatment
- translucency
- cleanability
- weight considerations

A clear topcoat finish may be applied to fabrics for added durability and cleanability. Acrylic was the first topcoat to be used. Later, polyvinyl fluoride (PVF) and polyvinylidene fluoride (PVDF) were introduced.

5. TEXTILES AS ROOFING MATERIALS

Single-ply and multi-ply materials are used in the roofing market to protect buildings. In traditional built-up roofing (BUR), alternating plies of felts, fabrics or mats are assembled in place and bonded together with layers of bituminous products such as asphalt or coal tar.

Single-ply roofing was introduced in the U.S. in the 1960s. It is gaining importance in commercial roofing applications (Figure 6.27). Single-ply roofing is made of single layer, watertight, weatherable membrane and sealed at the seams and edges. There are three types of single-ply membranes: elastomers or thermosets (rubber), thermoplastics and modified bitumens [13].

EPDM is the most commonly used single-ply membrane elastomer. Other elastomers used are neoprene or chloroprene rubber, chlorosulfonated polyethylene (CSPE) or Hypalon® by DuPont, chlorinated polyethylene (CPE), and polyisobutylene (PIB). Woven polyester or non-

FIGURE 6.27 Single-ply roofing (courtesy of Cooley, Inc.).

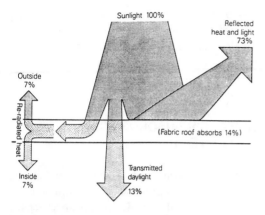

Sunlight 100%

Reflected
heat and light
73%

Outside
7%

(Fabric roof absorbs 14%)

Re-radiated heat

Transmitted
daylight

Inside
7%

13%

FIGURE 6.28 Typical light transmittance and reflectance levels of roofing fabric in hot climates (The Textile Institute [1]).

woven fiberglass materials are used as substrates to support these rubber sheets.

Polyvinyl chloride (PVC) is the major thermoplastic roofing material which may be used with reinforcing fabric or fibers. Modified bitumen is a rubber reinforced asphalt that consists of one or more prefabricated reinforced sheets, modified with either atactic polypropylene (APP) or a styrene-butadiene-styrene (SBS) to improve elasticity and durability.

Typical properties required in roof materials are toughness, non-wicking, delamination resistance, chemical resistance, UV resistance, flame resistance and fungus resistance. Table 6.6 shows properties of single-ply roof membrane.

Roofing fabrics are designed to provide different levels of sunlight transmittance and reflectance. Figure 6.28 shows typical light transmission and reflectance levels for hot climates.

6. STORAGE VESSELS

Flexible coated fabrics are used in collapsible storage containers. The major requirements for such storage systems are high tear and tensile strength with minimal weight. Examples of these types of structures include liners for transport containers and reservoirs, gas holders, marine dracones, and portable liquid (fuel, water) containment systems (Figure 6.29). Marine dracones are used as towing containers. They are made of woven nylon base fabrics that are coated

TABLE 6.6 Typical Properties of Single-Ply Roof Material (courtesy of Seaman Corp.).

Material Property	Test Methods (units)	Product Data
Thickness	ASTM D-751 (inch)	.036 nominal
Weight	(oz)	32 nominal
Tensile	ASTM D-751 (lbs)	375 × 350
Strength	ASTM D-882 (psi)	8,500
Elongation	ASTM D-751 (%)	20 warp × 30 fill
Tear strength	ASTM D-751 (lbs)	100 × 100
8″ × 10″ Sample puncture resistance	Fed. Std. 101 B Method 2031 (lbs)	250
Water vapor transmission	ASTM E-96 Proc. A (gm/m²/24 hours)	1.3
Water absorption, 14 days @ 70°F	ASTM D-471	1
Dimensional stability	ASTM D-1204	0.5
Low temperature flexibility	ASTM D-2136 (F)	−30
Factory seam strength	ASTM D-751, grab method (% of fabric strength)	100
Shore "A" hardness	ASTM D-2240	80
Accelerated weathering	Carbon arc with water spray	5,000 hrs/no cracking, blistering or crazing
Hydrostatic resistance	ASTM D-751 Method A. Proc. 1 (psi)	500 psi min
Flame resistance	Mil-C20696C Type II Class 2	Pass
Oil resistance	Mil-C-20696C	No swelling, cracking or leaking
Hydrocarbon resistance	Mil-C-20696C	No swelling, cracking or leaking

FIGURE 6.29 Collapsible fuel and water tank (courtesy of Cooley, Inc.).

on both sides. Neoprene is used for outer coating. Inner coating depends on the cargo, e.g., nitrile rubber for oil products and approved polyurethane for water. The total weight of the coated fabric is around 5 kg/m² and that of the base fabric is in the range of 600−700 g/m² [4].

Geomembranes, which are thin flexible sheets, are used as impermeable liners to hold liquids or to prevent leakage of toxic wastes into the soil. The base fabric, usually made of nylon, polyester or polypropylene, provides the necessary tear and puncture resistance to the geo-

membrane. Coated fabric liners are replacing concrete or compacted clay liners in reservoirs, canals, and irrigation channels. The inner coating must be resistant to weathering effects and chemicals.

7. FIBER REINFORCED CONCRETE AND CEMENT

One of the problems of a cement based matrix in civil engineering applications is the inherently brittle type of failure which occurs under tensile stress systems or impact loading. To improve the mechanical properties and decrease cracking of concrete, low volume of textile fibers is used in the matrix. Figure 6.30 shows the effect of fiber reinforcement on concrete failure strain. In the construction industry, a major reason for the growing interest in the performance of textile fiber in cement based materials is the desire to increase the toughness or tensile properties of the basic matrix.

The fibers used to reinforce concrete and cement can be divided into two main groups:

(1) Fibers with moduli lower than the cement matrix. These fibers include cellulose, nylon and polypropylene. If these fibers are used

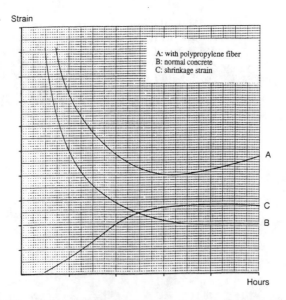

FIGURE 6.30 Effect of fiber reinforcement on concrete failure strain (courtesy of Polysilk).

to support permanent high stress cracked composite, considerable elongations or deflections can take place over time. Therefore, these fibers are generally used in situations where the matrix is expected to be uncracked, but where transitory overloads such as handling stresses, impacts or wind load is significant. Another concern is that they generally have large values of Poisson's ratio and this combined with lower moduli, means that if stretched along their axis, they contract sideways much more than the other fibers. This results in high lateral tensile stress at the fiber-matrix interface which is likely to cause a short aligned fiber to debond and pull out. Woven meshes or networks of fibrillated fibers are used to improve the situation.

(2) Fibers with moduli higher than the cement matrix. These fibers include glass, carbon and Kevlar®. High modulus short fibers may require mechanical bonding to avoid pull-out.

The elongations at break of textile fibers are two to three orders of magnitude greater than the strain at failure of the matrix. Therefore, the matrix would crack long before the fiber strength is reached. If the modulus of elasticity of the fiber is less than the cement matrix, due to low fiber volume fraction, the modulus of the composite is not greatly different from that of the matrix.

The maximum particle size of the matrix is important since it affects the fiber distribution and the quantity of fiber that can be included in the composite. The average particle size of cement paste before hydration is between 10 to 30 microns. Concrete which is intended to be used with fibers should preferably have particles between 10 mm and 20 mm for uniform fiber distribution.

Application areas of fiber reinforced concrete and cement include pavements, decorative cladding, timber substitute, thin shell roofs, internal waterproofing of cracked reinforced concrete tanks, permanent formwork (waffle floor pans, bridge deck formwork), water tanks, swimming pools, low cost and refugee housing, grain silo, pipes and noise barriers.

The commercial viability of fiber composites depends on the costs of the fibers because the matrix is very cheap. Therefore, it is a common practice to use the cheapest suitable fiber in the lowest volume required to fulfill the strength and durability requirements.

8. TEXTILES FOR ACOUSTIC AND HEAT INSULATION

Textile materials are used in the interior of buildings for mainly three acoustical purposes:

- to improve audibility
- to preserve the natural quality of sound
- to prevent transmission of undesired sound

Absorbent textile materials such as carpets, textile wall coverings and curtains are used to reduce the reverberation time. Reverberation time is defined as the time a sound takes to die away in a building. In an ordinary living room it is around 0.5 seconds and in large buildings it can be as long as 10 seconds. Complete elimination of reverberation is not desired. Too much absorption could attenuate sound and reduce audibility.

There are two types of noise in buildings: "airborne noise" and "impact noise." Airborne noise is transmitted from the air through openings in the building such as open windows and doors. Impact noise is transmitted directly through the structure of the building. Textiles such as curtains, felts and especially carpets offer excellent airborne sound absorption properties.

TABLE 6.7 Acoustical Properties of a Typical Fabric for Noise Reduction (courtesy of Chemfab).

Frequency (Hz)	124	250	500	1,000	2,000	4,000
Sound absorption (sabins)	43.0	49.7	38.7	50.4	52.1	54.2
Sound absorption coefficient (sabins/ft^2)	0.52	0.65	0.54	0.69	0.69	0.71
Noise reduction coefficient (NRC) = 0.64						

FIGURE 6.31 Tyvek® spunbond material for house insulation (courtesy of DuPont).

According to Carpet and Rug Institute of America, the main factors affecting sound absorption are as follows:

(1) Cut pile construction provides better absorption than loop pile construction (Section 4.7).
(2) In cut pile constructions, pile height and density increase absorption.
(3) In loop pile constructions, pile height is more important than pile density.
(4) Acoustic performance is independent of the pile fiber type.

Teflon® fiberglass composite materials are used for noise reduction. The structure is usually porous to facilitate the attenuation of sound within the fabric. Table 6.7 shows acoustical properties of a typical fabric used for noise control.

Housewrap materials are used in buildings for thermal insulation (Figure 6.31). They reduce the flow of air into and out of the house, cutting heating and cooling energy costs. Moisture vapor is allowed to pass through the wrap material which reduces damage from inwall condensation.

References and Review Questions

1. REFERENCES

1. Cumberbirch, R. J. E., "Textiles, Engineering and Architecture," *Textiles,* Vol. 16, No. 2, 1987.

2. Svedova, J., ed., *Industrial Textiles,* Elsevier, 1990.

3. Trevira® High Tenacity Polyester in Coated and Laminated Fabrics, Hoechst Celanese, October 1993.

4. Lomax, G. R., "Coated Fabrics, Part 2 – Industrial Uses," *Textiles,* Vol. 14, No. 2, 1985.

5. Blumberg, H., "Building with Coated Fabrics: The Present Position and World-Wide Trends" in *The Design of Textiles for Industrial Applications,* P. W. Harrison, ed., The Textile Institute, Manchester, 1977.

6. Perlstein, S., "Beam It Up," *Fabrics and Architecture,* May – June, 1994.

7. Armijes, S. J., "The Haj Terminal 10 Years Later," *Fabrics and Architecture,* November – December, 1991.

8. "Denver Airport Rises under Gossamer Roof," *The Wall Street Journal,* Nov. 17, 1992.

9. "Air, Tent and Tensile Structures: Design and Specification Guidelines," *Fabrics and Architecture,* Specguide 95, November – December, 1994.

10. Rebeck, G., "The Nomads at Rest," *Fabrics and Architecture,* IFAI, May – June 1994.

11. *Techtextil-Telegramm,* Messe Frankfurt GmbH, Jan. 24, 1994, No. 31, Issue E.

12. Swomley, G. A., "Fabric Selection, Brochure Ideas," *Industrial Fabric Products Review,* September 1988.

13. David, B., "Single-Ply Roofing: An Overview," *Industrial Fabric Products Review,* September 1990.

1.1 General References

Homan, M., "A Promising Outlook," *Industrial Fabric Products Review,* February 1993.

Dean, J., "A Changing Market," *Industrial Fabrics Product Review,* February 1988.

Rebeck, G., "Tension Structures," *Industrial Fabric Products Review,* June 1987.

Muhlberger, H., "Shading Seville – Expo '92," *Fabrics and Architecture,* July – August, 1992.

Gorman, H., "On a Cushion of Air," *Fabrics and Architecture,* Summer 1990.

O'Conner, T., "Put the Show on the Road," *Fabrics and Architecture,* November – December, 1993.

Lasko, S., "Large Tents for Events," *Fabrics and Architecture,* September – October, 1991.

Armijos, S. J., "Shelters of Fabric," *Fabrics and Architecture,* March – April, 1992.

Grossman, S., "The Topcoating Option," *Fabrics and Architecture,* January – February, 1991.

Smith, R., "The Electronic Awning," *Fabrics and Architecture,* Summer 1990.

Gorman, J., "Things That Glow in the Night," *Fabrics and Architecture,* July – August, 1992.

"U.S. Building Codes Summary: Air, Tent and Tensile Structures," *SpecGuide 1994, Fabrics and Architecture*, December 1993.

"U.S. Building Codes Summary: Awnings and Canopies," *SpecGuide 1994, Fabrics and Architecture*, December 1993.

Perlstein, S., "New Shades of Umbrellas," *Fabrics and Architecture*, July–August, 1994.

Perlstein, S., "Beam It Up," *Fabrics and Architecture*, May–June 1994.

Gorman, J., "Advanced Composites Advancing," *Fabrics and Architecture*, September–October 1992.

Gorman, J., "That's a Wrap," *Fabrics and Architecture*, January–February 1992.

Gorman, J., "There's a Fiber in My Concrete," *Fabrics and Architecture*, September–October 1991.

Gurian, M., "Acoustical Fabric Developments," *Fabrics and Architecture*, Winter 1990.

Daugherty, H., "Characterization of Architectural Fabrics," *Fabrics and Architecture*, January–February 1992.

Warshaw, R. I., "Designing Single-Ply Roofing Systems," *Fabrics and Architecture*, Summer 1990.

Kerrigan, S. and Warhshaw, R. I., *Fabrics and Architecture*, Summer 1989.

Mikko, C., "Fighting the Sun," *Fabrics and Architecture*, March–April, 1994.

"Awnings and Canopies: Design and Specification Guidelines," *Fabrics and Architecture*, December 1993.

Sudnik, Z. M., "Textiles in Buildings, Part 1 – Acoustics," *Textiles*, Vol. 2, No. 2, June 1973.

2. REVIEW QUESTIONS

1. Why is coating or laminating necessary for architecture and construction textiles?
2. What are the advantages of using textiles in buildings compared to concrete?
3. Discuss the major properties required for fabrics in buildings.
4. Define air and tension fabric structures. Explain the major differences between the two. Find out the names and locations of ten more (i.e., not mentioned in this chapter) large air and tensile fabric structures worldwide.
5. Explain the important fabric and structural considerations when designing buildings with industrial textiles.
6. If you were to choose a fabric construction type for a large shopping mall roof, would you choose an air or a tension structure? Explain your reasons.
7. What are the important test methods for fabrics in architecture and construction? List and explain each (hint: use Appendix 1).
8. What are the advantages of collapsible fabric storage vessels for the military?
9. What kind of fibers are used to reinforce concrete? Explain their effects.
10. Why are textile materials effective in sound and heat insulation?

TEXTILE STRUCTURAL COMPOSITES

Textile Structural Composites

S. ADANUR

1. INTRODUCTION

Whenever advancing technology creates a need for a combination of properties that no single material can provide, composites become the materials of choice. Composites can be defined as a combination of dissimilar materials to perform a task that neither of the constituent materials can perform alone. Since hardly any material is used in its pure form today, composites have a broad meaning and usage. Generally speaking, a composite is made of two components: reinforcement material and matrix (binder). The human body can be considered as a composite made of bones (reinforcement material) and flesh (matrix). There are several forms of synthetic composites such as metal-metal, metal-ceramic, metal-polymer, ceramic-polymer, ceramic-ceramic, and polymer-polymer combinations. A common example of synthetic composites is concrete which is reinforced with metal.

A textile composite is made of a textile reinforcement structure and a matrix material. Textile reinforcement structures can be made of fibers, yarns or fabrics (woven, braided, knit, nonwovens) and are generally flexible. These structures are called textile preforms. Textile preforms can be in various shapes and forms. Matrix materials can be thermoplastic or thermoset polymers, ceramic or metal. The consolidation of the textile structure with the matrix material produces textile structural composites.

Textile composites can be flexible or rigid. Examples of flexible textile composites are coated fabrics (Chapter 6.0), automobile tires (Chapter 15.0) and conveyor belts (Chapter 16.0). Rigid textile composites are the subject of this section.

Main characteristics of rigid textile composites are high stiffness, high strength and low density. Other characteristics include but are not limited to high temperature resistance, corrosion resistance, hardness and conductivity depending on the product design for a specific application. Textile structural composites have higher strength-to-weight ratio than metal composites. Well established textile manufacturing techniques allow near-net-shape manufacturing which reduces cost and material waste considerably. The application of traditional textile technology to organize high performance fibers for composite material applications has provided a route to combining highly tailored materials with enhanced processability. Braiding, weaving and knitting technologies have allowed fibers to be arranged locally in optimized configurations and globally into preforms for conversion processes such as resin transfer molding and pultrusion.

Another advantage of textile composites is that they can be made anisotropic. Many structural materials including metals have homogeneous and isotropic properties in general. In isotropic materials, strength, stiffness, thermal and other properties are equal in all directions and at all locations. With the use of oriented fibers in bundles or layers, textile composites can be made anisotropic so that they exhibit different properties along different axes. Strength, stiffness, thermal and moisture expansion coefficients can vary by more than ten times in different directions. As a result, by proper alignment of fibers with respect to loading direction, for example, the weight of the composite can be reduced further.

The strength and stiffness of the composite structure are functions of fiber and matrix properties. The functions of matrix materials are to bind the fibrous materials together and protect them from outside effects. Matrices transport the forces and stresses acting on the boundary of the composite to the fibers. They also help to strengthen the composite structure.

Application areas of composites are steadily expanding. They have replaced metals and metal alloys successfully in many applications including automotives, aerospace, electronics, military and recreation. Typical application examples are military and commercial aircraft parts, rocket and missile components, automotive parts, electronic boards, home appliances, construction and sporting goods.

2. TEXTILE COMPOSITE MATERIALS

2.1 Reinforcement Materials

For continuous fiber composites, fiber material has a great effect on composite properties. For primary and secondary load bearing applications, high modulus fibers and yarns must be used in textile reinforcement structures. High modulus materials have very low extensions under high stresses. The most commonly used fiber materials in textile reinforcement structures are fiberglass, carbon and aramid fibers which are high modulus materials. They are usually stiffer and more brittle than traditional textile materials which may require some modification in textile processing.

Glass Fibers

Glass fibers are widely used in textile composites due to relatively low cost and good per-

formance. The inherent properties of various types of glass fibers are shown in Table 7.1. Among the glass fibers, E-glass and S-glass are the most widely used materials in textile composites. E-glass is used in electronic boards because of its good electrical properties, dimensional stability, moisture resistance and lower cost. S-glass has higher tensile strength, high elastic modulus and better thermal stability but it is also more expensive. Therefore it is used in advanced composites where cost-performance benefits can be justified.

Carbon Fibers

Carbon fibers are produced by heat treatment of organic precursors such as rayon, polyacrylonitrile (PAN) and pitch. Carbon fibers have the highest modulus and strength of all reinforcing fibers at both room and high temperatures. Table 7.2 shows typical mechanical properties of commercially available carbon fibers. Their density is low; however, performance/cost ratio limits their use to high performance applications such as aerospace where stiffness/weight ratio is the primary concern. Carbon fibers are electrically conductive which requires special care when using them around electric motors and electronic controls. Since carbon fibers are difficult to wet with resins, surface treatments are used to increase the number of active chemical groups and roughen the fiber surface. In order to reduce fiber abrasion, improve handling and provide fiber-matrix compatibility, it is common practice to impregnate carbon fibers with epoxy size prior to shipment. These materials are called prepregs. Figure 7.1 shows various forms of carbon fibers.

TABLE 7.1 Properties of Glass Fibers [1,2].

	Specific Gravity	Tensile Strength (MPa)	Tensile Modulus (GPa)	Coefficient of Thermal Expansion, 10^{-6}/K	Dielectric Constant*	Liquid Temperature (°C)
E-glass	2.58	3,450	72.5	5.0	6.3	1,065
A-glass	2.50	3,040	69.0	8.6	6.9	996
ECR-glass	2.62	3,625	72.5	5.0	6.5	1,204
S-glass	2.48	4,590	86.0	5.6	5.1	1,454

*At 20°C and 1 MHz.

TABLE 7.2 Typical Properties of Commercially Available Carbon Fibers [3].

Product	Manufacturer	Precursor	Density (g/cm³)	Tensile Strength (GPa)	Tensile Modulus (GPa)
AS-4	Hercules, Inc.	PAN	1.78	4.0	231
AS-6	Hercules, Inc.	PAN	1.82	4.5	245
IM-6	Hercules, Inc.	PAN	1.74	4.8	296
T300	Union Carbide/Toray	PAN	1.75	3.31	228
T500	Union Carbide/Toray	PAN	1.78	3.65	234
T700	Toray	PAN	1.80	4.48	248
T-40	Toray	PAN	1.74	4.5	296
Celion	Celanese/ToHo	PAN	1.77	3.55	234
Celion ST	Celanese/ToHo	PAN	1.78	4.34	234
XAS	Grafil/Hysol	PAN	1.84	3.45	234
HMS-4	Hercules, Inc.	PAN	1.78	3.10	338
PAN 50	Toray	PAN	1.81	2.41	393
HMS	Grafil/Hysol	PAN	1.91	1.52	341
G-50	Celanese/ToHo	PAN	1.78	2.48	359
GY-70	Celanese	PAN	1.96	1.52	483
P-55	Union Carbide	Pitch	2.0	1.73	379
P-75	Union Carbide	Pitch	2.0	2.07	517
P-100	Union Carbide	Pitch	2.15	2.24	724
HMG-50	Hitco/OCF	Rayon	1.9	2.07	345
Thornel 75	Union Carbide	Rayon	1.9	2.52	517

Aramid Fibers

Perhaps the most widely known aramid fiber is Kevlar® produced by DuPont. There are several versions of Kevlar® used in textile structural composites as shown in Table 7.3. Kevlar® 29 is used for high toughness, good damage tolerance and ballistic protection. Kevlar® 49 has high modulus and is the most widely used aramid in composites. Kevlar® 149 has ultra-high modulus. There are several other aramid fibers in the market including Twaron® from Enka Corporation, Nomex® from DuPont and HM-50 from Teijin of Japan. Aramid fibers are especially suitable in applications that require high tensile strength-to-weight ratio such as missiles, pressure vessels and tension systems. Aramid fibers are available in various forms including continuous filament yarns, rovings, staple and textured yarns, woven and nonwoven fabrics.

Other Fibers

Boron fibers are used in epoxy, aluminum, or titanium matrices. The resulting composites have high strength-to-weight ratio, and good compressive strength. Spectra® from Allied offers very high tensile strength and stiffness and the lowest specific density; however, the melting temperature is relatively low.

2.2 Matrix Materials

Matrix material in a composite system serves several purposes. Matrices bind fibrous materials together and hold them in particular positions and orientations giving the composite structural integrity. They serve to protect fibers

FIGURE 7.1 Yarn, prepreg, fabric and chopped forms of carbon fibers (courtesy of Nippon Oil Company).

TABLE 7.3 Properties of Kevlar® Fibers [4].

Material	Density	Filament Diameter (μm)	Tensile Modulus (GPa)	Tensile Strength (GPa)	Tensile Elongation (%)	No. of Filaments
Kevlar® 29 (high toughness)	1.44	12	83	3.6	4.0	134−10,000
Kevlar® 49 (high modulus)	1.44	12	131	36−4.1	2.8	134−5,000
Kevlar® 149 (ultra-high modulus)	1.47	12	186	3.4	2.0	134−1,000

from environmental effects and handling. The matrix system transfers forces acting on the boundaries of composites to the fibers. Matrices also help to strengthen the composite structure.

The reinforcement material and resin must be compatible for good penetration and bonding. Both thermoset and thermoplastic resins are used as matrices in textile structural composites. Table 7.4 shows comparison of thermoset and thermoplastic resins for commercial aircraft composites.

Thermoset Resins

The most widely used thermoset resins are polyester and epoxy. Polyester resins are used in applications where high temperature is not a requirement. The advantages of polyester resins are low cost, low viscosity for good flow and fiber surface wetting, and low curing temperatures. The disadvantages are poor strength, poor impact performance and high shrinkage during the curing process.

Epoxy resins are ideal for higher temperature applications. They offer versatility, broad range of physical properties, mechanical capabilities and processing conditions. Depending on manufacturing conditions, epoxy resins can provide toughness, chemical and solvent resistance, flexibility, high strength and hardness, creep and fatigue resistance, good fiber adhesion, heat resistance and excellent electrical properties. Examples of commercially available resins are Tactic from Dow Chemical, Araldite from Ciba-Geigy and Epon HPT from Shell Chemical Company. A curing agent needs to be used with epoxy resins.

Thermoplastic Resins

Thermoplastic materials can be classified as commodity thermoplastics and engineering thermoplastics. Commodity thermoplastics such as polyethylene, polypropylene, polyvinyl chloride and polystyrene exhibit very little resistance to

TABLE 7.4 Comparison of Thermoset and Thermoplastic Resins for Commercial Aircraft Composites [5].

Property	Thermosets	Thermoplastics
Resin cost	Low	Low to high
Prepreg tack/drape	Excellent	None (revised lay-up techniques are required)
Volatile-free prepreg	Good	Good to excellent
Prepreg shelf life and out-time	Poor	Excellent
Prepreg quality assurance	Fair	Excellent
Prepreg cost	Good	High
Composite processing	Slow	Slow
Shrinkage	Moderate	Low
Composite mechanical properties	Good	Good
Interlaminar fracture toughness	Low	High
Resistance to fluids/solvents	Good	Poor to good
Crystallinity problems	None	Yes

TABLE 7.5 *Properties of Commonly Used Thermoplastics [5].*

Material	Supplier	T_g (°C)	T_m (°C)	Tensile Strength (MPa)	Tensile Modulus (MPa)	Fracture Toughness (J/m²)
Udel P-1700	Union Carbide	190	None	76	2,200	3,200
Radel	Union Carbide	220	None	–	–	5,500
Vitrx PES 200P	ICI	220	None	83	2,410	2,600
Utem	General Electric	220	None	110	3,300	3,700
Torlon	Amoco	275	530	193	4,800	3,400
Ryton	Phillips	85	285	65	3,800	210 (high crystallinity)
PEEK	ICI	143	343	–	–	–
Avimid K-II	DuPont	277	None	110	2,850	14,000

high temperatures. High temperature thermoplastic materials are relatively new and their temperature resistance can be superior to that of epoxy. They offer toughness and improved hot/wet resistance. Examples of engineering thermoplastic resins which have been considered for use in composites are PEEK (polyetheretherketone), PPS (polyphenylene sulfide) and PEI (polyetherimide). Table 7.5 shows properties of some commercially available thermoplastics that can potentially be used as matrix systems.

Polyimide resins are classified as thermoplastic polyimides that are derived from a condensation reaction and crosslinked polyimides that are derived from an addition reaction.

Commingling

Fibers and thermoplastic matrix materials such as polyetheretherketone (PEEK) in fiber form can be commingled together before fabrication of textile preform. The thermoplastic fibers melt under heat and pressure during consolidation and become the matrix. Commingling is suitable for dense three-dimensional fabrics where resin penetration may be difficult.

3. CLASSIFICATION OF TEXTILE REINFORCEMENT STRUCTURES

Textile reinforcements can be in various forms and shapes. Classification of textile reinforcement structures can be done in several ways depending on the preform structure's parameters. Ko lists several variables for classification of textile structures [6]. They are dimension (1, 2, or 3), direction of reinforcement (0, 1, 2, 3, 4, . . .), fiber continuity (continuous, discontinuous), linearity of reinforcement (linear, nonlinear), bundle size in each direction (1, 2, 3, 4, . . .), twist of fiber bundle (no twist, certain amount of twist), integration of structure (laminated or integrated), method of manufacturing (woven, orthogonal woven, knit, braid, nonwoven), and packing density (open or solid). Scardino classified various reinforcement systems as shown in Table 7.6. Due to fiber discontinuity, random fiber orientation and lack of fiber integration, chopped fiber constructions (Type 1) are not suitable for structural composites. Continuous filament yarn constructions (Type 2) are especially suitable for tensile loading applications. However, they are not well suited for

TABLE 7.6 *Textile Composite Systems [7].*

Type	Reinforcement System	Textile Construction	Fiber Length	Fiber Orientation	Fiber Entanglement
1	Suspended	Chopped fiber	Discontinuous	Uncontrolled	None
2	Linear	Filament yarn	Continuous	Linear	None
3	Laminar	Simple fabric	Continuous	Planar	Planar
4	Integrated	Advanced fabric	Continuous	3-D	3-D

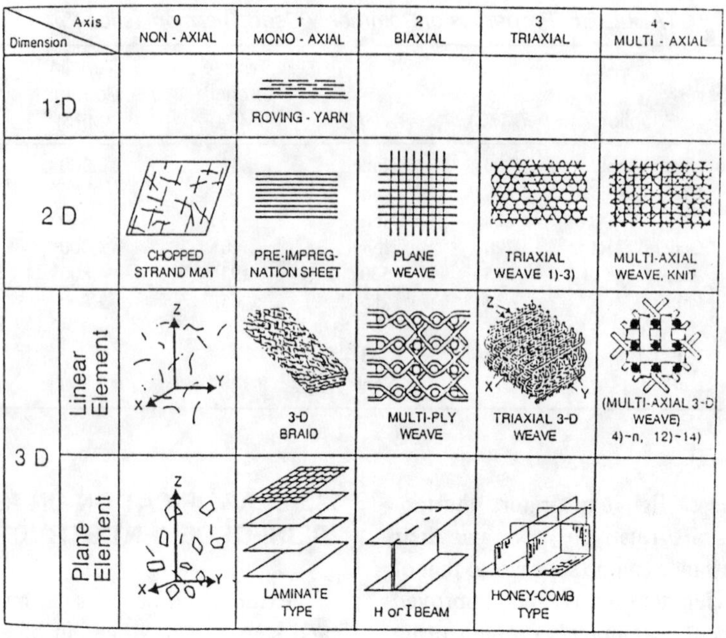

Axis / Dimension	0 NON - AXIAL	1 MONO - AXIAL	2 BIAXIAL	3 TRIAXIAL	4 ~ MULTI - AXIAL
1 ˙D		ROVING - YARN			
2 D	CHOPPED STRAND MAT	PRE-IMPREG-NATION SHEET	PLANE WEAVE	TRIAXIAL WEAVE 1)-3)	MULTI-AXIAL WEAVE, KNIT
3 D Linear Element		3-D BRAID	MULTI-PLY WEAVE	TRIAXIAL 3-D WEAVE	(MULTI-AXIAL 3-D WEAVE) 4)~n, 12)~14)
3 D Plane Element		LAMINATE TYPE	H or I BEAM	HONEY-COMB TYPE	

FIGURE 7.2 Classification of textile reinforcement structures based on axis and dimension [8].

general load-bearing applications because they are prone to splitting and delamination among filament layers. The laminar systems (Type 3) are made of woven, knit, braided or nonwoven fabric layers in which fibers are continuous, entangled and oriented in at least two directions. These structures are most suitable for load-bearing panels in flat form. However, they are prone to delamination between the fabric layers if not stitched properly. In integrated systems (Type 4) fibers are oriented in three-dimensional space. These structures give the best results for general load-bearing applications. Fibers are continuous and entangled throughout the structure. Splitting or delamination is not a concern. Their construction can be woven, knit, braided or laid-up with special manufacturing equipment.

Fukuta et al. classified textile structures based on fiber/yarn axis and dimension of the structure as shown in Figure 7.2.

The fabric requirements for composite applications are dimensional stability, conformability and moldability. Figure 7.3 shows behavior of various fabric structures under uniaxial stress in various directions.

Three dimensional (3-D) fabric structures were developed within the last two decades to withstand multi-directional mechanical stresses and thermal stresses. 3-D structures also improved interlaminar strength and damage tolerance significantly.

4. MANUFACTURE OF TEXTILE PREFORMS

The most common constituent material forms used in textile composites are fibers, fabrics, nonwovens and laminates.

Application of sizing agents is critical during textile reinforcement manufacturing. Sizing or finishing agents are surface coatings applied to a reinforcement to protect it from damage, to aid in processing or to enhance the composite mechanical properties. For example, sizing a carbon fiber tow, which may have several thousand fibers, protects individual filaments from contact damage or break during weaving. Sizing agents include film forming organics and polymers such as polyvinyl alcohol (PVA), polyvinyl acetate (PVAc), adhesion promoters such as silanes, interlayer agents such as elastomeric coating and chemical modifiers such as silicone carbide [10]. A sizing agent should be

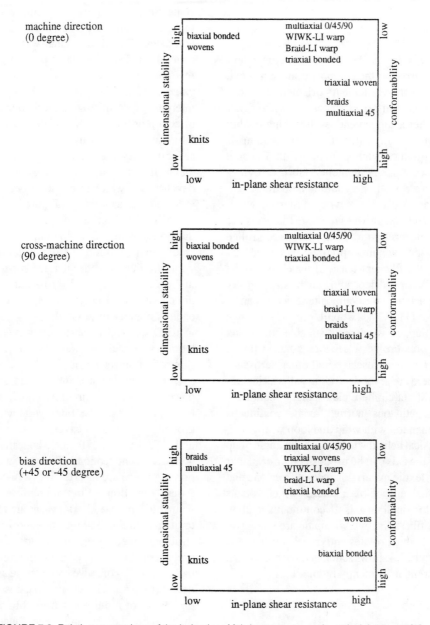

FIGURE 7.3 Relative comparison of the behavior of fabric structures under uniaxial stresses [9].

chemically compatible with the reinforcement material and should not deteriorate the mechanical properties of the interphase between reinforcement and matrix.

4.1 Fiber Forms

Short or chopped fiber reinforced composites are not as strong as the composite structures that allow controlled filament orientation. Continuous fiber reinforced structures with highly aligned fiber arrangement result in high quality and stronger composites. Filament winding is used to produce such structures. In filament winding, continuous filaments, fabrics, rovings or tapes are wound on a rotating mandrel.

There are two different types of filament winding: wet and dry. In wet winding, filaments are fed to the mandrel from a resin bath at a controlled angle. In dry winding, prepregs are used. By changing the filament tension, which is around 0.25 to 1 lb per end depending on the size, density and compaction of the composite structure can be controlled [11]. The wound preform is then cured. Filament winding is suitable for rocket engines, pressure vessels, storage tanks, pipes, helicopter blades, fuselage and other aerospace parts. There are several winding patterns including helical, circumferential, polar, continuous helical, continuous normal axial, continuous rotating mandrel with wrap and loop wrap.

In classical helical winding, a helix angle of up to 85° to the axis can be used. In polar or planar winding, textile material is wound from one end to the other end of the mandrel which rotates about its longitudinal axis. In continuous rotating mandrel, fibers and overwrapping are wound on a mandrel which rotates and travels through the oven. Figure 7.4 shows schematics of helical and circumferential filament winding.

4.2 Fabrics

Woven, braided, knitted, laminated, non-crimp and nonwoven fabric structures are used to reinforce composites.

Woven Fabrics

Woven fabrics that are used in composites can be grouped as two-dimensional (2-D) and three-dimensional (3-D) structures. In 2-D structures, yarns are laid in a plane and the thickness of the fabric is small compared to its in-plane dimensions. Single layer designs include plain, basket, twill and satin weaves which are used in laminates. Two-dimensional woven fabrics are generally anisotropic, have poor in-plane shear resistance and have less modulus than the fiber materials due to existence of crimp and crimp interchange. Reducing yarn crimp in the loading direction or using high modulus yarns improves fabric modulus. To increase isotropy, in-plane shear rigidity and other properties in bias or diagonal direction, triaxially woven fabrics are developed in which three yarn systems interlace at 60° angles as shown in Figure 7.5.

Unlike 2-D fabrics, in 3-D fabric structures, the thickness or Z-direction dimension is considerable relative to X and Y dimensions. Fibers or yarns are intertwined, interlaced or intermeshed in the X (longitudinal), Y (cross), and Z (vertical) directions. For 3-D structures, there may be an endless number of possibilities for yarn spacing in a 3-D space. 3-D fabrics are woven on special looms with multiple warp and/or weft layers. Figure 7.6 shows various 3-D woven structures.

In polar weave structure, fibers or yarns are placed equally in circumferential, radial and axial directions. The fiber volume fraction is around 50%. Polar weaves are suitable to make cylindrical walls, cylinders, cones and convergent-divergent sections. To form such a shape, prepreg yarns are inserted into a mandrel in the radial direction. Circumferential yarns are wound in a helix and axial yarns are laid parallel to the mandrel axis. Since the preform lacks the structural integrity, the rest of the yarns are impregnated with resin and the structure is cured on the mandrel. Polar weaves can be woven into near-net shapes. A near-net shape is a structure that does not require much machining to reach the final product size and shape. Since fibers are not broken due to machining, net shapes generally perform better than machined parts.

In orthogonal weave, reinforcement yarns are arranged perpendicular to each other in X, Y and Z directions. No interlacing or crimp exists between yarns. Fiber volume fraction is between 45 and 55%. By arranging the amount of yarn in

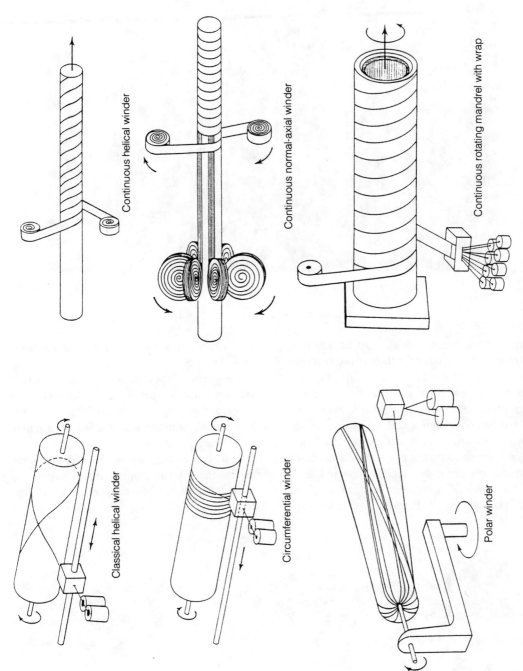

Continuous helical winder

Continuous normal-axial winder

Continuous rotating mandrel with wrap

Classical helical winder

Circumferential winder

Polar winder

FIGURE 7.4 Filament winding patterns (Richardson [11]).

FIGURE 7.5 Triaxial weaving.

each direction, isotropic or anisotropic preforms can be obtained. Except for the components that are fundamentally Cartesian in nature, orthogonal weaves are usually less suitable for net shape manufacturing than the polar weaves. Unit cell size can be smaller than polar weaves which results in superior mechanical properties. Since no yarn interlacing takes place in polar and orthogonal structures, they are also referred to as "nonwoven 3-D" structures in the composites industry. However, it is more proper to label these structures as woven structures with zero level of crimp.

In angle interlock type of structures, warp (or weft) yarns are used to bind several layers of weft (or warp) yarns together (Figure 7.7). In place of warp or weft yarns, an additional third yarn may also be used as binder. Stuffer yarns, which are straight, can be used to increase fiber volume fraction and in-plane strength. If the binder yarns interlace vertically between fabric layers, the structure is called orthogonal weave.

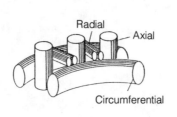

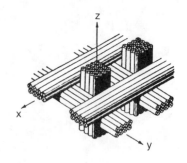

5-direction construction   polar weave   orthogonal weave

FIGURE 7.6 Schematics of various 3-D woven fabric structures for composites [12].

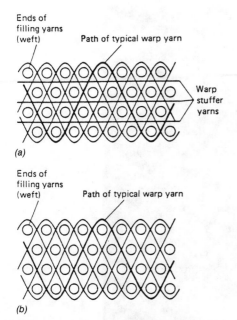

FIGURE 7.7 Angle interlock fabric; (a) with and (b) without added stuffer yarns [12].

Angle interlock or multi-layer fabrics for flat panel reinforcement can be woven on traditional looms, mostly on shuttle looms. The warp yarns are usually taken directly from a creel. This allows mixing of different yarns in the warp direction. Other more complex 3-D fabrics such as polar and orthogonal weaves require specialized weaving machines. Several weaving machines were developed to weave complex 3-D structures (Figure 7.8).

The most widely used materials in 2-D or 3-D weaving are carbon/graphite, glass, and aramid. Any material that can be shaped as a fiber can be woven into preforms. Woven preforms can be made of a single type of fiber material or different fiber and yarn materials can be used as a hybrid structure. Due to the nature of woven structure geometry and weaving process, when selecting a fiber for weaving or for any other textile manufacturing process, fiber brittleness and bending rigidity need to be considered. For example, carbon and graphite fibers, which account for 90% of all 3-D-woven preforms, are prone to break and fracture during weaving. Figure 7.9 shows preform and composite samples made of carbon fibers.

Adanur and Tam developed multi-layer fabric

structures using an on-loom stitching technique (Figure 7.10). In this method, a regular loom is used to stitch fabric layers together during weaving. Warp and/or fillling yarns can be used as binder.

Prepregs

A prepreg is a textile structure that is impregnated with uncured matrix resin. There are various forms of prepregs such as unidirectional and multi-directional tape prepregs and woven fabric prepregs. Common fibers that are used for prepregs are graphite, fiberglass and aramid. Figure 7.11 shows a schematic of a typical prepreg machine for unidirectional tape prepreg. Fibers are wound and collimated as a tape. The matrix resin is heated to reduce viscosity and dispersed on the fibers. The prepreg is calendered for uniform thickness.

Prepregs are suitable for hand and machine lay-up. Figure 7.12 shows uni- and multi-directional lay-ups. Increasing the number of oriented plies increases the isotropic strength. Four ply directions, i.e., $0°/90°/+45°/-45°$ orientations, are considered to be sufficient for isotropic properties.

Woven fabric prepregs are widely used in composite manufacturing. Hot melt or solvent coating processes are used to prepreg the fabrics. The hot melt process is similar to prepregging unidirectional tapes. In solvent coating, fabric is

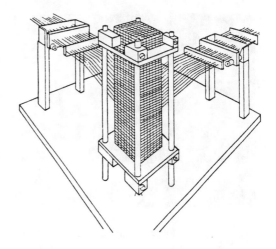

FIGURE 7.8 Schematic of King's 3-D machine [13].

FIGURE 7.9 Woven 3-D preform and composite samples made of carbon fibers (courtesy of Prof. Mansour H. Mohamed).

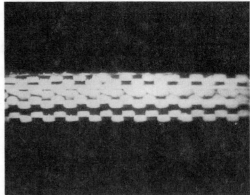

FIGURE 7.10 On-loom stitched, four layer fabric structure (Adanur and Tam).

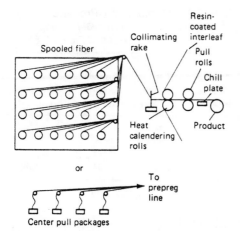

FIGURE 7.11 Schematic of a prepreg machine [14].

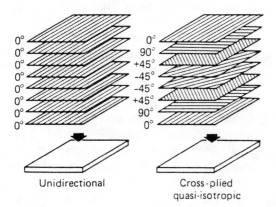

FIGURE 7.12 Uni- and multi-directional lay-ups [15].

TABLE 7.7 Graphite Woven Fabric Structures
for Prepregs [16].

Weave	Construction (tows/cm)	Weight (g/m²)	Thickness* (mm)
Plain	4.5 × 4.5	193	0.18
8-End satin	8.5 × 8.5	370	0.34
5-End satin	4.3 × 4.3	370	0.34
Crowfoot satin	4.1 × 4.1	185	0.17

*Cured ply thickness at 62 vol% fiber for high-strength, low modulus fiber-epoxy prepreg.

immersed in a bath containing 20 to 50% of a solvent and resin mixture. The fabric is then dried. Hot melt prepreg fabrics have less drape-ability and lower tack, better hot/wet mechanical properties and they are more expensive. Solvent coated woven prepregs offer better drape, have lower mechanical properties and are less expensive.

Most common fabric designs for prepregs are plain, basket, twill, crowfoot (1/3 harness satin) and 5 and 8 harness satin designs. Increasing the number of harnesses increases the composite strength. Woven prepregs offer flexibility during fabrication but they are more expensive than tape prepregs. Table 7.7 shows widely used woven fabric types for woven prepregs.

Fabric construction plays a critical role on composite properties as shown in Figure 7.13. The fabric density (number of warp and weft

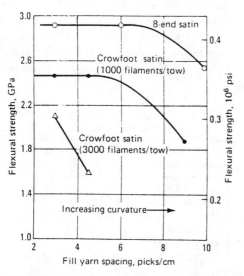

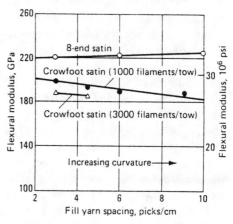

FIGURE 7.13 Effect of fabric construction on composite flexural strength (left) and flexural modulus (data normalized to 100 vol% warp fiber) [16].

yarns per inch as well as yarn densities) should be selected carefully to meet the required properties. Too dense fabric structures do not conform to complex shapes and may not allow resin penetration during consolidation. On the other hand, loose fabrics may not have the strength required in the reinforcing structure due to small amount of fiber. Tracer yarns (yarns with smaller denier and different color) are used in woven structures to control fiber or ply alignment.

A towpreg is either a single tow or a fiber strand that is impregnated with matrix resin. Towpregs are especially suitable for filament winding. They are used in the hard-to-form areas and joints of structural components. Most towpregs are made by a solvent coating process.

Hybrid Fabrics

Hybrid fabrics are made of more than one type of fiber or yarn. By combining different fiber properties in hybrid fabrics, the properties of reinforcing structure can be optimized while reducing the cost. Hybrid composites offer new possibilities for the designer as shown in Table 7.8.

Graphite, boron, glass and aramid fibers can be mixed in hybrid woven structures. Table 7.9 shows impact resistance of hybrid composites.

Braided Structures

Although braiding is not a major manufacturing process for traditional textiles relative to weaving and knitting, it is one of the major manufacturing processes for textile composite preforms. Braiding is done by intertwining yarns using an over and under sequence (Section 4.6). Due to several advantages including versatility, good mechanical properties and low manufacturing costs, the interest in braiding has grown for composite manufacturing in recent years. In comparison to woven structure, braided fabric structure usually has low shear resistance and therefore highly deferrable in the axial and radial directions. The reason for this is the lack of beat-up during braid formation. This characteristic of braided structure makes it particularly suitable to conform to surfaces of varying cross-

TABLE 7.8 Change in the Properties of Hybrid Fabric Composites Depending on the Mixture Ratio of Fibers [16].

Ratio of Aramid to Graphite Fiber	Tensile Modulus (GPa)	Tensile Strength (MPa)	Compressive Strength (MPa)
100/0	35.8	544	152
50/50	48.2	400	227
25/75	57.2	434	317
0/100	59.9	434	558

sectional shape such as cones and nozzles as shown in Figure 4.59. Braiding allows production of near-net shape structures. The yarns can also be deflected without discontinuity to make holes or cutouts on the preform.

A braided structure can be two- or three-dimensional. Two-dimensional bias braid structure is obtained with two sets of yarn carriers rotating in opposite directions which is the traditional braiding process (Figure 4.55). By inserting additional yarns in the axial direction, triaxial braided structure is obtained (Figure 4.57). Triaxial braids are particularly suitable for composites to be used as tension or compression members [17].

3-D braided structures are obtained by intertwining or orthogonal interlacing of multi-yarn systems (Figure 7.14). In the four-step braiding method, yarns are intertwined to form a multilayer 3-D structure (Figure 7.15). Some of the braiding yarns traverse the internal layers and bind the two exterior layers together. Some sort of a beat-up is needed to push the yarns into the

TABLE 7.9 Impact Resistance of Hybrid Epoxy Composites [16].

Hybrid Composite (weight %)	Izod Impact Strength Unnotched (J/m)
Graphite 100%	1,495
Graphite 75%, aramid 25%	1,815
Graphite 50%, aramid 50%	2,349
Aramid 100%	2,562
Graphite 100%	1,495
Graphite 75%, glass 25%	2,349
Graphite 50%, glass 50%	2,989
Glass 100%	3,843

FIGURE 7.14 Three-dimensional braiding (courtesy of Prof. A. El-Shiekh).

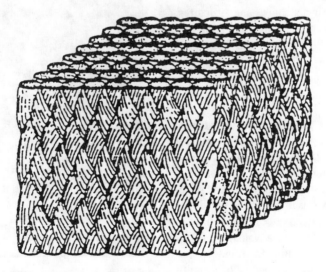

FIGURE 7.15 Three-dimensional braid produced by intertwining of multi-yarns.

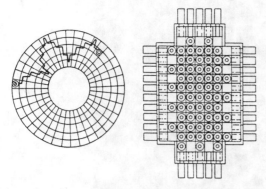

FIGURE 7.16 Schematic of circular and rectangular 3-D braiding [18].

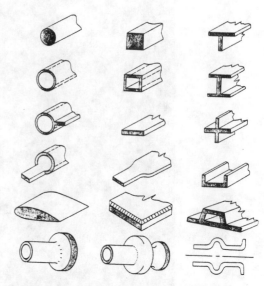

FIGURE 7.17 Various net-shape structures that can be produced by 3-D braiding [18].

fabric structure after each round of braiding. In a two-step system to make 3-D braids, axial yarns and braider yarns are used. The braider yarns move around axial yarns which are fixed parallel to each other. The resultant tubular structure has good reinforcement in axial direction but is weak in circumferential direction. The multi-layer interlock braiding method, which is similar to two-step braiding, allows greater circumferential reinforcement. In this method, yarns from adjacent layers are interlocked together.

Figure 7.16 shows two setups for circular and rectangular braiding in which the carriers move intermittently in X or Y direction. There are several research programs to completely automate the 3-D braiding systems in the United States. The braiding method is very versatile to produce various near-net shapes and structures. It is even possible to produce nuts and bolts with a braiding system (Figures 7.17 and 7.18).

Knit Fabrics

Both weft knit and warp knit structures can be used for composite reinforcement, especially for flexible composites. Weft knit structure can easily conform to different shapes. Weft or warp knitting is suitable to manufacture 3-D knit structures. By inserting yarns in the 0° and 90° direction, additional reinforcement can be obtained in the knit structure. One disadvantage of knit structures is low fiber volume fraction.

Figure 7.19 shows the structure of a multiaxial warp knit (MWK) fabric. The fabric is

FIGURE 7.18 Three-dimensional braiding of a cone-shape structure (courtesy of Hercules).

FIGURE 7.19 Multi-axial warp knit fabric [19].

made of warp (0°), weft (90°) and bias (α°) yarns that are stitched together by a chain or tricot stitch through the thickness of the fabric. The orientation of diagonal reinforcing yarns can be varied by a rotating turntable. The reinforcing yarns are held together by the loops. Latch needles with rounded hooks are used to reduce the damage to reinforcing yarns. Although the resultant multi-layer fabric is not a true 3-D structure, it offers strength in various horizontal directions. The reinforcing yarns can be stiff, such as carbon, but flexible yarns are used for knitting loops such as nylon, polyester or Kevlar®. Major

fabric structural parameters are: linearity of the bias yarns, the number of axes and the stitching technique.

In weft knitting of 3-D shapes, knitting is stopped on some needles while continuing on the other needles. Presser-foot technology applies an even downward force to loops on all needles which prevents tension buildup on the needles that are not knitting.

Malimo-type structures are weft inserted warp knit structures in which non-linear bias yarns are used. In weft inserted warp knitting, reinforcing yarns are added in the cross direction to increase structural integrity. In the Malimo stitch-bonding process, warp filaments (in machine direction) and a sheet of weft filaments (in cross-machine direction) are fed into the looping area of the machine (Figure 7.20). These two sets of filaments are then stitch bonded together with a knitting process.

Due to their low elongation to break, carbon fibers are not suitable for some knitting processes. Therefore, precursor yarns or fine filaments are knitted first and carbonized by heat treatment later.

Non-Crimp Fabric Structures for Composites

In a different variation of multi-axial warp knitting, called LIBA system, several layers of uncrimped yarns are stacked and stitched together along several axes by knitting needles

FIGURE 7.20 Malimo stitch bonding machine [20].

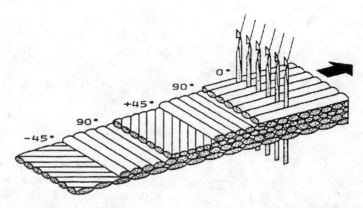

FIGURE 7.21 Warp knitted, multi-axial fabric structure, LIBA system [21].

piercing through the yarn layers. The layers can be oriented freely within the broad limits (0°, 45°, 90°) as shown in Figure 7.21. Nonwoven or non-oriented fibrous layers can be placed on top and bottom of the structure. Stitching yarns secure the orientation of the reinforcing fiber layers and prevent them from slipping. LIBA structural composites are generally used in the form of plastic laminates in boat construction, domes, etc. [21]. A special type of Malimo machine is used to manufacture these structures (Figure 7.22).

Non-crimp fabrics are a relatively new class of textiles. These fabrics offer a reinforcement form that has the potential to overcome the deficiencies of woven fabrics without introducing any apparent additional property reductions. It is reported that COTECH® non-crimp fabrics offer several advantages over crimped yarn structures [22]. In non-crimp structures, fibers are not subject to the kink stresses inherent in woven fabrics, thus their optimum properties are maintained. They have been used successfully in a wide variety of applications in the automotive, aerospace, construction, corrosive and marine industries.

Combination of fibers can be used within layers or from layer to layer, as engineered for optimum performance as shown in Figure 7.23. Reinforcing fibers are continuously placed in layers at different orientations to the next according to the desired properties. The layers are stitched together in a manner that permits high drapeability and conformance to complex shapes.

It is claimed that non-crimp fabrics are stronger and more predictable than other types of reinforcement structures. Other characteristics include less resin requirements, greater

FIGURE 7.22 Malimo stitch-bonding machine with multi-axial filling insertion [20].

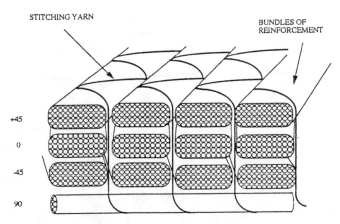

FIGURE 7.23 Schematic of multi-axial non-crimp fabric structure (courtesy of Tech Textiles USA).

impact resistance, less fatigue and reduced lay-up time. Fiber wet-out is also easier. Table 7.10 shows results of non-crimp COTECH® quadriaxial fabrics tested at various directions.

Nonwovens

Nonwoven unidirectional fabrics are used as reinforcing structures in composites. However, due to firmness of the binding material, they do not readily conform to complex shapes or contours. Several two-dimensional nonwoven fabrics produced with traditional methods can be stitched or knitted into multi-layer structures with considerable thickness or bulk. Due to lack of yarn interlacing, nonwoven structures are not as strong as fabric structures in which yarns are interlaced. Nonwoven preforms are easier to manufacture. Needlepunching and stitch bonding are the most widely used methods to produce nonwoven reinforcement structures. In stitch bonding, yarns are knitted through a nonwoven mat or array of yarns.

Laminates

Laminate preforms are made of two or more fiber and/or fabric layers bonded together. The performance properties of the laminate composites depend on the selection, orientation and composition of individual layers. Laminate construction can be quasi-isotropic or anisotropic (i.e., property depends on direction) as shown in Figure 7.12. Each individual layer may be of the same or different material performing a separate and distinct function. Fabric layers may or may not be stitched together. Laminar composites made of unstitched layers are prone to delamination. Stitching the layers together can be done on-loom or off-loom. Angle interlock structure is an example of on-loom stitching. Off-loom stitching can be done by sewing the layers together with a sewing machine. Laminate composites have good strength, low material and labor costs and are easier to manufacture. Although stitching improves delamination resistance, 3-D preforms of stitched layers are not as

TABLE 7.10 *Tensile and Flexural Data for Laminates Produced from Quadriaxial Fabrics at Different Orientations [23].*

Direction (degrees)	Ultimate Tensile Strength (MPa)	Tensile Modulus (GPa)	Flexural Strength (MPa)	Flexural Modulus (GPa)
0	295	19.6	374	15.2
−45	269	21.1	369	13
+45	258	21	344	11.8
90	239	19.4	302	12.4

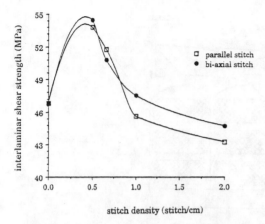

strong as 3-D woven preforms because the amount of yarn in the thickness direction (stitch yarn) is relatively small. In addition, it was shown by Adanur et al., that too dense stitching may damage the fibers resulting in lower mechanical properties than non-stitched samples as shown in Figure 7.24.

Finally in manufacturing of textile structural composites, it should be noted that there may be several options to produce a textile composite structure. As an example, several alternate ways to manufacture an I-beam are shown in Figure 7.25. This is, indeed, the beauty of the textile technology which offers unlimited possibilites to the engineer for design and manufacture of composites (Figures 7.26 – 7.28).

FIGURE 7.24 Effect of stitch density on interlaminar shear strength for seven layer off-loom stitched glass-epoxy composites [24].

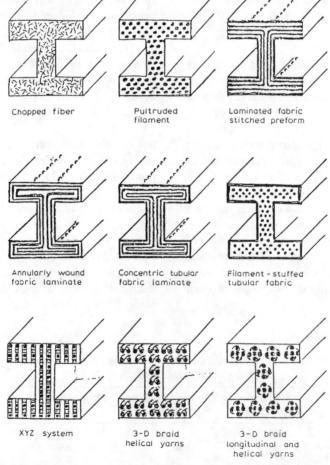

FIGURE 7.25 Various ways to manufacture a textile composite I-beam [9].

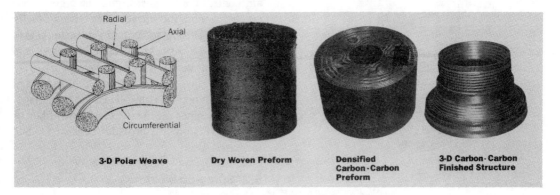

FIGURE 7.26 Construction stages of a composite part (courtesy of Hercules).

FIGURE 7.27 Various parts made of graphite-epoxy composites (courtesy of Hercules).

FIGURE 7.28 Graphite turbine wheel (courtesy of Nippon Oil Company).

5. COMPOSITE MANUFACTURING

Two major steps in manufacturing of composites are wetting of textile reinforcement structure with resin (matrix material) and curing which is three-dimensional formation of a polymer network. During curing, hardening of the resin takes place and bonding is formed between the resin and fibers in the reinforcement structure (consolidation). Curing can be done unaided or with application of heat and/or pressure for faster polymerization.

There are many methods in composite manufacturing that are well documented in the literature. Since this book is concerned mainly with textile manufacturing processes, a detailed discussion of composite manufacturing is out of the scope of this book. Table 7.11 shows the principal composite processing techniques. A brief explanation of the most widely used methods will be given below. For further information, the reader may refer to the references.

5.1 Hand and Machine Lay-up

This is the simplest way of manufacturing composites. Fiber or fabric layers are placed on a mold with resin applied to successive layers until the desired thickness is reached as shown in Figure 7.29.

A gel coat is applied on the mold for better quality surface. Prepregs are very suitable for hand lay-up technique to avoid any wet process. As mentioned earlier, prepregs are yarns and fabrics that are already impregnated or melted with resin. A roller is used to remove entrapped air, control the thickness, guarantee good wet-out and smooth the surface. The curing usually takes place at room temperature or under heat to speed up the process. Usually polyester and epoxy resins are used for hand lay-up.

Machine lay-up is the automated form of hand lay-up. Computer controlled automatic tape-laying machines are used to lay down fiber or fabrics. This process provides consistency and increased speed.

5.2 Spray-up Molding

Chopped fibers and resin are simultaneously deposited on a mold using spraying equipment as shown in Figure 7.30. Gel coat is applied by a spray gun. Curing takes place at room or elevated temperatures. Polyester and epoxy resins are used.

5.3 Vacuum-Bag and Pressure-Bag Molding

In vacuum-bag molding, suction is used to remove entrapped air, voids and excess resin in the composite structure. A film (cellophane, polyvinyl alcohol or nylon) is placed over the lay-up and sealed at the edges with a special sealing compound. Vacuum of approximately 25 inch Hg is drawn between the bag and the mold. Curing is done under heat in a chamber. Vacuum is usually maintained during curing and subsequent cooling process. Vacuum-bag molding is suitable for large complex parts and shapes. The resulting composite is usually dense. Honeycomb materials, mats, fabrics, chopped fibers and prepregs are used with vacuum-bag molding. Typical resins include epoxies and polyesters.

In pressure-bag molding, a rubber bag is placed over the lay-up and is inflated using air to remove voids, entrapped air and drive out excess resin.

5.4 Injection Molding

Injection molding is similar to metal die casting. It is a high pressure process to produce thermoplastic and thermoset parts. The matrix is melted and forced into the mold cavity, where it

TABLE 7.11 *Major Composite Processing Techniques (Richardson [11]).*

Process	Remarks
Autoclave	Modification of vacuum and pressure bag; low production; low void content, dense parts, limited to autoclave size; wet
Blow molding	Mainly closed-mold process; multi-layers; short fibers or particulates; high volume; small-to medium-sized products
Casting (simple)	Open-mold process; low production; little control of reinforcement orientation; monomers or polymers
Compression molding	Closed-mold process; preforms available; some control on orientation; dense products
Expanding	Mainly closed-mold process; low to high production rates; limited control of reinforcement orientation; small to large products; monomers or polymers
Extrusion	Closed mold process; continuous lengths; multiple layers; continuous long fibers possible; preforms possible; some control on orientation
Filament winding	Mainly open-mold process; low production; control of orientation; wet
Hand lay-up	Open-mold process; low production; control of reinforcement orientation; large parts; wet
Injection molding	Closed-mold process; short fibers or particulates; little control on reinforcement orientation; small to medium-sized projects
Laminating (continuous)	Mainly closed-mold process; medium to high production; control of reinforcement orientation; small to large products; continuous; monomers or polymers
Matched-die molding	Closed-mold process; low to high production; some control of orientation; preforms; medium-sized products; monomers or polymers
Mechanical forming	Mainly colsed-mold process; medium to high production; preforms; some control of reinforcement orientation
Pressure-bag molding	Open-mold process; low production; control of reinforcement orientation; preforms; wet
Pulforming	Mainly closed-mold process; continuous; some control of reinforcement orientation; preforms; monomers or polymers
Pultrusion	Mainly closed-mold process; continuous; some control of orientation; preforms; monomers or polymers
Reaction injection molding	Closed-mold process; small to medium-sized products; medium production; monomers or polymers
Rotational casting	Open-mold process; low production; little control of reinforcement orientation; small to large products; powders and wet
Spray-up	Open-mold process; low production; little control of reinforcement orientation; preforms; wet
Thermoforming	Mainly open-mold process; preforms; medium to high production; little control of reinforcement orientation; mostly short fibers or particulates
Transfer molding	Closed-mold process; high production; dense parts; little control of reinforcement orientation; small to medium products; short fibers or particulates; monomers or polymers
Vacuum-bag molding	Open-mold process; low production; control of reinforcement orientation; preforms; wet

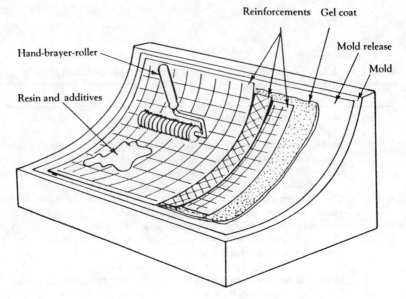

FIGURE 7.29 Hand lay-up technique (Richardson [11]).

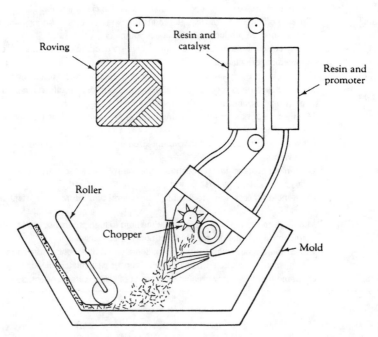

FIGURE 7.30 Spray-up technique (Richardson [11]).

freezes and is ultimately ejected as the finished part. The injection molding process permits finer part detail and can be easily automated. The part and mold can be designed such that near-net shape parts can be manufactured. There is a limit to the amount and types of fiber reinforcement that can be included in an injection molded part. The injection molding process of thermoplastic and thermosets is slightly different. In the thermoplastic injection molding process, a molten thermoplastic material is forced through an orifice into a cool mold where it solidifies. In thermoset injection molding, a reacting material is forced into a warm mold where the material further polymerizes or crosslinks to a solid part [25].

5.5 Autoclave

An autoclave is a large insulated cylindrical metal pressure vessel with a door at one or both ends. In autoclave, which is similar to vacuum-bag and pressure-bag molding, heat and pressure are applied on the lay-up from prepreg materials. Vacuum may be applied in the early stage to remove entrapped air. Air can be used for pressurizing and forced hot air is used for steam heat-up. Pressure, which is in the range of 345 to 690 MPa, is maintained during heating and cooling. Typical temperature range is between room temperature and 350°F.

In autoclave process, the lay-up is built on a mold plate which has the shape of the part to be produced. For a smooth surface, both sides of the laminate are usually covered with a fine polyester fabric which is peeled off after curing. The top surface of the lay-up is covered with a porous release film and "bleeder" or "breather" clothes if necessary. The whole assembly is then covered with a non-porous membrane and placed in the autoclave as shown in Figure 7.31. Autoclave produces high quality and consistent products but it is a rather slow process.

5.6 Pultrusion

Continuous reinforcement fibers or rovings of glass, carbon, aramid, etc., are drawn through a resin bath for impregnation and through a heated die to produce the desired shape and control the resin content in a continuous process as shown in Figure 7.32.

Pultrusion is especially suitable for unidirectional reinforcement for simple cross-sectional structures such as circular rods, tubes, channel and I-beams. Drawing velocity and the mold temperature are critical during the process. Polyester and epoxies are commonly used in pultrusion. The resulting composite has excellent mechanical properties due to good fiber alignment and high fiber fraction.

5.7 Compression Molding

In this method, prepregs or wet impregnated textile preforms are placed into an open mold. The mold is closed and heat and pressure are applied until the structure is formed and cured.

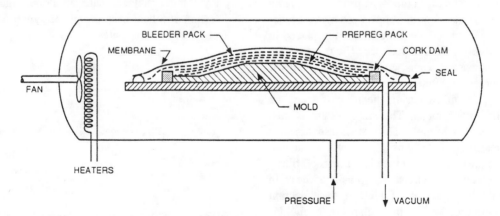

FIGURE 7.31 Schematic of autoclave molding [26].

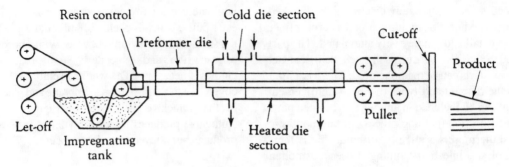

FIGURE 7.32 Pultrusion process (Richardson [11]).

The laminate thickness can be controlled by adjusting the final distance between press platens. Excess resin is usually allowed to escape. Both thermosetting and thermoplastic polymers are suitable for compression molding. The textile preforms may be vacuumed before molding to eliminate air bubbles in the composite. Compression molding is especially suitable for manufacturing flat or slightly curved panels or laminates. The advantages of compression molding include low cost, little material waste with close tolerances, part-to-part uniformity and reproducibility, good control of fiber-to-resin ratios and void content, and shorter cycle times. The compound can be heavily filled with little orientation of resin or additives.

5.8 Resin Transfer Molding (RTM)

Resin transfer molding, also called resin injection molding, is suitable for manufacturing of high fiber volume fraction (up to 70%) composite structures. In the RTM process, the resin system is transferred, at low viscosity and pressure, into the textile preform already placed in the closed mold (Figure 7.33). The resin system may consist of resins, curing agents, catalysts, promoters, inhibitors, etc., which may be premixed or mixed during the process using an on-line static mixer. The resin system is then injected into the mold. The resin is transferred at a pressure of 20−80 psi and/or with a vacuum in the range of 26−29 inch Hg. For good filling and wetting characteristics, viscosity of the resin should be less than 100 cP at the injection temperature. After filling with resin, the mold is sealed and heated for curing. Boron, Kevlar®,

glass, ceramic and graphite textile reinforcement structures are suitable for RTM. Woven, stitched, braided, knit and other preforms can be consolidated in the RTM process. Typical resin materials are polyester and epoxy. RTM offers high production rates, more consistent parts, and material and labor savings.

Comparison of resin transfer and compression molded glass-epoxy composites showed that the RTM process gives better product uniformity with less void volume [28].

6. COMPOSITE PROPERTIES AND TESTING

6.1 Predicting the Properties of Fiber Reinforced Composites—Rule of Mixtures

As stated earlier, the primary purpose of composites is to obtain properties that are not possible by any of the constituent materials alone. Figure 7.34 is an excellent example of the synergistic effect of textile composites.

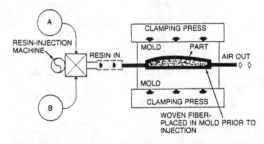

FIGURE 7.33 Schematic of RTM process (copyright 1989 Society of Manufacturing Engineers [27]).

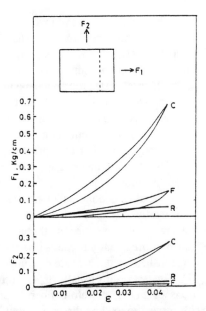

FIGURE 7.34 Synergistic effect in textile composites; F: woven fabric, R: silicon rubber, and C: their composites [29].

Continuous fibers, which have a high aspect ratio (l/d), give the best composite properties in terms of stiffness and strength. Discontinuous fibers with a large aspect ratio are also preferred. As volume fraction of fibers (VF) increases, strength and stiffness also increase. The upper limit of VF is approximately 80% which is determined by the geometry of the assembly. Process capability to surround the fibers with the matrix materials should also be considered when determining the VF during design stage.

In general, the rule of mixtures states that the value of the descriptive parameter of a fibrous composite denoted by P_c is given by

$$P_c = \sum_{i=1}^{n} f_i \times p_i$$

$$= f_1 \times p_1 + f_2 \times p_2 + \ldots + f_n \times p_n \quad (7.1)$$

where

n = number of components in the composite
f_i = the fraction of a component
p_i = the value of the same descriptive parameter for the individual component

If the fiber alignment is in more than one direction, the rule of mixture does not apply. Since textile structural composites are usually made of two components, i.e., fiber reinforcement and resin, $n = 2$. Therefore, some properties of fiber reinforced composites can be *roughly* estimated as follows.

Tensile Strength

According to the rule of mixtures, total stress in a continuous fiber composite is equal to the sum of the stresses in its constituents. For textile structural composites, the strength of the composite depends on the stress transfer from matrix to fibers. The amount of stress transferred is affected by the fiber length and the alignment of fibers with respect to loading direction. As the angle between the fiber axis and loading direction increases, the strength of a short fiber composite decreases. The strength of a textile structural composite can be calculated as

$$\sigma_c = f_f \left(1 - \frac{l_f}{2l} \right) \sigma_f + f_m \sigma_m \quad (7.2)$$

where

σ_c = ultimate tensile strength of composite
f_f = volume fraction of fiber
l_f = critical fiber length
l = fiber length
σ_f = fiber tensile strength
f_m = matrix volume fraction
σ_m = matrix tensile strength

This equation is valid when $l \geq l_f$. The critical fiber length is defined as the length at which fibers are stressed to the breaking stress and is given by

$$l_f = \sigma_f / 2\tau \quad (7.3)$$

where τ is the interfacial shear stress parallel to the fiber surface. Most of the time, the l_f can be taken as 2,000 μm [11]. If the aspect ratio is very large, then the equation above becomes

$$\sigma_c = f_f \sigma_f + f_m \sigma_m \quad (7.4)$$

Density

$$\varrho_c = f_f \varrho_f + f_m \varrho_m \qquad (7.5)$$

where ϱ_c, ϱ_f and ϱ_m are the densities of the composite, fiber and matrix materials respectively.

Modulus of Elasticity

When the direction of loading is parallel to fibers,

$$E_c = f_f E_f + f_m E_m \qquad (7.6)$$

where

E_c = elastic modulus of the composite
E_f = elastic modulus of the fiber
E_m = elastic modulus of the matrix

When loading direction is normal to the fibers,

$$1/E_c = f_f/E_f + f_m/E_m \qquad (7.7)$$

6.2 Test Methods

Composite behavior is different than that of isotropic and homogeneous materials which exhibit well-defined elastic and stress-strain properties. The composite properties largely depend on constituent materials, production technique and microstructure. Compared to single phase materials such as metals, plastics or ceramics, composite material properties are more variable. As a result, the spread of test data is much wider which requires a greater number of samples to be tested. The causes for the variation are due to relatively immature manufacturing processes, the variations inherent in textile structures and having two or more materials in the structure. The variation in the test data of composite structures often necessitates statistical analysis.

Composite test methods can be classified as destructive and non-destructive testing. It is important that the sample that is tested represent the composite structure. In destructive testing, the samples are damaged or destroyed while during non-destructive testing they are not. The most widely used destructive test methods include tension, compression, impact, three-point bending (long beam flexural test), four-point bending, interlaminar fracture toughness and short beam shear test. Nondestructive testing includes X-ray, ultrasonics, acoustic emission, thermography, holography, computer tomography, penetration, microwave, temperature differential and infrared scanning [11].

Composite testing is done either to measure physical properties or to detect defects. Composite testing is usually done in three areas: resins and reinforcement structures; composite parts; and finished products made of composite parts [27]. The test results usually depend on the orientations of fibers, therefore fiber orientations with respect to loading direction should be specified in the test results.

Testing of Reinforcement Structures

Several chemical, physical and mechanical tests are done on textile reinforcement materials without the matrix. The fiber surface characteristics are very critical for a good bonding with the matrix. Common methods to measure fiber surface characteristics are X-Ray Photoelectron Spectroscopy (XPS) and Electron Spectroscopy for Chemical Analysis (ESCA). In these methods, the fiber surface is bombarded with X-rays and the energy of the electrons knocked off the surface is measured. The fiber finish is also critical for bonding.

Physical tests of fibers include visual characterization, density and diameter measurements. In textile terminology, fiber density is usually expressed in denier (denier is the weight, in grams, of a 9,000 meter long fiber). Mechanical testing of fibers and tows includes tensile strength and bending rigidity. A typical tow includes several thousand individual filaments.

Textile preforms that are already woven, knit, braided, etc., are also tested for several properties. Number of yarns per unit length is critical for reinforcement. The reinforcement structure should be carefully inspected for defects, fiber breakage or damage, inclusion and fuzzing.

Testing of Matrix Materials

Curing characteristics of thermoset resins and the melting characteristics of thermoplastic resins are critical in composite manufacturing. Resin-fiber interaction affects the load bearing characteristics of composites. The completeness of the cure for thermoset resins is measured by Differential Scanning Calorimetry (DSC). Matrix materials affect the toughness and creep of the composites. Toughness is the ability of absorbing work and is measured as the area under the stress-strain curve. As shown in Figure 7.35, thermoplastic resins generally absorb more work than thermoset resins.

Two types of impact tests are used to measure the toughness of resins: pendulum test (Izod and Charpy methods) and the Falling Dart (Dart Impact) method (ASTM D 3029).

Creep behavior of a composite is influenced by the matrix. Creep is measured by loading the parts with specific weights and measuring the elongation over time at constant temperature.

Testing of matrix materials for environmental effects include resistance to heat, solvents, moisture and sunlight. Outer space and upper atmosphere applications require special testing.

Testing of Composite Parts

Testing of cured composites can be classified as quality testing, mechanical testing, environmental testing and field testing [27].

Quality Testing

Quality testing of cured composites includes fiber weight and volume fraction, void content, void detection and visual inspection. Fiber weight and volume fraction are determined with destructive tests. To measure the fiber weight, the composite specimen is weighed first. Then the resin is removed usually using a solvent; the remaining fibers are then dried and weighed. Fiber volume fraction is determined from the composite density, neat resin density and the bare fiber density. Fiber volume fraction of composite is one of the most important parameters which directly affects mechanical properties. Several

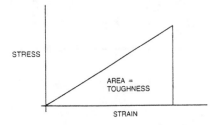

A) THERMOSET STRESS/STRAIN CURVE

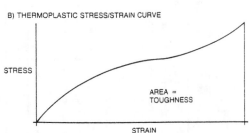

B) THERMOPLASTIC STRESS/STRAIN CURVE

FIGURE 7.35 Toughness of thermoset and thermoplastic composites (copyright 1989 Society of Manufacturing Engineers [27]).

methods are available to measure the density of the composite including the displacement method (ASTM D792) and density gradient method.

Non-destructive methods are used to indicate voids, the location and extent of defects for potential failure. The types of defects in composites include voids, delamination, debonds, wrinkles, moisture, inclusions, damaged reinforcements, reinforcement misalignment and matrix cracking. Major non-destructive methods include radiography, ultrasonics, acousto-ultrasonics, acoustic emission, thermography, optical holography and eddy current. Table 7.12 shows major non-destructive methods.

Mechanical Testing

Mechanical testing includes tensile test, compression test, flexural test, shear test and toughness test. Tensile test determines the tensile strength, tensile modulus and elongation of the composite. Anisotropic composites have different tensile properties in the fiber and cross-fiber directions. Tensile testing of composites is done according to ASTM D 3039. The test specimen is tabbed on the ends as shown in

TABLE 7.12 Non-Destructive Tests for Composites (copyright 1989 Society of Manufacturing Engineers [27]).

	Radiography	Ultrasonics	Acousto-Ultrasonics	Acoustic Emission	Thermography	Optical Holography	Eddy Current
Principle/characteristic detected	Differential absorption of penetrating radiation	Changes in acoustic impedance caused by defects	Uses pulsed ultrasound stress wave stimulation	Defects in part stressed generate stress waves	Mapping of temperature distribution over the test area	3-D imaging of a diffusely reflecting object	Changes in elec. cond. caused by material variations
Advantages	Film provides record of inspection; extensive database	Can penetrate thick materials; can be automated	Portable, quantitative, automated, graphic imaging	Remote and continuous surveillance	Rapid, remote measurement; need not contact part, quantitative	No special surface preparation or coating required	Readily automated; moderate cost
Limitations	Expensive; depth of defect not indicated; rad. safety	Water immersion or couplant needed	Surface contact, surface geometry	Requires application of stress for defect detection	Poor resolution for thick specimens	Vibration-sensitive if not coupled	Limited to elec. cond. materials; limited penetration depth
Defects Detected							
Voids	Yes	Yes	Yes	No	Yes**	Yes	Yes
Debonds	Yes*	Yes	Yes	Yes	Yes**	Yes	Yes
Delaminations	Yes*	Yes	Yes	Yes	Yes**	Yes	Yes
Impact damage	Yes**	Yes	Yes	Yes	Yes**	Yes	Yes
Density variations	Yes	Yes	Yes	No	No	No	Yes
Resin variations	Yes	Yes	Yes	No	No	No	Yes
Broken fibers	Yes†	Yes	Yes	Yes	No	Yes	Yes
Fiber misalignment	Yes	Yes	Yes	Yes	No	No	Yes
Wrinkles	Yes	Yes	Yes	Yes	Yes	Yes	Yes
Resin cracks	Yes**	Yes	Yes	No	Yes	Yes	Yes
Porosity	Yes	Yes	No	No	Yes**	Yes	Yes
Cure variations	No	Yes	Yes	No	No	No	No
Inclusions	Yes	Yes	Yes	Yes	No	No	Yes
Moisture	No	Yes	Yes	No	Yes	Yes	Yes

*Should be physically separated.
**Minor damages may not be detected.
†Might give problems.
Visual/optical techniques can be applied only to surfaces through openings or to transparent materials.
Liquid penetrants technique is applicable only to flaws open to surface; not useful on porous materials.

Figure 7.36 to ensure that it breaks between the tabs. The sample thickness is suggested to be between 5 to 10 mm (0.2″ to 0.4″) [27]. Due to stiffness of composite, strain gauges and extensometers are attached to the sample to measure the elongation.

The compression test is done according to ASTM D 3410 in which compression strength and compression modulus are determined. The same equipment as in the tensile testing is used except the cross-head moves in the opposite direction. In compression after impact tests, first, an impact test is done on the specimen. Then the specimen is subjected to compression test. As a result, the damage caused by the impact is measured.

The flexural test is a mixture of tensile and compression tests because the sample is subjected to compressive stresses on the top and to tensile stresses on the bottom as shown in Figure 7.37. ASTM D 790 specifies the flexural tests for composites. The span-to-thickness ratio is usually between 16:1 to 32:1 depending on the stiffness of the composite. The precise thickness measurement is very critical in flexural tests.

The short beam test is also called the interlaminar shear strength test and is done according to ASTM D 2344. This test is very similar to flexural tests except that the span-to-thickness ratio is approximately 4:1. The three-point test is more widely used. Specimens for other shear tests such as double notched shear test, double cantilever beam test, the 45° test and the rail shear test have different shapes.

Impact toughness tests can be pendulum-type (Izod and Charpy) and the falling dart-type as in the case of matrix tests.

Environmental and Field Testing

Some composite structures such as epoxy/carbon and epoxy/glass can be affected by moisture and water. To test the effects of moisture and heat, composite samples are placed in boiling water for 30 minutes. They are cooled down to room temperature for one hour and placed in boiling water again for 30 minutes. Then, the samples are tested for flexural strength.

In the thermal cycling test, the temperature is changed between −250°F to 232°F (−155°C to

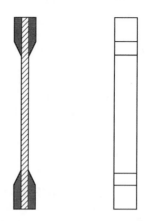

FIGURE 7.36 Composite specimen for tensile testing.

three point bending

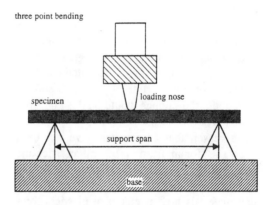

four point bending

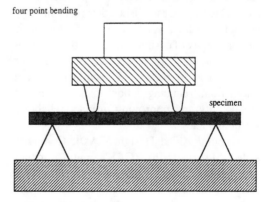

FIGURE 7.37 Flexural tests.

110°C). For composite engine parts, the range is 0°F to 450°F (−18°C to 230°C) [27]. The purpose of cycling tests is to determine the microcracks due to differences in the coefficients of thermal expansion of the fiber and matrix. Crimp tests are also done on composites similar to matrix testing.

Flammability tests are required for some applications such as inside panels for trains or airplanes. Vertical burn test is used to test flame resistance. A 1″ × 5″ sample is subjected to flame in vertical position and the rate of burning is measured.

7. DESIGNING WITH COMPOSITES

The major application areas of textile structural composites have been in aerospace and military applications as well as in some sporting goods. With the end of the cold war, the focus has been shifted towards increasing the use of composites in civilian applications such as automotives, construction and electronics in which composites may provide great advantages. As an example, it was estimated that for a reduction of every 100 lb in a 2,800 lb automobile, the savings would be 0.3 miles per gallon [11].

The reason that composites have not been widely used for applications other than space and defense is the cost associated with design and manufacturing of composites. Compared to other materials such as metals or plastics, composites lack a well established design technology which makes them costly. One reason for this is the complexity of design considerations for composites. The main areas that offer opportunity for improvement in composite manufacturing are automation, reducing labor costs, increasing productivity and improving product reliability and consistency.

Computer Aided Design (CAD), Computer Aided Manufacturing (CAM) and Computer Aided Engineering (CAE) systems are widely used for composite design, engineering and manufacturing. Figure 7.38 shows a typical production cycle for composite manufacturing.

Composite design and development requires careful consideration and selection of materi-

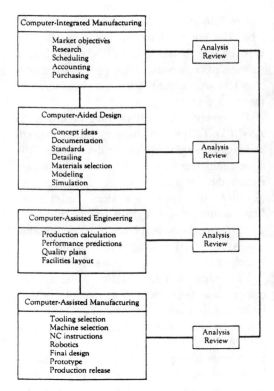

FIGURE 7.38 Typical composite design production cycle [11].

als, structural design and production technique. Proper selection of these components is a must to ensure success in application of composites.

Material selection is very critical to meet the design, economic and applications requirements. Considering the availability of a wide variety of materials, selection of materials with right properties require special attention. Most of the time manufacturer data sheets may not be sufficient to select material for a particular application. Material form may also be a determining factor for production technique. Functional property factors, processing parameters and economics need to be considered carefully when selecting a material. Functional property factors include tensile strength, creep, thermal expansion and resistance, impact resistance, density, service life, etc. Previous experience from similar successful (or unsuccessful) applications may be very valuable. Processing parameters to screen when selecting composite materials include product shape, tooling, production quantity and rate, capital investment for equipment and technology, reliability and quality. Material costs may be considerable in the total cost of high performance composites. Two factors that are often used for cost comparison of composites are apparent density (weight per unit volume) and bulk factors (ratio of the volume of loose molding powder to the volume of the same mass of matrix after molding). With the increased use of composites, recyclability of composite materials is gaining more importance.

The production method is determined to a large extent by the textile reinforcement structure, matrix formulation and form. Production technique considerations include tooling, capacity of the equipment, pressure, temperature, surface quality, postcuring cycles and production rates. For example, injection molding is suitable for short fiber reinforced composite manufacturing. Pultrusion, filament winding and laminating are more suitable for continuous fiber reinforced composites.

Composite design is a complex process. Composite structure can be designed to be isotropic (short fiber composites), quasi-isotropic and anisotropic as shown in Figure 7.39.

Anisotropic designs may be preferred since they allow the composite to have directional properties based on the orientation of reinforcement components. When designing composites, the end use or function, environmental conditions, reliability and safety requirements, and specifications must be considered very carefully. For design analysis, finite element method is used. Depending on the application areas, safety factor can vary between 1.5 and 10.

8. APPLICATIONS OF TEXTILE STRUCTURAL COMPOSITES

The advantages of textile composites can be summarized as follows [11]:

- Energy savings are important — low energy cost during manufacturing and long-term energy saving due to lighter components. The energy cost of producing composite automobile body is 40% less than that of a steel body.

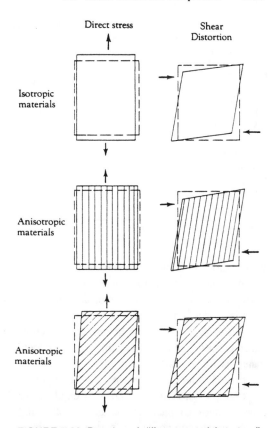

FIGURE 7.39 Reaction of different materials to tensile and shear forces [11].

- Due to anisotropic design, directional properties could be obtained.
- Good fatigue resistance and high degree of damage tolerance are seen.
- Corrosion resistance is an advantage.
- Composite parts can be designed with varying degrees of electrical and thermal conductivities.
- The availability of a wide range of fibers, yarns, fabrics, matrix materials and manufacturing processes for reinforcement fabrication and composite consolidation allows flexibility in designing various components and structures.

The current application of composites in automotives includes drive shafts, side rails, doors, cross members, oil pans, suspension arms, leaf springs, wheels, quarter panels, trunk

TABLE 7.13 *Applications of Composites in Aircrafts [30].*

Composite Component	F-14	F-15	F-16	F-18	B-1	AV-8B	727	757	767	Lear Fan
Doors	X			X	X	X		X	X	X
Rudder		X				X		X	X	X
Elevator							X	X	X	X
Vertical tail		X	X	X	X	X				X
Horizontal tail	X	X	X	X	X	X				X
Aileron						X		X	X	
Spoiler								X	X	
Flap				X		X		X		X
Wing box			X			X				X
Body						X				X

decks, hoods, hinges, transmission support, bumpers, seat frames, and wheels.

In the defense industry, composites are ideal materials for lightweight mobile, easily transportable vehicles for tactical shelters, ballistic combat and logistic applications. Composites are ideal materials for aircraft and aerospace applications where high strength-to-weight ratio is required. For example, glass fiber composite strength is five times higher than that of aluminum. Graphite and boron epoxy composite stiffness is five times higher than that of aluminum. Table 7.13 summarizes the application areas of composites in military and aerospace applications.

In aircraft applications, usually carbon, ara-

mid, glass fiber or filament reinforced epoxy based textile structural composites are used. Figure 7.40 shows application areas of composites on a Boeing 737-300 commercial airplane in which 3% of the total structural weight is made of composites. Carbon reinforced composites in aircrafts can provide up to 30 to 40% structural weight savings which can be used to carry more fuel or payload for the same takeoff weight.

In addition to structural body composites, composites are also used for interior parts such as overhead luggage compartments, sidewalls, ceilings, floors, galleys, lavatories, partitions, cargo liners, etc. Interior parts are usually made of fiber reinforced epoxy or phenolic resin (for

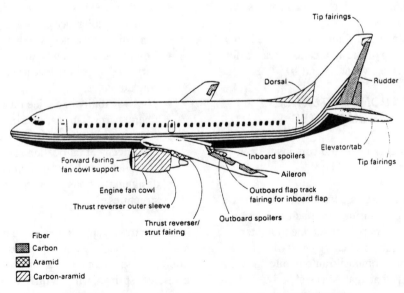

FIGURE 7.40 Fiber composite parts on the Boeing 737-300 [30].

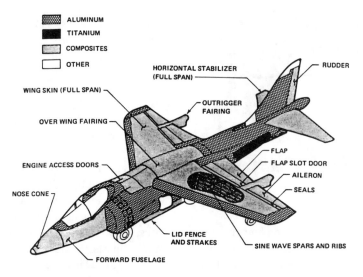

FIGURE 7.41 Graphite-epoxy composites on an AV-8B (courtesy of Mc-Donnell Douglas).

fire resistance, low smoke and toxic gas emissions) honeycomb sandwich structures. Important properties are impact resistance, stiffness and surface smoothness. Woven, knit, filament-wound and braided textile structures are used for reinforcement.

In aerospace applications, 40% of the composites is used for military applications. Carbon fiber reinforced composites constitute 26% of the structural weight of the U.S. Navy's AV-8B

(Harrier II) aircraft (Figure 7.41). The composite parts include the wing box, forward fuselage, horizontal stabilizer, elevators, rudder, over-wing surfaces, etc. On the F-18 fighter, 10% of the structural weight and 50% of the surface area are made of carbon fiber reinforced composites as shown in Figure 7.42. Other military aircrafts that use textile structural composites include the B-1B bomber, F-14A, F-16, and the Navy's V-22. Approximately 60% of the

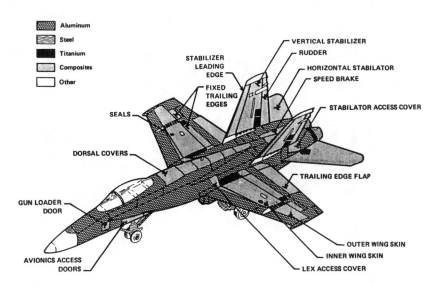

FIGURE 7.42 Composite distribution on an F/A-18 (courtesy of McDonnell Douglas).

FIGURE 7.43 Navy's tiltrotor V-22.

tiltrotor V-22 is composed of graphite composites (Figure 7.43) Almost 90% of the Voyager, which travelled around the world without refueling, was made of graphite fiber composites.

Composites are used in space structures such as missiles, rockets and satellites. Space structures require low weight, high stiffness, low coefficient of thermal expansion and dimensional stability which are the main properties offered by composites. The application areas of composites in missile systems include rocket motor case, nozzle, skirts and interstage structures, control surfaces and guidance structural components. Figure 7.44 shows schematic of BE-3 rocket motor in which rocket case and the nozzle are made of composites. Structural components used in space include trusses, platforms, shells, pressure vessels and tanks. Figure 7.45 shows a satellite truss.

E-glass, S-glass, aramid and carbon-graphite fibers are widely used in space and missile composite structures. In automotive applications, textile structural composites have to meet either structural or appearance requirements. A structural part performs a load bearing function in the structure.

Glass fiber composites have been used in boats very successfully. Corrosion resistance, light weight and low maintenance make composites ideal for marine applications such as hull structures, fairwaters, sonar domes, antennas, floats, buoys, masts, spars, deckhouses, etc. Composites are used in mine warfare vessels, tankers, trawlers, ferries, sonar domes, submarines, powerboats, racing yachts, pleasure boats, luxury yachts and laminated sailcloths.

Use of composites in sporting goods is also increasing at the expense of wood and metal. graphite-, boron- and Kevlar®-epoxy composites are used in golf carts, surf boards, hang-glider frames, javelins, hockey sticks, sail planes, sailboats, ski poles, playground equipment, golf shafts, fishing rods, snow and water skis, bows, arrows, tennis rackets, pole-vaulting poles, skateboards, bats, helmets, bicycle frames, canoes, catamarans, oars, paddles, etc. (Chapter 14.0, Sports and Recreation Textiles).

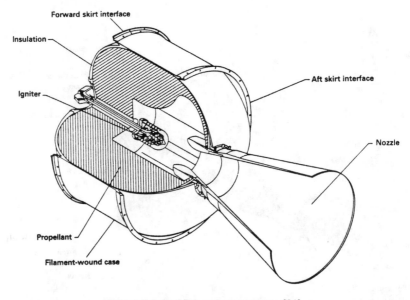

FIGURE 7.44 BE-3 racket motor case [31].

FIGURE 7.45 Satellite strut and truss made of graphite-epoxy composites (courtesy of Hercules Aerospace Products Group).

Composites have huge potential for mass-produced, modular manufactured composite homes, mobile homes, apartments, hotels and motels. Other application areas include musical instruments, appliance and furniture industries, medical implants such as artificial joints and organs, electronics, etc.

9. FUTURE OF TEXTILE STRUCTURAL COMPOSITES

Most of the composites in existence today were originally developed for aerospace applications. Detailed information and databases are available on the performance of these materials for aerospace applications. Although the results have been very successful, the cost factor has hindered the wide spread of composites into other application areas and industries. Other factors that hinder the faster growth of composite usage include capital investment, inspectability,

part consistency, lack of well-established manufacturing techniques and industry-wide specifications. Figure 7.46 shows the increased usage of composites in aerospace applications over the years.

With the end of the cold war, there are indications that the composite research in aerospace and defense areas is decreasing which leaves the transportation and construction to be the biggest potential areas for composite growth. As a result, due to different and probably less demanding applications, new low cost composites need to be developed. Great potential exists for the utilization of low cost composites in mechanical engineering, civil engineering, structural engineering and other areas such as automotives and electronics.

In automotives, composite bodies and parts will help by increasing the fuel efficiency of the vehicle due to light weight. By using composites, the number of parts in automobiles can be reduced from thousands to hundreds. Although

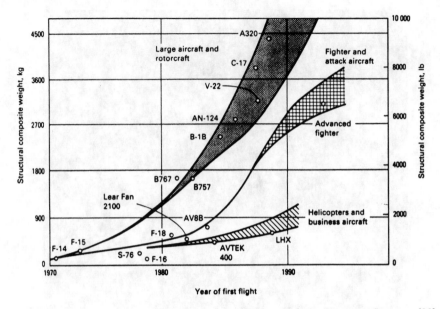

FIGURE 7.46 Usage of composites in aircraft applications within the last twenty-five years [30].

concept models of all-composite automobiles have been produced by the "big three" car manufacturers, at the present time, they are too costly for mass production.

The emphasis on smart or intelligent composites and materials has increased recently. A smart or intelligent material may be defined as a material that possesses intrinsic sensing, controlling and actuating capabilities. A smart or intelligent structure is a system that is capable of sensing, controlling and actuating. The growing use of composites in highly sophisticated systems such as aerospace, may necessitate the development of smart or intelligent composites. This, in turn, may require new developments in polymers, fibers and other textile structures. Although some intelligent materials can be found in naturally occurring biosystems, the development of synthetic intelligent materials is still at the conceptual stage.

7.2

References and Review Questions

1. REFERENCES

1. Watson, J. C. and Raghupathi, N., "Glass Fibers," in *Engineered Materials Handbook, Vol. 1, Composites,* ASM International, 1993.

2. PPG Industries, Lexington, NC.

3. Hansen, N. W. "Carbon Fibers," in *Engineered Materials Handbook, Vol. 1, Composites,* ASM International 1993.

4. Pigliacampi, J., "Organic Fibers," in *Engineered Materials Handbook, Vol. 1, Composites,* ASM International, 1993.

5. Gosnell, R. B., "Thermoplastic Resins," in *Engineered Materials Handbook, Vol. 1, Composites,* ASM International, 1993.

6. Ko, F. K., "Three-Dimensional Fabrics for Composites," in Chou, T., and Ko, F. K., eds., *Textile Structural Composites,* Elsevier, 1989.

7. Scardino, F., "Advanced Stitching Technology," *FIBER-TEX 1991, The Fifth Conference on Advanced Engineering Fibers and Textile Structures for Composites,* NASA Conference Publication 3176.

8. Fukuta, K., Onooka, R., Aoki, E, and Nagatsuka, Y., in Kawabata, S., ed., *15th Textile Research Symposium,* The Textile Machinery Society of Japan, 1984.

9. Scardino, F., "An Introduction to Textile Structures and Their Behavior," in Chou, T., and Ko, F. K., eds., *Textile Structural Composites,* Elsevier, 1989.

10. Bascom, W. D., "Fiber Sizing," in *Engineered Materials Handbook, Vol. 1, Composites,* ASM International, 1993.

11. Richardson, T., *Composites: A Design Guide,* Industrial Press Inc., 1987.

12. Magin, F. P., "Multidirectionally Reinforced Fabrics and Preforms," in *Engineered Materials Handbook, Vol. 1, Composites,* ASM International, 1993.

13. King, R. W., "Three-Dimensional Fabric Material," AVCO Corporation, United States Patent 4,001,478, Jan. 4, 1977.

14. Dominguez, F. S., "Unidirectional Tape Prepregs," in *Engineered Materials Handbook, Vol. 1, Composites,* ASM International, 1993.

15. Dominguez, F. S., "Multidirectional Tape Prepregs," in *Engineered Materials Handbook, Vol. 1, Composites,* ASM International, 1993.

16. Dominguez, F. S., "Woven Fabric Prepregs," in *Engineered Materials Handbook, Vol. 1, Composites,* ASM International, 1993.

17. Skelton, J., "Triaxially Braided Materials for Composites," in Chou, T. and Ko, F. K., eds., *Textile Structural Composites,* Elsevier, 1989.

18. Ko, F. K., "Braiding," in *Engineered Materials Handbook, Vol. 1, Composites,* ASM International, 1993.

19. The Karl Mayer Guide to Technical Textiles, Karl Mayer Textilmaschinenfabrik GmbH, 1988.

20. Malimo Stitch-Bonding Machines, Malimo Maschinenbau GmbH.

21. Horsting, K. et al., "New Types of Textile Fabrics for Fiber Composites," *SAMPE*

Journal, Vol. 29, No. 1, January/February, 1993.

22. Massey, D., Tech Textiles USA, Phenix City, AL, private communication.

23. "The Mechanical Properties of COTECH® Non-Crimp Fabric Based Composites," Technical Report, Tech Textiles USA, 1993.

24. Adanur, S., Tsao, Y. P. and Tam, C. W., "Improving Fracture Resistance in Laminar Textile Composites," *First International Conference in Composite Engineering (ICCE/1),* Aug. 28–31, 1994, New Orleans.

25. Murray, A. D., "Injection Molding," *Engineered Materials Handbook, Vol. 1, Composites,* ASM International, 1993.

26. Bader, M. G., "Molding Processes—An Overview," in Bader, M. G. et al., *Delaware Composites Design Encyclopedia, Vol. 3, Processing and Fabrication Technology,* Technomic Publishing Co., Inc., 1989.

27. Strong, A. B., *Fundamentals of Composite Manufacturing: Materials, Methods and Applications,* Society of Manufacturing Engineers, 1989.

28. Adanur, S. and Tsao, Y. P., "Stitch Bonded Textile Structural Composites," *SAMPE International Conference,* October 1994, Atlanta.

29. Kawabata, S., "Nonlinear Mechanics of Woven and Knitted Materials," in Chou, T. and Ko, F. K., eds., *Textile Structural Composites,* Elsevier, 1989.

30. Anglin, J. M., "Aircraft Applications," in *Engineered Materials Handbook, Vol. 1, Composites,* ASM International, 1987.

31. Policelli, F. J. and Vicario, A. A., "Space and Missile Systems," in *Engineered Materials Handbook, Vol. 1, Composites,* ASM International, 1987.

1.1 General References

Reinhart, T. J., "Introduction to Composites," in *Engineered Materials Handbook, Vol. 1, Composites,* ASM International, 1993.

Russell Diefendorf, "Carbon/Graphite Fibers," in *Engineered Materials Handbook, Vol. 1, Composites,* ASM International, 1993.

Clayton, A. M., "Epoxy Resins," in *Engineered Materials Handbook, Vol. 1, Composites,* ASM International, 1993.

May, C. A. and Tanaka, Y., eds., *Epoxy Resins, Chemistry and Technology,* Marcel Dekker, 1973.

Mohamed, M. H., "Three-Dimensional Textiles," *American Scientist,* Vol. 78, Nov.– Dec. 1990.

Chou, T. and Ko, F. K., eds., *Textile Structural Composites,* Elsevier, 1989.

Dominguez, F. S., "Prepreg Tow," in *Engineered Materials Handbook, Vol. 1, Composites,* ASM International, 1993.

Schwartz, M. M., *Composite Materials Handbook,* McGraw-Hill, 1984.

Summerscales, J., "Marine Applications," in *Engineered Materials Handbook, Vol. 1, Composites,* ASM International, 1987.

Kelly, A., ed., *Concise Encyclopedia of Composite Materials,* Pergamon, 1994.

2. REVIEW QUESTIONS

1. Define textile structural composites and list their basic characteristics. Why are composites used?
2. Explain the functions of reinforcement materials and matrices in textile composites.
3. What is the function of third direction (thickness or Z direction) fibers in a reinforcement structure?
4. Explain the following terms using sketches:
   - triaxial weave
   - angle interlock fabric
   - "5-D" weave
   - polar weave
   - orthogonal weave
   - prepreg
   - hybrid fabric
   - triaxial braid
5. How does resin transfer molding (RTM) machine work? What are the advantages and

disadvantages of RTM method compared to other composite manufacturing methods?

6. Explain the rule of mixture for composites.

7. The volume fraction of fibers in a fiberglass-epoxy composite is 50%. Glass fibers have a modulus of elasticity (parallel to the fibers) of 3.02×10^6 psi. If the elastic modulus of the epoxy is 410,000 psi, calculate the elastic modulus of the composite.

8. Explain the following:
   - aspect ratio
   - volume fraction of fibers (VF)

9. What is the difference between three-point bending and four-point bending test methods in terms of properties measured?

10. Define isotropic, quasi-isotropic and anisotropic textile structures. Explain the advantages and disadvantages of each.

FILTRATION TEXTILES

Filtration Textiles

S. ADANUR

1. INTRODUCTION

Filtration can be defined as the separation of one material from another. Therefore, filtration is basically a process of separation. The main purpose of filtration is to improve the purity of the filtered material. Sometimes filtration is used to recover solid particles. Textile filter materials are generally used in solid-gas separation and solid-liquid separation.

Filtration plays a critical role in our daily lives by providing healthier and cleaner products and environment. Textile materials are used in filtration of air, liquids, manufactured foods, and in industrial production. Air filters are widely used in air conditioning systems and in the engines of transportation vehicles. Filtration fabrics are used in vacuum cleaners, power stations, petrochemical plants, sewage disposal, chemicals and cosmetics. Another high volume filtration application is in cigarettes. Filtration is widely used in textile manufacturing. During man-made fiber manufacturing, molten and solution polymers are filtered to remove the foreign materials with metal media filter elements. Water filtration is the most important process in paper manufacturing that cannot be done without fabrics (Chapter 12.0).

Textile materials especially woven and nonwoven fabrics are particularly suitable for filtration because of their complicated structures and considerable thickness. Textile fabrics are in fact a three-dimensional network of fibers enclosing small pockets of void volume. Dust particles have to follow a ''tortuous'' path around textile fibers. Due to their structures, textile fabrics have high filtration efficiencies. The dust collection efficiency of fabric filters can range from 25 to 99.9% depending on the fabric type. High dust collection efficiencies and reasonable filter life (before plugging) can be obtained with woven and nonwoven fabrics. Textile materials do not restrict the flow of the fluid too much, yet they efficiently stop the particles [1,2]. Figure 8.1 shows examples of filtration products.

2. PRINCIPLES OF FILTRATION

The main objective of the filter medium is to maximize the possibility of collision and the subsequent retention of the suspended particles in the fluid (air/gas or liquid) stream with the media's fibrous structure while minimizing the energy loss of the stream of the fluid. The filtering element collects finer particles in a fluid stream while allowing fluid to flow through. Capturing these particles provides cleaner fluid [3,4].

A large portion of the total volume occupied by the fabric is in fact air space. The ratio of the volume of air or void contained in the fabric to the total fabric volume is defined as porosity. The amount and distribution of this air space influences a number of important fabric properties, including the efficiency of filtration in industrial fabrics. Knowledge of the air permeability of fabric is particularly important for many purposes such as suitability for use as vacuum cleaner bags and filters. Permeability is the capacity of a porous medium to transmit fluids. Air permeability of a fabric is a measure of the ease with which air can flow through the material. It is the rate of air flow under pressure differential across an area of the material. Gas and liquid permeability usually increases with

FIGURE 8.1 Various filtration products (courtesy of Fugafil® Saran GmbH & Co.).

the porosity of a fabric but the type of finish in a cloth has a considerable effect on its permeability [5]. As the porosity of the filter media increases, the pressure drop decreases as shown in Figure 8.2. Pressure drop should be minimized for higher filter efficiency. The figure also shows that increasing flow rate increases the pressure

drop. The relationship may not be linear depending on the system design and configuration.

Filtration takes place at different planes in the fabric. Therefore, there is a high probability that the particles in the fluid stream are captured by the fibers due to physical bonding forces between particles and fibers. If the dimensions of the particle are larger than the pore sizes in the textile material, the particle is stopped easily. In woven fabric filters, the pores for filtration are formed at yarn interstices. In nonwovens, the pores are formed by the small spaces that occur between the individual fibers. For the particles that are smaller than the pore sizes, there are five separate mechanisms of arrest as shown in Figure 8.3 [6,7]:

(1) Interception—When a particle tries to pass the fiber surface at a distance smaller than the particle's radius, it merely collides with the fiber and may be stopped or arrested.

(2) Inertial deposition—The velocity of a flow increases when passing the spaces of a filter because of the continuity equation. When a heavy particle is carried by the flow, it is thrown out of the flow streamlines due to its inertia (mass × speed). This may cause the particle to be caught by other fibers.

(3) Random diffusion (Brownian motion)—Due to Brownian-type motion, which can be described as random vibration and movement of particles in a flow, instead of trying to pass straight through the openings of the filter,

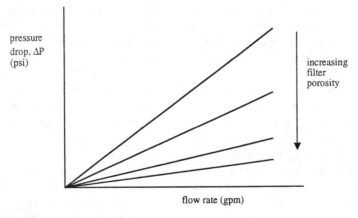

FIGURE 8.2 Relationship between flow rate and filter pressure drop for different filter porosities.

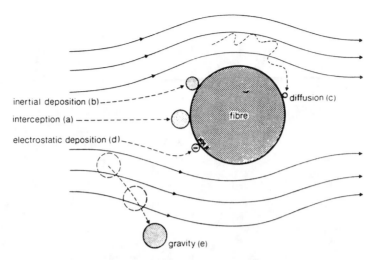

FIGURE 8.3 Filtration mechanisms [7].

particles follow a zig-zag route which increases the chance of being caught by the fiber material.

(4) Electrostatic deposition—Submicron particles are difficult to capture even with a combination of mechanical methods. It is well known that strong electrostatic forces of fibers attract the particles. Therefore, fibers may be given permanent electrical charges to attract small or medium sized particles. Charged fibers increase the filtration efficiency.

(5) Gravitational forces—Under the influence of gravity, a particle that is sinking, may collide with the fiber and get caught.

Depending on the process of separation, filtration is generally referred to as particle filtration, microfiltration, ultrafiltration, nanofiltration and reverse osmosis (hyperfiltration) as shown in Table 8.1. Modern filtration techniques can separate the particles as small as 1 Angstrom in size.

3. FILTRATION EQUIPMENT

Filter equipment and configurations vary and can be divided into several groups. The two major groups are filters for dry filtration and filters for liquid filtration. Depending on the application, different shaped industrial filters (e.g., bag, flat) are made from textile materials and are used on drum, disk, plate and frame, belt, vessel and other filters. Detailed discussion of filter equipment is out of the scope of this book. Therefore, only some examples of filtration equipment will be given below. The reader may refer to the references for in-depth information on filtration equipment.

Liquid filter bags are used in the processing of petroleum derivatives, chemicals, cutting oil, cleaning fluid, paints, pharmaceuticals, food processing, beverages, cosmetics, semiconductors, etc. Micron rated liquid filters are shown in Figures 8.4 and 8.5.

Figure 8.6 shows the view of a typical vessel filter for liquid filtration. The 3M company also developed a radial pleated filter element for vessel filters. The radial pleat design fills the entire filter cartridge with pleats that are stacked horizontally. According to the manufacturer, radial pleat design offers more surface area, better performance, longer life and less wasted space compared to vertically pleated filter media.

Rotary drum filters can be vacuum-type (rotary drum vacuum filters) or pressure-type (pressure drum filters). A rotary drum vacuum filter, which is used for slurry filtration, is a large drum (up to 20 ft in diameter and 20 ft long) covered by filtration fabric. The drum rotates continuously and slowly. The lower part of the drum is dipped in the container containing the

TABLE 8.1 The Filtration Spectrum (copyright by Osmonics, Inc., Minnetonka, Minnesota).

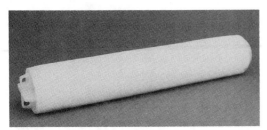

FIGURE 8.4 Liquid cartridge filter (courtesy of 3M Company).

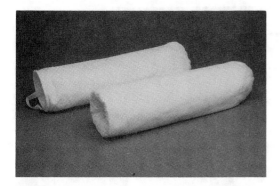

FIGURE 8.5 Multiple layer bag filters for high efficiency (courtesy of 3M Company).

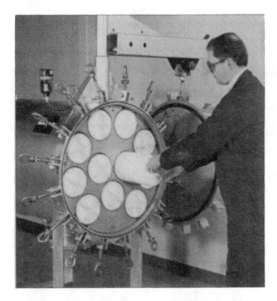

FIGURE 8.6 Multi-cartridge horizontal filter vessel (courtesy of 3M Company).

slurry. Vacuum is used inside the drum to suck the fluid leaving a layer of solid on the outside of the fabric. As the drum rotates, the solid layer is lifted and using a knife, it is separated from the fabric. In a rotary pressure filter, the rotating drum, which is covered with fabric, is contained in a sealed chamber. Treatment media is fed to the chamber under pressure. The pressure drum filter is used for continuous mechanical separation of solid-liquid mixture. It can be used at high pressure and temperatures.

In rotary disk filters, a number of disks, which are covered with filter cloth, are mounted on a horizontal shaft. The disks are rotated in the suspension container and vacuum is applied through the core of the shaft. Cake collected on the emergent sector of the disk is removed before reentering the suspension. There are various types of rotary disk filters.

The filter press, which is used for discontinuous liquid filtration, consists of many pairs of plates that are covered with fabric and pressed together. Slurry is pressed inside the compartment under high pressure; the liquid is filtered through the fabric leaving the solid layer on the fabric surface. Then the plates are separated, the solid layer falls off which completes the cycle.

In belt filters, a suction tray supports a rotating filter cloth. There is a continuous cycle of vacuum stroke/return stroke. When suction is applied, filtration is produced down on to the vacuum tray which is carried along by the belt. Then the vacuum is released and the cloth continues moving forward while the tray is pulled back to the original position which completes the cycle (Figure 8.7).

Figure 8.8 shows a cylindrical filter configuration which is used in the textile industry for filtering out separately coarse particles/fibers and fine dust out of the return air from the machines. Air flows from inside to outside.

Bag filters are used for dust removal in industrial atmospheres. Depending on the application, dust can be captured on the internal or external surface of the bag. Large numbers of woven or nonwoven fabric tubes or bags are used in baghouse systems. Suction created by a fan is used to pull the dusty air through the bags which collect the dust on the fabric surface. The filtration efficiency must be around 99.9% for fine

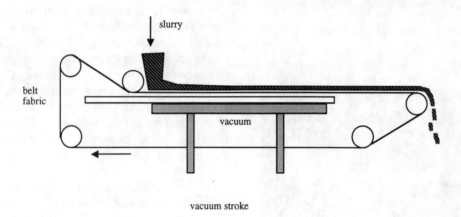

slurry

belt
fabric

vacuum

vacuum stroke

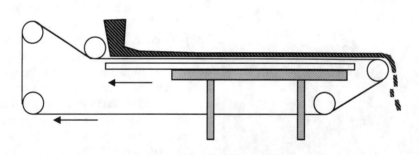

return stroke

FIGURE 8.7 Principle of belt filter.

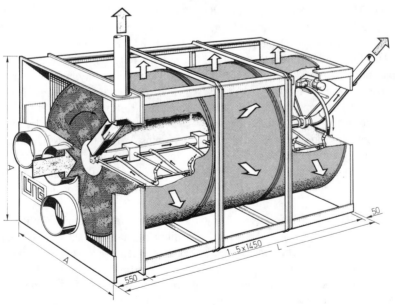

FIGURE 8.8 Cylindrical drum filter used in the textile industry to separate coarse particles, fiber and dust from the air flow (courtesy of LTG).

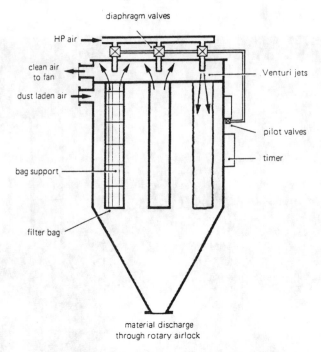

FIGURE 8.9 Schematic of a pulse jet baghouse [1].

dust collection. Collected dust must be removed from the fabric surface periodically to allow efficient filtration. This can be done in several ways including pulse jet, reverse air and shaking. In older systems, the filter bags in one compartment are shaken while the air flow is diverted to another compartment. In new systems, a blast of compressed air is blown into the bags to shake them briefly without interrupting the filtration process. Dust is collected in a container below the bags and removed periodically. Figure 8.9 shows the schematic of a pulse jet baghouse.

4. TEXTILES IN DRY FILTRATION

Application areas for dry filtration include the mining industry, chemical, iron and steel industries, cement, aggregate plants, utilities, the primary non-ferrous smelting industry, lime, clay, kaolin and ceramic works, feed, grain and food industries, woodworking and furniture industries, paper-related industries, and textile plants. Filter media is also used in conveying, milling, mixing and shipping of bulk materials [8].

4.1 Fibers and Fabrics

The most widely used fiber in dry filter media is polyester (approximately 70% of all materials). Other fibers for dry filtration include nylon, polypropylene, acrylic, aramids and polytetrafluoroethylene, glass and sulfar (PPS). Table 8.2 lists the properties of textile fibers used in dry filtration.

Cotton fabrics are used in older shaker type systems. Woven fiberglass is currently used for the removal of fly ash in utilities. Microglass blanket media is used for high ASHRAE (American Society of Heating, Refrigerating and Air-conditioning Engineers) efficiency, extended surface filters in HVAC (Heating, Ventilation and Air Conditioning) applications. Polyester felt fabric is used for reverse jet or pulse jet systems [9].

Polyester is the most widely used fiber in gas filtration because of its strength, relatively high temperature resistance (up to 150°C) and low cost (Figure 8.10). The disadvantages of polyester for filtration applications are low resistance to alkalis, acids and steam. Homopolymer acrylic is used when chemical conditions and temperatures up to 130°C require a better fabric.

TABLE 8.2 Properties of Textile Fibers for Dry Filtration (copyright 1989 Filter Media Consulting, Inc.).

*All Information is based on data of the different fiber manufactures, verified and acknowledged.

FIBER — GENERIC NAME / TRADE NAME	COTTON	WOOL	POLYAMID NYLON 66	POLYPROPYLENE HERCULON®	POLYESTER DACRON®	ACRYLIC COPOLYMER ORLON®	HOMOPOLYMER ACRYLIC DRALON T	AROMATIC ARAMID NOMEX®	AROMATIC ARAMID TEIJINCONEX®	POLYTETRAFLUORETHYLENE TEFLON®	POLYTETRAFLUORETHYLENE TOYOFLON®
Recommended continuous operation temperature (dry heat)	180° F / 82° C	200° F / 94° C	200° F / 94° C	200° F* / 94° C	270° F / 132° C	248° F / 120° C	257° F / 125° C	400° F / 204° C	392° F / 200° C	500° F* / 260° C	500° F / 260° C
Water vapor saturated condition (moist heat)	180° F / 82° C	190° F / 88° C	200° F / 94° C	200° F / 94° C	200° F* / 94° C	230° F / 110° C	260° F / 125° C	350° F / 177° C	356° F / 180° C	500° F* / 260° C	500° F / 260° C
Maximum (short time) operation temperature dry heat	200° F / 94° C	230° F / 110° C	250° F / 121° C	225° F / 107° C	300° F / 150° C	248° F / 120° C	302° F / 150° C	465° F / 240° C	482° F / 250° C	550° F / 290° C	550° F / 290° C
Specific density	1.50	1.31	1.14	0.9	1.38	1.16	1.17	1.38	1.37-1.38	2.3	2.3
Relative moisture regain in % (at 68° F & 65% relative moisture)	8.5	15	4.-4.5	0.1	0.4	1.0	1.0	4.5	4.5	0	0
Supports combustion	Yes	No	Yes	Yes	Yes	No	Yes	No	No	No	No
Biological resistance (bacteria, mildew)	No. If not treated	No, if not treated	No Effect	Excellent	No Effect	Very Good	Very Good	No Effect	No Effect	No Effect	No Effect
Resistance to alkalies	Good	Poor	Good	Excellent	Fair	Fair	Fair	Good	Good	Excellent	Excellent
Resistance to mineral acids	Poor	Good	Poor	Excellent	Fair+	Good	Very Good	Fair	Fair	Excellent	Excellent
Resistance to organic acids	Poor	Good	Poor	Excellent	Fair	Good	Excellent	Fair+	Fair	Excellent	Excellent
Resistance to oxidizing agents	Fair	Fair	Fair	Good	Good	Good	Good	Poor	Poor	Excellent	Excellent
Resistance to organic solvents	Very Good	Very Good	Very Good	Excellent	Good	Very Good	Very Good	Very Good	Very Good	Excellent	Excellent
Comments:				*250° F for Type 154						*Not Recommended	

(continued)

TABLE 8.2 (continued).

*All information is based on data of the different fiber manufactures, verified and acknowledged.

FIBER — GENERIC NAME	EXPANDED PTFE	POLYETHERIMIDE	SULFAR (PPS)			POLYKETONE	POLYIMIDE	GLASS	METAL*	CERAMIC	
TRADE NAME	RASTEX*	AKZO* PEI	RYTON®	TEIJIN PPS	BAYER PPS	ZYEX*	P 84*	FIBERGLASS*	BEKINOX®	NEXTEL 312*	FIBERFAX*
Recommended continuous operation temperature (dry heat)	500° F 260° C	338° F 170° C	375° F 190° C	375° F 190° C	375° F 190° C	460° F 240° C	500° F 260° C	500° F 260° C	1020° F 550° C	2100° F 1150° C	2300° F 1260° C
Water vapor saturated condition (moist heat)	500° F 260° C	338° F 170° C	375° F 190° C	375° F 190° C	n.y.e.*	460° F 240° C	383° F 195° C	500° F 260° C	1020° F 550° C	2100° F 1150° C	2300° F 1260° C
Maximum (short time) operation temperature dry heat	550° F 290° C	392° F 200° C	450° F 232° C	446° F 230° C	446° F 230° C	570° F 300° C	580° F 300° C	550° F 290° C	1110° F 600° C	2600° F 1427° C	3260° F 1790° C
Specific density	1.6	1.28	1.38	1.34–1.35	1.37	1.30	1.41	2.54	7.9	2.7	2.7
Relative moisture regain in % (at 68° F & 65% relative moisture)	0	1.25	0.6	0.24–0.25	<0.6%	0.1	3.0	0	0	0	0
Supports combustion	No	No	No	No	Self Quenching LOI 39–41%	No	No	No	No	No	No
Biological resistance (bacteria, mildew)	No Effect	No Effect	No Effect	No Effect	n.y.e.*	No Effect	No Effect	No Effect	No Effect	No Effect	No Effect
Resistance to alkalies	Excellent	Good ph<9	Excellent	Excellent	Excellent	Excellent	Fair	Fair	Excellent	Good	No Effect
Resistance to mineral acids	Excellent	Good	Excellent	Excellent	Excellent	Very Good	Very Good	Very Good	Good	Very Good	Very Good
Resistance to organic acids	Excellent	Excellent	Excellent	Excellent	Excellent	Excellent	Very Good	Very Good	Very Good	Very Good	Excellent
Resistance to oxidizing agents	Excellent	Good	Fair*	Fair	Fair**	Good	Very Good	Excellent	Excellent	Excellent	Excellent
Resistance to organic solvents	Excellent	Good*	Excellent	Excellent	Excellent	Excellent	Excellent*	Very Good	Excellent	Excellent	Excellent
Comments:		*PEI fiber is dissolved by partially chlorinated hydrocarbons	*PPS fiber is attacked by strong oxidizing agents. For example at 200° F for 7 days.		*n.y.e. Not Yet Examined **Depending on Concentration		*Soluble only in strong polar solvents (DMF, DMAC, DMSO, NMP)		*INCONEL 601™		

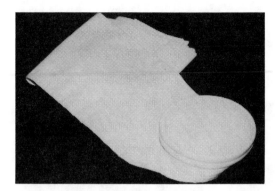

FIGURE 8.10 Filtration bag made of 100% polyester (courtesy of Wellington Sears Company).

Ryton® (polyphenylene sulfide, PPS), commercialized in 1983 by Phillips Fibers, has high-performance filtration qualities. It is used as high-temperature filtration fabric up to 190°C in continuous use. Nomex® (aramid) is used for temperatures up to 200°C (Figure 8.11). Teflon (polytetrafluoroethylene or PTFE) can withstand up to 260°C maximum. Woven glass filter bags are currently used up to 250°C.

Fiberglass has a high particle holding capacity. Recently, there has been an increased concern in regard to the safety of fibrous glass and health effects from its use in HVAC applications [10]. The installation of materials containing fiber glass may cause skin irritation. This is caused by a reaction to broken ends of fibers that become embedded in the outer layer of the skin. Irritation does not normally persist for a long period of time if the skin is gently washed. If large amounts of airborne fiberglass are released, upper respiratory problems occur including scratchiness and burning of the throat. Overexposure to glass fibers may cause a transitory lung condition involving persistent coughing and wheezing. Fiberglass has also been found to aggravate existing conditions such as asthma and bronchitis. Studies have been conducted to monitor the effect of the release of glass fibers from air filter media [11]. Fiber levels in the indoor air of public buildings equipped with glass fiber filters were found to be low. In addition, the possible risk of lung cancer from exposure to fiberglass was found to be insignificant.

DuPont developed a Teflon®-fiberglass blend nonwoven filter fabric called Tefaire®. It is claimed that the opposite triboelectric character of the two fibers in the blend results in the generation of electrostatic charges during operation which improves the performance and efficiency in submicron particulate collection. Dust is collected on the Teflon® fiber; the glass fiber bundle appears cleaner (Figure 8.12). According to the manufacturer, Teflon® fibers improve the filter bag life, while the small amount of fine glass fibers increase the total available fiber surface area and reduce the porosity of the felt which results in improved dust collection efficiency.

Polypropylene textile substrates have been used for the manufacture of industrial filter fabrics. The reason for selecting polypropylene as raw material for filter fabrics lies in the fact that polypropylene has excellent acid resistance, alkali resistance and abrasion resistance. It is

NOMEX Type 450 felt (Approx. 500 g/m²)

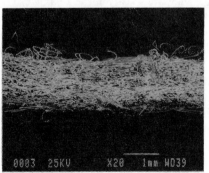

NOMEX high density felt without support fabric (Approx. 300 g/m²)

FIGURE 8.11 Cross section of nonwoven Nomex® fabric (courtesy of DuPont).

FIGURE 8.12 Electromicroscopic photograph of Tefaire® nonwoven filter fabric (courtesy of DuPont).

inexpensive and can be melt blown. Polyimide and Gore-Tex® membrane filters are also used in dry filtration.

Ceramic fibers are suitable for hot gas filtration around 1,000°C. The major supplier of such fabrics is currently 3M with a material called Nextel®. There are other high temperature ceramic materials being offered by European manufacturers [9].

For filtration of gases, nonwoven fabrics are mainly used. Figure 8.13 shows several filter products for air filtration. Needlepunched filter fabrics provide good dimensional stability, excellent particle retention and freedom from plugging. They are mostly made of polyester, homopolymer acrylic, nylon, polypropylene, Nomex®, Ryton®, P84® and PTFE.

Melt blown nonwovens are also growing at a fast rate. Melt blown and microfiber nonwovens are used primarily in the higher efficiency ASHRAE and Eurovent classes with efficiencies predominantly between 80–95%. Many melt blown microfiber nonwovens currently used in high efficiency ASHRAE/Eurovent filtration are often electrostatically charged and referred to as electret nonwovens. An electret is a non-conductive polymeric material that maintains a long-lived electrostatic charge. They can be divided into different fields of manufacturing: electrostatically charged from a polymer solution, split fiber corona charged and post media charged materials. Special electret nonwoven fabrics are manufactured in different weight ranges and thickness and consequently, different efficiencies up to 99.9996% according to DOP (DiOctyl Phthalate liquid) conditions of 0.1 micron at air flow rates of 5.3 cm/s [12]. Figure 8.14 shows the comparison of electrostatic capture of particles and microfiberglass filters. Electrically charged media captures particles such as NaCl particles throughout its depth, while still permitting air to flow freely through the filter. Microfiberglass filters require a more closed construction and capture NaCl particles primarily on the filter surface which results in higher airflow resistance and greater pressure drops.

4.2 Applications

Needlepunched filters are used in baghouses and fabric filters for dust filtration. Nomex® (aramid) filter bags are used in hot mixed asphalt plants. Thermobonded, molded and melt blown nonwovens are used in industrial and medical

FIGURE 8.13 Various filter products for air filtration (courtesy of 3M Company).

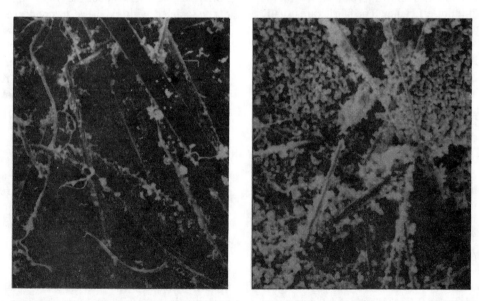

FIGURE 8.14 Electrostatic capture of NaCl particles on fibers (left) versus mechanical capture of NaCl particles with a microfiberglass filter (courtesy of 3M Company).

face masks. High loft glass and polyester filters are common in ordinary heating and air conditioning systems in buildings, offices and factories. Carded rayon, cotton and polyester blends are used in the low-to medium efficiency range. Medium- to high-efficiency microglass blanket media is used in critical applications such as pharmaceutical, medical, nuclear, and electronic industries. Nonwoven air filters are also used in automotives, lawn mowers, chain saws and portable gas-powered blowers. Other application areas of nonwoven filters include air-intake filters for the protection of industrial ventilation, blower engines, turbines, pneumatic controls, compressors, silencers and vacuum pumps [8].

In industrial applications such as coal burning, cement manufacture, steel making, etc. large amounts of dust are produced which should not be allowed to escape to the atmosphere. Other particles such as carbon black, fine pigments, etc., must also be filtered. Fabric filters are very suitable for dust collection in baghouse systems. Nonwoven fabrics allow more air flow with the same filtration efficiency. The fabrics are usually made dense with low air permeabilities. Depending on the application conditions, the filters may become useless in several months due to abrasion, chemical attack or blocking by dust. Some conditions may be very harsh including high gas temperatures, steam, oxides of sulfur, acid, and alkaline.

4.3 Filters in Air Conditioning

An important area of filter application is in air conditioning units. The most important reason for air filtration is to clean the air. Two applications of melt blown nonwovens are HVAC (Heating, Ventilation, and Air Conditioning) and HEPA (High Efficiency Particulate Air filter) filtration. The HVAC industry requires higher standards for air filters because of the importance of indoor air quality. The efficiencies of HVAC systems are regulated through ASHRAE in North America and under the designation Eurovent 4/5 in Europe. Almost all currently used HEPA filters worldwide are based on microglass, wet-laid paper with minimum efficiencies according to the DOP test of 99.95% for

0.3 micron particles. The next generation of filters is referred to as ULPA (Ultra Low Penetration Air filter) with required efficiencies of 99.9995% for particles of 0.1 micron. Such efficiencies can be obtained normally at the expense of relatively high differential pressure. Seed particles, pollen, fibrous particles, dust, carbon black, smoke and aerosols in the atmospheric air must be filtered for air conditioning systems not to exceed 4 mg/m^3. The size of these particles may be less than 1 micrometer. Nonwoven filters are ideal for this purpose. For many years, the HVAC industry has used air filters to protect equipment and prevent discoloration of ceilings and other equipment. Recently, consumers are becoming aware of the detrimental health effects in relation to exposure of airborne contaminants [13]. Particles in the air that have human health implications are now given increased attention.

HVAC filters are used as part of a forced air heating and cooling system to reduce indoor concentrations of hazardous airborne particles. Some particle filters reduce the risk of lung cancer associated with exposure to condensed tobacco smoke vapors. Microbiological particles such as fungi, bacteria and insects are also removed by effective air filtration. Other airborne particles have health effects that result from their composition and substances that are absorbed into their composition [14]. The typical amounts of dust concentrations in the countryside and in industrial areas are 0.1 mg/m^3 and 5 mg/m^3 respectively. A level of 0.2 mg/m^3 dust concentration is considered to be acceptable. Therefore, filters in air conditioning units are expected to hold back 90% of the atmospheric dust.

Filters for residential air conditioning systems are usually made of a thick low-density mat of polyester or glass fibers. They are designed to allow high air flow rate and to have good dust collection capacity. The filter should not build up pressure across its thickness. Sometimes, fibers in the mat are covered with a sticky film to help with particle capture. The latest developments include an electrostatically charged material from 3M.

Clean air is a must for many applications. In hospital operation rooms, clean air which is free of harmful bacteria is needed (HEPA filtration). Bacteria size can range from less than a micron to 10 microns. A particle free environment is also critical for clean rooms in microcircuit manufacturing where the distance between parts may be a few microns apart. For this type of critical application, usually a second filtration stage is used. The second filter is usually made of a denser fabric or a glass fiber paper with high filtration efficiency but low air permeability. As a result, a large filter area is needed which may be obtained by a folded structure. The typical filtration efficiency of the second filter is around 99.9 % (i.e., dust penetration of 0.15).

5. TEXTILES IN LIQUID FILTRATION

5.1 Fibers and Fabrics

Polypropylene fibers are widely used in woven and nonwoven structures in liquid filtration to improve filtration properties because of their resistance to chemical breakdown. Polypropylene represents approximately 50 % of all materials in liquid filtration. Polyester in woven and nonwoven structures is used in liquid filtration. Viscose rayon fibers are also suitable for liquid filtration.

Woven fabrics are common in liquid filtration. The most widely used designs include plain, basket and satin weave. Warp knit fabrics are used for relatively coarse filtration. The use of nonwovens in liquid filtration is also increasing. Needlepunched, adhesive bonded, impregnated and calendered nonwoven fabrics are used in liquid filtration. Needlepunched fabrics are used for plate and frame presses, belt filters and sometimes drum and disk filters.

Rayon or polyester needled felts are used in micron-rated (between 3 μm and 100 μm) vessel bags which are used as disposable filter elements in single- and multi-cartridge vessels. Nylon and polypropylene felts are also used for this purpose. Spunbonded filter media is used to filter and clean industrial fluids and oils extensively used in machinery manufacturing plants. Wet-laid nonwovens are used in food filtration especially for dairy products such as milk. Nonwovens are also used in medicine for blood filtration and dialysis [8].

Forming fabrics in papermaking, which are used to drain water and chemicals from the slurry, are made of polyester monofilaments (Chapter 12.0). A small percentage of nylon is also used to improve abrasion resistance.

5.2 Applications

Depending on the particle concentration, liquid filtration can be divided into two types: low particle concentration and high particle concentration in which solid concentration is more than 10 %. Another type of classification for liquid filtration is continuous and discontinuous for which drum (rotary) filter and plate press are used respectively.

When the particle concentration is below 10 %, the liquid is pumped through a filter fabric that has small pores. Usually a filter aid is used to prevent the blockage of small pores in the fabric. The filter aid, which consists of a powder such as diatomite, forms a porous layer on top of the fabric and increases the solid holding capacity of the filter and removes undesirable impurities from the liquid.

In the second type of filtering application, such as collection of china clay, particulate chemical production, dewatering of sewage, and separation of coal dust, sand and clay, the solid concentration is usually more than 10 % which is called "slurry." Rotary vacuum drum filters are used for some of the slurry filtration applications. Filter presses are used in discontinuous filtration processes. Usually polypropylene woven fabric is used for rotary filters and filter presses where abrasion is critical.

For some applications, a woven, stitchbonded or spunbonded backing may be used. For high temperature applications, the fibrous web may be heat-shrunk. To increase the smoothness of the surface, a surface finish may be applied. Smooth surface is necessary for easy removal of the sludge or cake. Polypropylene fibers offer good chemical resistance, durability, low moisture absorption and a hydrophobic surface which improves the cake removal.

In sewage treatment plants, belt filter presses, drum, plate or belt filters have been used to dewater sewage sludges. Recently, needlepunched fabrics are being used instead of woven fabrics in these applications. Comparison of

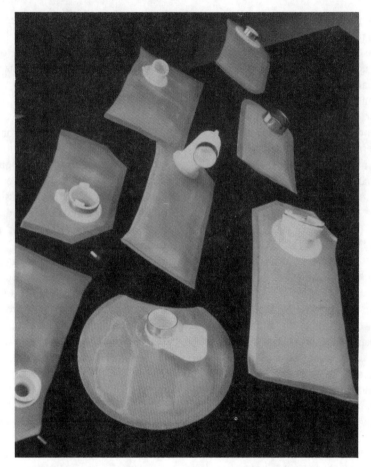

FIGURE 8.15 Filters for oil pumps (courtesy of Fugafil® Saran GmbH & Co.).

woven nylon and needlepunched polypropylene filters in these applications favored the use of nonwovens.

Petroleum and its products are filtered using filter candles. Filter candles are made into a tube shape with typical dimensions of 250 mm length, 30 mm inside diameter and 60 mm outside diameter. Nonwoven filter cloth is used for filtration of oil. For high purity such as aerospace hydraulics, spunbonded filter cloth made of extremely fine micron-sized glass, polyester or polyvinyl chloride fibers is used. The new sensitive high-performance petrol pumps in injection systems set high requirements for the filtration characteristics of the petrol filter with respect to impurities and air intake. Figure 8.15 shows filters used in oil pumps.

Filter products are widely used in the food industry. Water cartridge filters are used to purify water. Modern milk filtration is done in a continuous fashion in automatic filter systems. The filter cloth must have good tensile strength.

Woven and nonwoven fabrics made of metal fibers or yarns are used for filtration of impurities in melt and spinning of textile fibers.

6. DESIGNING FOR FILTRATION

When designing a textile structure for filtration applications, several factors should be considered that include flow velocity, pressure inside the filtering system, size and concentration of the particles, and the nature and components of the suspension filtered. When constructing a filter for a specific application, the fiber must be selected such that it can withstand the harsh environmental conditions such as temperature, abrasion, chemical conditions, etc. As in the case of most industrial applications, use of synthetic fibers is increasing due to several advantages including reduced fabric weight because of the higher strength of man-made fibers, easier handling and cloth replacement, easier separation of the filter cake from the woven cloth, resistance to rot, higher rate of filtration, improved fatigue resistance, good dimensional stability, high temperature resistance, better abrasion, corrosion and chemical resistance.

Filter fabrics used in filter presses are calendered to obtain a smooth fabric surface which improves the removal of filter cakes. Calendering is done on a machine with electrically heated calender rollers.

Natural randomness of the textile structures increases the probability of the particles being caught by the fibers. Therefore, nonwoven fabrics, in general, provide higher filtration efficiencies than woven or knit fabric. Woven and knit fabrics are considered to have two-dimensional structures while nonwoven fabrics may give three-dimensional structures with larger thickness which increases the distance for the particle to travel. It is also easy to construct nonwovens in layers. A coarse open fabric removes larger particles. As the fluid progresses through the filter, the fibers and pores become finer to trap finer particles.

The advantages of nonwoven filter media over woven filter media include [3]:

- high permeability
- higher filtration efficiency
- less blinding tendency
- no yarn slippage as in the case of woven media
- good gasketing characteristics
- good cake discharge

Thomas and Kebea [15] investigated the filtration properties of wet-laid polyester nonwoven filters. Their results showed that as the fiber diameter increases, filtration properties become less efficient.

In order to reduce the energy requirements, the pressure drop across the filter must be kept low which can be achieved by materials with large void volume. Since textile materials have a large number of air passages, they are particularly suitable for filtration applications. The void volume (the volume in the fabric that is not occupied by the fibers) is around 70% for woven and knit fabrics and can be as high as 98% for some nonwovens. Nonwovens offer the possibility of disposable filters.

All air filters change the direction of air flow and induce turbulence. This causes an unavoidable pressure drop and loss of energy. The pressure drop causes the compressed air to perform less work. Excessive pressure drop requires more work from the fan to give the desired flow of air although air flow is not usually changed. Low air flow causes poor heat transfer between the coils and the air in an HVAC system. This causes the compressor to have a higher operating pressure and shortens the life of the system. It also results in increased energy costs in operating the system [16]. A lower pressure drop gives higher air flow and better heat transfer. Thus, reduced pressure drop provides more energy efficiency and overwhelming savings in power usage. In many cases, more efficient filtration results in a higher pressure drop. Finer filtration, however, does not necessarily cause a decrease in air flow and higher pressure drop. A balance must be reached between the efficiency of heating/cooling and filtration efficiency.

7. TESTING

Filtration tests are conducted to measure the filtering capacity of fabrics for the intended suspension in liquid filtration. In a filtration test, the relationship between filter delivery and filtration time is determined using different pressures and liquid suspension temperatures. Other factors that need to be considered for commercial liquid filtration systems include the rate of filter choking, service life of the filter cloth, filtrate purity and cake removal.

The efficiency of a filter has a direct relation with the particle size. The larger the particle size for a particular filter, the higher the efficiency. The high efficiency of fibrous filters is obtained by the design of the fiber system such as size of fiber, orientation and packing [3].

Important test characteristics with fabrics used

in gas and dust filtration include permeability, differential pressure, efficiency, strength and chemical resistance. These parameters are all tested, particularly if fabrics are failing [9].

Efficiency of air filters measures the ability of the filter to remove particulate matter from an airstream. Arrestance is one type of efficiency measurement that determines the amount of particulate matter that is retained in a filter when introduced into the airstream [17]. The ASHRAE Standard Test Method 52-76 is widely used for arrestance measurements. In this method, a clean filter sample is placed into a wind tunnel.

A known amount of ASHRAE Synthetic Dust is then introduced into the airstream through a funnel at the top of the duct upstream of the filter. The dust is fed at a controlled rate until a uniform distribution of particles covers the filter. The filter remains in the airstream for five minutes to allow any particles not trapped in the filter media to pass through. The filter is removed and weighed to determine the increase in weight due to the added dust. Finally, the percentage of dust held (arrestance) is calculated by dividing the difference in weight of the filter before and after testing by the amount of dust injected.

References and Review Questions

1. REFERENCES

1. Morris, W. J., "Filtration Fabrics," *Textiles*, Vol. 14, No. 3, 1985.

2. Chatterjee, K. N., "Polyester Needle Punched Nonwoven Dust Filter for Controlling Air Pollution," *Man-Made Textiles in India*, Vol. 35, January 1991.

3. Sinha, S. K., "Use of Textiles in Protecting Environment," *Textile Trends*, Vol. 20, November 1992.

4. Petty, W. C., "Air Filtration: An Often Overlooked Productivity Factor," *Hydraulics and Pneumatics*, Vol. 38, June 1986.

5. Chatterjee, K. N., "Air Permeability of Nonwoven Filter Fabrics," *The Indian Textile Journal*, Vol. 102, April 1992.

6. 3M Filtration Products, 1992.

7. Floyd, K. L., "Textiles in Filtration," Textiles, Vol. 4, 1975.

8. Bergmann, L., "Nonwovens for Filtration Media," *TAPPI Journal*, January 1989.

9. Bergmann, L., private communication.

10. Castillo, J. E., "Fiberglass: Information You Should Know," *Professional Safety*, Vol. 37, November 1992.

11. Bunn, W. B. et al., "An Update on the Health Effects of Man-Made Vitreous Fibers and Potential Fiber Exposures," *TAPPI Journal*, Vol. 74, June 1991.

12. Bergmann, L., "Melt Blown Nonwovens in Filtration," *Nonwovens Industry*, Vol. 23, February 1992.

13. Montgomery, K. D., "Comparison of ASH-RAE Dust Spot and Fractional Aerosol Efficiencies of Air Filters," *TAPPI Journal*, February 1994.

14. Offerman, F. J. et. al., "Performance of Air Cleaners in a Residential Forced Air System," *ASHRAE Journal*, Vol. 34, September 1992.

15. Thomas, B. P. and Kebea, R., "Filtration Properties of Wet-Laid Polyester Non-wovens," *TAPPI Journal*, November 1988.

16. Decowsky, G. M., "Energy Conservation Idea: Air Filter Pressure Drop Costs Money," *Hydraulics and Pneumatics*, Vol. 36, September 1983.

17. "Test Performance of Air Filter Units," Underwriters Laboratories, UL Standard 99-77, 1977.

1.1 General References

Broughton, R. M., private communication.

Rothwell, E., "Design and Operating Characteristics of Reverse-Jet Assemblies for Pulse-Jet Dust Collectors," *Filtration and Separation*, July/August 1988.

Malchesky, P. S., "Applying Filtration and Separation Techniques for Improved Health," *Filtration and Separation*, March/April 1989.

Reisch, M. S., "High-Performance Fibers Find Expanding Military, Industrial Uses," *Chemical and Engineering News*, February 2, 1987.

Svedova, J., ed., *Industrial Textiles*, Elsevier 1990.

Golesworthy, T. J. and Pragnell, R. J., "Filter Fabric Selection for Coal-Fired Flue Gases," *Filtration and Separation*, November and December 1990.

Dickenson, C., *Filters and Filtration Handbook,* 3rd Edition, Elsevier, 1994.

Rothwell, E., "Who Needs Better Filter Media," *Filtration and Separation,* March/April 1987.

Passant, F. H., "Gas Cleaning in the Nuclear Industry," *Filtration and Separation,* May/June 1987.

Cheremisinoff, N. P. and Azbel, D. S., *Liquid Filtration,* Ann Arbor Science, 1983.

Hunter, J. S. III, "Recent Developments in Buoyant Media Liquid Filtration," *Filtration and Separation,* November/December 1987.

Molter, W. et al., "Fast, Automated Testing of Different Air Filter Media," *Filtration and Separation,* January/February 1992.

Callis, R., "Practical Application of High Temperature Filters," *Filtration and Separation,* July/August 1991.

Versen, R. A., "Clearing the Air about Fiber-Glass Air Filters," *Filtration and Separation,* March/April 1994.

Morgan, H., "Cost-Effective Clean Room Designs," *Filtration and Separation,* January/February.

Matteson, M. J., and Orr, C., eds., *Filtration Principles and Practices,* 2nd Edition, Marcel Dekker, Inc., 1987.

2. REVIEW QUESTIONS

1. Why are textile structures effective in filtration? Explain.

2. What are the main fabric requirements for dry and liquid filtration?

3. What are the advantages and disadvantages of woven and nonwoven fabric structures for filtration?

4. Make a table showing application areas of textile filter media along with fibers and fabric types that can be used in each application.

5. What are the major test methods to measure the efficiency of filters?

GEOTEXTILES

Geotextiles

S. ADANUR

1. INTRODUCTION

Geotextiles are a member of a larger family called geosynthetics. Other members of the geosynthetics family are geogrids, geonets, geomembranes, and geocomposites. Figure 9.1 shows samples of various geosynthetic materials. Geotextiles, which are true textile structures, are the subject of this section. Brief definitions of other geosynthetic family members are given at the end of this chapter.

Geotextiles are permeable textile structures made of polymeric materials and are used mainly in civil engineering applications in conjunction with soil, rock or water. The Romans used a primitive form of geotextiles for their road construction. The first modern-day commercial geotextiles, which also were known as "filter fabrics," were used for erosion control in the 1950s. The first geotextiles were woven monofilament fabrics with a high percentage open area. Nonwoven needlepunched fabrics were introduced in Europe in the late 1960s. The first nonwoven (thermally bonded and needled) geotextiles were introduced in the United States in 1972.

Geotextiles are the largest group of geosynthetics in terms of volume. They are used in geotechnical engineering, heavy construction, building and pavement construction, hydrogeology, and environmental engineering. Geotextiles are designed for one or more of the following functions (Figure 9.2):

- separation
- reinforcement
- filtration
- drainage
- protection

When impregnated with polymeric sealing materials, geotextiles act as moisture barriers as well. Table 9.1 shows a summary of the application of geosynthetics.

Properties of commercially available woven and nonwoven geotextiles are given in Appendix 3.

2. GEOTEXTILE MATERIALS AND MANUFACTURING

Geotextiles are traditional textile products such as woven and nonwoven fabrics. Knit fabrics are hardly used as geotextile materials. Recently, special structures were developed with the same applications as geotextiles. These new types of materials are generally very coarse compared to conventional-type geotextiles. Examples of these new structures are webbings, mats and nets.

The most important factors in manufacturing of geotextiles are polymer-type, fiber-type, fabric design and type of bonding (for nonwovens). Polypropylene and polyester are the two most widely used polymers in geotextiles. Polyamide (nylon) and polyethylene are used to a much lesser extent. Monofilament, multi-filament, staple and slit film yarns are used in geotextiles. Table 9.2 summarizes geotextile materials and manufacturing methods.

Common weave types for geotextiles are plain, basket and twill weaves. Plain weave is the most common. Geotextiles are woven on wide industrial looms. Figure 9.3 shows woven monofilament and slit film fabrics.

Nonwoven manufacturing of geotextiles includes fiber production (spinning), fiber preparation, web formation, web bonding, and

FIGURE 9.1 Geosynthetic samples (courtesy of IFAI Geotextile Division).

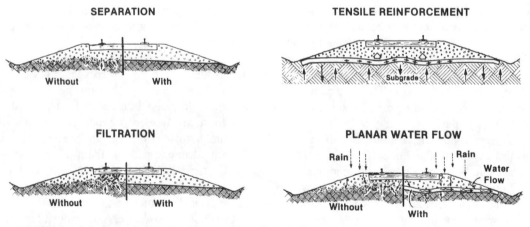

FIGURE 9.2 Schematic of railroad bed construction with and without geotextile (courtesy of Hoechst Celanese).

TABLE 9.1 Geosynthetics Application Summary [1].

Application	Primary Function	Products
Subgrade stabilization	Separation/reinforcement/filtration	Geotextile/geogrid
Railroad track bed stabilization	Drainage/separation/filtration	Geotextile
Sedimentation control silt fence	Sediment retention/filtration/separation	Geotextile
Asphalt overlay	Waterproofing/stress-relieving layer	Geotextile
Soil reinforcement		
Embankments	Reinforcement	Geotextile/geogrid
Steep slopes	Reinforcement	Geogrid/geotextile
Vertical walls	Reinforcement	Geogrid/geotextile
Erosion control filter	Filtration/separation	Geotextile
Subsurface drainage filter	Filtration	Geotextile
Geomembrane protection	Protection/cushion	Geotextile
Subsurface drainage	Filtration/fluid transmission	Prefabricated drainage composites
Surficial erosion control	Turf reinforcement	Erosion control mats

TABLE 9.2 Geotextiles and Manufacturing Methods [2].

	Structure	
	Nonwoven	Woven
Raw material	Polypropylene, polyester, polyethylene, polyamide (nylon), etc.	
Fiber type	Continuous filament	Monofilament Multi-filament Slit film Fibrillated
	Staple filament	Fibrillated
Bonding process	Needlepunching Heat bonding	Weaving

Mass per unit area: a basic identifier typically ranging from 3 oz/yd^2 to 16 oz/yd^2.

FIGURE 9.3 Woven geotextiles (courtesy of Hoechst Celanese).

finishing. Filaments or short fibers are first arranged into a loose web, then bonded together by chemical, thermal or mechanical bonding processes (Section 4.8, Nonwoven Fabrics).

In a chemical bonding process, which is used the least, a cementing medium such as glue, latex or resin is added to bind the filaments or short fibers together. The impregnated web is then cured and/or calendered.

In the thermal bonding (also called melt or heat bonding) process, filaments or short fibers are subjected to heat and melted at their crossover points (Figure 9.4). Thermal bonding can be achieved with or without additives. If the melting temperature of the fibers in the web is relatively low, they can be easily melted at their crossover points under heat without additives. Alternately, if the melting temperature of the web fibers is high, bonding additives such as powder and granules of fibers can also be used which melt at lower temperatures and bond the web fibers together. However, this method is rarely used. The most common fusing methods are through air or steam heating and calender bonding. In air heating, hot air is used to melt the fibers which produces high loft, low density fabrics in general

[3]. Hot air is either blown through the web on a conveyor belt or sucked through the web over a porous drum. In calender bonding, the web is drawn between cylinders under heat and pressure. Calendering gives strong, stable and low loft fabrics. Binder materials can be homofil (monocomponent) and heterofil (bicomponent or sheath/core spun). The heat bonded nonwoven fabrics are normally relatively thin.

Mechanical bonding is done by a needlepunching process. In this method thousands of barbed needles, attached to a board, are punched through the loose web and withdrawn, leaving filaments entangled. The needlepunched nonwoven geotextiles are relatively thick (Figure 9.5).

Spunbonding, which combines fiber spinning, web forming, web bonding and finishing in a continuous process, is widely used to produce geotextiles from polymers. Spunbonding process may be followed by needlepunching process for some geotextiles.

Special fabrics such as webbings, mats and nets are generally coarser than classical geotextiles. Webbings are very coarse woven fabrics made of strip-like fibers. A webbing looks like a very coarse slit film woven fabric. Mats are

FIGURE 9.4 Heat bonded nonwoven geotextile (courtesy of Hoechst Celanese).

FIGURE 9.5 Needlepunched nonwoven geotextile (courtesy of Hoechst Celanese).

made of coarse and rather rigid filaments with a tortuous shape, bonded at their intersections. Some mats look like a very coarse and open nonwoven fabric. Nets consist of two sets of coarse, parallel extruded strands (using rotating spinnerettes) intersecting with a constant angle. Strands of one setting are connected to strands of the other set by partial melting at the intersections. When designing and manufacturing geotextile fabric for a particular civil engineering application, construction considerations need to be evaluated very closely.

3. GEOTEXTILE PROPERTIES AND TESTING

Numerous geotextile fabric and fiber properties are critical for the performance of geotextiles in the field. All the properties of geotextiles fall under three major categories.

- intrinsic properties, such as physical and mechanical properties, that are called properties of a geotextile in isolation
- geotextile properties that influence soil-geotextile interaction
- endurance properties

Geotextile tests can be two types: index tests and performance tests. An index test generally does not produce an actual property value but provides a value or indicator from which the property of interest can be qualitatively assessed. Index tests can be used as a means of product comparison and for specifications and quality control evaluation. They are usually rapid and efficient to perform. Performance tests require testing of the geotextile with soil to obtain a direct assessment of the property of interest. They provide a direct measure of the soil influence on the particular geotextile property and vice versa. Performance tests should be done in conjunction with the site soils under specific design conditions.

3.1 Physical Properties

The important physical properties of geotextiles are thickness, mass/unit area and flexibility. Specific gravity may become an important parameter for some applications involving water [4]. If the specific gravity of the geotextile is less than one, the fabric may float in water. Specific gravity of a geotextile fiber is almost equal to the specific gravity of the base polymer. Specific gravities of some geotextile materials are: polypropylene $= 0.90$, polyester $= 1.38$, nylon $= 1.14$, polyethylene $= 0.92$ (Table 17.2).

The geometry of geotextiles is essentially characterized by the fact that the thickness is much smaller than the two other dimensions, the length and the width. In other words geotextiles are bidimensional products. Thickness of a geotextile is measured at a specified pressure of 2.0 kPa [5]. Thickness is important for in-plane water flow.

The mass per unit area (oz/yd^2 or g/m^2) characterizes the quantity of material in the direction perpendicular to the plane of a geotextile [6]. In general, fabric cost and some mechanical properties are related to mass per unit area.

Most geotextiles are very flexible in nature with very low bending resistance or stiffness as a result of the mechanical property of the fibers used. In stiffness or flex test [7], the geotextile gravitationally bends under its own weight. Flexibility of geotextiles enables them to be produced and shipped into rolls and installed easily. Functionally, reinforcement may be important in reinforcement applications.

3.2 Mechanical Properties

Important mechanical properties of geotextiles are compressibility, tensile strength, seam strength, tear, burst, impact and puncture strength.

Compressibility of woven and thermally bonded nonwoven geotextiles is low. Needlepunched geotextiles have high compressibility. Compression of a geotextile reduces its transmissivity (in-plane flow) and permittivity (cross-plane flow) since it changes the pore structure of the fabric.

Tensile properties are important properties of geotextiles. Adequate tensile strength is necessary for all types of geotextile functions and installation. In general, maximum tensile stress, elongation at break, toughness and modulus values are reported from tensile tests. Toughness is the work done per unit volume before fabric failure which is the area under the stress-strain curve. Figure 9.6 shows typical load-elongation curves of woven and nonwoven geotextiles. Various specimen sizes for tensile testing of geotextiles are indicated in ASTM D 4595 (wide

width strip method) and ASTM D 4632 (grab method). Mostly $4''$ and $8''$ wide samples are used for tensile testing ($20''$ wide samples may also be used). Figure 9.7 shows the wide-width tensile test of a nonwoven geotextile sample. A wide specimen is especially needed for nonwovens because of their necking during testing.

In confined tensile tests, the specimen is surrounded by soil to simulate the in situ conditions. Figure 9.8 shows a plot of confined and unconfined load-elongation curves for a woven and needlepunched nonwoven geotextile. Confinement significantly increased the load values for the nonwoven geotextile.

Geotextile fabric ends are seamed together for property continuity either in a shop or in the field. Various types of seams are used for different fabrics. Seaming is usually done by sewing the ends together. Other methods of seaming include epoxy, resin, and mechanical seaming. Several sewn seam types are shown in Figure 9.9.

Although 100% seam efficiency is desired, it is hardly possible. 90% seam efficiency is considered to be good. Seam efficiency (SE) is defined as

$$SE\ (\%) = (T_s/T_g) \times 100 \qquad (9.1)$$

where T_s is the tensile strength of the seam area and T_g is the tensile strength of the geotextile without a seam. Seam efficiency decreases with increasing fabric strength. In testing of sewn

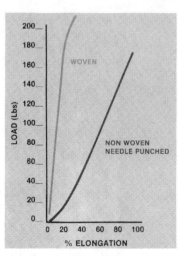

FIGURE 9.6 Typical tensile load-elongation diagrams of woven and nonwoven geotextiles [8].

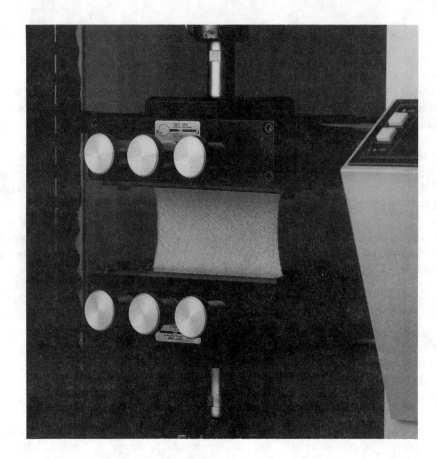

FIGURE 9.7 Wide-width tensile test of a geotextile [9].

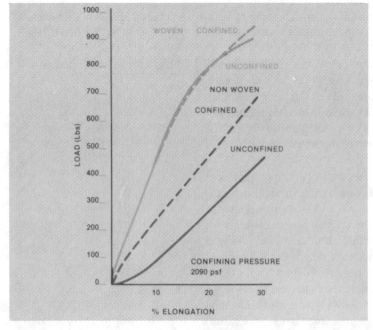

FIGURE 9.8 Load-elongation curves of confined and unconfined geotextiles [8].

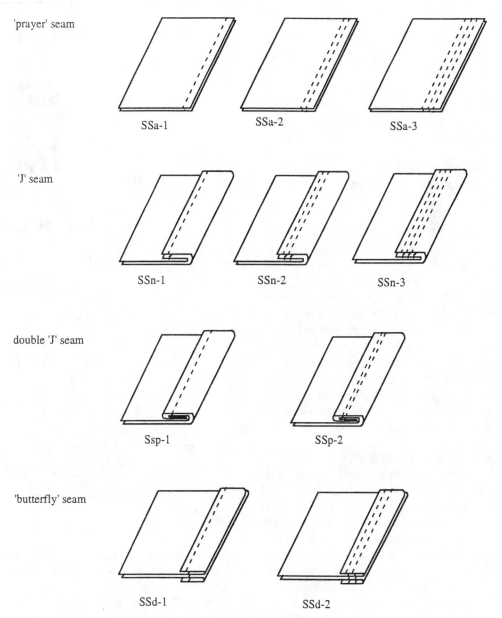

'prayer' seam

SSa-1   SSa-2   SSa-3

'J' seam

SSn-1   SSn-2   SSn-3

double 'J' seam

Ssp-1   SSp-2

'butterfly' seam

SSd-1   SSd-2

FIGURE 9.9 Sewn seam types for geotextiles [10].

geotextile seam strength (ASTM D 4884), an eight inch wide seam is gripped in its entire width with clamps of a tensile testing machine. A longitudinal (perpendicular) force is applied at a specified rate of extension until the seam or the fabric ruptures.

Properties related to the ability of a geotextile to withstand damage are typically the resistance to tear and puncture. ASTM D 4533 is used for trapezoidal tear test of geotextiles. The puncture test is done according to ASTM D 4833. Figure 9.10 shows schematics of some of the mechanical tests for geotextiles.

3.3 Properties Influencing Soil-Geotextile Interaction

Two kinds of interactions are possible between soil and geotextile: mechanical and physical. The mechanical interaction is characterized by the shear strength developed between soil and geotextile. Soil to geotextile friction is very critical in reinforcement. A high contact shear strength is required when geotextile is used to reinforce soil and a low contact shear strength is required when soil and geotextile are designed to move against each other. The shear strength is governed by the angle of internal friction developed between soil and geotextile. Two different tests are done to test the friction behavior of geotextiles: direct shear and pullout tests (Figure 9.11).

In direct shear (ASTM D 5321), the geotextile is fixed to the bottom half of the shear box which moves against the other half that contains soil under normal stress. Shear forces versus normal forces are plotted from which shear strength parameters can be obtained. The pullout test is carried out with the geotextile embedded in the soil. This test is usually performed with a normal stress or surcharge, which represents normal stress in the field, applied above the topsoil layer. The horizontal force applied to pull the geotextile out of the soil is recorded as shown in Figure 9.12. Pullout resistance is calculated by dividing the maximum load attained by the test specimen width. Plot of pullout resistance versus deformation at different points along the geotextile can be obtained for different applied normal stresses. From this information, response curve interac-

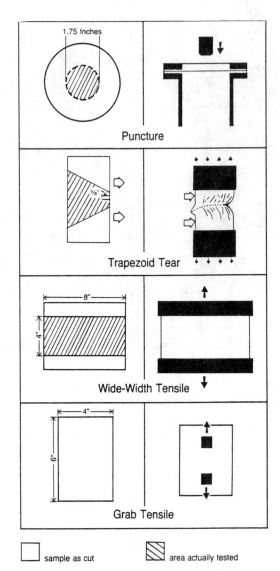

FIGURE 9.10 Schematics of some mechanical tests for geotextiles [1].

tion coefficients can be generated. ASTM is currently developing a standardized pullout test method.

Swan [11] studied the effect of geotextile geometry on the shear strength parameters of a geotextile. He showed that the fabric geometry greatly influences the shear strength parameters of a geotextile for a given soil type. His study also showed that the geotextiles having similar strength, construction and yarn geometry can have different shear strength in a particular soil due to difference in geotextile geometry.

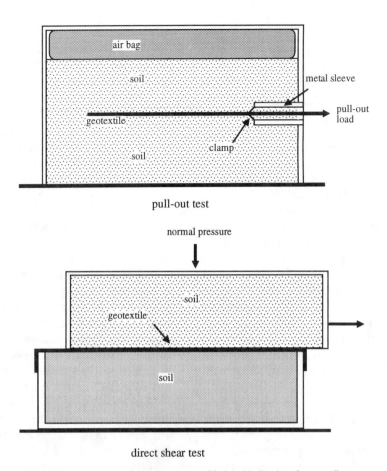

FIGURE 9.11 Test methods to measure frictional behavior of geotextiles.

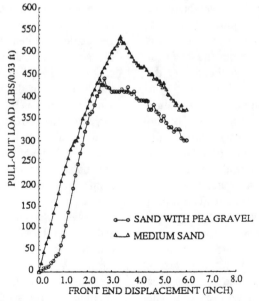

FIGURE 9.12 Typical pullout load versus displacement
curves for woven geotextiles (Adanur, Mallick and Zhai).

307

The physical interaction is characterized by the hydraulic properties of the geotextile. These properties are porosity, percent open area (POA), apparent opening size (AOS), permittivity and transmissivity of a geotextile. Porosity, which is the ratio of void volume to total volume, indicates the total void volume of the geotextile available for establishing effective filtration with the in situ soil. Pore sizes of some geotextiles may change under normal load in the field. POA is the ratio of open area to total fabric area which is a valid measurement for woven monofilament geotextiles only. Monofilament geotextiles have a POA range of $2-60\%$. Apparent opening size or Equivalent Opening Size (EOS) is measured for woven and nonwoven geotextiles. To determine the AOS with ASTM D 4751, a geotextile sample is placed in a sieve frame. Sized glass beads are placed on the surface of the geotextile. Several sizes of glass beads are used. The sample is tested with the smallest bead first. The geotextile, glass beads and frames are shaken laterally. The jarring motion causes the beads to pass through the sample. The beads remaining on the geotextile surface as well as the ones that fall off as a result of turning the specimen over and tapping the rims of the sieves are weighed. The beads that pass through the sieve are also weighed. The procedure is repeated using succeedingly larger beads until the weight of the beads passing through the sample is 5% or less. A plot of percent passing versus particle size is obtained. The AOS of each specimen is the U.S. sieve number which represents the size of the beads of which 5% or less pass. The AOS for the test is determined by averaging the AOS values for five specimens. This test has some limitations. For example, beads may stick to the geotextile due to the development of static electricity. Bead clumping may also occur.

Permittivity or cross-plane water flow is an important index test for filtration. The mechanism by which water moves through the geotextile is characterized by the permittivity of geotextile. ASTM test method D 4491 is used to determine the water permeability of a geotextile as the permittivity. The method consists of a falling head test and a constant head test. The constant head test is used when the flow rate of water through the geotextile is so great that it is

difficult to obtain a reading of head change versus time in the falling head test. In the falling head test, a column of water is allowed to flow through the geotextile. Readings of head changes versus time are recorded. The flow rate in this test must be slow enough to obtain the accurate readings. In the constant head test, a head of 5 cm of water is maintained on the geotextile throughout the test. The quantity of flow versus time is measured.

Since permittivity may change for some geotextiles under load, a test method for geotextile permittivity under load is being developed by the ASTM. This method provides a means of measuring the water flow in the normal direction through a known cross section of a geotextile sample. The test is conducted under specified constant hydraulic head over various applied normal compressive stresses. This procedure assumes that Darcy's law is valid, which implies saturated conditions and laminar flow. This procedure is normally performed using deaerated water.

Filtering efficiency and flow rate of a geotextile for silt fence applications using site specific soil are determined using ASTM D 5141. This test procedure is useful in determining the spacing between silt fences or barriers. It provides a way to evaluate geotextiles with different soils under a variety of conditions that simulate silt fence or barrier installation. This method may be used to simulate several storm events on the same geotextile specimen. The test procedure consists of placing a geotextile sample across a flume and passing sediment laden water through it. The time it takes for the water to flow through the sample as well as the amount of soil passed by the geotextile is recorded. This information is used to calculate the amount of soil retained, filtering efficiency, and the flow rate. This test method should be carried out with soil obtained from the construction project site.

Transmissivity of a geotextile characterizes the flow of water in the plane of the geotextile and is generally very small. Transmissivity of a geotextile depends on the cross section of the fabric. Woven and nonwoven heat bonded geotextiles are relatively thin and as a result have negligible transmissivity, while transmissivity of needlepunched fabrics is slightly higher. ASTM D 4716 covers constant head hydraulic transmissivity (in-plane flow) of geotextiles and geotextile related products. It is used to determine the hydraulic transmissivity of a sample. This is determined by measuring the quantity of water that passes through a test specimen in a specific time interval under a specific normal stress and a specific hydraulic gradient. The hydraulic gradient and the specimen contact surfaces are selected to simulate a given set of field parameters. Measurements are repeated under normal stresses ranging from 1.45 psi to a maximum specified value. Figure 9.13 shows the schematic of the test.

3.4 Endurance Properties

These properties are related to the resistance of a geotextile to progressive deterioration. Durability of a geotextile depends on the stress exerted by the external media and the base material used to make the geotextile, which is usually polymer. The geotextile can be subjected to two types of external actions, mechanical and physicochemical (degradation).

Abrasion is defined as the wearing away of any part of a material by rubbing against another surface. Abrasion can be caused by large soil particles such as gravel or the failure of soil-geotextile system. The abrasion resistance test of geotextiles is done using the sliding block method (ASTM D 4886). In this method a sample is mounted on a stationary platform. The sample is rubbed by an abradent which has specific surface characteristics. The abradent is rubbed with a controlled amount of pressure and abrasive action on a horizontal axis using uniaxial motion.

The resistance to abrasion is determined and is expressed as a percentage of the breaking load before abrading.

Physicochemical degradation of geotextiles can be caused by long term confinement under sustained load or exposure to the atmospheric conditions such as ultraviolet ray, moisture, different chemicals (ASTM D 5322), etc. When a geotextile is subjected to a constant long-term load, the base polymer in the geotextile suffers from long-term creep deformation. Creep of a geotextile is an important property for reinforcement application because, as the geotextile undergoes creep deformation, its thickness decreases with time. In tension creep testing of geotextiles (ASTM D 5262), the geotextile sample is maintained at a specific temperature and humidity. Unless otherwise mentioned, a temperature of $21 \pm 2°C$ and relative humidity of $50\% - 70\%$ shall be used. A sustained load is applied to the sample in one step, and the total elongation of the sample is measured as a function of time. From this information, the tension creep behavior of a geotextile can be measured. The creep behavior of geotextiles is greatly influenced by soil confinement conditions.

Another important performance property of a geotextile is its long-term drainage capability or its resistance to clogging. The soil-geotextile system clogging potential is determined using the ASTM D 4716 hydraulic transmissivity test method or ASTM D 5101 gradient ratio test method. The latter test method requires setting up a cylindrical clear plastic permeameter with a geotextile and soil. Water is passed through this system with varying differential heads and site-

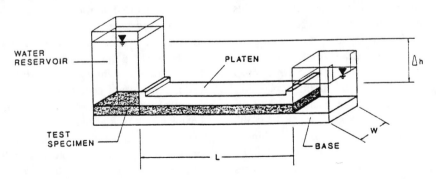

FIGURE 9.13 Constant head hydraulic transmissivity testing device (copyright ASTM, reprinted with permission [12]).

TABLE 9.3 ASTM Standard Test Methods for Geosynthetics [12].

D 1987	Test method for biological clogging of geotextile or soil/geotextile filters
D 3020	Specifications for polyethylene and ethylene copolymer plastic sheeting for pond, canal and reservoir lining
D 3083	Specification for flexible poly(vinyl chloride) plastic sheeting for pond, canal, and reservoir lining
D 4354	Practice for sampling of geosynthetics for testing
D 4355	Test method for deterioration of geotextiles from exposure to ultraviolet light and water (xenon arc-type apparatus)
D 4437	Practice for determining the integrity of field seams used in joining flexible polymeric sheet geomembranes
D 4439	Terminology for geotextiles
D 4491	Test methods for water permeability of geotextiles by permittivity
D 4533	Test method for index trapezoid tearing strength of geotextiles
D 4545	Practice for determining the integrity of factory seams used in joining manufactured flexible sheet geomembranes
D 4594	Test method for effects of temperature on stability of geotextiles
D 4595	Test method for tensile properties of geotextiles by the wide-width strip method
D 4632	Test method for grab breaking load and elongation of geotextiles
D 4716	Test method for constant head hydraulic transmissivity (in-plane flow) of geotextiles and geotextile related products
D 4751	Test method for determining apparent opening size of a geotextile
D 4759	Practice for determining the specification conformance of geosynthetics
D 4833	Test method for index puncture resistance of geotextiles, geomembranes, and related products
D 4873	Guide for identification, storage, and handling of geotextiles
D 4884	Test method for seam strength of sewn geotextiles
D 4885	Test method for determining performance strength of geomembranes by the wide-strip tensile method
D 4886	Test method for abrasion resistance of geotextiles (sandpaper/sliding block method)
D 5101	Test method for measuring the soil-geotextile system clogging potential by the gradient ratio
D 5141	Test method to determine filtering efficiency and flow rate of a geotextile for silt fence application using site-specific soil
D 5199	Test method for measuring nominal thickness of geotextiles and geomembranes
D 5261	Test method for measuring mass per unit area of geotextiles
D 5262	Test method for evaluating the unconfined tension creep behavior of geosynthetics
D 5321	Test method for determining the coefficient of soil and geosynthetic or geosynthetic and geosynthetic friction by the direct shear method
D 5322	Practice for immersion procedures for evaluating the chemical resistance of geosynthetics to liquids
D 5323	Practice for determination of 2% secant modulus for polyethylene geomembranes

specific soils. Measurements of differential heads and flow rates are taken at different time intervals, which are used to determine hydraulic gradient.

Common causes of degradation of geotextiles are temperature, chemical, hydrolysis, biological, sunlight degradation and polymer aging. Deterioration of geotextiles from exposure to ultraviolet light and water can be determined according to ASTM D 4355. In this procedure, geotextile specimens for the machine and cross-machine directions are exposed for 0, 150, 300 and 500 hours of ultraviolet exposure in a xenon-arc device. The exposure consists of 120 minute cycles. The cycles are made up of 102 minutes of light only, followed by 18 minutes of water spray and light. Five samples are tested for each

total exposure time for each direction. After the samples have been exposed, they are subjected to a strip tensile test. These test results are compared to the test results for five unexposed control samples to determine the deterioration that has occurred due to ultraviolet exposure. Results are reported as percent strength retained.

ASTM D 4594 determines the effects of temperature on stability of geotextiles. This procedure consists of conditioning geotextile samples at selected temperatures in an environmental chamber that is attached to a tensile testing machine. The temperatures selected are the temperatures at which the geotextiles will perform or are typically exposed in the field. A two inch strip tensile test is performed on the samples as specified in ASTM D 5035, while the selected temperatures are maintained. Control samples are also tested under standard laboratory test conditions. Tensile strength and percent elongation properties are recorded at various test temperatures. Strength changes due to the effects of temperature are determined. Chemical degradation is a major concern since it may potentially cause the failure of the geotextile. Table 9.3 summarizes the geotextile test methods approved by the ASTM.

4. GEOTEXTILE FUNCTIONS

The main functions of geotextiles are separation, reinforcement, stabilization, filtration, drainage, waterproofing and protection. Table 9.4 shows geotextile application areas versus geotextile functions.

4.1 Separation Function

Geotextiles are used to separate two dissimilar materials such as two layers of soil with different properties. The purpose of separation is to maintain or improve the integrity and performance of both materials. Figure 9.14 shows examples of separation function where the geotextile keeps the soil particles from migrating and mixing with the coarse aggregate. The geotextile ensures that the aggregate maintains its load-bearing ability

(Figure 9.15). Without geotextile, subsoil and aggregate base intermix and load-bearing capacity are reduced. With geotextile as a separator between subsoil and aggregate base, load-bearing capacity is maintained.

In almost every application, geotextiles play a separation role. Important geotextile property requirements for separation are tensile strength, puncture, tear, AOS and permittivity. If the geotextile is locked into position by the material above and below, it may be subjected to a lateral in-plane tensile stress. This tensile stress is caused by the stones just above the fabric. Table 9.5 shows the parameters of influence for separation function at different stages of construction. Table 9.6 lists the major geotextile strength requirements for separation.

4.2 Reinforcement and Stabilization Function

Geotextiles are high tensile strength materials and soils, in general, are low tensile strength (but high compression strength) materials. Therefore, geotextiles are ideal materials to increase soil quality and thus to increase soil structural stability. Figures 9.16 and 9.17 show reinforcement applications of geotextiles.

Geotextile increases load-bearing capacity by providing tensile mechanism to the soil. This is

TABLE 9.4 Geotextile Application Areas versus Geotextile Functions [2].

Areas of Application	Separation	Filtration	Drainage	Reinforcement	Protection	Waterproofing
Paved and unpaved roads						
Wet, soft subgrade	X	X	X	O		
Film subgrade	X	O	O	O		
Repaving				O	O	X
Drainage	O	X	O			
Sports fields	X	X				
Erosion control/hydraulic construction	O	X				
Railroads	X	X				
Containments	O	X	O	O	X	O
Embankments on soft soils	X	X	X	O		
Reinforced soil walls and slopes			O	X		
Tunnels			X		X	

Symbols—X: primary function, O: secondary function.

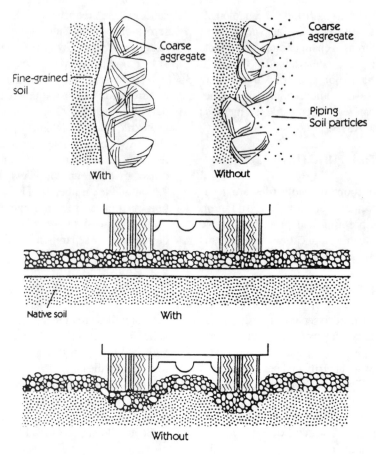

FIGURE 9.14 Separation function of geotextiles (courtesy of Amoco).

FIGURE 9.15 Effect of geotextile as a separator (courtesy of Reemay).

TABLE 9.5 *Parameters of Influence for Separation Function [2].*

Application Condition	Parameters		
	Mechanical	Hydraulic	Long-Term Performance
During installation	Impact resistance	Effective opening size	UV resistance
	Elongation at break	Mass per unit area	
During construction	Puncture resistance	AOS	Chemical stability
	Burst pressure resistance	Mass per unit area	UV resistance
	Elongation at break		
After completion of	Puncture resistance	AOS	Chemical stability
construction	Burst pressure resistance	Mass per unit area	Resistance to decay
	Tear propagation resistance		
	Elongation at break		

TABLE 9.6 *American Association of State Highway and Transportation Officials (AASHTO) M288 Separation Strength Requirements [13].*

Property	Test Method	High Survivability Level	Medium Survivability Level
Grab strength (lbs)	ASTM D 4632	270/180	180/115
Elongation (%)	ASTM D 4632	< 50%/ > 50%	< 50%/ > 50%
Seam strength (lbs)	ASTM D 4632	240/160	160/105
Puncture strength (lbs)	ASTM D 4833	100/75	70/40
Trapezoid tear (lbs)	ASTM D 4533	100/75	70/40

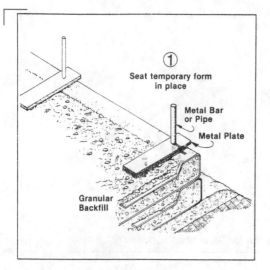

① Seat temporary form in place

Metal Bar or Pipe

Metal Plate

Granular Backfill

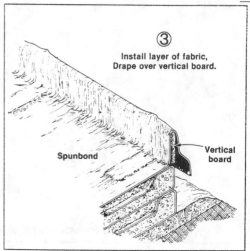

③ Install layer of fabric, Drape over vertical board.

Spunbond

Vertical board

② Place vertical form board.

Hand Holes or Handles 3 to 4 Ft. Centers

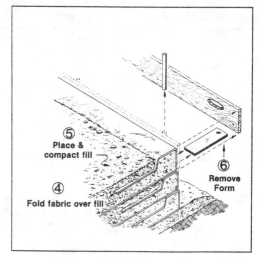

⑤ Place & compact fill

④ Fold fabric over fill

⑥ Remove Form

FIGURE 9.16 Reinforcement function of geotextiles: geotextile wall construction steps (courtesy of Hoechst Celanese).

FIGURE 9.17 Geotextile used to stabilize a steep highway embankment (courtesy of Polyfelt).

TABLE 9.7 *Parameters of Influence for Reinforcement Function [2].*

Application Condition	Parameters		
	Mechanical	Hydraulic	Long-Term Performance
Slope failure	Geotextile tensile strength	Hydraulic boundary	Creep of the geotextile-soil
	Geotextile-soil friction	conditions	system
	Confined strength-elongation	In-plane permeability	Chemical and decay
			resistance
Base failure	Tensile strength	Hydraulic boundary	Chemical and decay
	Geotextile-soil friction	In-plane permeability	resistance

achieved through a bonding mechanism to the geotextile-soil system. Three different mechanisms can play a role during a geotextile reinforcement function: membrane-type, shear-type and anchorage-type. In membrane reinforcement, a vertical load is applied to a geotextile on a deformable soil. In shear reinforcement, a geotextile that is placed on a soil is loaded in the normal direction and two materials are sheared at their interface, similar to direct shear tests. In anchorage reinforcement, a tensile force is applied to the geotextile that is surrounded by soil on both sides (similar to pullout tests).

Geotextile requirements for reinforcement function are mechanical (shear strength, tensile strength), hydraulic and durability (chemical and decay resistance) as shown in Table 9.7.

4.3 Filtration Function

In filtration function, geotextile acts as a filter by allowing free liquid flow through its plane and by retaining soil particles on the upstream side. Therefore, the geotextile must meet two contradictory demands simultaneously: open structure for flow of water and tight fabric structure for soil retention. Another important factor is the long-term soil-fabric system flow compatibility

that will not clog during the operation period. As a result, important geotextile properties for filtration function are cross-plane permeability (permittivity), soil retention and compatibility over an indefinitely long period of time. Table 9.8 shows geotextile requirements for filtration function.

Typical application areas of geotextiles for filtration are pipe underdrains, drainage for retaining walls and erosion control structures as shown in Figures 9.18 and 9.19.

4.4 Drainage Function

The drainage function of a geotextile involves transmission of liquid in the plane of fabric without soil loss. Figure 9.20 shows an earth dam with geotextile chimney drain. The major difference between filtration function and drainage function is the direction of flow which makes in-plane permeability (transmissivity) critical for the drainage function. Drainage refers to planar flow as opposed to filtration which refers to flow across the geotextile. The soil retention and long-term compatibility requirements are similar to filtration [14].

Table 9.9 shows geotextile requirements for drainage function. Fabrics must be relatively

TABLE 9.8 *Parameters of Influence for Filtration Function [2].*

Application Condition	Parameter		
	Mechanical Filter Stability	Hydraulic Filter Stability	Long-Term Performance
Permanent filter	Minimum and maximum apparent geotextile opening size AOS	Geotextile permeability	Porosity AOS (minimum) Chemical properties of water and soil
Temporary filter	AOS (min. and max.)	Geotextile permeability	

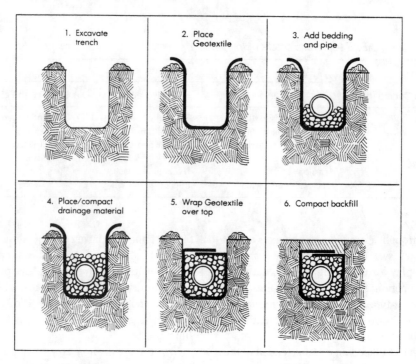

FIGURE 9.18 Sequential procedure for geotextile-lined underdrain construction (courtesy of Reemay).

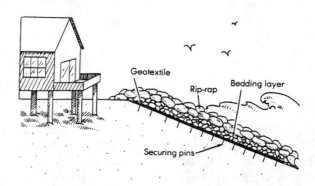

FIGURE 9.19 Geotextile for erosion control (courtesy of Reemay).

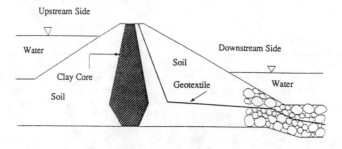

FIGURE 9.20 Drainage function of geotextiles [2].

316

TABLE 9.9 Parameters of Influence for Drainage Function [2].

Application Condition	Parameters		
	Mechanical	Hydraulic	Long-Term Performance
Permanent drainage	Influence of normal over-burden pressure	In-plane permeability Mass per unit area Apparent opening size AOS	AOS Porosity Chemical properties of water and soil Chemical stability decay resistance
Temporary drainage	Influence of normal over-burden pressure	In-plane permeability Mass per unit area AOS	

thick to provide significant transmissivity. The only materials of geotextiles with in-plane flow are needlepunched geotextiles.

Drainage can be classified as gravity flow and pressure flow. For gravity drainage, a geotextile is placed on a slope that creates the necessary driving force for drainage. Typical application areas for gravity drainage are chimney drains in dams, pore water dissipaters behind retaining walls, and flow interceptors as in fin drains [4]. In pressure flow, water flows from locations of higher pressure to locations of lower pressure regardless of the geotextile's orientation. Examples of pressure drainage are reinforced earth walls, earth embankments, dams, and beneath surcharge fills.

4.5 Waterproofing Function (Moisture Barrier)

Geotextiles can act as waterproof materials when impregnated with bitumen or polymeric sealing materials. After impregnation, the water and vapor permeability of the fabric become very low, in both cross-plane and in-plane flow. For the waterproofing function, geotextiles must be homogenous and have affinity for the sealing material. Fabric shrinkage which may be caused by temperature variations should be low. Adequate elongation is desired to compensate for stresses due to temperature fluctuations. Physical and mechanical requirements of geotextiles for waterproofing function are shown in Table 9.10.

4.6 Protection Function

In a variety of structures, geotextiles are used with geomembranes. Geotextiles can provide long-term protection of geomembranes against mechanical damage, such as perforation and abrasion, during and after installation. Required

TABLE 9.10 Parameters of Influence for Waterproofing Function [2].

Application Condition	Parameters	
	Physical	Mechanical
Repaving	Bitumen saturation level Shrinkage Product homogeneity	Geotextile tensile strength Geotextile elongation Inherent stiffness
Liquid membranes	Saturation capacity Shrinkage Product homogeneity	Geotextile tensile strength Inherent stiffness Flexibility
Membrane Encapsulated Soil Layers Method (MESL)	Saturation capacity Product homogeneity	Burst pressure resistance Puncture resistance Elongation at break

FIGURE 9.21 Application of geotextile for protection.

TABLE 9.11 Parameters of Influence for Protection Function [2].

Application Condition	Parameters	
	Mechanical	Long-Term Performance
Landfill and reservoir geo-membrane construction	Puncture resistance Burst pressure resistance Friction coefficient	Chemically stable: pH $= 2-13$ Decay resistant Creep resistance
Stress relief for crack retardation	Mass per unit area	Creep resistance
Roof construction	Puncture resistance	Chemical compatibility
Tunnel construction	Burst pressure resistance Puncture resistance Abrasion resistance	Chemically stable: pH $= 2-13$ Decay resistance

properties of geotextiles for protection function are primarily puncture resistance and abrasion resistance. They also must be chemically stable and decay resistant for long-term performance. Typical application areas are highway tunnels, landfills, water and sewage tunnels, railroad and subway tunnels and reservoirs. Figure 9.21 shows geomembrane protection with geotextiles. First geotextile, then synthetic liner is installed.

Table 9.11 shows geotextile requirements for protection function. A significant increase in the puncture resistance of the geomembrane was reported when geotextile was used as a protective layer underneath the membrane. In this regard, the laboratory test estimated that the thickness of the geomembrane could be reduced by a factor of at least two when a geotextile is used.

4.7 Combined Functions

Most of the time, geotextiles must serve more than one function (Table 9.4). For example, in road construction separation, filtration and reinforcement functions are involved (Figure 9.22). Beneath railroad ballast, the geotextile must

FIGURE 9.22 Geotextile in road construction (courtesy of Polyfelt).

meet the requirements for separation, reinforcement, filtration, and drainage. In these situations, in addition to the primary function, additional functions must all be satisfied. Since the required factor of safety for each function must be satisfied, the cumulative factor of safety will increase progressively.

5. DESIGNING WITH GEOTEXTILES

Table 9.12 shows the important criteria and principal properties required to evaluate a geotextile against a specific application. A geotextile designer can select a geotextile for a particular application using one or a combination of the following methods [4]:

(1) Design by cost and availability—In this method, which is not practiced much any more, design decision of selecting a geotextile fabric for a particular application is based on the price and availability of geotextile fabrics.

(2) Design by specification—This method is common with public agencies. In this method fabric properties are specified by the agency for each application category. Method of specifying geotextiles may vary from state to state. American Association of State Highway and Transportation Officials (AASHTO) has published manuals for the specification guidelines. It should be noted that these specifications generally list minimum required fabric properties [15].

(3) Design by function—In this method, the required numerical value of the geotextile

TABLE 9.12 *Important Criteria and Principal Properties Required for Geotextile Evaluation [14].*

Criteria and Parameter	Property	Application			
		Filtration	Drainage	Separation	Reinforcement
Design requirements					
Mechanical strength					
Tensile strength	Wide-width strength				X
Tensile modulus	Wide-width modulus				X
Seam strength	Wide width				X
Tension creep	Creep		*		X
Soil-fabric friction	Friction angle				X
Hydraulic					
Flow capacity	Permeability	X	X	X	X
	Transmissivity		X		
Piping resistance	Apparent opening size (AOS)	X		X	X
Clogging resistance	Porimetry	X			
	Gradient ratio	X			
Constructability requirements					
Tensile strength	Grab strength	X	X	X	X
Seam strength	Grab strength	X	X	X	
Puncture resistance	Rod puncture	X	X	X	X
Tear resistance	Trapezoidal tear	X	X	X	X
Longevity (durability)					
Abrasion resistance**	Reciprocating block abrasion	X			
UV stability†	UV resistance	X			X
Soil compatibility††	Chemical	X	X	?	X
	Biological	X	X	?	X
	Wet-dry	X	X		
	Freeze-thaw	X	X		

*Compression creep.
**Erosion control applications where armor stone may move.
†Exposed fabrics only.
††Where required.

TABLE 9.13 Geotextile Design Chart [2].

(1) Project analysis
- Problem addressed
- Feasibility of geotextiles (yes/no?)

(2) Determination of geotextile primary and secondary functions
- Separation
- Filtration
- Drainage
- Reinforcement
- Cushion
- Waterproofing

(3) Determination of required safety factor: $f_{s,required}$

(4) Determination of on-site parameters
- Soil: grain size distribution
            soil strength
            permeability
- Water: rate of flow
            flow conditions
- Stress: static
            dynamic

(5) Calculation of minimum required values: $X_{required}$

(6) Determination of actual geotextile minimum values: X_{actual}
- Mechanical: puncture resistance
                burst strength
                tear resistance
                tensile strength
- Hydraulic: in-plane permeability
                cross-plane permeability
                effective opening size (EOS) of
                   the geotextile
                apparent opening size (AOS)
                flow rate

(7) Calculation of actual safety factor: $f_{s,actual}$

(8) Comparison of safety factors: $f_{s,actual}$ versus $f_{s,required}$

property is calculated for the primary function that the geotextile will serve. Then, a factor of safety (FS) is calculated as follows:

FS = allowable property/required property

where allowable property is a value based on a laboratory test method that models the field conditions and required property is a value based on a design method that models the field conditions. Allowable property, P_a, is calculated as

$$P_a = P_t \{1/(f_1 + f_2 + f_3 + \ldots)\} \quad (9.2)$$

where P_t is the test property and f_1, f_2, f_3, etc.

are partial safety factors to account factors such as installation damage, creep deformation, degradation effects, soil clogging, etc. These partial factors of safety values can be obtained by direct experimentation and measurement. If the calculated safety factor is greater than the required FS, then the geotextile is accepted for that particular application. Table 9.13 shows steps for designing with geotextiles.

6. APPLICATION EXAMPLES OF GEOTEXTILES

Geotextiles play several important roles in hazardous or sanitary landfills, roads, retaining walls, earth embankments, dams, coastline slope protection, streambed lining, sand and dune protection, foundation reinforcement, temporary walls, pipelines, erosion control structures, clay liners, reservoirs and tanks [16−22]. There are around eighty specific application areas for geotextiles [2,21]. Figures 9.23−9.25 show some applications of geotextiles. The roles of geotextiles in major applications are explained below.

Filtration geotextiles are used to retain erodible soil on slopes and within subsoils for protection against subsurface erosion. Geotextiles retain soil particles in place while allowing seeping groundwater to freely drain, which prevents fine soil particles from clogging drainage systems. It also prevents the buildup of hydrostatic pressures in protected slopes and enhances slope stability for erosion control. Thick drainage geotextiles have limited capability of allowing liquids or gases to flow within the fabric's plane. Biodegradation of landfill waste produces large quantities of carbon dioxide and methane. Geotextiles are used as part of the landfill cover system to collect and vent the gases to prevent random gas migration, buildup and possible explosion. Geotextiles used as silt fences and silt curtains remove sediments from flowing water. Figure 9.23 shows application of geotextiles for filtration.

Reinforcing geotextiles provide stability for water-sensitive structures that can be weakened by forces of water. Mixing forces of water can be

FIGURE 9.23 Applications of geotextiles for filtration (courtesy of Hoechst Celanese).

controlled using separation geotextiles to isolate aggregate from soil. Geotextiles are used to prevent soils from moving in riverbeds. Reinforcing geotextiles enhance the stability of earth structures through their tensile strength and physical interaction with soil. Cost-effective retaining wall systems, steepened slopes and embankments over soft foundations are possible with reinforcement of soil using geotextiles. Figure 9.24 shows an example of geotextile application for reinforcement.

Separation geotextiles preserve the strength and drainage characteristics of coarse road base aggregate by preventing the aggregate from mixing with fine subgrade soils. Stabilization geotextiles provide separation, confinement and reinforcement to enhance subgrade load-bearing capacity.

Landfill liners are designed to intercept, collect and discharge the leachate produced by the waste in the landfill. A composite liner is absolutely necessary for any type of landfill, even if it is municipal and non-hazardous type. A state-of-the-art double liner system has four major segments: (1) leachate collection system, (2) primary liner, (3) leak detection system, and (4) secondary liner. Figure 9.25 shows a longitudinal section of a landfill liner. As shown in the figure, a geotextile filter is an essential part of the geocomposite drainage layer along side slope. Without it, waste would plug the open area of the drainage zone and render the drainage core

FIGURE 9.24 Geotextile reinforced wall (courtesy of Hoechst Celanese).

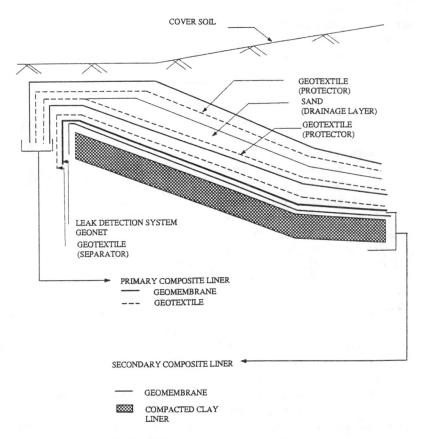

COVER SOIL

GEOTEXTILE
(PROTECTOR)

SAND
(DRAINAGE LAYER)

GEOTEXTILE
(PROTECTOR)

LEAK DETECTION SYSTEM
GEONET
GEOTEXTILE
(SEPARATOR)

PRIMARY COMPOSITE LINER
—— GEOMEMBRANE
– – – GEOTEXTILE

SECONDARY COMPOSITE LINER

—— GEOMEMBRANE

▨ COMPACTED CLAY
LINER

FIGURE 9.25 Cross section of a landfill liner.

ineffective. Geotextile filter should be resistant to long-term clogging. The second layer is the drainage layer. It can be made of geotextile or geocomposite. Geocomposite drainage layer is more preferred than the natural soil drains because it can be installed more easily and could be anchored at the top which will prevent shear failure in future. The third layer is the protecting layer where geotextile is used. This protection layer is required if drainage layer is made of sand or gravel. The protecting layer prevents the drainage layer material from puncturing or abrading the primary geomembrane liner. Geotextile, or different types of polymer foams can be used as a protection layer. Next is the primary liner. A barrier and a separator together make up the primary liner system.

The use of a nonwoven geotextile and asphalt for the lining for reservoir floors and canal linings is applied worldwide. The geotextile supplies high mechanical strength and asphalt provides adequate waterproofing. Geotextiles have also found applications in dams. The Tucuri dam is situated in the Amazon Region of the north of Brazil, in the valley of the Tocantins river. The dam was inaugurated at the end of 1984. It was one of the greatest hydroelectric power generating projects in Brazil [23]. Soil investigation in the metabasic strata revealed that soil in this zone had very high permeability which was totally unacceptable for the safety of dam. Geotextile was introduced in between compacted fill and the metabasic layer to prevent possible piping of fine soil. Geotextiles are also used for the restoration of dams as was the case with the McMicken dam (length 15 km, height 10 m) in Phoenix, Arizona [24].

The development and use of geosynthetic clay liners (GCLs) are relatively new. There are four currently available commercial products, each of which consists of essentially dry bentonite clay agglomerates placed on a geotextile or geomembrane carrier layer. Often the whole sandwich is needled throughout. An important property of the clay in the liner is its hydration. The natural affinity of dry clay to water is strong enough to naturally fill the soil's void and to continue to absorb until 100% saturation is achieved. Literature reports values of free swell of bentonite clay up to 1,200%. Overburden pressure can keep this volume change at minimum and at low overburden stress the volume change was reported not to exceed 200%. The second property is permeability. Different techniques are used for different products to measure the permeability. The permeability of the prefabricated clay liners is extremely low varying from 10^{-8} to 10^{-9} cm per second.

7. GEOTEXTILES MARKET

Geotextiles have shown tremendous growth within the last twenty years as shown in Figure 9.26. The market activity of geosynthetics in the geotechnical area and environmentally related areas has been very strong for the past decade. Available data on the geosynthetic market in North America shows that geotextile demand is steadily increasing with the increase in application areas. With the continuous development of different types of synthetic fibers, fabrics and composites, the application range is expected to increase in the future.

8. GEOSYNTHETICS

Brief descriptions of other members of the geosynthetics family are given for the reader's convenience. Geomembranes are impervious sheets of rubber or plastics and are always used

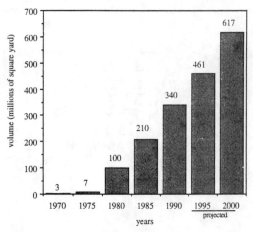

FIGURE 9.26 Market growth of geotextiles (courtesy of IFAI).

as a moisture or vapor barrier. They are the second largest group of geosynthetics after geotextiles. They are used for linings and covers of liquid or solid storage facilities. Geonets are formed by continuous extrusion of polymeric ribs placed at acute angles to one another. When the ribs are opened relatively large apertures are formed into a net-like configuration. Geonets are used to convey fluid and their applications are related to drainage applications. Geogrids are plastic materials, formed into a very open gridlike configuration having very large apertures. This is a more technical reinforcement segment of geosynthetics. Geogrids have two applications: separation and reinforcement. Geocomposites are usually composed of two geosynthetics, such as geotextile and geogrid, geotextile and geomembrane, or geomembrane and geonet. The application areas of geocomposites are numerous and are growing steadily. Functions of geocomposites vary greatly. Major functions are separation, drainage, filtration, moisture barrier and protection.

9.2

References and Review Questions

1. REFERENCES

1. Geosynthetics, Use with Confidence, IFAI Geotextile Division, 1991.

2. *Polyfelt Geotextiles, Design and Practice Manual,* Polyfelt Inc., Oct. 1994.

3. *The Nonwovens Handbook,* INDA, Association of the Nonwoven Industry, 1988.

4. Koerner, R. M., *Designing with Geosynthetics,* Third Edition, Prentice Hall, 1994.

5. ASTM D 1777-64 (1975), *1993 Annual Book of ASTM Standards,* ASTM, Philadelphia, PA.

6. ASTM D 5261-92, *ASTM Standards on Geosynthetics,* Third Edition, Philadelphia, PA, 1993.

7. ASTM 1388-64 (1975), *1993 Annual Book of ASTM Standards,* ASTM, Philadelphia, PA.

8. Geotextile Selection Guide, Hoechst Celanese, 1988.

9. Polyfelt Geotextiles, Technical Data/Samples, 1992.

10. Diaz, V. and Myles, B., *Field Sewing of Geotextiles, a Guide to Seam Engineering,* IFAI, St. Paul, 1990.

11. Swan, R. H, Jr., "The Influence of Fabric Geometry on Soil/Geotextile Shear Strength," *Geotextiles and Geomembrane,* Vol. 6, 1987.

12. *ASTM Standards on Geosynthetics,* Third Edition, Philadelphia, PA, 1993.

13. LINQ Tech Note #9, LINQ Industrial Fabrics, Inc., 1994.

14. DeBerardino, S. J., "Drainage Principles and the Use of Geosynthetics," *Geotextiles and Geomembranes,* Vol. 11, 1992.

15. Geotextile Design and Construction Guidelines, National Highway Institute, April 1992.

16. EPA/600/2-86/085 September 1986, "Geotextiles for Drainage, Gas Venting, Erosion Control at Hazardous Waste Sites."

17. Daniel, D. E. and Koerner, R. M., "Landfill Liners from Top to Bottom," *Civil Engineering,* December 1991.

18. Delmas, P. H., Matychard, Y., "Landsliding Confrontation with Geotextile Reinforced Earth Work," *Third International Conference on Geotextiles,* 1986, Vienna, Austria.

19. Englemann, H. R., "Nonwoven/Geotextile/Asphalt Composite in Water Reservoir Construction," *Third International Conference on Geotextiles,* 1986, Vienna, Austria.

20. Kysala, Z. "Geotextile as a Structural Member for Load Redistribution," *Third International Conference on Geotextiles,* 1986, Vienna, Austria.

21. Eith, A. W., Boschuk, J. and Koerner, R. M., "Prefabricated Bentonite Clay Liners," *Geotextiles and Geomembranes,* Vol. 10, 1991.

22. Trauger, R. "Geosynthetic Clay Liner Installed in a New Municipal Solid Waste Landfill," *Geotechnical Fabrics Reports,* November 1991.

23. Dib, P. S. and Aguir, P. R., "Tucuri Dam: Non-Woven Geotextile as One of the Anti-Piping Barriers," *Third International Conference on Geotextiles,* 1986, Vienna, Austria.

24. Deatherage, J. D., Beckwith, G. H., and Hansen, L. A., "Restoration of McMicken Dam," *Third International Conference on Geotextiles,* 1986, Vienna, Austria.

2. REVIEW QUESTIONS

1. List the members of the geosynthetic family and describe each. What are the main differences among them?
2. What are the most common fibers and fabric designs for geotextiles? Why are they suitable for geotextiles?
3. Explain the pullout test for geotextiles. What is the significance of this test?
4. Explain the major manufacturing methods of nonwoven geotextiles.
5. Define the following properties of geotextiles and explain how they are measured:
   - seam strength
   - percent open area (POA)
   - apparent opening size (AOS)
   - permittivity
   - transmissivity
   - long-term drainage capability
6. Summarize the geotextile functions. What geotextile properties are the most important for each function?
7. Explain the major steps when designing with geotextiles.

MEDICAL TEXTILES

10.1

Medical Textiles

S. ADANUR

1. INTRODUCTION

Textile materials in the medical field gradually have taken on more important roles. As more research has been completed, textiles have found their way into a variety of medical applications. In addition to protective medical apparel, textiles in fiber and fabric form are used for implants, blood filters, and surgical dressings. Woven and knitted materials in both synthetic and natural form play a part in the biotextile field, but non-woven materials also have proven to be effective and cost-efficient.

Medical apparel must protect the healthcare professional from any contact with possibly contaminated fluids. This has become more important with the growing risks of HIV and other viruses. However, with this protection, healthcare garments also must be comfortable and not restrict movement. There is also the decision as to whether it is better to disinfect the used garment in order to reuse it or dispose of the garment completely.

Textile structures in implantation are identified by structure, material composition, and behavior of fiber surface and degradation. A major concern with artificial implants is the reaction the body will have towards the implant. A biotextile in implantation must meet mechanical requirements and it must be biocompatible. Biocompatibility testing evaluates the response of the host system to the medical textile. Results of this testing must be viewed along with the risks and benefits of the device.

Since there is such a broad range of properties in textile materials, the required properties of a specific medical device usually can be acquired by modification efforts. That is, specialists from physicians to textile chemists and technologists can work as a team utilizing specific knowledge of their field to create an appropriate product.

Due to recent advancements in medical procedures and textile engineering, use of textile materials in the medical industry is growing. Between 1980 and 1990, the medical textile products grew at a compound annual rate of 11% in the United States. It is estimated that the growth rate will be 10% annually until the end of the century. In the United States, medical textile product sales grew from $11.3 billion in 1980 to $32.1 billion in 1990. This figure is expected to reach $76 billion by the year 2000 [1]. Approximately 70% of medical textiles are disposable and 30% are reusable.

2. MATERIALS USED IN BIOTEXTILES

The term *biomaterials* is defined as materials that are used in contact with tissue, blood, cells, protein and any other living substance. Biomaterials include metals, ceramics, polymers, natural fibers and their composites [2]. Biomedical polymers, natural fibers and medical textiles are within the scope of this book on industrial textiles.

Advancing medical technology necessitated interdisciplinary research by polymer scientists, textile engineers and medical doctors that began in the 1960s and has increased since the 1970s. Biomedical polymers have been used successfully in the research and development of artificial organs that simulate functions of natural organs. Table 10.1 shows major application areas of biomedical polymers.

TABLE 10.1 *Biomedical Application of Polymeric Materials (reprinted by permission of CRC Press [3]).*

Field	Purpose	Examples
Therapy	Repair or replacement of injured tissue	Prosthesis of bone, joint, tooth, etc.; artificial heart valve; patch grafting; artificial blood vessel; artificial shunt; contact lens; intraocular lens; artificial skin
	Assist or temporary substitution for physiological functions of a failed organ	Artificial heart and lung, ventricular assist system, artificial heart, artificial blood, artificial kidney, artificial liver, artificial pancreas, apheresis therapy, biological response modifier
	Disposable articles in daily medical treatment	Tubing, catheters, syringe, suture, etc.
Drug formulation	Novel drug delivery systems for amelioration of pharmacokinetics	Devices for controlled release of drugs, targeting design of drugs, pulsatile release devices
Diagnostic examination	New items in clinical laboratory tests	Reagents and tools with quick response, high accuracy, and high sensitivity; clinical test for new marker of disease; cell labeling
Bioengineering	New technology in tissue culture in vitro	Synthetic substrate or carrier particles for cell culture, additives for cell fusion, hybrid organ
	Separation of blood components	Plasma separation, cell separation, removal of virus and bacteria

2.1 Characteristics of Materials for Medical Use

The major requirements for biomedical polymers are non-toxicity (e.g., non-pyrogenicity, non-allergenic response, non-carcinogenesis), the ability to be sterilized (radiation, ethylene oxide gas, dry heating, autoclave), mechanical properties (strength, elasticity, durability), and biocompatibility (bioinert, bioactive) [4,5]. If an implanted material in the living body causes temperature rise, chronic tissue inflammation, allergic reaction, carcinoma or deformity, the material is considered toxic. As biomedical materials may be contaminated with bacteria, sterilization is important for biomedical polymers. The sterilization techniques can be physical (dry heat, steam, ionizing radiation) or chemical. However, sterility does not guarantee absence from pyrogenic bacterial products. Non-toxicity is not enough for biocompatibility which includes bioinertness and bioadhesiveness. The parameters for biocompatibility tests include toxicity, thrombogenicity, hemolysis, teratogenicity, mutagenicity, carcinogenicity, infection behavior and sensitivity [6]. Diffusion properties are critical in controlled drug delivery systems and membranes in artificial kidneys [7], optical properties are important for contact lens ma-

terials [8], and bulk and surface properties must be suitable for interfacial agreement with host tissues and fluids [9,10]. An analysis of structure-property relationships of biomedical polymers is provided in Reference [4].

Polyurethanes are widely used in biomedical applications. Thermoplastic polyurethane elastomers are suitable for hemodialysis sets, blood oxygenation tubing, blood bags, gas therapy tubing, heart assist devices, pacemaker lead insulators, and connectors [11]. Biomer®, a segmented polyether polyurethane with long fatigue life and high tensile strength, was used for blood contacting surfaces in artificial heart pumps and left ventricular devices [12]. However, it has recently been discontinued due to its propensity to degrade in the body. Ultra-high molecular weight polyethylene is suitable for joint replacement while porous high density polyethylene permits tissue ingrowth. Low density polyethylene is used for non-load-bearing applications. Polypropylene is suitable for steam sterilization due to a high melting point of 170°C. Although polytetrafluoroethylene has poor load-bearing properties, its derivatives as fiber, expanded, pore sponge, and composites cause minimal tissue reaction [13]. Silicone rubber polymer has been used successfully for a wide range of in vivo (internal) applications. It has thermal and oxidative stability, good flexibility and elasticity, tissue and blood compatibility and is inert and relatively non-toxic. It has been used successfully for tubing and shunts, mammary augmentation, capping nerve ends, vascular applications, pacer lead insulators, intraocular lenses, catheters, plastic and reconstructive surgery, replacement of cartilage or bone, bladders, sphincters, testicles and penile struts. Polymethylmethacrylate (PMMA) has been used for dentures, repair of cranial defects, jaw contour correction, spinal fixation, penile inserts, testicular implants, and as bone cement. Polyurethane has been used for aortic patch grafts, arterial venous shunts, pacer lead insulators and percutaneous devices. Hydrophilic polymers (hydrogels) are used for temporary and permanent prostheses. Insoluble gel networks can be formed which offer softness and mechanical comfort. End uses include contact lenses, breast and other substitutes, bone ingrowth sponges, drug delivery devices and catheters. Polyethylene terephthalate (PET) or polyester is widely used for fibrous prostheses. It is relatively inert, flexible and resilient. PET can be sterilized by all methods.

Use of composite or bicomponent materials such as carbon fiber reinforced polylactic acid polymer, hydrogel coatings on various polymers, parylene coated polypropylene, metallized epoxy, sucrose or CMC and bone cements is increasing.

Textile materials that are used in medical applications include fibers, yarns (monofilament and multi-filament), fabrics (woven, knit, nonwoven) and composites. Depending upon the application, the major requirements of medical textiles are absorbency, tenacity, flexibility, softness and at times biostability or biodegradability. Medical textile materials can be natural or synthetic, biodegradable or non-biodegradable.

Cotton and silk are the most common natural materials for medical textiles. These, along with regenerated cellulosic fibers (viscose rayon), are used extensively in non-implantable materials and healthcare/hygiene products. Man-made materials used in medical textiles include polyester, polyamide, polypropylene, polytetra-fluoroethylene (PTFE), carbon and glass [14].

There are several other specialty fibers used in medical textiles. Collagen, which is a biodegradable material obtained from bovine skin, is a protein that can be in either fiber or hydrogel (gelatin) form. Collagen fibers, which are used as sutures, are as strong as silk. The hydrogel is formed by crosslinking collagen in 5−10% aqueous solution. Calcium alginate fibers, which are produced from Laminariae seaweed, are effective in wound healing. It is non-toxic and biodegradable. Chitin is a polysaccharide obtained from insect skin, crab and lobster shells. Fibers and fabrics made of chitin are relatively non-thrombogenic, can be absorbed by the body and have good healing characteristics. Non-woven chitin fabrics are used as artificial skin dressings. They adhere well to the body and stimulate new skin formation while promoting healing rate with less pain. Alkali treatment of chitin yields chitosan which can be spun into filaments and used for slow drug release membranes. Research has been done on polylactics and polygalactics for sutures and possibly other products.

2.2 Biodegradability and Bioabsorbability

Biomedical fibers can also be classified according to biodegradability. Materials that are absorbed by the body two to three months (or longer) after implantation are considered as biodegradable and include polyamide, some polyurethanes, collagen and alginate. Although cotton and viscose rayon are biodegradable, they are not used as implants. Materials such as polyester, polypropylene, polytetrafluoroethylene (PTFE) and carbon are not absorbed by the body and are considered non-biodegradable.

Biodegradable and bioabsorbable materials are generally called bioabsorbable materials. They are decomposed in the body and their decomposition products are metabolized and excreted from the body. They may also be used to temporarily replace human organs. Biomedical application of bioabsorbable polymers are absorbable surgical sutures, synthetic skin, adhesives and joints, DDS (Drug Delivery Systems) and biohybrid organs as shown in Table 10.2.

Bioabsorption of polymeric materials has two stages: decomposition and absorbtion. During

TABLE 10.2 *Biomedical Applications of Bioabsorbable Polymers (reprinted by permission of CRC Press [15]).*

Application	Shape	Example
Binding materials	Suture, clip	Poly(α-hydroxy acids), poly(1,4-dioxan-2-one), polyglyconate, collagen (catgut)
	Adhesive	Poly(α-cyanoacrylate)
Bone-setting materials	Plate, screw, rod, pin, splint	PLLA, polyglactin, hydroxyapatite
Hemostatic materials	Wool, dressing, powder, spray	
Antiadhesion materials	Sheet, jelly, spray	Gelatin
Matrix for tissue culture	Sponge, mesh, nonwoven fabrics, tube	
Artificial tendon and ligament	Fiber	PLLA, polyglactin (combined with carbon fiber)
Artificial blood vessel	Fiber, porous material	PLLA, polyglactin
Wound cover attributes	Fiber	Collagen, chitin, polyglactin, poly-L-leucine
Synthetic skin	Fiber	Collagen
Drug delivery system	Microcapsule, microsphere, needle, hollow fiber	All degradable polymers

decomposition, main chain bonds break, molecular weight decreases and non-toxic monomeric and oligomeric compounds are produced. Depending upon the polymer, decomposition may or may not require an enzyme. Polymers decomposed by the enzyme-specific reaction are called enzymatically degradable polymers and those decomposed by non-specific hydrolysis or oxidation in contact with body fluids are called non-enzymatically degradable polymers. By definition, only the enzymatically degradable polymers are called "biodegradable." In the absorption stage, the decomposition products are rendered innocuous by bioprocesses and excreted from the kidney and sudoriferous glands. In addition to degradability, the bioabsorbable materials must satisfy other requirements such as biocompatibility, mechanical and chemical properties, and must also be non-toxic. The degradation and absorption rate should be compatible with the healing rate of biotissues and organs. The healing rate is different for various human biotissues: it takes three to ten days for dermal tissues, one to two months for internal organs and two to three months for hard tissues and at least six months for regeneration of large organs. Bioabsorbable implant materials should maintain their mechanical properties and functions until the biotissues are completely cured. After complete healing of biotissues, the implanted materials should be degraded and absorbed quickly

to minimize their side effects. Therefore, it is important to know the degradability properties of materials. Factors that affect degradability include chemical structure, crystallinity, hydrophilic/hydrophobic balance, shape and morphology.

Enzymatically degradable materials include polypeptides, polysaccharides, polyesters, nucleic acids and non-enzymatically degradable polymers are polyesters, poly(ester-ethers), poly(amide-esters). A comprehensive list of enzymatically and non-enzymatically degradable biopolymers is given by Kimura [15].

2.3 Blood Compatibility

Blood compatibility is necessary in biomedical materials, especially blood contacting devices, such as artificial hearts and artificial vessels as well as blood purification devices and blood catheters. Blood compatibility implies that a material should not cause thrombosis, which is the clotting of blood formed within a blood vessel. In addition to thrombosis, materials used in long-term applications should not cause alteration of the plasma proteins, destruction of the enzymes, depletion of electrolytes, adverse immune responses, damage to adjacent tissue, cancer, toxic and allergic reactions, or destruction of the cellular elements of the blood such as the red blood cells, white blood cells, and platelets.

Although several materials have been used successfully in conjunction with blood, at present, there is no polymer that is truly blood compatible.

The molecular designs of blood compatible polymers are classified in four categories. These are based on the approach for regulation of the blood materials interaction and are summarized in Table 10.3.

Polyurethanes and polyurethane ureas are used for the construction of intraaortic balloons due to favorable reports on their blood compatibility and good mechanical properties [17]. Because of their viscoelasticity, polyurethanes have been widely researched by device designers. However, there have been reports of their lack of long-term stability and their use is very limited as permanent implants. Lycra® is a well-known example of a segmented polyurethane urea that was developed by DuPont for elastic fiber applications. It has good physical properties and good biological performance, but it is difficult to dissolve, which makes it difficult to be used in the fabrication of complex medical devices. More soluble segmented polyurethanes have been prepared based on methylene and either polypropylene glycol or polytetramethylene glycol [18].

Carbon fibers have good blood compatibility and are appropriate for applications with rigid structures, but they are not suitable for structures that are thin or that require being flexed. Coatings of polyamides, polyimides, cellulose acetate, silicone rubber, and segmented polyetherurethane elastomers have been used for making composites of carbon. These composites make carbon more flexible.

Microfibers in composite form are often blood compatible. They are able to anchor viable fibroblastic and pseudo-endothelial cells, thus producing a living surface. This surface is considered blood compatible. Typical microfibers include those made of polyesters, and parylene-coated polypropylenes. Other fibers include polytetrafluoroethylene (PTFE) which is used as artificial arteries and arteriovenous shunts as well as a tip material for shunts. Polyethylene is often used as tubing to carry blood and for intravascular catheters.

3. CLASSIFICATION OF MEDICAL TEXTILES

Depending on the application areas, medical textiles can be classified as follows:

(1) Surgical textiles
- textiles for implantation (e.g., sutures, vascular grafts, fabrics for heart valves and repair, artificial joints, fabrics for hernial repair, surgical reinforcement meshes, fibrous bone plates, etc.)
- non-implantable textiles (e.g., bandages, wound dressings, plasters, etc.)

TABLE 10.3 *Blood Compatible Polymers (reprinted by permission of CRC Press [16]).*

Synthetic polymer	Hydrophobic surface	Polytetrafluoroethylene (PTFE)
	zero critical surface tension theory	Polydimethylsiloxane
		Polyethylene
	Hydrophilic surface	Poly(2-hydroxymethyl methacrylate)
	zero interfacial free energy concept	Poly(acrylamide)
	Heterogenic surface	Block-type copolymer
	microdomain concept	Graft-type copolymer
	Molecular cilia mechanism	Poly(ethylene glycol)
	Negative charge surface	Sulfonated polymer
Synthetic polymer + biologically active polymer	Pseudomembrane formation	Expanded PTFE
		Poly(ethylene terephthalate)
	Heparinization	Heparin releasing polymer
		Heparinized polymer
	Immobilization of fibrinolysis enzyme	
Synthetic polymer + biological molecules	Albumin adsorbed surface	Polyurethane with alkyl group
	biomembrane-like surface formation	

(2) Textiles in extracorporeal devices (e.g., artificial kidney, liver, lung, etc.)

(3) Healthcare and hygiene products (e.g., bedding, protective clothing, surgical gowns, cloths, wipes, etc.)

4. TEXTILES FOR IMPLANTATION

Although the natural way to replace a defective body part would be transplantation, this is not always possible due to several reasons including availability, performance requirements etc. Therefore, physicians often have to use an artificial substitute (biomaterials) such as biotextiles. A foreign or synthetic material or part used to replace a body part is referred to as a prosthesis. Table 10.4 shows major implantable textile materials.

Although it depends on specific application, in general, the biological requirements for a satisfactory artificial implant may be stated as follows [1,19]:

(1) A suitable artificial surface for the body cells to easily adhere to and grow on

(2) Porosity, which determines the rate at which tissue will grow and encapsulate the implant; Implant material should be sufficiently porous.

(3) Fiber diameter, in general, smaller than the cells for their adherence. Although human tissue is capable of encapsulating objects much larger than fibers, it can better encapsulate the small circular fibers than large irregular cross-sectional fibers.

(4) Biodegradability or biostability depending on the application.

(5) Non-toxicity, where fiber polymer or fabrication techniques must be non-toxic and fibers should be free of contaminants

The implantable material must meet mechanical requirements for the particular application. Biocompatibility requires that the biotextile must interact with the host in a controlled and predictable way. In addition to these conditions, the biomaterial must not damage blood cells or cause formation of destructive blood clots. Therefore, it is essential that the structure-property-function relationships of implant materials and body organs are well understood. For example, the primary function of hard tissue such as bones and teeth is load-carrying. On the other hand, collagen-rich tissues such as skin, tendon, and cartilage act as load-bearing elements. Elastic tissues, such as blood vessels, ligamentum nuchae, and muscles exhibit a large deformation with small cyclic load. Applications of textiles as implants include abdominal wall, artery, bio-

TABLE 10.4 Implantable Textiles [1].

Product Application	Fiber Type	Manufacture System
Sutures		
Biodegradable	Collagen, polylactide, polyglycolide	Monofilament, braided
Non-biodegradable	Polyamide, polyester, Teflon®, polypropylene, polyethylene	Monofilament, braided
Soft tissue implants		
Artificial tendon	Teflon®, polyester, polyamide, polyethylene, silk	Woven, braided
Artificial ligament	Polyester, carbon	Braided
Artificial cartilage	Low density polyethylene	
Artificial skin	Chitin	Nonwoven
Eye contact lenses/ artificial cornea	Polymethyl methacrylate, silicone, collagen	
Orthopaedic implants		
Artificial joint/bones	Silicone, polyacetal, polyethylene	
Cardiovascular implants		
Vascular grafts	Polyester, Teflon®	Knitted, woven
Heart valves	Polyester	Woven, knitted

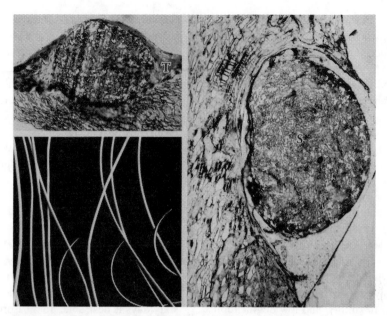

FIGURE 10.1 Expanded PTFE monofilament suture (courtesy of W. L. Gore and Assoc.). Upper left: suture retrieved with a temporopopliteal graft at two months. Tissue (T) is attached to the suture. Right: suture (S) retrieved with a soft tissue patch at four years. Bottom left: unused sutures.

hybrid organs, bone, heart valve and wall, hip joint, ligament, tendon, trachea and vein.

4.1 Sutures

Sutures are mono- or multi-filament threads that are used to close wounds, join tissue and tie off bleeding vessels. In the thirties, silk and linen were used as sutures until the development of nylon. BASF then developed a "pseudo-mono-filament," which was a combination monofilament and multi-filament material. Following polyester in the suture market in the late sixties was polyglycolic and polylactic acid, synthetic absorbable suture materials. Dexon® and Vicryl®, resorbable materials, are presently used in a braided configuration.

Suture materials can be classified as absorbable and non-absorbable sutures. Absorbable sutures include collagen (catgut), polyglycolic acid, polyglycolic acid-polylactic acid copolymer, polydioxanone and polyglycolic acid polycarbonate copolymer. Absorbable sutures are used internally. Non-absorbable sutures are made of "expanded" polytetrafluoroethylene (ePTFE), polypropylene, polyester and silk. They are generally used for closing cutaneous or oral incisions where the suture can be easily removed.

These absorbable sutures are used to stop internal bleeding, and are degraded by the body fluids after the healing process. Non-absorbable suture materials are unresorbed or considered to be unresorbed in the body for a long period of time.

Catgut, silk and linen are traditional suture materials. Virgin silk is used for ophthalmic sutures due to its thin size. However, its application for other medical areas is limited. The surgical suture of silk fibroin is not bioabsorbable. Linen and steel have been used in the past but have proven less appropriate due to tissue reaction and poor handling properties. Recently, absorbable surgical suture has been developed from chitin. Its tensile strength is better than that of the other absorbable sutures such as Dexon® and catgut. Figure 10.1 shows application of expanded PTFE monofilament suture.

Manufacturing Processes

There are advantages and disadvantages to the

various methods in which suture materials can be manufactured, whether it be a monofilament, a braid, a pseudo-monofilament or a twisted strand. It is up to the user to determine the type of material and the form that will be the most beneficial. Each process is discussed below.

A monofilament with its smooth surface can only be made from synthetic material by extrusion. One property of the monofilament is a minimal tissue drag with low tissue reaction. This is because of the smooth surface. It is also easy to make or place a knot in the depth of the body. This advantage is especially critical in intestine closure. Disadvantages of the monofilament include low knot security and less flexibility. Typical monofilament materials used are polyester, polyamide, polypropylene, and polydioxanone. Polyester has a high knot-pull tensile strength, good flexibility, and low degradation. Polypropylene has excellent tissue drag and stability. Polyamides have a similar chemical structure to that of collagen and degrade within the body over a period of time depending on the mass of the implant and the site within the body it is used. Polydioxanone and polyglycolic acid-polycarbonate copolymer are monofilaments with long resorption time. The tensile strength is low and the material can be found in the body for about 180 days.

Various yarns of different multi-filament fibers can be braided together to form a suture material. There are anywhere from eight to sixteen yarns in a braided material for this purpose. Obviously, due to the manufacturing method, the braid is rougher on the surface, which causes tissue drag to be high. A lubricant can coat the braided suture material to lower the tissue drag and allow better knotability. Braids are also flexible and easy to handle, which are also beneficial properties of sutures. Polyesters, polyamides, and silks are commonly used for braids. The synthetic resorbable materials, polyglycolic acid and the polyglycolic acid-polylactic copolymer are resorbed in the body within twenty to twenty-eight days [20].

The pseudo-monofilament has a core of several twisted materials coated with an extrusion of the same material as the core. Because of its low coefficient of friction, it has low tissue drag, good knotability, low knot security, and fair flexibility. The twisted suture consists of twisted yarns in order to achieve a desired diameter and strength. The rough surface causes high tissue drag, but has good knotability and security. The flexibility is about the same as the braid.

Selection of the appropriate suture material is likely the decision of the surgeon. The various properties have to be considered depending on what the use will be. However, resorbable monofilaments have become popular suture materials.

Test Methods

The tested properties of sutures include knot-pull tensile strength and diameter, bending stiffness, surface roughness, knot run-down, and knot security. Degradation time between thirty to sixty days is desirable. To measure suture properties, exact standardized testing methods and devices should be developed with the aims of specifying needed properties to help the surgeon to choose the right suture for a specific purpose to minimize operation and healing risk.

The knot-pull tensile test is based on the surgeon's knot and the shape of the clamps that fix the suture at jaws. The methods of measuring diameter are not always accurate, so exceptionally strong sutures are often used to be safe in regard to yarn strength. Bending stiffness determines how easily a suture is handled. A slack suture is easier to knot and is more secure. A stiffness tester presses a yarn loop onto measuring bolts of a precision scale by the circling movement of the clamp that fixes the suture ends. This determines bending stiffness.

A rough suture can cause damage to biological tissue and can damage a surgeon's operating gloves. However, a rough surface allows for better knot stability and security due to the friction between the surfaces. The Hommel roughness tester has a light-weighted, rounded measuring tip that runs along the moving thread, the yarn causes an up and down motion of the tip which is recorded on-line with a plotter.

Knot run-down, knot security and knot snugging-down are important parameters that depend on suture stiffness, surface roughness, and coating material. Knot security means whether the knot stays under steady tension in its determined place. When a slack or smooth suture is used there is less force to form the knot, and the tension on the biological tissue is small. Table 10.5 shows knot-pull tensile strength of some sutures.

Bacteria on the Degradation of Absorbable Sutures

Bacteria in surgical wounds can cause difficulties in the healing process. Complications can be especially critical with absorbable sutures that degrade by bacteria or bacterial enzymes.

Testing was performed by Elser et al. on Dexon® and Maxon® suture material for bacterial effects [21]. Dexon® is a coated and non-stained braided, absorbable multi-filament made from polyglycolic acid. Maxon® is an absorbable monofilament made from polyglycolate. The incubation liquid for the test was a bacterial suspension based on nutrient broth inoculated with *Staphylococcus aureus* ATCC 6538P (106/ml). The results were that both Dexon® and Maxon® degradation was accelerated significantly due to the presence of the bacteria. With Dexon®, 50% degradation occurred within 10 days in sterile broth and within 8 days in the presence of bacteria. Maxon®'s degradation was 50% within 24 days in sterile broth and 22 days in the presence of bacteria.

4.2 Soft Tissue Implants

Biomedical materials are used in applications such as soft tissue compatible artificial prostheses, artificial skin patches, artificial tendon, and artificial corneas. Important properties that affect cell attachment and tissue growth are chemical structure, electric charge, hydrophi-

licity and hydrophobicity, roughness of surface, microheterogeneity and material flexibility. There is an inverse relationship between the percentage of initial cell attachment and the relative rate of cell growth. As the affinity of the cell to the material increases, the cell growth decreases. It was reported that cell attachment and growth are promoted on the positively charged substrate and are restrained on the negatively charged substrate [22,23]. Hydrophilic material surfaces promote cell attachment. As the surface roughness increases, cell attachment increases, but cell growth decreases. For the cell attachment and growth on the material surface, a hard footing in molecular order is needed for the cells to anchor to the surface.

Soft tissue compatible biological polymers are collagen, silk protein, cellulose, chitin and chitosan. Collagen, which is a main component of soft tissue in the human organism, is a protein which forms connective tissue in vivo. Therefore, artificial implants with collagen modified surfaces easily adhere to soft tissue. Collagen is also used to make artificial skin dressing which has a similar constitution to the living skin and is regarded as a living skin-equivalent hybrid skin [24]. Silk protein is made of two types of protein, silk fibroin and silk sericin. Important characteristics of cellulose include chemical stability, hydrophilicity, good mechanical properties and chemical functionality.

Soft tissue compatible artificial materials include silicone rubber, polyurethane, hydrogels and carbon fiber. Silicone rubber is a cross-linked polymer of poly(dimethyl siloxane). It has been used in artificial breasts, ears and noses [25]. Hydrogels are insoluble water-containing materials. They are made by introduction of either crosslinks, hydrophobic groups, or crys-

TABLE 10.5 Knot-Pull Tensile Strength of Various Sutures (courtesy of W. L. Gore and Assoc.).

Average Diameter (mm)	Knot-Pull Tensile Strength (kg)		
	Monofilament Polypropylene	Braided Polyester	Braided Silk
0.044	0.09	–	0.09
0.084	0.32	0.27	0.32
0.174	1.05	1.27	0.95
0.320	2.59	2.95	2.27
0.450	5.00	5.23	4.00

talline domain into water-soluble polymers. High water content inhibits cell attachment and growth, but gains good oxygen permeability. Carbon fibers are used to repair joints in the arm, shoulder, knee and ankle.

Tendons and Ligaments

Textile materials offer the strength and flexibility to replace tendons, ligaments and cartilages in both reconstructive and corrective surgery. In fact, the structure of ligaments, which connect bone to bone, and tendons, which connect muscle to bone, is similar to polymeric textiles because they are composed of a multiple of fibrils and fiber bundles.

Man-made tendons have been made of woven and braided porous structures coated with silicone. The major requirements for artificial tendons are tissue compatibility, tensile strength, fatigue resistance, porosity and flexibility. Common pore sizes of polyethylene tendons may vary between 100 microns square to 500 microns square.

Nonwovens are not suitable as ligament prosthesis materials because of their lack of strength. Woven and knitted structures are used as artificial ligaments. Braided fabrics with a stress strain behavior similar to a natural tendon or ligament, are the most suitable structures.

Braided composite textile structures made of carbon and polyester are particularly suitable for knee ligament replacement (Figure 10.2). Braided polyester ligaments have good strength and creep resistance to cyclic loads. The requirements for ligament and tendon implants are both biological (biocompatibility, long-term stability, supporting tissue proliferation) and biomechanical (physiological progressive stress-strain behavior, low creeping, high sheer strength). Bioabsorbable polymers are preferable for manufacturing of ligaments and tendons. A braided polypropylene implant usually consists of several strands that are braided in a tape-like structure. A braided carbon fiber implant is typically made up of thirty-two strands of 3,000 fibers each. These strands are braided at an angle of approximately 45° (Figure 10.3). There have been braided ePTFE textiles used as ligament as well.

Low density polyethylene (LDPE) has compatible properties with natural cartilage and

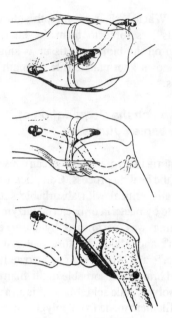

FIGURE 10.2 Schematic of an artificial anterior cruciate ligament implanted in a right knee (courtesy of Sulzer Ruti [26]).

therefore is used for facial, ear, nose, and throat cartilage replacement [1]. Carbon ligaments are not widely used due to splintering of the carbon filament with flexion.

Skin Dressings

In order to conform to irregular surfaces, elastic and flexible materials are used as skin dressings. They promote skin regeneration. There are two opposite requirements for skin dressing to meet: it should prevent dehydration of the wound and obstruct bacterial entry, but it should also be permeable enough to allow the passage of discharge through pores or cuts.

For flexibility and absorbability of body flu-

FIGURE 10.3 Braided ligament structure without load (left) and under load (courtesy of Sulzer Ruti [26]).

TABLE 10.6 Properties of Materials Used for Wound Cover Attributes (reprinted by permission of CRC Press [30]).

Polymer	Strength (MPa)	Elongation (%)
Poly-L-leucine	200	13
Chitin	750	12
Collagen	360	20

ids, skin dressings are made of woven and non-woven fabrics as well as microporous layers. Most of the commercially available skin dressings use absorbable enzymatically degradable polymers, such as collagen, chitin and poly-L-leucine whose properties are shown in Table 10.6 [15]. In hybrid skin dressing, synthetic polymers and cell cultures may be combined to form a biological/synthetic composite.

Even today there is no single biological or synthetic material that can meet the requirements of a skin substitute exactly. The long list of necessary properties for skin substitutes includes tissue compatibility, water vapor transmission similar to normal skin, impermeability to exogenous microorganisms, inner surface structure that permits ingrowth of fibrovascular tissue, prevention of wound contraction, flexibility and pliability to permit conformation to irregular wound surface, elasticity to permit motion of underlying body tissue, resistance to linear and shear stresses, low cost and indefinite shelf life.

Hernia Repair

Textile mesh fabrics are used in hernia repair. Polypropylene mesh such as the Meadox Trelex Natural® mesh is an example of fabrics used for hernia repair. Polypropylene is resistant to infection and it is not allergenic. Gore-Tex® soft tissue patch, which is used in hernia repair, is made of expanded PTFE (Figure 10.4).

4.3 Hard Tissue Implants

Hard tissue compatible materials must have excellent mechanical properties compatible to the hard tissue. Table 10.7 shows the mechanical properties of hard tissues and some artificial materials for hard tissue replacement.

Typical characteristics of polymers related to hard tissue replacement are good processability, chemical stability, and biocompatibility. Applications include artificial bone, bone cement, and artificial joints. The current practice is to combine bioactive ceramics with polymers or metals to improve interfacial properties as shown in Figure 10.5. In recent years, biodegradable and adhesive bone cements were developed.

Orthopedic implants are used to replace bones and joints, and fixation plates are used to stabilize fractured bones. Textile structural composites (Chapter 7.0) such as carbon composites are replacing metal implants for this purpose. A nonwoven fibrous mat made of graphite and

FIGURE 10.4 Cross-sectional view of Gore-Tex® soft tissue patch and its use in vetral hernia repair (courtesy of W. L. Gore and Assoc.).

TABLE 10.7 Mechanical Properties of Hard Tissue Materials (with permission from CRC Press [27]).

		Tensile Strength (MPa)	Tensile Modulus (GPa)	Compressive Strength (MPa)	Bending Strength (MPa)
Bone	Cortical bone	70	16	140	150
Teeth	Dentin	53	19	300	–
	Enamel	11	84	390	–
Ceramics	Hydroxyapatite	115	90	520	130
	Carbon	200	12	800	170
Metals	SUS 316	518	200	–	–
	Titanium	406	110	–	–
Polymers	Poly(methyl methacrylate)	46	2	77	100
	Ultra-high molecular weight polyethylene	200	0.7	–	–
	Poly(L-lactic acid)	50	6	–	–
	Silicone	7	60	–	–

Teflon® is used around the implant to promote tissue growth.

4.4 Vascular Implants

Artificial veins or arteries are used to replace segments of the natural cardiovascular system that are blocked or weakened. A typical example is to replace a section of the aorta where an aneurism has occurred. Another example is the arteries in the legs of diabetic patients that have a tendency to be blocked. Grafts are inserted to bypass the blockages and restore circulation.

Grafts usually have a tube shape. The most

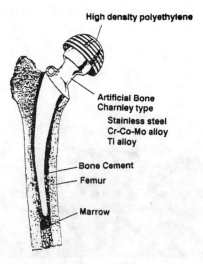

High density polyethylene

Artificial Bone Charnley type

Stainless steel Cr-Co-Mo alloy Ti alloy

Bone Cement

Femur

Marrow

FIGURE 10.5 Replacement of artificial bone with bone cement (reprinted by permission of CRC Press [27]).

widely used fibers are polyester and expanded PTFE. Polyester vascular grafts can be woven, knit or extruded. Expanded polytetrafluoro-ethylene grafts are rendered porous by expanding sintered particles of Teflon®. Straight or bifurcated grafts are produced with warp knitting or weaving. The structure in warp knit grafts is reverse lock knit and full tricot. Tubular woven grafts are made on narrow label looms. Mostly plain weave is used with yarns between 50 – 150 denier. In some woven vascular grafts, a leno weave is used for every few ends to minimize unravelling and fraying. Heat setting creates a crimped structure and improves the handling characteristics of polyester vascular grafts.

The most important property of a graft is its porosity. Woven graft structures are typically less porous than knitted structures which diminishes blood leakage through the interstices but at the same time can hinder the tissue growth. Knitted vascular grafts are more porous which allows tissue growth but leakage may be a concern. Typical water permeabilities of woven grafts are 50 – 500 cc/cm²/min and of knitted grafts are 1,000 – 2,000 cc/cm²/min [28]. In order to fill the interstices of the graft and prevent leakage, knitted and woven grafts with internal and/or external velour surfaces have been developed (Figures 10.6 and 10.7). Most knitted and woven grafts are preclotted prior to transplantation. Blood is intentionally soaked into the textile wall allowing clot formation to occur. The proclotting steps reduce or eliminate blood loss through the

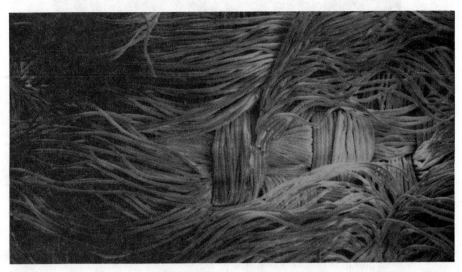

FIGURE 10.6 Outer surface of plain weave, double velour graft (courtesy of Meadox Medical, Inc.).

textile wall once the graft is implanted. Presealed prostheses (also called composite grafts) are impregnated with collagen (e.g., Hemashield® grafts and fabrics from Meadox Medical) or gelatin first to seal pores (Figure 10.8). Approximately two weeks (gelatin) to three months (collagen) after implantation, the seal degrades allowing tissue ingrowth. Polyester fabrics are used as suture cuffs on artificial heart valves to allow suturing the valve to the surrounding tissue. The fabric also helps secure the valve with the tissue growth.

Figure 10.9 shows various expanded PTFE grafts. In addition to porosity, good handling and suturing characteristics, satisfactory healing (rapid tissue growth), mechanical and chemical stability (good tensile strength and resistance to deterioration) are the major requirements of vascular grafts. Since vascular grafts are subjected to static pressure and repeated stress of pulsation in application, they should have good dilation and creep resistance. Application examples of vascular grafts are shown in Figure 10.10.

Crimp in grafts is for handling and length

FIGURE 10.7 Inner surface of full tricot knit, double velour (courtesy of Meadox Medical, Inc.).

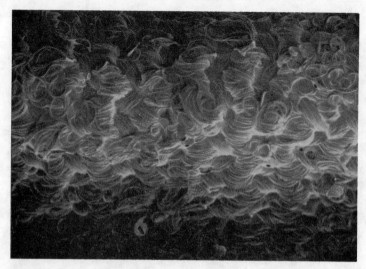

FIGURE 10.8 Inner surface of full tricot knit, double velour, collagen impregnated graft (courtesy of Meadox Medical, Inc.).

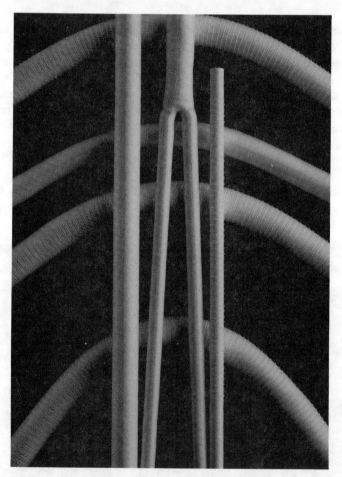

FIGURE 10.9 Various vascular grafts (courtesy of W. L. Gore and Assoc.).

FIGURE 10.10 Application examples of grafts (courtesy of W. L. Gore and Assoc.).

sizing. Crimped geometry improves the graft's bending ability without kinking. Crimp is removed at implantation by the surgeon. The elasticity of the graft depends on the elastic modulus of the material, its thickness and radius. To determine the circumferential deformation under internal pressure (dilatability), a parameter called "compliance" is used. Compliance is the percent change in the external diameter per unit change in pressure assuming a constant vessel length and neglecting wall thickness variation. It was reported that knit grafts showed more compliance than woven grafts, and warp knit graft dilated less than the weft knit graft as internal pressure was increased [29]. Due to their crimped structure, all the grafts showed high longitudinal distensibility, knit grafts being the highest. Composite graft structures, which are obtained by impregnation with biodegradable compounds, resulted in stiffer structures with less elasticity. Externally supported grafts exhibited good dimensional stability.

Figure 10.11 shows a different type of vascular graft structure before and after implantation. The wall of the graft is composed of 1,400–1,600 layers of 8–12 micron diameter fibers.

Vascular grafts have been successfully implanted in individuals during the last thirty years [30]. The manufacturers of vascular grafts in the

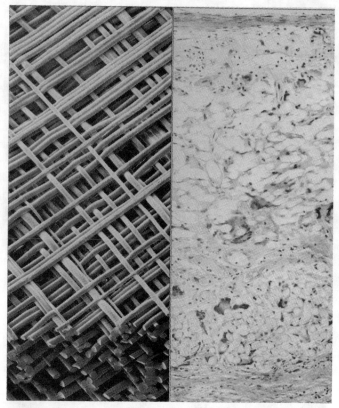

FIGURE 10.11 A solvent bonded vascular graft before and eighteen months after implantation (courtesy of Corvita Corporation).

United States include Meadox Medical, Inc. (New Jersey), Corvita Corporation (Florida) and C. R. Bard (Massachusetts).

4.5 Biomaterials in Ophthalmology

Natural and synthetic hydrogels physically resemble the eye tissue and hence have been used in ophthalmology as soft corneal contact lenses. Soft contact lenses are made of transparent hydrogel with high oxygen permeability. Hard contact lenses are made of poly(methyl methacrylate) and cellulose acetate butyrate. Flexible contact lenses are made from silicone rubber.

Hard contact lens materials can be hydrophilic or hydrophobic. The lens material should have high surface energy and be easily wettable by tears. The permeability of the lens to oxygen is important.

4.6 Dental Biomaterials

Since regeneration of tooth tissue is not possible, biopolymers are used in dental treatments to substitute for defects in tissue. Major requirements of dental polymers include translucence or transparency, stability, good resilience and abrasion resistance, insolubility in oral fluids, non-toxicity, relatively high softening point and easy fabrication and repair. The most widely used polymer for dental use is poly(methyl methacrylate) (PMMA) and its derivatives. Other materials for denture base polymers are polysulfone and polyethersulfone. Crosslinked acrylic resin teeth were produced to improve wear resistance [31,32].

5. NON-IMPLANTABLE TEXTILES

Non-implantable materials are used for external applications on the body with or without skin contact. They include surgical dressings, bandages, release fabrics etc., as shown in Table 10.8.

5.1 Surgical Dressings

Wound dressings are used for many purposes including protection against infection, absorption and exudation of blood and excess fluids, healing and application of medication (Figure 10.12). Ideally, surgical dressings should be soft, pliable, pad the wound to protect from further injury, be easily applied and removed, be sterile, lint-free and non-toxic. Wound care products are usually made of three layers: wound contact layer, absorbent layer and base material. Wound dressing should not adhere to the wound allowing easy removal without disturbing new tissue growth. The middle layer absorbs blood or liquids while providing a cushioning effect to protect the wound. The base material is coated with an acrylic adhesive to hold the dressing in place. The tightly woven cloths provide a smoother dressing pad and probably absorb more quickly, but looser structure will give more bulk for greater protection.

The collagen, alginate and chitin fibers are proven to be effective in healing of wounds. The interaction between the alginate fibers and exuding liquids creates a sodium calcium alginate gel

FIGURE 10.12 Wound dressing (courtesy of Johnson & Johnson).

which is hydrophilic, permeable to oxygen, impermeable to bacteria, and helps with the formation of new tissue [29].

Gauze and paraffin-coated gauze are the most common dressings used. Most gauze is made from cotton in the form of a loose plain weave. The typical yarn densities per inch are $12-19$ for warp and $8-15$ for filling (Figure 10.13). 44's cotton count of warp yarns and 54's weft carded yarns are common sizes for gauze manufacturing. Gauze is used mostly as a dressing for direct wounds, but it is also used in internal pads and general swabbing applications.

Perhaps the most difficult property to achieve is dressing removal because when a gauze is placed in contact to the wound the exudate, the discharge from a wound, adheres to the fibers of the dressing. Burns and skin graft sites must have their dressings changed frequently. When the dressing is removed, it is not only painful, but it

TABLE 10.8 Non-Implantable Textile Materials [1].

Product Application	Fiber Type	Manufacture System
Wound care		
Absorbent pad	Cotton, viscose	Nonwoven
Wound contact layer	Silk, polyamide, viscose, polyethylene	Knit, woven, nonwoven
Base material	Viscose, plastic film	Nonwoven, woven
Bandages		
Simple inelastic/elastic	Cotton, viscose, polyamide, elastomeric yarns	Woven, knit, nonwoven
Light support	Cotton, viscose, elastomeric yarns	Woven, knit, nonwoven
Compression	Cotton, polyamide, elastomeric yarns	Woven, knit
Orthopaedic	Cotton, viscose, polyester, polypropylene, polyurethane foam	Woven, nonwoven
Plasters	Viscose, plastic film, cotton, polyester, glass, polypropylene	Knit, woven, nonwoven
Gauzes	Cotton, viscose	Woven, nonwoven
Lint	Cotton	Woven
Wadding	Viscose, cotton linters, wood pulp	Nonwoven

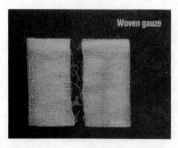

FIGURE 10.13 Woven gauze (courtesy of Johnson & Johnson).

can also destroy the regenerating tissues. This can delay the healing process, cause scarring, and reopen the wound for possible bacteria entrance. The paraffin-coated gauze, which is usually multi-layered, is a little easier to remove than dry gauze, but it can still stick and it does not rapidly absorb wound liquids. It is also possible that this petroleum based coating can liquefy at body temperature and can introduce foreign materials into the wound. It is used to treat burns and scalds. Gauze may be impregnated with plaster-of-paris. When applied wet in several layers, the plaster will harden to make a cast. Rayon has been researched as a gauze, but when wet it often forms a tangled mass and becomes slimy. Operating room applications require sterilization which is done in many ways including steam, dry heat, ethylene oxide and irradiation sterilization. Finishing agents, such as wetting agents and optical whiteners are not added to gauze fabrics because of the possibility of irritation and possible carcinogenic effects.

Wadding is an absorbent material. To prevent adhesion to wound or fiber loss, it is covered with a nonwoven fabric. Lint is a plain weave cotton fabric that is frequently used in treatment of mild burns.

Because sterilization is a major concern for surgical dressings, nonwovens can be considered. Nonwovens can be smooth, and lint free for the most part. This allows for a lesser chance for debris to be left in the wound. Nonwovens can be made softer and more absorbent by latex or thermal calendering. For postoperative dressing, sophisticated nonwoven structures such as perforated film on absorbent base, polymer/nonwoven welded laminate and metallized nonwoven fabric are used. Nonwoven fabrics made

of atelocollagen filaments are used as wound dressing for burns.

Polypropylene is promoted as "the most unwettable of all fibers," but it can also be quite wettable if converted into a fibrous web by melt blowing. In meltblowing a stream of molten polymer is subjected to blasts of air which form tiny fibrils that fall randomly as a web on a conveyor belt. This gives polypropylene a new role as an absorbable dressing material [33]. Polypropylene in fiber and fabric form can take on a more traditional role as a facing material coupled with an absorbent substrate. Common uses would be in burn dressings and bandages. The polypropylene fabric would be in skin contact and would transmit liquid to the absorbent material. The resulting product would be comfortable, harmless, and able to keep the skin dry [34].

Research has also been performed on carbonized rayon fabric used in dressings. It forms a type of charcoal which is well known for absorbing unwelcome smells.

5.2 Bandages

Bandage fabrics are used for various medical applications that include keeping the dressing over the wound. Bandages can be woven, knitted, nonwoven and elastic or non-elastic. Elastic yarns provide support and comfort in the bandage fabric structure. Warp and weft knitting can be used to produce tubular structures.

Elastic crepe bandages, which are used for sprained wrist and ankle support, are woven from cotton crepe yarns. These are spun with a high twist so the bandage will maintain its tension and recover from stretch. Elasticity can also be obtained by using two warp beams during weaving, one under normal tension and one under high tension. Stretch and recovery properties apply sufficient tension to support the sprained limb. Bandages can also be made in tubular form by a weft knit. This bandage type, available in various diameters, is applicable for holding a dressing on a finger or limb.

Compression bandages are used to exert a certain amount of compression for the treatment and prevention of deep vein thrombosis, leg ulceration and varicose veins. Depending on the compression they provide, compression bandages are classified as light, moderate, high and extra-high compression bandages. They can be woven, warp or weft knitted with cotton and elastomeric yarns.

Nonwoven orthopaedic cushion bandages are used under plaster casts and compression bandages for padding and comfort. They are made of polyester or polypropylene, with blends of natural and other synthetic fibers. Polyurethane foam may also be used. Light needlepunching gives bulk and loft to the structure for greater cushion effect.

6. TEXTILES FOR EXTRACORPOREAL (BIOMEDICAL) DEVICES

Textile materials are used in mechanical organs such as artificial kidney (hemodialysis units), artificial liver and mechanical lung (blood oxygenator) for blood purification (Table 10.9).

In the artificial kidney, blood is circulated through a membrane that retains unwanted waste material. The membrane may be a flat sheet or a bundle of hollow regenerated cellulose fibers in the form of cellophane. Multi-layer fabrics, made of several layers of needlepunched fabrics with different densities are also used in artificial kidneys.

Blood purification is an effective therapy for incurables such as end-stage renal failure. It is used to correct the abnormality of blood quality and quantity in treating sickness. Principles of blood purification therapies are dialysis, filtration and adsorption. Separation membranes and adsorbents are used in blood purification devices. The membrane separation depends on membrane pore size. Purification methods are hemodialysis (dialysis, membrane pore size $1-8$ nm), hemofiltration (filtration, membrane pore size $3-60$ nm), plasma exchange (filtration, porous membrane pore size $0.2-0.6$ micrometer) and hemoadsorption. Hemodialysis accounts for more than 90% of the blood purification treatment in the world which corresponds to 30 million treatments per year keeping 300,000 patients alive. Hemodialysis includes removal of metabolic substances, adjustment of electrolytes and pH and removal of excess water by ultrafiltration and dialysis. During the process, blood contacts with dialysate through a membrane (Figure 10.14).

The materials used in dialysis membranes are regenerated cellulose, cellulose triacetate, acrylonitrile copolymer, poly(methyl methacrylate), ethylene-vinyl alcohol copolymer, polysulfone and polyamide which can be grouped as cellulose and synthetic polymer systems. Eighty percent of the dialyzers use cellulose materials which have excellent permeability for low molecular substances. Pore sizes of membranes vary between $1-3$ nm for conventional membranes and $4-8$ nm for large pore membranes. Today, high efficient hollow fibers have replaced coil or laminate in dialyzer devices. The cross section of a hollow fiber-type dialyzer is shown in Figure 10.15, which consists of 4,000 to 20,000 hollow filaments having an external diameter of 200 to 300 micrometers. Blood flows inside of the fibers and the dialysate flows outside of the fibers. Almost the same materials are used for hemofiltration.

Mechanical lungs use microporous membranes that provide high permeability for gas flow and low permeability for liquid flow which is similar to the natural lung where oxygen and blood come into contact. In these devices,

TABLE 10.9 Materials Used in Extracorporeal Devices [1].

Product Application	Fiber Type	Function
Artificial kidney	Hollow viscose Hollow polyester	Remove waste products from patient's blood
Artificial liver	Hollow viscose	Separate and dispose patient's plasma, and supply fresh plasma
Mechanical lung	Hollow polypropylene Hollow silicone Silicone membrane	Remove carbon dioxide from patient's blood and supply fresh oxygen

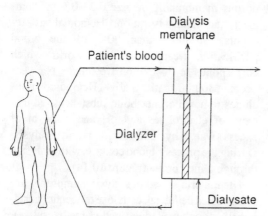

FIGURE 10.14 Schematic of hemodialysis treatment (reprinted by permission of CRC Press [35]).

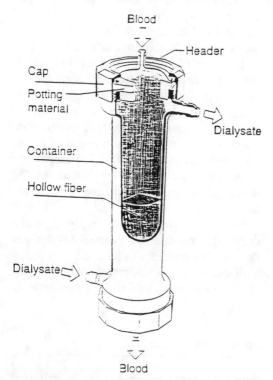

FIGURE 10.15 Cross section of a hollow fiber-type dialyzer (reprinted by permission of CRC Press [35]).

oxygen flows around hollow fibers at extremely low pressure. Blood flows inside of the fiber. The oxygen permeates the micropores of the fiber and comes in contact with the blood. The pressure gradient between the blood and oxygen is kept near zero to prevent mixing of oxygen and blood. Red blood cells capture oxygen by diffusion process.

Hollow fibers are also used in artificial livers and pancreases. Organ cells are placed around the fibers and blood flows inside the fiber. Blood nutrients pass through the fiber wall to the organ cells and enzymes pass from the cells to the blood.

7. HEALTHCARE AND HYGIENE PRODUCTS

As shown in Table 10.10, these products cover a wide range of textile materials for various applications.

7.1 Protective Healthcare Garments

The purpose of protective healthcare garments is to protect healthcare professionals from contamination by blood and other infectious fluids. Protective healthcare textiles include operating and emergency room textiles, barrier products, breathable membranes, surgeon and nurse's caps and masks, footwear, coats, etc. (Figure 10.16). Because of the "unknown" factor, professionals in the emergency room may be at greater risk than professionals in operating rooms. With the spread of HIV and other viruses, the importance of protecting medical personnel is growing which demands development of more effective protective textiles.

Biological protective garments are defined by the Occupational Safety and Health Administration (OSHA) as follows:

Personal protective equipment will be considered appropriate only if it does not permit blood or other potentially infectious materials to pass through to or reach the employee's work clothes, street clothes, undergarments, skin, eyes, mouth or other mucous membranes under the normal conditions of use and for the duration of time which the protective equipment will be used.

TABLE 10.10 *Healthcare and Hygiene Textile Materials [1].*

Product Application	Fiber Type	Manufacture System
Surgical clothing		
Gowns	Cotton, polyester, polypropylene	Nonwoven, woven
Caps	Viscose	Nonwoven
Masks	Viscose, polyester, glass	Nonwoven
Surgical covers		
Drapes	Polyester, polyethylene	Nonwoven, woven
Cloths	Polyester, polyethylene	Nonwoven, woven
Bedding		
Blankets	Cotton, polyester	Woven, knitted
Sheets	Cotton	Woven
Pilowcases	Cotton	Woven
Clothing		
Uniforms	Cotton, polyester	Woven
Protective clothing	Polyester, polypropylene	Nowoven
Incontinence diaper/stock		
Coverstock	Polyester, polypropylene	Nonwoven
Absorbent layer	Wood fluff, superabsorbents	Nonwoven
Outer layer	Polyethylene	Nonwoven
Cloths/wipes	Viscose	Nonwoven
Surgical hosiery	Polyamide, polyester, elastomeric yarns, cotton	Knit

According to this definition, there are two basic requirements for a protective textile garment: it should prevent infectious materials from passing through the skin and it should last long enough.

Recently, two hydrostatic test methods have been adopted by the ASTM as emergency standards:

(1) ES-21-1992 Test Method for Resistance of

Protective Clothing Materials to Synthetic Blood—In this test method synthetic blood with a surface tension of 40 dynes/cm (which is the surface tension of human blood) is used. The fabric specimen is wetted with the synthetic blood for five minutes; then it is kept in a pressurized cell at two psi for one minute. With no pressure, the penetration of blood to the downstream side is visually observed for the next fifty-four minutes.

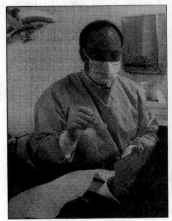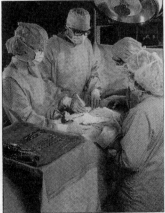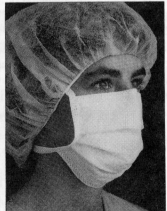

FIGURE 10.16 Various textile products to protect healthcare professionals (courtesy of Fiberweb North America, Inc.).

(2) ES-22-1992 Test Method for Resistance of Protective Clothing Materials to Penetration by Blood-Borne Pathogens Using Viral Penetration as a Test System.

National Fire Protection Association (NFPA) issued a standard in August 1992 addressing biological protection: NFPA 1999 Standard on Protective Clothing for Emergency Medical Operations.

Protective apparel in the medical field should be affordable, breathable, comfortable, dependable, and effective. According to OSHA, "Fluid-resistant clothing shall be worn if there is a potential for splashing or spraying of blood or other potentially infectious materials" and "Fluid-proof clothing shall be worn if there is a potential for clothing becoming soaked with blood or other potentially infectious materials" [36,37].

The impact of AIDS (Acquired Immune Deficiency Syndrome) and other contagious diseases have made hospitals very cautious about protective fabrics. The medical profession was once concerned mainly with protecting patients from germs, now they must protect themselves. These precautions have caused an increase in demand for medical products ranging from gloves and face masks to complete gowns. Manufacturers are also responding to higher demands of protection by producing products with increased barrier levels. Industry is also having to consider the environmentally conscious population when deciding between recycling potentially contaminated products or disposing of the products in landfills.

In the United States, fabric based products used by medical personnel in patient care are considered medical devices. Therefore, their manufacture and sale is regulated by the Federal Drug Administration (FDA). Some of the test methods to determine antimicrobial properties of medical textiles are as follows:

- fungal tests: ASTM 6-21 and AATCC 30 Parts III and IV
- bacterial test methods: AATCC 147, AATCC 100

The protective material has to be waterproof but breathable, i.e., it must allow transmission of moisture vapor. Two types of materials are used to satisfy these contradicting demands [38]:

(1) Monolithic membranes—They are made of polyurethane, polyether/polyester or other copolymers. They have solid layer(s) to provide protection. These film membranes contain polymers that are made of "hard" and "soft" segments in the molecular chain. Due to structural differences between these segments, there is a thermodynamic intolerance between them that causes a separation at their interfaces. Semipermeability takes place at these points of microseparation. The phenomenon is driven by the difference in relative humidity between the surface of the skin and the air.

(2) Microporous membranes—Microporous membranes provide comfort by allowing body perspiration to be transmitted from the skin surface to the air through a fabric. When humidity on the surface of the skin is greater than the atmosphere, moisture can pass through the pores of a microporous membrane by gaseous diffusion and convection. The dimensions of the pore are about $0.1-0.2$ micron which excludes liquid water compared to water vapor which has been estimated to be 0.0004 micron.

Equipment used in an operating room is usually Class I Medical Device. Therefore, in order to show commitment to standards and quality, some textile companies that produce medical textiles register as Medical Device Manufacturers [39].

Needlesticks and sharp objects are the main causes for infection of medical professionals by HIV. Current focus is on development of protective materials that will prevent penetration of needles and sharp objects. Proper sealing of seam areas is critical.

Healthcare garments can be woven, knit or nonwoven. Vinyl garments are also used. Healthcare garments could be washable (and therefore recyclable) or disposable.

Woven cotton fabrics are traditionally used in some surgical gowns because cotton does not produce static electrical charges that can build up and produce electric sparks. However, it may release particles from the surgeon and also generate high levels of dust (lint) which is the source of contamination. Bar-Bac® Plus is a patented, piled, 50/50 polyester/combed cotton blended and tightly woven into fabric. It has a durable, moisture resistant and static control finish. It breathes and has proven to be comfortable to wear. Other properties include abrasion resistance, good tensile strength, fast drying, and reusability. ASEPR Surgical and Precautionary Apparel is made up of high construction continuous filament polyester fabrics. This product has been very successful where high bacteria filtration is required. It is also moisture resistant, lightweight, comfortable, and reusable.

Gore-Tex® is a two layer laminated fabric produced by W. L. Gore & Associates as a barrier material. It is reported that it keeps water and bacteria-proof properties even after fifty laundry autoclave cycles. Other properties reported include good moisture, vapor transmission, light weight and softness. Gore-Lite® is a coating material similar to Gore-Tex film. It is used in nonwovens for medical and surgical purposes. This coating makes the fabric breathable yet liquid-proof. The coating has been used on spun laced, wood pulp/polyester with various reinforcements, woven polyester fabric, and Gore-Tex barrier fabric.

A high performance acrylic fiber, Courtek M, can prevent the buildup of bacteria in cloth and medical equipment. Developed by Courtaulds, the fiber contains a combination of antimicrobial compounds, based on metallic salts. The compounds are bound into the fiber matrix [40]. This makes them more effective because they are not harmed by wear or washing. An aseptic chlorofiber developed by Rhovyl is designed to fight against the proliferation of microorganisms where hygiene is a concern. It contains the Mercurobutol compound, an antiseptic well known in the medical world. This compound is trapped within the basic polymer molecule during the manufacturing process. It is able to remain unaffected through repeated washings. Tests on various bacteria revealed a rate of reduction greater than 90% in the development of bacteria on aseptic chlorofiber compared to traditional chlorofiber. This fiber is also anallergenic and non-toxic, as well as comfortable.

Nonwoven surgical gowns are used to prevent sources of contamination and are composed of nonwoven fabrics and polyethylene films. Nonwovens are known for their disposability, but their performance has proven to be better for protecting medical professionals from infectious liquids. One way the nonwovens industry is marketing itself is by using the term "single use" instead of "disposable" products. Money is the largest problem because many hospitals have much capital invested in laundry facilities, which does not allow for disposables.

Surgical drapes are used in the operating room to cover patients and cover cloths are used to cover the area around the patient. The choice of material is nonwovens which are used as backing material on one or both sides of a film. While the film is impermeable to bacteria, nonwoven backing is highly absorbent to both body perspiration and secretions from the wound. Hydrophobic finishes may also be used as a bacteria barrier. Back to back laminated loop-raised warp knitted polyester fabrics, containing a microporous PTFE film in the middle, provide permeability and comfort as well as resistance to microbiological contaminants.

Surgical masks are made of three layers: a middle layer of extra fine glass fibers or synthetic microfibers covered on both sides by acrylic bonded parallel-laid or wet-laid nonwoven. The performance requirements for surgical facemasks are high bacterial filtration capacity, high air permeability, lightweight and non-allergenic. Nonwoven disposable surgical caps are made of cellulosic fibers with the parallel-laid or spunlaid process. Hydroentanglement is gaining popularity in producing operating room disposable products and garments.

The general requirements for surgical drapes and gowns include liquid repellency and bacterial barrier properties, aesthetics (including conformability, tactile softness, comfort), strength, fiber tie-down properties (lint propensity and abrasion resistance), flame resistance, static safety and toxicity. For surgical drapes, stiffness is very critical because barrier performance may be affected by conformability to patient or equipment. For gowns, comfort and stiffness may affect perspiration and movement. Strength requirements include tensile, tear, burst and puncture resistance. Linting is not wanted

FIGURE 10.17 Various incontinence textile products (courtesy of Viola).

because particles from gown or drape may complicate the wound healing process. Good abrasion resistance is a necessity for the safety of barrier materials. Flame resistance is needed especially for laser applications and oxygen administration. The Consumer Product Safety Commission (CPSC) requires 3.5 seconds burn time on CS-191-53 for gowns, head coverings and surgical drapes. For static discharge propensity, the National Fire Protection Association requires noncellulosic fabrics to be incapable of discharging a spark with sufficient energy to ignite flammable anesthetic gases. Test methods to measure surface resistivity or static discharge are used. Non-toxicity is needed to prevent skin irritation; allergic reaction; tissue or cell damage; and carcinogenic, mutagenic or tetragenic reactions [41].

Nonwoven medical textile products are manufactured using various processes including wet and dry forming, spunbonding, spun lacing and melt blowing.

Laboratory tests for healthcare garments include water repellency, launderability (if recyclable), burst strength and tear strength. The guidelines for operating room materials have been defined by several institutions including the Centers for Disease Control (CDC), Association of Operating Room Nurses (AORN), and the American College of Surgeons' Committee on Operating Room Environment [42,43].

7.2 Other Medical Textiles

Other textile products used in hospitals include bedding, clothing, shoe covers, mattress covers, incontinence products, cloths and wipes. Cotton leno woven blankets have replaced traditional woollen blankets in order to reduce the risk of cross-infection. Nurse's apparel is made of conventional fabrics since no specific requirement is needed other than comfort and durability. In isolation wards and intense care units, composite fabrics are used for protective clothing. These fabrics consist of tissue reinforced with a polyester or polypropylene spun-laid web.

Incontinence textile products can be either diapers or bedding sheets (Figure 10.17). The disposable diapers have three layers: inner covering layer (coverstock), an absorbent layer and an outer layer. The inner covering layer consists of a hydrophilic finish treated polyester web or a spun-laid polypropylene nonwoven.

Tissue paper or nonwoven bonded fabrics are used in cloths and wipes which usually have an antiseptic finish. Cloths and wipes are used to clean wounds prior to wound dressing or to treat rashes and burns. Surgical hosiery materials possess compression characteristics and are used for various applications including support to the limb, treatment of venous disorders, protection in physical activities, etc.

References and Review Questions

1. REFERENCES

1. Rigby, A. J. et al., "Medical Textiles, Textile Materials in Medicine and Surgery," *Textile Horizons*, May 1994.

2. Szycher, M., ed., *Biocompatible Polymers, Metals and Composites*, Technomic Publishing Co., Inc., Lancaster, PA, 1983.

3. Tanzawa, H., "Biomedical Polymers: Current Status and Overview," in Tsuruta, T. et al., eds., *Biomedical Applications of Polymeric Materials*, CRC Press, 1993.

4. Hayashi, T., "Interactions between Polymers and Biosystems," in Tsuruta, T. et al., eds., *Biomedical Applications of Polymeric Materials*, CRC Press, 1993.

5. Fung, Y. C., *Biomechanics: Mechanical Properties of Living Tissues*, Springer-Verlag, Berlin, 1981.

6. Planck, H. et al., "Fact Database: Mediplast," in Planck, H. et al., *Medical Textiles for Implantation*, Springer-Verlag, 1989.

7. Stancell, A. F., "Diffusion through Polymers," in *Polymer Science and Materials*, Tobolsky, A. V. and Mark, H. F., eds., John Wiley & Sons – Interscience, NY, 1971.

8. Blaker, J. W., "Ophthalmic Optics," in *Applied Optics and Optical Engineering*, Shannon, R. R. and Wyant, J. C., eds., Academic Press, Orlando, FL, 1979.

9. Hench, L. L. and Ethridge, E. C., *Biomaterials – An Interface Approach*, Academic Press, NY, 1982.

10. *Biomedical and Dental Applications of Polymers*, Gebelein, C. G. and Koblitz, F. F., eds., Plenum Press, NY, 1979.

11. Ulrich, H. et. al, "Synthesis and Biomedical Applications of Polyurethanes," in *Synthetic Biomedical Polymers, Concepts and Applications*, Szycher, M. and Robinson, W. J., eds., Technomic Publishing Co., Inc., Lancaster, PA, 1980.

12. Phillips, W. M. et al., "The Use of Segmented Polyurethane in Ventricular Assist Devices and Artificial Hearts," in *Synthetic Biomedical Polymers, Concepts and Applications*, Szycher, M. and Robinson, W. J., eds., Technomic Publishing Co., Inc., Lancaster, PA, 1980.

13. Wagner, J. R., "Synthetic Polymer Body Prosthesis Materials" in Durso, D. F., ed., *The Technical Needs: Nonwovens for Medical/Surgical and Consumer Uses*, TAPPI Press, 1986.

14. Cook, J. G. in *Handbook of Textile Fibers – Man-Made Fibers*, 5th Edition, Merrow Pub. Co. Ltd, Durham, 1984.

15. Kimura, Y., *Biodegradable Polymers in Biomedical Applications of Polymeric Materials*, Tsuruta, T. et al., eds., CRC Press, 1993.

16. Ishihara, K., "Blood Compatible Polymers," in Tsuruta, T., *Biomedical Applications of Polymeric Materials*, CRC Press, 1993.

17. Planck, H., "General Aspects in the Use of Medical Textiles for Implantation," in Planck, H., *Medical Textiles for Implantation*, Springer-Verlag, 1989.

18. Bruck, S., *Blood Compatible Synthetic*

Polymers, Charles C Thomas Publisher, Springfield, IL, 1974.

19. Tittmann, F. R. and Beach, W. F., "Parylene Coated Polypropylene Microfibers as Cell Seeding Substrates," in *Synthetic Biomedical Polymers, Concepts and Applications,* Szycher, M. and Robinson, W. J., eds., Technomic Publishing Co., Inc., Lancaster, PA, 1980.

20. Kniepkamp, H., "The Historical Development of Sutures Comparing the Manufacturing Process, Handling Characteristics and Biocompatability," in Planck, H. et al., *Medical Textiles for Implantation,* Springer-Verlag, 1989.

21. Elser, C., Renardy, M., and Planck, H., "The Effect of Bacteria on the Degradation of Absorbable Sutures," in Planck, H. et al., *Medical Textiles for Implantation,* Springer-Verlag, 1989.

22. Macieira-Coelho, A. et al., "Properties of Protein Polymers as Substratum for Cell Growth in Vitro," *J. Cell Physiol.,* Vol. 83, 1974.

23. Rembaum, A. and Senyei, A. E., "Interaction of Living Cells with Polyionenes and Polyionene-coated Surfaces," *J. Biomed. Mater. Res. Symp.,* Vol. 8, 1977.

24. Bell, E. et al., "Living Tissue Formed in Vitro and Accepted as Skin-Equivalent Tissue of Full Thickness," *Science,* Vol. 211, 1981.

25. Gifford, G. H. et al., "Tissue Reactivity of Silicone Rubber Implants," *J. Biomed. Mater. Res.,* Vol. 10, 1976.

26. Freudiger, S., "Requirements for Artificial Cruciate Ligaments," *Technical Textiles International,* June 1994.

27. Ishihara, K., "Hard Tissue Compatible Polymers in Biomedical Applications of Polymeric Materials," Tsuruta, T. et al., eds., *Biomedical Applications of Polymeric Materials,* CRC Press, 1993.

28. Moreland, J., "An Overview of Textiles in Vascular Grafts," *International Fiber Journal,* October 1994.

29. Debille, E. et al., "Dilatability and Stretching Characteristics of Polyester Arterial Prosthesis. Evaluation of the Elastic Behavior," in Planck, H. et al., *Medical Textiles for Implantation,* Springer-Verlag, 1989.

30. Goggins, J., Meadox Medicals, Inc., private communication.

31. Nakabayashi, N., "Dental Biomaterials," in *Biomedical Applications of Polymeric Materials,* Tsuruta, T. et al., CRC Press, 1993.

32. Phillips, R. W., *Skinner's Science of Dental Materials,* W. B. Saunders, Philadelphia, 1991.

33. Lennox-Kerr, P., "Medical Textiles—Poised for Massive Growth," *Textile Technology International,* 1988.

34. Latham, R. H., "Polypropylene—A New Concept in Medical and Hygiene Fabrics," in *The University of Leeds Conference on Medical Applications of Textiles,* July 1981.

35. Nishimura, T., "Polymer Materials for Blood Purification" in *Biomedical Applications of Polymeric Materials,* Tsuruta, T. et al., CRC Press, 1993.

36. Greenwald, E. and Zins, H., Medical Textiles, IFAI, Nashville, 1991.

37. *Federal Register.* "Occupational Exposure to Bloodborne Pathogens: Proposed Rule and Notice of Hearing," 30 May 1989, Part II, Department of Labor, Occupational Safety and Health Administration, 29 CFR Part 1910, Vol. 54, No. 102: 23042-23139.

38. Martz, J., "Protective Health Care Garments; The Challenge to Meet New Chemical and Technical Standards at an Affordable Cost," *IFAI Second International High-Performance Fabrics Conference,* Nov. 12–13, 1992, Boston, MA.

39. Stan Littlejohn, Medical Textiles, *IFAI 91 Expo,* Nashville.

40. Owen, P., "Acrylic Kills Bacteria," *Textile Monthly,* July 1990.

41. National Fire Protection Association, "Standard for Facilities," NFPA-99-1984, Batterymarch Park, MA.

42. "Association of Operating Room Nurses

Standards for Draping and Gowning Materials," in *AORN Standards of Practice,* 1982.

43. "American College of Surgeons Committee on Control of Surgical Infections," *Manual on Control of Infection in Surgical Patients,* J. B. Lippincott Co., 1976.

2. REVIEW QUESTIONS

1. Why are textile materials, in general, suitable for medical applications?
2. Define biocompatibility. What are the requirements for biocompatibility?
3. What are biodegradability and bioabsorbability? Explain the major characteristics of these types of materials.
4. Why is blood compatibility important for medical textiles?
5. What are the major requirements of textiles for implantation? Explain.
6. What is an extracorporeal device? How are textiles used in these devices?
7. What properties are important for sutures? What kinds of test methods are used for sutures?
8. What are the differences between artificial tendons and ligaments?
9. What types of mechanical properties are important for hard tissue implants? Why?
10. What types of fabric structures are used for vascular implants? Explain their properties.
11. What is the structure of fibers used in extracorporeal devices?
12. Explain how monolithic and microporous membranes work.

MILITARY AND DEFENSE TEXTILES

Military and Defense Textiles

S. ADANUR

1. INTRODUCTION

"Of the last 3,422 years, only 268 have been free of armed conflict somewhere in the world" [1]. This fact indicates the significance of military and defense textiles. Textiles has long been considered one of the most essential industries—"second only to steel in essentiality" —for the armed forces [2].

The U.S. Defense Department has in its inventory approximately 10,000 items made partially or entirely from textiles. Some 300 of these items are considered "combat essential" which include uniforms, protective clothing, parachutes, sweaters, socks, gloves, coveralls, sandbags, sheets, blankets and hospital supplies [3].

Although the military strategies of the world powers are changing direction as a result of the end of the cold war, there will be a need for military textiles as long as there are military personnel. The recent trend in war fighting is towards quick deployment of agile forces made of fewer troops with increasingly sophisticated gear and weapons [4]. The military textiles should meet the requirements of these new trends.

Soldiers are exposed to environmental conditions more than civilians. The differences between estimated outdoor-exposure lifetimes for civilian office workers, outdoor staff and peacetime soldiers are 3%, 8% and 20%, respectively, which explains why civilian clothing is not adequate for military use. Due to extremely strict requirements, the design, development and acceptance of a new military uniform takes a lot longer than it does in civilian clothing.

Textiles in military and defense can be grouped into two categories [5]:

- protective clothing and individual equipment, including battle-dress uniforms, ballistic protective vests and helmets, chemical protective uniforms, field packs, equipment belts and suspenders, mountain climbing ropes, etc. (Figure 11.1)
- textiles used in defense systems and weapons, including tents, shelters, parachutes, harnesses, cords, tarpaulins, and textile composites (Figure 11.2)

Typical military textile and apparel items include [3,6]:

- service uniforms
- camouflage and utility uniforms
- high performance uniforms such as those used by aircraft crews, parachutists and explosive ordnance disposal specialists
- chemical and biological protective suits, masks, gloves, coveralls, boots
- liners for protective clothing
- anti-mine boots
- mine removal chutes
- sweaters, socks, underwear, hats, gloves, ponchos, helmets
- rainwear
- scarves, handkerchiefs
- helicopter dust covers
- rope, netting, cord, shoelaces
- cockpit ejection and weapons delivery chutes
- flags
- tents, tent liners, other shelters
- strong, lightweight fabrics for aircraft components (Chapter 7.0)
- ballistic blankets
- wiping cloths for gun cleaning kits

FIGURE 11.1 MIlitary protective clothing (courtesy of the U.S. Army).

FIGURE 11.2 Military tents (courtesy of the U.S. Army).

- military hospital supplies, gauze, bandages, slings
- towels
- aircraft fuel cells
- sandbags
- equipment packs
- gas mask bags
- duffle bags
- cargo and flare parachutes
- mail bags
- bomb curtains for Air Force and Navy
- draperies, carpets
- sheets, blankets
- cots
- sewing threads for protective apparel
- load-carrying equipment
- electrical insulation
- ammunition pouches, gun covers

2. PROTECTIVE CLOTHING AND INDIVIDUAL EQUIPMENT

In the past, the main objective of military protective clothing was to protect the soldier from environmental effects such as rain, snow, cold, heat and wind as well as to give him freedom to maneuver. With the development of chemical, biological, thermonuclear and more effective fragmentation weapons, small arms surveillance and sensor systems, the requirements for military protective clothing have increased dramatically. Textiles used in military clothing have to satisfy strict, quantifiable requirements.

In modern military establishments the individual soldier is treated as a "system" (Figure 11.3). The clothing of a soldier consists of uniforms, hosiery, knitwear, underwear, shirtings, etc. The purpose of military protective clothing is to help maximize the survivability, sustainability and combat effectiveness of the individual soldier system against extreme weather conditions, ballistics, and nuclear, biological, chemical (NBC) warfare [4,5].

The major requirements for an advanced integrated combat clothing system can be grouped into four categories [1]:

- physical requirements: durability to prolonged exposure to inclement weather

FIGURE 11.3 A soldier and his equipment is considered a "system" (courtesy of the U.S. Army).

and heavy wear, good tensile and tear strength and abrasion resistance
- environmental requirements: water repellency, windproofness, snow-shedding and insectproofness
- physiological requirements: low weight, easy to wear, minimum heat stress, air, moisture and vapor permeability and comfort
- battlefield requirements: ballistic protection, flame resistance, resistance to chemical and biological agents, resistance to long range thermal effects of nuclear weapons, good camouflage properties and low noise generation

In addition to these requirements, uniforms for special duties may require other properties such as resistance to propellant fuels, strong

TABLE 11.1 Recent Advances in Military Clothing [5].

Technology	Materials	Items
High performance fibers/fabrics	Kevlar® Spectra®	Lightweight combat helmet
Fine denier fibers	Nylon (staple and filament)	Hot weather battle-dress uniform arctic overwhite
Microdenier fibers	Synthetic down (Primaloft®)	Cold weather clothing and sleeping bags
Stitchless seaming	Synthetic fabrics	Chemical protective clothing Rainwear Sleeping bags Cold weather clothing

acids, and liquefied gases (oxygen, nitrogen, ammonia).

Some of these requirements may be in conflict. For example increasing environmental protection may create physiological concerns such as heat stress and fatigue. The requirements on the military uniform dictated by battlefield conditions may impair the operational efficiency of the soldier. Physical properties must give the clothing durability for extended periods of use. Low weight and bulk are usually preferred for operational efficiency but in the case of ballistic vests, the level of protection would probably be reduced. Tightly woven fabrics may provide improved water resistance, but reduce air permeability.

Recent advances in military clothing include high performance fibers/fabrics, flame resistant materials, microdenier fibers, bioengineered fibers, blended yarns and stitchless seaming. These new technologies have led to the materials and end products that are shown in Table 11.1. Table 11.2 shows integrated protection provided by recently improved battle-dress uniform fabrics.

2.1 Ballistic Protection

Principles of Ballistic Protection

Ballistic protection of a soldier involves protection of the body and eyes against projectiles with various shape, size, and impact velocity. The principle of ballistic protection by cloth armor is to dissipate the energy of the fragment/shrapnel through stretching and breaking the yarns in the many layers of high performance woven fabrics in the ballistic vest (Figure 11.4). Each fabric layer reduces the energy of the projectile. In each layer, impact energy from the struck fibers is absorbed and dispersed to other fibers in the weave of the fabric. Their transfer occurs at crossover points where the fibers are interwoven. In general, a plain balanced, woven fabric is preferred for ballistic protection to maximize the number of crossover points. Since the woven fibers work together to dissipate the impact energy, a relatively large area of the fabric becomes involved in preventing the bullet's penetration [7].

Two conflicting requirements have to be met

TABLE 11.2 Integrated Protection Offered by the New, Improved Battle-Dress Uniform Fabrics [5].

Fabric Characteristics	Nylon, Cotton	Nomex®, Kevlar®, P-140	Cotton, Kevlar®, Nylon, P-140
Flame resistance	N	Y	Y
Liquid chemical agent resistance	Y	N	Y
Electrostatic resistance	N	Y	Y
Day/night camouflage	Y	Y	Y
Durability	Y	Y	Y+
Weight (oz/yd²)	7.0	5.5	6.5
Cost ($/linear yd)	4.00	20.00	11.00 (estimate)

Y: yes, N: no.

FIGURE 11.4 Multiple microflash photograph of a Kevlar® yarn being impacted by a 1.1 gram fragment simulating projectile at 250 m/s (courtesy of U.S. Army Natick Research, Development and Engineering Center).

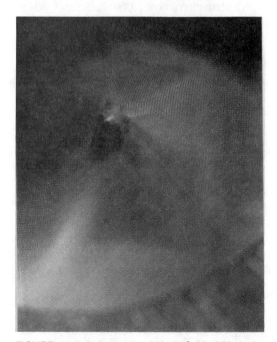

FIGURE 11.5 A single ply of Kevlar® 29, 200 denier fabric being impacted by a 1 gram right circular cylinder projectile at 180 m/s, 83 microseconds after impact (courtesy of U.S. Army Natick Research, Development and Engineering Center).

in ballistic protective clothing: (1) ballistic performance which generally requires high mass and bulk, and (2) lightweight and comfortable clothing. Eye protection requires light transparent materials that resist ballistic penetration, are scratch-resistant, and have optical clarity [8]. Factors that influence the energy absorption characteristics of ballistic protective systems are properties of constituent materials, fabric design parameters, number of fabric layers, fabric density, and impact conditions such as projectile mass, striking velocity and geometry [9].

Mechanics of ballistic impact has been the subject of extensive research since World War II [10−16]. Upon impact by a projectile, a single yarn develops a V-shaped transverse deflection with the projectile at the apex of the V (Figure 11.4), whereas a layer of cloth develops a tent-shaped transverse deflection with the projectile at the apex of the tent (Figure 11.5). At the impact moment, a strain wave develops and propagates away from the impact point with the speed of sound. In the strain wave's wake, the textile yarn develops a motion toward the impact point which feeds the developing transverse deflection. In single yarns, the longitudinal wave speed is equal to the square root of the material's specific modulus. Similarly, transverse and longitudinal waves develop and propagate away from the impact point in each layer of a multiple-layer fabric armor.

Figure 11.6 shows the typical energy absorption characteristics of a single-ply fabric made of Kevlar® 29. Fabric energy absorption decreases rapidly when the projectile velocity just exceeds the system's ballistic limit, then decreases gradually and levels out at high velocities. Ballistic limit is defined as the projectile velocity at which penetration by the projectile starts. The location of strain front in nearby yarns depends on the transverse wave velocity and the fabric warp and weft count. The longitudinal wave speed depends on woven fabric material and construction parameters. It was reported that the ballistic response of fabrics is closely related to the ballistic response of constituent yarns. Due to yarn slippage at the impact point, loose woven structures or fabrics with low yarn-to-yarn friction have low ballistic resistance. Balanced weaves provide better ballistic resistance than unbalanced weaves.

Fibers and Fabrics for Ballistic Protection

Vests made of layers of "ballistic" nylon fabric ("fragmentation vest") were used in the past for protection against fragmentation weapons and low-velocity small arms. The disadvantage of fragmentation vests was the heavy load caused by layers of high tenacity fibers. Wetting reduced their ballistic performance and created an impermeable cover which increased heat stress by reducing moisture-vapor permeability [1]. It is important that the armor does not lose its protective properties when wet.

New requirements for protection against more dangerous weapons have spurred the development of high performance fibers and flexible composites. Today the most widely used materials for ballistic protection are aramid (Kevlar® by DuPont, Twaron® marketed by Akzo Industrial Fibers, Technora® by Teijin of Japan), ultra-high molecular weight polyethylene (Spectra® by AlliedSignal, Dynema® by DSM of Netherlands, Tekmilon® by Mitsui Petro Chemical of Japan) and liquid-crystal polymer-based (Vectran® by Hoechst Celanese) fibers [17,18].

Kevlar® offers drastically improved ballistic protection while reducing the weight and bulk of the vest. As a result, helmets and vests made of Kevlar® have replaced the old "steel pot" helmet and nylon vest. The vest for the U.S. Army Personnel Armor System for Ground Troops (PASGT) is mainly made of Kevlar® 29 with a light nylon shell fabric. It was reported that ballistic protection was improved up to 40% by using Kevlar®. All of the projectile energy is absorbed by the fabric armor system up to the ballistic limit. When the projectile velocity exceeds the ballistic limit, the energy absorption capacity of the fabric armor system decreases rapidly [9,19]. Strain in the fabric decreases with distance from the impact point.

Kevlar® KM2, which is the latest Kevlar® from DuPont for ballistic protection, has a tensile modulus of 28 g/den versus 22 g/den for Kevlar® 29. Kevlar® is up to five times stronger than steel on an equal weight basis and significantly stronger than nylon (Figure 11.7). The energy dissipation properties are also reported to be improved. The performance of Twaron®, which

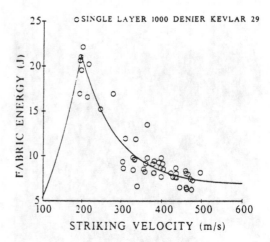

FIGURE 11.6 Energy absorption characteristic of single-layer Kevlar® 29 fabric [9].

is another aramid fiber from Akzo, is similar to Kevlar® 29 and 129 fibers. Recently introduced Twaron® 2000, which is the second generation of Twaron®, is reported to be superior in ballistic performance and comfort. Twaron® 2000 is offered in densities of 840, 1,000 and 1,500 denier with a filament count of 1,000. Technora®, which is a para-aramid fiber, offers good tensile strength and modulus, chemical and hydrolytic resistance, fatigue resistance and dimensional stability. Stiff and highly oriented molecular

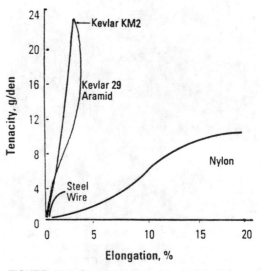

FIGURE 11.7 Stress-strain curves of Kevlar® versus steel and nylon (courtesy of DuPont).

structure causes creep and stress relaxation to be low [18].

Spectra® fibers are produced using a patented gel-spinning process. Spectra® has very high tensile strength and good chemical resistance. Spectra® 1000 has higher strength and modulus than Spectra® 900. Spectra® Shield is a composite armor structure made of nonwoven ballistic armor material in which two layers of unidirectional fibers are impregnated and bound together with a polymeric resin in perpendicular directions. Multiple layers of Spectra® Shield can be pressed together to the desired thickness and weight for different ballistic requirements. It can be made flexible, thin and lightweight. Spectra® Shield provides good protection against high-velocity rounds, multiple hits and shots fired at an angle. Vests made of flexible composites are being used more and more because they offer many advantages over woven fabrics. Flexible composite body armor weighs around 3 to 4 lbs compared to 5 to 6 lbs of traditional nylon vests. Their ballistic performance is also better. In addition to intercepting the projectile, composite armors may also minimize blunt trauma which is the punching effect on the body caused by the arrested projectile. A ballistic vest also provides a measure of protection against vehicle accidents and other impact injuries.

Dyneema® fiber from DSM, which is made of ultra-high molecular weight polyethylene, is produced using a gel-spinning process. Parallel orientation of the molecules and high crystallinity give the fiber ultra-high strength. Dyneema® SK60 is used for general applications and Dyneema® SK66 is used in fabrics for ballistic applications. In UD66, which is a relatively new construction, the yarns are laid unidirectionally, cross-plied and impregnated with a matrix. Tekmilon® is a high modulus, high strength, high molecular weight polyethylene fiber with low weight, and good water, abrasion and chemical resistance. Its properties are similar to Spectra® and Dyneema® [18].

Vectran® from Hoechst Celanese is a thermoplastic monofilament made of liquid crystal polymer. Its properties include high strength and rigidity, low moisture absorption, negative coefficient of thermal expansion and good chemical resistance. It is reported to be five times stronger than steel and ten times stronger than aluminum [18].

The use of hybrid structures is increasing in ballistic protection systems (Figure 11.8). Aramid fabrics blended with ultra-high molecular weight polyethylene unidirectional cross-plied nonwoven material which is impregnated with matrix offers the flexibility and comfort of fabric and the high performance of composite nonwoven material.

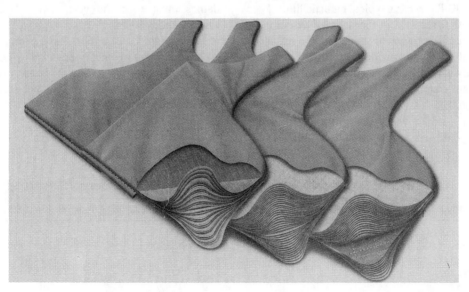

FIGURE 11.8 Ballistic panel configurations; all Kevlar® 129, all Spectra® Shield and hybrid (Spectra Shield and Kevlar) package (courtesy of IFAI).

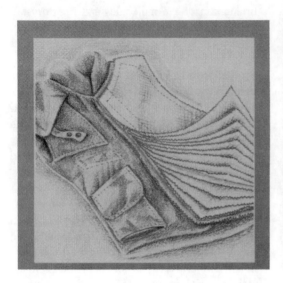

FIGURE 11.9 The PASGT vest (courtesy of DuPont).

Ballistic Vests

Each PASGT vest contains 13 layers of MIL-C-44050 Type II, 14 oz/sq. yd ballistic fabric of Kevlar® covered by an 8 oz/sq. yd. camouflage ballistic nylon fabric (Figure 11.9). The helmet/vest system is designed to work together as a system to protect the head and vital body parts. The PASGT vest has a typical ballistic limit, V_{50}, of 1,650 ft/sec (V_{50} is the average velocity at which half the fragments penetrate the material and half are stopped within a specified velocity spread; determined using a 0.22 caliber, Type 2, 1.1 gram fragment simulating projectile; MIL-STD-662E).

Table 11.3 shows typical properties of U.S. military ballistic fabrics made of Kevlar®. Type I fabric is used for Combat Vehicle Crewman (CVC) vests (ballistic limit, V_{50}, 1,625 ft/sec, Figure 11.10), and Type III fabric is used for the Ranger Vest (ballistic limit, V_{50}, 2,100 ft/sec, Figure 11.11). In addition to ballistic protection, vests made of Kevlar® offer excellent flame protection. The non-melting and self-extinguishing properties of Kevlar® help keep the heat and flames outside of the vest, away from the wearer's body [6].

Lightweight body armors have been widely used by law enforcement personnel within the last twenty-five years. As a result, officer homicides in the United States have been reduced from 130 per year in the early 1970s to 70 per year since the early 1980s [20].

The protective clothing for Explosive Ord-

TABLE 11.3 Military Ballistic Fabrics of Kevlar®, MIL-C-44050 (courtesy of DuPont).

Property	Type I* (Kevlar® 29)	Type II** (Kevlar® 29)	Type III† (Kevlar® KM2)
Denier	1,000	1,500	850
Weave	Plain	2 × 2 Basket	Plain
End × picks per inch	31 × 31	35 × 35	31 × 31
Weight, oz/yd²	8.3	14.0	6.8
Thickness (mil)	14	25	10

*Available scoured (Class 1) or scoured and treated for water repellency (Class 2) for use in fragment- or bullet-resistant vests.
**Available scoured (Class 1) for use in helmets or scoured and treated for water repellency (Class 2) or use in fragment-resistant vests.
†Available with adhesion modified treatment (Class 3) for use in helmets.

nance Disposal (EOD) specialists requires far better protection. The EOD suit's panel is made of sixteen layers of Kevlar®. The suit is available in five sizes and weighs between 54 and 64 lbs. The jacket is designed to provide all-around protection while the trousers provide protection in the front. The EOD helmet is similar to the basic troop helmet with a bonnet of twelve more layers of Kevlar®. It was reported that the EOD suit can stop a 2 lb pipe bomb at close range [18].

Today, a variety of personal body armor is available which includes concealable vests and T-shirts to wear under a uniform, ''executive'' vests and overshirts for plainclothes persons, windbreakers, raincoats, briefcases, etc. [7].

Ballistic Helmets

The use of composites allowed the development of different helmet sizes for the men and women of the U.S. Army. The PASGT helmet is a one-piece design made of nineteen layers of Kevlar® 29 MIL-C-44050 Type II fabric and polyvinyl butyral (PVB) and phenol-formaldehyde (PF) resins (Figure 11.12). The helmet is manufactured using a compression molding technique [19]. Figure 11.13 shows the evolution of performance versus weight for the ballistic protective helmet. The PASGT helmet replaced the old ''steel pot'' helmet in the late 1970s. It weighs approximately 3.25 lbs and provides excellent ballistic and flame protection. Typical ballistic limit, V_{50}, of the helmet is around 2,000 ft/sec. The lightweight PASGT helmet is made of MIL-C-44050 Type III fabric. It provides ballistic protection equal to Kevlar® 29 but with a 15% reduction in weight. The ''improved'' Combat Vehicle Crewman (CVC) helmet utilizes the same material system as the lightweight PASGT helmet. It provides PASGT-level ballistic protection and superior fragmentation resistance at a lower weight. Both helmet systems have a typical ballistic limit (V_{50}) of 2,150 ft/sec. The effective number of plies is thirty-four for both helmets.

Designing for Ballistic Protection

The important design considerations for ballistic garments are the selection of ballistic-resistant material, the required degree of protection

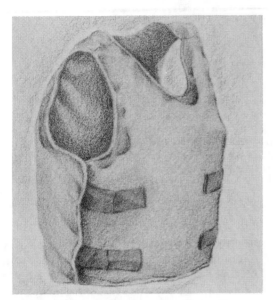

FIGURE 11.10 The CVC vest for protection against spall and fragmentation as well as against heat and fire (courtesy of DuPont).

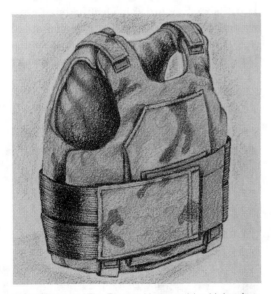

FIGURE 11.11 The Ranger Vest provides higher fragmentation resistance at a lighter weight (courtesy of DuPont).

FIGURE 11.12 The old M-1 steel helmet (left) compared with the U.S. current Personnel Armor System for Ground Troops (PASGT) Kevlar® helmet (right). Both helmets are the same weight, but the PASGT helmet provides substantially increased protection (courtesy of U.S. Army Natick Research, Development and Engineering Center).

(projectile or bullet type, caliber, impact velocity), final weight of the uniform, wearer comfort and ease of movement. Each fiber and fabric type has unique ballistic properties. Depending on the specific threat to be addressed, composite ballistic material can improve the performance of the protective system. Coated ballistic fabrics may also be used. Some manufacturers use a non-ballistic layer to increase blunt-trauma protection. There are different ways of assembling ballistic panels into a single unit such as bias stitching around the edge of the panel, tack-stitching at several locations, biaxial stitching, or quilting the entire panel. Ballistic protective clothing is generally worn beneath the shirt in civilian life, but the reverse is true for the majority of the military [19].

It is important to note that an armor which is effective against a projectile at a certain impact speed may not be effective against the same projectile if the impact speed is increased. The National Institute of Justice (NIJ) Standard-0101.03, which was issued by the National Bureau of Standards' Law Enforcement Standards Laboratory in 1987, identifies six formal armor classification types and a special armor type: Type I, IIA, II, IIIA, III, IV and Special Type Armor. Each armor type provides either multiple-hit or single-hit ballistic protection against a specified bullet at a specified velocity.

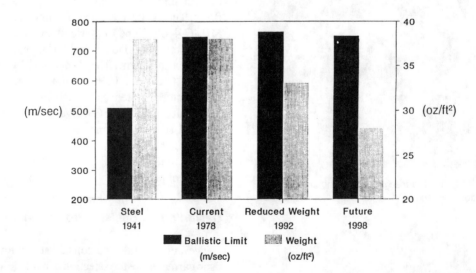

FIGURE 11.13 Ballistic protective helmet performance versus weight [5].

For example, Type III "provides protection against 7.62 mm FMJ bullets (U.S. military designation M-80) with nominal masses of 150 gr, impacting at a velocity of 2,750 feet per second. It also provides protection against such threats as a .223 Remington (5.56 mm FMJ), .30-carbine FMJ 12-gauge rifle slug, as well as Type I through Type IIIA threats." The National Institute of Justice Technology Assessment Program Information Center (TAPIC) has developed a voluntary national testing program for ballistic items. With this program, first the armor is identified against the NIJ Standard-0101.03 for compliance. Then the armor is sent to an independent testing lab for performance tests [19]. Special type armor covers any other protection level that is not covered in one of the six armor types.

2.2 Chemical and Biological Protection

Modern chemical warfare started during World War I. Since then, various chemical agents including asphyxiants, irritants, vesicants, blood agents, and nerve agents have been used in battlefields. To protect the soldier against chemical and biological warfare agents, permeable (activated carbon systems), semipermeable material systems and impermeable barrier materials are used for the U.S. Army (Figure 11.14).

Currently, the permeable chemical protective fabric systems use activated carbon liners. These fabric systems allow evaporative cooling for heat dissipation and excellent chemical agent vapor protection. The supplemental finish on their shell fabrics provides adequate liquid protection. Permeable chemical barrier fabrics consist of a nylon tricot base fabric with polyurethane foam that is impregnated with activated charcoal. An outer fabric is laminated with this system.

Semipermeable materials are produced in two forms: coating and films. These materials have nonporous, ultraporous or microporous structures which can affect their qualifications for the end uses. Non-porous materials offer good chemical/biological (CB) and environmental protection because of their solid structure. However, they exhibit low moisture vapor transmission rate. Microporous and ultraporous materials offer the most desirable properties; high moisture vapor transmission rate, high hydrostatic resis-

FIGURE 11.14 Chemical and biological protective military suit (courtesy of the U.S. Army).

tance and excellent CB protection. The tests required to assess these structures include moisture vapor transmission rate, hydrostatic resistance, aerosol resistance and chemical resistance. There are many semipermeable materials which are potentially applicable for chemical/biological protective clothing; however, a few have balanced properties. Presently, a microporous semipermeable membrane is being used by the U.S. Army in combination with an active carbon-containing foam in the Air Crew Uniform Integrated Battlefield for CB protection [21].

Selectively permeable materials are a type of semipermeable materials. They are thin, lightweight, flexible, and allow selective permeation of water vapor molecules, but not chemical and biological warfare agent molecules in liquid, aerosol and vapor forms. Selectively permeable materials prevent chemical agent vapor permeation without the need for adsorptive materials. Integrated protective fabric systems using selectively permeable membranes will also provide environmental protection, i.e., wind, rain, snow and other wet-weather environments. Research in selectively permeable materials is being done by the U.S. Army Natick Center.

The impermeable barrier materials, which are

made of rubber and coated fabrics, act as a physical barrier to chemicals including petroleum, oils and lubricants (POL). The impermeable fabric systems provide the best CB agent protection; however, they will cause heat stress to the wearers in hot, moderate climates and body chill in cold climates because of their extremely low moisture vapor transmission rates. Completely sealed, impermeable clothing cannot be used for long periods of time. Therefore, these materials require the use of a heavy, bulky and expensive microclimate cooling system.

The Army Natick R&D Center developed air-conditioned undergarment clothing systems to be used where heat stress is likely such as in a tank. Figure 11.15 shows the air-cooled undergarment assembly. The system is connected to a refrigerant source in the tank that distributes conditioned air which makes operation possible up to 12 hours in very hot conditions. The conditioned air (around 70° F) is passed into a tube in the vest. The vest directs the cool air to the upper torso to cool the soldier's vital organs. The air that reaches the vest is controlled by a switch.

Waterproof, breathable, sorptive materials (WBS) which contain activated carbon or super-activated carbon have been identified by the U.S. Army as potential alternatives for the permeable chemical protective fabric liner and impermeable materials as they offer the performance of permeable materials and the protection of impermeable materials.

Vests with microclimate cooling systems have been developed. In these systems, an antifreeze solution, which is cooled by a device carried in the backpack, is circulated inside the vest's tubing. The vest's temperature is controlled by the liquid temperature. The backpack includes a motor to drive the circulator system, a battery to drive the motor and a reservoir for the coolant [19].

Chemical protective (CP) gloves are made of butyl rubber. The CP gloves are produced in three thicknesses: 7, 14 and 25 mil. The CP gloves provide chemical warfare (CW) agent resistance. However, these gloves do not provide flame retardance and POL resistance. Research in this area is continuing.

In a separate application, a totally encapsulated chemical-protective suit was developed by

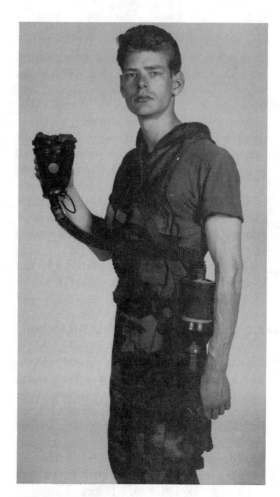

FIGURE 11.15 Standard microclimate cooling air vest for M1A2 main battle tank (courtesy of U.S. Army Natick Research, Development and Engineering Center).

the U.S. Coast Guard for protection during chemical spill response (Figure 11.16). The suit is made of fluoropolymer (Teflon®)/aramid composite material.

2.3 Nuclear Protective Fabrics

The need for protection from nuclear flash appeared in the 1940s. The major effects of nuclear weapons are blast, short-range radiation (gamma and neutron radiation), and longer-range high intensity thermal radiation. There is hardly anything that can be done to protect a person from blast and short-range radiation close to the nuclear burst. Therefore, the purpose of military protective clothing is to provide high-intensity

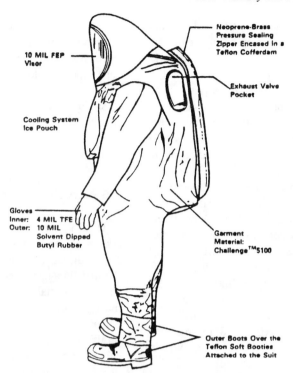

Neoprene-Brass
Pressure Sealing
Zipper Encased in a
Teflon Cofferdam

10 MIL FEP
Visor

Exhaust Valve
Pocket

Cooling System
Ice Pouch

Gloves
Inner: 4 MIL TFE
Outer: 10 MIL
        Solvent Dipped
        Butyl Rubber

Garment
Material:
Challenge™5100

Outer Boots Over the
Teflon Soft Booties
Attached to the Suit

FIGURE 11.16 Chemical response suit developed by the U.S. Coast Guard [22].

thermal radiation and flame resistance. Non-melting, non-flammable textiles with good resistance to intense thermal spikes even for short periods of time are used for this purpose [23].

It is well known that for a material exposed to radiation of a particular wavelength, the total of the relative proportional values of absorption, transmittance and reflectance is one. To protect against high-intensity thermal radiation, absorption and transmittance should be low and reflectance should be high. Fabrics would be ideal for this purpose, but they might create a problem for camouflage in certain areas.

From the 1940s to the 1980s, all nuclear protective fabrics were made of 100% cotton twill or sateen. However, cotton has some shortcomings for nuclear protection which include irregular surface (not desirable for decontamination) and high linting, which reduces the protection. The second generation of nuclear protective fabrics utilized a 65/35 polyester/cotton blend which accounts for 80% of all nuclear protective wear in the U.S. Poly/cotton fabric has a longer life than all-cotton fabrics because of

higher yarn strength and ease of decontamination. Nuclear protective clothing are decontaminated at special protective clothing laundries. Recently, new and better nuclear protective fabrics have been developed including nylon/polyester and nylon/carbon blends. Fabric construction of 99/1 nylon/carbon fabric offers improved barrier properties, reduced heat stress, reduced linting, better contamination resistance and easier decontamination [24].

2.4 Environmental Protection

The U.S. Army requires different uniforms for use in five basic climates: hot-wet, hot-dry, temperate, cold-wet, and cold-dry. In hot-wet, hot-dry and temperate environments battle-dress uniforms are worn that are made from a blend of 50% cotton, 49% high tenacity nylon, and 1% static dissipative fiber. The balance of durability and comfort are critical performance factors. The temperate battle-dress uniform is made from a 7 ounce per square yard twill fabric. The hot weather and desert battle-dress uniforms are made from a 6 ounce per square yard ripstop poplin fabric. Woodland camouflage patterns consisting of the four colors of light and dark green, brown and black, are applied using roller screen printing [25]. The three color desert pattern consisting of light tan, light brown and light khaki are applied using the same technique. The sewing threads are cotton, rayon or polyester with a polyester core. For flame resistance, plain and oxford weave aramid cloth is specified. The aramid fiber blend consists of 92% meta-aramid, 5% para-aramid and 3% static dissipative fiber. The fabrics are woodland or desert camouflage printed. Research is continuing for improved flame-resistant fabrics for military use.

Protection from extremely cold weather is another aspect of military clothing. The U.S. Army Natick Research, Development and Engineering Center developed the Extended Cold Weather Clothing System (ECWCS) for this purpose. It is made of synthetic fibers which transport the moisture away from the body. The system, which provides insulation, waterproofness and moisture vapor permeability, consists of polypropylene underwear, cold weather trousers, field jacket and trouser made with a semi-permeable film laminated fabric. This system provides warmth and dryness without excess weight. This system performs well at −25°F. By adding a polyester fiber pile shirt and bib overalls, the range can be extended to −60°F [19].

Protection against liquid water penetration is achieved by polymer coating fabrics using rubber, PVC, neoprene, acrylics, or polyurethanes. Waterproof fabrics may allow moisture-vapor permeability even though it may be a small amount, e.g., up to 200 g/m² per 24 hr for a lightweight polyurethane coated nylon fabric. For an effective moisture-vapor permeable clothing a minimum of 2,000−2,500 g/m² per 24 hour permeability is required [1].

Although fabrics treated with water-repellent finishes can withstand light showers, they are not waterproof. The U.S. Army treats various tightly woven nylon and cotton blend fabrics with a fluorocarbon-based water and oil repellent finish. The combination of the dense fabric construction and durable fluorocarbon treatment results in an air and moisture vapor permeable fabric that provides durable water and oil repel-

lency for the life of the garment. Since the fabrics will also repel perspiration they are used in garments that are not worn next to the skin such as caps, hoods, cold weather field coats and trousers.

Smooth fabric surfaces are good for snow-shedding. Continuous filament nylon or polyester fabrics give smooth fabric surfaces. Silicon treatment also reduces snow or ice adhesion.

2.5 Surveillance and Camouflage

Textiles for military uniforms, individual equipment and tents must reduce visibility to sensor threats such as image intensifiers, radar, thermal imagers, and multi-spectral sensors. Textile materials used in military applications must provide protection against infrared image detection by blending in with the terrain. The traditional camouflage colors against a subtropical background are black, brown and green (Figure 11.17). In deserts, the camouflage colors become different tones of brown. In snowy areas, the obvious choice for camouflage is a white color garment which may be a light overgarment [26].

2.6 Other Textile Materials for Individual Equipment

Special high altitude flight suits are used by the pilots in the U.S. Air Force. The flight jackets are generally made of Nomex® IIIA. The fabric is designed to protect the body against pressure fluctuations. It also provides extra fire protection.

Antifragmentation trousers made of Kevlar® are effective against low velocity fragments and blasts. Parachutist's rough terrain suits made of Kevlar® offer increased puncture protection from branches. Inclusion of Kevlar® in parachute fabrics also increases flame resistance, tear strength and cut resistance. Ballistic blankets increase fragmentation resistance for ordnance and cargo. Gloves made of Kevlar® increase flame resistance and provide improved cut and wear resistance for working with razor wire.

Safety and protective fabrics for non-military applications are covered in Chapter 13.0.

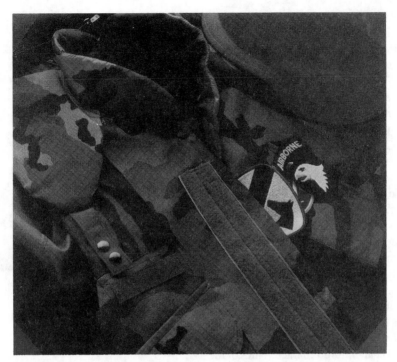

FIGURE 11.17 Camouflage fabrics (courtesy of Highland Industries).

2.7 Physiological Requirements in Military Clothing

The three most common properties required in military clothing are low weight and bulk, protection against high-intensity thermal radiation and moisture-vapor permeability.

The load carried by the combat soldier is agreed to be 18.5 kg in peacetime and anywhere between 39 and 57 kg under combat conditions (Figure 11.3). The mass per unit area of fabric is therefore a very important factor. Most NATO countries use a 305 g/m² 100% cotton fabric for the required durability during combat. 50/50 cotton/nylon blends and 65/35 polyester/cotton blends are also used at 245 and 195 g/m² [1].

By increasing the mass of fabric or bulk, transmission of heat to the skin can be reduced. Untreated natural fibers such as cotton are not flame resistant. Wool would burn and form a char. Thermoplastic fibers such as nylon and polyester melt under heat and may cause additional hazard. Therefore, the best way to protect against high-intensity thermal radiation and flames is to use fire retardant finishes or inherently flame resistant materials. Nomex®, Kevlar®, Kermel®, PBI are inherently flame resistant. However, these fibers are more expensive.

Low mass and bulk decreases insulation efficiency which is related to thickness. Recent development of microdenier fibers provided satisfactory insulation while reducing the thickness and bulk.

High moisture-vapor permeability is necessary to prevent condensation of sweat which may result in insulation loss and increased heat loss that creates cold chill and hypothermia.

3. TEXTILES USED IN DEFENSE SYSTEMS AND WEAPONS

A very broad range of textile products is used in military equipment including tents, tactical shelters, rigid walls, tarpaulins, vehicle and gun covers, equipment webbings, sleeping bags, camouflage nets, inflatable boats and other structures, life rafts, pillow (portable) tanks, fuel cells, parachutes and other air-drop equipment (Figures 11.18–11.20).

The major requirements for tentage are mobility, low weight and ease of erection and deployment. For shelters, resistance to dimensional changes under wet conditions is critical. Other properties required in tents and shelters include environmental protection, chemical/biological (CB), infrared and radar avoidance techniques, waterproofness, rot and mildew resistance and non-flammability [8]. Duck fabrics made of 100% cotton with a variety of fire, weather, water and mildew resistant treatments have been used in the U.S. Army for tents. Fabrics made of synthetic and blend yarns are also used for tentage [27]. Camouflage net structures, usually made of spun nylon, are used to cover military vehicles and weapons such as tanks and aircraft. Polyurethane-coated nylon fabric pieces (pigment coated) are patched on the net. The colors and pigments are chosen to match the visual and near-infrared characteristics of the background. Different camouflage screens are used against ultraviolet, far-infrared and radar detection.

Inflatable structures are used for technical or rest purposes. There are several ways to build inflatable structures: using pressurized tubes built into the structure, pressurizing the space between double fabrics or slightly pressurizing the whole structure with continuous pumping.

Fabrics for fuel cells are used in helicopters and fighter planes against crush damage. They are made of nylon yarns in plain or basket weave and coated with neoprene. The fabrics are tied into the military aircraft and have the same shape as the inside of the aircraft. The major requirements for these fabrics are bursting resistance, crunch and ballistic protection. These fabrics also protect cables, wires, etc. in the aircraft.

The main requirements for cargo parachutes are high strength and low weight. Personnel parachutes must maintain high strength at high temperatures.

3.1 Textile Composites in Military Use

Textile structural composites (Chapter 7.0) are widely used in military equipment and hardware.

FIGURE 11.18 Fluoropolymer laminated composite fabric used in self-contained, chemically protective mobile medical station for the U.S. Army Natick Center (courtesy of Chemfab).

For example, glass fiber reinforced composites are used to make hulls of mine sweepers and the conning towers of submarines. Missile casings and nose cones are manufactured by filament winding (Figure 11.21). Composites made of carbon and aramid reinforcement fibers, which have high strength-to-weight ratio, are replacing metals such as steel and aluminum in military

FIGURE 11.19 Air-drop equipment (courtesy of Highland Industries).

FIGURE 11.20 A command post area camouflaged with textiles (courtesy of U.S. Army).

FIGURE 11.21 Filament winding of graphite fibers for lightweight, high strength racket motor case for the small ICBM mobile missile (courtesy of the U.S. Department of Defense).

FIGURE 11.22 F/A-18A (courtesy of U.S. Army).

hardware including fighter planes and armored vehicles. Boron, silicon carbide and aluminum oxide fibers are also being used. A boron/epoxy skin on the F-14 fighter plane is 20% lighter than titanium which reduces plane weight by several hundred pounds. On the F-18 fighter, 10% of the structural weight and 50% of the surface area are made of carbon reinforced composites (Figure 11.22).

In naval applications, composite propellers are used for the Mark 46 torpedo in the U.S. Navy. Composite propellers performed better than forged aluminum and decreased the production costs by 55%. Other advantages offered by the composite propellers are weight reduction, chemical inertness and acoustic properties such as transparency to electronic detection. Compression molding technique is used to manufacture the composite propellers [28].

Due to cabin pressurization cycles, cracks may develop around fasteners and windows in older aircrafts. Boron-epoxy composites are used to patch cracks in aircrafts. It was reported that boron-epoxy patches increase the fatigue life

2.5 times and inspection interval five times compared to aluminum patches. Boron-epoxy patching is also easier to do in less time than metal patching. Boron-epoxy reinforcements have been used for the U.S.'s C-141 fleet for many years [29].

4. OTHER APPLICATIONS

Fiber-optic systems are finding more and more usage in military applications including guidance systems for aircraft, missiles, and remotely piloted vehicles. They are also used under water to detect the enemy submarines [30].

Smart textile materials offer good potential for use in military applications (Chapter 24.0). New polymers and composites that can detect and respond to environmental stimuli (temperature, light, humidity) can be used to reduce heat build-up and stress. Textile materials that will change their color or properties as the background changes can provide good dynamic camouflage. Research in these areas is continuing.

11.2

References and Review Questions

1. REFERENCES

1. Morris, J. V., "Protective Clothing for Defense Purposes," in *The Design of Textiles for Industrial Applications,* P. W. Harrison, ed., The Textile Institute, 1977.

2. Weinberger, C., Former Secretary of Defense, in *Textiles and National Defense,* American Textile Manufacturers Institute, Washington D.C.

3. *Textiles and National Defense,* American Textile Manufacturers Institute, Washington DC.

4. Chamberlain, G. and Joyce, M., "The Modular Soldier," *Design News,* August 20, 1990.

5. Tassinari, T., "Advances in Military Protective Clothing," *IFAI Second International High-Performance Fabrics Conference,* Nov. 12 – 13, 1992, Boston, MA.

6. Kevlar® Aramid Fiber, DuPont, 1994.

7. Dress for Survival, Personal Body Armor Facts Book, DuPont, 1992.

8. Broad Agency Announcement, DAAK60-93-T-0001, The U.S. Army Natick Research, Development and Engineering Center, October 1993.

9. Cunniff, P. M., "An Analysis of the System Effects in Woven Fabrics Under Ballistic Impact," *Textile Research Journal,* Vol. 62, No. 9, 1992.

10. Cunniff, P., "Dimensional Analysis of Woven Fabric Body Armor under Ballistic Impact," Fiber Society Fall Meeting, Nov. 1989.

11. Cunniff, P., "Constitutive Properties of Liquid Crystalline Polymers under Ballistic Impact," *NASA Fiber-Tex 1989,* Greenville, SC, Oct. 1989.

12. Jamison, J. W. et al., "Dynamic Distribution of Strain in Textile Materials under High Speed Impact, Part III," *Textile Research Journal,* Vol. 32, 1962.

13. Montgomery, T. G., The Effects of Geometric Shape of an Impacting Projectile on the Energy Absorption Process in Woven Fabric Systems, Ph.D. thesis, North Carolina State University, 1980.

14. Morrison, C., The Mechanical Response of an Aramid Textile Yarn to Ballistic Impact, Ph.D. thesis, University of Surrey, March 1984.

15. Winson, J. R. and Zukas, J. A., "On the Ballistic Impact of Textile Armor," *Journal of Applied Mechanics,* Vol. 6, 1975.

16. Wilde, A. F. et al., "Photographic Investigation of High Speed Impact on Nylon Fabric," *Textile Research Journal,* Vol. 43, 1973.

17. Young, S. A., "The Anatomy of Body Armor," *Safety and Protective Fabrics,* Industrial Fabrics Association International (IFAI), September 1992.

18. Park, A., "Materials are Integral to Armor Effectiveness," *Safety and Protective Fabrics,* Industrial Fabrics Association International (IFAI), August 1993.

19. Chamberlain, G., "Dressed to Fight," *Design News,* Sept. 8, 1986.

20. Park, A., "Issues Complex in Soft Body

Armor Selection,'' *Safety and Protective Fabrics,* Industrial Fabrics Association International (IFAI), May 1993.

21. Truong, J. Q., "Semipermeable Materials, Fact Sheet," U.S. Army Natick Research, Development and Engineering Center.

22. Stull, J. et al., "Hydrogen Fluoride Exposure Testing of the U.S. Coast Guard's Totally Encapsulated Chemical Response Suit," in *Performance of Protective Clothing: Second Symposium, ASTM STP 989,* S. Z. Mansdorf, R. Sager, and A. P. Nielsen, eds., American Society for Testing and Materials, Philadelphia, 1988.

23. Seidel, L. E., "Defense Needs Offer Textile Opportunities," *Textile Industries,* November 1983.

24. Bugge, G., "Cleaning Nuclear Protective Wear," *Safety and Protective Fabrics,* May 1994.

25. Military Specification, MIL-C-44031D, U.S. Army Natick Research, Development and Engineering Center, August 22, 1989.

26. Military Specification, MIL-C-83429B, U.S. Army Natick, 26 September 1990.

27. Weil, E. D., "Recent Progress in Flame Retardant Fabric for the Armed Forces," *American Dyestuff Reporter,* February, 1987.

28. Pegg, R, and Reyes, H., "Advanced Composites: Now They've Joined the Navy," *Design News,* September 8, 1986.

29. Lynch, T. P., "Composite Patches Reinforced Aircraft Structures," *Design News,* April 22, 1991.

30. Robinson, G. M., "Fiber Wins over the Military," *Design News,* April 22, 1991.

2. REVIEW QUESTIONS

1. What are the major requirements for military/defense textiles for soldier protection?
2. What are the typical materials used in battledress uniforms?
3. Explain the principle of ballistic protection using textiles.
4. What is blunt-trauma effect? How could it be reduced?
5. What are the requirements for camouflage fabrics for different climate and landscape conditions?

PAPER MACHINE CLOTHING

Paper Machine Clothing

S. ADANUR

1. INTRODUCTION

Large quantities of high value-added, high performance textile fabrics and felts are used in the paper industry. Although the average cost of paper machine clothing is only two cents for a dollar's worth of paper sold, paper machine clothing is one of the most critical factors in papermaking. In theory, paper can be made without a paper machine (which may cost several hundred million dollars) or without many of the chemicals involved, but it cannot be made without fabrics that are used for forming, pressing and drying of paper. Therefore, although the volume of paper machine clothing may be small, it is a stable and secure market due to the obvious need as long as the papermaking process is not revolutionized. The United States, which constitutes 5% of the world's population, has 12% of the world's paper and paperboard mills that produce 30% of the world's paper and paperboard [1].

Paper machine fabrics' design, properties and applications have tremendous effects on paper properties and papermaking process. Proper fabrics can increase productivity, decrease paper machine downtime and increase the bottom line. Therefore it is crucial that the textile engineer develops the right fabric and that the papermaker knows a good deal about papermaking fabrics. The successful fabric supplier must be able to listen and understand the papermaker.

2. PAPER MACHINES AND PAPERMAKING PROCESS

A basic overview of the paper machine would be helpful for understanding the application of paper machine clothing. Figure 12.1 shows a paper machine. There are three main sections in a typical paper machine: forming section, press section and dryer section. Figure 12.2 shows these three sections in a conventional Fourdrinier paper machine. Figure 12.3 shows the longitudinal cross section of a Fourdrinier paper machine. Consistency of paper is represented as percent solids in the figure.

2.1 Forming Section

The fabrics that are used in this section are called forming fabrics. The term "forming" comes from the formation of paper sheet. Water and wood fiber mixture (slurry) is pumped onto rotating forming fabric. Typical consistency of slurry at the headbox exit is 0.5% fiber and 99.5% water. Water is filtered through the forming fabric and wood fibers are retained on the top surface, thus the sheet is formed as shown in Figure 12.4. There are three types of hydrodynamic processes during formation: drainage, oriented shear, and turbulence [3]. Oriented shear is the result of the difference between fabric speed and headbox exit speed. At the end of the forming section, sheet, which has a consistency of approximately 20% fiber and 80% water, is transferred from forming fabric to press section by open draw or suction pickup.

There are mainly four types of machines for the wet end: Fourdrinier, twin wire, multi-ply and multi-cylinder formers. Schematic of a Fourdrinier forming machine is shown in Figure 12.2. Figure 12.5 shows a twin wire (fabric) former and Figure 12.6 shows a two-ply former. Forming fabrics rotate in the directions shown in the figures. Use of multi-cylinder machines is

FIGURE 12.1 Overall view of a high-speed publication grades paper machine with a flat "wire" Fourdrinier and top "wire" former. "Wire" is an old term from metal fabric days that is still used for today's paper machine fabrics (courtesy of Beloit Corporation).

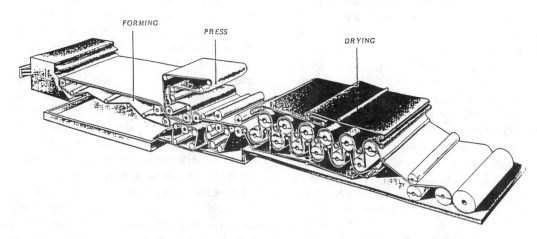

FIGURE 12.2 Schematic of a Fourdrinier paper machine.

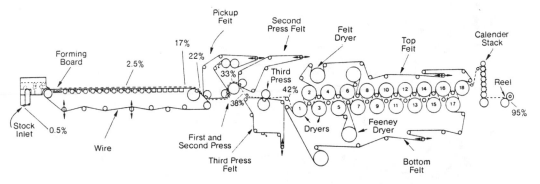

FIGURE 12.3 Longitudinal cross section of a typical Fourdrinier paper machine (Albany International [2]).

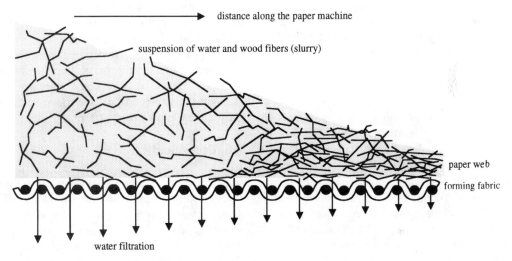

FIGURE 12.4 Schematic of sheet formation.

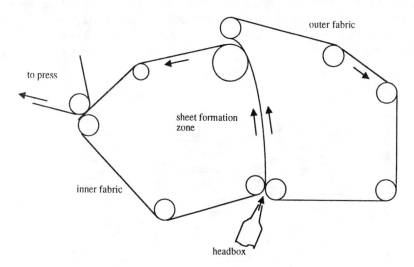

FIGURE 12.5 Schematic of a twin wire (fabric) former.

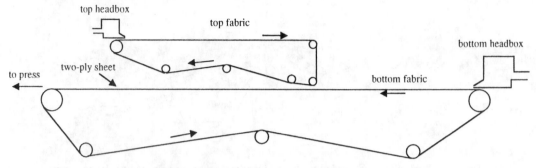

FIGURE 12.6 Schematic of a two-ply former.

decreasing. There are various machine elements to improve and control the drainage and formation which are outside of the scope of this book [4−7].

2.2 Press Section

The function of pressing is to continue the water removal process which started in the forming section, consolidate the sheet, give texture to the sheet surface, support and transfer the sheet. The fabrics in this section are called press fabrics or press felts. Although it depends on sheet grade and paper machine, typical sheet consistency at the end of press section is 40% fiber and 60% water. At the end of the press section, sheet is transferred to the dryer section.

During pressing, the sheet is compressed between one or two felts and two rolls in the press nip to squeeze water from inside the web and out of the felt fibers. Figure 12.7 shows this process in a plain press nip. Increased compression increases water removal [8−10].

There are different types of pressing mechanisms (suction press, grooved roll transverse press, blind-drilled press, fabric press, shrink sleeve press, double-felted and double-vented nip press, extended nip press, impulse drying and press drying) and press configurations (straight-through press, transfer press, twinver press and no-draw press) [2,5,6].

2.3 Dryer Section

After pressing, the residual water in the sheet is removed by evaporation (mass transfer) in the dryer section by using steam (heat transfer). The fabrics in this section are called dryer fabrics or dryer felts. The consistency after this section is on the order of 95% fiber and 5% water which is the typical consistency of regular paper. The sheet is then wound on a roll.

The drying zone can be considered as a "black box" into which wet paper, steam and air enter. The dried paper, the moist and heat-laden air, and the condensate leave the box [2]. Figure 12.8

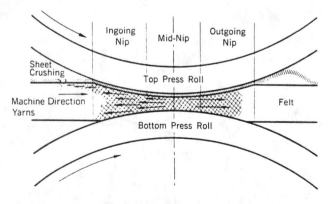

FIGURE 12.7 Plain press nip [2].

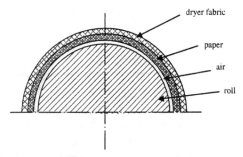

FIGURE 12.8 Drying configuration.

shows the arrangement in the drying zone. Dryer fabric presses the sheet tight against the cylinder. Most of the energy used to make the sheet is consumed in the dryer section [11].

Conventional cylinder drying, which uses a series of steam-heated cylinders, is the most common drying method. Drying configurations include single-felt with double-tier dryer cylinders, single-felt with single-tier (Unorun) dryer cylinders and double-felt with double-tier dryer cylinders which are shown in Figure 12.9. In single-felted systems, a single fabric wraps both the top and bottom dryer cylinders and supports the sheet continuously. However, with this system, the bottom dryers are not utilized to their full extent because dryer fabric is sandwiched between the cylinder and the sheet. To improve the efficiency, the bottom dryers were replaced with smaller diameter rolls that led to single-tier configuration. In the double-felted configuration, separate fabrics are used for the top and bottom dryer cylinders. The disad-

vantage is that the sheet is not supported by the fabric between the cylinders.

3. FORMING FABRICS

Among the three types of paper machine clothing, forming fabrics are probably the most critical to design. They affect the final sheet properties more than press or dryer fabrics. If the sheet structure is not formed correctly in the forming section, it would hardly be possible to correct it later on the paper machine. Therefore, forming fabric design requires careful attention and a high level of engineering. Forming fabrics are used as the support, drainage and transportation means in papermaking.

3.1 Materials

Until the 1950s, forming fabrics were being made of phosphor bronze and stainless steel. There were very few design variations. The average fabric life was one to two weeks. Fabric stability was good, however the forming fabric was prone to damage due to crease and breaks. Metal fabrics had high drainage rates and poor fiber retention and sheet release. The first synthetic forming fabric was used in 1957. Although fabric life was not improved initially, sheet quality was better. By the mid-1970s, more than half of all paper machine clothing had mono- or multi-filaments. Multi-layer forming fabrics were introduced in the mid-1970s. In 1982, three

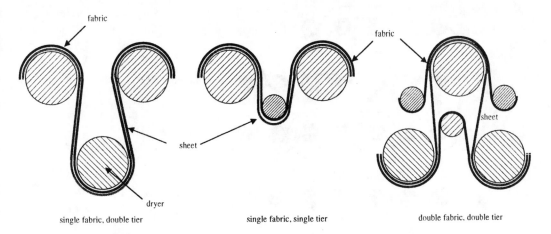

FIGURE 12.9 Configuration of dryer fabric with respect to cylinders and sheet.

layer fabrics were introduced. Today almost all of the forming fabrics are made with high molecular weight and high modulus monofilament polyester. For wear resistance, alternating nylon yarns are used in the bottom layer of multilayer fabrics. However, nylon is not dimensionally stable under wet conditions which limits its use in forming fabrics. With the increase of alkaline paper making, more and more nylon yarns are being used along with polyester. Polyester has good stress-strain properties and it is resistant to acid, but is subject to hydrolysis under alkaline conditions.

Today, with the use of recycled fibers, contaminants in the pulp such as pitch, wax, latex, etc., may create problems during paper manufacturing. Therefore yarns with contaminant resistance properties are in high demand. There are several ways to achieve contaminant resistance such as yarn coating, by adding contaminant resistant additives to the polymer and fabric coating. Fabric coating also improves stability. Generally, fluorocarbon based coatings are used.

3.2 Fabric Designs and Structures

There are basically three types of forming fabric design: single layer, two layer and three layer.

Single Layer Forming Fabrics

Figure 12.10 shows schematics of single layer forming fabric designs. There is one layer of warp and one layer of weft yarns. In practice, single layer fabrics are woven with two to eight harnesses. Figure 12.11 shows a photomicrograph of a five harness single layer forming fabric.

Two Layer Forming Fabrics

Two layer designs have one layer of warp yarns and two layers of filling yarns. There are two variations of two layer fabrics: standard two layer and two layer extra designs. Figure 12.12 shows a schematic of a standard two layer design and Figure 12.13 shows a photomicrograph of a standard two layer fabric. In this design, the ratio of the number of top weft yarns to bottom filling yarns is 1 to 1. The second type of two layer forming fabric is called "two layer extra" in which the ratio of the top filling yarns to bottom filling yarns is 2:1 (in paper machine fabrics industry, a filling yarn is often called a "shute").

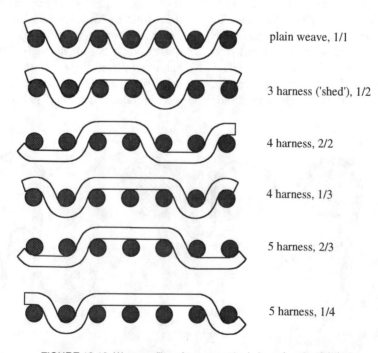

plain weave, 1/1

3 harness ('shed'), 1/2

4 harness, 2/2

4 harness, 1/3

5 harness, 2/3

5 harness, 1/4

FIGURE 12.10 Warp profiles of common single layer forming fabrics.

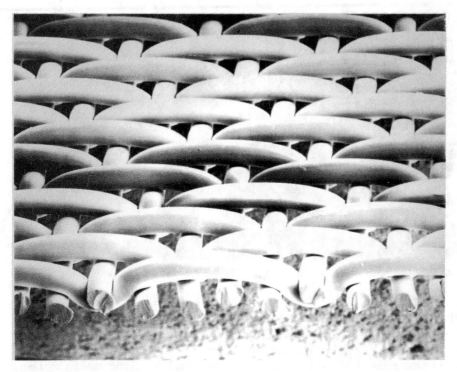

FIGURE 12.11 Five harness single layer forming fabric (courtesy of Asten Forming Fabrics, Inc.).

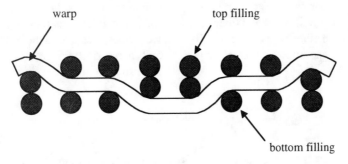

FIGURE 12.12 Warp profile of a standard two layer fabric design.

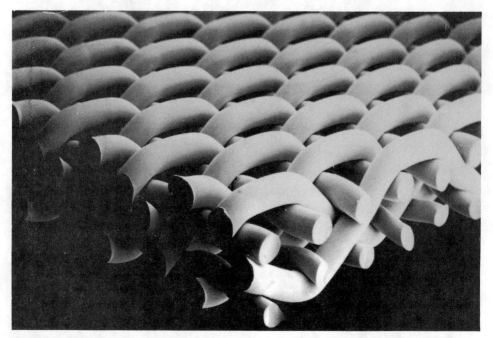

FIGURE 12.13 Photomicrograph of a standard two layer fabric (courtesy of Asten Forming Fabrics, Inc.).

The warp strand profile of a seven harness two layer extra design is shown in Figure 12.14. Two layer extra design provides higher fiber support compared to standard two layer designs. Therefore, their main application area is in the fine paper market. Figure 12.15 shows the photomicrograph of a two layer extra design.

The advantages of two layer fabrics include improved stability, resistance to wear and damage, improved sheet formation, greater water drainage capacity, more uniform wood fiber distribution, higher wood fiber support, less two-sidedness in the sheet and improved retention of fines and fillers.

Three Layer Forming Fabrics

These fabrics have two separate fabric layers (top and bottom) connected with a binder strand (stitch). As a result, there are two sets of warp yarns and two sets of filling yarns as shown in Figure 12.16. In general, the top layer is finer than the bottom layer. Advantages of three layer designs are better formation and sheet quality due to the fine top layer, improved retention of fines and fillers, increased stability, improved wear resistance due to the coarser bottom layer, and increased drainage rate, drainage capacity, fiber support and sheet smoothness.

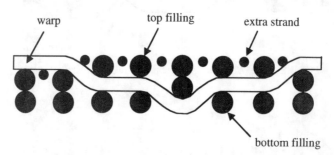

FIGURE 12.14 Warp profile of two layer extra design.

FIGURE 12.15 Photomicrograph of a two layer extra fabric design (courtesy of Asten Forming Fabrics, Inc.).

Figure 12.17 shows a photomicrograph of a typical three layer design. The stitch strand is usually finer than the other yarns. Therefore, it is well protected against wear due to paper machine elements.

The forming fabric designer can select and control yarn materials, warp and beat count, weave pattern, number of harnesses and yarn diameters in order to achieve the desired fabric properties. These variables have effects on several fabric properties (air permeability, fabric stability, fabric modulus, open area, fabric caliper) and paper machine operations (drainage, formation, fiber retention, sheet release, wire mark, fabric cleaning and life).

3.3 Manufacturing of Forming Fabrics

Design and manufacturing of forming fabrics require careful engineering. Manufacturing equip-ment for forming fabrics is considerably heavier than that for standard textile goods. The following are the major steps in manufacturing of forming fabrics.

Raw Material Testing and Evaluation

Today, almost all forming fabrics are made of monofilaments. The tolerance range for the monofilament yarn diameter is $\pm 2\%$. Therefore it is crucial that the yarn diameters are checked before weaving. Some of the other tests on monofilament yarns are modulus, elongation and shrinkage evaluations.

Warping and Weaving

Section warping is the general type of warping for forming fabrics. Only shuttle looms are used for forming fabric manufacturing. Their construction is very heavy due to the high forces involved, especially during beat-up. Today, looms are manufactured at widths well over 500 inches. For three layer fabrics, two separate warp beams are used. There are two types of weaving forming fabrics—flat weaving and endless weaving. In flat weaving, warp yarns on the loom become machine direction yarns on the paper machine. In endless weaving, filling yarns on the loom become machine direction yarns on the paper machine.

Preliminary Heat Setting

After weaving, forming fabrics are heat set at high temperatures and under tension in warp and filling directions. The maximum heat set tem-

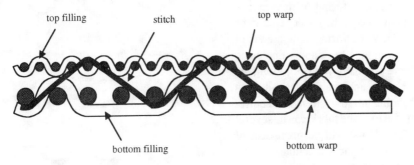

FIGURE 12.16 Schematic of a three layer design.

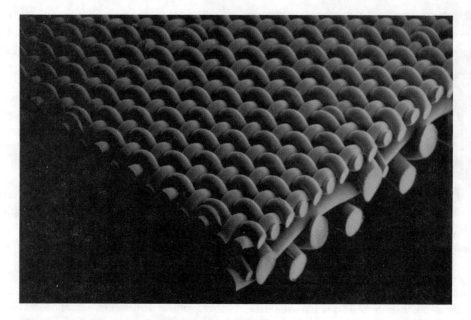

FIGURE 12.17 Photomicrograph of a three layer fabric (courtesy of Asten Forming Fabrics, Inc.).

perature depends on the fabric material, however it is usually around 475°F for standard polyester fabrics. The purpose of heat setting is to relieve the internal stresses in the fabric after weaving in order to have a stable fabric structure. Fabric modulus is drastically increased by heat setting. Fabric structure becomes more stable after heat setting due to crimp interchange. Warp crimp decreases and weft crimp increases during heat setting (Figure 12.18).

Seaming

Flat woven fabrics need to be seamed after heat setting to form an endless belt. Seaming is the most time-consuming, costly and tedious part of forming fabric manufacturing. Weft yarns are raveled out on both ends of the fabric, creating a warp fringe. Then, warp yarns on both ends of the fabric are woven together one by one using filling yarn. There are many ways of joining the warp ends together to achieve the desired seam strength. Depending on the strength require-ment, the seamed area of the fabric can be several inches long. There is naturally a great deal of research going on in the paper machine clothing industry in order to completely automate the seaming process. There is no need to seam end-

less fabrics since they are woven as an endless belt on the loom.

Final Heat Setting and Finishing

After seaming, the fabric is given a final heat set to relieve the stresses in the seaming area. The final heat setting temperature is lower than the preliminary heat setting temperature in order not to alter the properties of the fabric that were set during preliminary heat set. After final heat set-ting, fabric edges are trimmed to the paper

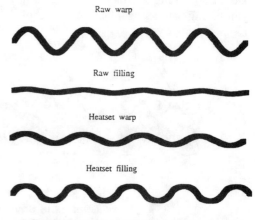

FIGURE 12.18 Crimp transfer during heat setting.

machine width. An adhesive is applied on the edges to prevent yarn unraveling during operation.

Forming fabrics that will be used for finer paper grades are sometimes sanded with an abrasive roll. Sanding removes the acute knuckle points on the fabric and provides a smoother surface.

3.4 Physical Properties of Forming Fabrics

Some physical properties of forming fabrics are crucial to its successful performance on paper machine.

Mesh and Beat Count

Mesh is the number of warp yarns per inch and beat is the number of filling yarns per inch. These numbers are usually used to distinguish forming fabrics and they affect the other fabric properties. Lower mesh and beat numbers usually mean a coarser fabric.

Air Permeability

This is one of the most important fabric properties. Air permeability is the measurement of air flow through the fabric at a pressure difference of 0.5 inch of water. It is measured as the air volume (cubic feet) passing through a unit area (square feet) of fabric per minute (cfm).

Modulus

Modulus is another important property of forming fabrics. With the increasing speeds of paper machines, tension on fabrics has increased significantly. In order to prevent excess stretching of fabric beyond the take-up limit of the machine under high tension, fabrics should have enough resistance to stretching in the machine direction. In practice, modulus is expressed in terms of load per lineal unit length across the fabric width.

Fiber Support Index

Fiber support is critical in sheet formation.

Especially with the increased use of recycled fibers, that are generally shorter, fiber support capability of forming fabrics has become more important. A statistical model was developed by Beran to calculate the fiber support of various forming fabrics [12]. In its simple form, the fiber support index is calculated as follows:

$$FSI = 2/3 \, [a \cdot W + 2 \cdot b \cdot S] \quad (12.1)$$

where

a = support factor for warp yarns
b = support factor for weft yarns on sheet side
W = number of warp yarns per inch
S = number of filling yarns per inch
a = (number of weft yarns under warp knuckle +1)/number of harnesses
b = (number of warp yarns under filling knuckle +1)/number of harnesses

Drainage Index

Drainage index is a tool for rating fabrics based on relative drainage rates. Drainage index was developed by Johnson [13] and is given by the following formula:

$$DI = (b \cdot S \cdot P)/1{,}000 \quad (12.2)$$

where

b = support factor from weft strands
S = number of filling yarns per inch
P = air permeability (cfm)
b = (number of strands under filling knuckle +1)/number of harnesses

Fabrics with higher DI values have higher drainage rates.

Caliper (Fabric Thickness)

Low caliper is usually preferred especially in multi-layer fabrics since it reduces the water carry-back and increases drainage.

Open Area

Open area in a fabric is an indication for straight-through drainage. Some two layer designs do not have any open area at all. In these designs, warp coverage is usually more than 100%, therefore, water drainage path is not straight. The open area of a fabric is calculated as follows:

Open Area (%)

$$= [1 - (\text{warp count})(\text{warp diameter})]$$

$$\times [1 - (\text{weft count})(\text{weft diameter})]$$

$$\times 100 \qquad (12.3)$$

Void Volume

Void volume is the volume that is not occupied by yarns in fabric volume. Void volume affects drainage of forming fabric. As the void volume increases, drainage also increases.

3.5 Applications of Forming Fabrics

Selection of Forming Fabric Design

Different paper grades require different forming fabric designs. When selecting a fabric design for a specific application, the following paper and paper machine parameters need to be considered:

(1) Paper properties
- grade (tissue, fine paper, newsprint, kraft, nonwovens)
- type of furnish and recycled fiber content
- paper weight
- wire mark considerations
- formation type

(2) Paper machine parameters
- machine type
- machine length and width
- speed
- type of pickup
- paper machine elements (rolls, foils, vacuum elements)
- operating tension

The drainage path of forming fabric is very important. Figure 12.19 shows drainage paths of single layer and multi-layer fabrics. In single layers, there is straight-through drainage. In multi-layer fabrics, there may be straight-through drainage and angular drainage at the same time. Drainage path determines the orientation of fibers in the sheet which in turn affects the physical properties of sheet such as tensile strength ratio and bulk. Figure 12.20 shows filling yarn profiles of various forming fabric designs.

Fabric surface topography also plays an im-

single layer

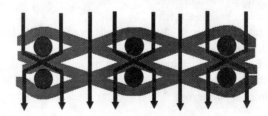

two layer

FIGURE 12.19 Drainage paths of single and two layer fabrics.

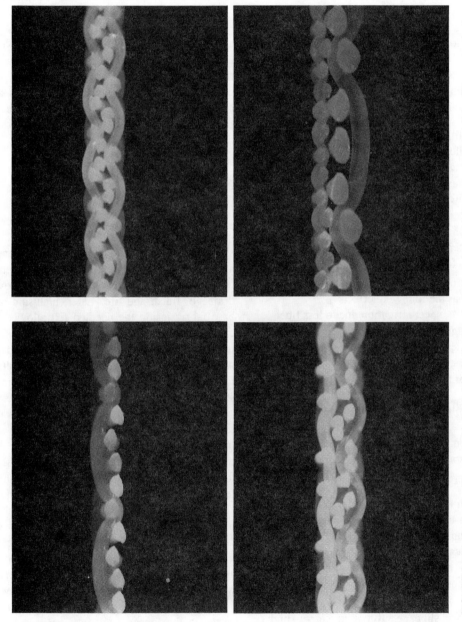

FIGURE 12.20 Cross direction view of forming fabrics. Top left: single layer; top right: two layer; bottom left: two layer extra; bottom right: three layer (courtesy of Lindsay Wire).

393

portant role. Figure 12.21 shows the paper side surfaces of typical single layer, two layer, two layer extra and three layer forming fabrics.

Hole size and shape determine wood fiber bridging between the yarns. They also affect fiber orientation. For a given warp count, hole size can be altered by changing the number of weft yarns per unit length.

Running attitude of forming fabrics are dictated by the paper grade and wood fiber types. Running fabrics with long filling knuckles up or down affects the paper properties and fabric running conditions (Figure 12.22). There are two fabric running attitudes:

(1) Long filling knuckle up (to the sheet side) has the following effects
   • better sheet formation
   • easier fabric cleaning
   • reduced load
   • better fiber bridging
   • good sheet release
   • better drainage
   • less fiber bleeding
(2) Long filling knuckle down (to the machine side) has the following effects
   • poorer sheet formation due to less fiber bridging
   • difficulty in cleaning
   • higher load due to larger wear surfaces
   • poor release

3.6 Forming Fabric Operating Problems

Common problems affecting life and performance of forming fabrics can be summarized as follows:

(1) Stretch—Fabrics with low modulus may stretch too much under high operating tensions on paper machine. Proper use of high modulus warp material and heat setting techniques can reduce fabric stretch during operation.
(2) Fabric wear—The main causes for fabric wear are slippage, fillers in slurry, high vacuum, abrasive roll surfaces, and flatbox cover materials.
(3) Contamination—Various contaminants may fill the fabric and make it useless after a

while. The contaminants can be organic (pitch, size, starch, defoamers, ink, wax, latex), inorganic (calcium carbonate, calcium sulfate, calcium silicate, talc, titanium dioxide, clay) or biological (aerobic, anaerobic, fungi, algae). Several shower methods are used in order to keep fabrics clean.

Edge unraveling and ridges/creases also create problems on paper machine. Paper machine misalignment, guide malfunction, and roll damage are among the causes for ridges and creases. Proper alignment and maintenance of paper machine reduce the risk of ridges and creases.

4. PRESS FELTS

Until the mid-1940s, wool was the choice of material for press felts. Introduction of staple synthetic fibers drastically improved the strength and life of the felt. In the early 1960s, needle-punched felts were developed with improved drainage, finish and life. Later, inclusion of a base fabric inside the needled felt allowed increased press loading since it provided greater incompressible void volume and improved the water handling capacity of the felt. The resulting fabric was the batt-on-mesh felt that was much stronger, easier to clean, and had longer life.

The main functions of a press fabric are water removal from the sheet, sheet support and transportation, providing uniform pressure distribution, and imparting proper surface finish to the sheet. The felt should provide proper cushion for the sheet to resist crushing, shadow marking and groove marking. Other functions include transferring the sheet from one position to another in the case of closed draws, driving the undriven felt rolls and cylinder formers. The amount of water that the felt can absorb and the water flow resistance are affected by the void volume (volume that is not occupied by fibers or yarns) in the fabric. Low flow resistance and ability to maintain void volume under load are important during operation. Important felt properties include strength, adequate void volume, required permeability, low compressibility,

(a)

(b)

FIGURE 12.21 Forming fabric surfaces: (a) single layer, (b) two layer (courtesy of Tamfelt, Inc.).

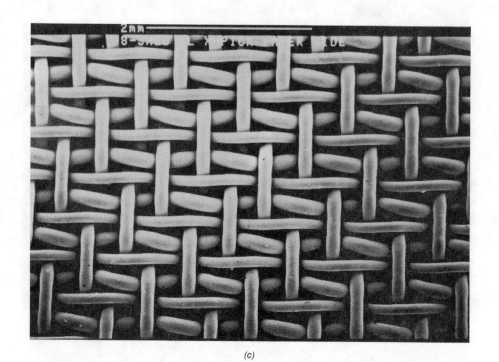

(c)

(d)

FIGURE 12.21 (continued) Forming fabric surfaces: (c) two layer extra, (d) three layer fabrics (courtesy of Tamfelt, Inc.).

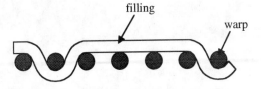

long filling knuckle up

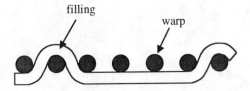

long filling knuckle down

FIGURE 12.22 Fabric running attitudes.

batt/base ratio, compaction resistance, abrasion resistance, heat and chemical resistance.

4.1 Press Felt Designs

Conventional woven felts, which were woven with spun yarns have become obsolete. Other major felt designs are as follows.

Batt-on-Base Felt

A woven fabric, usually of spun yarns, is used as the base fabric on which fiber batt is needle-punched. Batt-on-base felts are made of either a combination of wool and synthetic fibers or 100% synthetic fibers (mostly nylon), which are more common. The base fabric, which can have a synthetic content between 50–100%, can be woven as endless belt or can be woven flat and seamed later. A web is formed with carded fibers and needlepunched into the base fabric. The batt web has a synthetic content between 20 and 100%. After washing and chemical treatment, the felt is stretched on the dryer to the operating dimensions. Singeing is usually done to prepare the surface and remove the loose fibers. Finally, the felt is dried at an overstretch length for easy installation. Figure 12.23 shows the random nature and structure of the top surface of a needled felt.

Batt-on-Mesh Felt

Needlepunched batt-on-mesh felt structure is similar to batt-on-base felt structure. Batt-on-mesh fabrics are generally made of 100% synthetic materials. The base structure may be a single- or multi-layer fabric woven with 100% monofilament, a combination of mono- and multi-filament, all high twisted multi-filament, all high twisted multi-filament treated with resin for rigidity, or any combination of these yarns. Since these yarns are stiffer than spun yarns, batt-on-mesh fabrics have higher resistance to compression and compaction than batt-on-base felts. The base mesh is woven endless and the batt is applied on needle-punching machines. Figure 12.24 shows a batt-on-mesh felt with single layer monofilament base fabric. Installation of batt-on-mesh fabrics may be more difficult than batt-on-base fabrics due to stiffer base structures.

Multi-filament batt-on-mesh fabrics are suitable for suction, grooved and shrink sleeve presses. Resin treated multi-filament fabrics are applicable on first and second suction presses for coarse paper grades such as kraft. Monofilament or mono- and multi-filament combination base structural felts have excellent stability, resistance to wrinkles, and good finish for fine paper applications on suction pickup positions, straight suction presses, and grooved or shrink sleeve presses.

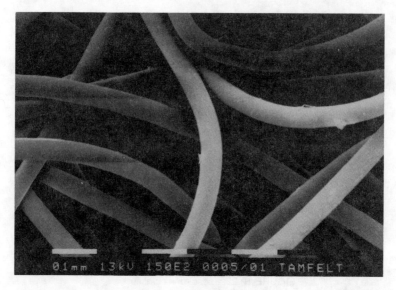

FIGURE 12.23   Top surface of a needle-pressed felt (courtesy of Tamfelt, Inc.).

Baseless Felt

In this type of structure, no base fabric is used; felt is made by needle punching only. The absence of base yarns reduces marking and provides a uniform pressure distribution in the press nip. Nonwoven felts are suitable for suction, grooved roll and shrink sleeve presses mainly for fine paper and board machines for a smooth surface.

Felts with No-Crimp Base-Fabric (Knuckle-Free or Fillingless)

The base fabric is not a woven fabric but is formed by one, two, or more, completely separate uncrimped yarn layers. In case of a one layer structure which is also called fillingless felt, only warp yarns are used without any filling yarns. However, in some variations, very fine filling yarns may be used. Figure 12.25 shows the two layer no-crimp felt structure. It is reported that the no-crimp base fabric structure provides improved sheet dewatering, optimum resistance to compaction and dimensional changes, resistance to filling and improved micro-uniformity of pressing. Eliminating knuckles in the base structure reduces the possibility of paper fines and dirt being trapped at the yarn crossover points. Moreover, it reduces

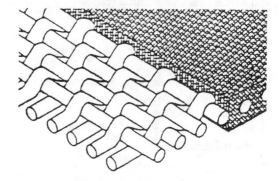

FIGURE 12.24   Batt-on-mesh press felt structure [14].

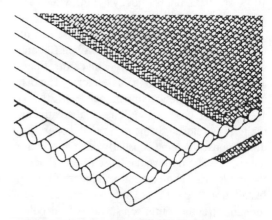

FIGURE 12.25   Two layer no-crimp base press felt [14].

sheet marking. Other advantages of fillingless felts include improved abrasion resistance, easy cleaning, better drainage and smoother surface. However, due to lack of interlacing of yarns, fabric stability may suffer.

Laminated Press Felts

In this relatively new press felt structure, one or more different base fabric structures are included in the fabric. The top base fabric is usually a single layer. The bottom base fabric can be one, two, or three layer integrally woven fabric (Figure 12.26). Laminated fabrics allow a wider range of base fabric pressure uniformity and low batt-to-base ratio which is critical for open and clean operation. As the machine speeds increase, the nip resistance time decreases and better surface contact between the sheet and press fabric becomes a requirement.

Continuous research is being done to develop new and improved felt designs. The latest developments include utilization of round, hollow yarns in cross-machine direction and 100% monofilament (no batt) press fabrics. Hollow yarns flatten under pressure and immediately rebound after exiting the nip. Flexing of yarns provides better sheet contact which in turn improves water removal and smoothness. It is reported that machine runnability and printability on the sheet are also improved [14].

4.2 Manufacturing of Press Felts

Base fabrics are woven (except no-crimp base fabrics) similar to forming fabrics. They can be woven as endless belt or flat. Some looms for endless weaving can be as wide as 30 meters (Figure 12.27). Flat fabrics are seamed later to form a continuous belt. Today, approximately 25−30% of the press felts are seamed. Usually pin seaming is used which makes installation of the fabric on the paper machine a lot easier. Figure 12.28 shows a press fabric seam. Monofilament warp yarn ends are looped and woven back into the fabric at two ends of the fabric. A monofilament pin passes through the loops connecting the two ends together after the installation of the fabric on the paper machine.

After base fabric manufacturing, carded fiber web is needle-punched into the base structure to form the batt. In the past, the same length and diameter batt fibers were being used in the fabric. In modern press felts, different length and diameter fibers are used to obtain stratified batt layers, the coarsest fibers being next to the base fabric and the finest fabrics being on the surface. Heat-resistant fibers or other special purpose fibers such as fusible fibers can be used to improve bonding.

The remaining major manufacturing steps are heat-setting for stabilization, washing, chemical treatment and singeing. Chemical treatment im-

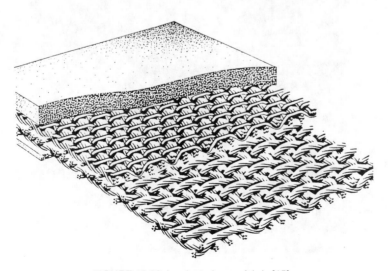

FIGURE 12.26 Laminated press fabric [15].

FIGURE 12.27 Weaving machine for endless base fabric manufacturing (courtesy of Steele Heddle).

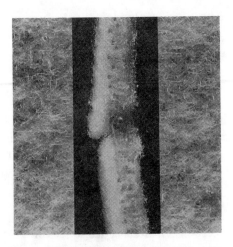

FIGURE 12.28 Pin seamed press fabric (courtesy of Albany International).

proves start-up characteristics, abrasion, bacterial and chemical resistance, compaction resistance, and contaminant resistance. Singeing is done to remove loose fibers and properly prepare the surface.

4.3 Properties of Press Fabrics

Based on their structures (number and dimensions of base yarns, size of batt fibers, smoothness of the surface finish), press felts are classified as superfine, fine, medium and coarse. The important properties of press felts are given below.

Felt Mass and Thickness

Felt mass is defined as the fabric weight per unit area. Press felt should have a uniform mass distribution across the width. Uneven mass or thickness (caliper) may cause press bounce or vibration. Thickness is important for felt wear and compaction characteristics that influence void volume and drainage of the felt. The rate of thickness change during the operation depends on the felt design. The thickness decreases quite rapidly during the early days of the operation. After a certain thickness loss, the fabric has to be removed from the machine.

Batt-to-Base Ratio

Batt-to-base ratio is defined as the mass of the batt fibers divided by mass of the base fabric. As this ratio increases, compactness of the felt also increases, which is not very desirable for dewatering capability. However, a small batt-to-base ratio may cause marking on the sheet due to base yarns.

Air Permeability

Similar to forming fabrics, it is measured as the air volume (cubic feet) passing through unit area (square feet) of fabric per minute (cfm).

Void Volume

Void volume is the volume that is not occupied by the yarns or fibers in the fabric. It is an indication of the amount of water that felt can absorb.

Flow Resistance

Properly designed felt should provide adequate drainage and filtration of water and therefore should have minimum resistance to flow. Resistance to water flow in the felt is measured in three directions using water permeability testers—machine direction, cross-machine direction and vertical direction. Low flow resistance indicates higher water acceptance capacity of the felt.

Compressibility and Resiliency

Compressibility is a measure of compactness under load and resiliency is a measure of rebounding capability of the felt after compaction. Compaction resistance is a desired property in press felts. A compaction resistant fabric can be either incompressible or compressible but resilient.

Uniformity of Pressure Distribution

Pressure from the rolls should be transferred to the sheet uniformly by the felt in order to prevent uneven sheet moisture profiles. Press felts with fine base fabrics and fine batt fibers provide more uniform pressure distribution. Too coarse base yarns, improper weave design or insufficient batt may also cause sheet marking.

The major structural requirements of modern press felts are dimensional stability, abrasion resistance, contaminant and chemical degradation resistance. The felt should be dimensionally stable throughout its operation. Some felts may get narrower after running some time due to heat from furnish and steam boxes. Smooth sheet transfer also requires dimensional stability. Abrasive inorganics in alkaline papermaking and papermaking chemical additives for mechanical conditioning have increased the importance of abrasion resistance in press felts. Increased use of secondary fibers has made contaminant resistance a necessity in today's press felts. Papermaking chemical additives such as bleaching, oxidizing and cleaning agents require chemical degradation resistance.

5. DRYER FABRICS

Major functions of dryer fabrics are aids in drying, sheet support and transportation, cockle prevention and shrinkage control. Air permeability and hydrolysis resistance are the two most important properties of dryer fabrics.

The early dryer fabrics were made of cotton/asbestos materials. Polyester monofilament fabrics and multi-filament fabrics that included Nomex® replaced cotton/asbestos fabrics by the early 1970s. Additives such as staboxyl are added to polyester to increase hydrolysis resistance. Hydrolysis resistance is the resistance of a material to breakdown at high moisture and temperatures which is the operation environment in the dryer section of the paper machine. Starting around the mid 1980s, dryer fabric manufacturers began using monofilament yarns with superior hydrolysis and high temperature properties such as PCTA and PPS (polyphenylene sulfide) in dryer fabrics which offered clean/dry running characteristics. Other materials that have been used in dryer fabrics are acrylic, aramid and polyamide. Figure 12.29 shows comparison of several materials for hydrolysis resistance. The tensile strength was measured after exposing yarns to high temperature and moist conditions in an autoclave for a certain period of time.

5.1 Dryer Fabric Designs

Paper grade and operating environment in the paper machine are the most important factors that affect the design of dryer fabrics. Hydrolysis resistant yarns have become standard materials in dryer fabrics due to high performance requirements of today's paper machines. In addition to the requirement for hydrolysis yarns in the structure, the dryer fabric has to maximize both heat and mass transfer. Heat transfer is increased by increasing the fabric contact surface and mass transfer is increased by increasing the fabric permeability. Increasing fabric tension on the machine also increases heat transfer.

Standard dryer fabric designs can be classified as monofilament and multi-filament designs. Needlepunched dryer fabrics, which are similar to press felts in construction (i.e., batt fibers are needled into a base fabric on both sides) are not used much any more. Most of the dryer fabrics are made of two or three layers. However, use of single-ply fabrics is increasing. The current trend in dryer fabrics is 100% monofilament fabric with low permeability, low caliper, soil release properties and a seam that is compatible with the fabric structure.

Monofilament Designs

Monofilament fabrics are made of hydrolysis resistant polyester monofilament yarns. A stuffer yarn may be used to control air permeability. A stuffer can be monofilament, in

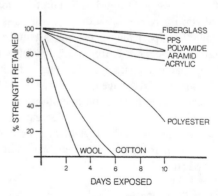

FIGURE 12.29 Hydrolysis resistance of materials (CPPA [16]).

which case the fabric becomes 100% monofilament, or spun yarn. Therefore, there are mainly three variations of monofilament dryer fabrics:

- monofilament weave
- monofilament weave with monofilament stuffer for low air permeability
- monofilament weave with spun stuffer

Monofilament designs are especially suitable for finer paper grade applications such as newsprint and fine paper due to their clean and dry running characteristics, high sheet contact and good heat transfer, resistance to sheet mark and low air-carrying characteristics. The monofilament yarns used can be round (0.4 mm to 0.5 mm in diameter) or flat (0.27 mm × 0.54 mm to 0.57 mm × 0.88 mm). The predominant machine direction (MD) yarn size is 0.38 mm × 0.60 mm. Round MD yarns generally require increased ends per inch. Round monofilament stuffer usually has a diameter of around 0.45 mm. Spun stuffer yarn can be spun polyester, continuous filament polyester, spun acrylic, or sheath/core type structure. The air permeability depends on the fabric density and stuffer yarn. Figure 12.30 shows monofilament dryer fabrics.

In another variation of monofilament fabrics, which are called low caliper and low permeability monofilament fabrics, usually single or at the most one-and-a-half layer designs are used. They can be 100% monofilament or cross-machine direction yarns can be a combination of monofilament and sheath/core (cabled) yarn for low air permeability.

Hollow yarns are now being used in the cross-machine direction to lower air permeability.

Spiral Fabric

Spiral fabric is a special type of structure in which helical formed loops of monofilament yarns are joined together to form an endless fabric. The joining yarn, which is a straight monofilament yarn, is inserted through the "eye" that is formed when the right- and left-hand spiral yarns are merged into each other as shown in Figure 12.31.

Polyester is the most common yarn in spiral fabrics. PPS yarns are also being used to con-

struct spiral fabrics. Spiral fabrics have air permeabilities in the range of 100 – 1,000 cfm. Permeability can be altered by inserting yarns in the cross-machine direction. This type of fabric is resistant to stretching, bowing and narrowing. Spiral fabric is especially suitable for brown paper applications.

Multi-Filament Designs

These are open mesh fabrics made of multi-filament yarns. Their designs are varied to control air permeability. There are two major types of multi-filament designs. The first type uses Nomex® aramid yarn in the warp or machine direction which is the load-bearing direction (Figure 12.32). Nomex® has excellent resistance to thermal and hydrolytic degradation. The second type of multi-filament design uses blends of yarns such as nylon, polyester or acrylic. The mechanical properties of these fabrics are comparable to Nomex®-content fabrics, however hydrolysis resistance is not as good as Nomex®-content fabrics. Therefore, Nomex®-content fabrics are used in harsher environments such as brown paper and corrugating applications.

5.2 Manufacturing of Dryer Fabrics

Manufacturing processes are similar to forming and press fabrics. Dryer fabrics are woven flat and seamed later. Both monofilament and multi-filament fabrics are heat set for dimensional stability. Multi-filaments may be chemically treated. The seam is the weakest area in the fabric. Improper seam design may mark the sheet.

There are various types of seam designs, depending on the fabric design and application area. The clipper seam, in which steel hooks are used to join the seam ends, is being used less, only for very coarse sheet grades (linerboard, bag and medium). This seam style is used mostly for multi-filament fabrics. In a multi-filament seam, the seam is made of a webbing with woven loops which is woven into the body of the fabric. A spiral seam is used where marking is not very critical. It is constructed by raveling out warp

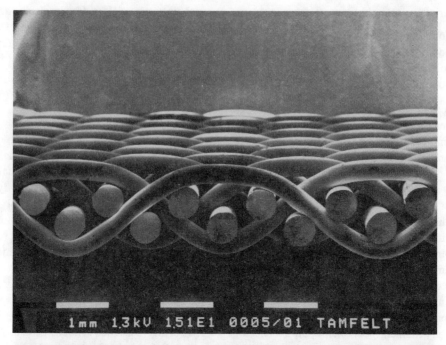

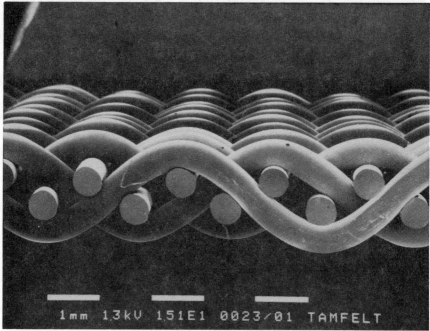

FIGURE 12.30 Monofilament dryer fabrics (courtesy of Tamfelt, Inc.).

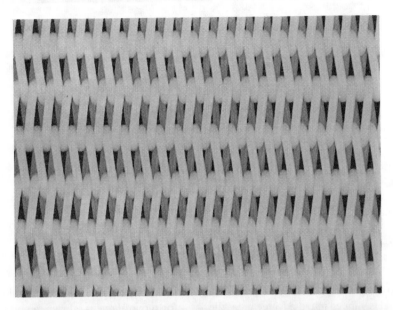

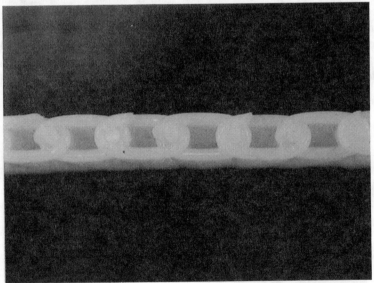

FIGURE 12.31 Spiral fabric (courtesy of Asten Dryer Fabrics, Inc.).

FIGURE 12.32 Multi-filament dryer fabric with Nomex® in MD and wrapped fiberglass in CD direction (courtesy of Asten Dryer Fabrics, Inc.).

yarns and creating a fringe. Then, a precoiled synthetic spiral is slipped into the fringe area, the fabric is folded over and sewn [16]. The spiral loops engage a similar spiral on the other end of the fabric. The sewing thread should be buried in the fabric to avoid abrasion and seam failure. This seam style is used for mono- and multi-filament fabrics. In the joined spiral seam, which is for monofilament fabrics, a spiral with loops is used as well. However, instead of sewing, the area next to the spiral is woven into the fabric, thus the spiral is anchored into the fabric (Figure 12.33). The woven seam area is around 40−60 mm in the warp direction. In a pin seam, all the

FIGURE 12.33 Spiral coil seam in monofilament fabric; the spiral yarn is PEEK (courtesy of Asten Dryer Fabrics, Inc.).

FIGURE 12.34 Pin seam in monofilament dryer fabric. The larger white area is a sheet of paper placed for contrast. The joining wire is hidden, but is located between the two white areas (courtesy of Asten Dryer Fabrics, Inc.).

warp yarns are woven back into the fabric but only some warp yarns participate in loop formation. Then, the loops on each end are connected by a pin in the weft direction. The pin seam is used for monofilament fabrics only (Figure 12.34).

Monofilament dryer fabrics may be sanded to obtain a smoother surface especially for the finer paper grades.

5.3 Application of Dryer Fabrics

The factors that affect dryer fabric selection are paper grade, cylinder temperature and air moisture, operating life and economic considerations, paper machine speed, pocket ventilation, fabric drying equipment, operation tension, and dryer section configuration. Dryer fabric design should not mark the sheet.

Since monofilament fabrics are resistant to filling and can be cleaned more easily, they are suitable for recycled paper manufacturing. Teflon® content monofilaments are used in the construction of fabrics used with recycled pulp. Open-mesh multi-filament and monofilament fabrics provide better moisture profile of the web across the machine width. With the increased machine speeds, fabric operating tension has

reached the $1.5-2.5$ kN/m range which should be a fabric design factor especially in the seam area.

6. CURRENT AND FUTURE TRENDS IN PAPER MACHINE CLOTHING

Paper machine clothing design, properties and applications have tremendous effects on paper properties and papermaking process. Proper paper machine clothing improves product quality, increases productivity, decreases paper machine downtime and increases the bottom line. Advances in raw materials and paper machine clothing design usually follow the advances in paper machine design, paper making processes, paper grades and properties very closely. As a result, a more "engineered" fabric is demanded.

6.1 Trends in Papermaking

- increased recycling: In 1991, over 35% of all paper consumed in the United States was recovered for recycling. In 1993, the percentage reached 40%.
- easy fabric installation, less downtime, better runnability and improved fabric efficiency

- wider paper machines
- higher paper machine speeds resulting in higher tensions on fabrics
- higher running temperatures for through-air dryer, press, and dryer fabrics
- more alkaline pulping in fine paper
- better formation and sheet quality
- lower basis weight in all paper grades especially in tissue to achieve high bulk
- increased strength/weight ratio of paper products
- increased forming consistency
- more complicated chemical environments

6.2 Trends in Raw Materials

For forming fabrics:

- contaminant resistance
- hydrolysis resistance for through-air dryer fabrics
- high modulus and stretch resistance for high machine speeds
- abrasion resistance
- chemical resistance (peroxide, chlorine, ozone, acid, alkaline)
- shaped monofilaments (flat, square, oval, hollow)
- improved loop and knot strength
- low moisture nylon

For press fabrics:

- low moisture regain nylon and new polymers
- chemical resistance (chlorine bleaching, peroxide and ozone bleaching)
- compaction resistance for fatigue and impact
- shaped monofilaments
- high temperature polyamides for high temperature pressing (impulse drying)

For dryer fabrics:

- high temperature resistance yarns (PPS, PCTA, polyamides, PEEK)
- monofilaments for permeability control (hollow, sheath)

- contaminant resistance
- conductive yarns for static electricity
- improved loop and knot strength
- abrasion, stretch and chemical resistance

6.3 Current and Future Developments in Forming Fabrics

- higher fiber support for increased retention, and improved formation
- improved fabric stability to increase crease resistance for improved weight profiles
- greater and controlled drainage in multi-layer designs
- improved contaminant resistance fabrics
- reduced wire mark on the sheet by reducing the number of crossovers and knuckle height using smaller yarns and surfacing
- better wear and abrasion resistance (including extruded edge wear treatment) to deal with abrasive inorganics in alkaline papermaking and for protection from increased high pressure showering when running recycled furnish
- forming fabrics to reduce tissue basis weight
- higher temperature, better hydrolysis fabrics for through-air dryer
- improved seaming techniques for easy installation — completely automated seaming
- specialized fabrics for controlled fiber orientation
- anti-static fabrics for better formation and release in nonwovens
- new designs for high consistency forming

6.4 Trends in Press Felts

- smoother needlepunching to reduce sheet marking
- increased and uniform drainage capacity
- batt fiber abrasion resistance due to alkaline papermaking and chemical additives

- superior resistance to compaction and dimensional changes
- stratified or laminated fabrics for better pressure uniformity and superior sheet smoothness
- seamed felts rather than endless weaving—Seamed felts offer faster and safer installation with minimum machine adjustment.
- machine direction oriented batt fibers for smooth surface (advantages: high sheet finish, low flow resistance for faster drainage, reduced hydraulically induced vibrations)
- contaminant resistance due to recycling and chemical additives
- chemical degradation resistance due to bleaching, oxidizing and cleaning agents
- 100% monofilament construction for improved drainage

- foam to replace batt fibers partially or totally
- felts with three or four layers
- felts for higher temperature running conditions to enhance dewatering

6.5 Trends in Dryer Fabrics

- contaminant release fabrics
- ultra-high contact, smooth surface
- better resistance to heat and hydrolysis
- design to reduce sheet stretching and sheet-to-fabric friction for smoother, more uniform sheet, reduced fabric wear and fewer sheet breaks
- maximum drying and ventilation efficiency with less energy consumption
- increased use of monofilament fabrics rather than multi-filaments which stay clean longer and carry less moisture

12.2

References and Review Questions

1. REFERENCES

1. Storat, R. E., "The U.S. Pulp, Paper and Paperboard Industry: A Profile," *TAPPI Journal,* March 1993.

2. Paper Machine Felts and Fabrics, Albany International, 1976.

3. Parker, J. D., *The Sheet Forming Process,* TAPPI Press, January 1990.

4. Hansen, V. E., "Water Management for the Wet End," *1991 Wet End Operations,* TAPPI Press.

5. Smook, G. A., *Handbook of Pulp & Paper Technologist,* TAPPI Press, 1982.

6. *Pulp and Paper Manufacture, Volume 7: Paper Machine Operations,* TAPPI Press, 1991.

7. Banks, R., "Clothing Performance Evolution Has Led to Paper Machine Advances," *Pulp and Paper,* May 1988.

8. Reese, R. A., "Pressing Operations," in Thorp, B. A. and Kocurek, M. J., eds., *Pulp and Paper Manufacture, Volume 7: Paper Machine Operations,* TAPPI Press, 1991.

9. Wicks, L. D., "Press Section Water Removal Principles and Their Applications," *Southern Pulp & Paper Manufacturer,* Jan. 1978.

10. Bliesner, W. C., "Sheet Water Removal in a Press: Time to Review the Fundamentals," *Pulp & Paper,* Sept. 1978.

11. Hill, K. C., "Paper Drying," in *Pulp and Paper Manufacture, Volume 7: Paper Machine Operations,* TAPPI Press, 1991.

12. Beran, R. L., "The Evaluation and Selection of Forming Fabrics," *TAPPI Journal,* Vol. 62, No. 4, 1979.

13. Johnson, D. B., "Retention and Drainage of Multi-Layer Fabrics," *Pulp & Paper Canada,* Vol. 87, p. 5, 1986.

14. Antos, D., "Advantages of Nonwoven Press Felts," *Paper Age,* April 1993.

15. Sassaman, S., "Pressing: High Speed Tissue Press Fabrics," *Paper Age,* June 1993.

16. Stowe, B. A. et al., "Paper Machine Clothing," in *Pulp and Paper Manufacture, Volume 7: Paper Machine Operations,* TAPPI Press, 1991.

2. REVIEW QUESTIONS

1. Explain how paper is made. What is the role of fabrics in papermaking?

2. Compare the major structural properties of forming, press and dryer fabrics. What kind of materials are used in each?

3. What are the requirements of recycled pulp on forming fabric properties? Explain.

4. What are the advantages of two layer extra forming fabrics?

5. What is the purpose of heat setting paper machine fabrics?

6. Define the following terms:
   - mesh count
   - beat count
   - air permeability
   - fiber support index (FSI)
   - drainage index
   - void volume

7. What are the factors to consider when select-

ing a forming fabric design for a particular application?

8. Explain the major press felt structures.

9. What are the important properties of press felts?

10. Define hydrolysis resistance. Which fibers are the best for hydrolysis resistance?

11. Explain how a pin seam is formed using a schematic.

SAFETY AND PROTECTIVE TEXTILES

Safety and Protective Textiles

S. ADANUR

1. INTRODUCTION

Safety and protective textiles refer to garments and other fabric-related items designed to protect the wearer from harsh environmental effects that may result in injury or death. It may also be necessary to protect the environment from people as in the case of clean rooms. Safety and protective materials must often withstand the effects of multiple harsh environments.

Industrial textiles are used to protect from one or more of the following:

- extreme heat and fire
- extreme cold
- harmful chemicals and gases
- bacterial/viral environment
- contamination
- mechanical hazards
- electrical hazards
- radiation
- vacuum and pressure fluctuations

Ballistic protection, which is a form of mechanical protection, and low visibility (camouflage) textiles are covered in Chapter 11.0, Military and Defense Textiles. Bacterial/viral protective textiles are included in Chapter 10.0, Medical Textiles. Other categories of safety and protective textiles are the subject of this chapter.

The General Duty clause of the federal OSHA Act, paragraph 5a1 states that "to identify risks and hazards in the workplace and seek out appropriate protective garments and equipment for the protection of workers is the responsibility of the employer" [1]. OSHA's new Personal Protective Equipment (PPE) rule 1910.123 went into effect on October 5, 1994. As a result of the rule, employers must [2]:

- conduct a hazards assessment in the work area
- ensure that employees wear safely designed and constructed PPE
- establish, complete and certify an employee training program for the correct application of PPE in each specified area

OSHA implemented this new standard to ensure the safety of an estimated 12 million U.S. workers at risk of eye, face, head, foot and hand injuries. OSHA holds employers responsible for the implementation of the tasks above. Failure to comply may result in substantial penalties. It should be noted that OSHA regulations generally do not provide specifics about the personal protective clothing to be used, leaving that to the end user, manufacturers, trade groups or standard setters to design and make the equipment, develop standards, certification and selection procedures.

2. HIGH TEMPERATURE TEXTILES

In the past, asbestos, a mineral fiber, was an important high temperature resistant textile material. Within the last thirty years, many high heat and temperature resistant fibers have been developed as shown in Table 13.1. Asbestos, on the other hand, has been largely superseded since it causes health hazards that may lead to cancer. However, its properties are often compared with newly developed high temperature materials.

A high temperature textile material is defined as a material that can be used continuously at temperatures over 200°C (~400°F) without decomposition and without losing its major physical properties. Although some fibers can be

TABLE 13.1 General Timeline for Development of High Temperature Fibers
(courtesy of Industrial Textile Associates).

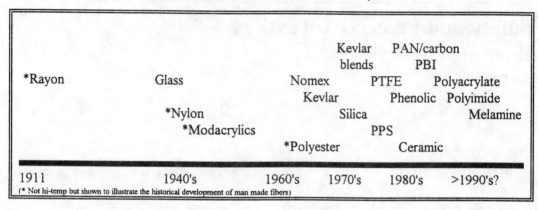

			Kevlar blends	PAN/carbon PBI	
*Rayon	Glass		Nomex Kevlar	PTFE Phenolic	Polyacrylate Polyimide
	*Nylon		Silica		Melamine
	*Modacrylics			PPS	
		*Polyester		Ceramic	
1911	1940's	1960's	1970's	1980's	>1990's?

(* Not hi-temp but shown to illustrate the historical development of man made fibers)

used for insulation and retain some strength above 2,000°F, they generally lose their textile characteristics at that temperature range [3,4].

Table 13.2 lists the major high temperature materials along with their manufacturers. Important properties of some high temperature fibers are given in Table 13.3. Table 13.4 shows the application areas of some high performance-high temperature fibers.

Thermal resistance is the ability of the fiber to remain relatively unchanged when exposed to radiant, conductive and convective heat. If heat

TABLE 13.2 High Temperature Fibers (courtesy of Industrial Textile Associates).

ARAMID
Meta-Aramid:
  Apyeil (Unitika)
  Nomex (DuPont)
  Conex (Teijin)
  Fenilon (Russian)

Para-Aramid:
  Kevlar (DuPont)
  Technora (Teijin)
  Twaron (Akzo)
  Hoechst para-aramid (pilot)
  Kolon para-aramid (pilot)
  SVM (Russian)

GLASS (various producers)
  E-glass
  S-glass
  Cardable glass
  Miraflex—bicomponent
    (Owens Corning)

MELAMINE
  Basofil (BASF)

PHENOLIC (also known as
Novaloid)
  Kynol (Gun Ei Chem—Nippon Kynol)
  Philene (St. Gobain)

POLYBENZIMIDAZOLE
  PBI (Hoechst Celanese)
  Togilen (Russian)

POLYAMIDE
  P84 (Lenzing)
  Arimide (Russian)

POLYAMIDE IMIDE (sometimes
referred to as meta-aramid)
  Kermel (Kermel—Joint venture
    Amoco and Rhône Poulenc)

POLYPHENYLENE-SULFIDE
(PPS or sulfar)
  Ryton (Amoco—formerly Phillips)
  Procon (Toyobo)
  PPS (Teijin, Hoechst)

POLYETHERETHERKETONE
  PEEK (Shakespeare, H-C)

POLYACRYLATE
  Inidex (Courtaulds)

PTFE (fluorocarbon)
  Teflon (DuPont)
  Rastex (W. L. Gore)
  Polyfen (Russian)
  Tetratex (Tetratec)

PAN/CARBON (preoxidized
polyacrylonitrile fiber)
  Panox (RK Carbon)
  Panotex (Universal Car.)
  Pyromex (Toho Rayon)
  Celiox (CCF/Toho)
  Fortafil (Great Lakes)
  Uglen, Gralen, Evlon (Russian)

SILICA
  Leached glass (various)
  Quartzel (Quartz Prod. and Quartz
    & Slice)
  Silica, Quartz (Russian)

CERAMIC
  Nextel (3M)
  Taikalon (Noai Bussan Japan/China)

Note: This is not an exhaustive list. Others are being developed; ©1995, Industrial Textile Associates, Greer, SC 29650.

TABLE 13.3 Properties of Selected High Temperature Fibers (courtesy of Industrial Textile Associates).

	Fiber										
Properties	Aramid[3] M = meta P = para	PAN/ Carbon	Glass	Melamine	PBI	Novoloid (Phenolic)	PPS	Polyacrylate	PTFE	Polyimide	Polyamide Imide
	Fiber Properties										
Tenacity—dry g/d	4.0–5.3 M 21–27 P	24	15.3	1.8	2.6–3.0	1.3–2.4	3.5	1.3–1.7	0.9–2.0	3.7	3.4
Tensile—000 psi	90 M 400 P	31	500	30	50	20–30	61	25–33	40–50	67	75–80
Elongation at break %	22–32 M 2.5–4.0 P	19	4.8	12	25–30	30–60	40	20–30	19–140	19–21	16–21
Moisture regain %	6.5 M 4.0 P	9	<0.10	5	15	6–7.3	0.6	12	0	3	4–4.5
Specific gravity	1.38 M 1.44 P	1.4	2.5	1.44	1.43	1.27	1.37	1.5	2.1	1.41	1.34
Avg. toughness—g/d	0.30 M 0.85 P	N/A	0.37	N/A	0.4	N/A	N/A	N/A	0.15	N/A	N/A
Abrasion resistance	Good M Poor P	Poor	Poor	Fair	Good	Poor	Good	Fair	Good	Good	Good
Resilience	Excel	Poor	Poor	Fair	Excel	Fair	Good	Good	Poor	Fair	N/A

(continued)

417

TABLE 13.3 (continued).

						Fiber					
Properties	Aramid[3] M = meta P = para	PAN/ Carbon	Glass	Melamine	PBI	Novoloid (Phenolic)	PPS	Polyacrylate	PTFE	Polyimide	Polyamide imide
Chemical Resistance											
Solvents	Excel M / Good P	Good	Excel	Excel	Excel	Excel	Excel	Good	Excel	Excel	Excel
Acids dilute:	Good M	Good	Excel	Good	Excel	Excel	Excel	Excel	Excel	Excel	Good
concentrated:	Fair P	Poor	Excel	Poor	Excel	Poor[1]	Good	Excel	Excel	Excel	F-Good
Alkalies dilute:	Good M	Good	Fair	Excel	Excel	Excel	Excel	Excel	Excel	Poor	Good
concentrated:	Good P	Good	Fair	Excel	Good	Excel	Excel	Excel	Excel	Poor	Good
Ultraviolet (UV)	Poor	Good	Excel	Good	Good	Excel	Excel	Excel	Excel	Good	Good
Thermal Properties											
Limiting oxygen index (LOI)	30 M / 29 P	55	>100	32	38	33	34	43	>100	40	32
Thermal conductivity (BTU-in/hr^2 per °F)	0.26 M / 0.30 P	<.03	7.2	0.20	0.26	0.28	0.3	0.31	0.2	N/A	0.08
Usable temp. °C Short term:	315–370	V-high	V-high	260–370	>595	Decomp. >400	260	[2]	Decomp. >430	<485	>420
Continuous:	230	220	315	>200	315	205	205	160	288	<260	250
Smoke emission— density:	1.0	N/A	Low	Low	Trace	<0.30	N/A	Trace	Low	<1.0	<2.0
Relative Cost											
1 = Low 4 = High	3	3	2	2	4	2	3	3	4	3	3

[1]Nitric and sulfuric.
[2]Auto ignition @ 450°C and 100% O_2.
[3]M—meta-aramid (as Nomex); P—para-aramid (as Kevlar).
Source: Producers literature and ITA database. © 1995—ITA, Greer, SC.

TABLE 13.4 *Typical Applications of High Temperature Fibers (courtesy of Industrial Textile Associates).*

Market	Desired Properties	Fibers Used
Construction		
Hi-temp pipe wrap	Tensile, thermal, hydrolytic properties	Aramids, glass
Inflatable roofs	Hi-strength, fire resist., low creep	Glass
Coated Fabrics		
Hi-temp silicone and similar coatings, seals, diaphragms, inflatables	Strength, fatigue resist., low creep	Aramids, glass, PBI, others
Filtration		
Hot air/gas and corrosive liquids	Hi-temp and chemical resistance retention of properties	Aramids, PPS (sulfar), glass polyimide, other
Friction Products/Packing, Seals, Gaskets		
Brake linings, clutch facings	Modulus, thermal stability, non-toxic, low wear	Aramids, polyacrylate, carbon glass, phenolic
Gaskets, packing, seals	Compressibility, thermochemical resistant, recovery, sealability	PTFE, PBI, polyimide, phenolic
Protective Clothing		
Garments	Thermal, absorption, dyeable, good textile properties	Aramids, PBI, polyimides, PAN, others
Gloves/aprons/sleeves	Thermal, abrasion and cut resistance	Aramids, glass
Reinforcement, Plastic, Advanced Composites		
Aerospace, interior	Strength, low density, thermal	Aramid, carbon
Aerospace, structural	+ compression, vibration dampening hi-temp, ablative, electrical prop.	Carbon, quartz, glass, aramid
Automotive, structural	Tensile, compression and flexural prop.	Carbon, glass
Reinforcement/Rubber		
Belts/mechanical	Strength, stiffness, hydrolytic stability, temperature resistance	Aramid, other
Tires-radial/hoses	Strength, stiffness, adhesion, low fatigue, oxidation resist, thermal	Aramid, other
Ropes/Cordage		
Rescue/fire resistant/escape chutes/slides/slings	Strength, thermal, low elongation compressed storage properties	Aramid, other
Miscellaneous Industrial		
Hi-temp fire blankets, welding and fire barriers	Thermal resistance, extreme hi-temp, reasonable cost	Aramid, glass, PBI Leached glass
Sewing threads	Strength, weatherability, thermal + extreme hi-temp	Aramid, PTFE Leached glass, quartz, ceramic
Fire block seating	Thermal, low toxicity, light weight	Aramid, polyimide
Wire insulation	Thermal, electrical, strength	PPS, PEEK, aramid, glass

© 1995, Industrial Textile Associates, Greer, SC 29650.

dissipation is desired, high thermal conductivity and/or high emissivity materials are used. Another measure of high temperature performance is the Limiting Oxygen Index (LOI) which is used to rate the flammability of fibers. LOI is the percent oxygen level that must be present in the oxygen/nitrogen mixture of air before the fiber would ignite and burn when exposed to flame. Therefore, higher LOI number generally indicates better flame resistance. Fibers with LOI in the mid-20s and above are considered flame resistant and self extinguishing. For example, when exposed to flame at normal atmosphere, cotton, which has an LOI of 18, would ignite and continue to burn even after the flame is removed. The oxygen content of atmosphere is about 16%. Figure 13.1 and Table 13.3 show LOI of selected high temperature fibers.

In general, fabrics made of high temperature fibers have a weight range of 2−50 ounces per square yard with typical FR apparel fabrics being in the 6−8 oz range. Twill and sateen weave designs are common. Batting and nonwovens are used extensively for thermal insulation. Compound structures with more than one layer of different materials are effective in certain applications. The price of some newer high temperature fibers are considerably higher than traditional non-FR textile materials such as cotton, nylon or polyester.

Several factors should be considered when selecting a high temperature fiber or fabric for a particular application [5].

- the nature of application (i.e., personnel or machinery protection, insulation, welding protection, stress relieving, gasketing, fire curtains, filtration, etc.)
- temperature range and heat generation (maximum continuous temperature, exposure time, direct or indirect heat source, radiant, conduction or convection)
- environmental conditions (types of chemicals present, direct contact or airborne, wet and dry environment, abrasion and strength requirements, direct contact with skin, etc.)
- special needs (special test requirements, special coatings, etc.)

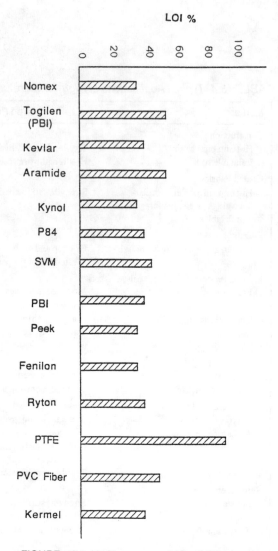

FIGURE 13.1 Limiting oxygen index (LOI) of various high temperature fibers [4].

The effect of increased temperature on fabric strength and other properties is important for performance reasons. For optimum performance, fabric should not lose its mechanical strength under heat. Figure 13.2 shows the strength of several materials at different temperatures. In general, fabric properties that are temperature dependent are affected by the amount of heat absorbed, not necessarily by the mechanism of heat absorption. By increasing the fabric weight strength loss can be delayed. However, fabric weight decreases with increasing exposure time as shown in Figure 13.3.

Material	Weave Type	Weight (g/m²)
PBI	plain	153
Kynol	2/1 twill	159
95/5 Nomex/Kevlar	cloque	180
Kevlar	cloque	175
Cottin	oxford	217
Polyester	plain	220

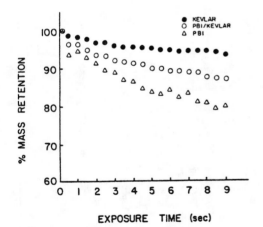

STRENGTH RETENTION (%)

1	PBI
2	KYNOL
3	NOMEX 95/5
4	KEVLAR
5	COTTON
6	POLYESTER

TEMPERATURE (°C)

FIGURE 13.2 Strength retention of fabrics at various temperatures (copyright ASTM [6]).

% MASS RETENTION

EXPOSURE TIME (sec)

FIGURE 13.3 Mass retention of PBI, Kevlar and PBI/ Kevlar blend in thermal protective performance tests (copyright ASTM [7]).

3. FLAME RESISTANT PROTECTIVE CLOTHING

National Burn Information Exchange data indicate that the most devastating burn injuries occur when a person's clothing ignites. The greatest factor affecting whether a person survives a flash fire and how rapidly he or she recovers, is the percentage of second and third degree burns on the person's body. The lower percentage burn of the body increases the chance of survival. Wearing flame resistant (FR) protective apparel reduces burn injury and increase the chance of surviving a flash fire.

Flame resistant fabrics are designed to resist ignition and self-extinguish when ignited. In general, a properly designed flame resistant fabric should prevent the spread of flames when subjected to intensive heat or flame. It should also self-extinguish immediately as soon as the ignition source is removed. Resistance to both flame and associated heat transfer through the garments is defined as thermal protection.

Primary and secondary protective clothing for flame resistance are defined by ASTM in Standard F1002. Primary protective clothing is designed to be worn for work activities where significant exposure to molten substance splash, radiant heat, and flame, etc., is likely. Secondary protective clothing is designed for continuous protection from possible intermittent exposure to molten substance splash, radiant heat and flame. Examples of primary protective clothing are firefighter turnout gear and fire entry suits. Continuously worn flame resistant uniforms are considered as secondary protective clothing.

Flame resistance in fabrics can be achieved in two ways:

(1) By using inherently flame resistant materials and fibers

(2) By using special treatments of fibers or fabrics

In the first group, flame resistance is an essential or inherent characteristic of the fiber. In the second case, chemical treatments or compounds are incorporated into or onto the fiber or fabric to achieve flame resistance.

Flame resistant fabrics can be produced from

Table 13.5 Commercially Available Flame Resistant Fibers for Work Apparel Fabrics [1].

Generic Name		Fiber	Manufacturer	Moisture Regain (%)	Tenacity (g/den)	Comments
Aramid	(meta)	Nomex	DuPont	6.5	4.0–5.3	Fiber forming substance is a long chain synthetic polyamide. Excellent durability and chemical resistance. Relatively poor colorfastness to laundering and light exposure.
	(para)	Kermel	Rhône-Poulenc (France)	3.4	4.0–4.5	Blended with Nomex for strength. Difficult to dye. Excellent strength and abrasion resistance. Sensitive to bleach, light, and strong mineral acids.
		Kevlar	DuPont	4.3	21–27	
		Twaron	Akzo (Holland)	4.0	22.6	
Modacrylic		SEF Kanecaron	Monsanto Kaneka (Japan)	2.5	1.7–2.6	Fiber forming substance is a long chain synthetic polymer containing acrylonitrile units. Excellent chemical and abrasion resistance. High thermal shrinkage.
PBI		pbi	Hoechst Celanese	15.0	2.8	Polymer is a sulfonated poly(2,2-*m*-phenylene-5,5-dibenzimidazole). Will not ignite, does not melt. Excellent chemical resistance. Dyeable in dark shades only.
Polyimide		P84	Lenzing (Austria)	3.0	3.7	Fiber forming substance is a long chain synthetic polyimide. High thermal shrinkage. Thermal properties inferior to Nomex.
FR Viscose		PFR Rayon	Lenzing (Austria)	11.5	2.6–3.0	Man-made cellulosic, properties similar to cotton.
FR Cotton		FR Cotton	Natural fiber	8.0	2.4–2.9	Flame resistant treated in fabric form. Poor resistance to acids. Abrasion resistance only fair. Relatively poor colorfastness to laundering and light exposure. Wear properties similar to untreated cotton.
Vinal		Vinex FR9B	Westex	3.0	3.0	Fiber composed of vinyl alcohol units with acetal crosslinks. Sheds aluminum splash, but very sensitive to shrinkage from wet and dry heat.
FR Polyester		Trevira FR polyester	Hoechst Celanese	0.4	4.5	Polyester with proprietary organic phosphorus compound incorporated into the polymer chain. Properties similar to regular polyester except for inherent flame resistance. Melt point 9°C lower than regular polyester.

a single type of fiber (e.g., 100% SEF Modacrylic) or from blends of two or more fibers (e.g., DuPont's Nomex® IIIA which is made of Kevlar®, Nomex® and carbon). Blending of fibers is a common practice in industry to optimize the desired flame resistance and other properties, including cost. By mixing natural fibers such as cotton and wool with synthetic fibers, fabrics with high comfort and good aesthetic properties can be engineered to meet the specific thermal requirements. However, caution should be exercised when mixing FR fibers that may react differently to flame. Future FR garments will withstand longer exposure times in hazardous conditions, reduce the potential for body burns, and enhance the ergonomical aspect for better job efficiency.

3.1 Flame Resistant Fibers and Fabrics

Treated cotton is widely used in apparel for flame resistance. The most commonly used synthetic flame resistant fibers include aramids, PBI and PFR rayon. SEF acrylic is used in some chemical resistance requirements and in children's sleepwear. Table 13.5 shows the commercially available flame resistant fibers for work apparel fabrics. Many blending combinations are possible using fibers in the table. Brief descriptions of flame resistant fibers and fabrics are given below. For more detailed information the reader may refer to Reference [1]. Table 13.6 shows the FR fabric consumption in the United States.

Flame Resistant (FR) Cotton

Various flame resistant treatments have been applied to cotton goods over the years. The two most widely used treatments for cotton today are:

(1) A precondensate of tetrakis (hydroxymethyl) phosphonium chloride (THPC) reacted with urea and cured with gaseous ammonia to form a polymeric molecule, e.g., Proban from Albright and Wilson

(2) A heat cured dialkylphosphonamide, e.g., Pyrovatex CP from Ciba Geigy

Depending on the treatment, flame retardant finishes for cotton may be temporary or permanent. While FR treated cotton is suitable for foundries, flame cutting, welding, chemical environments, oil, gas and petrochemicals, it is not recommended for protection against molten white metals such as aluminum (molten metals will stick to the fabric) and in critical static control operations. Although cotton is resistant to alkalis, many acids will completely destroy FR and non-FR cotton. Chlorine bleach destroys the flame retardant polymer in treated cotton fabric. Starches and fabric softeners should also be avoided. Life expectancy of the FR cotton is similar to 100% cotton.

FR-8 cotton fabric is treated with a non-phosphorus flame resistant finish. It is exclusively intended for protection against aluminum splash. A system of antimony oxide and bromide compounds is bound to the fiber by a special latex binder. The disadvantages include stiffness, little moisture absorbency, and short treatment life.

Fabric weight is an important parameter for thermal insulation with FR cotton materials. It was reported that increasing fabric weight increased protection time as shown in Figure 13.4. Twenty-eight FR cotton fabrics of weights ranging from 4.5 oz/sq. yd to 13.5 oz/sq. yd were exposed to a mixture of convective and radiant energy from a thermal source of 2.0 cal/cm^2 sec. It should be noted that the fabric itself, not the flame retardant treatment provides the insulative protection.

TABLE 13.6 Estimated Consumption of FR Fabrics in the United States (millions of square yards [8]).

Fabric Type	1985	1987	1990	1991	1996
FR Cotton	11.5	11.5	11.5	12.5	16.5
Nomex III	7.5	8.5	10.0	11.0	14.5
PBI/Others	2.5	3.0	1.5	1.0	1.5
Total	21.5	23.0	23.0	24.5	32.5

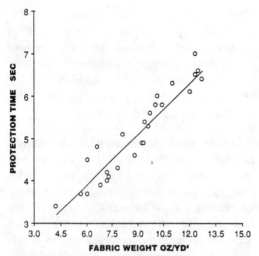

FIGURE 13.4 Protection time versus fabric weight for FR cotton fabrics [9].

Table 13.7 shows the effect of laundering on the performance of FR treated cotton fabric. Testing was conducted in the new condition as well as after 25, 50, 75 and 100 heavy soil industrial launderings. The results showed that char length increases with increasing number of washes.

Nomex®

Nomex®, which was introduced in 1961 by DuPont, is an inherently flame resistant material. It can be dyed in fiber or fabric form with various colors. Nomex® offers excellent thermal stability and does not melt, but decomposes at 700°F. Nomex® is recommended in petrochemicals, utilities and fire fighter uniforms but should not be used for protection against molten substances, welding operations and in critical static

control applications. Although 100% Nomex® fabric offers good FR properties, Nomex® is usually blended with other fibers to optimize properties. Nomex® III is a blend of 95% Nomex® and 5% Kevlar which has higher strength. Nomex® IIIA, which consists of 93% Nomex®, 5% Kevlar and 2% carbon core nylon, was developed to improve static electricity properties. Nomex® III and IIIA are extremely durable.

Nomex® Delta A is designed to reduce the risk of explosions sparked by static electricity prevalent in the petroleum and chemical industries. Nomex® Delta T is suitable for protection from flash fires or other sources of intense heat for extended periods of time. Nomex® Delta FF incorporates fine fibers for hot gas filtration applications.

Vinex®

Vinex® FR9B (Westex, Inc.), an inherently flame resistant fabric, is a blend of 85% vinal (composed of vinyl alcohol units with acetal crosslinks) and 15% rayon. Unlike many other FR fabrics, it offers good protection against molten aluminum and other "white" metal splash. Vinex® can be piece dyed. It is resistant to most acids, alkalis and organic solvents. The thermal stability is relatively poor and moist heat above 140°F will cause excess shrinkage.

SEF Modacrylic®

SEF Modacrylic® (Monsanto) is an inherently flame resistant synthetic, acrylonitrile fiber. It is suitable for chemical plants, tire manufacturing plants, paint and varnish manufacturing plants.

TABLE 13.7 Effect of Laundering on FR Treated 100% Cotton Fabric [10].

Fabric Weight (oz/yd²)	Average* Char Length (inch)**									
	Initial		25 Washes		50 Washes		75 Washes		100 Washes	
	Warp	Fill	Warp	Fill	Warp	Fill	Warp	Fill	Warp	Fill
7.0	2.2	2.1	2.2	2.2	2.5	2.5	2.7	2.7	3.0	3.4
9.0	2.6	2.3	2.5	2.6	2.6	2.5	2.5	2.3	2.0	2.8

*Average of five specimens in each direction.
**FTMS 191A Method 5903 Vertical Flammability Test.

However, it is not recommended against molten substances or welding exposure. SEF offers good colorfastness to laundering. It is resistant to most acids, weak alkalis, bleaches and most solvents. Since SEF garments shrink considerably over 140°F, their use in FR clothing has been declining.

Flamex II®

Flamex II®, a treated fabric, is a 70/30 cotton/polyester blend produced by Galey & Lord Inc. It is dyeable in various colors with good colorfastness to laundering, which makes it suitable for professional fire fighters and utilities personnel. Its use is not recommended against molten substances and welding. Flamex II® is sensitive to chlorine bleach. Flamex®, which was an earlier version, had 50/50 cotton/polyester ratio.

PBI (Polybenzimidazole)

PBI from Hoechst Celanese, is an inherently flame resistant fiber. First used in manned spacecraft applications, PBI is mostly used for military and aerospace applications as well as professional fire fighters. PBI, which has a natural color of gold, can be dyed in dark colors. It has good resistance to most acids, alkalis and solvents, however, chlorine bleach reduces PBI's strength. PBI is thermally very stable and retains fiber integrity after flame exposure. Garment life is claimed to be compatible with Nomex®. Kombat 450, which is a blend of 60/40 Kevlar/PBI fabric produced by Southern Mills, is suitable for career apparel applications. PBI is used in high temperature industrial gloves, foundry applications, firemen's protective apparel and as an aircraft seat fire blocker.

Kermel®

Kermel® (Rhône-Poulenc) is a polyamide imide aramid fiber which is offered only in blended form with other fibers. Kermel®/wool blend is used for dress uniforms, sweaters, and underwear. A blend of 50/50 Kermel®/FR viscose rayon is suitable for petrochemical, professional fire fighter and work apparel. Its use for molten materials, welding or in critical static control operations is not recommended. Kermel® is solution dyed, i.e., dyed in fiber form. It is resistant to most acids and alkalis.

FireWear®

FireWear® is an inherently flame resistant yarn which is a blend of 55% FFR-acrylic (Fibrous Flame Retardant Fiber) and 45% combed cotton produced by Springs Protective Fabrics. It may be used in knitted fabrics (T-shirts, polo shirts, sweatshirts, protective hoods) and in woven fabrics (shirts, pants, etc.). FireWear® is dyeable in fabric form. Application areas include petrochemicals, utilities, and fire fighter uniforms. It offers excellent thermal stability.

Upon exposure to flames, certain molecular components of the FFR fiber emit a non-combustible gas—released through the tiny pores in the fiber—that smothers the fire similar to a fire extinguisher. These gases shut off the oxygen feeding the flames, thereby stopping further burning. Figure 13.5 shows FFR fibers and FireWear® woven and knit fabrics before and after exposure to flames.

Trinex®

Trinex® (Westex, Inc.) is a blend of modacrylic, polyester and rayon. The ratio is 40/30/30 for pants fabric and 50/30/20 for shirt fabrics, respectively. It is inherently flame resistant. Trinex® is used for fire fighter and utility worker uniforms, and airline flight personnel. It is not recommended for molten materials, welding operations, or in critical static control operations. Other properties include relatively poor thermal stability, good resistance to chemicals, aging, sunlight and abrasion.

P84

P84 (Lenzing) is an inherently flame resistant long chain polyimide fiber. It is usually blended with other fibers. In the United States, Springs Protective Fabrics produces Reliant P84, which is a 40/60 P84/Kevlar blend for fire fighter turnout gear and A.L. International produces 40/60 P84/viscose rayon for knit protective hoods. P84

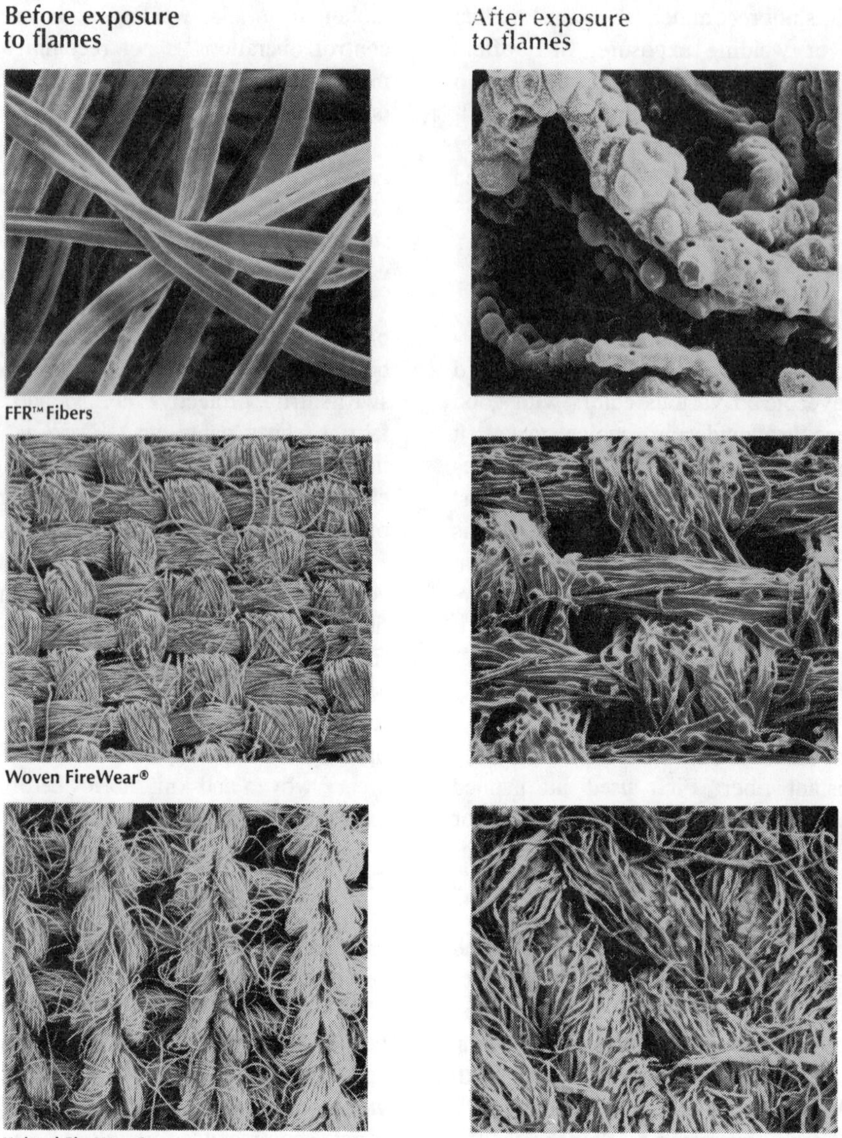

Before exposure
to flames

After exposure
to flames

FFR™ Fibers

Woven FireWear®

Knitted FireWear®

FIGURE 13.5 FFR fibers and FireWear® fabrics before and after exposure to flames (courtesy of Springs Protective Fabrics).

is solution dyed. It offers good resistance to organic solvents, acids and bleach. P84's thermal stability is low.

Kynol®

Kynol®, which is produced by Gun Ei Chemical in Japan and distributed by American Kynol, Inc., is an inherently flame (and chemical) resistant novoloid fiber and can be made in fiber, yarn, fabric and felt forms. It can also be used in aluminized and rubberized fabrics. Kynol® is a non-melting and non-shrinking organic fiber which offers light weight and soft hand. The practical temperature limits for long-term application are 150°C in air and 200−250°C in the absence of oxygen [11].

Aflammit®

Aflammit® is a group of FR products developed by Thor Chemicals that is durable to repeated washings. Aflammit® KWB can be applied to cellulosics to meet required standards of flame retardancy. It has been successfully applied to cotton single jersey fabric for children's sleepwear. Aflammit® PE imparts a durable FR to polyester. Latest developments include FR finishing of cotton-rich polyester/cotton blends without significantly affecting fabric hand [12].

Glo-Tex®

Glo-Tex® from TBA Textiles, is a protective lining material for gloves designed to handle hot, dry objects during industrial manufacturing. Glo-Tex is a double layered needled felt that is strong enough to withstand electrical sparks and molten metal droplets. One side is carbon and aramid. The other has hygroscopic FR viscose fibers that absorb moisture and perspiration.

Solar Alpha®

Solar Alpha® is an endothermic, heat storage fiber that efficiently absorbs sunlight and helps retain body heat. It consists of a nylon sheath and a core containing zirconium carbide. Zirconium carbide absorbs radiation with a wavelength below 2 μm, comprising 98% of the energy from sunlight. Future application areas could include space suits and vehicles.

Reliant®

Reliant® P is a 40/60 blend of Lenzing P fiber and Kevlar from Springs Industries. It will not burn in air or melt. It resists chemicals, acids, alkalis, solvents, abrasion, high temperature and direct flame. Hydropel finishes can be applied for water repellency.

Synergy®

Synergy®, from Springs Industries, is created by combining the inherent flame resistance of Nomex® III with Freedom stain release finish, Wickable comfort finish and Hydropel water repellent finish. It is used for heat and flame protection. These fabrics exhibit low thermal shrinkage, shorter char length and resistance to embrittlement.

AraMax®

AraMax® is a combination of 60% Kevlar® and 40% Nomex® and is developed by Springs Industries. Thermally stable, high strength Kevlar is wrapped with abrasion and fire resistant Nomex® to yield a strong, durable fabric.

Requirements for FR Textiles

Important parameters when selecting flame-resistant protective apparel are flame resistance, ergonomical design for wearer comfort, durability, easy maintenance, and aesthetics. Although flame resistance is essential for the protection function, it is not sufficient by itself to meet the requirements for flame resistant clothing. Other requirements are equally important in today's market. Comfortable, lightweight garments are the current trends in FR clothing. For example, moisture vapor transport ability of the fabric is critical for the effectiveness of a fire fighter. In addition, there may be special requirements such as chemical or molten substance resistance depending on the workplace. Furthermore, the garments should withstand a certain number of washings without losing color, appearance or physical properties. Dimensional stability is important after repeated washings.

3.2 Testing for Flame Resistance

Flame resistant fabrics for work and fashion apparel are required to meet the federal flammability standard, 16 CFR-1610 which is administered by the Consumer Products Safety Council (CPSC). Using this standard, ease of ignition and flame spread time are measured. A test specimen at 45° angle is exposed to a one-second ignition and the time to burn five inches of the specimen is measured. Based on burning characteristics, three types of flammability are identified [1]:

(1) Class 1: normal flammability (flame spreads in 4 seconds or more)
(2) Class 2: intermediate flammability (applies to napped fabrics)
(3) Class 3: rapid and intense burning (spread time is less than 4 seconds)

All wearing apparel sold in the United States must be made of fabrics in Class 1 or 2.

The most widely used test methods to evaluate the performance of thermal protective fabrics are as follows:

(1) The vertical flame resistance test, Method 5903.1 of Federal Test Standard 191A, which is sponsored by the General Services Administration (GSA), Office of Federal Supply Services. In this method, a twelve inch long fabric, which is suspended vertically, is subjected to a controlled flame on the cut edge. After twelve seconds, the flame is removed and three characteristics are reported (Figure 13.6):

- afterflame: the time (in seconds) that visible flame remains on the fabric
- afterglow: the time (in seconds) that glow remains on the fabric
- char length: fabric length that is destroyed by the flame (inches)

Five samples are tested in each direction (warp and filling) of the fabric and the average of each group is reported. It should be noted that Method 5903.1 is only a test method and it does not establish requirements. Table 13.8 shows typical vertical flame test results using this method. Some manufacturers and professional agencies have established more specific standards. For example, California OSHA specifies a

TABLE 13.8 Typical Vertical Flame Test Results Using FTM 5903A (DuPont [13]).

	Char Length (inch)	After-Flame (seconds)
Flame Resistant Fabrics		
Aramid (6 oz/yd²)	3	0
FR cotton (10 oz/yd²)	3	0
Non-Flame Resistant Fabrics		
100% Cotton	12	36
65/35 Polyester/cotton		
blend	12	48

maximum of 2.0 seconds after-flame and five inches char length for fire fighter uniforms. Heat resistance of thread materials is commonly measured using Federal Test Method 191A, 1534.

(2) ASTM Standard Test Method for Thermal Protective Performance (TPP) of Materials for Clothing by Open-Flame Method, D 4108. This test method is used to rate textile materials for thermal resistance and insulation when exposed to a convective heat flux. It is suitable for testing of woven, knit, non-woven fabrics and batting materials. The material is exposed to heat from a standard flame on one face. The amount of heat energy, which would be expected to cause a second degree burn on human tissue in contact with or close to the opposite face, is measured. ASTM D 4108 specifies two methods for TPP testing: one for single layer fabrics (such as those used in the shirts, pants and coveralls worn by industrial workers) and one for multi-layer fabrics (such as those used in three layer firefighter turnout coats). During single layer fabric testing, 0.25 inch spacer is placed between the fabric sample and the heat sensor. In multi-layer fabric testing, the fabric sample and the heat sensor is in intimate contact, without the spacer.

Other ASTM standards related to flame resistance clothing include:

- D 1230, Test Method for Flammability of Apparel Textiles
- F 1358, Test Method for the Effects of Flame Impingement on Protective Clothing Materials Not Designated for Primary Flame Resistance

FIGURE 13.6 Vertical flame test of FR cotton (courtesy of American Kynol, Inc.).

- F 1449, Guide for Care and Maintenance of Flame Resistance and Thermal Protective Clothing
- Method for Thermal Protective Performance of Materials for Clothing by Convective and Radiant Heat Method (New Standard)
- Test Method for Thermal Protective Performance of Materials for Clothing by Radiant Heat Method (New Standard)

(3) The National Fire Protection Association (NFPA) has issued several standards on fire fighter protective clothing.
- NFPA 701, Fire Tests for Flame Resistant Textiles and Films
- NFPA 702 (for clothing except hats, gloves, footwear and interlinings)
- NFPA 1971, Standard on Protective Clothing for Structural Fire Fighting
- NFPA 1973, Standard on Gloves for Structural Fire Fighting
- NFPA 1975, Standard on Protective Clothing for Station Work Uniforms. In this standard, two new requirements are established: garments must not melt or drip when exposed in a forced air oven at 500°F for five minutes; and garments must be approved by a third party organization.
- NFPA 1976, Standard on Protective Clothing for Proximity Fire Fighting
- NFPA 1977, Standard on Protective Clothing and Equipment for Wildlands Fire Fighting

In 1992, NFPA issued NFPA 1999, Standard on Protective Clothing for Emergency Medical Operations which includes design, performance and labeling requirements for emergency medical protective gloves, garments and facewear.

Table 13.9 summarizes flammability test methods for various applications.

In Europe, the European Standardization Committee (CEN) is responsible for standard development. CEN standards specific to protective clothing are developed by Technical Committee (TC) 162 [14].

Finally, it should be noted that extreme care should be taken when conducting the tests. The test results may vary slightly from laboratory to laboratory although the same standard test method is used as shown in Table 13.10. One reason for this variation may be the fact that every flame or fire is different.

3.3 Performance of FR Textiles

Baitinger and Konopasek examined a variety of FR cotton fabrics for thermal protective performance [15]. Fabrics were subject to convective and radiant exposure (ASTM D 4108). The results are shown in Table 13.11.

Barker et al. measured the ability of FR fabrics to block heat transfer when in contact with a hot surface [16]. They found that protection time drops off exponentially with surface temperature. Applied pressure increased the contact between the fabric and hot surface and decreased the effective thickness of compressible fabrics resulting in increased heat transfer rate. Figure 13.7 shows their results. Table 13.12 shows tolerance time of tested fabrics at different surface temperatures.

Thermal comfort of the protective garment is very important for the wearer. At high outside temperature and humidity, the evaporation of sweat becomes a factor. Therefore, to improve comfort, water vapor permeability of protective clothing must be high.

Special attention should be given to sewing thread when manufacturing flame resistant garments. Sewing thread failure can lead to downfall of the entire protective system. Table 13.13 shows properties of the some sewing threads for high temperature applications.

Some companies and research institutions developed instrumented mannequin testing procedures to predict the relative protective performance of actual garments in a flash fire. Figure 13.8 shows a schematic of such a system. These tests use a controlled laboratory simulation of a flash fire exposure on a life-size mannequin to predict estimated burn injury (Figure 13.9). Variables include fabric type, fabric weight, garment style, use of undergarments, heat flux and exposure time.

TABLE 13.9 Major Characteristics of Various Fabric Flammability Test Methods (courtesy of U.S. Testing Co. Inc.).

Sponsor	Name of Test	Procedure I.D. No.	Sample	Specimen Size (inches)	No. of Specimens	Angle of Specimen	Ignition Source	Conditioning	Properties Measured
1a CPSC	Flammability of Clothing Textiles	CS 191-53 16 CFR 1610.4	Fabrics Apparel	2 × 6	5 original 5 DC & W	45°	Butane burner	Oven-dry 1/2 hr @ 221°F	Time of flame spread, ease of ignition, flame intensity
1b ASTM	Flammability of Clothing Textiles	ASTM D 1230	Fabrics Apparel	2 × 6	5 original 5 DC & W	45°	Butane burner	Oven-dry 1/2 hr @ 221°F	Time of flame spread, ease of ignition
2 CPSC	Children's Sleepwear	FF 3-71/5-74 16 CFR 1615.4 16 CFR 1616.4	Sleepwear sizes 0–6X sizes 7–14	3.5 × 10	10 fabrics 15 seams	Vertical	Methane burner	Oven-dry 1/2 hr @ 221°F	Char length
3 GSA	Flame Resistance of Clothing, Vertical	FTMS 191 Method 5903.1	Fabrics	2.75 × 12	5W, 5F	Vertical	Methane burner	70°F, 65% RH	After-flame, afterglow, char length
4 GSA	Flame Resistance of Clothing, Vertical	FTMS 191 Method 5905.1	Fabrics	2.75 × 12	5W, 5F	Vertical	Butane burner	70°F, 65% RH	After-flame and % consumed
5 GSA	Burning Rate of Cloth, 45° Angle	FTMS 191 Method 5908	Fabrics	2 × 6	5	45°	Butane burner	Oven-dry 1/2 hr @ 221°F	Ease of ignition, rate of burning
6 GSA	Burning Rate of Cloth, 30° Angle	FTMS 191 Method 5910	Fabrics	1 × 6	3W, 3F	30°	Match	4 hrs @ 140°F	Rate of burning
7a CPSC	Mattresses	FF-4-72 16 CFR 1632.4 16 CFR 1632.6	Mattress and mattress pads ticking	Finished item 20 × 20	6 surfaces 6 surfaces	Horizontal	Cigarette	48 hrs @ 65–80°F and 55% RH max.	Char length and ignition
7b State of CA Bureau of Home Furnishings	Mattresses	Tech Bulletin 106	Mattress and mattress pads ticking	Finished item 20 × 20	6 surfaces 6 surfaces	Horizontal	Cigarette	48 hrs @ 65–80°F and 55% RH max.	Char length and ignition

430

TABLE 13.9 (continued).

Sponsor	Name of Test	Procedure I.D. No.	Sample	Specimen Size (inches)	No. of Specimen	Angle of Specimen	Ignition Source	Conditioning	Properties Measured
7c State of CA Bureau of Home Furnishings	Mattresses	Tech Bulletin 121, 129	Mattress	N/A	1	Horizontal	Propane burner	70°F, 50% RH	Heat release, weight loss, smoke density
8a NFPA	Fire Tests for Flame Resistant Textiles and Films	NFPA 701, large scale	Fabrics and films	Installed: (1) in sheets 5 × 84 (2) in folds 25 × 84	5W, 5F 2W, 2F	Vertical	Burner	Oven-dry 1 hr @ 140°F	After-flame, afterglow, char length
8b NFPA	Fire Tests for Flame Resistant Textiles and Films	NFPA 701, small scale	Fabrics and films	3.5 × 10	5W, 5F	Vertical	Natural gas	1 hr @ 140°F	After-flame, flaming residues, char length
8c NFPA	Fire Tests for Flame Resistant Textiles and Films	NFPA 701 curtain and drapery under 3.0 oz/sq yd	Fabrics	3.5 × 10	5W, 5F	Vertical	Natural	1 hr @ 140°F	Weight loss
9 City of Boston	Fabric Classification	BFDIX-1	Treated fabrics	4 × 12	3L, 3W	Vertical	Propane 45° torch	Ambient	After-flame, afterglow, char length
10 City of New York	Board of STDs. Appeals Bylaw	Calender No. 294-40-SR	Fabrics	2 × 12.5	3L, 3W	Vertical	Burner	Ambient	After-flame, afterglow
11a State of CA	Fire Code Title 19	Par. 1237.1, small scale	Fabrics	2.5 × 12.5	3W, 3F	Vertical	Burner	1 hr @ 140°F	After-flame, char length
11b State of CA Fire Marshal's Office	Fire Code Title 19	Par. 1237.3, large scale	Fabrics	5 × 84	3W, 3F	Vertical	Burner	1 hr @ 140°F	After-flame, char length
12a CPSC	Surface Flammability of Carpets and Rugs	FF 1-70/FF 2-70 16 CFR 1630.4 16 CFR 1631.4	Carpets and rugs 24 sq. ft.	9 × 9	8	Horizontal	Methanamine burning tablet	Oven-dry 2 hrs @ 221°F	Area of flame spread (greatest diameter)

(continued)

TABLE 13.9 (continued).

Sponsor	Name of Test	Procedure I.D. No.	Sample	Specimen Size (inches)	No. of Specimens	Angle of Specimen	Ignition Source	Conditioning	Properties Measured
12b ASTM	Surface Flammability of Carpets and Rugs	ASTM D 2859	Carpets and rugs	9 × 9	8	Horizontal	Methenamine burning tablet	Oven-dry 2 hrs @ 221°F	Area of flame spread (greatest diameter)
13a State of CA Bureau of Home Furnishings	Upholstery Materials	Tech Bulletin 117	Fabrics and filling materials	Specimens and procedures vary with materials		Specimens and procedures vary with materials	Specimens and procedures vary with materials		
13b UFAC	Upholstery Materials	UFAC-1990	Fabrics and filling materials						Char length
13c NFPA	Upholstery Materials	NFPA-260	Fabrics and filling materials						Char length
13d BIFMA	Upholstery Materials	BIFMA	Fabrics and filling materials						Char length
13e Brit Stnds	Upholstery Materials	BS 5852	Fabrics and filling materials						Char length, weight loss
13f NFPA	Upholstered Furniture or Mock-ups	NFPA 261	Component materials	Varies with procedure	1	Horizontal	Cigarette	48 hrs @ 73°F, 50% RH	Char length
13g State of CA Bureau of Home Furnishings	Upholstered Furniture	Tech Bulletin 116	Complete upholstered	N/A	3	Horizontal	Cigarette	65–80°F, 55% RH	Char length
13h State of CA Bureau of Home Furnishings	Seating Furniture	Tech Bulletin 133	Complete upholstered	N/A	1	N/A	Propane burner	70°F, 50% RH	Heat release, weight loss, smoke density
14a NFPA	Protective Clothing	NFPA 1971	Component materials	2.75 × 12	5W, 5F	Vertical	Bunsen burner	70°F, 65% RH	After-flame, afterglow, flaming drips, char length
14b NFPA	Station Work Uniforms	NFPA 1975	Fabrics	2.75 × 12	5W, 5F	Vertical	Bunsen burner	70°F, 65% RH	Melt and drip, after-flame, char length

TABLE 13.9 (continued).

Sponsor	Name of Test	Procedure I.D. No.	Sample	Specimen Size (inches)	No. of Specimens	Angle of Specimen	Ignition Source	Conditioning	Properties Measured
15a Federal Aviation Authority	FAA Regulations for Compartment Interiors	FAR. 25.853 Appendix F, Pt. 1	Compartment materials	4 × 14	3L, 3W	Horizontal	Burner	70°F, 50% RH	Time of flame spread
15b Federal Aviation Authority	FAA Regulations for Compartment Interiors	FAR. 25.853 Appendix F, Pt. 1	Compartment materials	2.75 × 12	3L, 3W	Vertical	Burner	70°F, 50% RH	After-flame, burn length, flaming drips
15c Port Authority of NY and NJ	Upholstery of Seating	Same as FAA 25.853	Fabrics	2.75 × 12	3W, 3F	Vertical	Methane	70°F, 50% RH	After-flame, char length, flaming drips
16 Dept. of Transportation and State of CA	Flammability of Interior Materials of Cars, Trucks, Multipurpose Passenger Vehicle Buses	FMVSS 302	Interior materials	4 × 14	3L, 3W	Horizontal	Burner	70°F, 50% RH	Flame spread
17 Fed. Test Method STD.	Mackey	FTMS 191 Method 5920	Cloth, related materials	2 × 12 minimum	2 min.	N/A	N/A	4 hrs @ 60°C	Tendency of material to undergo self-heating at moderate temp.

(continued)

TABLE 13.9 *(continued).*

Sponsor	Name of Test	Procedure I.D. No.	Sample	Specimen Size (inches)	No. of Specimens	Angle of Specimen	Ignition Source	Conditioning	Properties Measured
18a Canvas Products Assoc. Int'l	Camping Tentage	CPAI-84 ASTM D 4372	Tenting materials	Floors 9 × 9 Walls 2.75 × 12	8 8W, 8F	Horizontal Vertical	Pill burner	70°F, 65% RH Leached & weathered	Area of flame spread, after-flame, char length
18b NY State	Tents	Dept. of Labor Bd. STDS. and Appeals Code 36, Sec. 3.8C	Fabrics	2.75 × 12	5W, 5F	Vertical	Bunsen burner	70°F, 65% RH	After-flame, char length
19 Canvas Products Assoc. Int'l	Sleeping Bags	CPAI-75	Sleeping bags	Entire bag	5 original 5 laundered	Horizontal	Bunsen burner	70°F, 65% RH	Burning rate
20 ASTM	Blankets	ASTM D 4151	Fabric	2.75 × 2.75	5 face 5 back	Horizontal	Butane burner	Oven-dry 1/2 hr @ 220°F	Pass/fail
21 ASTM	Apparel Fabrics Semi-Resistant	ASTM D 3659	Fabric	6 × 15	5W	Vertical	Fan burner	Oven-dry 1/2 hr @ 220°F	4 burn time, weight loss
22 U.S. Forestry Depart.	Modified USTC– Comparative Inhalation	5100-1A	Fire tent laminate	10 × 4.5 to 10 × 36	1 min.	45°	Radiant panel	Ambient	Inhalation toxicity

TABLE 13.10 TPP Tests of Single Layer Fabrics (courtesy of DuPont).

	TPP Value	
	Manufacturer Results	Independent Test Results
Typical Shirt Weights		
Aramid (4.5 oz/yd^2)	10.6	9.1
FR Cotton (5.5 oz/yd^2)	–	5.5
FR Cotton (6.0 oz/yd^2)	6.7	–
Typical Pants Weights		
Aramid (6.0 oz/yd^2)	13.1	11.6
FR Cotton (7.5 oz/yd^2)	8.8	7.3
FR Cotton (9.0 oz/yd^2)	9.8	9.2

TABLE 13.11 Protection Times for FR Cotton Fabrics Exposed to Radiant and Convective Heat (copyright ASTM [15]).

Fabric Type	Color	Weight (g/m^2)	Thickness (mm)	Protection Time, s		
				Radiant Heat, 0.3-cal/cm^2/s Contact	Convective and Radiant Heat, 1.0-cal/cm^2/s Contact	Convective and Radiant Heat, 2.0-cal/cm^2/s Contact
Westex – breeze tone	Blue	190.0	0.330	50.5	6.5	3.7
Graniteville – plain jean shirting	Blue	203.5	0.330	30.1	6.4	3.7
Ameritex – flannel	Print	223.9	0.508	49.7	10.0	4.8
Graniteville – supershirting	Green	230.6	0.457	54.0	7.4	3.9
Westex – shirting	Navy	244.2	0.483	49.8	7.4	4.2
Graniteville – supershirting	Blue	244.2	0.406	50.0	7.3	4.0
Graniteville – supershirting	Navy	247.6	0.457	59.7	6.7	4.1
Graniteville – supertwill	Green	312.1	0.533	65.3	10.2	4.9
Westex – sateen	Gray	318.8	0.584	63.0	9.8	5.4
Westex – twill	Green	373.1	0.711	88.7	10.6	6.3
Westex – denim	Navy	417.2	0.762	73.0	10.8	7.0
Graniteville – bull denim	Green	417.2	0.711	65.7	11.4	6.5
Westex – whipcord	Green	424.0	0.686	77.0	10.8	6.6
Graniteville – mallard duck	Black	407.0	0.686	47.4	11.7	6.1
Graniteville – mallard duck	Yellow	430.8	0.711	61.6	11.9	6.4

TABLE 13.12 Insulation of FR Fabrics in Surface Contact (copyright ASTM [16]).

Fabric Composition	Tolerance Time,* s, at Surface Temperature, °C				
	100	150	200	300	375
Nomex® III (6 oz)	23.0	6.6	13.9	1.8	1.2
Nomex® III (10 oz)	>30.0	12.3	5.8	2.8	2.2
Kevlar (8 oz)	23.3	7.8	4.4	2.4	1.8
Aluminized Nomex® III (10 oz)	12.1	4.1	2.5	1.3	1.3
FR Cotton (13 oz)	20.2	5.8	3.4	1.8	1.6
FR Cotton (7.3 oz)	6.6	2.2	1.3	0.0	0.0
Aluminized FR Cotton (14 oz)	16.2	4.8	2.9	1.7	1.4
FR Wool (18 oz)	>30.0	>30.0	15.9	5.3	3.0
FR Rayon (9 oz)	12.0	3.3	2.0	1.2	0.0
FR Modacrylic (7 oz)	23.9	6.9	3.5	1.6	1.3
Novoloid (9 oz)	14.2	4.5	2.5	1.4	1.2
Aramid/glass (18 oz)	>30.0	>30.0	19.9	10.0	7.4
Silica (35 oz)	>30.0	>30.0	27.1	14.3	11.6
PAN (17 oz)	>30	16.3	8.1	4.2	2.7
Aluminized PAN (18 oz)	27.3	6.5	3.4	2.0	1.6

*Tolerance time: time to second degree burn injury predicted using the Stoll criterion.

TABLE 13.13 Properties of Selected Sewing Threads for FR Clothing [17].

Fiber	Yarn Type	Sewability
Nomex®	Spun filament	Generally good
Kevlar®	Spun filament	Spun: excellent, filament: can pull apart
Fiberglass	Filament	Breaks easily, PTFE coating often used
PTFE	Filament	Slick, machine adjustment necessary
Carbon	Spun filament	Slower speed and machine adjustments needed
Quartz	Filament	Fibers may fracture

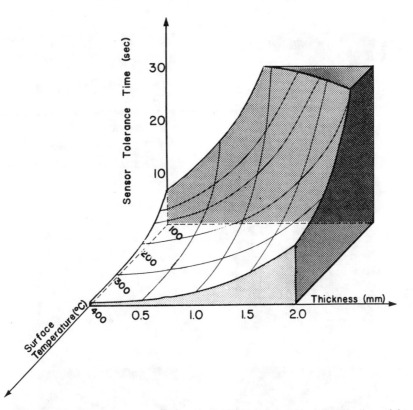

FIGURE 13.7 Effect of plate temperature and fabric thickness on insulation time (copyright ASTM [16]).

Garments exposed directly to the propane gas flame (at 1000°C) for 5 seconds

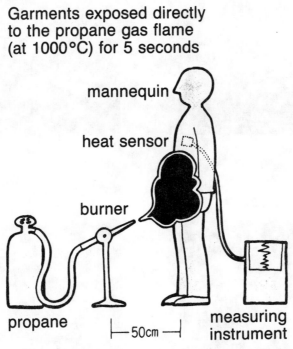

FIGURE 13.8 Schematic of a typical mannequin test setup (courtesy of Teijin Limited).

FIGURE 13.9 Mannequin test results of protective garments made of 100% meta-aramid fabric (top row) and polyester/cotton fabric (bottom row) (courtesy of Teijin Limited).

3.4 Applications of FR Textiles

FR textiles are used in a wide variety of applications including fire fighter uniforms, furniture manufacturing, airline seats, children sleepwear, etc.

Fire Fighter Protective Clothing

Among the FR protective garments, fire fighter suits probably deserve some special attention. A fire fighter's protective clothing system includes coat, pants, underwear, helmet, boots, gloves, stationwear uniforms and breathing devices. In addition to flame resistance, all clothing layers should have characteristics such as resistance to cuts and punctures, mobility, waterproofness, fit and durability.

Protective clothing for structural fire fighting is made of three layers: the outer layer, the moisture barrier and the thermal barrier (Figure 13.10). The minimum performance requirements for each component are found in NFPA 1971.

The outer layer must be heat and flame resistant as well as abrasion resistant. It must be suitable for wet, dry, hot and cold conditions. The two most common materials for the outer layer are 60/40 Kevlar® (or para-aramid)/PBI blend and Nomex III. Other blended materials used for the outer shell are P-84/Kevlar®, Kevlar®/Nomex® and neoprene coated Nomex® [18].

The main purpose of the moisture barrier is to keep the fire fighter dry. Therefore, the material of this layer must prevent water penetration. Woven, knit, spunbonded or needlepunched substrates coated or laminated with neoprene or polytetrafluoroethylene (PTFE) are commonly used as the moisture barrier.

Although the first two layers are flame resistant, the inner or thermal barrier layer provides the main protection against heat. The thermal barrier layer can be made in several ways. Fibers can be needlepunched into a batting and quilted to a face cloth for stability and strength. They also can be made into a light, airy batting that can be quilted to a face cloth. Different fibers can be blended together to form a composite with unique characteristics.

FIGURE 13.10 Fire fighter protective clothing (courtesy of Fyrepel).

It is crucial to the performance of the fire fighter that these layers are combined together in a way that does not affect the movement of the fire fighter. Table 13.14 gives the major properties and thermal protective performance of typical fire fighter protective ensembles.

There are various types of FR protective garments depending on the purpose of application. Figures 13.11–13.13 show typical FR protective clothing for different purposes.

Furniture Manufacturing

Increasingly strict fire safety standards require the use of FR fabrics in furniture manufacturing. Figure 13.14 shows flammability comparison

TABLE 13.14 TPP Ratings for Fire Fighter's Protective Ensembles at 2.0 cal/cm²/s, 50% Radiant/50% Convective Load (copyright ASTM [19]).

Ensemble Description	Area of Application	Weight (g/m²)	Thickness (mm)	TPP Rating
OS Neoprene on 7.5 oz Nomex III	Coat or pants body	407	0.5	
VB Neoprene on Nomex ripstop		258	0.3	
TL Nomex quilt		373	4.5	
Total		1,038	5.7	67.4
OS 7.5 oz Nomex III	Coat or pants body	244	0.6	
VB Neoprene on TL Nomex	OS Yoke design	–	–	
Needlepunch		654	2.8	
Total		899	3.4	47.9
OS 7.5 oz Nomex III	Coat or pants body	244	0.5	
VB Neoprene on polycotton		288	0.3	
TL Nomex quilt		336	4.1	
Total		868	5.4	38.6
OS 7.5 oz Nomex III	Coat or pants body	241	0.6	
VB Gore-Tex on Nomex ripstop		126	0.2	
TL Nomex quilt		353	4.2	
Total		719	5.4	46.5
OS 7.5 oz Nomex III	Coat or pants body	251	0.6	
VB Gore-Tex on E-89 Nomex		119	0.6	
TL Nomex quilt		359	4.8	
Total		729	6.8	54.3
OS 7.5 oz Nomex III	Coat or pants body	241	0.6	
VB Neoprene on Nomex ripstop		251	0.2	
TL Nomex quilt		342	4.2	
Total		834	5.4	41.8
OS 7.5 oz Nomex III	Coat or pants body	261	0.6	
VB Neoprene on Nomex ripstop	Ensemble after sweating	258	0.3	
TL Nomex quilt		315	4.0	
Other: 1 g water added				
Total		834	4.9	23.0
OS 7.5 oz Nomex III	Coat or pants body	254	0.6	
VB Neoprene on Nomex ripstop	Ensemble after sweating	251	0.3	
TL Nomex quilt		353	4.9	
Other: 2g water added				
Total		858	6.3	32.6

TABLE 13.14 (continued).

Ensemble Description	Area of Application	Weight (g/m²)	Thickness (mm)	TPP Rating
OS 7.5 oz Nomex III				
VB Neoprene on polycotton	Collar design	244	0.6	
TL Not applicable		302	0.3	
Other: FR corduroy		–	–	
Total		909	1.7	28.9
OS 7.5 oz Nomex III		244	0.6	
VB Neoprene + TL Nomex	Collar design	–	–	
Needlepunch		685	2.7	
Other: FR corduroy		363	0.8	
Total		1,292	4.3	56.6
OS 7.5 oz Nomex III		244	0.5	
VB Neoprene on polycotton	Storm flap design	288	0.3	
TL Not applicable		–	–	
Other: 7.5 oz Nomex III		244	0.5	
Total		776	1.5	26.3
OS 7.5 oz Nomex III	Facing or storm flap design	241	0.6	
VB Neoprene + TL Nomex		–	–	
Needlepunch		624	2.8	
Other: 7.5 oz Nomex total		241	0.6	
Total		1,105	4.1	53.2
OS 7.5 oz Nomex III	Fly OS design	241	0.6	
VB Neoprene on polycotton		298	0.3	
TL–none				
Total		539	0.9	19.6
OS 7.5 oz Nomex III	Shoulder and fly OS design	244	0.6	
Other: 7.5 oz Nomex III		244	0.6	
Total		488	1.2	17.0
Outer leather	Elbow, knee, shoulder	1,200	1.7	
OS 7.5 oz Nomex III	OS design	244	0.6	
VB Not applicable				
TL Not applicable				
Total		1,444	2.2	22.3

OS: outer shell, VB: vapor (moisture) barrier, TL: thermal liner.
Total thickness may exceed the sum of parts because of air entrapment between layers.
Each TPP rating is an average of three separate tests. A minimum TPP rating of 35.0 is being considered as a standard for coats or pants.

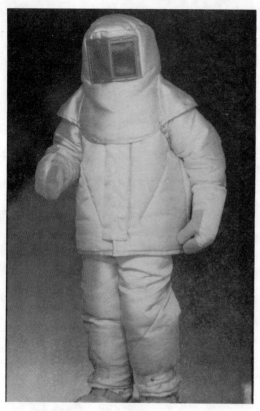

FIGURE 13.11 Fire entry suit. This suit is designed for short duration entry into total flame engulfment or furnace. It consists of separate coat and pants constructed from multiple layers of glass insulation, aluminized glass, and non-aluminized glass surface (courtesy of Fyrepel Products).

FIGURE 13.12 Kiln entry suit for worker who must function in kiln or other extreme heat situations that do not involve total flame, but require high quality heat protection. It is made of multiple layers of glass and an extra layer of aluminized glass between the wearer and heat or non-ferrous splash (courtesy of Fyrepel Products).

FIGURE 13.13 Proximity suit (left) and proximity coverall. These suits are designed for performance of maintenance and repairs in high heat areas. They are made of multi-layer construction, with the outer layer composed of high temperature aluminized glass. A moisture/steam barrier lining provides protection in areas where exposure to hot liquids, steam or hot vapor is a possibility (courtesy of Fyrepel Products).

tests of chairs made with and without FR materials. The chair on the left is protected with a spun laced Kevlar® fire blocking sheeting. Both chairs were exposed to 80 seconds of scorching flame.

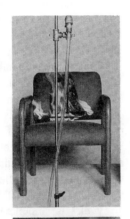

FIGURE 13.14 Flammability comparison of chairs with (left) and without FR textiles (courtesy of DuPont).

4. CHEMICAL PROTECTIVE CLOTHING (CPC)

The purpose of chemical protective clothing is to isolate parts of the body from direct contact with a hazardous chemical. Considering the number of hazardous chemicals used today, the importance of protective clothing is obvious. Recently, the application of chemical protective clothing has widened; e.g., to protect farmers from insecticides, truck drivers from toxic/flammable materials during loading/unloading, and fire fighters from chemical substances, etc.

Chemical protective clothing should be considered as the last line of defense [20]. Every CPC item has limitations. Fabrics used in CPC can protect against only a limited number of specific hazards for certain chemicals. No matter how effective the protective clothing, the worker's exposure to the hazardous chemicals should be minimized through engineering controls (isolation) or administrative controls (avoidance).

The Environmental Protection Agency (EPA) has defined four levels of protection [20]:

(1) Level A offers the highest protection from skin and respiratory exposure to chemical hazards. Level A protection must be gas and vapor tight and prevent any external contaminant from reaching the wearer. Level A ensembles include a self-contained breathing apparatus (SCBA).

(2) Level B protection must prevent the passage of any liquid into the ensemble, but may not prevent the penetration of gases or vapors. SCBA is used for respiratory protection.

(3) Level C is the same clothing as Level B, but with an air-purity respirator. Level C systems provide limited protection from minor contamination. Level C protection is not usually used by emergency-response personnel.

(4) Level D is the normal work clothing to keep off dirt, grease, limited spill protection, etc., and is not considered protective clothing by emergency-response personnel.

Chemical protective clothing may be categorized as encapsulating and non-encapsulating. Encapsulating ensemble covers the whole body including the respiratory protection equipment. The non-encapsulating system is assembled from separate components and the respiratory apparatus may be worn outside the suit. Encapsulating garments usually provide better chemical protection. All Level A and Level B protective clothing is encapsulating. Non-encapsulating systems usually apply to Level B or below.

4.1 Chemical Protective Materials

In general, porous fabrics are used for particulate protection and coated/laminated fabrics are used for liquid and gas protection. Table 13.15 gives some of the currently used materials for chemical protection.

Particulate Protective Textiles

Examples of materials for particulate protection are spunbonded olefins (e.g., Tyvek®), hydroentangled spun lace (e.g., Sontara®), cellulosic materials (e.g., Kaycel®) and spunbond-meltblown-spunbond composites (e.g., Kleenguard®).

Tyvek® is a spunbonded polyethylene material made by DuPont. It has high strength, low permeability and meets the federal flammability requirements. Tyvek® is formed by a continuous process from very fine $0.5-10$ micrometer fibers. These non-directional fibers are first spun and then bonded together by heat and pressure, without binders or fillers. Figure 13.15 shows the fiber structure of Tyvek®.

Tyvek® is suitable for cutting and sewing. However, garments made of Tyvek® may cause heat stress. Perforated Tyvek® allows through air flow due to tiny air holes opened in the structure. Usually antistatic treatment is applied on the surface of Tyvek®, which is opaque.

Kaycel® of Kimberly-Clark is made of cellulosic fabric reinforced with a plastic scrim. Hydraspun FR®, produced by Dexter Corpora-

TABLE 13.15 Materials for Chemical Protective Clothing [21].

Material	Material Type	Applications
Butyl/nylon	Supported elastomer	Splash, encapsulating suits
Butyl rubber	Elastomer	Gloves
Challenge® 5200 (Teflon®/fiberglass)	Plastic laminate	Encapsulating suits
Chemrel®	Plastic laminate	Splash suits
Chlorinated polyethylene (CPE)	Plastic film	Splash, encapsulating suits
Fluorinated ethylene propylene (Teflon®)	Plastic film	Visors
Natural rubber	Elastomer	Gloves, suit gaskets
Neoprene®	Elastomer	Gloves, suit gaskets
Neoprene® nylon	Supported elastomer	Splash suits
Nitrile rubber	Elastomer	Gloves
Polycarbonate	Plastic film	Visors
Polyethylene coated Tyvek®	Supported plastic	Splash suits
Polymethacrylate	Plastic film	Visors
Polyvinyl chloride (PVC)	Plastic film	Gloves, visors
PVC/nylon	Supported plastic	Splash suits
Responder®	Plastic laminate	Encapsulating suits
Saranex® coated Tyvek®	Supported plastic	Splash suits
Silvershield®	Plastic laminate	Gloves
Viton®	Elastomer	Gloves
Viton®/Nomex®/chlorobutyl rubber	Elastomeric laminate	Encapsulating suits
Viton®/nylon/neoprene	Elastomeric laminate	Encapsulating suits

tion, is made of wood pulp and hydrocarbons with excellent FR properties and moderate barrier properties. Spun laced Nomex® offers fair barrier protection, however, FR properties are identical to Nomex®. Spunbonded polypropylene is a lightweight, breathable fabric. In Kleenguard® (Kimberly-Clark), meltblown polypropylene is heat bonded between two layers of spunbonded polypropylene: the outside layer gives strength and durability while the inner layer provides barrier protection. Kleenguard has good breathability which makes it suitable for closed areas. Sontara is a spun laced nonwoven fabric (polyester and cellulose) that is treated to be flame retardant. It is suitable for garment making, however barrier properties are moderate. DuPont also developed aramid versions of Sontara available in 100% Kevlar, 100% Nomex®, Nomex®/Kevlar blend and other combinations.

Contaminant resistant garments such as Tyvek® and Kleenguard® are used during asbestos removal to protect the worker (Figure 13.16). To remove asbestos, the area is usually sealed off which creates heat stress [22].

Liquid and Gas Protection

Fabrics which are coated or laminated with special films are suitable for chemical protection involving liquids and gases. They are used in hazardous or toxic material handling (Figure 13.17).

Saranex® is a multi-layer film that can be laminated to Tyvek®. Saranex 23P® is a multi-layer structure that consists of low density polyethylene (outside layer), Saran (second layer), ethylene vinyl acetate (EVA) copolymer (third layer) and Tyvek® (inner layer). The EVA acts as an adhesive. Saranex is very suitable for garment manufacturing with various seam types. By laminating 1.25 mil of polyethylene with Tyvek®, a tough fabric with good barrier properties is produced that is suitable for garment manufacturing with different seam types.

Barricade®, another DuPont product with good barrier qualities, is a multi-layer film laminated to a substrate of polypropylene. For better protection, total encapsulation suits made

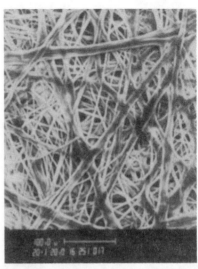

FIGURE 13.15 Fiber structure of Tyvek® (courtesy of DuPont).

of special coated fabrics such as DuPont's Teflon® onto Nomex® or fiberglass fabrics may be needed. There are other products in the market for chemical protective clothing. Various nylon, cotton, or polyester fabrics coated with butyl, PVC and neoprene also offer good barrier properties. In addition, there are several proprietary laminates in the marketplace.

FIGURE 13.16 Tyvek®, in uncoated form, is used in industrial garmets for asbestos removal (courtesy of DuPont).

FIGURE 13.17 Garments of coated/laminated Tyvek® are used for hazardous or toxic material handling (courtesy of DuPont).

4.2 Requirements and Testing

Several parameters are of interest in protection against hazardous chemicals such as degradation, permeation, penetration and breakthrough time. Degradation is the change in physical properties (weight, dimensions, tensile strength, elongation, swelling, delamination, deterioration) of the material exposed to chemicals. Permeation is the movement of a chemical through a protective clothing material on a molecular level. Permeation takes place in three stages [23].

- absorption of chemical molecules by the outside surface of the protective material
- diffusion of the molecules into the material
- desorption of the chemical molecules by the inner surface into the collecting medium

Penetration is the flow of a chemical through porous material, seams, pinholes, closures or other imperfections of the protective material on a non-molecular level. Breakthrough time is the time the chemical takes to travel from outside to the inside surface of the protective material. Table 13.16 shows the methods to measure chemical resistance.

Contamination can be on the surface or internal. Generally water, detergent or solvent washing is enough to remove surface contamination. Contamination throughout the material is more difficult to clean without altering the fabric characteristics.

ASTM specifies fifteen liquid and six gaseous chemicals to use in the evaluation of protective clothing materials as shown in Table 13.17 (ASTM F 1001). Table 13.18 gives breakthrough times and permeation rates of several chemical protective materials.

Breakthrough time and permeation rate measurements give good indication about the performance of protective materials. However, these measurements alone are not enough and they should be complemented by other physical property tests. ASTM Committee F-23 on Protective Clothing developed the test method F 739, Test for Resistance of Protective Clothing Materials to Permeation by Liquids or Gases

TABLE 13.16 Methods for Measuring Chemical Resistance of Materials
(Texas Research Institute [21]).

Method	Measurements	Uses/Limitations

Degradation	Physical property changes
	  Weight
	  Dimensions (optional)
	  Tensile strength (optional)
	  Elongation (optional)
	Visual observations (swelling, delamination,
	  discoloration, deterioration)
		Allows material screening
		Assesses material-chemical compatibility
		Indicates inappropriate materials
		Does not completely evaluate material barrier
		  effectiveness

Penetration	Time to penetration (liquid penetration first
	  observed)
	Other visual observations
		Assesses material ability to hold out liquids and
		  liquid-tight integrity of seams
		Can be used to recommend materials for liquid
		  splash protection
		Does not show vapor penetration or permeation

Permeation	Breakthrough time
	
	Steady state or maximum permeation rate
	
	Minimum detection limit
	Visual observations
		Completely assesses material barrier effective-
		  ness against chemical liquids, vapors and gases
		Can be used to recommend materials for vapor
		  protection and liquid splash protection
		Breakthrough time depends on detector sensitivity

TABLE 13.17 ASTM F 1001 Recommended Chemicals to Evaluate Protective Clothing Materials
(Texas Research Institute [21]).

Liquids	Gases
Acetone	Ammonia, anhydrous
Acetonitrile	1.3-Butadiene
Carbon disulfide	Chloride
Dichloromethane	Ethylene oxide
Diethylamine	Hydrogen chloride
Dimethylformamide	Methyl chloride
Ethyl acetate	
Hexane	
Methanol	
Nitrobenzene	
Sodium hydroxide	
Sulfuric acid	
Tetrahydrofuran	
Tetrachloroethylene	
Toluene	

TABLE 13.18 Breakthrough Times and Permeation Rates of Selected
Chemical Protective Clothing Materials (copyright ASTM [24]).

Chemical	Clothing Material	Thickness (mm)	Breakthrough Time (h)	Permeation Rate (mg/min/m²)
Cl₂				
0.1%	Natural rubber latex	0.46	>8.0	ND
	Neoprene	0.41	>8.0	ND
0.002%	PE Tyvek	0.15	>8.0	ND
HCN (liquid) 100%	PVC	0.79	0.5	2.9
	Butyl	0.38	1.0	0.15
	PE Tyvek	0.15	1.0	1.1
CH₂O				
37%	Natural latex	0.15	0.1	33.3
	Nitrile	0.28	>6.0	ND

The values are mean values of triplicate tests.
ND: No detectable permeation during the test run.

under Conditions of Continuous Contact. This method quantifies three parameters: the normalized breakthrough time, the permeation time and the cumulative amount of chemical that has permeated at any time. In this method, the outside surface of the protective clothing is exposed to a challenging chemical. The inside surface is kept in contact with a collection medium that is monitored for the chemical content [25]. ASTM developed other test methods related to chemical protective clothing that include:

- F 903 Test Method for Resistance of Protective Clothing Materials to Penetration by Liquids
- F 0955 Test Method for Evaluating Heat Transfer through Materials for Protective Clothing upon Contact with Molten Substances
- F 1001 Guide for Selection of Chemicals to Evaluate Protective Clothing Meterials
- F 1002 Performance Specification for Protective Clothing for Use by Workers Exposed to Specific Molten Substances and Related Thermal Hazards
- F 1052 Practice for Pressure Testing of Gas Tight Totally Encapsulating Chemical Protective Suits
- F 1060 Test Method for Thermal Protective Performance of Materials for Protective Clothing for Hot Surface Contact
- F 1154 Practices for Qualitatively

Evaluating the Comfort, Fit, Function, and Integrity of Chemical Protective Suit Ensembles
- F 1291 Test Method for Measuring the Thermal Insulation of Clothing Using a Heated Mannequin
- F 1194 Guide for Documenting the Results of Chemical Permeation Testing on Protective Clothing Materials
- F 1296 Guide for Evaluating Chemical Protective Clothing
- F 1301 Practice for Labeling Chemical Protective Clothing
- F 1359 Practice for Determining the Liquid-Tight Integrity of Chemical Protective Suits or Ensembles under Static Conditions
- F 1383 Test Method for Resistance of Protective Clothing Materials to Permeation by Liquids or Gases under Conditions of Intermittent Contact
- F 1407 Test Method for Resistance of Chemical Protective Clothing Materials to Liquid Permeation-Permeation Cup Method
- F 1461 Practice for a Chemical Protective Clothing Program
- F 1494 Terminology Relating to Protective Clothing
- Test Methods for Measuring the Performance Characteristics of Exhaust Valves Used in Chemical Protective Suits (New Standard)

- Practice for Determining the Resistance of Chemical Protective Coveralls, Suite or Ensembles to Inward Leakage to Liquids under Dynamic Conditions (New Standard)
- Practice for Evaluating Resistance of Protective Clothing Materials to Permeation by Liquid Chemical Warfare Agent (New Standard)

NFPA has issued three standards on chemical protective clothing:

- NFPA 1991, Standard on Vapor Protective Suits for Hazardous Chemical Emergencies
- NFPA 1992, Standard on Liquid Splash Protective Suits for Hazardous Chemical Emergencies
- NFPA 1993, Standard on Support Function Protective Clothing for Hazardous Chemical Operations

All three standards address a complete ensemble of chemical protective clothing and provide an extensive list of performance requirements related to suit integrity, material chemical and flame resistance, material strength and durability and component function.

Important physical property requirements of protective clothing are abrasion resistance, flex fatigue, tear resistance and weight. Since the main function of chemical protective clothing is to protect the wearer against chemicals, the clothing is not necessarily designed for protection against fire or heat. However, the possibility of a flash fire or explosion is quite high during cleanup of hazardous material (haz-mat) spill. For this reason, haz-mat garment materials are made or treated with flame resistant materials. FR rayon, FR polyester and FR treated olefins can be used for mild protection. Fabrics made of aramid fibers such as Nomex® provide good protection against high temperatures [26]. Chemical protective clothing may cause heat stress which is a concern for the wearer. By using microporous fabrics such as Gore-Tex® the situation can be improved. For vapor-impermeable barriers, usually a personal cooling system is required in certain applications.

Goldman identified six parameters that should be taken into account in predicting the heat stress of the garments: air temperature, ambient air motion, ambient vapor pressure, mean radiant heat temperature, task work load and the type of clothing worn by the worker [27]. These parameters affect the body temperature, heart rate and skin dampness which are important for comfort and health.

4.3 Applications

Simple chemical protective clothing is gloves, boots, face shields, and aprons. Hoods, coveralls and splash protective suits provide additional body protection. Highly sophisticated systems such as totally encapsulating suits provide a "gas-tight" envelope around the wearer.

Chemical protective clothing is customarily classified as durable, disposable or "limited use." Durable protective clothing can be used many times but may need costly decontamination treatments between uses. It is also difficult to guarantee decontamination. Durable protective clothing is usually made of woven or knit fabrics and is coated. Laminated nonwovens are also used. Disposable protective clothing is made of nonwovens and can be used only once. It is typically used in the medical field (Chapter 10.0). Limited use garments may be used several times depending on the degree of contamination [28].

Protective garment manufacturing requires careful attention to fiber and fabric selection, cutting, sewing, and inspection. The seams may be especially critical depending on the end-use requirements. Typical seam types are glued seams, serged, single needle seams and bound seams. Sealed and bound seams and sealed and taped seams are also used. The sealed seams can be ultrasonic, thermal or RF sealed. Sealed seams provide the greatest resistance to toxic material penetration [23,28]. To protect the worker and to protect the manufacturer from the product liability suits, strict quality assurance should be practiced during manufacture, inspection and packaging. Proper labeling is also important for this purpose. OSHA or EPA standards must be followed where applicable. Third party certification is becoming important for this purpose.

Hazardous material response, toxic waste dump cleanup and manufacture of some chemicals require complete encapsulation of workers. A typical totally encapsulating chemical protective (TECP) suit is made up of several components: self-contained breathing apparatus, lens, suit closure, vent valves, suit membranes, seams, gloves and boots.

Factors to consider when selecting a protective garment include type of application, level of protection and type of chemicals, productivity, cost, acceptance and durability (decontamination versus reuse). Different garments provide different degrees of protection, have different sealing properties, etc. Their selection also depends on whether respiratory equipment is used or not. The goal is to provide just the right amount of protection in order to minimize cost and maximize production by acceptability.

5. MECHANICAL PROTECTION

Industrial textiles are used to protect the wearer from mechanical hazards which include cut, tear, puncture, splash, impact and abrasion. Textiles for mechanical protection can be manufactured from high strength textile fibers used alone or with metallic fibers (see section on Metallized Fabrics).

According to the U.S. Labor Department, each year more than one million workers suffer job related injuries, 25% are to the hands and arms [29]. Cut resistant textiles are used for many applications including food preparation, construction, forest services, utility works, metal works and tree trimming among others. The use of protective gloves, aprons, sleeves and chaps reduce the number of injuries, the severity of the injury and lost time as well as insurance and legal costs.

The types of materials used in mechanical and physical protective clothing include para-aramids, high density polyethylene (HDPE), nylon/cotton blends in woven and knit configurations, as well as specialized fabrics with metal yarns.

Cut resistance can be achieved in several ways: increasing fabric weight, using inherently cut resistant materials and using composite yarns. Heavier fabrics provide better cut resistance, however increased weight may adversely affect the user performance. Materials such as extended chain polyethylene and para-aramid offer high inherent cut resistance. By wrapping cut resistant fibers around steel or fiberglass yarns, composite yarns with excellent cut resistance can be obtained [30].

5.1 Gloves

Glove selection must be done according to the end-use requirements. The requirements for a surgeon's glove and sheet metal worker's glove are different. Gloves should be comfortable and should not hinder the wearer's manual dexterity in performing his or her job. Gloves made of continuous filament yarns like nylon and polyester are suitable for clean rooms and paint shops since they are lint-free.

Standard work gloves made of cotton or leather provide little cut resistance. A 14-ounce para-aramid glove provides more cut resistance than a 30-ounce cotton glove [31]. Gloves made of composite yarns combining para-aramid or extended-chain polyethylene with steel wire or fiberglass provide excellent cut protection. They provide better dexterity and comfort with the protection levels of metal mesh gloves.

Gloves with puncture resistance are needed when dealing with small, pointed objects. For this type of application, the glove construction becomes very important. Loosely knit gloves are not suitable for puncture resistance. Tightly woven fabrics made of strong fibers increase puncture protection. Leather also provides puncture resistance because of its dense, tough surface. Nonwovens made of metallic and high strength textile fibers provide excellent puncture resistance. After forming, the web is highly consolidated and coated. The resulting fabric is suitable for puncture resistant gloves, protective vests and outdoor apparel.

For protection against abrasion, glove thickness becomes important. Leather gloves provide the best abrasion resistance followed by synthetic fibers.

Thermal protective gloves must have insulating properties to prevent heat transfer through the glove to the wearer's hands. Gloves made of leather, FR cotton and aramid fibers provide various levels of thermal protection. Leather is not recommended for severe exposure. FR cotton gloves decompose at $600-650°F$. Gloves or glove liners made of aramid fibers provide excellent thermal protection. Meta-aramids decompose at $700-800°F$ and para-aramids decompose at $800-900°F$. Aramid fibers can also be aluminized for protection against molten metals.

For hazardous chemicals and liquids, nonporous gloves are used. Gloves made of rubber and other elastomeric materials, such as neoprene can prevent liquids from penetrating through the skin. Fabrics made of natural or synthetic fibers also can be coated for liquid protection.

5.2 Chain Saw Clothing

Loosely woven or knit fabrics comprising a large number of long yarns are used in chain saw clothing. Although these fabrics may be easy to cut in conventional tests, when they are hit by the fast moving chain (as fast as 3,300 feet per second), they "shred out" and, in less than 0.15 seconds, jam the drive sprocket of the chain saw and bring it to a dead stop (Figure 13.18) [32]. In the past, chain saw chaps were made with ballistic nylon. They are now usually manufactured using alternating layers of woven and needlefelt para-aramid. Para-aramid fabrics require fewer layers for adequate protection, making it less bulky and more comfortable. Chaps constructed of ballistic nylon typically are made with eight to fourteen layers, while para-aramid chaps require only six layers [33].

ASTM and NFPA have been working to establish viable test methods for measuring the cut resistance of protective clothing. AlliedSignal developed the BetaTec procedure to measure cut resistance. In this test, a weighted pivoting beam with a cutting edge (a standard razor blade) is brought into contact with a revolving finger-like mandrel. The number of revolutions made by this action is electronically counted until cut-through is detected. DuPont altered this method to minimize the blade wear (The Modified Beta-

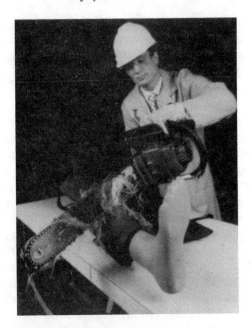

FIGURE 13.18 Chain saw chaps provide effective cut and slash protection (courtesy of DuPont).

Tec approach) [30]. ASTM test methods related to mechanical protection are as follows:

- F 1342 Test Method for Protective Clothing Material Resistance to Puncture
- F 1414 Test Method for Measurement of Cut Resistance to Chain Saw in Lower Body (Legs) Protective Clothing
- F 1458 Test Method for Measurement of Cut Resistance to Chain Saw of Foot Protective Devices

6. ELECTRICAL PROTECTIVE CLOTHING

There are mainly two types of protection from electrical hazards: electromagnetic protection and electrostatic protection.

6.1 Electromagnetic Protection

Utility workers that work close to power lines and electrical equipment may be exposed to electric shocks and acute flammability hazards.

To protect the workers in these situations, rubber gloves, dielectric hard hats and boots, sleeve protectors, conductive Faraday Cage garments, rubber blankets, non-conductive hot sticks, and live line buckets are used.

In January 1994, OSHA issued a new standard, 29 CFR Part 1910, addressing the work practices to be followed during the operation and maintenance of electric power generation, transmission and distribution facilities. In the same document, the electrical protective equipment requirements contained in the electric power generation standard are also revised [34].

Conductive protective clothing is necessary for people who work in the vicinity of very high-voltage electrical equipment. In addition to conductivity, these garments must be flame resistant, wear resistant and comfortable. Humans are sensitive to electrical energy because of their sophisticated nervous system. Conductive protective materials are made by closely weaving metal and natural or synthetic fibers. These garments (called "Live Line" garments) were introduced in the early 1970s and are still in use today. A typical protective fabric is woven from spun yarns containing a mixture of fire retardant textile fibers and stainless steel fibers (8 – 12 micron diameters). It was proved that "Faraday suits" made of fabrics with 25% stainless steel fiber/75% wool blend or 25% stainless steel fiber/75% aramid fiber can be effective at voltages up to 400 kV [35]. It should be emphasized that the garment protects the wearer from the effects of electromagnetic field generated by high voltage in the vicinity of the power line, i.e., the garment cannot protect the individual if there is direct contact with the high voltage line.

King et al. measured the thermal protective performance of single layer and multi-layer fabrics exposed to electrical flashovers [36]. They exposed fabric specimens to a 12 s (720 cycle) arc at a distance of 57 mm from the electrodes which created an incident energy flux density of 211 kW/m². Table 13.19 shows the results of some of the single layer fabrics they tested. All of the fabrics were flame resistant. They showed that those parts of the body covered by only one fabric layer were exposed to the highest risk of burn injury, whereas for those covered with six or more layers, the risk of burn injury was negligible. They concluded that more protection is obtained by using two or more fabric layers (Table 13.20).

Figure 13.19 shows the response of shirts made from 100% ordinary cotton, Nomex® III and FR treated cotton to electric arc blast. The shirts were exposed to an average arc energy of 96 cal/cm² for 1/6 seconds.

Radiation from electromagnetic fields (EMF) generated by power lines is another public concern. There have been mixed reports about the relation between exposure to electromagnetic fields and health hazards like leukemia and brain cancer. It is suggested that the "Live Line" technology could be readily adapted for protection against EMF radiation with slight modifications [29].

In April 1994, ASTM issued F 1506, Standard Performance Specification for Use by Electrical Workers Exposed to Momentary Electric Arc and Related Thermal Hazards [37].

6.2 Electrostatic Protection

Static electricity is an electrical charge at rest that is generated by the transfer of electrons. The transfer of these electrons between objects with different electrostatic potentials or to the ground can cause electrostatic discharge. The human body, which can be considered a capacitor, may get electrostatically charged. The body transfers its charge to a metallic object upon contact, thus creating an electrostatic discharge. This discharge can damage an electronic part, cause an explosion in a hazardous atmosphere, cause a computer glitch through signal misinterpretation and can cause an involuntary reflex leading to injury. Antistatic garments and materials are also critical for the space programs [29,38].

The generation of static electricity caused by rubbing two materials together is called the triboelectric (static) effect. As a result of the triboelectric effect, materials can be ranked in order of positive to negative charging, which is called triboelectric series as shown in Table 13.21. When any two substances in the list contact each other and are separated, the material highest on the list gains a positive charge.

Table 13.22 shows typical voltage generation during daily life. Since electric parts can be

TABLE 13.19 The Thermal Protective Performance (TPP) of Single Layer Fabrics (copyright ASTM [36]).

Fiber(s)	Name	Structure	Finish	Thickness (mm)	Mass/Unit (g/m²)	Lightness (%)	Protection Time (s)	TPP Rating (MJ/m²)
Cotton	Shirting	2/1 twill	Proban®	0.51	177	34.1	2.4	0.46
Cotton	Drill	3/1 twill	Proban	0.65	288	32.7	4.7	0.92
Cotton	Satin	4/1 satin	THPOH-NH₃	0.73	328	52.6	4.7	0.83
Cotton	Denim	3/1 twill	Proban	0.81	348	28.9	4.7	0.89
Cotton	Brushed	4/1 twill	Proban	1.12	414	89.3	5.5	1.26
Cotton	Satin	4/1 satin	P44	0.66	429	87.2	4.4	0.95
Cotton	Denim	3/1 twill	Proban	0.93	491	84.1	5.6	1.12
15/85 P/C	Denim	2/1 twill	Proban	0.75	361	28.2	4.1	0.84
15/85 P/C	Denim	2/1 twill	P44	0.66	453	31.2	5.2	1.29
Wool	Suiting	2/2 twill	FR	0.68	262	14.3	4.6	0.95
60/40 wool/Cordelan®	Serge	2/2 twill		0.61	243	55.2	3.9	0.97
60/40 Cordelan®/wool	Suiting	2/2 twill		0.60	274	23.0	4.4	0.98
Aramid	Nomex® III	1/1 plain		0.58	178	87.9	4.1	0.72
50/50 Kevlar®/Nomex aramid		1/1 plain		0.64	250	83.7	4.1	0.77
Aramid	Nomex III	1/1 plain		0.89	293	88.3	4.7	0.90
Aramid (reclaimed)		Needlepunched felt		3.65	353	26.7	6.1	1.41
Aramid	Norfab®	1/1 plain		2.45	747	77.2	9.4	1.77
Glass		1/1 plain		0.28	205	92.1	4.5	0.77
Glass		1/1 plain		1.53	646	68.0	4.9	0.84
Glass and Al foil		1/1 plain		1.42	677	95.9	15.7	2.80
Novoloid	Kynol®	Needlepunched felt		2.50	157	49.2	3.8	0.69
Novoloid	Kynol	1/1 plain		0.39	163	46.7	2.4	0.45
Novoloid/aramid	Kynol	1/1 plain		0.54	201	46.6	2.4	0.44
Novoloid/glass	Kynol	1/1 plain		2.11	561	52.4	5.1	1.07
Polybenzimidazole	PBI	2/1 twill		0.95	308	27.5	4.1	0.59
Oxidized acrylic	Pyron®	2/1 twill		1.28	419	14.3	3.9	0.73
Modacrylic	SEF®	2/1 twill		0.52	217	24.7	2.3	0.43

TABLE 13.20 The TPP Ratings of Two Layer Assemblies (copyright ASTM [36]).

	Outer Fabric			Inner Fabric			Two Layer Assembly		
Fiber/Finish	Mass/Unit Area (g/m²)	TPP (MJ/m²)	Fiber/Finish	Mass/Unit Area (g/m²)	TPP (MJ/m²)	Mass/Unit Area (g/m²)	Protection Time (s)	TPP (MJ/m²)
15/85 Polyester/ FR cotton	361	0.84	FR Cotton	177	0.46	538	6.7	1.67
15/85 Polyester/ FR cotton	361	0.84	Aramid	178	0.72	539	7.3	1.81
15/85 Polyester/ FR cotton	361	0.84	Novoloid	163	0.45	524	7.2	1.56
FR Wool	262	0.95	FR Cotton	177	0.46	439	7.6	1.55
FR Wool	262	0.95	Aramid	178	0.72	440	6.3	1.31
FR Wool	262	0.95	Novoloid	163	0.45	425	6.3	1.31
Aramid	293	0.90	FR Cotton	177	0.46	470	11.8	2.19
Aramid	293	0.90	Aramid	178	0.72	471	6.5	0.97
Aramid	293	0.90	Novoloid	163	0.45	456	7.3	0.98
PBI	308	0.59	FR Cotton	177	0.46	485	8.1	1.20
PBI	308	0.59	Aramid	178	0.72	486	10.3	1.47
PBI	308	0.59	Novoloid	163	0.45	471	8.5	1.32

Before Arc Blast

NOMEX® III

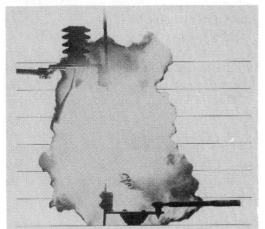

During Arc Blast

Flame-Retardant Treated Cotton

After Arc Blast

FIGURE 13.19 Response of shirts made of different fabrics to an electric arc blast. Ordinary 100% cotton shirt is shown before, during and after arc blast. Nomex® III and FR treated cotton shirts arc blast are shown on the right (courtesy of DuPont).

TABLE 13.21 Triboelectric Series [29].

Positive (+)	
	Acetate
	Glass
	Human hair
	Nylon
	Wool
	Fur
	Aluminum
	Polyester
	Paper
	Cotton
	Wood
	Steel
	Acetate fiber
	Nickel, copper, steel
	Brass-stainless steel
	Rubber
	Acrylic
	Polystyrene foam
	Polyurethane foam
	Saran
	Polyethylene
	Polypropylene
	PVC (vinyl)
Negative (−)	Teflon

damaged by electrostatic discharge voltages of 100 volts or less, static control is extremely important for the electronics industry.

Using textile fabrics made of intimately blended metal and textile fibers is an effective way to control static buildup. Garments made of these fabrics dissipate the static generated by the wearer before it reaches dangerous levels. Continuous multi-filament metal fibers can also be woven into fabrics at intervals to control static. High concentration spun yarns are also used for this purpose.

Metal fiber content is usually varied between 1% to 5% depending on the hazard level of application. A typical antistatic garment is a blend of 65% polyester/32.5% cotton/2.5% stainless steel fabric. Antistatic garments were

introduced in the medical market in the late 1960s. They were later adapted to electronics manufacturing in clean rooms and other areas including emergency chemical response and cleanup, fuel or ordnance handling operations and general chemical process industry.

In the late 1960s, NASA developed a method to test the static charge generation and discharge rates on flat plastic and textile materials. Recently, an ASTM F-23 committee has developed a new standard for evaluating triboelectric charge generation on protective clothing materials [Draft ASTM F 23.20.05, Resistance of Protective Clothing Materials to Triboelectric (Static) Charge Generation]. This method measures the maximum charge accepted and rate of discharge for static produced on materials by rubbing friction. Measurement of material surface resistivity and electrostatic decay has been by the common test methods used to evaluate material static electric hazards. AATCC Test Method 76 is used to measure surface resistivity. Federal Test Method Standards (FTMS) 191A, 5931 and 101B, 4046 provide methods to measure electrostatic decay of fabrics and plastics. Table 13.23 shows the ranking of some chemical protective clothing materials in the market for triboelectric charge.

7. CLEAN ROOM TEXTILES

Clean room (anticontaminant) textiles protect the environment from the wearer. The human body sheds one billion skin cells every day. The body and its clothing also carry a good amount of dust, ions, hair, textile lint, cosmetics, perfume, and tobacco smoke, all of which are incompatible with high-tech production methods [39]. Clean room items are used in clean rooms to keep the atmosphere clean and prevent con-

TABLE 13.22 Typical Voltage Generations [29].

Action	65–95% Relative Humidity	10–20% Relative Humidity
Walking on carpet, sliding on auto seat	1,500 volts	35,000 volts
Wearing synthetic fiber apparel	1,500 volts	30,000 volts
Sitting on vinyl upholstered chair	1,500 volts	18,000 volts

TABLE 13.23 *Representative Triboelectric Charge Measurements for Chemical Protective Clothing Materials [38].*

Material	Peak Voltage	Voltage after 5 Seconds
Aluminized aramid	< -5	< -5
Aluminized fiberglass	< -5	< -5
Aramid (untreated)	2,010	895
Chlorobutyl rubber/aramid	450	200
Elastomer/plastic laminate	$-9,500$	$-1,300$
Teflon/aramid	$-12,500$	$-12,100$
Teflon/aramid (optimized)	-300	< -10

tamination of items in the room which may be food, pharmaceuticals, microelectronics, aerospace components, optics and automotive components. Each market has its own requirements regarding fabric containment capability.

Anticontaminant textiles should protect the wearer from hazardous materials in the clean room as well. Exposure to high purity pharmaceuticals or the solvents used in the semiconductor industry, for example, can be harmful to a worker. Clean room textiles should be lint-free, antistatic and resistant to human contamination such as from hair or dead skin.

7.1 Clean Room Classification and Required Garment Types

There are various garment types for clean rooms, as recommended by the Institute of Environmental Services (IES). The lower the number associated with the class, the cleaner the environment [39].

- A Class 100,000 room usually requires a hat or hood with full hair cover, beard cover, coveralls or zippered frock, footwear and gloves (100,000 represents the number of 1.0 micron diameter or larger particles per 1 cubic foot of air). Facilities rated higher than Class 100,000 generally are not considered clean rooms.
- For Class 10,000 rooms, a hoop replaces the hat from the Class 100,000 ensemble.
- Class 1,000 rooms require full coveralls in place of a zippered frock.
- A Class 100 environment requires full facial cover.

- A Class 10 environment necessitates a complete facial enclosure.

7.2 Fabrics for Clean Rooms

Woven, nonwoven and laminate fabrics are used in clean room garments. The fabrics used to make clean room textiles are designed to contain particles as small as 0.2 microns or less in diameter [40]. Table 13.24 lists general fabric recommendations for clean room applications.

Woven fabrics are made of continuous multifilament yarns. Fabrics made of staple and natural yarns are never used in clean room applications because of the excessive amount of lint they generate. Important characteristics of woven fabrics for clean room applications are containment (filtration) efficiency, air permeability, moisture vapor transmission rate, abrasion resistance and static decay properties. The typical air permeability range is $1-45$ cfm and typical filtration efficiency can vary from 15% to 60% for particles measuring 0.3 to 0.5 microns. Woven fabrics can also be used to control electrostatic discharge (ESD).

Continuous filament polyester is widely used in woven anticontaminant garments. These garments can readily be sterilized through gamma radiation. Other major materials are Tyvek®, coated and laminated fabrics such as Gore-Tex, and inherently antistatic fabrics. Tyvek, a spunbonded polyethylene, is a lint-free product with good strength and abrasion resistance properties that make it suitable for clean room applications (Figure 13.20). It is easy to sterilize and has excellent particle containment properties. Another material for clean room application is Kleenguard® from Kimberly-Clark, which is a

TABLE 13.24 General Fabric Recommendations for Clean Room Applications (Mission Clean Room Services [40]).

| Industry | Wovens | | | | Nonwoven | Laminates | |
	Herringbone	Taffeta	Twill	High-Density	Spunbonded	PTFE Polyolefin	Urethane
Integrated circuit			X	X		X	X
Disk drive			X	X		X	X
Pharmaceutical	X	X	X	X	X		
Aerospace		X	X				
Electronic assembly		X	X				
Optics			X	X		X	X
Food	X	X		X	X		
Automotive			X				

spunbond/meltblown/spunbond heat-fused polypropylene laminate.

Conductive yarns such as carbon, nickel or aluminum can be woven into clean room fabrics for static control which is required in semiconductor and microchip manufacturing. These fabrics also prevent sparks in flammable environments. Bicomponent fibers with a conductive core can provide the required antistatic properties in the fabric. In general, inherently flame retardant fabrics are required for clean room garments. Chemical protection ability of the garment is also important.

During fabrication of clean room garments, loose fabric edges, holes, gaps and other means, by which particles could escape into the air from the wearer or garment, must be avoided. Cut edges should be preserged, or burned and turned under, then lock stitched, or bound and sealed.

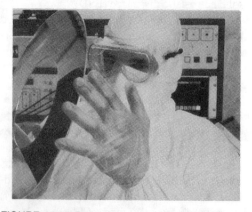

FIGURE 13.20 Clean room garment of Tyvek® (courtesy of DuPont).

Particle entrapment regions such as pockets, belts, pleats, fold-over collars or cuffs, emblems, logos and pen-tabs should be avoided. Neckline and collar edges should be over-edged before setting. Clean room garments are usually sewn using low-texture, continuous filament sewing threads. Some companies use ultrasonically bonded seams to eliminate the "blow through" associated with needle holes.

8. RADIATION PROTECTION

Protective clothing is used in radioactive environments to prevent the transfer of radioactive materials, such as dirt containing radioactive elements, from directly contaminating human skin. The dirt emits radiation and the skin absorbs the radiation energy becoming irradiated (damaged) in proportion to the dose received.

Radiation protection is necessary for nuclear plant employees, X-ray professionals, workers in cancer treatment centers and other places subject to ionizing radiation.

For alpha- and some beta-radiation producing matter, irradiation injury can be prevented by simply keeping the radioactive dirt off the skin, out of the eyes, nose and mouth. As long as the radiative materials do not contact human tissue, no damage is done. In these cases, goggles, respiratory protection, gloves and lightweight protective clothing may be enough for protection. However, in case of gamma radiation, the radiation energy will penetrate deeply into the suit and wearer. Dirt that emits gamma radiation irradiates and damages uncontaminated tissue in its proximity. The radiant energy absorbed by tissue from a distance will cause damage even if the radioactive dirt does not touch the human tissue. The techniques used for protection from gamma and some beta radiation-emitting radioactive materials are contamination control, time, distance and shielding. Shielding is done by placing a dense (heavy) radiation barrier such as lead between the radioactive dirt and the worker [41].

Nuclear protective wear fabrics are made of cotton, polyester/cotton or nylon/polyester [42]. Until the 1980s twill or sateen cotton fabrics were the main material for all nuclear protective wear. Recently, polyester/cotton (65/35) has

been the fabric of choice for nuclear protective wear. Poly/cotton fabrics represent 80% of all nuclear protective wear in the United States. More recently other synthetic fabrics have been introduced to the nuclear protective wear market. One example is 99% nylon/1% carbon fabric which offers better performance over poly/cotton fabrics.

There are no regulatory requirements for specifications of radiation protection clothing. Each institution is responsible for protecting their employees. As an example, Table 13.25 lists the types of protective garments used at Duke Power Co.'s (Charlotte, NC) nuclear sta-

TABLE 13.25 *Typical Examples of Clothing for Radiation Protection (Duke Power Co. [43]).*

Dry/light work	
15,000 – 150,000 dpm*/100 cm²	Hood, disposable or cloth
	Shoe covers, disposable and rubber
	Gloves, cloth and rubber
	Coveralls, disposable or cloth
1,000 – 15,000 dpm/100 cm²	Hood, disposable or cloth
	Shoe covers, rubber
	Gloves, cloth and rubber
	Coveralls, disposable
Wet/light work	
15,000 – 150,000 dpm/100 cm²	Hood, disposable and wet suit
	Shoe covers, disposable and rubber
	Gloves, cloth and rubber
	Coveralls, disposable or cloth, and wet suit
1,000 – 15,000 dpm/100 cm²	Hood, disposable
	Shoe covers, rubber
	Gloves, cloth and rubber
	Coveralls, disposable
< 1,000 dpm/100 cm²	No protective clothing required
Dry/heavy work	
15,000 – 150,000 dpm/100 cm²	Hood, disposable or cloth
	Shoe covers, disposable and rubber
	Gloves, cloth and heavy rubber
	Coveralls, disposable and cloth
1,000 – 15,000 dpm/100 cm²	Hood, disposable or cloth
	Shoe covers, disposable and rubber
	Gloves, cloth and heavy rubber
	Coveralls, cloth
Wet/heavy work	
15,000 – 150,000 dpm/100 cm²	Hood, disposable and wet suit
	Shoe covers, disposable and rubber
	Gloves, cloth and heavy rubber
	Coveralls, cloth and wet suit
1,000 – 15,000 dpm/100 cm²	Hood, cloth or disposable
	Shoe covers, disposable and rubber
	Gloves, cloth and heavy rubber
	Coveralls, disposable and cloth
< 1,000 dpm/100 cm²	No protective clothing required

*Disintegrations per minute, a unit of radiological contamination.

TABLE 13.26 Some Commercially Available Insulation Materials (44).

Products	Manufacturer	Characteristics	End Use
Drylete®	Hind	Combines hydrofil nylon and hydrophobic polyester to push moisture away from the body and then pull it through fabric for quick drying	Skiwear, running and cycling apparel
Gore-Tex®	W. L. Gore	The original waterproof/breathable laminated fabric with pores large enough for water to escape, small enough to block rain	Outerwear, gloves, hat, footwear, running and cycling apparel
Hydrofil®	Allied	A new super-absorbent hydrophilic (water-loving) nylon that sucks moisture away from skin	Linings, long underwear, cycling apparel
Silmond®	Teijin	A durable water repellent and windproof fabric made from polyester microfibers	Outerwear
Supplex®	DuPont	A strong, quick drying nylon fabric, in smooth or textured weaves, that feels like cotton	Outerwear, skiwear, running and cycling apparel
Synera®	Amoco	A strong, lightweight polypropylene fabric that transports moisture away from skin	Long underwear, jacket linings
Tactel®	DuPont	Extremely soft, quick drying nylon in smooth or textured versions	Outerwear, skiwear, running and cycling apparel
Thermolite®	DuPont	An insulation made of polyester fibers that have been coated to make them slippery and more drapeable.	Skiwear, climbing apparel
Thermoloft®	DuPont	A semithick insulation that contains hollow polyester fibers that trap warm air	Skiwear, outdoor apparel
Thinsulate®	3M	Thin insulation made of polyester and polypropylene fibers	Skiwear, gloves, footwear
Thintech®	3M	A waterproof/breathable laminated fabric that is highly resistant to contamination by determination by detergents or perspiration	Skiwear, outerwear

tions. Protective garments for radiation include coveralls, head covers and hoods, gloves, shoe covers, rain suits, lab coats, modesty tops and gym shorts. Respiratory protection equipment may be necessary when contamination levels (loose) are greater than 25,000 disintegrations per minute (dpm)/100 cm^2 for heavy work and 100,000 dpm/cm^2 for light work.

Coveralls provide general protection for the whole body, including the arms, legs and torso. Cotton coveralls are used for normal to heavy work conditions. Disposable coveralls may be worn over personal clothing for low-level contamination. Hoods are worn to protect the head

and hair from contamination. Cotton and disposable hoods are used. Gloves can be rubber or cotton. Rubber shoe covers are worn over disposable shoe covers. Plastic rain suits are used when water may be involved. They are also used in highly contaminated areas to provide maximum protection from contamination. Lab coats are used where the chance of contamination is low. Modesty tops and gym shorts are worn under protective clothing [43].

Non-ionizing radiation hazards may be encountered around microwave relay towers, high-voltage transmission lines (see the section on Electrical Protective Clothing), and even among those who work around computers and radiating electronic equipment. Metallized materials that provide dissipation of radio frequency and electromagnetic radiation are being used for non-ionizing radiation protection.

9. THERMAL INSULATION (PROTECTION FROM COLD)

Table 13.26 lists some of the insulation materials on today's market. The list includes the product name, the manufacturer of the product, different characteristics of the fiber and the product's end uses.

One new advancement in fiber technology involves the invention of a unique, patented thermal insulation of a fine bonded polyester called Primaloft®, manufactured by Albany International. Primaloft has been proven to be superior to fine goose down. It is used for insulating such things as sleeping bags and snow skiing clothes because it is lightweight, durable and maintains its insulating properties even when wet. Primaloft® carries no more than 20% of its own weight in moisture. This is minute when compared to 108% − 1,000% for synthetics. Primaloft's insulating value when wet is also better than that of goose down. Figure 13.21 shows the comparison of several insulation material diameters.

The figure of the duck down shows a branch that comes from waterfowl down quill. The larger branch fibers give loft and compressional resistance while the smaller fibrils exemplify many micro-spaces of trapped air that give the down its insulating abilities. The figure of the Primaloft® fibers and air spaces shows how it is very similar to that of the duck down in terms of size, number and distribution. Dead air space is an effective insulator; more important than the fiber material itself. Its unique characteristics result in giving similar control of convective and radiant heat transfer, overall insulating performance that is just as good or maybe even better than that of down and other synthetic insulators.

Primaloft's microscopic air pockets help it to resist the passage of water while allowing body moisture, in the form of water vapor, to escape. There are millions of these tiny air pockets that cause the Primaloft® to entrap heat and minimize convective and radiant heat transfer. Its smaller fibers help to create a soft hand, exceptional drape and great compressibility.

Albany International has shown with a Toothpick Theory that there can be a perfect balance between density and loft (Figure 13.22). The Toothpick Theory consists of imagining heat traveling around one pencil to escape. The heat would have to travel from distance A to B, but would have an easy path to travel between these

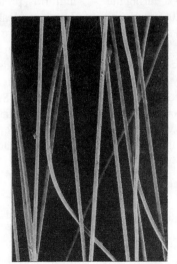

FIGURE 13.21 SEM of various insulation materials: duck down (left), Primaloft® (center), and synthetic insulation (courtesy of Albany International).

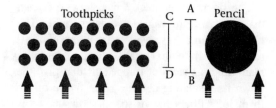

FIGURE 13.22 Toothpick theory (Albany International).

two points. Then imagine the heat having to travel through 15−20 toothpicks from distance C to D. It is more effective for an insulator that the heat travel the shorter distance through 15−20 times more fibers.

Insulation textiles for military applications are explained in Chapter 11.0, Military and Defense Textiles.

10. HIGH VISIBILITY TEXTILES

Studies have shown that reduced visibility contributes to fatal pedestrian accidents. It was reported that night time vehicles strike and kill more than 4,000 pedestrians and injure more than 30,000 pedestrians annually in the United States [45]. It is believed that high visibility materials (HVM) assist in avoiding worker and pedestrian deaths or serious injuries. According to the National Safety Council, the work-related deaths have declined more than 40% during the past ten years, from 12 to 7 per 100,000 workers. The use of HVM has at least doubled during the same period [46].

Dark clothes reduce and white clothes increase visibility at night. For safety purposes, it is necessary to increase visual contrast. The primary technique to increase conspicuity is through the use of high visibility materials. There are three major types of high visibility products:

- reflective materials (they shine when struck by light)
- photoluminescent materials (give yellow light in dark)
- fluorescent materials (i.e., red-orange is visible during the day)

Textile substrates are used in reflective per-

sonal safety products. Fabrics are covered with reflective microprism sheeting for high visibility personal safety items. These items are used by pedestrians, highway workers, cyclists, joggers, hikers, police, firemen and other professionals. State Departments of Transportation, the Occupational Safety and Health Administration (OSHA) and the National Fire Protection Association (NFPA) require the use of reflective material on workers' garments. Airlines, law enforcement and emergency medical services also use reflective materials.

In a recent development, 3M's Scotchlite® reflective transfer film was micro-slitted to form narrow retro-reflective yarns. These yarns can be woven, knit or braided into fabrics. 3M's Scotchlite® 8710 retro-reflective transfer film has 50,000 tiny, highly reflective glass beads per square inch. These glass beads, due to their round shape, reflect light back to its source [47].

Photoluminescent or phosphorescent materials have the ability to absorb daylight or artificial light, store its energy and emit a green-yellow glow in darkness. High quality photoluminescent safety fabrics remain visible to darkness-adapted eyes for up to six hours [48]. Zinc-sulfide crystals, which are non-radioactive and non-toxic pigments, provide the visibility in dark. They may be recharged repeatedly when exposed to light for a few minutes. Photoluminescent fabric-backed vinyl can be sewn on uniforms, protective apparel, back supports, etc. They are ideal for emergency exits and escape routes (Figure 13.23).

In some applications, a combination of the three methods are used for optimum safety during the day and night.

ASTM Subcommittee E-12.08, High Visibility Materials for Pedestrian Safety, is working to develop standards for both manufacturing and grading materials.

11. METALLIZED FABRICS

As it has been mentioned throughout the chapter, metallized products are used in industrial, specialty and protective clothing applications. There are various ways to combine metals with textile materials for specific applications.

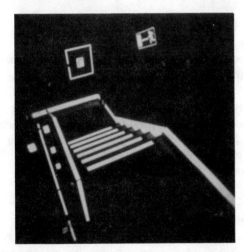

FIGURE 13.23 Photoluminescent extruded plastic profiles for electrical emergency lighting (courtesy of American Permalight, Inc.)

Metallized fabrics provide good abrasion resistance, reflectivity over extended time, wear resistance and molten metal splash resistance.

11.1 Coating/Laminating

Textile fabrics are used as substrates in metallized protective materials. Woven, knit and nonwoven fabrics may be coated or laminated with metal surfaces. Substrate fabrics can be made of aramids, carbon based fibers, PBI, glass, cotton, rayon and others. Aluminum is widely used in metallized fabrics (Figure 13.24).

In aluminized fabric, aluminum molecules are deposited on a PET film. Examples are Mylar®

from DuPont and Hostaphan® from Hoechst. The aluminized film can reflect up to 90% of radiant heat. Gold can be used for reflection of up to 100%, but it is expensive.

Laminated metallized fabrics can be made of several layers of materials. A typical five-ply dual mirror aluminized fabric has the following layers: aluminum, protective film, a second layer of aluminum, heat stable adhesive and fabric.

11.2 Blending

Metal sliver can be blended with synthetic or natural fibers to produce conductive textiles. Stainless steel sliver used for this purpose usually has 4. 8 or 12 micron fiber diameters and weighs approximately 1, 2 or 4 grams. The fiber length may vary from 1.5 to 6 inches. There are several methods to produce metal fibers including bundle drawing (most common), wire drawing, shaving, shearing, melt spinning, melt extracting and stretch casting. Figure 13.25 shows typical cross sections of metal fibers from each process. For maximum conductivity, the steel fiber is introduced at the last drawing operation.

Protective fabrics made from metal-based blended fibers are suitable to protect individuals from the hazardous effects of electrostatic discharge and electromagnetic radiation.

11.3 Composite Yarns

Multi-filament metal fiber yarns can be twisted or wrapped with textile yarns to produce composite yarns. These yarns are suitable for cut resistant apparel items, antistatic brushes for business machines, lightning strike protection and antistatic filter bags. The most widely used metal yarns are 12 microns/91 filaments, 25 microns/91 filaments and 8 microns/812 filaments.

11.4 Nonwoven Metal Based Fabrics

Chopped metal fibers can be air or wet layered with textile fibers to form nonwoven textiles. For air layering, 1 inch fiber length and 4-38 micron fiber diameters are used. For wet layering, fiber lengths of 0.125 to 0.5 inch have been success-

FIGURE 13.24 Aluminized textile garment (courtesy of Gentex).

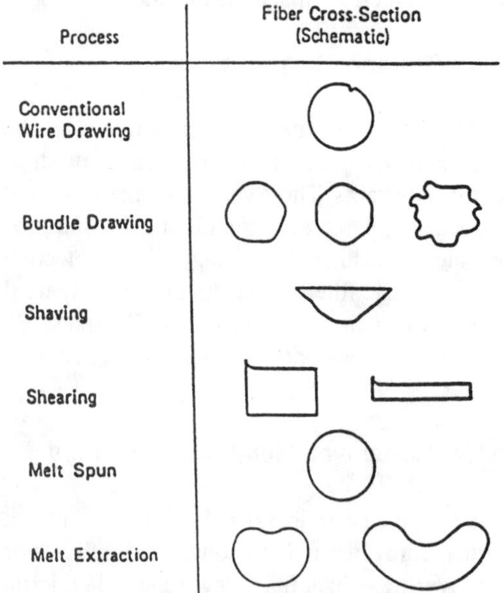

Process	Fiber Cross-Section (Schematic)
Conventional Wire Drawing	
Bundle Drawing	
Shaving	
Shearing	
Melt Spun	
Melt Extraction	

FIGURE 13.25 Typical metal fiber cross sections [49].

fully used. Binders or sintering may be used for stabilization. During sintering, the organic binder fibers are burned off, leaving a 100% metallic fiber structure. In general, fiber diameters of 4−15 microns in 0.125−0.250 inch lengths are suitable for this process [49].

Test methods and characteristics to evaluate the metallized products include the following:

- Military Specification, MIL-C-87076A, for Aluminized, Twill Weave, Aramid, Coated Cloth
- MIL-C-24924A Class I (fire proximity garments)

12. SPACE SUITS

Space suits used by the astronauts during space shuttle missions represent the ultimate protective clothing in existence today. They protect the astronauts from heat, cold, chemicals, micrometeoroids and pressure fluctuations among other things. The description of the space suit below is excerpted with permission from several sources [50−53].

12.1 Suits for Outside the Space Shuttle (in Vacuum)

Since there is no atmospheric pressure or oxygen in space to sustain life, astronauts wear space suits assembly (SSA) outside the space shuttle, in the vacuum. The space suit is also called Extravehicular Mobility Unit (EMU). It protects astronauts from the harsh environment of space outside the Space Shuttle's crew cabin. The EMU includes many individual components that, when assembled, form a single space suit (Figure 13.26).

When fully assembled, the EMU becomes a complete short-term, "soft" spacecraft for one person. It provides pressure, thermal, and micrometeoroid protection, oxygen, cooling water, drinking water, food, waste collection (including carbon dioxide removal), electrical power and communications. Space suits for space shuttle missions are pressurized at 4.3 pounds per square inch (psi). On earth, the suit and all its parts weigh about 112 kilograms. In

FIGURE 13.26 Space suit (courtesy of NASA).

orbit, they have no weight at all, but do retain their mass, which astronauts feel as a resistance to a change in motion.

There are several layers of the space suit. Table 13.27 lists the shuttle space suit assembly materials and functions. Figure 13.27 shows the upper parts of the space suit.

(1) Micrometeoroid/tear protection layer – This is the outermost layer of the space suit. It comprises the outer layer of the Thermal Micrometeoroid Garment. This layer is of Ortho Fabric®, a blend of woven Nomex® and Teflon® with Kevlar® rip stops (a rip stop is a fabric design in which strong yarns or fibers are placed to prevent tear growth). It protects subsequent layers of the space suit, namely, the pressure restraint layer and the pressure bladder, from abrasions and tears.

Protection of the astronaut from micrometeoroid penetration is one of the unique functions of the space suit. The average micrometeoroid has a density of 0.5 g/cm^3 and a speed of 20 km/s.

(2) Super insulation layers (aluminized PET, such as Mylar® from DuPont and Hostaphan® from Hoechst) – The next five layers of the Thermal Micrometeoroid Garment provide the astronaut with thermal protection. The material in these layers is aluminized Mylar® film, which is reinforced with PET scrim.

The space suit protects the astronaut from extreme outside temperatures ranging from $-180°$F to $235°$F. Inside the SSA, the temperature may range from $50°$F to $110°$F.

(3) Second micrometeoroid layer (Neoprene®)

TABLE 13.27 *Shuttle EMU Space Suit Assembly Materials and Application Functions (courtesy of NASA).*

Material	Function 1	Function 2	Function 3
Ortho fabric (woven Nomex and Teflon with Kevlar rip stop)	Abrasion and tear protection for pressure restraint and bladder	Thermal control	Micrometeoroid protection as part of Thermal Micrometeoroid Garment
Aluminized Mylar film reinforced with Dacron scrim	Thermal insulation	Micrometeoroid protection as part of Thermal Micrometeoroid Garment	
Neoprene coated nylon cloth	Micrometeoroid barrier layer	Inner-liner for Micrometeoroid Garment	
Fiberglas 8 ply layup	Pressure/load structure for hard upper torso assembly	Life support and display module attachment structure and fluids and gas connections	Helmet, arm, and lower torso attachment structure
Dacron cloth	Pressure restrain/load structure for arms, gloves, and lower torso assembly	Single axis joint flexure for elbow, wrist, finger, waist, knee, and ankle mobility	
Polyurethane-coated nylon cloth	Pressure bladder for arms and lower torso assembly		
Dipped polyurethane film	Pressure bladder for gloves		
Multi-filament stretch nylon	Attachment restraint for coolant tubing and ventilation gas ducting	Attachment restraint for biomedical instrumentation and radiation dosimeters	Attachment restraint for body comfort interlining
Ethylene vinyl acetate tubing	Water coolant distribution for body thermal control		
Nylon chiffon	Body comfort lining		

—As the inner lining for the Thermal Micrometeoroid Garment, this layer is the final barrier of micrometeoroid protection. The material is Neoprene® coated nylon cloth.

(4) Pressure restraint layer (PET)—This layer is composed of synthetic fabric placed over the pressure bladder fabric to give it additional support and shape.

(5) Pressure bladder (polyurethane coated Nylar®)—Pressure bladders are much like today's tubeless tires, composed of rubber (urethane) to seal in the air pressure and nylon to restrain expansion. For the bladder in the space suit, nylon is dipped in polyurethane a minimum of six times to create an impermeable barrier between the pressure of the pure oxygen inside the space suit and the vacuum of space on the outside.

(6) Liquid cooling and ventilation garment (ethylene vinyl acetate tubing)—This garment covers the astronaut's upper and lower body. The garment itself contains ethylene vinyl acetate tubing, through which water is circulated to control the temperature.

(7) Body comfort lining (nylon chiffon)—This is the first layer of fabric touching the astronaut's skin. It is made of nylon chiffon, a lightweight material, to be as comfortable as possible. It is positioned between the astronaut and the liquid cooling and ventilation garment.

(8) Pressure restraint system—The materials used for the EMU pressure restraint system are a composite of items. Airtight linings with comfort fabrics are a significant part, but the space suit must also be flexible and retain its shape as well as inside pressure.

(9) Gloves—The restraint layer of the EMU glove is placed over the bladder (Figure 13.28). The bladder is formed when a ceramic cast of a hand is dipped seven or eight times into a vat containing urethane. Then the hand-shaped bladder is stripped from the ceramic cast and inflated to check

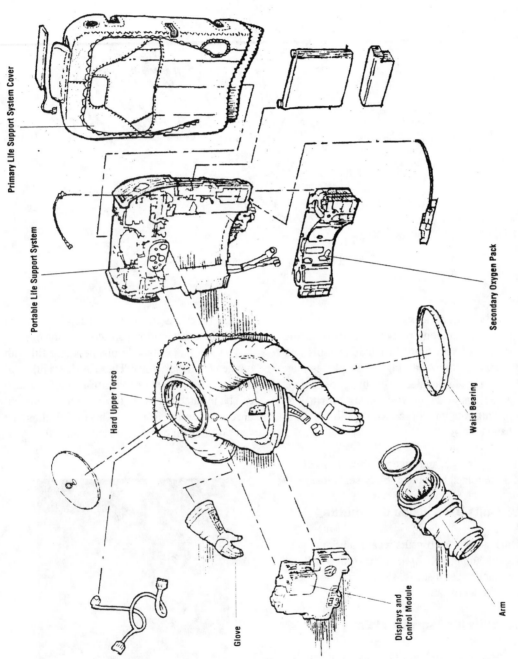

Primary Life Support System Cover

Portable Life Support System

Secondary Oxygen Pack

Hard Upper Torso

Waist Bearing

Glove

Displays and
Control Module

Arm

FIGURE 13.27 Extended view of the upper part of the EMU (courtesy of NASA).

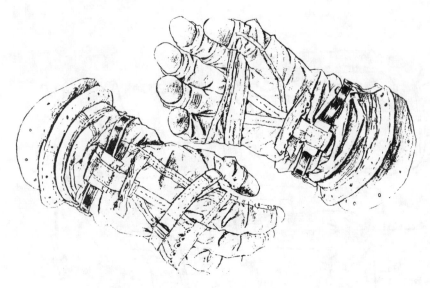

FIGURE 13.28 The EMU glove (courtesy of NASA).

for defects and leaks. The gloves have fingertips of silicone rubber that permit some degree of sensitivity in handling tools and other objects. Metal rings in the gloves snap into rings in the sleeves of the upper torso. The rings in the gloves contain bearings to permit rotation for added mobility in the hand area.

The upper torso and lower torso sections of the EMU are joined by a metal body-seal closure.

12.2 Suits Used inside the Shuttle

During orbit, astronauts wear traditional garments including FR flight suits, trousers, lined zipper jackets, knit shirts, sleep shirts, soft slippers and underwear.

12.3 Suits for Ascent and Entry

During ascent and entry of the space shuttle, each crew member wears a different clothing system than the one worn in space (in vacuum). The special equipment consists of a partial pressure suit, a parachute harness assembly and a parachute sack. The suit, which consists of a helmet, communication assembly, torso, gloves and boots, provides counterpressure and antiex-

posure functions in an emergency situation in which the crew must parachute from the orbiter.

The suit has inflatable bladders that fill with oxygen from the orbiter. These bladders fill automatically at reduced cabin pressure. They also can be inflated manually during entry to prevent the crew member from blacking out. Without the suit pressing on the abdomen and legs, the blood

FIGURE 13.29 Static-free fabric for the space shuttle's cargo bay (courtesy of NASA).

would pool in the lower part of the body and cause a person to blackout as the space craft returns from microgravity to the Earth's gravity.

12.4 Textiles for Space Station

The next generation of space suits is being designed for the space station program. The Mark III suit, a combination of soft and hard elements, is being developed at the NASA Lyndon B. Johnson Space Center (JSC) in Houston.

The AX-5, a hard, all-metal suit, is being developed by the NASA Ames Research Center (ARC) in California. Both suits are being designed to operate at a pressure of 8.3 psi.

Specially designed static-free fabrics will be used in the space station. They have been used for the space shuttle program. Figure 13.29 shows static-free woven coated fabric liner for the space shuttle's cargo bay. The liner was in service on Discovery's Hubble Space Telescope mission.

13.2

References and Review Questions

1. REFERENCES

1. Flame Resistant Protective Work Apparel, An Industry Update, Red Kap Industries, June 1993.

2. Henry, N. W., "Complying with OSHA's New PPE Standard," *Safety and Protective Fabrics*, January/February 1995.

3. Smith, W. C., "High Temperature Fibers, Fabrics, Markets – An Overview," in *High Temperature Fabrics, 78th Annual IFAI Convention*, San Francisco, October 1990.

4. Suda, J., "New Products and Niche Markets in High-Temperature Clothing," *IFAI 1993 Industrial Fabric and Equipment Exposition*, Denver, CO.

5. Graham, W., "High Temperature Fabrics from the Weavers Point of View," in *High Temperature Fabrics, 78th Annual IFAI Convention*, San Francisco, October 1990.

6. Schoppee, M. M. et al., "Protection Offered by Lightweight Clothing Materials to the Heat of a Fire," in *Performance of Protective Clothing, ASTM STP 900*, R. L. Barker and G. C. Coletta, eds., American Society for Testing and Materials, Philadelphia, 1986.

7. Shalev, I. and Barker, R. L., "Predicting the Thermal Protective Performance of Heat-Protective Fabrics from Basic Properties," in *Performance of Protective Clothing, ASTM STP 900*, R. L. Barker and G. C. Coletta, eds., American Society for Testing and Materials, Philadelphia, 1986.

8. Jackson, D., "Demand Heats Up," *Safety and Protective Fabrics*, September 1992.

9. Technical Brief, Proban®/FR-7A® Flame Resistant 100% Cotton Fabrics, Westex Inc.

10. Technical Brief, Indura® Flame Resistant Cotton Fabric, Westex Inc.

11. Hayes, J. S., Jr., Novoloid Fibers, American Kynol, Inc.

12. Shanley, L. S. and Carroll, T. C., "New Developments in Thermal Protection," *Industrial Fabric and Equipment Exposition*, October 1994, Indianapolis, IN.

13. Rappaport, J. B. and Kopf, M. K. S., "Testing Protects Workers from On-the-Job Fires," *Safety and Protective Fabrics*, May 1993.

14. Stull, J. O., "Setting the Standards," *Safety and Protective Fabrics*, September 1992.

15. Baitinger, W. F. and Konopasek, L., "Thermal Insulative Performance of Single-Layer and Multiple Layer Fabric Assemblies," in *Performance of Protective Clothing, ASTM STP 900*, R. L. Barker and G. C. Coletta, eds., American Society for Testing and Materials, Philadelphia, 1986.

16. Barker, R. L. et al., "Measuring the Protective Insulation of Fabrics in Hot Surface Contact," in *Performance of Protective Clothing: Second Symposium, ASTM STP 989*, S. Z. Mansdorf, R. Sager, and A. P. Nielsen, eds., American Society for Testing and Materials, Philadelphia, 1988.

17. Tannenbaum, H., "Hanging by a Thread," *Industrial Fabric Products Review*, September 1988.

18. Ponnwitz, J. H., "Firefighter PC Materials: Their Function, Application," *Safety and Protective Fabrics*, November 1993.

19. Veghte, J. H., "Functional Integration of Fire Fighters' Protective Clothing," in *Performance of Protective Clothing, ASTM STP 900*, R. L. Barker and G. C. Coletta, eds., American Society for Testing and Materials, Philadelphia, 1986.

20. Schroll, R. C., "Emergency-Response CPC Is Last Line of Defense," *Safety and Protective Fabrics*, May 1993.

21. Stull, J. O., "Suitable Suits," *Industrial Fabric Report Review*, September 1988.

22. Smith, W. C., "The Protective Clothing Market," *Industrial Fabric Products Review*, September 1988.

23. Goldstein, L., "Limited Use Protective Clothing," in *Protective Clothing: An Industry Perspective, 77th Annual IFAI Convention*, Orlando, October 1989.

24. Henry, W. N. III, "How Protective Is Protective Clothing?" in R.L. Barker and G. C. Coletta, eds., *Performance of Protective Clothing, ASTM STP 900*, 1986, p. 57.

25. Winter, J. E., "Protective Clothing Permeation Testing: Calculations and Presentation of Data," in R. L. Barker and G. C. Coletta, eds., *Performance of Protective Clothing, ASTM STP 900*, 1986.

26. Smith, W. C., "Thermal Resistance in Limited Use Garments," in *Protective Clothing: An Industry Perspective, 77th Annual IFAI Convention*, Orlando, October 1989.

27. Goldman, R. F., "Heat Stress in Industrial Protective Encapsulating Garments," in *Protecting Personnel at Hazardous Waste Sites*, Butterworth, Stoneham, MA, 1985.

28. Honnigford, L., "Safety and Protective Clothing, Limited-Use Sees Growth," in *Protective Clothing: An Industry Perspective, 77th Annual IFAI Convention*, Orlando, October 1989.

29. Toon, J. J., "Metal Fiber Fabrics for Mechanical, Electromagnetic and Electrostatic Protection," *Textile Technology Forum '94*, IFAI, Indianapolis, IN.

30. Phillips, C. and Bledsoe, E., "An Update in Cut-Resistance Technology," *Safety and Protective Textiles*, May 1993.

31. Rivet, E. and Blocker, R., "Factors Affecting Glove Selection," *Safety and Protective Fabrics*, September 1992.

32. Turner, R., "PPE for Europe: A Guide on How to Succeed," *Safety and Protective Fabrics*, May 1994.

33. Jagielski, K., "Market Report: Cut and Slash Protection," *Safety and Protective Fabrics*, February 1994.

34. *Federal Registrar*, Rules and Regulations, Vol. 59, No. 125, June 30, 1994.

35. Davies, J., "Conductive Clothing and Materials," in *Performance of Protective Clothing: Second Symposium, ASTM STP 989*, S. Z. Mansdorf, R. Sager, and A. P. Nielsen, eds., American Society for Testing and Materials, Philadelphia, 1988.

36. King, M. W., "Thermal Protective Performance of Single-Layer and Multiple-Layer Fabrics Exposed to Electrical Flashovers," in *Performance of Protective Clothing: Second Symposium, ASTM STP 989*, S. Z. Mansdorf, R. Sager, and A. P. Nielsen, eds., American Society for Testing and Materials, Philadelphia, 1988.

37. Baitinger, W. F., "PC Issues Are Hot in the Electrical Utility Business," *Safety and Protective Fabrics*, January/February 1995.

38. Stull, J., "Measuring Static Charge Generation on PC Materials," *Safety and Protective Fabrics*, August 1994.

39. Ravnitzky, M., "Toward Cleaner Clean Rooms," *Safety and Protective Fabrics*, March 1993.

40. Spector, R., "A Study in Cleanroom Fabrics," *Safety & Protective Fabrics*, August 1994.

41. Hardy, B., "Protection Methods Reduce Contamination," *Safety and Protective Fabrics*, November 1993.

42. Bugge, G., "Cleaning Nuclear Protective Wear," *Safety and Protective Fabrics*, IFAI, May 1994.

43. Courtney, G. L., "Protective Clothing Specification for Radiation-Contaminated Environments," *Safety and Protective Fabrics*, March 1993.

44. Walzer, E., "Dressed for Chill," *Health*, November 1990.

45. DeMaio, J., "See and Be Seen," *Safety and Protective Fabrics*, September 1992.

46. Lesley, G., "Correlating HVM Usage with Worker Accidents," *Safety and Protective Fabrics*, August 1994.

47. Textile Protection, America's Textile International, January 1995.

48. Batzke, M., "See, and Be Seen, in the Dark," *Safety and Protective Fabrics*, IFAI, May 1994.

49. Toon, J. J., "An Overview of Metal Fiber Production," *Safety and Protective Fabrics*, August 1994.

50. Spacesuit Guidebook, NASA, July 1991.

51. Spacesuit Wall Chart, NASA, January 1990.

52. Gomes, C. A., "Protected in Space," *Safety and Protective Fabrics*, September 1992.

53. "Wardrobe for Space: From Shuttles to Stations," *Safety and Protective Textiles*, March 1993.

2. REVIEW QUESTIONS

1. Summarize the application areas for safety and protective textiles. State the major requirements for each application area.

2. Define high temperature textiles. What are the major requirements for high temperature textile materials?

3. What is the significance of limiting oxygen index (LOI)?

4. What are the factors to consider when selecting a high temperature fiber for a specific application?

5. What are the two ways to achieve flame resistance in textile materials? Explain. Discuss the advantages and disadvantages of both methods.

6. Which is more critical for fire resistance: fiber material or fabric structure? Why?

7. List the major tests for flame resistance. Explain three of them.

8. Explain the structure of the fire fighter suit? What type of fiber materials are used?

9. What is the purpose of chemical protective clothing? What are the differences between flame resistant clothing and chemical protective clothing?

10. What are the types of chemical protection? Explain the textile material requirements for each.

11. How does the cut resistant clothing work? Explain the ways to achieve cut resistance.

12. What are the requirements for electromagnetic and electrostatic protection? Explain the textile materials used for each type of protection.

13. Define a clean room textile. What kinds of fabrics are used for clean rooms?

14. What are the characteristics of textile materials for radiation protection?

15. What is the most important factor for protection from cold? Explain the Toothpick Theory.

16. What is the significance of high visibility textiles? How is high visibility achieved?

17. List the application areas of metallized fabrics. Explain the major methods to produce metallized fabrics.

18. Explain the major characteristics of a space suit.

SPORTS AND RECREATION TEXTILES

Sports and Recreation Textiles

S. ADANUR

1. INTRODUCTION

Textile materials are used in virtually every sport from exercising to camping to football. High performance textile fibers and fabrics are used in uniforms, equipment and sport facilities.

Textile manufacturers today are giving themselves an edge in the increasingly competitive sports and recreation market by developing their own signature products. Creating new fabrics is usually a lengthy and expensive process. Companies put their fibers and fabrics through rigorous tests to measure strength, abrasion resistance and breathability among other things. Several companies even have environmental labs where they can alter temperature and humidity conditions while athletes work out wearing the prototype garments.

Use of textile structural composites in sporting goods is increasing. Their applications include composite roller blades, bike frames, golf clubs, tennis rackets, ski and surf equipment, etc. Sports such as golf, baseball and tennis rely heavily on composites for their essential equipment. These sports would be drastically different if composite materials were not used. Due to their high strength and durability, composites have won favor in the sporting industry. Figure 14.1 shows some of the sporting goods made of composites.

2. SPORTS UNIFORMS

Today's fast-moving and innovative technology has forced manufacturers to go to extreme measures in order to keep up with market demands and competition when it comes to ac-

tivewear for athletes and exercisers. With activewear now a booming $3.1 billion a year industry, manufacturers are subjecting fabrics to strenuous tests and "workouts" to ensure that people can enjoy sports and other recreational activities without having to worry about getting wet or chilled.

Sport garments that are used next to the skin are usually made of plain or satin woven fabrics. Warp knits made from continuous filament fibers and weft knits made of spun yarns are widely used in shirts for active sports like football (Figure 14.2). Nylon, cotton, polyester/cotton and polyester/viscose are common fibers in sport shirts. Nylon and polyester warp knit fabrics offer light weight, stretchability and drapeability. Warp knit shirts made from mixtures of acetate, nylon and polyester can be piece-dyed [1].

For very active sports, synthetic fibers are preferred because they do not retain moisture and therefore do not get heavy upon sweating like cotton does. Synthetic sports uniforms also have better dimensional stability. Synthetic fibers offer the three major requirements in today's high technology sports uniforms:

- warmth, wind resistance, moisture wicking and lightness
- comfort and feel of natural fibers
- style and a variety of colors

With the advanced technology, however, natural fibers like cotton and wool are making a comeback in high performance, outdoor activities. For example, cotton now can be made windproof, breathable and water resistant.

Spandex is a superfine polyurethane fiber that can stretch up to five times its original length and

475

FIGURE 14.1 Examples of sporting goods made of composites (courtesy of Nippon Oil Company).

FIGURE 14.2 Knit fabrics provide strength and stretch for active sports (courtesy of Auburn University Photographic Services).

recover immediately. Lycra® Spandex, an elastane fiber, is widely used in swimwear, skiwear and gymnastic uniforms. It offers close-fitting, stretching and non-restrictive properties. A blend of Spandex and other fibers such as cotton, silk, rayon, wool, nylon, Supplex®, Tactel®, polypropylene, merino wool, angora and even cashmere are used in a variety of activewear.

For heavier fabrics, such as track suits and jogging suits, nylon, polyester, acrylics, and their blends with acetate, cotton and wool are used. These fabrics may be brushed inside for warmth and are cut loosely for comfort.

Dedicated exercisers who were no longer willing to abandon fresh-air workouts in the off-season, needed clothes that would protect them from the elements but would let the heat of the sport escape. Cotton, wool and goose down tend to become saturated with sweat – hardly ideal conditions considering damp fabrics chill the body twenty-five times faster than dry ones. So wool gave way to polypropylene, which is a petroleum-based fiber that absorbs virtually no moisture. Polyester replaced goose down. Nylons that feel like cotton were unveiled in the early 1980s. Fabric manufacturers have engineered sophisticated synthetic fibers that use a combination of mechanisms to wick away wetness by using an inner hydrophobic (water-hating) layer that pushes perspiration away from the skin to an outer hydrophilic (water-loving) layer where the moisture easily evaporates.

In 1976, W. L. Gore & Associates developed an innovative fibrous material important to sports and exercise lovers – Gore-Tex®. It is a film-like membrane, or laminate, with the consistency of Saran Wrap and billions of micropores per square inch that made regular fabric both waterproof and breathable. Each micropore is smaller than a drop of water; therefore, the rain cannot penetrate. However, the micropore is bigger than a molecule of water vapor, so perspiration can penetrate the garment and evaporate outside. Gore-Tex is bonded to the inside surface of the fabric by adhesive dots, rather than continuously. Its pores will not clog up from body oils.

In 1983 a subsidiary of Thoratec Laboratories Corp. in Berkeley, California developed a fabric called Bion® II which differs from Gore-Tex® in that it is a continuous film through which perspiration diffuses to evaporate on the outside surface. Bion II can also be laminated directly to the cloth and is thin, thus making it able to produce fabrics with good feel and drapeability.

Flame and abrasion resistant textiles such as Nomex® are used to protect race car drivers in case of an accident. Hoods, undergarments, socks, suits, shoes and gloves made of high performance textiles are used for protection (Figure 14.3).

Fabrics made of 100% acrylic fibers are used in hunting. Fabrics are usually treated with rain and stain repeller. Fabric colors and patterns can vary according to the terrain. Depending on the purpose, high-visibility or low-visibility (camouflage) fabrics are used (Figures 14.4 and 14.5).

The use of textile fabrics in the upper part of training and jogging shoes is increasing at the expense of leather. The fabrics in footwear are generally made of three layers: woven nylon outside, foam in the middle and warp knit fabric inside. Textile fabrics offer many advantages over leather including consistent and more regular quality, lightweight when wet and washability for easy care. Unlike leather, fabrics dry-up soft after wetting. Fabrics can be made in various colors which is important from a fashion point of view. By using mesh fabrics, ventilation can be increased. Nonwovens, such as melt-bonded nylon, are also used as inner fabrics in footwear [1].

2.1 Winter Activewear

New fibers are being created to replace the old, traditional synthetic fibers for outdoor sports. For example, the classic cotton windbreaker has now been replaced by an all-weather garment made of high filament nylon treated with a waterproof/breathable microporous membrane. Desscente, a Japanese sportswear firm, is manufacturing top-of-the-line skiwear of Solar A, a cloth which contains particles known as zirconium carbide that absorb solar energy and convert it into thermal power. Ski fabrics have been designed which change their porosity depending on the temperature. In another garment development, ski fabrics can change their

FIGURE 14.3 Race car driver suits are made of flame resistant textiles (courtesy of DuPont).

color as a result of the intensity of sunlight. The fibers used for winter activewear are listed in Table 13.26, Chapter 13.0, Safety and Protective Textiles.

Gentec Corporation developed a laminate that incorporates an impermeable water barrier to prevent any wetness from penetrating from the exterior. Figure 14.6 shows the schematic of this principle. It provides protection against rain, snow, sleet and wind. The perspiration is transferred to the exterior, keeping the wearer dry and comfortable. Good abrasion resistance, flex-

FIGURE 14.4 High visibility apparel for hunting (courtesy of Highland Industries).

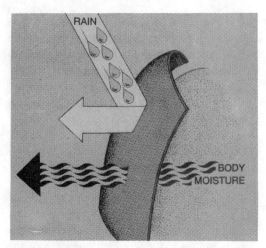

FIGURE 14.6 Schematic of the principle of a water-proof, breathable laminate fabric (courtesy of Gentec Corporation).

ibility and strength combined with the light weight and breathability provides good wet weather protection.

3. CAMPING AND HIKING

Various tent structures and materials have been developed for camping purposes. General tent structures are covered in Chapter 6.0, Architecture and Construction Textiles.

Recently, tent fabrics have been developed with insecticide built in the fabric. For example, Expel®, developed by Graniteville Co., is treated with the effective insecticide permethrin, in formulation with other active ingredients. It was reported that Expel® bug repellant fabric keeps its repellency for two years [2].

Sleeping bags are made of synthetic microdenier fibers such as Primaloft® from Albany International [3], which is a 0.5 denier polyester fiber (Chapter 13.0). A fiber is considered microfiber if it is less than one denier. Microfibers can be as small as 0.001 denier. They are effective in heat insulation and are widely used in leisurewear and sportswear (Figure 14.7). Microfibers have also been used in tents for very cold weather. Polyester and polyamide are widely used. Another major application area for microdenier fibers is fashionwear. Table 14.1 shows typical properties of sleeping bags.

FIGURE 14.5 Low visibility uniform for hunting (courtesy of Milliken & Company).

FIGURE 14.7 Sleeping bag (courtesy of IFAI).

4. BASEBALL

Baseball, though an extremely traditional sport, has not escaped the invasion of high technology. It, along with many other sports, has had the privilege of experiencing today's innovative and advanced scientific inventions. The technology in baseball involves the ever-important baseball bat.

White ash has been the overriding choice of bat manufacturers for decades because of its high impact resistance and strength-to-weight ratio. However, the decline in quality and availability of ash in northeastern forests has led the bat manufacturers to look elsewhere for more useful, quality products from which to develop bats.

This is why the aluminum bat has cut into the wooden bat market. But the aluminum bat is not the preferred bat of baseball players. They choose wooden bats over aluminum bats because they are not cold to the hand, have a more natural feel and make the sound people are used to hearing when a baseball is hit.

Engineers have worked to develop a composite baseball bat that delivers the look, feel, sound and performance of traditional hardwood bats with the durability of metal bats. Two important types of composites have been developed—the graphite composite bat and the wood composite bat. The graphite composite bat is made on automated filament winding machines. These machines precisely position strong

TABLE 14.1 Typical Properties of Sleeping Bags (courtesy of Albany International).

	Length		Elbow Girth	Knee Girth	Fill Weight		Total Weight	
	Regular	Long			Regular	Long	Regular	Long
Three season (20°F)	5'11"	6'6"	61"	44"	1 lb 11 oz	1 lb 12 oz	2 lb 13 oz	2 lb 14 oz
Winter (−5°F)	5'11"	6'6"	61"	44"	2 lb 7 oz	2 lb 9 oz	3 lb 12 oz	3 lb 15 oz
Expedition (−30°F)	5'11"	6'6"	63"	46"	3 lb 9 oz	3 lb 11 oz	4 lb 14 oz	5 lb 0 oz

graphite and glass fibers that are subsequently bonded together with epoxy resin to create a hollow bat structure.

The wood composite bat is made of a high strength inner core fabricated from resin impregnated synthetic fibers and yarns, integrated with an ashwood outer surface. The wood composite bat has performed remarkably like a traditional hardwood bat but is far less likely to break than a solid wooden bat. One of the latest versions of this bat was tested and it withstood more than 3,000 hits with no drop in performance. Some of the bats even lasted for thousands of hits.

Bat design requires extensive testing on the field and in the laboratory. An engineer must look at several different aspects of the dynamics of the bat, such as the weight distribution, the placement of center of percussion, the cosmetics, the handle size, the baseball organization rule constraints, the sound of the bat and the impact of the ball on the bat. The bats are subjected to numerous trials and repeated impacts along their lengths in order to confirm structural integrity, to find the bat's bending frequencies and to develop data regarding the behavior of balls as they rebound off the bat.

The new and exciting technology within the sport of baseball is promising to the textile industry. There are still many challenges facing the engineer in the area of the composite-style bats and they are making innovative discoveries every day.

5. TENNIS

Today, tennis rackets are going through some remarkable changes in the design of the racket. Recently, sporting goods manufacturers began combining computer aided design techniques with breakthroughs involving materials, racket shapes and control of string tension. However, the most significant of these changes is material selection. Yamaha International put the first composite fiberglass tennis racket on the market in 1974. Since then there has been tremendous growth in the use of many different textile composites in racket design. In fact, manufacturers combined fiberglass with other fibers — graphite and ceramics — in an attempt to increase the strength and durability of the rackets.

These improvements led to the use of a new resin bonded matrix of woven ceramic, graphite and boron fibers together with fiberglass and Kevlar®. The way composites are constructed allows them to be more important in the racket's performance than any one material such as wood or aluminum. Fibers crosslink with the resin when bonded together to form the racket's durable frame. The various composite materials are pressed into a resin film and then heated. As this mix warms up, the different fibers begin to crosslink making them stronger than any previous material used for this application. These rigid materials produce both the lowest weight and high strength while simultaneously reducing the vibrations of the racket.

Graphite is the current top choice material that combines both lightweight and durability increasing both power and control of the racket. The drawback to graphite is simply the cost of the material. Slazenger's first graphite racket arrived in 1976, made in a German factory that normally produced graphite composite skis. Graphite fiber rackets are up to five times stronger than aluminum ones and thirty times stronger than the traditional wooden racket.

Other space aged fibers such as Twaron®, and boron are combined with high performance carbon producing an excellent racket; again the cost is high especially for the recreational player. Boron fiber composites are six times stiffer than steel and five times stronger than aluminum.

High performance industrial textiles allow special features to be molded into the equipment rather than being added later, and this cuts the price of production while maintaining the reliability of composites. In the highly competitive tennis racket industry, manufacturers are constantly searching for better, stronger and more durable fibers to mold into a new and improved racket.

Tennis balls are made of woven and needled tennis felts. Slit film fabrics are used for wind break screening for outdoor courts. Tennis net is made of very durable polyethylene monofilament fibers.

6. FOOTBALL

Manufacturers of footballs realized that there was a problem with catching the traditional ''pigskin'' ball during cold weather. For that

FIGURE 14.8 Textile materials are used in footballs (courtesy of Johnston Industries).

reason, engineers began to make the football softer using a butyl rubber bladder which in turn would yield making it easier to grasp in the colder weather. However, this softer ball was not without its share of problems. It was much less durable than the traditional "pigskin" leather.

The design of football was revolutionized with the use of textile composites. The bladder of the football was made of a multi-layer arrangement of polyurethane. This increased the strength and durability. The lining is made of a specially engineered 3-ply, high performance spun polyester. The seams of the football are made of Kevlar® 29. The polyurethane does not deform like the conventional butyl rubber bladder and is currently the choice of the National Football League, NFL (Figure 14.8).

7. GOLF AND HOCKEY EQUIPMENT

In the 1980s, golfers discovered high performance clubs made of a graphite composite. Engineers are adjusting the amount and orientation of graphite fibers to change the dynamics of the golf club. With these improvements in the use of graphite, the flexibility has been greatly increased. Different combinations of graphite are used to manufacture clubs with specific properties for the amateur to the professional.

Some prominent clubs consist of a mixture of boron and graphite shafts with a high modulus of elasticity. Specific stiffness, a measure of the modulus of elasticity with respect to weight, is more than seven times that of the traditional component, steel. The high modulus and high strength make for a lighter, stiffer shaft. These are very desirable qualities which composites offer to improve the sport of golf.

Hockey sticks are made of textile structural composites (Figure 14.9).

8. SKATES

Looking for materials that would provide both impact strength and stiffness, manufacturers of roller skates turned to composites for the answer.

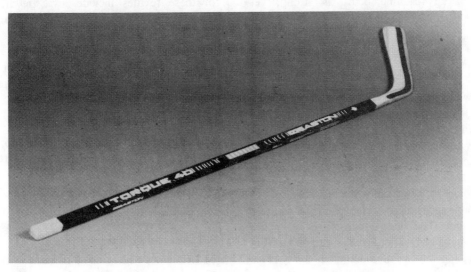

FIGURE 14.9 Composite hockey stick (courtesy of Johnston Industries).

Not wanting to use steel, manufacturers of these "in-line" skates used a glass reinforced thermoplastic resin. This resin met both the very rigid requirements for impact strength and stiffness.

The in-line skate frame experiences particularly high stresses because of the way it is designed. Unlike conventional skates, which distribute the force of the person equally among four wheels, the in-line skates have concentration of stresses on centered wheels. This is a considerable amount of stress to contend with. Rollerblade engineers originally used steel for this application. The steel provided large flexural and tensile capabilities along with the needed impact resistance. The drawback to using steel is obviously the weight. Switching to the composite thermoplastic resin reduced the weight of each skate 50%.

The composite also helped to simplify the skate's assembly. By reducing the number of bolts and washers, the manufacturers were able to produce the rollerblades faster. Since composite material can form a more complex structure than the steel, the reduction of pieces can take place.

Along with these improvements over the steel version of the skate, chemical resistance and abrasion resistance have increased. Composites also lend themselves well to acquiring bright colors that steel could not produce.

Structural cores of skateboards are made with layers of reinforcement fabrics impregnated with resins. The resulting board is strong, yet lightweight.

9. BIKES

Bicycling enthusiasts believe that bike frames should be as rigid as possible to prevent the loss of energy and unstable steering. Going completely against this theory, bicycles have been produced consisting of completely "elastic" material. The flexible frame allows for the absorption of the shocks of a mountain bike.

A cable and spring system replaces the frame's down tube. The cable connects the top of the frame and a conventional coil spring to the bottom. A separate spring made of an epoxy resin impregnated composite attaches to the frame in front of the seat. This in turn compresses to absorb the shock when the front wheel of the bicycle hits a bump. The elastic properties of the composite enable the frame to elongate and shorten, while the coil spring compresses and extends. An epoxy impregnated continuous filament glass fiber composite forms the inside of the system. The composite attached to the frame of the bicycle acts as a spring to permit frame elongation and energy storage. This energy storage means the cyclist can exert less power and receive better benefits. In fact, all the energy put into the bicycle is returned.

High performance composite materials which have both flexural strength and fatigue resistance increase the durability of bikes. These composites are now used widely in mountain bikes and racing bikes.

10. MARINE PRODUCTS

Textile materials are used in various marine products for function and fashion purposes including mooring covers, boat tops, shading, sail covers, etc. Table 14.2 lists major properties of textile marine products. Figures 14.10 and 14.11 show typical uses of textile materials in marine applications. Table 14.3 shows typical specifications of an awning marine fabric.

Industrial textiles made their way into the sailboat industry a long time ago. The requirements for these textiles include low stretch, high strength failure along with good resistance to weather, aging and chemicals. Of course, the material must be waterproof as well.

Four types of sailcloth account for more than ninety percent of the textiles manufactured for sailcloth in the United States [4].

(1) Woven polyester fabric—Has low stretch achieved with elimination of crimp; high tenacity and high initial modulus yarns are used with a plain weave

(2) Woven nylon fabric—Main properties are lightweight, limited porosity, good breaking and tear resistance using a plain weave and can be coated to control porosity

(3) Laminates with oriented polyester film

TABLE 14.2 Major Characteristics of Marine Products (courtesy of
Marine Fabricators Association, IFAI).

Product	Advantages	Disadvantages
Acrylic	Best color selection, excellent colorfastness, UV and mildew resistant	Poor abrasion resistance, subject to slight stretch, requires occasional water repellency treatment
Undercoated fabrics	UV and mildew resistant, excellent colorfastness, waterproof, good abrasion resistance	Costly, not breathable
Acrylic polyester twill	Excellent colorfastness, good abrasion resistance, lightweight	Dark under color, limited colors
Cotton	Inexpensive, good abrasion resistance	Severe shrinkage, limited colors, subject to fungus and UV degradation
Cotton/polyester blends	Less shrinkage than cotton, initially stronger than cotton	UV degradation, moderate shrinkage, limited colors
Acrylic coated polyester	High strength, good water and mildew resistance	Colors rub off, poor colorfastness
Vinyl laminated polyester	Waterproof, dual color combinations, good cleanability, good dimensional stability	Hardens and cracks over time, leakage at seams is possible, not breathable
Fabric-backed vinyl	Waterproof, high strength	Heavy, shrinks if cotton backed, hardens and cracks over time
Dacron cover cloth	Lightweight, dimensionally stable	Poor UV resistance
Vinyl coated polyester	Lightweight, waterproof, easily cleanable, excellent UV resistance	Expensive, not breathable, condensation may be a problem
Roll vinyl 20−40 mil	Inexpensive	Poor optical clarity, yellows, hazes, poor dimensional stability
Polished sheet vinyl 20−40 mil	Good optical clarity, rollable, good dimensional stability	Stiff in low temperatures, yellows, hazes, scratches
Polycarbonate sheet 25−60 mil	More rigid, excellent dimensional stability, excellent optical clarity	Should not be rolled, scratches, yellows
Polyester thread	Inexpensive, excellent strength	Poor colorfastness, good UV resistance
Teflon thread	Colorfastness, excellent UV resistance, warranted	Very expensive, poor sewability

TABLE 14.3 Typical Characteristics of an Awning Marine Fabric (courtesy of
Sunbrella, Glen Raven Mills, Inc.).

Property	Test Method	
Fiber content		100% solution dyed acrylic
Hydrostatic pressure	AATCC 127-1985	39 cm
Oil resistance	AATCC 118-1983	5
Construction	ASTM D 3775-85	75 (warp) × 35 (filling)
Weight	ASTM 3376-85	9.00 oz/sq yd
Breaking strength	ASTM D 1682-64	250 lbs (warp), 151 lbs (filling)
Tear strength	ASTM D 2262-83	13.5 lbs (warp) × 8.7 lbs (filling)

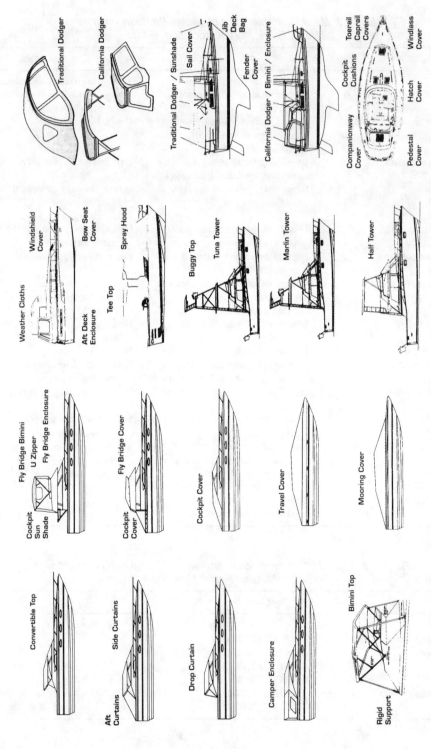

FIGURE 14.10 Schematics of typical uses of marine fabrics (courtesy of Marine Fabricators Association, IFAI).

Traditional Dodger

California Dodger

Traditional Dodger / Sunshade

Sail Cover

Jib
Deck
Bag

Fender
Cover

California Dodger / Bimini / Enclosure

Toerail
Caprail
Covers

Companionway
Cover

Cockpit
Cushions

Hatch
Cover

Windlass
Cover

Pedestal
Cover

Weather Cloths

Windshield
Cover

Bow Seat
Cover

Aft Deck
Enclosure

Tee Top

Spray Hood

Buggy Top

Tuna Tower

Marlin Tower

Half Tower

Fly Bridge Bimini

U Zipper

Fly Bridge Enclosure

Cockpit
Sun
Shade

Fly Bridge Cover

Cockpit
Cover

Cockpit Cover

Travel Cover

Mooring Cover

Convertible Top

Side Curtains

Aft
Curtains

Drop Curtain

Camper Enclosure

Bimini Top

Rigid
Support

FIGURE 14.11 Application of textiles in boats (courtesy of Marchem Coated Fabrics, Inc.).

(4) Exotics such as carbon fiber and layered composites – The common characteristic of these fibers is the relatively high cost.

Spectra®, a high performance polyethylene fiber from AlliedSignal, is widely used in sailcloth due to its high strength and low creep under constant use. It is also suitable in rope and cordage uses. Spectra® has good abrasion and fatigue resistance. Carbon fibers are used in high performance sailboats such as the Stars and Stripes which won the America's Cup in 1988. Kevlar is also applied to sailcloth for high performance quality for yacht sails.

Coated fabrics are used in inflatable boats and other marine applications (Figure 14.12).

Diving Suit

A typical scuba diving suit is made of laminated elastic products. The composite is made up of two stretch fabrics each adhesively laminated to both sides of a urethane film. The fabrics are composed of nylon and spandex and

FIGURE 14.12 Inflatable boats for leisure purposes (courtesy of Reeves International).

Dense mat of textured pile → fibres.

Base fabric to which pile → fibres are attached.

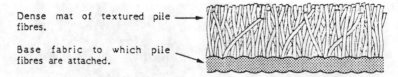

FIGURE 14.13 Cross section of artificial grass fabrics [5].

the adhesive utilizes a dot matrix to allow for as much elasticity as possible. The product is waterproof and breathable since it passes moisture vapor.

11. TEXTILES IN SPORTS SURFACES

Artificial sports surfaces were developed to replace natural grass which is expensive to establish and maintain. Compared to natural grass, textiles offer the advantages of uniformity, all-weather playing surfaces and constant performance with little deterioration [5].

Basically two types of textile products have been developed to replace natural grass: textile pile ''carpet'' (artificial grass) and felt-like textile structures. Pile carpet-type surface was developed first. In the early 1960s, Monsanto produced AstroTurf artificial playing surface. The pile fibers, which are tape-like fibers made of nylon 6,6 for wear resistance, are knitted into a tough backing fabric as shown in Figure 14.13. The base fabric, which is generally made of polyester tire cord yarns with high strength, is laid on a foam underpad that provides cushioning and provides safety against hard falls. The pile fibers are crimped to improve resiliency and increase density for uniform and consistent bouncing of the ball. The whole assembly of carpet and cushion is placed on a firm foundation

such as concrete. If used outdoors, the foundation should provide water drainage. Artificial playing surfaces can be permanent or roll-unroll type to accommodate different sports.

Artificial fabrics with felt-like surfaces consist of a dense nonwoven top layer and a shock absorbing base as shown in Figure 14.14. The felt is usually made of densely needlepunched synthetic fibers such as polypropylene. The fibers are fusion-welded to prevent fiber loss. In some applications, dense acrylic fibers are needled into a woven polypropylene base fabric which is laid on a polyester or polyamide fiber needled polypropylene base fabric.

Sometimes, soil under the natural grass is reinforced by a synthetic fiber web. This can be done in two ways: (1) natural grass turf is rolled back and a fibrous web is placed under it (2) the web is laid under the soil and the soil is seeded. The fibrous web is flexible, rot-resistant and porous which allows grass roots to develop within and through the web. The fibrous web stabilizes the soil, prevents grass from being uprooted during play, reduces surface damage and improves natural grass growth and turf recovery.

The major requirements of artificial playing surfaces are durability, wear resistance, resistance to damage, stain resistance and cleanability. For outside use, the fibers must be resistant to photochemical degradation, and biodegradation (resistance to fungi and insects).

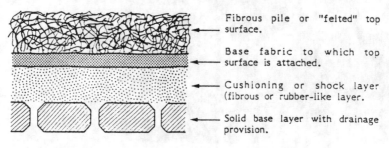

Fibrous pile or "felted" top ← surface.

Base fabric to which top ← surface is attached.

Cushioning or shock layer ← (fibrous or rubber-like layer.

Solid base layer with drainage ← provision.

FIGURE 14.14 Cross section of felt-like artificial surface [5].

Low moisture regain is also needed for outdoor applications. Polypropylene, polyamide, polyester and acrylic fibers are widely used in artificial playing structures. To have colorfastness on the top surface, colorants are added into molten polymer before extrusion of the fibers. These colorants are inert and unaffected by light and various weather conditions. Although green is predominant, artificial fiber surface can be made in any color. Artificial surfaces are suitable for painting to mark the field lines. Artificial fiber surfaces also allow relatively easy replacement of worn out or damaged parts. The damaged section is cut out and a new piece is seamed or bonded in.

When designing an artificial sports surface, the characteristics of that particular sport should be considered very carefully. For example, the artificial surface should cause the bouncing of a ball similar to the natural surface, otherwise the character of that particular sport would be altered. Surface must be uniform for even bouncing, and comfortable for the players. Especially important, the artificial fibers should not cause damage to player's skin upon contact during sliding.

Polymeric all-weather surfaces are used on track fields. These surfaces are around 10 mm thick and are made of solid PVC, polyurethane and other polymers with a non-slip surface. They are bonded to a concrete or asphalt substrate as shown in Figure 14.15.

Field Covers

Textile field covers are used to protect the playing surface from rain, snow, blistering sunshine and freezing winds (Figure 14.16). Professional football and baseball teams are required by league rules to have one [7].

Fabric weight is critical for field covers. The cover must be light enough to move around but heavy enough to withstand severe weather conditions. Baseball covers can weigh as much as 2,500 lbs. Polyvinyl chloride is widely used as a coating and laminate. PVC provides waterproofness, mildew resistance and protects the fabric from the sun's unltraviolet rays. Plasticizers in the resin protects the fabric from cracking at $-40°F$. Both laminated and coated fabrics are used as field covers. Polyester and nylon are the most common materials as supporting fabrics in field covers. Polyester offers good dimensional stability at different temperatures and good UV resistance. Nylon is considered to be more suitable for field covers than the polyester but it is prone to shrinkage.

12. HOT-AIR BALLOONING

Rip-stop woven nylon fabrics in the range of $30-60$ g/m^2 weight are used for hot-air ballooning. Rip-stop design is suitable to prevent the

tough-resilient tufted nylon fibers

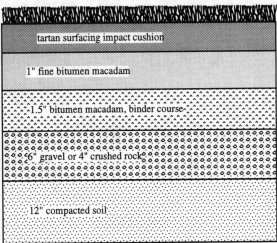

tartan surfacing impact cushion

1" fine bitumen macadam

1.5" bitumen macadam, binder course

6" gravel or 4" crushed rock

12" compacted soil

FIGURE 14.15 Typical cross section of a tartan turf [6].

FIGURE 14.16 Field cover used to protect the playing surface in a stadium (courtesy of Cooley, Inc.).

FIGURE 14.17 Hot-air balloon (courtesy of Queen Light Electronic Ind., Ltd.).

FIGURE 14.18 The Crocky-Woc water slide (courtesy of Airspace USA, Inc.).

FIGURE 14.19 Educational Softplay materials (courtesy of Airspace USA, Inc.).

FIGURE 14.20 "Super" Superman (courtesy of IFAI).

FIGURE 14.21 Application examples of Cordura® nylon (courtesy of DuPont).

propagation of fabric tear. The fabric is coated with a thin layer of polyurethane to reduce porosity. A typical balloon is made of around 100 panels, totaling 1,500 m² of fabric (Figure 14.17). The lift of the balloon is provided by heating the air inside the balloon with propane burners.

13. FABRICS FOR CHILDREN'S FUN AND PLAY

Coated and laminated fabrics are used in a variety of applications for children's fun and play. Figures 14.18, 14.19 and 14.20 show some examples.

14. OTHER FIBERS AND FABRICS

There are several other fabrics for leisure products produced by various companies. For example, DuPont developed the Cordura® family of fibers in the mid-1970s which are suitable for a variety of applications. Cordura® is an air-textured, high tenacity nylon fiber. It can be blended with Supplex® nylon, Taslan® air-jet textured yarn, Lycra® spandex or Antron® nylon which results in high durability and/or quiet movement. Applications of Cordura® blends include apparel, backpacks, boots, camera bags, golf bags, handbags, horse blankets, hunting apparel, hunting gear, indoor/outdoor furniture, soft-sided luggage, protective covers, shoes, ski covers, boot bags, sport-bags, totes, duffles, upholstery, video and computer bags and wallets (Figure 14.21). Recently, DuPont introduced Cordura® Plus for the leisure fabric market. It increased the performance for soft luggage, backpacks, camera bags and other leisure applications.

References and Review Questions

1. REFERENCES

1. Hill, R., "Fibers and Fabrics in Sports," *Textiles*, 1985, Vol. 14, No. 2.
2. Introducing Expel, Information Kit, Graniteville Company.
3. Albany International, "Primaloft," 1992.
4. Doyle, B., "Low-Stretch High-Strength Fabrics for Sailmaking and Other Applications," *IFAI Second International High Performance Fabrics Conference*, Boston, MA, November 1992.
5. Cumberbirch, R. J. E., "Textiles in Artificial Sports Surfaces," T1985, Vol. 14, No. 1.
6. Goldberg, I. J., "Synthetic Sports Surfaces," *Textiles*, Vol. 3, No. 2, June 1974.
7. "The Big Ballfield Cover-Up," *Industrial Fabric Products Review*, November 1986.

1.1 General References

Ashley, S. "High Tech up at Bat," *Popular Science*, May 1992, Vol. 240.
Ashley, S. "Wood–Composite Baseball Bats Take the Field," *Mechanical Engineering*, August 1992, Vol. 113.
Bradford, C. "Action Softwear," *Health*, September 1988, Vol. 20.
Kindel, S. "Dry and Cool," *Forbes*, August 1989, Vol. 134.
Murray, C., "New Materials Take the Field," *Design News*, November 19, 1990, Vol. 46.
Sparrow, D., "The Racket Revolution," *Health*, June 1989, Vol. 21.
Walzer, E., "Dressed for Chill," *Health*, November 1990, Vol. 22.
Wenddland, M., "Tennis Rackets Enter the Space Age," *High Tech.*, April 1986, Vol. 6.

2. REVIEW QUESTIONS

1. Summarize the applications of textiles in sports and recreation.
2. What are the advantages of textile structural composites in sporting equipment?
3. What are the main fibers and fabrics used in marine products?
4. What is the most important factor to consider when replacing a natural sports field with textiles?

TRANSPORTATION TEXTILES

Transportation Textiles

S. ADANUR

1. INTRODUCTION

Industrial textiles are widely used in transportation vehicles and systems including cars, trains, buses, airplanes and marine vehicles. Transport conveyors (Chapter 16.0) reinforced with textile substrates are used for passenger transportation at airports and shopping centers.

Approximately 50 square yards of textile materials is used in an average car for interior trim (seating areas, headliners, side panels, carpets and trunk), reinforcement, lining, underlay fabrics, tires, filters, belts, hoses, airbags, sound dampening and insulation [1]. In 1991, the U.S. automobile industry consumed more than 500 million square yards (418 million square meters) of fabric [2]. Body cloth, headliner and carpet were the main applications. For these applications, the United States consumes approximately 200 million square yards (167 million square meters) of nonwovens. Figure 15.1 shows major textile applications in automotives. Table 15.1 lists the fiber type and manufacturing methods of major nonwoven products used in automotives.

2. TIRES

The tire industry consumes approximately 75% of the textile fibers used in the rubber industry (Figure 15.2). In 1993, 165 million tons of fibers (55% polyester, 43% nylon, 2% rayon) were used in North America for tire reinforcement. Worldwide, the figure is 744 million tons (57% nylon, 24% polyester, and 19% rayon) [1].

Woven fabric made of Irish flax was used in the first tire which was produced by J. B. Dunlop in the late 1800s. Cotton soon replaced the flax and remained the main reinforcement material until World War II when viscose rayon and nylon were introduced. Later, polyester, fiberglass and more recently aramid and steel fibers were introduced as reinforcement materials. The important physical properties of textile cord reinforcement materials for tires are strength, low elongation, low moisture regain, thermal stability, fatigue strength, flexibility, and adhesion to rubber.

There are three major types of tire designs as shown in Figure 15.3:

(1) Diagonal (cross-ply) tire in which the cord yarns make a biased angle with the tire rolling direction. The layers of cords (plies) run diagonally from bead to bead at opposite bias directions. The cords need to be evenly tensioned to prevent tire growth.

(2) Belted bias tire (semiradial tire). These tires are similar to a diagonal design but with the addition of reinforcement plies (belts) beneath the tread. The belts extend across the crown area of the tire, have lower angle than the carcass plies, with opposed bias directions to the casing plies and to each other. Polyester cords and glass belts were introduced in semiradial tires. The belts increase the stability of the tread area. Semiradial tires were manufactured mainly in the United States. This construction was obsolete by 1990.

(3) Radial design also uses belts, but the carcass cords under the belts are arranged perpendicular (90°) to the tire rolling direction. The design provides a very flexible sidewall which contributes to long life.

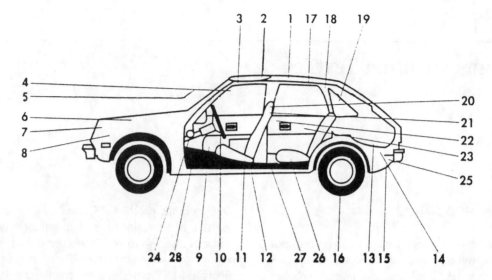

FIGURE 15.1 Application of textiles in cars [3]: 1. saloon roof, 2. sunroof, 3. backing material for interior roof, 4. padding for sunvisors, 5. covering material for sunvisors, 6. carburetor filters, 7. battery separators, 8. belts and hoses, 9. door trim, 10. airbags, 11. covering for seat belt anchorage, 12. seat belt, 13. boot trunk liners, 14. boot floor covering, 15. silencer (muffler) wraps, 16. tire reinforcement, 17. inside roof lining, 18. bodywork parts, 19. window frames, 20. covering for molded seats, 21. upholstery, 22. insulation and sound-proofing, 23. decorative fabric, 24. filters, 25. molded fuel tanks, 26. polyurethane coated backing, 27. carpeting, 28. backing for tufted carpeting.

Other advantages are better ride at high speed, better cornering, improved wear resistance and road-holding ability. Glass fiber, aramid and steel are used in the belt, but steel became the dominant material by 1990. Radial tires were developed after diagonal and belted bias tires and have taken over the passenger and truck tire markets.

For example, in 1993, radial passenger tire production totaled 193 million units in the United States and other constructions were 5 million units. Light truck tire production for 1993 was 24 million radials and 4 million other constructions. Medium and heavy duty radial truck tires are mainly steel carcass and belt. The 1993 production

TABLE 15.1 *Fiber Product and Formation/Bonding Methods for Automotive Applications (Chrysler Motor Corp. [2]).*

Application	Fiber Type	Manufacturing Method
Transmission oil filter	Polyester	Needled, bonded
Dash insulator	Reprocessed	Dry-laid, bonded
Seat cover slit sheet	Nylon	Spunbonded
Seat foam reinforcement	Polypropylene	Needled
Headlining substrate	Glass	Melt blown, air-laid
Shelf panel cover	Polypropylene	Needled
Landau vinyl backing	Polyester	Needled
Trunk liners, floor covering	Polyester	Needled
Door trim panel padding	Polyester	Dry-laid, bonded
Door trim panel carpet	Polypropylene	Needled
Floor carpet underpad	Reprocessed	Needled
Hood panel insulator	Glass	Melt blown, air-laid
Carpet tufting fabric	Polyester	Spunbonded

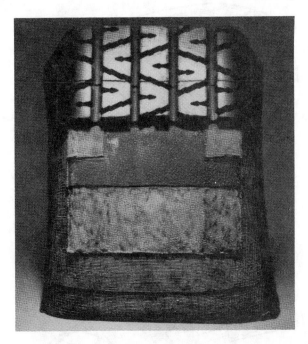

FIGURE 15.2 Textile cord fabrics in a tire (courtesy of Akzo).

of these radials was 10 million units compared to less than 2 million units of bias construction.

Table 15.2 lists the angles between the cord yarns and tire circumferential centerline for all three designs. In all tire designs, carcass plies made of textile materials or steel give the tire its strength. Polyester is the predominant yarn used in the United States today as the carcass reinforcement for radial passenger cars and light trucks. A small percentage of rayon is also used.

2.1 Manufacturing

Figure 15.4 shows the schematic of the major steps in tire manufacturing. Carcass plies made

of textile materials are woven cord fabrics made of cords and light filling yarns.

A cord is produced by plying several twisted filament yarns together. The twist direction of each constituent yarn and the cord are opposite. The three major systems used to form tire cord are ring twisting, direct cabling and two-for-one twisting. The twist level is usually high (10 twists per inch). The cord yarns are then woven as the warp of the cord fabrics with a low filling count of one to two picks per inch. "Combs" or "thread bars" are used to aid in warp spacing (ends per inch) uniformity. Usually rapier, air-jet and shuttle looms are used for cord fabric weaving. Filling yarns are very fine compared to warp cords. The major function of the filling yarn is to maintain the warp spacing during han-

TABLE 15.2 Typical Cord Angles in Tire Designs [4].

Tire Condition	Cross-Ply		Belted Bias		Radial	
	Casing	Belt	Casing	Belt	Casing	Belt
Raw	54°–63°	–	58°–61°	52°	80°–90°	16°–21°
Vulcanized	29°–36°	–	32°	26°	75°–90°	12°–18°

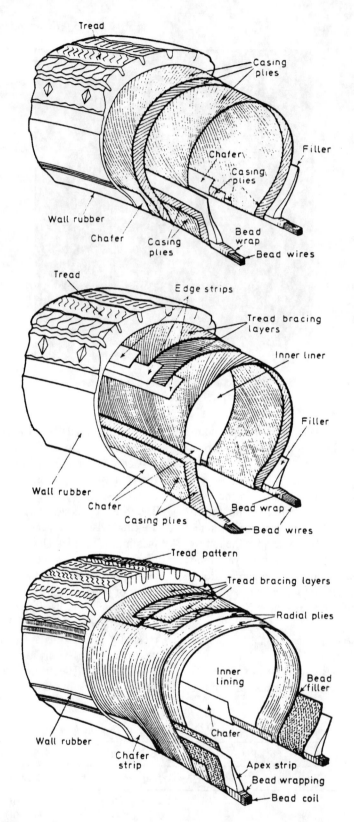

FIGURE 15.3 Major tire designs. Top: diagonal tire (four-ply construction), tubed. Middle: belted bias tire (two-ply construction), tubeless. Bottom: radial tire (two-ply construction), tubeless [4].

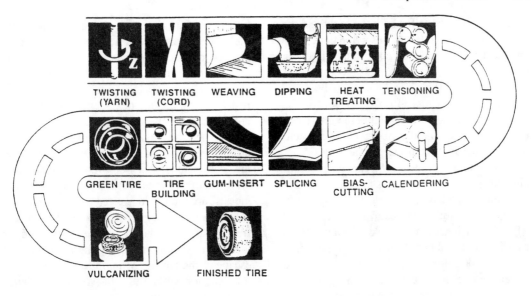

TWISTING (YARN)   TWISTING (CORD)   WEAVING   DIPPING   HEAT TREATING   TENSIONING

GREEN TIRE   TIRE BUILDING   GUM-INSERT   SPLICING   BIAS-CUTTING   CALENDERING

VULCANIZING   FINISHED TIRE

FIGURE 15.4 Tire production process (courtesy of Crain Communications, Inc.).

dling of the fabric. The filling yarns do not contribute to the performance of the ply and therefore the tire. Relatively fine filling yarns are used to avoid misalignment of the warp cords during treating and calendering with rubber. The cord fabric is treated with adhesives and calender coated with rubber to a thickness of approximately 1 mm [4]. Treating the cord also heat sets for physical properties. Once the warp yarns are firmly placed in the rubber compound, the need for filling yarns ceases, and in subsequent stages of tire manufacturing the cord spacing changes as the tire is shaped. In fact, the very existence of filling yarns may adversely affect the uniform distribution of the tire forces and cord geometry. To avoid this, either the filling yarns are broken, or highly extensible filling

yarns are used. Radial tire designs in particular need the elimination of the filling influence. Extensible filling yarns consist of a core (undrawn nylon or polyester filament with up to 200% elongation) and a cotton sheath. The superextensible core ensures a uniform partition of warp ends as the tire is shaped, and the sheath provides sufficient rigidity to enable the fabric to be processed through treating and calendering, but then breaks during shaping of the tire. Steel cords and some treated textile warp cords are fed directly from a creel through the rubber calender, thus eliminating the need for filling yarns. Table 15.3 shows the properties of polyester tire yarns.

Rayon or polyester cords and steel belts are extensively used in radial passenger tires. Vis-

TABLE 15.3 Typical Physical Properties of High Tenacity Polyester Tire Yarns (courtesy of Hoechst Celanese).

Denier	Filament Count	Tenacity (g/d)	Breaking Load (N)	Elongation (%) At Break	Elongation (%) At 45 N	Initial Modulus (g/d)	% Hot-Air Shrinkage at 177°C
1,000	192	8.9	88.1	12.6	5.3	115	13.4
1,000	200	7.9	77.3	8.5	4.7	115	3.9
1,000	328	8.0	78.7	9.8	5.0	115	9.5
1,300	192	8.9	114.7	12.5	4.4	115	13.4
1,300	200	7.9	101.0	9.0	4.0	115	3.9
1,500	279	8.9	132.2	13.0	4.1	115	12.8
1,500	300	7.4	109.0	10.5	3.6	115	3.9

cose rayon is suitable for both the body and the belt. In bias car and truck tires nylon cord is the choice of material. Nylon cord is used in radial truck tires with a steel cord belt. Polyester and glass cords are traditionally used in bias and semiradial car tires, with polyester cord in the body, and glass cord in the belt of the tire. However, by the late 1980s, fiberglass had essentially disappeared from the tire industry.

Various adhesive and heat treatments are applied to synthetic cord fabrics to optimize adhesion and performance. The adhesive is typically an aqueous solution of resin and latex called RFL. Steel wire is brass coated which increases its adhesion to rubber. Thermoplastic cord fabrics made of nylon and polyester are subjected to heat setting under controlled time, load and temperature to improve properties and dimensional stability.

Fatigue resistance of cord fabric materials is critical in tire applications. Nylon fibers have the highest resistance to dynamic stresses followed by polyester and viscose rayon [5].

The use of steel, polyester and aramid fibers in tires is increasing at the expense of nylon and rayon, particularly in passenger tires. Compared to steel belts, the disadvantage of textile belts are poorer self centering (more difficult steering), shorter life due to higher internal abrasion, and lower uniformity due to greater tire growth.

In the 1980s, the high performance, high speed radial tire was developed for high speed sports cars and performance luxury touring cars. They utilize a shoulder guard or cap ply of cord fabric wound over the shoulders of the steel belts or over the entire belt. Most frequently these are nylon cord fabric, but other fibers including aramid have been used. The cords are placed in the rolling direction of the tire, and act as a compressing force to resist the centrifugal force of the steel belt as the tire performs at high speed and high rpm.

Some smaller volume, but important textile contributions are made to tire building and tire performance in the bead area.

2.2 Chafer

Square woven (but bias cut and applied) chafer fabrics are built into the surface of the bead area of the tire where it makes contact with the wheel rim. The fabric aids in maintaining the desired shape of the bead area as the tire is cured, and protects the tire during mounting and dismounting on the rim. The chafer also resists abrasion and degradation of the bead area during the life of the tire.

Historically cotton and rayon were the fibers used in chafers, but in the 1970s nylon became the dominant fiber. Both monofilament and multi-filament yarns are used. In all cases the fabrics are adhesive treated (RFL) both for adhesion to the rubber and to impregnate the fiber bundles (yarns) so that the pressurized air in the tire cannot wick along the filaments and escape to the outside or into the sidewall of the tire. The former causes a flat tire and the latter leads to tire failure.

2.3 Bead Wrap

The bead wire bundles are generally wrapped with treated nylon cords, or woven or warp knitted nylon fabrics to maintain the integrity of the wire bundle during cure. These fabrics are treated for adhesion and tack to uncured rubber, and are slit to width by the textile supplier for direct application at the tire building machine.

Standard test methods for tire cords and tire cord fabrics are given in ASTM D 885, Standard Methods of Testing Tire Cords, Tire Cord Fabrics, and Industrial Filament Yarns Made from Man-Made Organic Base Fabrics (standard units) and in ASTM D 885 M (metric units).

3. AIRBAGS

An airbag is an automatic safety restraint system that has gained significant importance within the last decade. Airbags are built into the steering wheel and instrument panel. They are not alternative but supplemental to seat belts because airbags provide protection only against head-on collisions while seat belts provide protection regardless of the crash direction (Figure 15.5).

Airbags are made to inflate from collisions that occur within a 60% arc in front of the vehicle, and with collisons equivalent to about 12

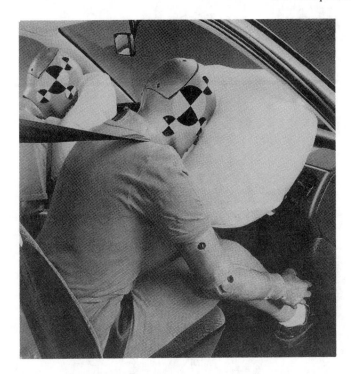

FIGURE 15.5 Seat belts and airbags (courtesy of Sulzer Ruti).

miles per hour into a fixed wall. Upon crash, sensors set off an igniter in the center of the airbag inflator. Sodium azide pellets in the inflator ignite and release gases that primarily consist of nitrogen. The gas then passes through a filter, which removes ash or any particles, into the bag, causing it to inflate (Figure 15.6).

Figure 15.7 shows the operation sequence of an airbag. Since almost all collisions occur within 0.125 second, the airbag is designed to inflate in less than 0.04 second or 40 milliseconds (ms). In a collision, the airbag begins to fill within 0.03 second. By 0.06 second, the airbag is fully inflated and cushions the occupant

from impact. The airbag then deflates 0.12 second after absorbing the forward force. The entire event, from initial impact to full deployment, takes about 55 milliseconds—about half the time to blink an eye [6].

Data show that more than half of all severe injuries and deaths in automotive accidents are the result of frontal collisions. The proven effectiveness to prevent death and decrease certain types of injuries in accidents has made airbags almost a standard item in vehicles by legislative mandates. A recent study by the Insurance Institute showed that airbags are much more effective than seat belts alone for safety. Compared

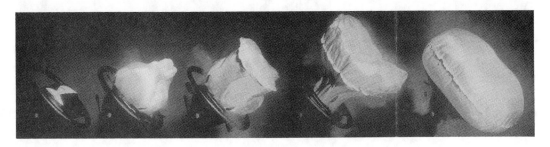

FIGURE 15.6 Deployment of an airbag (courtesy of Sulzer Ruti).

FIGURE 15.7 Operation sequence of an airbag (courtesy of Sulzer Ruti).

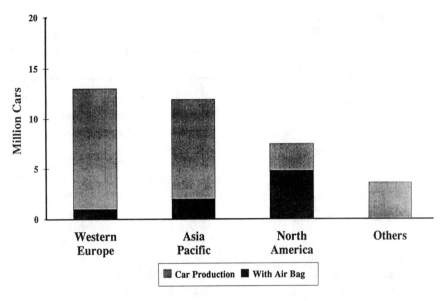

FIGURE 15.8 The 1992 world car production with airbags (*Automotive News* [8]).

to seat belts alone, airbags have reduced death by 28%, serious injury by 29% and hospitalization by 24%.

The first airbags were introduced in the early 1970s by General Motors but were not readily accepted by the consumer. The market for airbags was assured in the United States when the Department of Transportation (DOT) implemented the Federal Motor Vehicle Safety Standard (FMVSS) 208 in 1984 [7]. Because of this law, the United States leads the commercialization of the airbag as shown in Figure 15.8.

3.1 Materials and Properties of Airbags

The main requirements in airbag fiber materials are high strength, heat stability, good aging characteristics, energy absorption, coating adhesion and functionality at extreme hot and cold conditions. At present, coating is used only for driving side airbags. Tear strength, toughness and fog resistance of bag materials are important. Packageability and strength are critical for bag performance. Softness for reduced skin abrasion is desired.

The most widely used yarns in the airbag market are 315, 420, 630 and 840 denier Nylon 6,6 yarn. Nylon 6 is also used in a small percentage of U.S. made airbags. DuPont, Allied-Signal, Akzo and Toray are the major fiber suppliers of airbags. Table 15.4 shows physical properties of Nylon 6,6 airbag fibers. PET, which has good dimensional stability even at humid environmental conditions and good compaction, is beginning to be used in certain applications as airbag material.

Driver and passenger side bags differ slightly in their properties as shown in Table 15.5. Passenger side fabrics have different properties and are uncoated because the gas that inflates the airbag is dispersed over a larger area. Fabrics must be free of knots, splices, spots and broken ends.

3.2 Manufacturing

Airbags are made of woven fabrics. In the United States, airbag fabrics are manufactured

TABLE 15.4 Typical Properties of Nylon 6,6 Airbag Yarns (courtesy of AlliedSignal).

Denier	420	840
Filaments	68	140
Tenacity (g/d)	7.9	8.4
Elongation (%)	21	21
Free shrinkage (% at 177°C)	6.1	6.5
Melting temperature (°C)	256	256

TABLE 15.5 Typical Characteristics of Driver
and Passenger Side Fabrics Used
in Airbags (courtesy of AlliedSignal).

Driver Side	Passenger Side
25 × 25 plain weave 840 denier Nylon 6,6 Scoured, heat set, coated	25 × 25 rip stop 840 denier Nylon 6,6 Scoured, heat set
46 × 46 plain weave 420 denier Nylon 6,6 Scoured, heat set, coated	41 × 41 plain weave 630 denier Nylon 6,6 Scoured, heat set
	49 × 49 plain weave 420 denier Nylon 6,6 Scoured, heat set

by Milliken, JPS Automotive, Takata/Highland Industries, Clark-Schwebel, and Stern & Stern. Reeves International of Italy is a major manufacturer in the world. Figure 15.9 shows a typical dense woven fabric for airbags.

After weaving, the driver side airbag fabric is coated with black neoprene rubber or silicon rubber. Major requirements for coating are good adhesion, antiblocking, long term flexibility, resistance to cyclic temperature change (from −40°F to 250°F), ozone resistance, long term stability, low air permeability and low cost. Table 15.6 shows typical fabric properties of an 840 denier nylon airbag fabric. Coating companies in the United States include Reeves, Takata/Highland, Milliken, Alpha and Bradford.

Cutting and sewing of airbag fabrics demand careful attention. Dimensional tolerances are very small. Sewing thread fiber type, weight, construction and coating need to be selected properly. Nylon 6,6, Polyester and Kevlar® aramid fibers are used for sewing threads. Sewing patterns and stitch types are critical to the performance of the airbags.

An airbag module consists of airbag, inflator, mounting hardware and molded cover. Crash sensors and diagnostics are used as part of the system. Major module manufacturers are located in the United States (American Bag Co. division of Milliken, Reeves, AlliedSignal Safety Restraints, Takata/Highland Industries, Morton International, TRW Inc., GM-Inland Fisher Guide, Breed), Europe (Electrolux Autoliv, TWR Repa) and Japan (Takata Industries). The module manufacturing companies are responsible for designing, testing, constructing, certifying and supplying the car man-

FIGURE 15.9 A dense woven airbag fabric (courtesy of Sulzer Ruti).

TABLE 15.6 *Typical Properties of 840 Denier,
Nylon 6,6 Airbag Fabric (courtesy
of AlliedSignal).*

	Uncoated	Coated
Fabric construction		
(epi × ppi)	25 × 25	25 × 25
Thickness (inch)	0.013	0.107
Weight (oz/yd²)	5.69	8.30
Tensile strength (lbs)		
Warp	533	450
Filling	549	478
Elongation (%)		
Warp	33.6	28
Filling	35.3	38
Tongue tear (lb)		
Warp	199.4	85
Filling	192.5	83

ufacturers with a total unit ready to install in the car. It should be noted that each airbag module contains non-textile parts as well.

Inflators are primarily made with sodium azide which upon combustion produces hot nitrogen gas. The face of the bag moves toward the occupant at speeds near 100 mph and full inflation typically occurs in about 40 milliseconds, depending on design and size of the airbag, at about 5 to 10 psi. It is important that the bag be designed to begin deflation just prior to contact so the occupant "rides down" the bag for maximum safety.

As new types of pyrotechnic or other inflator systems are developed, fabric of lighter weight and lower bulk becomes more important.

The major considerations for airbag assemblies are reliable crash protection, reduced space requirements, reduced unit weight and low cost.

3.3 Testing

One of the unique characteristics of the airbag market is the potential legal liability that may lead to lawsuits for failures or injuries incurred. As a result, testing requirements are very demanding: over fifty tests are performed on airbags for physical, chemical, environmental properties and grading. Since there are only a few established industry standards, any change in manufacturing or materials will require performance validation which may exceed $50,000.

Full certification by an automaker for a complete airbag module may cost $100,000. Several organizations such as The American Society for Testing and Materials (ASTM), The Society of Automotive Engineers (SAE), and the Automotive Occupant Restraints Council (AORC) are working on standardizing the various specifications and test methods. ASTM has developed the following practices for inflatable restraints:

- D 5427-93a Accelerated Aging of Inflatable Restraint Fabrics
- D 5428-93a Performance of Inflatable Restraint Modules (Evaluating)
- D 5446-94 Physical Properties of Fabrics in Inflatable Restraints, Determining
- D 5426-93 Visual Inspection and Grading Fabrics Used for Inflatable Restraints

The two standards developed by the SAE are SAE J1538 APR88 (Glossary of Automotive Inflatable Restraint Terms) and SAE J1856 MAY89 (Identification of Automotive Airbags). Table 15.7 gives the list of typical inspection parameters for airbags. Traceability from the completed unit back to the raw materials is very important in airbag manufacturing.

3.4 Airbag Market and Future Trends

The market growth of airbag market chain (from fiber to module, ready to install) has been phenomenal in recent years—from $800 million in 1990 to an expected $6.6 billion by the year 2000 worldwide. The 75 million linear yards of fabric (60-inch wide) needed by the year 2000 will have a market value of about $500 million. Typically, it takes about 1.7 square yards of fabric to make a driver side airbag and about 3.7 square yards to make a passenger side airbag. Almost the same amount of fabric is used for light trucks and van airbags [6,8]. Chrysler, Ford and General Motors (GM) have driver side airbags on all of their domestic cars. Airbag installation is also increasing in Europe and the Far East.

Polyester fibers will most probably be used in airbag manufacturing in the future. Fabrics may have higher cover factors (low porosity) and

TABLE 15.7 Inspection Parameters for Airbag Fabrics (courtesy of AlliedSignal).

Parameter	Uncoated Fabric	Coated Fabric
Abrasion (chafe mark)	X	X
Baggy (weave) cloth	X	X
Baggy (slack, loopy) selvage	X	X
Bias (skewed) fabric	X	X
Blotch (oil spot)	X	X
Bow	X	X
Broken (missing) end	X	X
Broken (cut) pick	X	X
Coarse (heavy) end		X
Coarse pick		X
Crease (hard wrinkle)	X	X
Curled selvage	X	X
Cut selvage		X
Double pick (mispick)		X
Drag-in (jerk-in)	X	X
End-out (missing end)	X	X
Excessive talk		X
Fine end	X	X
Fine filling	X	X
Float	X	X
Foreign material	X	X
Fuzz balls		X
Harness skip	X	X
Holes (not pinholes)	X	X
Incomplete coating		X
Kink (loop)		X
Knot	X	X
Lump		X
Mixed end	X	X
Mixed filling	X	X
Orange peel		X
Pick-out mark	X	X
Pills		X
Pinholes (pinheads)		X
Scalloped edge (misclip)	X	X
Seam mark		X
Selvage mark		X
Slough off		X
Slub (cockle, stripback)		X
Snag	X	X
Splice	X	X
Thin (weak) place	X	X
Tight end(s)		X
Tight pick(s)		X
Tight edge (tight selvage)	X	X
Wrong draw	X	X

may be calendered. Current development work is concentrated on 35 × 35 plain weave made of 650 denier nylon yarns for driver side and 41 × 41 plain weave made of 440 denier polyester yarns for passenger side. Driver side bag will probably be scoured, heat set and coated while the passenger side bag may be scoured, heat set and calendered. Coating of the driver side bag may also be eliminated in the future. Lighter denier fibers may be used for smaller fabric packages.

Future opportunities for airbag use include side impact bags, rear seat bags, trucks, airplanes and buses.

4. SEAT BELTS

The purpose of seat belts is to prevent the forward movement of the wearer in a controlled manner during sudden deceleration of the vehicle. Seat belts are an easy to use, effective and inexpensive means of protection in an accident. There are various types of seat belts depending on the vehicle. In passenger cars, the seat belt fits across the lap and diagonally across the chest which is the most widely used type in the world. In passenger airplanes, the simple lap strap is used that does not restrain the upper part of the body. Racing drivers wear a full harness which consists of a strap over each shoulder and a lap belt. In general, automobile seat belts are made of polyester and aircraft seat belts are made of nylon.

The seat belt should be able to carry a static load of around 3,300 lb (1,500 kg) with a maximum extension of 25 to 30% [9]. Other major requirements are abrasion, heat and light resistance, lightweight and flexibility for ease of use. To meet these requirements, high tenacity polyester or nylon continuous filament yarns are used in the warp and weft direction. Polyester has the majority of the market by far and the use of nylon is declining. A typical yarn for seat belts is made of 320 ends of 1,100 dtex each. Warp direction in the belt is more critical since the load is applied mostly in that direction during an accident. Twill or satin-type weaves are used for seat belts as shown in Figure 15.10.

In this type of design, long warp knuckles on

FIGURE 15.10 Two/two twill seat belt design [10].

both sides of the belt provide the necessary strength in the loading direction. This configuration also provides a lightweight, slim and flexible fabric with a smooth surface that is comfortable to wear and easy to use. The woven fabric is shrunk during finishing to improve the energy-absorption properties. As a result of the shrinkage, the weight would increase typically from 50 g/m to 60 g/m. Proper dyestuff should be used to provide fastness to light, rubbing and perspiration. No dye should be transferred from the seat belt to garment by rubbing even in wet conditions.

A properly designed seat belt should provide non-recoverable extension or stretch during the collision to reduce the deceleration forces on the body (elastic stretch is not wanted since it may cause whiplash damage). A typical seat belt should allow the passenger to move forward around 12 inches during an accident by controlled extension of the belt. The extension and reduced recovery from stretch properties are given to seat belts by controlled heat relaxation during finishing. Since the stretch on the belt is not recovered, seat belts need to be replaced after a major accident. Otherwise, a seat belt should last as long as the life of the car.

In a serious accident, a well-designed seat belt provides a deceleration of 20 *g* (*g* is gravitational acceleration) or more, which means that there is a considerable amount of force exerted on the body by the seat belt. Although this force may cause some damage to the body, this damage is far less compared to the damage (i.e., death or serious injury) without a seat belt.

The Society of Automotive Engineers has developed the following standards for seat belts:

- Seat Belt Assembly Webbing Abrasion Performance Requirements, SAE J114 MAR86

- Dynamic Test Procedure, Type 1 and Type 2 Seat Belt Assemblies, SAE J117
- Seat Belt Assembly Webbing Abrasion Test Procedure, SAE J339 MAR86

5. AUTOMOTIVE INTERIOR TRIM

Various woven and nonwoven fabrics are used for seat facing, seat bolsters, headliners, doors and side panels. Fabrics provide comfort and better appearance in car interiors while still meeting the performance and consistency requirements. Automotive interiors have an influence on new car buyers.

Each application requires different fabric properties. The overall main trends in automotive industry's fabric requirements are:

- softness, luxury, hand and aesthetics
- patterning, design flexibility
- dyeability
- "moldable" fabrics
- stretch 5–200%

The Society of Automotive Engineers has developed the following test methods and standards for general automotive textile characteristics and interior trim materials. The SAE interior trim and textile standards were developed through the IFAI Transportation Division before they were published as SAE standards.

- Felts, Wool and Part Wool, SAE J314 MAY81
- Test Method for Measuring Thickness of Automotive Textiles and Plastics, SAE J882 FEB85
- Test Method for Determining Dimensional Stability of Automotive Textile Materials, SAE J883 FEB86
- Test Method for Wicking of Automotive

Fabrics and Fibrous Materials, SAE J913 FEB85

- Test Method for Determining Resistance to Abrasion of Automotive Bodycloth, Vinyl and Leather and the Snagging of Automotive Bodycloth, SAE J948 FEB86
- Test Method for Determining Resistance to Abrasion; Beading; and Fiber Loss of Automotive Carpet Materials, SAE J1530 JUN85
- Method of Testing Resistance to Scuffing of Trim Materials, SAE J365 FEB85
- Test Method for Measuring Weight of Organic Trim Materials, SAE J860 JAN85
- Test Method for Determining Blocking Resistance and Associated Characteristics of Automotive Trim Materials, SAE J912a
- Hot Odor Test for Insulation Materials, SAE J1351 APR87
- Accelerated Exposure of Automotive Interior Trim Components Using a Controlled Irradiance Water Xenon-Arc Apparatus, SAE J1885 MAR92
- Accelerated Exposure of Automotive Interior Trim Components Using a Controlled Irradiance Air-Cooled Xenon-Arc Apparatus, SAE J2212 MAR93
- Accelerated Exposure of Automotive Interior Trim Materials Using an Outdoor Under Glass Variable Angle Controlled Temperature Apparatus, SAE J2229 FEB93
- Accelerated Exposure of Automotive Interior Trim Materials Using Outdoor Under Glass Controlled Sun-Tracking Temperature and Humidity Apparatus, SAE J2230 FEB93
- Test Method for Determining Window Fogging Resistance of Interior Trim Materials, SAE J275 FEB85
- Non-metallic Trim Materials—Test Method for Determining the Staining Resistance to Hydrogen Sulfide Gas, SAE J322 DEC85
- Flammability of Polymeric Interior

Materials—Horizontal Test Method, SAE J369 JUN89

5.1 Seat Fabrics

Textile fabrics and leather have replaced PVC for car seats. However, vinyl is still used with either a polyester/cotton knit or a polyester non-woven substrate as trim on the back and in the skirt (boxing) of the seat. Major seating fabric structures are:

- flat woven fabrics
- pile woven fabrics
- warp knits
- circular knits
- double needle bar raschel

The use of flat woven fabrics is decreasing while the use of all-pile fabrics is increasing. The dominant fiber used in car interiors is polyester. Nylon was used in early seat fabrics. Polypropylene is used successfully in accent colors and some bolster fabrics. Major yarn structures are flat continuous filament, false twist, air-jet textured and knit-de-knit. Regularity of texture is critical. The main requirements of seat facing fabrics are soft handle, design color and pattern, piece dyeability and mold-ability via stretch (Figure 15.11).

Seat fabrics can be yarn dyed (for spun or textured yarns) or piece dyed. Figure 15.12 shows the major process steps for both methods. Very little solution dye is used for woven fabrics. Color and shade reproducibility are critical for seat fabrics which require computer control on dyeing machines. Setting of thermoplastic fibers is usually necessary. Finishes such as lubricants and antistats are usually applied to seat fabrics. After finishing, the fabric is laminated. Flame lamination, which is widely used, produces a trilaminate of face fabric, polyurethane foam and lining. The foam lining has a thickness of 2–10 mm. Testing of seating fabrics includes heat stability, light degradation and pile distortion. Ovens with temperature and humidity control are used for the tests. Fogging tests are also performed for safety concerns.

Recently, Inland Fisher Guide (IFG), a components division of General Motors, has developed a new technology for a seamless seat

FIGURE 15.11 Seat fabrics (courtesy of Stork).

Yarn Dyed

Piece Dyed

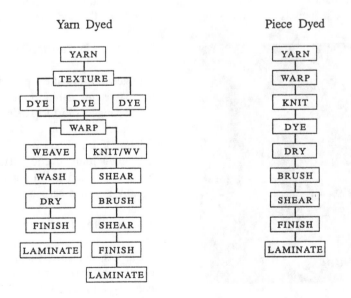

FIGURE 15.12 Manufacturing steps for seat fabrics.

cover. Using 3-D knitting, a fully integrated, three-dimensional seating fabric is produced while substantially reducing the need for cut-and-sew operations. The process is centered around a fully integrated CAD/CAM system [11].

Seating fabrics are becoming lighter, thinner and more contoured. Therefore, stretch is needed to maintain contours of the design. The Society of Automotive Engineers has developed the following standards for seating:

- Dynamic Flex Fatigue Test for Slab Polyurethane Foam, SAE J388
- Load Deflection Testing of Urethane Foams for Automotive Seating, SAE J815 MAY81
- Urethane for Automotive Seating, SAE J954
- Dynamic Durability Testing of Seat Cushions for Off-Road Work Machines, SAE J1454 JAN86

5.2 Liners and Panels

Warp knit, flat woven, circular knit and non-woven fabrics are used as headliners. Headliner fabrics must be easily moldable while retaining textile aesthetics (Figure 15.13). Nylon is the prevalent fiber in headliners. It offers a good hand and good resilience.

For door and side panels, usually a back

FIGURE 15.13 Headliner (courtesy of Freudenberg).

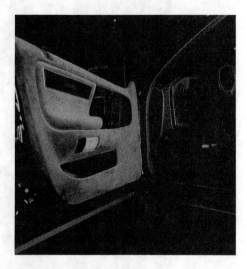

FIGURE 15.14 Door panel (courtesy of Freudenberg).

coating is applied to fabrics. Needlepunched fabrics dominate applications like seat backs and panels, trunk liners, lower door panels, package trays and kick panels (Figure 15.14). Spunbonded polyester fabric may be used as a primary and secondary backing substrate for tufting precision-molded and ready-to-be-fitted automotive elements such as carpets, trunk linings and door panels.

Textile materials and substrates on door panels and molded headliners are tested together for adhesion in environmental chambers with humidity and temperature control.

5.3 Floor Coverings

In general, tufted cut pile carpeting is used in the passenger cabin floor. Increasingly finer tufted gauges, short nap heights and more dense structures are being used in the United States (Figure 15.15). Loop pile fabrics may also be used in the future due to warranty requirements. Nylon BCF and nylon staple are used almost equally in tufted carpets [12].

5.4 Textiles for Sound Dampening

Carpets, nonwovens and other fibrous materials are effective for sound-deadening in cars. Foam-type materials and solid polymeric sheets filled with minerals increase sound damping.

FIGURE 15.15 Floor coverings (courtesy of Freudenberg).

Sound insulators in cars can be spacer-type (needled cotton, resinated cotton, slab foam, cast foam, fiberglass), barrier-type (ethyl vinyl acetate, polyvinyl chloride), or damping-type (viscoelastic materials).

Hoodliners are used as part of the acoustical package design to make the cars quieter by preventing noise from reaching the passenger compartment (Figure 15.16). It is estimated that a hoodliner is installed in over 10 million ve-

FIGURE 15.16 Hoodliner for sound dampening (courtesy of Freudenberg).

hicles each year in North America. Most of the hoodliners used in North America are made with a bonded mat of glass fiber as a sound absorbing element. These fibers are bonded together using a crosslinking thermosetting chemical binder system. Research is being done to replace fiberglass in hoodliners with other materials such as a thin spunbond nonwoven fabric [13].

The Society of Automotive Engineers developed the following standards for insulation:

- Acoustical and Thermal Materials Test Procedures, SAE J1324 OCT89
- Test Method for Measuring the Relative Drapeability of Flexible Insulation Materials, SAE J1325, FEB85
- Test Method for Measuring Thickness of Resilient Insulating Paddings, SAE J1355 APR87

6. AUTOMOTIVE EXTERIOR TRIM

Any fabric for the exterior of a car needs a coating reinforcement such as polyvinyl chloride or rubber. Weather resistance is the most important consideration for exterior fabrics. Exterior fabrics are usually considered a fashion statement expressing individuality.

6.1 Convertibles

In the 1920s and 1930s convertibles were being made of dyed cotton canvas duck or sateen. The modern convertibles started in 1951 with a PVC coated double texture butyl rubber fabric. Orlon, rayon and nylon were also considered along with hypalon exterior coatings. Production of convertibles in the United States was discontinued in 1977 due to a government regulation which required greater protection in the event of a rollover. In 1982, Chrysler developed a new design and body construction that met the safety requirements and reopened the convertible market (Figure 15.17) [14].

Conventional convertibles are made of dyed (usually jet black), vinyl coated cotton sateen fabric. This product, which has been around for some thirty years, is both functional and eco-

FIGURE 15.17 Convertible topping fabric (courtesy of Haartz Corp.).

nomical. More recently, solution dyed acrylics and acrylic polyester blends are being used as the outer fabric combined with another plain drill or jacquard lining fabric to produce the soft convertible topping fabric. These lining fabrics may be cotton, polyester, or nylon or a blend of a combination.

The soft topping for sport utility vehicles is a vinyl coated heavy-duty polyester-cotton sateen. Another variation is a low count square weave polyester scrim coated with vinyl on both sides. The vinyl top in the United States has traditionally been the symmetrical small diamond or pinpoint texture. The finish in Europe has a rougher animal leather grain.

6.2 Vinyl Roof Treatments

The "Landau" roof originated as an option in 1964 on luxury cars to give the vehicle an expensive leather look (Figure 15.18). The substrate under the vinyl is a nonwoven needle-punched polyester backing. With the use of a foam pad underneath, this enabled the top to assume a luxurious padded look and feel. Use

FIGURE 15.18 Automotive roof treatment (courtesy of Haartz Corp.).

of roof treatments is all but over on current production. A small but profitable aftermarket restyling market remains.

6.3 Other Automotive Exterior Fabrics

Other exterior fabrics include sports car front end covers, truck covers, rear window curtains for hatchbacks, binding fabrics and tire covers. Front end covers are used to protect the body finish from gravel and insects. These fabrics are made of vinyl on nonwoven polyester or expanded vinyl or polyester knit. The light utility truck cover is used to protect the cargo. The backing fabric is a weft inserted loop sheeting or a conventional sheeting made of polyester. Although the rear window curtains for hatchbacks are used inside the vehicle, they are positioned directly beneath the rear window glass and subject to magnified ultraviolet rays. Binding fabrics are used to cover either a raw edge or provide a bridge to fill in a narrow gap. Vinyl coated polyester/cotton binder fabrics are used for vinyl products. For the soft acrylic tops, a matching acrylic duck is used with or without a light vinyl or adhesive back-coating. A variety of fabrics are used as tire covers ranging from heavy cotton knitted fabrics to nonwoven 6–10 oz polyesters coated with vinyl.

The following standards have been developed by the Society of Automotive Engineers for exterior trim:

- Accelerated Exposure of Automotive Exterior Materials Using a Controlled Irradiance Water-Cooled Xenon-Arc Apparatus, SAE J1960 JUN86
- Accelerated Exposure of Automotive Exterior Materials Using a Solar Fresnel-Reflective Apparatus, SAE J1961 DEC88
- Outdoor Weathering of Exterior Materials, SAE J1976 JUN89
- Accelerated Exposure of Automotive Exterior Materials Using a Fluorescent UV and Condensation Apparatus, SAE J2020 JUN89
- Test Method for Determining Visual Color Match to Master Specimen for Fabrics, Vinyls, Coated Fiberboards and Other Automotive Trim Materials, SAE J361 MAR85
- Method of Testing Resistance to Crocking of Organic Trim Materials, SAE J861
- Instrumental Color Difference Measurement for Exterior Finishes, Textiles and Colored Trim, SAE J1545 JUN86
- Retroreflective Materials for Vehicle Conspicuity, SAE J1987 MAR91

7. TRUCK AND CAR COVERS

Cars with low drag coefficients are more fuel efficient. Textile fabrics (tarpaulins) are used as truck covers to reduce the drag and therefore improve truck aerodynamics as shown in Figure 15.19. The small arrows indicate turbulence, giving rise to drag. This is especially effective for trucks that carry open cargoes. It was shown that when properly tied down, tarpaulins give fuel savings of up to 7% [15]. Tarpaulins are coated fabrics made of high strength polyester and nylon fibers. Tarpaulins are also excellent means to protect cargoes against the weather. They prevent cargo debris from flying. Tarpaulins are a legal requirement in some states for certain types of cargo.

The Industrial Fabrics Association International (IFAI) estimated that in 1991, 15 million square meters of fabric was consumed in the truck cover market. Replacement covers are estimated to have 80–85% of the market, of which 30% are open tops, and 70% are load covers. This translates into roughly 350,000 average-sized truck covers. There are several types of truck covers as shown in Figure 15.20. Soft side trailers and open-top trailer covers such as the one shown in Figure 15.21 remain more popular in Europe than in North America [16].

Textile fabric covers are used to protect cars from sun, rain, hail, snow and other harsh environmental effects (Figure 15.22). All-season car covers are made of polyester and impact-absorbing padding.

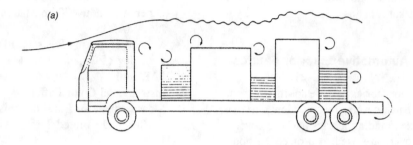

(a)

(b)

FIGURE 15.19 Typical air flows over trucks (a) without, and (b) with, tarpaulins [15].

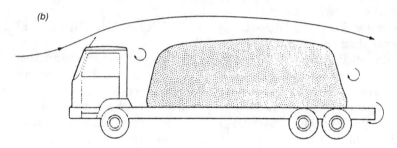

Fitted Load Cover

Retractable Cover

Retrofit Side Kit

Fitted Cover Over Frame

FIGURE 15.20 Schematics of truck covers (courtesy of Hoechst Celanese).

FIGURE 15.21 Trailer cover (courtesy of Sue Zorichak, IFAI).

8. BELTS, HOSES AND FILTERS IN CARS

Textile material reinforced hoses and belts are essential parts of car engines (Figure 15.23). Polyester and rayon reinforcements are used in radiator hoses, heater and air-conditioning, hydraulic lines, power steering lines and vacuum brake hoses. Heater and radiator hoses have to withstand the near-boiling temperature of water and other chemicals and radiant heat from the engine itself. The recent trend in

FIGURE 15.22 Car cover (courtesy of ATI).

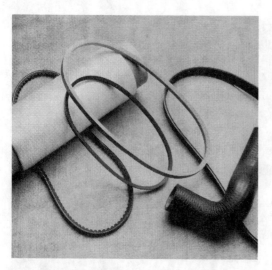

FIGURE 15.23 Belts and hoses used in cars (courtesy of Amoco Chemical Co.).

radiator hose is towards aramids. They should be resistant to vibration, oils and other solvents. The required properties of textile reinforcing materials for hoses and belts are strength, flexibility, temperature resistance and strong adhesion to elastomers.

Recently, high temperature thermal barrier materials such as fiberglass, ceramic and silica fibers, aramids, molded mineral fibers, metallized fabrics and silicone rubber coated fabrics are being used in modern cars for insulation of hydraulic, fuel, air-conditioning wires, engine mounts, wiring harnesses, exhaust gas recirculating tubes and electronic components.

Drive belts require similar requirements in textile reinforcement materials. V-belts are used to drive the alternator and water pump. With the inclusion of stronger and lighter textile reinforcements, toothed belts for timing drives have replaced chain drives completely. Due to its inextensibility, polyester is the main reinforcement material for toothed belts.

Textile reinforced belts offer low noise, low cost and light weight. Since the textile belts are more flexible than chains, smaller pulleys can be used for power transmission which allow compact designs. Another advantage of a textile reinforced belt is that it does not require lubrication. Belts and hoses for industrial applications are covered in Chapter 16.0.

General filtration textiles are covered in Chapter 8.0. Air filters are used in cars to clean the carburetor air and to provide clean air to the passengers. Oil and fuel filters clean the oil and fuel from impurities. Air filters are made of dry-laid nonwovens or paper. Oil and fuel filters are made of paper. Some dry-laid nonwovens are made of two or three sheets of nonwovens that are then needlepunched and sometimes combined with web construction, such as thermal bonding, spunbonding and hydroentangling.

9. TEXTILES FOR AIRCRAFTS

The Federal Aviation Administration (FAA) has stringent rulings concerning the flammability requirements for aircraft seat cushions. In order to comply with their specifications, the materials have to withstand burning and also not give off any toxic fumes. In the past, polyurethane foam alone was inserted into the aircraft seat cushions. When burned, these cushions released dense amounts of toxic fumes, causing airplane passengers to suffocate.

In airplane upholstery, a type of structure known as blocking layers is used to help pass flame resistance regulations. These layers of fabric consist of aluminized Zirpro wool and glass fiber or a blend of chemically treated wool and flame retardant synthetic fibers. Zirpro wool is a chemically treated flame resistant wool that has recently been developed. Seat covers with one of these kinds of layers help increase the chances of surviving a postcrash fire. The aluminized wool fire blocking layers are placed next to the foam filler in the seat cushion. The glass fiber is placed on the other side of the wool to give extra protection from fire. When this type of procedure is used, fire safety dramatically increases. However, these layers proved to be rather inflexible so for that reason they were better suited for critical locations such as flight attendant and pilot seats. The other type of fire blocking layer (consisting of chemically treated wool and man-made fibers) are more flexible and therefore are used for passenger seating.

A new fabric, known as Firotex® has been developed which is placed around the cushions as an inner cover. The Firotex® fabric is made of viscose rayon with a reinforced aramid inside to give needed abrasion resistance. The fabric can withstand burning with kerosene for up to two minutes.

Wool is another flame resistant material that is often used in aircraft furnishings. Due to the chemical and morphological structure of the fiber, the wool has a high nitrogen and moisture content. Therefore the wool will not usually support a flame or be susceptible to burning. If the wool does burn, it does not melt away. Instead, it leaves a foam-like ash which has good insulating properties and can be handled right after burning. This ash can also be easily removed so that major clean-ups are not required if part of the aircraft is burned. Wool does not shrink when burned so it will not leave any areas exposed to the fire. Wool is often used because it has a good appearance, and is easily maintained. Planes have curtains, carpets, upholstery, and blankets made from wool. Wool fabrics usually undergo a flame resistant treatment.

For carpeting used in airplanes, one flammability specification is known as the National Bureau of Standards radiant panel test. This test is considered to be more reliable for carpets than the vertical test. In this test, the carpet is positioned horizontally under an angled radiant panel with a small igniting flame located at the hottest area closest to the radiant panel. To be able to pass this test, a flame resistant latex must be applied to the carpet backings to keep it from burning. Cellulose fibers may also be used in this backing.

Other areas of the plane that require the use of textiles include seat belts and shoulder harnesses. These fabrics undergo tests in both the warp and the weft direction. Samples are taken at random from the item as it is installed in the aircraft. If this is not possible, simulated samples are constructed for testing purposes. The main test used for these fabrics is known as a horizontal test. This test uses a fabric length of 6 cm. All materials are tested as closely as possible to the form in which they are used in the airplane.

Protective clothing plays a major role in the aerospace industry. Protective garments must be worn when working in aerospace industry clean rooms. These rooms are where flight hardware is processed and they differ from the usual clean-room environment. The facilities at Kennedy Space Center for example, are large environmentally controlled high bays and are similar to airplane hangars. As air enters these facilities, it is filtered and monitored for airborne particulate, particulate fallout, and hydrocarbons. In order to maintain this level of cleanliness, workers must wear protective clothing. Garments are sometimes changed on a daily basis. The garments are made to be launderable, flexible, and comfortable. This will allow for longer life of the garment, thus resulting in a higher cost effectiveness. These garments are made to meet several conditions. They are made to be flame resistant, and have to withstand exposure to hazardous wastes. The components that make up the garments are 99% continuous filament polyester material with 1% raised grid carbon/polyester filament yarn. Nomex material is used in some of the garments that require extreme resistance to hazardous materials.

Another important factor to consider in aerospace is the space requirements. Workers have to be able to crawl and climb into various spaces. They also had to be very careful when working with orbiters since certain areas of the orbiter could not be stepped on. For this reason, garments with snaps, buckles, and belts are not used. Instead, the garment is made to be zipped up close to the wearer's body. This type of design makes it much easier to move around when wearing these garments.

Textiles have also been used as actual components of the space shuttle. The thermal protective system is aided by the placement of outer ceramic tiles and Nomex®-strain isolation pads. Composites have been used in forgings and castings of the shuttle. These composites are lightweight and provide for easier handling of the area into which they are installed (Chapter 7.0).

In a rather unique application, polychloroprene-coated 2-ply nylon, one-ply biased fabric is used for emergency slides (FAA slides) in airplanes (Figure 15.24). Critical properties of these fabrics are weight, breaking strength,

FIGURE 15.24 Airplane evacuation slide made of coated fabric (courtesy of Reeves).

tearing strength, flame resistance and air retention.

10. TEXTILES AS STRUCTURAL ELEMENTS IN TRANSPORT VEHICLES

The use of textile structural composites (Chapter 7.0) is increasing in manufacturing of transport vehicles including automotives, trucks and aircrafts. In commercial aircrafts, composite materials provide long-term fuel and chemical resistance. Honeycomb composite structures made of Nomex® and impregnated with phenolic resin are used for aircraft flooring due to their low density, light weight and chemical and fire resistance. In automotives, many body components are made of textile reinforced composites. In shipbuilding, graphite composite can reduce structural weight to increase payload capability, and improve ship stability. Boron reinforced aluminum was used in the space shuttle which saved around 300 lb of weight in the shuttle's cargo structure compared to conventional metals [17].

Textile fibers can be used in clutch and brake parts replacing asbestos. These parts require good abrasion and high temperature resistance. Clutch facings are reinforced with 20% by weight Kevlar® short fibers, combined with inorganic fillers and binder. Properly constructed disc brake pads, drum brake shoes, truck block

brakes and railroad brake shoes require less than 5% short fibers of Kevlar® by weight as the reinforcement material. Sheet gasketing can be reinforced with less than 6% Kevlar® pulp by weight combined with other fibers, fillers and rubber binders. Figure 15.25 shows various automotive parts that include Kevlar® as the reinforcement material.

Automotive Polymer Composites

Automobiles account for approximately two-thirds of the gasoline consumption and one-third of total energy use in the United States. Approximately 75% of a vehicle's fuel consumption is directly related to weight-related factors, the other 25% is related to air drag. A small weight reduction of current vehicles would result in tremendous gasoline savings as well as reduction in CO_2 emissions [18]. Auto companies have made an effort to reduce the weight of vehicles over the past twenty years. The average weight of a vehicle has been reduced nearly 25% while doubling fuel economy without sacrificing other desirable performance features.

Polymeric materials are expected to replace steel in many automotive application areas. Steel is heavy and subject to corrosion. Polymer composites are lighter, more flexible and can offer better dent resistance. Examples of substituting metals with polymer products include

FIGURE 15.25 Various automotive parts reinforced with Kevlar® aramid chopped fibers and pulp (courtesy of DuPont).

plastic bumpers, front and rear ends, radiator tanks, fenders, wheel covers and door handles (Figure 15.26). Table 15.8 shows the major areas for plastic and polymer composite body panels. It is estimated that polymeric materials accounted for 12% of the body weight in 1990.

Polyurea resins processed by reaction injection molding (RIM) are the current front runners in fender applications. These resins have been used on the Chevrolet Camaro four-seat sport coupe. Sheet molding compound (SMC) is the leading material for hoods. SMC is a composite of thermoset polyester modified with

FIGURE 15.26 Fender and wheel cover made of thermoplastic polyester alloy (courtesy of Hoechst Celanese).

TABLE 15.8 Plastic Automotive Parts and Polymers Used [18].

Automotive Parts	Polymers Used
Interior parts	
Trim	ABS, ABS alloys, modified PBT, acetal
Instrument panel	RIM polyurethane (PU), RIM polyurea
Instrument panel skin	ABS/PVC alloy, PPO
Console	ABS
Functional interior parts	
Radiator header tanks	Nylon
Brake reservoirs	Nylon, acetal
Electrical distributor caps	PBT, PPO blends
Fuel tanks	Acetal, nylon, modified PET, PBT
Gears	Acetal, nylon
Seat backs	PPO
Seat cushions	PU
Engine fans and shrouds	Nylon
Air cleaner housing	Nylon
Exterior parts	
Bumper fascia	RIM-PU, RIM-Polyurea, SRIM-PU, PPs, toughened PPs, fiber reinforced PPs, and other thermoplastic polyolefins (TPOs), PC/PBT Alloys, thermoplastic elastomers
Body panels	SMC, BMC, PET/PBT blends, TPOs, nylon, nylon/PPO blends
Exterior trim	PPO/HIPS blend, PC, PC/PET blend, PC/PBT blend
Hood	SMC
Functional exterior parts	
Headlamp assemblies	Nylon, PC, modified polyesters
Wheel hubs, covers	Nylon, PC, modified polyesters
Mirror housing	Nylon

RIM = reaction injection molding, PPO = polyphenylene oxide, ABS = acrylonitrile-butadiene-styrene, SMC = sheet molding compound, PC = polycarbonate, PVC = polyvinylchloride, PET = polyethyleneterephthalate, PP = polypropylene, PBT = polybutylene terephthalate.

FIGURE 15.27 Fabrics for coast guard boats (courtesy of Reeves).

fillers like calcium carbonate and reinforced with chopped glass fibers. The roof, doors, hatch, and spoiler of Camaro sport coupe and those of the fourth-generation Pontiac Firebird are made of chopped glass/polyester SMC. The material cost of SMC is very attractive to automakers. SMCs have found applications in specialty roofs for light-duty vehicles such as the Ford Blazer and Jeep Wranglers. SMC lift gates are utilized on Jeep Cherokees and Ford Broncos. SMC is being incorporated in side panels on the box of GM support trucks.

The Society of Automotive Engineers has developed the following standards for structural polymeric materials:

- Test Method for Determining Cold Cracking of Flexible Plastic Materials, SAE J323
- Test Method of Stretch and Set of Textiles and Plastics, SAE J855 MAR87
- Chemical Stress Resistance of Polymers, SAE J2016 JUN89

11. INFLATABLE PRODUCTS USED IN TRANSPORTATION

Coated fabrics are used in inflatable life-saving aids such as life jackets, rafts and aircraft emergency chutes. Modern life rafts, which are generally designed for 1–30 people, have multiple buoyancy tubes to keep the raft afloat even if a tube is punctured. Usually canopies are also included to protect against severe weather. In case of emergency, a small canister of compressed carbon dioxide, which is at $-78\,^{\circ}C$, is used to inflate the life raft in a few seconds. To prevent flex cracking due to cold gas during inflation and unfolding the raft, polymer coatings with good low-temperature flexibility are used. Civilian rafts are generally made of woven nylon fabrics and coated on both sides with natural rubber or natural rubber/SBR (styrene-butadiene rubber) blend. The coating contains daylight fluorescent pigments for maximum visibility. Naval life rafts are usually made of nylon laminates with an interply of butyl rubber which is less permeable to carbon dioxide than natural rubber providing longer use [19]. Leisure boats are inflated to a pressure of 15–25 kPa by compressed air rather than carbon dioxide.

Instead of strengthened fabric floors, naval and coast guard vessels have a rigid hull encircled with buoyancy tubes (Figure 15.27). These amphibious craft are suspended above the surface of water by a cushion of air that is generated by fans and airscrews. The coated-fabric "skirt" is fitted around the hull and contains the air cushion. The fabric is inflated to a pressure around 3 kPa. Heavy polyester or nylon fabrics coated with thick layers of Neoprene, Hypalon, or polyurethanes are used. Similar coated fabrics are also used to surround and contain oil slicks, for marker buoys and in flotation equipment for offshore oil and construction industries. Buoyancy bags are used for supporting pipelines and platform sections. Deflated bags can be attached to sunken objects and then inflated to raise them to the surface.

References and Review Questions

1. REFERENCES

1. Smith, W. C., "Automotives: A Major Textile Market," *Textile World*, September 1994.

2. DataTextile, *IFAI*, March 1992.

3. "EDANA Officials Report Automotive Nonwovens Gain Ground," *Southern Textile News*, March 22, 1993.

4. Blow, C. M. and Hepburn, C., eds., *Rubber Technology and Manufacture*, 2nd Ed., Butterworth Scientific, 1982.

5. Svedova, J., ed., *Industrial Textiles*, Elsevier, 1990.

6. Jagielski, K., "A Driven Market," *Industrial Fabric Products Review*, August 1992.

7. "Use of Dual Airbags Triples in 1993," *Industrial Fabric Products Review*, August 1994.

8. Nelb, G. W., "Automotive Airbags—What Is the Status?" E. I. du Pont de Nemours & Co., 1994.

9. Morris, W. J., "Seat Belts," *Textiles*, Vol. 17, No. 1, 1988.

10. Stirk, E. M., "High-Strength Webbings," *Textiles*, Vol. 11, No. 2, 1982.

11. Smith, T. L., "3-D Knitting Adds New Dimension to Interiors," *Automotive & Transportation Interiors*, November 1994.

12. Kalogeridis, C., "Auto Companies Focus on Textiles," *ATI*, February, 1994.

13. Wooten, H. L., "A New Era in Hoodliner Design," *Hi-Tech Textile Conference*, July 1993, Greenville, SC.

14. Alling, P. S., "The Evolution of Automotive Convertible Topping and Other Exterior Decorative Trim Fabrics," *The Conference on Textiles in Automotives*, Atlanta, GA, Nov. 9–11, 1993.

15. Shields, M., "Textiles for Road Vehicles," *Textiles*, Vol. 13, No. 2, 1984.

16. Rebeck, G., "A Hard Road," *Industrial Fabrics Products Review*, August 1992.

17. Reisch, M. S., "High-Performance Fibers Find Expanding Military, Industrial Uses," *Chemical and Engineering News*, February 2, 1987.

18. "Materials for Lightweight Vehicles," Office of Transportation Materials, U.S. Department of Energy, Washington, D.C., 1992.

19. Lomax, G. R., "Coated Fabrics, Part 2—Industrial Uses," *Textiles*, Vol. 14, No. 2, 1985.

2. REVIEW QUESTIONS

1. List the major applications of textiles in transportation vehicles.

2. What are the major tire designs? How are textiles used in each design?

3. Explain how a tire cord fabric is made. What are the properties of fibers used?

4. Explain how an airbag module works in a crash.

5. What is the predominant fiber used in airbags? Why?

6. Why is testing especially important in air-bag manufacturing? List the major tests for airbags.

7. Explain the structure of seat belts. What are the major requirements for seat belts?

8. What are the general requirements for automotive interior trim? Explain.

9. Why are truck covers used? What are the materials?

10. Discuss the use of textiles in aircrafts.

GENERAL INDUSTRIAL TEXTILES

General Industrial Textiles

S. ADANUR

This chapter includes examples of textile materials used in various applications that are not covered in the previous chapters.

1. TEXTILES IN AGRICULTURE

The main applications areas of textiles in agriculture include farming, animal husbandry and horticulture. The volume of special textiles that are manufactured for agricultural applications only is small compared to other areas of technical textiles. This does not mean that the use of textiles in agriculture is not significant. On the contrary, a wide variety of textile products that are designed for general industrial applications are used in agriculture in great quantities. These products include hoses, conveyor belts, tires, composites, filters, textiles for hydraulic applications, etc. With the apparent exception of protective clothing for insecticides, farmers usually make use of existing fabrics to fit their needs. This section will cover specific applications of textiles in agriculture.

The most important requirements of textiles for agricultural applications are weather resistance and resistance to microorganisms. Therefore, synthetic fibers are the choice of material for agricultural products. The use of textiles in horticulture (fruits, vegetables, trees, flowers) is increasing more rapidly than any other area in agriculture. The use of nonwovens, especially spunbonded fabrics, is increasing in agricultural applications at the expense of woven fabrics.

The following areas are some of the end uses of textiles in agriculture.

1.1 General Applications

Clothing, mud control, water hoses, windscreens, cordage, transportation, drainage, crop covers, landscaping and moisture retention. For example, sacks made of PVC coated polyester fabrics are used to transport agricultural products (Figure 16.1).

1.2 Animal Husbandry

Nylon and polyester identification belts are used in cows. Textile nets are used to support the large udders. Nonwoven fabrics are used to filter the milk in automatic milking systems. Nonwoven fabrics are used as an underlay to reduce mud on cattle paths and trails.

1.3 Horticulture

Application of textile materials in horticulture is growing. Nets, nonwoven mats, movable screens for glass houses, nonwoven sheets, mixed bed for mushrooms, cordage and strings are used in horticulture.

Trees are covered by nets to protect the growing expensive fruits from birds (Figure 16.2). Nets are also used for protection against hailstorms.

Light resistant woven and nonwoven polyester fabrics are used in the inside of greenhouses to protect the plants from extreme hot or cold. They are also used on the outside of the greenhouses as screens to control sunlight. Fabric protective greenhouses provide virus-free cultivation of young plants. Nonwoven sheets are also used in the field to protect young plants

FIGURE 16.1 Knitted sacks for agriculture product storage and transportation (courtesy of Karl Mayer).

FIGURE 16.2 Net providing protection against birds (courtesy of Karl Mayer).

FIGURE 16.3 Nonwoven fabric for plant protection (courtesy of Reifenhauser).

such as strawberries, potatoes and lettuce from extreme cold, night frost and viruses (Figure 16.3).

Nylon fabric beds are used to grow mushrooms. A mixture of horse manure and compost is laid on the fabric which is wrapped around rollers to easily remove the mixture.

1.4 Protective Clothing in Agriculture

Dermal and respiratory pesticide exposure of agricultural workers involved in spray operations may be significant. Spray operations include field or row crop spraying, orchard (air blast) spraying, spraying in greenhouses, aircraft spraying and mixing and handling. Pesticides can move through the fabric (wicking), wet inner layers of fabric and skin, and be dermally absorbed. Moreover, farm workers are involved in other tasks such as cultivation, harvesting, irrigation, scouting and fruit thinning. Protective clothing is used to reduce exposure to workers that are directly or indirectly involved in pesticide applications.

The risk factor (R) of personal exposure to pesticide spraying and handling depends on the possible routes of exposure and toxic characteristics of pesticide. This factor is defined as [1]:

$$R = \frac{\text{inhalative exposure}}{\text{relevant tox data}} + \frac{\text{dermal exposure}}{\text{relevant tox data}}$$

If the risk factor is greater than 1, use of pro-

tective clothing is recommended. Figure 16.4 shows the general guidelines for requirements of protective measures and recommended protective means. Figure 16.5 shows a typical protective garment and hood design for pesticide protection. Gore-Tex® and Saranex-laminated Tyvek® are the typical materials used.

Wearing of protective clothing such as rubber aprons, waterproof outer garments and face masks for a long period is difficult especially in hot climates. Therefore, selection of protective clothing materials is extremely important. The major requirements of the materials are [3]:

- protection against a range of pesticide formulations; aqueous organic solvent, or oil based
- lightweight
- low price
- readily washable without losing protective properties
- durable
- good air exchange to maintain wearer comfort
- acceptable to wearer

The penetration of pesticide through the fabric is influenced by capillary forces. Pesticide protective clothing can be made of traditional textiles such as cotton, polyester (or their blend) and treated with repellent finishes. However, special woven and nonwoven protective materials such as Tyvek® and Gore-Tex® offer better protection. Protective clothing should not cause heat stress or impair the worker's movement. The effectiveness of protective clothing is influenced by garment properties (fiber content, yarn and fabric geometry, textile finish), environmental conditions and pesticide characteristics. Fluorocarbon finishes provide better protection against pesticide penetration than non-fluorocarbon-finished fabrics and reduce wicking. However, hydrophobic fluorocarbon soil-repellent (SR) finishes may cause redeposition of soil in laundering [4,5].

Hobbs et al. tested several woven and nonwoven fabrics as barriers to aerosol spray penetration by methylene blue dye solutions [6]. The aerosol spray test was based on ASTM Method of Salt Spray (Fog) Testing, B 117-73, 1979 and was designed to assess the fabric resistance to

requirements of protection measures

application handling	maximum penetration rate								
	hand protection LD50 derm. [mg/kg]			body protection LD50 derm. [mg/kg]			respiratory protection LC50 [mg/l] 4 h		
	400–4000	50–400	< 50	400–4000	50–400	< 50	0,5–5	0,1–0,5	< 0,1
(tractor)	30	20	10	50	25	10	–	–	30
(spraying tractor)	10	5	5	10	5	5	–	20	5
(person spraying)	20	10	5	20	10	5	30	20	5
(pouring)	10	5	5	20	10	5	–	20	10

and therefrom derived protective means

(tractor)	+	+	+	A	A	B	–	–	⟨filtering face piece⟩
(spraying tractor)	+	++	++	B	B	B	–	⟨filtering⟩	⟨half mask⟩
(person spraying)	+	++	++	A	B	B	⟨filtering⟩	⟨filtering⟩	⟨half mask⟩
(pouring)	+	++	++	(A, B) + S			–	⟨filtering⟩	⟨half mask⟩

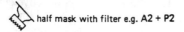

symbols

+	protective gloves with short term resistance
++	protective gloves with long term resistance
A	protective suits e.g. cotton coverall
B	protective suits (permeability < 5 %) e.g. Goretex, Propylen-Kimberly-Clark, Tyvek
S	apron of good resistance

⟨ filtering face piece

⟨ half mask with filter e.g. A2 + P2

FIGURE 16.4 Protection required during the application and mixing of pesticides of various toxicity, represented (top) by the maximum permissible penetration rate *P* in % and (bottom) by symbols or personal protection gear satisfying the penetration requirements (copyright ASTM [1]).

penetration of aerosol sprays. The results were recorded as pass or fail. The presence of any methylene blue dye on the fabric backing was considered as "fail." They concluded that the presence of fluorocarbon finish increased the resistance to aerosol penetration. In another study, it was found that water-repellent and soil-release finishes increase the fabric effectiveness for protection [7].

The USDA, EPA and NIOSH (National Institute for Occupational Safety and Health) recommend the use of protective apparel during handling and application of most pesticides [8]. The information about the proper personal protective equipment for the safe use of pesticides is included in the Federal Insecticide, Fungicide, and Rodenticide Act (FIFRA) and the Code of Federal Regulations, Title 40, Section

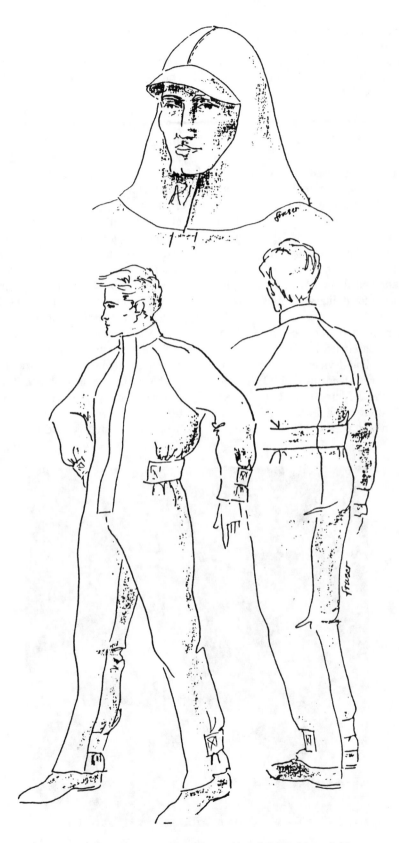

FIGURE 16.5 Typical protective garment and hood design for pesticide protection (copyright ASTM [2]).

162.10, Labeling Requirements of Pesticide Products issued by the Office of Pesticide Programs (OPP) of the EPA.

2. TEXTILES IN ELECTRONICS

Kevlar® aramid can be used as a high density, leadless electronic chip carrier. The carrier can be made of a printed copper surface and laminated with a woven Kevlar® structure which allows placement of more electrical connections in less space and restrains copper and epoxy thermal expansion. Kevlar® is also more resistant to vibration and failure, which makes the chip very suitable for military missiles and aircraft [9].

Nextel alumina-boria-silica ceramic fibers developed by 3M are used in electrical equipment, thermocoupling devices and wire insulation. N-312 can withstand up to 1,205°C and N-440 can withstand up to 1,370°C. These fibers can be woven into a fabric.

Polyester film laminates and Nomex® aramid paper are used as insulation in motors and generators. Nomex® paper is used in slot liners, phase insulation, wedges, coil insulation and high voltage insulation (Figure 16.6).

Thermount® from DuPont is a nonwoven aramid reinforcement material for printed wiring boards. It is suitable for high-density interconnects such as fine-pitch surface mount, direct-chip attach, or in multi-chip modules. Thermount® provides a low coefficient of inplane thermal expansion (CTE), which reduces strain on solder joints during thermal cycling and a low dielectric constant, which allows faster signal speeds and good dimensional stability for improved registration yields.

PTFE composites are utilized in low loss, low dielectric constant, printed circuit boards for microwave applications as shown in Figure 16.7.

Textiles are used as planar antenna covers as shown in Figure 16.8. The permanent hydrophobic surface of the modified PTFE composite provides a durable, weather resistant construction.

Kevlar® aramid is used in fiberoptic cables,

FIGURE 16.6 Nomex® as insulation material in a motor (courtesy of DuPont).

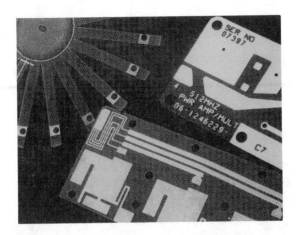

FIGURE 16.7 PTFE composites in printed circuit boards (courtesy of Chemfab).

FIGURE 16.8 RAYDEL® microwave transmissive fabric as planar antenna cover (courtesy of Chemfab).

FIGURE 16.9 Use of Kevlar® in cables (courtesy of DuPont).

data and sonabuoy mooring cables, air and sea towed antennae and acoustic arrays (Figure 16.9). The principal function of the reinforcement in fiberoptic and electromechanical cables is to limit the strain on the optical and/or electrical conductors.

Nonwoven fabrics are used in computer diskettes for cleaning purposes. Textiles are also used in professional and auto audio markets. Nonwoven and knit fabrics are used to cover boxes that house speakers and amplifiers.

3. BANNERS AND FLAGS

Banners and flags are used for decoration, architectural enhancement and signage. Banners can be grouped as interior and exterior banners. Exterior banners can be street lamp banners, across-street banners, draping down banners or across the sides of buildings banners. These distinctions are important when selecting fabrics and anchoring systems.

A flag is a banner that is "connected to a cable or pole at two points on one side of the fabric or with the support running through a sleeve attached to the fabric" [10]. Although flags can be used for aesthetic purposes, they usually have a symbolic purpose that represents governments, corporations or institutions (Figure 16.10).

Mildew resistance, UV resistance, tensile strength, abrasion resistance and tear resistance are critical for flags and banners for outside use. Longevity considerations, water repellency, color and appearance are important when selecting a fabric for banner or flag applications. Flame resistance may also be required especially for internal use. The most widely used materials are vinyl laminates, vinyl coated mesh, polyester, nylon, spunbonded polyolefin, unsupported films, solution-dyed coated acrylic, canvas and drill cloth. Banner or flag substrates can be woven, knitted, nonwoven or weft-inserted warp knit construction. Scrims are either woven or weft inserted.

Vinyl laminates are constructed by sandwiching polyester, nylon or fiberglass scrims between two layers of vinyls. The scrim provides dimensional stability, tensile and tear strength while the vinyl provides a smooth, uniform surface for printing graphics. Laminates are easy to process—they can be sewn, heat sealed or radio frequency (RF) welded. Vinyl coated meshes have open weave designs and are used where wind loads are a major factor. Polyester fabrics can be woven or knitted. They are coated or treated for durability, smoothness and printability. Polyester and nylon fabrics are widely used for flags. Spunbonded polyolefin fabrics offer the look and feel of paper and they are used for short-term banners. Unsupported films, such as polyethylene or vinyl sheeting are light and inexpensive plastics without any fabric reinforcement. Canvas and drill cloth are made of natural fibers (all the others are made of synthetics). They are less expensive but also less UV-resistant and tough. Table 16.1 lists the major properties of banner and flag fabrics.

4. TEXTILE REINFORCED PRODUCTS

The rubber industry uses textile yarns and fabrics extensively for reinforcement. The major rubber product items in which textiles are used as reinforcement are tires, conveyor belts, and power transmission belts. Tires are explained in Chapter 15.0, Transportation Textiles. Uses of textile materials in other rubber products are discussed in this section.

Major fibers used in the rubber industry are cotton, viscose rayon, nylon, polyester, glass and aramids. The major fiber properties required in rubber reinforcement are:

- high tenacity and modulus
- good dimensional stability
- moisture resistance
- resistance to high temperatures
- fatigue resistance
- good bonding with adhesives and rubber

4.1 Transport Conveyors

Transport conveyors are used in mineral works and transportation, material handling systems and for passenger transportation at air-

FIGURE 16.10 Flags combine two unique properties of textiles: flexibility and strength (courtesy of Wellington Sears Company).

TABLE 16.1 Major Properties of Banner and Flag Fabrics (courtesy of IFAI).

	Nylon	Polyester	PVC, Mesh, Laminates, Coated Nylon	Cotton, Polyester/Cotton	Acrylic, Modacrylic	Polyolefin
Description	Indoor/outdoor woven material, 200–400 denier most common, coatings/finishes available	Spun, woven or knit polyester fibers, used for indoor/outdoor applications. Finishes are for texture, performance characteristics	Indoor/outdoor materials, includes extruded woven, laminated, and coated materials	Woven, all-natural fabrics and polyester cotton blends, including bunting, canvas, drill cloth, and ducks	Acrylic fibers dyed in the formulation stage, giving woven material colorfastness. Modacrylic inherently fire-resistant	Spunbonded polyolefin fibers create a light-weight, weatherable material similar in touch to wax paper. Typical usage: short-term indoors, outdoors
Weight/sq yd	3–6 oz	4–12 oz	10–18 oz most common	8–18 oz	8–10 oz	3–5 oz
Colors	Wide range	Wide range	Wide range	Limited in natural state, some come dyed	Wide range	Limited
Graphic versatility	Applique, screenprint solvent, screenprint ink, acid dye, pressure-sensitive vinyl on coated materials	Applique, screenprint ink, dye process, sublimation print	Screenprint ink, pressure-sensitive vinyl, heat transfer film, hand paint	Screenprint ink, dye, applique, hand paint	Screenprint ink, applique, heat transfer film, heat-sealed in-set, hand paint	Screenprint ink, hand paint
Transparency	Transparent light colors, translucent dark colors	No	Opaque (some blocked with ink	No	White is translucent	Translucent
Maintenance	Yes, consult manufacturer	Yes, consult-manufacturer	Easily cleaned, consult manufacturer	Consult manufacturer	Yes	N/A
Life expectancy	6 mo average, some treated for UV resistance	6–12 mo	1 yr	6 mo	1 yr	2 wks outdoors
Fire-resistance (from manufacturer or treatable)	Yes, treatable	Some are FR	Many are; consult manufacturer	Some are; converted by manufacturer	Yes	No
UV resistance	Yes	Yes	Yes	Some are	Yes	No

FIGURE 16.11 Transport conveyor in mineral works (courtesy of Akzo).

ports, shopping centers, etc. (Figure 16.11). Textile or steel cords are used to reinforce rubber or plastic in conveyor belts and to provide longitudinal strength. In return, rubber or plastic covering protects textile materials from damage, moisture, etc. Depending on the application area, the major property requirements of conveyor belts are high strength and flexibility; low stretch in service; abrasion, impact and tearing resistance; temperature and fire resistance; long life; resistance to moisture, oils, and chemicals. Good adhesion of the reinforcing plies is essential.

A number of rubberized textile reinforcement plies are laminated, consolidated and vulcanized in special presses and curing machines to form conveyor belts (Figure 16.12). Usually three to six plies are used for reinforcement. Increasing the number of plies unnecessarily reduces the ratio of total ply strength to belt strength. The main textile reinforcement fibers are rayon, polyester, nylon and aramid. Cotton is used for relatively low strength requirements.

High modulus materials, such as aramid and polyester, are required in the longitudinal (warp) direction for long-haul high tensile belting. Aramid fibers are as strong as steel and they are non-corroding, fire resistant with a low thermal conductivity [2,3].

Simple fabric designs such as plain or basket weave are widely used in conveyor belts. Spe-

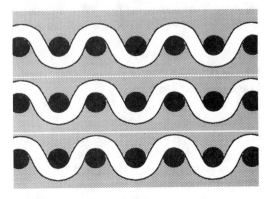

FIGURE 16.12 Schematic of multi-ply fabric reinforced conveyor belts.

cial weaves may be used to improve physical characteristics such as stretch, edge wear and tear. Two- or three-layer solid or hybrid fabrics provide dimensional stability. Blending different fibers in fabrics offers many advantages including improved adhesion, strength, impact resistance, fatigue resistance, stretch resistance, temperature and chemical resistance. Typical fiber combinations include aramid or polyester warp and nylon weft, nylon warp and weft. For bulk and adhesion, cotton and synthetic staple fibers are added. Fabrics are dried to avoid blowing of the laminate during the vulcanization. Fabric is passed over a multiple steam-heated drum drier or hot plate. Next, frictioning is done while the fabric is still hot. Frictioning is a coating process in which the large difference between the fabric and the rubber sheet at the nip of the calender forces the rubber into the fabric weave. In calendering, both sides of the full width fabric are coated simultaneously. Then, fabrics are cut to the width desired. The cut widths are passed through a doubling machine to adhere the layers together until the desired number of plies is obtained. Vulcanization is carried out in a press or continuous drum curing machine. To make the belt endless, the ends of the belt are brought together either by mechanical fastening or splicing. Splicing is done by vulcanizing the belt ends together. Fire resistance requirements for passenger and package carrying belts are increasing. PVC belting is used for fire resistance.

Belts are identified according to reinforcement fiber type: "E" indicates polyester, and "P" indicates nylon. "EE" designates a belt with polyester warp and weft, and "PP" indicates a nylon warp and weft belt. A belt with polyester warp and nylon weft is designated as "EP."

4.2 Textiles in Power Transmission Belting

Driving belts transfer the torque from a driving shaft to a driven shaft. The reinforcement material of the power transmission belt is made of textiles. Driving belts can be flat or V-shaped. Plain weave cotton fabrics (duck or cord) of $1.05-1.15$ kg/m² are used in flat transmission belts. Polyester and rayon are also used.

The fabrics are coated with rubber. The number of plies is usually between three and ten. Single-ply or compound weave fabrics are also common. The requirements for flat driving belts are similar to conveyor belts. Rayon, polyester, glass and aramid cord materials are used in endless V-belts which may have trapezoidal or wedge shape.

Textile reinforcement in the form of a cord fabric or yarns is placed at the neutral axis or center of the V-belt in between two layers of rubber. Then the whole structure is covered by fabric. Raw edge V-belts do not have the outside protective jacket and therefore the tension members are exposed. These belts are suitable for smaller diameter pulleys. The trend in driving belts is towards higher power transmission and smaller pulley diameters which results in steadily growing use of polyester fibers in V-belt manufacturing. In compact cars, this trend is very obvious. Besides, the diminished motor space is responsible for higher heat stresses to which V-belts are also exposed.

By joining several belts together, multi-V and poly-V belts are made. The major fiber requirements for V-belts are high strength and dimensional stability, resistance to impact and periodic bending stresses, resistance to high temperature and chemicals and good adhesion to rubber.

Positive drive belts or toothed belts are used in applications where no-slip precision drive and timing are required. Since these belts should have very low stretch, steel, glass or aramid cords are used. The teeth are molded with the belt and covered by tough nylon fabric for abrasion resistance.

4.3 Hoses

General use rubber hoses usually consist of three layers: lining or tube (inner layer), reinforcement or carcass (middle layer) and cover (outer layer) for protection as shown in Figure 16.13. Figure 16.14 shows a flexible hose made of coated fabric.

Depending on the requirements, aramid, carbon, cotton, flax, linen, rayon, nylon, polyester, polypropylene and glass fibers can be used as reinforcement materials. Coatings include neo-

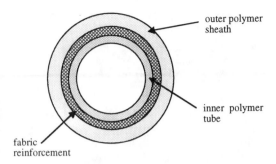

FIGURE 16.13 Schematic cross section of fabric-reinforced hose.

prene, pigmented silicone, fluoro silicone, PVC and laminated vinyl. The choice of the material generally depends on the operating pressure of the hose. The reinforcement yarn can be knitted, braided, spiral wound and circularly woven on the lining as shown in Figure 16.15. Reinforcement can also be achieved by wrapping fabrics straight or on the bias on the lining.

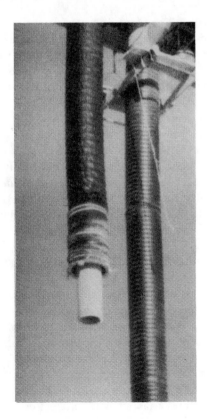

FIGURE 16.14 Flexible hose made of coated fabric (courtesy of Cooley).

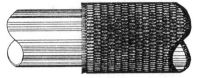

Circular Woven Hose

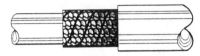

Knit Hose

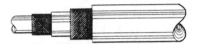

Spiral Wound Hose

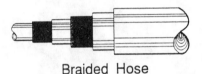

Braided Hose

FIGURE 16.15 Types of hose constructions (courtesy of Hoechst Celanese).

If more than one layer of reinforcement is used, a layer of insulating rubber is inserted in between. According to manufacturing technique, there are five major hose types: molded hose, hydraulic hose, machine made wrapped hose, handmade hose and circular woven hose. In molded hoses, the lining (tube) is formed first by extrusion. Then the lining is reinforced by braiding or spiraling the textile around it. Next, the rubber cover is applied to the reinforced hose by a cross-head extruder. Vulcanizing and molding follow.

In manufacturing of hydraulic hoses, the lining tube is extruded, blown onto a steel mandrel and then the braiding and/or spiraling of reinforcement is done. After covering and cloth wrapping, vulcanization takes place. Flexible hydraulic hoses are used in aircraft, automobiles, earth moving, materials handling and mining. In machine made wrapped hoses, woven fabrics (duck) at a bias angle of 45° are employed. These hoses are used for water, compressed air and steam delivery, for welding and shot blasting equipment and for conveying drinks and food [11].

Large size handmade hoses that are used in the oil industry are made by hand on large lathes. Lining, carcass and cover are hand applied and rolled into position. Steel wire helices are inserted by spiraling to prevent collapse under suction. In circular woven hoses, fabrics woven on circular looms are used. Warp threads are placed in the longitudinal direction. Some oil-pump hoses are produced by weaving around an oil-resistant lining tube. Polyester and nylon yarns are used in oil-pump hoses because of their high burst strength under pressure, low moisture retention and complete resistance to mildew (Figure 16.16). Wire

FIGURE 16.17 Fire hose (courtesy of AlliedSignal).

helices are included in the weft. Fire hoses are made by weaving a jacket (cover) on vertical looms (Figure 16.17). The fabric is treated for mildew resistance. The lining is extruded and inserted into the woven jacket. Consolidation and vulcanization follow.

In all hose designs, the angle of textile yarn cord or fabric relative to the hose lining is very critical. This angle (braid angle) ultimately affects the bursting pressure. Burst pressure is twice the hoop force divided by the mean diameter of the reinforcement and is calculated for different textile reinforcement designs.

4.4 Other Textile Reinforced Products

Textiles are used as reinforcement materials in a variety of rubber products including collapsible boats, air mattresses and cushions, raincoats, synthetic leather, industrial fabrics for outdoor and deep sea applications and containers for liquid transportation and storage. The textile structure in most of these applications is woven fabrics with different constructions and weaves. The factors to consider in selecting the type of textile reinforcement are

FIGURE 16.16 Hoses for oil-pumps (courtesy of Allied-Signal).

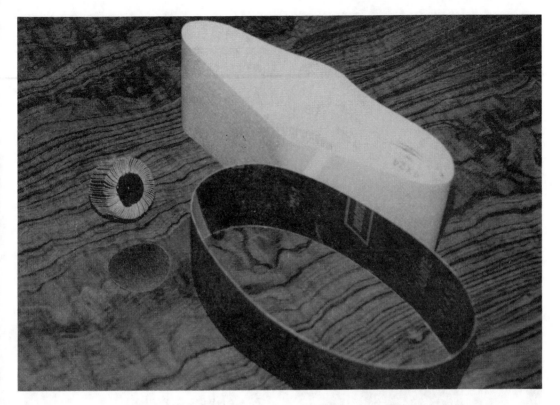

FIGURE 16.18 Abrasives (courtesy of Wellington Sears Company).

product quality, production technique and type of rubber. The basic fiber requirements for rubber products are strength, dimensional stability, and good adhesion to rubber.

Figure 16.18 shows abrasive products in which fabrics are used for structural integrity.

5. TRANSPORT BAGS AND SHEETS

Large bags made of coated textile fabrics are very convenient to transport loose materials of one ton or more (Figure 16.19). The coated fabric is made of a woven nylon or polyester backing coated on both sides with PVC. The coating provides a waterproof and airtight structure. Huge transport bags are widely used in civil engineering, agriculture, and the food industry.

Coated textile fabrics are also used as transport handling sheets. Due to their strength, they can handle various products such as hay, bricks, prefabs, etc.

6. FABRICS TO CONTROL OIL SPILLS

Oil spills are dangerous and not uncommon. The Oil Pollution Act of 1990 (OPA '90), which was formed as a direct result of the 1989 Exxon Valdez oil spill in Alaska, requires facilities that handle oil products near lakes, rivers, oceans or other waterways to maintain oil spill containment and cleanup capabilities [12].

Oil booms set out on the water contain damaging spillage (Figure 16.20). Most oil booms are made of PVC coated polyester, PVC coated nylon or urethane coated polyester. Sections of some booms are constructed from chloroprene coated fabric or nitrite rubber coated fabric. Polyurethane coated aramid fabrics are used to control oil spills in seas.

Rapid deployment booms are usually made of lightweight 12 to 18 ounce PVC coated polyester. The weight can be up to 100 to 110 ounces for conveyor belting materials for open-ocean barriers.

The fabrics must be oil, water and abrasion

FIGURE 16.19 Flexible containers for material storage and transportation (courtesy of Taiyo Kogyo).

FIGURE 16.20 Oil spill control with coated fabric.

resistant with good tear and tensile strength. For durability, some fabrics have modified plasticizers and coatings that stay flexible in the presence of hydrocarbons.

Boom Types

There are three main types of booms:

- Fence booms have a rigid or semi-rigid material that acts as a vertical screen against oil.
- External tension booms are controlled with an external tension bridle.
- Curtain booms have a flexible skirt held down by ballast weights or tension lines. Curtain boom types include internal foam, external foam, self-inflatable and pressure inflatable.

Other special purpose booms include tidal seal booms, fireproof booms for in situ oil burning, ice booms and sorbent booms.

7. CANVAS COVERS AND TARPAULINS

Canvas covers are made from heavy duty cotton fabrics of denim, twill, sateen and duck construction that are chemically treated for water and mildew resistance. These covers are breathable and used to protect objects such as heavy farm equipment, trucks and trailers. They may consist of several layers of fabric.

Tarpaulins are coated or laminated textile fabrics used in a variety of applications to protect materials from the effects of weather. In contruction (Chapter 6.0), building supplies, timber and wet concrete are covered with tarpaulins for protection. In transportation (Chapter 15.0), tarpaulins are used to cover cargoes that are transported by rail, road, and sea. Tarpaulins are generally waterproof. The most widely used base woven fabrics include polyester, nylon and polyethylene. Vinyl and polyethylene film are used for coating and laminating. More expensive coatings, such as neoprene and polyurethane are used for covers or containers that will be used in contact with fuels or oils. Coated fabrics for foodstuff and medical products have to be made of polymers that

TABLE 16.2 *Cordage Institute Rope Specifications.*

Fiber	Nominal Diameter (inch)	Nominal Circumference (inch)	Average Linear Density (lb/100)	Tensile Strength (lb min, new)
Sisal	4	12	435	84,000
Manila	4	12	434	94,500
Polyester	2.25	7	157	96,500
	2	6	128	106,000
Polypropylene	3	9	153	103,000
Nylon	2.125	6.5	109	95,500
	2	6	100	103,000
Polyester/ polypropylene	3	9	174	103,000
Kevlar® aramid	1.25	4	54	112,000

are FDA (Food and Drug Administration) approved.

Important general properties of canvas and tarpaulins are strength, tear resistance, puncture resistance, weather resistance, flexibility and durability. Flame resistance or UV resistance treatments may be applied. Due to synthetic materials used, tarpaulins are usually lighter and stronger than canvas covers.

8. ROPES AND NETS

Natural and synthetic fibers are used in ropes for various purposes. The main natural fibers are sisal and manila. Synthetic materials include polyester, polypropylene, polyester/ polypropylene composite, nylon and aramid. Strength, elongation, weight and abrasion characteristics of ropes are important. Kevlar® aramid is used in deep-sea work-system cables and balloon tethers. Table 16.2 shows the Cordage Institute Rope Specifications. Fishing nets and lines are made of nylon fibers (Figure 16.21).

9. HOME AND OFFICE FURNISHINGS

The home and office furnishing market is an important part of the textile industry. Home furnishing textiles include carpets and rugs, draperies and curtains, upholstery, blankets, bedspreads, sheets, pillowcases and tablecloths.

Home and office furnishing textiles form a bridge between consumer (traditional) textiles and industrial textiles. In many ways, they have the characteristics of both technical textiles and traditional textiles. For example, functions of carpets in homes and offices include heat insulation and sound deadening. Tufted and nonwoven mats are used in homes and offices for abrasion resistance to protect the original floor where the traffic is high (Figure 16.22). Awnings are used internally in offices and homes for aesthetic purposes (Figure 16.23). Wall coverings and partition panel fabrics in homes or offices are used for insulation, privacy and aesthetics (Figure 6.24). Ceiling tiles made of fabrics provide sound absorption and decoration. Drapery and upholstery fabrics in homes and offices play an important role in creating a warm leisure or friendly business environment (Figures 6.25 and 6.26). Figure 16.27 shows a tablecloth that may have been coated for leakproofness.

In addition to functional requirements, the aesthetic criteria, i.e., color, texture and appearance, are of primary importance to the user in home and furnishings. This is one of the differences between a typical industrial textile and a home furnishing textile. Functional requirements now include flame resistance in many home and office applications.

With the change in lifestyle and behavior of consumers, more exact or tailored requirements are demanded from furnishing textiles. There-

FIGURE 16.21 Fishing net made of nylon (courtesy of AlliedSignal).

FIGURE 16.22 Nonwoven mat (courtesy of Foss Manufacturing Co., Inc.).

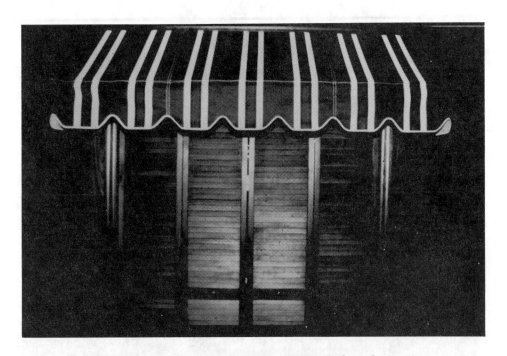

FIGURE 16.23 Internal awning (courtesy of Wellington Sears Company).

FIGURE 16.24 Wall covering and panel fabrics (courtesy of Foss Manufacturing Co., Inc.).

FIGURE 16.25 Home furnishings (courtesy of Wellington Sears Company).

FIGURE 16.26 Upholstery fabrics (courtesy of Wellington Sears Company).

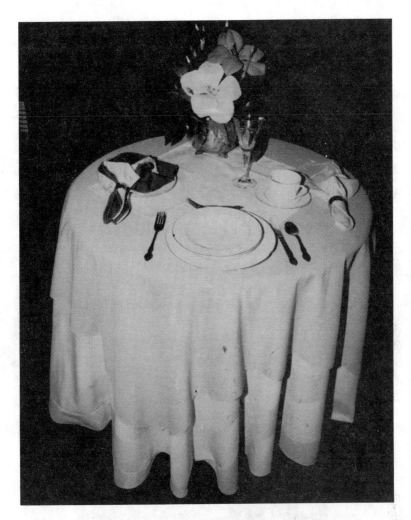

FIGURE 16.27 Tablecloth (courtesy of Wellington Sears Company).

FIGURE 16.28 Footwear components made of textiles
(courtesy of Foss Manufacturing Company, Inc.).

FIGURE 16.29 Three-dimensional fabric structure for athletic shoes (courtesy of Synthetic Industries).

FIGURE 16.30 Luggage lining fabric (courtesy of Reifenhauser).

fore, many times the physical properties of each product are custom tailored which is characteristic of technical textiles. As a result, design, manufacturing and testing of home and office furnishings are getting more sophisticated.

10. MISCELLANEOUS APPLICATIONS

Woven glass fabrics, which are coated with a protective vinyl, are used for window and door insect screening. The fabric is non-combustible and does not rust, corrode or stain. The typical mesh is 18 × 16 with colors of usually charcoal, silvery gray and bronze.

Another common application area of textiles is footwear. Structural components such as box toes, counters, midsoles, insoles, doublers, heel tucks and linings are made of textiles (Figure 16.28). Three-dimensional fabrics are used in athletic shoes. Figure 16.29 shows a spring coil fabric for high performance athletic shoes.

Substrates are used for coated fabrics for seating materials, footwear upper materials, vertical blinds and car wash fabrics. Luggage is another application area of textiles (Figure 16.30).

References and Review Questions

1. REFERENCES

1. Batel, W. and Hinz, T., "Exposure Measurements Concerning Protective Clothing in Agriculture," in *Performance of Protective Clothing: Second Symposium, ASTM STP 989,* S. Z. Mansdorf, R. Sager, and A. P. Nielsen, eds., American Society for Testing and Materials, Philadelphia, 1988.

2. Fraser, A. J. and Keeble, V. B., "Factors Influencing Design of Protective Clothing for Pesticide Application," in *Performance of Protective Clothing: Second Symposium, ASTM STP 989,* S. Z. Mansdorf, R. Sager, and A. P. Nielsen, eds., American Society for Testing and Materials, Philadelphia, 1988.

3. Litchfield, M. A., "A Review of the Requirements for Protective Clothing for Agricultural Workers in Hot Climates," in *Performance of Protective Clothing: Second Symposium, ASTM STP 989,* S. Z. Mansdorf, R. Sager, and A. P. Nielsen, eds., American Society for Testing and Materials, Philadelphia, 1988.

4. Laughlin, J. M., "Fabric Parameters and Pesticide Characteristics That Impact on Dermal Exposure of Applicators," in *Performance of Protective Clothing, ASTM STP 900,* R. L. Barker and G. C. Coletta, eds., American Society for Testing and Materials, Philadelphia, 1986.

5. Branson, D. H. et al., "Effectiveness of Selected Work Fabrics as Barriers to Pesticide Penetration," in *Performance of Protective Clothing, ASTM STP 900,* R. L. Barker and G. C. Coletta, eds., American Society for Testing and Materials, Philadelphia, 1986.

6. Hobbs, N. E. et al., "Effects of Barrier Finishes on Aerosol Spray Penetration and Comfort of Woven and Disposable Nonwoven Fabrics for Protective Clothing," in *Performance of Protective Clothing, ASTM STP 900,* R. L. Barker and G. C. Coletta, eds., American Society for Testing and Materials, Philadelphia, 1986.

7. Leonas, K. K. and DeJonge, J. O., "Effect of Functional Finish Barriers on Pesticide Penetration," in *Performance of Protective Clothing, ASTM STP 900,* R. L. Barker and G. C. Coletta, eds., American Society for Testing and Materials, Philadelphia, 1986.

8. "Apply Pesticides Correctly, A Guide for Commercial Applicators," U.S. Department of Agriculture and Environmental Protection Agency, U.S. Government Printing Office, Washington, D.C., 1976.

9. Reisch, M. S. "High-Performance Fibers Find Expanding Military, Industrial Uses," *Chemical and Engineering News,* February, Vol. 2, 1987.

10. "Banners and Flags, Design and Specification Guidelines," *Fabrics and Architecture,* November-December 1994.

11. Blow, C. M. and Hepburn, C., eds., *Rubber Technology and Manufacture,* 2nd Ed., Butterworth Scientific, 1982.

12. Ravnitzky, M., "A Boom in Oil Booms?" *Industrial Fabric Products Review,* April 1994.

2. REVIEW QUESTIONS

1. What are the application areas of textiles in agriculture?

2. Explain the advantages that textiles offer in electronics.

3. What are the main requirements for banner and flag fabrics? Explain.

4. What are the functions of textiles in rubber products?

5. Explain the common characteristics of industrial and home furnishing textiles.

FIBER PROPERTIES AND TECHNOLOGY

Fiber Properties and Technology

S. ADANUR

The major properties of textile fibers that have important industrial textile usage are included in this chapter. There are many synthetic fibers of the same generic class, manufactured by different companies. For those fibers which are reasonably well known or widely used, properties are listed generically and by trade names. Fibers with lesser known trade names are grouped under their proper generic class, with ranges of values being given wherever possible. The continual introduction of new man-made fibers makes it a genuinely impossible task to present a tabulation which will remain up-to-date for any reasonable length of time. Fiber names, types, and properties are constantly changing. Therefore, caution should be exercised in using the numerical data presented in the following tables. This chapter also discusses the significance of fiber properties as they influence the performance characteristics of yarns and fabrics.

Appendix 4 lists the characteristics, industrial uses and manufacturers of generic man-made fibers and trade names. Figure 17.1 shows longitudinal and cross-sectional photographs of selected industrial fibers.

1. REFRACTIVE INDEX

The refractive index (n) of a substance is defined as the ratio of the velocity of light in a vacuum to the velocity of light through the substance. Because of the molecular orientation within fibers, the speed of light along the fiber axis is usually different from that transverse to the fiber axis. The greater the amount of orientation and degree of crystallinity, the greater will be the difference between the refractive in-

dices which are parallel with, and transverse to, the fiber axis. The difference is called the "birefringence." For all fibers except Saran® the parallel index n_ϵ is greater than the transverse index n_ω, and so the birefringence is always positive. Refractive indices are useful for identifying fibers and, in research, indicating the nature of molecular order in experimental fibers. While refractive index may influence fiber luster, the effect of fiber crimp and the state of aggregation of fibers in yarns and yarns in fabrics are of far greater significance in obtaining or preventing fabric luster. Except as it is a measure of molecular order, which in turn may influence the tensile and elastic properties of fibers, refractive index is of no special significance in the industrial textile picture. Table 17.1 lists refractive indices and birefringences of natural and man-made fibers.

2. DENSITY (SPECIFIC GRAVITY) OF FIBERS

Table 17.2 lists densities of fibers. Density is defined as the ratio of a substance's unit volume weight to that of water at 4°C.

3. FIBER FINENESS AND DIAMETERS

Table 17.3 lists fiber diameters (μm) based on filament deniers for selected fibers.

4. FIBER STRENGTH AND TENACITY

While it might be desirable to catalog the absolute breaking strengths of individual fibers,

Name	Photomicrographs
Asterisk denotes trademark	Cross section, 500X Longitudinal, 250X

Polyester

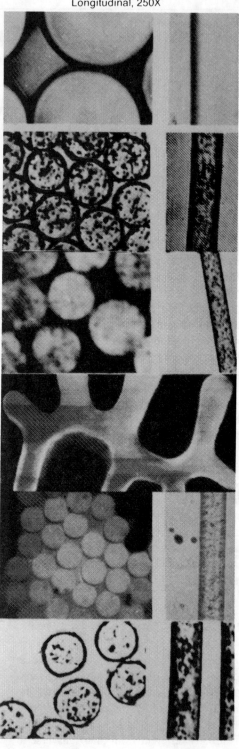

*A.C.E.
*Compet
Allied

* Dacron
Du Pont

*Fortrel
Wellman

4DG
Eastman

*Tairilin
Nan Ya Plastics Corp.,
America

*Trevira
Hoechst Celanese

FIGURE 17.1 Photomicrographs of man-made fibers (courtesy of McLean Hunter Publishing Co.).

Carbon

       *Thornel
       Amoco

Nylon

       Nylon 6

       Nylon 6,6

Lyocell

       * Tencel
       Courtaulds

Rayon

       *Fibro
       Courtaulds

       Cuprammonium

Photomicrographs

Cross section, 500X
Longitudinal, 250X

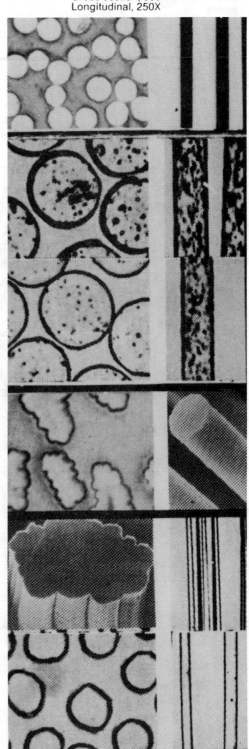

FIGURE 17.1 (continued) Photomicrographs of man-made fibers (courtesy of McLean Hunter Publishing Co.).

Name
Asterisk denotes
trademark

Photomicrographs
Cross section, 500X
Longitudinal, 250X

Rayon
North American
Rayon Corp.

*Saran
Pittsfield Weaving

Acetate

Acetate

Sulfar

*Ryton
Amoco
(Licensed by Phillips)

Acrylic

* Acrilan
Monsanto

*Creslan
*MicroSupreme
 Cytec

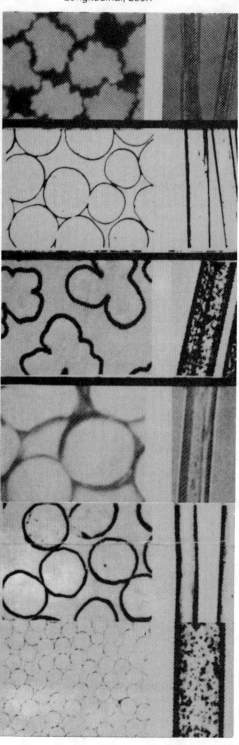

FIGURE 17.1 (continued) Photomicrographs of man-made fibers (courtesy of McLean Hunter Publishing Co.).

Name	Photomicrographs

Name
Asterisk denotes·
trademark

Photomicrographs
Cross section, 500X
Longitudinal, 250X

Modacrylic

    * SEF
      Monsanto

Olefin

    Polyethylene
    Hercules

    * Herculon
      Hercules

    *Marvess
    *Alpha
    Amoco

    *Essera,
    *Marquesa Lana,
    *Patlon III
    Amoco

    *Spectra 900
    *Spectra 1000
    Allied

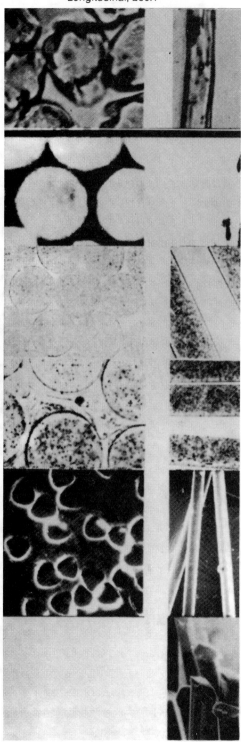

FIGURE 17.1 (continued) Photomicrographs of man-made fibers (courtesy of McLean Hunter Publishing Co.).

Name Asterisk denotes· trademark	Photomicrographs Cross section, 500X Longitudinal, 250X

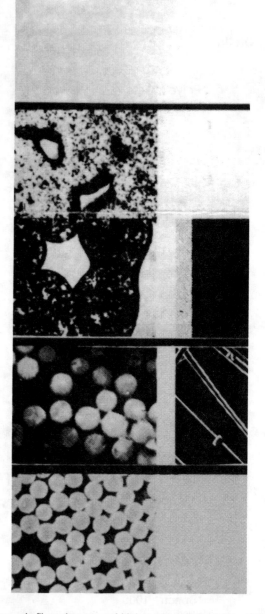

*Fibrilon
Synthetic Industries

Spandex

***Glospan/
Cleerspan, S-85**
Globe

* Lycra
Du Pont

Glass

Glass

Aramid

* Kevlar
Du Pont

FIGURE 17.1 (continued) Photomicrographs of man-made fibers (courtesy of McLean Hunter Publishing Co.).

Name

Asterisk denotes
trademark

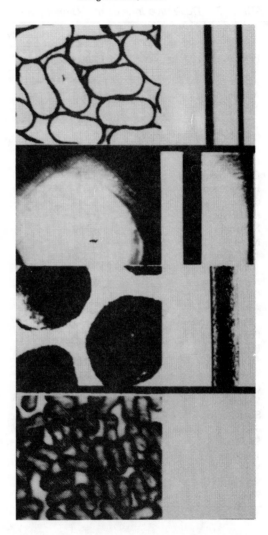

* **Nomex**
  Du Pont

Fluorocarbon

* **Gore-Tex**
  W. L. Gore

* **Teflon**
  Du Pont

Polybenzimidazole

PBI
Hoechst Celanese

FIGURE 17.1 (continued) Photomicrographs of man-made fibers (courtesy of McLean Hunter Publishing Co.).

TABLE 17.1 Refractive Indices and Birefringences of Fibers ([1], Fiber Manufacturers' Data).

Fiber	Refractive Index		Birefringence $(n_\epsilon - n_\omega)$
	Parallel (n_ϵ)	Transverse (n_ω)	
Acetate	1.479	1.477	0.002
Acrylic	1.525	1.520	0.004
Aramid			
Nomex®	1.790	1.662	0.128
Kevlar®	2.322	1.637	0.685
Asbestos	1.5–1.57	1.49	0.01–0.08
Cotton, raw	1.580	1.533	0.047
Cotton, mercerized	1.57	1.52	0.05
Fluorocarbon	1.37	–	–
Glass	1.547	1.547	0.000
Modacrylic	1.536	1.531	0.005
Novoloid	1.650	1.648	0.002
Nylon			
Nylon 6	1.568	1.515	0.053
Nylon 6,6	1.582	1.519	0.063
Nytril	1.484	1.476	0.008
Olefin			
Polyethylene	1.556	1.512	0.044
Polypropylene	1.530	1.496	0.034
Polycarbonate	1.626	1.566	0.060
Polyester			
Dacron®	1.710	1.535	0.175
Kodel®	1.632	1.534	0.098
Cuprammonium rayon	1.548	1.527	0.021
Viscose rayon	1.547	1.521	0.026
Saran	1.603	1.611	−0.008
Silk	1.591	1.538	0.053
Spandex	1.5	–	–
Triacetate	1.472	1.471	0.001
Vinal	1.543	1.513	0.030
Vinyon (PVC)	1.541	1.536	0.005
Wool	1.556	1.547	0.009

TABLE 17.2 Density of Fibers [1,2].

Fiber	Density (mg/mm³)
Polyester	
ACE®	1.38
Compet®	1.38
Dacron®	1.34–1.39
Fortrel®	1.38
Tairilin®	1.38
Trevira®	1.38
Thornel® carbon	
High strength	1.77
High modulus	1.77
Ultra-high modulus	1.96
Nylon 6	1.14
Nylon 6,6	1.13–1.14
Tencel® lyocell	1.56
Viscose rayon	1.52
Cuprammonium rayon	1.53
Saran	1.62–1.75
Acetate	1.32
Ryton® sulfar	1.37
Acrylic	1.17
Modacrylic	1.35
Nytril	1.20
Polyethylene, low density	0.92
Polyethylene, high density	0.95
Polypropylene	0.90
Spectra® olefin	0.97
Lycra® spandex	1.2
E-glass, single filament	2.54–2.69
S-glass, single filament	2.48–2.49
Glass, multifilament	2.5
Aramid	
Kevlar®	1.44
Nomex®	1.38
Fluorocarbon	
Gore-Tex®	0.8–2.2
Teflon®	2.1
Polybenzimidazole	1.43
Asbestos	2.1–2.8
Flax	1.54
Cotton	1.54
Novoloid	1.28
Polycarbonate	1.21
Silk	1.35
Vinyon (PVC)	1.40
Wool	1.31

cross-sectional shapes and areas of fibers can vary widely. When subjected to a tensile force, the total strength of a fiber is dependent upon both its intrinsic ability to remain intact and upon its dimensions. The absolute value of a fiber's strength may be meaningless until it is related to its cross-sectional area or its linear density, i.e., weight per unit length. As is the custom with all engineering materials, fiber breaking strengths can be listed on a pounds per square inch (psi) basis. More commonly, however, the textile trade uses the term "tenacity" to describe strength on a grams per denier (gpd) basis. This is because it is easier to determine a fiber's or yarn's weight per length than its weight per cross-sectional area, because yarn weight is an important textile physical and economic factor. Since denier is based upon weight per unit length, it is obvious that tenacity is influenced by the specific gravity of the fiber, while strength per unit area is not. The relationship between these properties is:

tensile strength (lb per sq. in. or psi)

$$= 12{,}800 \times \text{tenacity (grams per denier or gpd)}$$

$$\times \text{ specific gravity} \qquad (17.1)$$

In Table 17.4 tensile strengths are listed in psi and grams per denier for various fibers. The relationship between tex and denier is:

$$\text{tex} = \frac{\text{denier}}{9} \qquad (17.2)$$

Textile fibers are relatively light, therefore they can be highly efficient on a strength-to-weight ratio basis. High tenacity nylon, for example, with a specific gravity of 1.14 may have a tenacity of 9 grams per denier (81 grams per tex) as compared with a high strength steel (500,000 psi) or about 3.5 gpd. On a cross-sectional area basis, high tenacity nylon has a strength exceeding 100,000 psi.

While textile fibers may have high tenacities, often the corresponding strength of the textile fabric is disproportionately lowered, due to the inefficient translation of fiber strength into yarn strength and yarn strength into fabric strength. This is particularly true with staple fibers. Because of non-uniform fiber lengths, interfiber slippage, random fiber orientation within the yarn, and the helical path of fibers due to yarn

TABLE 17.3 Fiber Diameters of Various Textile Fibers [3].

Denier	Viscose Rayon	Acetate and Vinyon	Nylon	Polyester	Olefin
1	9.6	10.3	11.1	10.1	12.5
2	13.6	14.5	15.7	14.3	17.7
3	16.7	17.8	19.3	17.5	21.7
4	19.3	20.6	22.3	20.2	25.0
5	21.6	23.0	24.9	22.6	28.0
6	23.6	25.2	27.3	24.8	30.7
7	25.5	27.3	29.5	26.8	33.1
8	27.3	29.1	31.5	28.6	35.4
9	28.9	30.9	33.4	30.4	37.5
10	30.5	32.6	35.2	32.0	39.6
12	33.4	35.7	38.6	35.1	43.3
14	36.1	38.5	41.7	37.9	46.8
16	38.6	41.2	44.5	40.5	50.1
18	40.9	43.7	47.3	42.9	53.1
20	43.1	46.1	49.9	45.3	56.0

twist, only a part of the theoretically available strength, based upon the number of fibers in the yarn cross section, will actually result in the yarn.

Cotton has a tenacity of about 3.0 to 4.9 gpd (27 to 44 gpt). Because it exists only as staple, its fiber strength is not completely convertible into equivalent yarn strength, the efficiency of the conversion being on the order of magnitude of 50%. Thus, cotton yarns usually have tenacities not in excess of about 3 gpd (27 gpt). The competitive advantages of multi-filament high tenacity rayon, nylon, or other synthetic yarns become obvious.

It should be apparent that strength per se is not the only criterion by which the performance of textiles should be judged. Textile yarns and fabrics are normally used because they have flexibility and moderate strength, coupled with the ability to deform or strain under load, thereby absorbing energy, and the ability to recover or retract when the load is removed. Many other chemical and physical properties will influence the ultimate selection of the fiber and the yarn and fabric into which it is converted.

5. FIBER BREAKING ELONGATION

Table 17.5 lists breaking elongations or strains which fibers can undergo up to the point of fail-

ure. As a general rule it can be stated that an inverse relationship exists between fiber strength and elongation. For natural cellulosic fibers, linen and ramie are stronger but have lower elongations than cotton; the animal fibers are much weaker but have greater elongations than the cellulosics. Glass fibers are very strong but very brittle, i.e., they have low elongations. In the manufacture of man-made fibers, for a particular generic type, a range of inverse strength-elongation values is normally obtainable. Thus nylon can be made with a tenacity of 9 gpd (81 gpt), an elongation of 12% or 5 gpd (45 gpt) and an elongation of 30%.

6. FIBER LOAD-ELONGATION DIAGRAMS

Figure 17.2 shows fiber tenacity (gpd) versus percent elongation diagrams. Often such diagrams are popularly but somewhat erroneously called "stress-strain" curves. Actually the tenacity scale should be converted to a pounds per square inch (psi) scale in order to produce a true stress versus strain diagram. A variety of fibers with a wide range of tenacity-elongation diagrams is available. The selection of a fiber for a particular purpose, of course, is dependent upon many property requirements. The shape of a fiber's load-elongation diagram, initially and after the fiber has been cyclically loaded and

TABLE 17.4 Tensile Properties of Fibers [2].

Fiber	Breaking Tenacity (gpd)		Tensile Strength (psi)
	Standard	Wet	
Polyester			
A.C.E® and Compet®			
Filament–HT	8.3–9.0	9.0	135,000–160,000
Dacron®			
Staple and tow	2.4–7.0	2.4–7.0	39,000–106,000
POF	2.0–2.5	2.0–2.5	33,000–42,000
Filament–RT	2.8–5.6	2.8–5.6	50,000–99,000
Filament–HT	6.8–9.5	6.8–9.5	106,000–168,000
Fortrel®			
Staple series 400–RT	4.8–6.0	4.8–6.0	85,000
Staple series 300–HT	6.0–6.8	6.0–6.8	102,000
PO filament	2.0–2.5	2.0–2.5	80,000–88,000
Tairilin®			
POY	2.0–3.0	2.0–3.0	33,000–42,000
Staple	4.5–7.0	4.5–7.0	80,000–102,000
Trevira®			
Staple	3.5–6.0	3.5–6.0	41,000–105,000
POF	2.0–2.5	2.0–2.5	33,000–42,000
Filament–HT	7.2–8.2	7.2–8.2	118,000–140,000
Thornel® carbon			
High strength	24.1	24.1	550,000
High modulus	21.3	21.3	500,000
Ultra-high modulus	10.8	10.8	270,000
Nylon			
Nylon 6			
Staple	3.5–7.2		62,000–98,000
Monofil. and filament–RT	4.0–7.2	3.7–6.2	73,000–100,000
BCF–RT	2.0–4.0	1.7–3.6	
Filament–HT	6.5–9.0	5.8–8.2	102,000–125,000
Nylon 6,6			
Staple and tow	2.9–7.2	2.5–6.1	
Monofil. and filament–RT	2.3–6.0	2.0–5.5	40,000–106,000
Filament–HT	5.9–9.8	5.1–8.0	86,000–134,000
Tencel® lyocell–HT	4.8–5.0	3.8–4.2	
Rayon			
Fibro®			
RT–Multi-lobed	2.3	1.1	
IT–0.9 den or 0.25 in and up	3.0	1.5	
Cuprammonium, filament	1.95–2.0	0.95–1.1	
Rayon			
Filament–RT	1.9–2.3	1.0–1.4	
Filament–HT	4.9–5.3	2.8–3.2	
Saran®, monofilament	1.2–2.2	1.2–2.2	15,000–27,000
Acetate, filament and staple	1.2–1.4	0.8–1.0	20,000–24,000
Ryton® sulfar, staple	3.0–3.5	3.0–3.5	35,000–40,000
Acrylic			
Acrilan®, staple and tow	2.2–2.3	1.8–2.4	30,000–40,000
Creslan®, staple and tow	2.0–3.0	1.6–2.7	30,000–45,000
MicroSupreme, staple	2.0–3.0	1.6–2.7	30,000–45,000
SEF® modacrylic, staple	1.7–2.6	1.5–2.4	29,000–45,000

(continued)

TABLE 17.4 *(continued).*

Fiber	Breaking Tenacity (gpd)		Tensile Strength (psi)
	Standard	Wet	
Olefin			
Polyethylene			
Monofilament, low density	1.0–3.0	1.0–3.0	11,000–35,000
Monofilament, high density	3.5–7.0	3.5–7.0	30,000–85,000
Herculon®			
Staple	3.5–4.5	3.5–4.5	41,000–52,000
Buff filament	3.0–4.0	3.0–4.0	35,000–47,000
Marvess® and Alpha®			
Staple and tow	2.0–5.0	2.0–5.0	60,000–100,000
Multifilament	2.5–5.5	2.5–5.5	20,000–50,000
Essera®, Marquesa Lana®, Pation® III			
BCF	2.5–3.5	2.5–3.5	30,000–40,000
Staple	2.5–4.0	2.5–4.0	30,000–45,000
Spectra® 900	30	30	375,000,000
Spectra® 1000	35	35	425,000,000
Fibrilon®			
Staple	2.5–5.5	2.5–5.5	12,000–60,000
Fibrillated	3.5–5.0	2.5–5.0	40,000–60,000
Multifilament—RT	2.5–5.5	2.5–5.5	20,000–50,000
Spandex			
Glospan®/Cleerspan, S-85, multifilament	0.7		11,000
Lycra®			
Type 126, 127	1.0		11,000–14,000
Type 128	0.8		
Glass			
Single filament—E-glass	15.3	15.3	450,000–550,000
Single filament—S-glass	19.9	19.9	650,000–700,000
Multifilament	9.6	6.7	313,000
Aramid			
Kevlar®			
Kevlar 29/Kevlar 49	23	21.7	425,000
Kevlar 149	18		340,000
Kevlar 68	24.0		425,000
Kevlar HT (129)	26.5		490,000
Nomex®, staple and filament	4.0–5.3	3.0–4.1	90,000
Fluorocarbon			
Gore-Tex®	3.0–4.0	3.0–4.0	85,000–115,000
Teflon®			
TFE multifil., staple, tow, flock	0.9–2.0	0.9–2.0	40,000–50,000
FEP, PFA monofilament	0.5–0.7	0.5–0.7	14,000–20,000
PBI, staple	2.6–3.0	2.1–2.5	50,000
Asbestos	2.5–3.1		80,000–300,000
Cotton	3.0–4.9	3.3–5.39	59,500–97,000
Silk	2.4–5.1	1.75–4.01	38,500–88,000
Wool	1.0–1.7	0.85–1.44	16,500–28,000
Steel	3.5	3.5	512,500

HT: high tenacity.
POF: partially oriented filament.
POY: partially oriented yarn.
RT: regular tenacity.
IT: intermediate tenacity.

TABLE 17.5 Breaking Elongation, Elastic Recovery, Stiffness and Toughness of Fibers [2].

Fiber	Breaking Elongation (%)		Elastic Recovery (%)	Average Stiffness (gpd)	Average Toughness (gpd)
	Standard	Wet			
Polyester					
A.C.E.® and Compet®					
Filament—HT	13–22	13–15	77 at 5%; 75 at 10%	55–56	0.7–0.9
Dacron®					
Staple and tow	12–55	12–55	82 at 3%	12–17	0.2–1.1
POF	120–150	120–150			1.3–1.8
Filament—RT	24–42	24–42	76 at 3%	10–30	0.4–1.1
Filament—HT	12–25	12–25	88 at 3%	30	0.5–0.7
Fortrel®					
Staple series 400—RT	44–55	44–55			1.3–1.8
Staple series 300—HT	24	24			
PO filament	120–150	120–150			
Tairilin®					
POY	120–170	120–170			1.3–1.8
Staple	28–6	28–55	88 at 3%	10–30	0.3–1.5
Trevira®					
Staple	18–60	18–60	67–86 at 2% 57–74% at 5%	7–31	0.28–1.50
POF	130–145	130–145			1.3–1.8
Filament—HT	10–20	10–20	99 at 1%	54–77	0.35–0.55
Thornel® carbon					
High strength	1.6	1.6	100	1,500	
High modulus	1.0	1.0	100	2,300	
Ultra-high modulus	0.38	0.38	100	3,000	
Nylon					
Nylon 6					
Staple	30–90	42–100	100 at 2%	17–20	0.64–0.78
Monofil. and filament—RT	17–45	20–47	98–100 at 1–10%	18–23	0.67–0.90
BCF—RT	30–50	30–60			
Filament—HT	16–20	19–33	99–100 at 2–8%	29–48	0.75–0.84
Nylon 6,6					
Staple and tow	16–75	18–78	82 at 3%	10–45	0.58–1.37
Monofil & filament—RT	25–65	30–70	88 at 3%	5–24	0.8–1.25
Filament—HT	15–28	18–32	89 at 3%	21–58	0.8–1.28
Tencel® lyocell—HT	14–16	16–18		30	0.34
Rayon					
Fibro®					
RT—Multi-lobed	18–22				
IT—0.9 den or 0.25 in and up	18–22				
Cuprammonium, filament	8–14	24–28			
Rayon					
Filament—RT	20–25	24–29			
Filament—HT	11–14	13–16			
Saran®, monofilament	15–25	15–25	95 at 5–10%	5–10	0.165–0.265
Acetate, filament and staple	25–45	35–50	48–65 at 4%	3.5–5.5	0.17–0.30
Ryton® sulfar, staple	35–45		100 at 2%; 86 at 10%	10–20	
Acrylic					
Acrilan®, staple and tow	40–55	40–60	99 at 2%; 89 at 5%	5–7	0.4–0.5
Creslan®, staple and tow	35–45	41–50		6–8	0.62
MicroSupreme, staple	30–40	30–40		6–8	0.62
SEF® modacrylic, staple	45–60	45–65	100 at 1%; 95 at 10%	3.8	0.5

(continued)

TABLE 17.5 (continued).

Fiber	Breaking Elongation (%)		Elastic Recovery (%)	Average Stiffness (gpd)	Average Toughness (gpd)
	Standard	Wet			
Olefin					
Polyethylene					
Monofilament, low density	20–80	20–80	Up to 95 at 5%	2–12	0.3
			88 at 10%		
Monofilament, high density	10–45	10–45	Up to 100 at 1–10%	20–50	
Herculon®					
Staple	70–100	70–100	96 at 5%; 90 at 10%	20–30	1–3
Buff filament	80–100	80–100	96a at 5%; 90 at 10%	20–30	1–3
Marvess® and Alpha®					
Staple and tow	60–100	80–98	90–93 at 5%; 83 at 10%	3–10	1.5–4
Multifilament	20–50	85–99	92–96 at 5%; 87 at 10%	12–25	0.75–3.00
Essera®, Marquesa Lana®, Pation® III					
BCF	40–60	40–60	90 at 5%; 65–90 at 10%	5–10	0.9–1.05
Staple	30–450	30–450			
Spectra® 900	3.6	3.6		1,400	
Spectra® 1000	2.7	2.7		2,000	
Fibrilon®					
Staple	30–150	30–150	93 at 5%; 85 at 10%		
Fibrillated	14–18	14–18	85 at 5%; 75 at 10%		
Multifilament–RT	30–100	30–100	95 at 5%; 85 at 10%	12–25	0.75–3.00
Spandex					
Glospan®/Cleerspan, S-85, multifilament	600–700		99% at 50% (S-7)	0.17 (S-7)	
			98% at 200% (S-5)	0.05 (S-5)	
Lycra®					
Type 126, 127	400–625		97 at 50%	0.13–0.20	2.00
Type 128	800		99 at 200%		
Glass					
Single filament–E-glass	4.8	4.8	100	320	0.37
Single filament–S-glass	5.3–5.7	5.3–5.7	100	380	0.53
Multifilament	3.1	2.2	100	310	0.15
Aramid					
Kevlar®					
Kevlar, Kevlar 29/Kevlar 49	4.0/2.5	4.0/2.5	100 at 1%, 2%, 3%	500/900	
Kevlar 149	1.5		100 at 1%	1,110	
Kevlar 68	3.0		100 at 1%, 2%	780	
Kevlar HT (129)	3.3		100 at 1%, 2%	755	
Nomex®, staple and filament	22–32	20–30		70–120	
Fluorocarbon					
Gore-Tex®	5–20	5–20			
Teflon®					
TFE multifil., staple, tow, flock	19–140	19–140		1.0–13.0	0.15
FEP, PFA monofilament	40–62	40–62		7.0	0.10–0.12
PBI, staple	25–30	26–32		9–12	0.40
Cotton	3–7		74 at 2%; 45 at 5%	60–70	0.15
Silk	10–25		92 at 2%; 51 at 10%	60–116	0.4–0.8
Wool	25–35	25–50	99 at 25%; 63 at 20%	4.5	0.35

HT: high tenacity.
POF: partially oriented filament.
POY: partially oriented yarn.
RT: regular tenacity.
IT: intermediate tenacity.

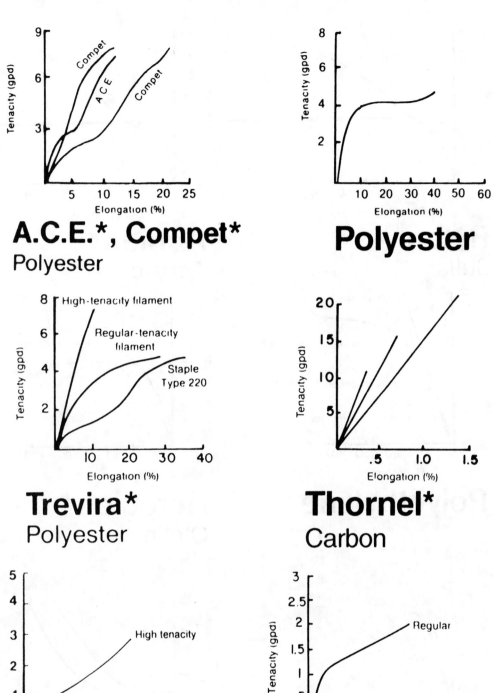

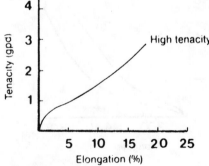

A.C.E.*, Compet*
Polyester

Polyester

Trevira*
Polyester

Thornel*
Carbon

Fibro*
Rayon

Cuprammonium

FIGURE 17.2 Typical tenacity versus elongation diagrams of fibers (courtesy of McLean Hunter Publishing Co.).

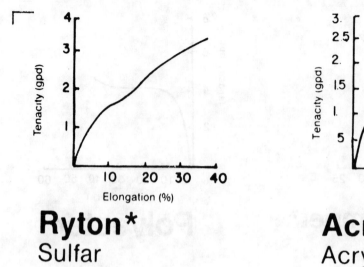

Ryton*
Sulfar

Acrilan*
Acrylic

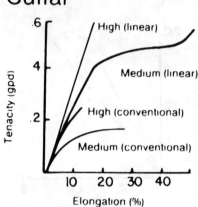

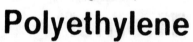

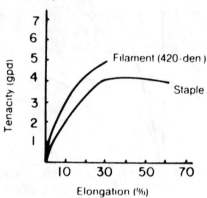

Polyethylene

Herculon*
Olefin

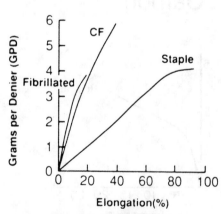

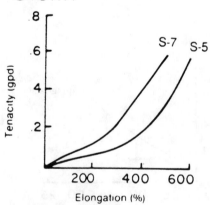

Fibrilon*
Olefin

Glospan*
Spandex

FIGURE 17.2 (continued) Typical tenacity versus elongation diagrams of fibers (courtesy of McLean Hunter Publishing Co.).

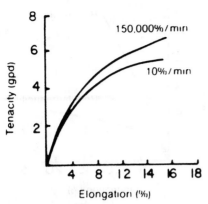

Nomex*
Aramid

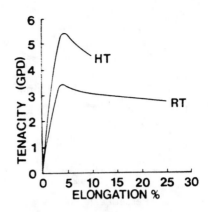

Gore-Tex*
Fluorocarbon

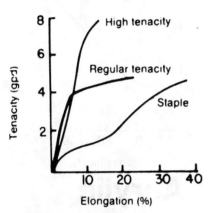

Dacron*
Polyester

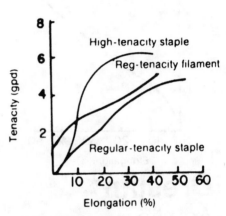

Fortrel*
Polyester

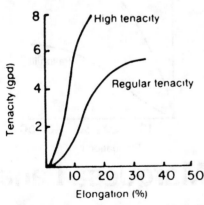

Nylon 6

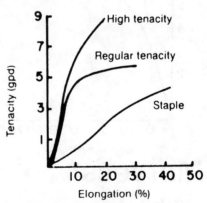

Nylon 6,6

FIGURE 17.2 (continued) Typical tenacity versus elongation diagrams of fibers (courtesy of McLean Hunter Publishing Co.).

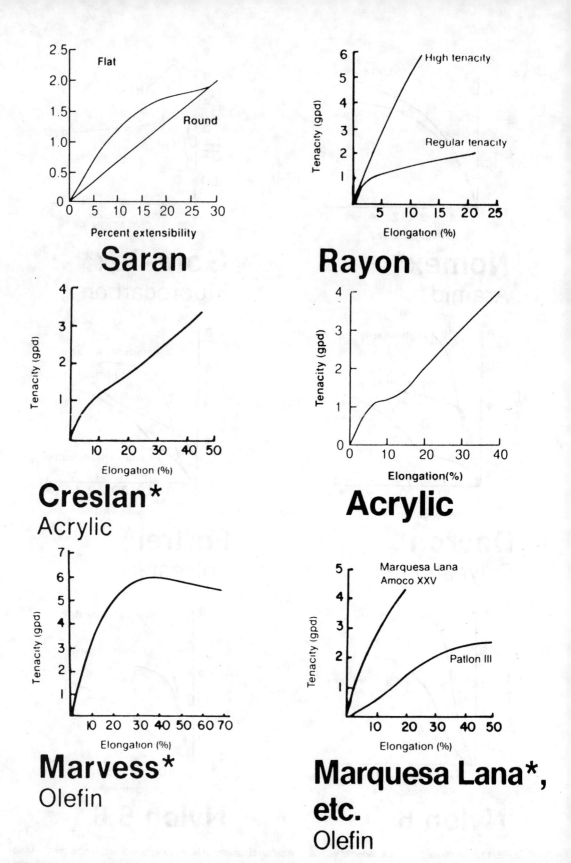

FIGURE 17.2 (continued) Typical tenacity versus elongation diagrams of fibers (courtesy of McLean Hunter Publishing Co.).

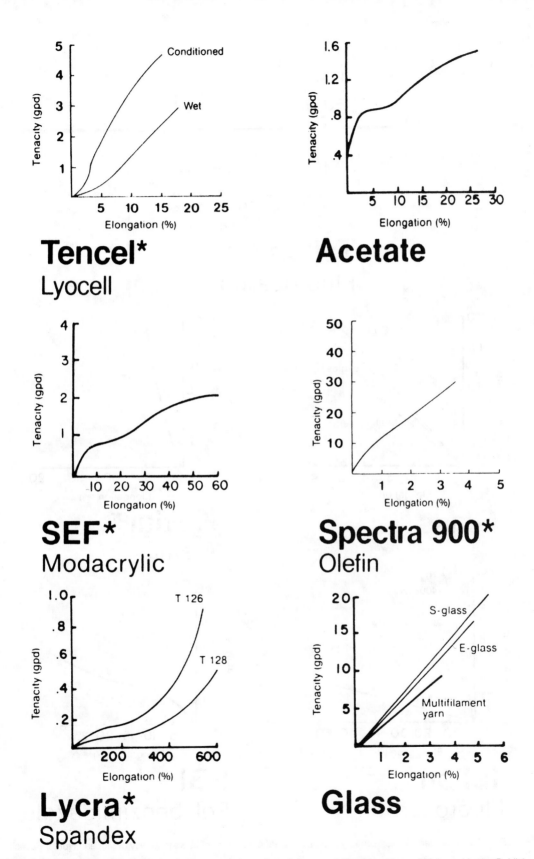

FIGURE 17.2 (continued) Typical tenacity versus elongation diagrams of fibers (courtesy of McLean Hunter Publishing Co.).

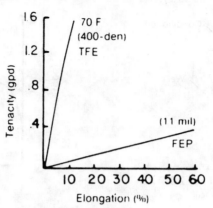

Teflon*
Fluorocarbon

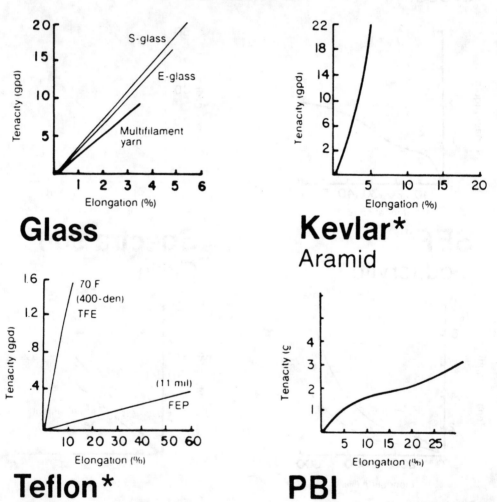

Glass

Kevlar*
Aramid

Teflon*
Fluorocarbon

PBI
Polybenzimidazole

FIGURE 17.2 (continued) Typical tenacity versus elongation diagrams of fibers (courtesy of McLean Hunter Publishing Co.).

unloaded at loads below rupture, is important in terms of the fiber's influence on such fabric properties as breaking and tear strength, energy absorption, dimensional stability, and abrasion resistance.

7. VISCOELASTICITY

Examination of fiber load-elongation diagrams shows that they are normally "non-Hookean," i.e., they do not obey Hooke's law which states that the strain (elongation) is linearly proportional to the applied stress (load). Figure 17.3 for a typical polyester, for example, shows an original region OA which is approximately linear. Point A is called the "proportional limit." Beyond this point, region AB develops where additional incremental loads produce proportionately larger increases in elongation, due to the propensity of the fiber to "creep," or slowly elongate under load. A slope change occurs at point B, followed by a stiffening region BC where additional loads beyond B cause smaller changes in elongation. Finally there is a second flow region CD.

Textile fibers and other high polymers such as rubbers, elastomers, films, and plastics which exhibit such varying "stress-strain" properties are stated to be "viscoelastic." The molecular and structural nature of these materials is such that during selected periods of progressive loading, they may function in a perfectly "elastic" or "spring-like" manner. During other periods of loading a "viscous flow" or creep type of deformation occurs. These two mechanisms most often act in varying contributory amounts at the same time and thus produce a viscoelastic type of stress-strain response.

The above description pertains to the application of a tensional force in one time loading to rupture. If a specimen is strained (elongated) to a value below rupture and the strain producing force is then removed, a strain recovery will usually take place. Repeated load applications and removals produce strains and recoveries which are dependent upon the material's viscoelastic properties. The phenomenon becomes complex, giving rise to hysteresis [Figure 17.10(b)]. The elongation recovery or "repeated stress" characteristics of fibers is discussed later in the chapter. They are of great significance because they influence such properties as fabric dimensional stability, wrinkle resistance, abrasion resistance, and energy absorption.

8. ELASTIC MODULUS

The ratio of stress to strain is called the "modulus" of a material. High modulus materials are stiff; they show small elongation under

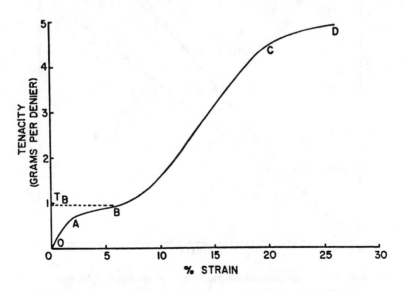

FIGURE 17.3 Analysis of a typical polyester tenacity-elongation diagram.

load. Low modulus materials are extensible; they show high elongation under load. "Young's modulus" is defined by ASTM as the ratio of change in stress to change in strain within the elastic limit of the material. The ratio is calculated from the stress, expressed in force per unit cross-sectional area, and the strain, expressed as a fraction of the original length. Young's modulus is of great value to the engineers utilizing wood, structural steel, and concrete since these materials are used below their elastic limits, and small deformations under load can be calculated with a reasonable degree of precision. The textile engineer and technologist also make use of this ratio, but because load-elongation diagrams of textiles are not usually linear over their entire use range, the "tensile modulus" of a fiber is not a constant and should be used only under conditions where

it is properly defined. ASTM also lists the following moduli definitions for textile materials:

- initial modulus: the slope of the initial straight-line portion of the stress-strain curve
- chord modulus: the ratio of the change in stress to the change in strain between two specified points in the stress-strain curve

8.1 Determination of Initial Modulus

In the case of a fiber exhibiting a region that obeys Hooke's law (i.e., has a linear region), a continuation of the linear region of the curve is constructed as shown in Figure 17.4. The intersection point B is the zero-strain point from which strain is measured. The initial modulus is

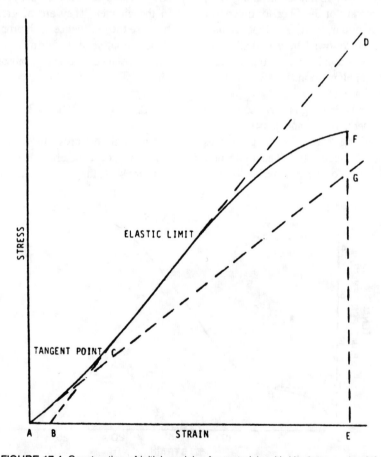

FIGURE 17.4 Construction of initial modulus for materials with Hookean region [4].

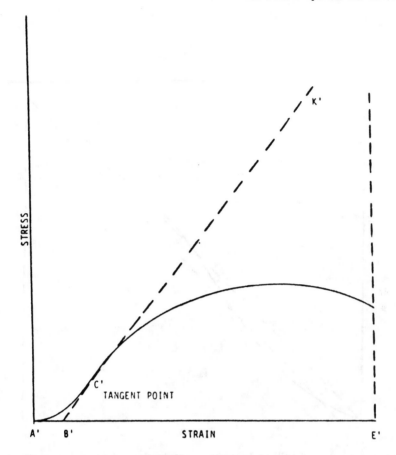

FIGURE 17.5 Construction of initial modulus for materials with no Hookean region [4].

determined by dividing the stress at any point along the line BD by the strain at the same point (measured from point B). Point C, the point where line BD first touches the stress-strain curve is the tangent point.

In the case of a fiber that does not exhibit any linear region (Figure 17.5), a tangent K'B' is constructed to the maximum slope. Point B' is the zero point from which strain is measured. Point C', the point where line K'B' first touches the stress-strain curve, is the tangent point. The initial modulus is determined by dividing the stress at any point along line B'K' by the strain at the same point (measured from point B').

8.2 Determination of Chord Modulus

In a typical stress-strain curve, a straight line is constructed through the zero-stress axis, such as zero strain point A" and a second point, such as 10% strain, point M" (Figure 17.6). The chord modulus is determined by dividing the stress at any point along line A"M" (or its extension) by the strain at the same point.

Figure 17.6 also represents a straight line constructed through any two specified points, point Q" and point R", other than zero and 10% strain. In this case, the line extends through the zero stress axis at point B". This intersection is the zero strain point from which strain is measured. The chord modulus can be determined by dividing the stress at any point along line Q"R" (or its extension) by the strain at the same point (measured from point B").

The modulus is also called the "stiffness" of the material since the higher the modulus the greater the resistance to deformation.

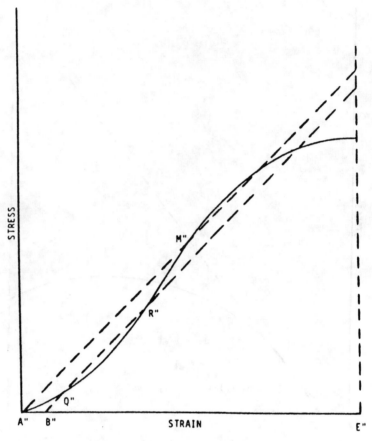

FIGURE 17.6 Construction of chord modulus [4].

Since engineering stress is given in pounds per square inch, modulus units for engineering materials normally are reported as psi per unit strain, or simply psi. Again, because convenient textile nomenclature uses the denier system, Young's moduli of fibers are reported as grams per denier per unit strain, i.e., the hypothetical tenacity required to strain the filament 100%, presuming strain extended at a rate equal to that of the initial slope of the stress-strain curve. The modulus (stiffness) is usually reported simply as "grams per denier." Moduli of major textile fibers are given in Table 17.5. Extensible fibers such as wool, acetate and acrylic have low textile moduli. Nylon and polyester can be drawn to varying degrees so that their moduli may have varying ranges (on the average, polyester slightly higher than nylon). Glass at 300 has a high modulus. However, carbon,

Spectra® olefin, and Kevlar® aramid fibers have moduli far above the range of other textile fibers. This is why these high modulus fibers, yarns, or fabrics are used in plastic laminates where minimum distortion under load is required. However, because of their high modulus and inability to deform under load (i.e., brittleness), excessively high stress concentrations may be induced in these fibers, and they may fail. In other low modulus fibers, the high strains resulting from the application of the non-uniform loads reduce the stress concentration, and the fiber can remain intact. This, among other reasons, is why high modulus fibers have not found wide use in textile fabrics where repeated flexing and abrasion, both being examples of non-uniform loading, are required. In uses where deformations are necessary, particularly under impact conditions, low modulus materials are advantageous, for the material will then extend rather than permit the buildup of an excessive force.

9. TOUGHNESS OR ENERGY ABSORPTION

The load-elongation diagram is a measure of work done or energy absorbed by the filament. This property is also called "toughness." Units normally are grams per denier (or tex). Table 17.5 lists energy absorption capacities of the various fibers.

Figure 17.7 shows that equal areas (i.e., energies) under a load-elongation diagram can be obtained via many different paths. For example, the area bounded by OBC is equal to OB′C′; but for Case I it is obtained via high strength and low elongation, while for Case II the strength is lower but the elongation is higher.

Except in those cases where energy must be absorbed with a minimum amount of deformation or elongation in the specimen, in order that the material maintain its dimensional stability, a combination of load bearing and elongation capacities is usually desirable. For example, glass is strong but its elongation is too low to be a good energy absorber. Wool is weak but is quite extensible, and so can absorb significant amounts of energy. However, because its load supporting capability is low and because the wool yarn, cord or fabric would become highly distorted, normally it is not used for industrial purposes. Cotton, high tenacity rayon, and especially polyester and nylon are tough, and hence have been widely used in ropes, conveyor belts, bagging, webbings, tents, awnings, and a wide array of industrial uses where a combination of lightness, strength, and toughness is required. Such end items usually are used many times, and therefore a further requirement is that they can repeatedly deform and absorb energy under load, and retract when the load is removed.

10. ELONGATION-RECOVERY PROPERTIES

If a vertically suspended textile fiber is dead-loaded with a weight equal, for example, to about half of its breaking load, the fiber will immediately elongate a fixed amount (Figure 17.8, part A). This elongation from 0 to 1 is called the "instantaneous elastic deflection" (IED). If the weight is then left hanging on the fiber for many minutes or hours, the fiber will continue to elongate or "creep" at a slow rate from 1 to 2 (part B). If the weight is then removed, the fiber will immediately and instantaneously recover or spring back by an amount approximately equal to the original IED. The recovery (from 2 to 3, part C) is called the "instantaneous elastic recovery" (IER). If the weight is then kept off, the fiber will tend to "creep back" and return to its original length. The amount which it recovers (3 to 4, part D) is called "primary creep" or "creep recovery." The amount by which it fails to return to its original length (4 to 0), is called "secondary creep" or "permanent set." Diagrammatically, the elongation versus time curve is depicted in Figure 17.9.

The instantaneous elastic deflection obeys Hooke's law, which states that the strain (elongation) is linearly proportional to the applied stress. Thus, if progressively increasing weights are hung on the fiber, the resulting IED's will be proportional to the applied loads and a linear load-IED diagram results. The creep portion of the elongation is the viscous effect referred to earlier. While, in practice, textiles may be dead loaded under a constant force, more often they are subjected to a heterogeneous combination of repeated below-rupture stresses and strains. When a specimen is progressively loaded to a

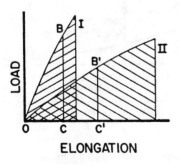

FIGURE 17.7 Energy absorption diagram.

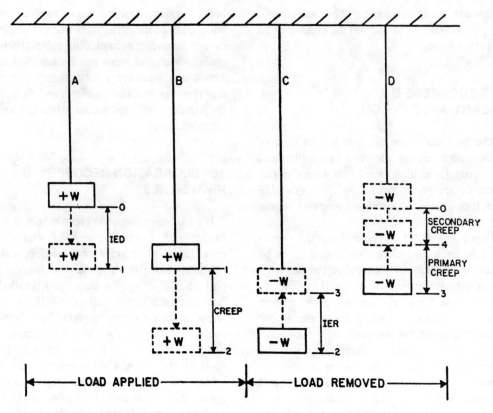

FIGURE 17.8 Elongation properties of a fiber under constant load.

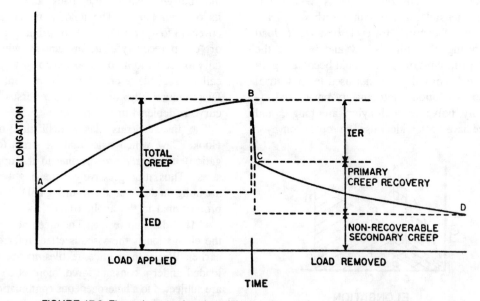

FIGURE 17.9 Elongation as a function of time under constant load and load removal.

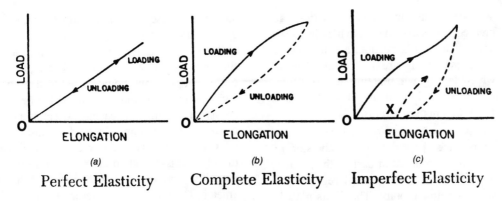

FIGURE 17.10 Three types of elasticity.

level below its rupture point, and then the load is progressively removed, one of the three types of load-elongation-recovery diagrams shown in Figure 17.10 will result.

Figure 17.10(a) depicts a truly Hookean material where the strain (elongation) is proportional to the applied stress (load) over the entire curve. Only instantaneous elastic deflection and recovery exist, thus producing a straight line for both loading and unloading, the two lines being superimposed at all points. Such a condition represents "perfect elasticity."

Figure 17.10(b) represents a condition where the strain is not proportional to the load, thus creep causes additional elongation. However, all the creep is primary, and upon load removal, complete recovery takes place, but via a path different from the loading curve, thus producing a "hysteresis" area. The primary creep recovery, while complete, is delayed and requires a longer recovery than the original loading time to be completed. Such a condition represents "complete elasticity." Figure 17.10(c) also represents a condition where the strain is not proportional to the load, thus creep causes additional elongation. Here, however, secondary as well as primary creep exists. Upon load removal recovery is incomplete, and a "permanent set" represented by O-X results. Such a condition represents "imperfect (and incomplete) elasticity."

Significance of Elongation-Recovery Properties

The abilities to deform upon load application and recover upon load removal are two of the most important properties which textile fabrics must have. Several pertinent examples are cited.

(1) Ability to absorb energy repeatedly—In order to be repeatedly useful, pump diaphragms, aircraft arrestation cables, conveyor belts, tire cords, fire hoses, crane slings, and the like must be capable of retracting absorbing energy. This is accomplished by the materials' elongation under load. If a fabric is to be used more than once, it must be capable of retracting to its original dimension after use in order that it again will have the ability to absorb an equal amount of energy on second or subsequent uses. If most of the elongation is non-recoverable secondary creep, the energy absorption capacity on a subsequent use will be significantly reduced to the point where failure may result.

(2) Ability to maintain dimensional stability—A tarpaulin, truck cover, auto seat cover, or tent may be subjected to deforming forces resulting from handling, or wind and weather. Its ability to deform or "roll with the punch" on load application, and return to its original dimension on load removal is necessary if the fabric is to be capable of absorbing stresses and strains repeatedly without becoming distorted and out-of-shape. In the case of conveyor belts, while high modulus-low elongation fabrics are usually demanded to keep "stretch" to a minimum, nevertheless the ability of the material to "give" slightly under load and recover upon load removal is a prerequisite.

(3) Wrinkle resistance—When fabrics are creased, yarns and fibers are bent in a manner such that some portions are in compression, and others are in tension. Creases which develop will be removable as a function of the ability of the fibers to recover from the imposed strains.

(4) Abrasion resistance—As will be discussed later in the chapter, a fabric's abrasion resistance is governed in part by the capacity of its fibers to absorb energy repeatedly at stresses below rupture. The fibers must be able to deform under the force produced by the abrasive action and to recover when the force is removed. Then, upon a subsequent abrasion stroke, the fibers again can deform and absorb energy without failure.

Table 17.5 shows the percent recovery of fibers which have been strained by varying amounts. After being strained 2%, staple Nylon 6, upon load removal, will recover to its original unstrained dimension. The Nylon 6 is thus said to recover 100% from the 2% strain. After 3% strain, regular tenacity Nylon 6,6 recovers 88%. This recovery capability, coupled with its high energy absorption permits it to repeatedly absorb large quantities of energy, thus accounting for nylon's excellent abrasion resistance.

11. RATE OF RECOVERY PROPERTIES

The more quickly and completely a fiber recovers from an imposed strain, the more nearly perfectly elastic it is. The ratio of the instantaneous elastic deflection to the total deflection may be used as a criterion of elasticity. The integrated closeness from a theoretical graph of perfect elasticity versus elongation is also used as a criterion under the term "elasticity index."

12. BENDING, COMPRESSION, SHEAR, AND TORSION STRAINS

In addition to tensile forces and strains, there are four other types of deformations to which engineering materials including textiles, may be subjected. These are simple compression, shear, torsion, and bending (Figure 17.11).

One of the outstanding attributes which a textile fiber must have is flexibility. In fact, this requirement probably should be included as an integral part of the definition of a textile fiber. The ability of the fiber repeatedly to bend and recover without failure is vital for most fabric uses. The stiffness of a textile fiber or any structural beam, is directly proportional to its elastic modulus, and to the fourth power of its diameter. The modulus is governed by the intrinsic nature of the fiber substance, i.e., the arrangement of atoms in the molecule, the amount of crystallinity, and the molecular orientation.

Organic fibers, in general, have stiffness moduli in the range of 10 to 200 grams per denier (Table 17.5). Rubbery or elastomeric materials have moduli of about 0.1 to 0.5 gpd. Glass, metals, and ceramics are in the range of 250 to 350. Conventional organic fibers have sufficient flexibility for most textile uses when their diameters are in the 10 to 70 micron range. In order for glass, ceramic, and metallic fibers to enjoy equal flexibility, they must be of even smaller diameter because of their higher moduli. Furthermore, the stiffness increases in direct proportion to the fourth power of the diameter, and so it becomes obvious that high modulus fibers must be drawn to very fine diameters if they are to be sufficiently flexible to perform as textiles. Furthermore, the specific gravities of glass, ceramics, and metals are considerably higher than the organic polymers. Therefore, inorganic fiber diameters must be small, not only to attain flexibility, but also to reduce fiber weight.

In simple compression, forces act in opposition to the direction they assume for tension. In the case of textiles, unless the fiber, yarn, or fabric is very short; true and complete compression does not exist, for the fiber will buckle and bending will result. This produces tension on the outside portion of the fiber and compression on the inside. Similarly, torsion and shear forces are negligible in flexible textile fibers, since the initial development of such forces produces strains in the material of such a magnitude that the forces immediately become tensional.

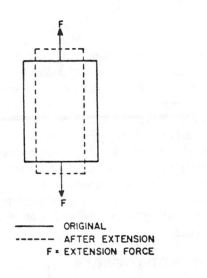

TENSION

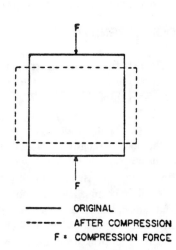

COMPRESSION

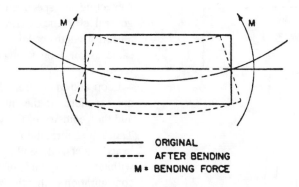

BENDING

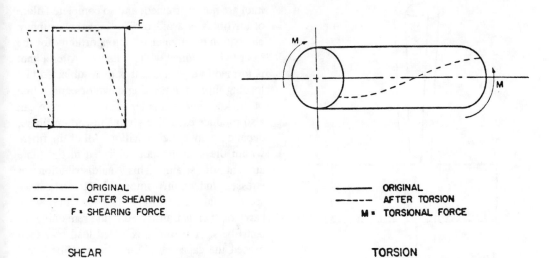

SHEAR

TORSION

FIGURE 17.11 Types of forces and resulting strains.

13. LOOP STRENGTH

Table 17.6 lists fiber loop and knot strengths. The loop strength test consists of looping one length of a fiber through another loop as shown in Figure 17.12, and elongating the loops until rupture occurs. As the rupture elongation increases, the loop strength efficiency also increases. The reason being that in low elongation brittle fibers, high stresses are quickly built up and tensile failure occurs. High elongation fibers can extend rather than fail, and a more efficient break ultimately results. When a fiber is bent around itself, a tension strain develops on the outside and a compression strain on the inside.

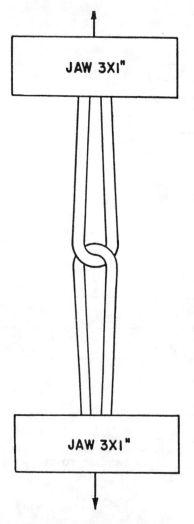

FIGURE 17.12 Loop strength test.

The greater the rupture elongation, the greater is the ability of the fiber to bend without failing.

High modulus fibers have low loop strength efficiencies and weak, extensible fibers have higher efficiencies. Remember, however, that a less efficient, but originally stronger, fiber may have a higher absolute loop strength than a more efficient weaker fiber. Cotton has a remarkably high loop strength accompanying its nominal elongation. Aramids, nylon and polyester being tough, strong fibers, exhibit high absolute loop strength.

14. KNOT STRENGTH

Knot strengths of several fibers are listed in Table 17.6. The knot test consists of making a simple overhand loop (Figure 17.13) and then subjecting the specimen to a tensile pull. The general efficiencies (i.e., the ratio of knot to tensile strength) of knot strengths are about the same as for loop strengths, and for the same reasons.

Loop and knot strengths of yarns and cords are functions of the intrinsic fiber properties and the geometry of the yarn or cord structure. The actual formation of a knot can cause selected fibers, depending upon their position within the yarn, cord, or rope, to be bent into a configuration so that the resulting strain exceeds the fiber's rupture elongation, and therefore some fiber breakage occurs. Other fibers in the knot are not so strained, and so complete failure of all fibers is avoided. The higher its rupture elongation the greater is the opportunity for the fiber to be strained without failure. After a knot is formed in a yarn and it is then subjected to a loading force, the necking down or compacting of the knot into a smaller, denser structure can also produce excessive strains in certain fibers, depending upon their position. All of the fibers within the knot are not subjected to the same amount of strain. This maldistribution of stresses and strains among the fibers causes some to fail prematurely. The load they were carrying then is transferred to the remaining intact fibers. However, this added load may then exceed the capacity of the adjacent fibers and they, too, fail. Thus, the failure proceeds from fiber to fiber and the entire cord or rope fails.

TABLE 17.6 Fiber Loop and Knot Strengths [2].

Fiber	Loop Strength (gpd)	Knot Strength (gpd)
Polyester		
A.C.E® and Compet®, filament—HT	6.0–7.0	5.0–6.0
Dacron®		
Staple and tow	2.1–6.4	2.1–6.4
Filament—RT	2.5–5.2	
Filament—HT	5.6–5.8	
Fortrel®		
Staple series 400—RT	4.4–5.6	
Staple series 300—HT	5.6–6.0	
Tairilin®, staple	2.0–5.6	
Trevira®		
Staple	1.2–5.6	2.0–5.6
Filament—HT	4.4–6.7	4.0–6.3
Nylon		
Nylon 6		
Monofil. and filament—RT	3.8–5.6	3.8–5.5
BCF—RT	2.0–3.0	2.0–3.0
Filament—HT	5.1–10.1	4.8–6.7
Nylon 6,6		
Staple and tow	3.7–5.9	3.7–5.9
Monofil. and filament—RT	2.0–5.1	2.0–5.1
Filament—HT	5.0–7.6	5.0–7.6
Tencel® lyocell—HT	2.2–2.6	2.1–2.3
Cuprammonium rayon, filament	2.15–2.25	
Saran®, monofilament	0.7–1.1	1.0–1.7
Acetate, filament and staple	1.0–1.2	1.0–1.2
Acrylic		
Acrilan®		1.9–2.6
Creslan®, staple and tow	1.9–2.3	
MicroSupreme, staple	2.5–3.0	
SEF® modacrylic, staple		1.6–2.5
Olefin		
Polyethylene		
Monofilament, low density		1.0–2.5
Monofilament, high density	2.5–4.0	2.5–4.5
Herculon®		
Staple	3.0–4.0	
Bulk filament	2.5–3.5	
Glass, multifilament	4.0	0.9
Aramid		
Kevlar®		
Kevlar 29/Kevlar 49	10.5	7.6
Nomex®, staple and filament	4.0–5.0	
Fluorocarbon		
Gore-Tex®	2.5–3.3	2.5–3.3
Teflon® TFE multifil., staple, tow, flock	0.8–1.4	0.8–1.4
Cotton, mercerized	2.5	

HT: high tenacity.
POF: partially oriented filament.
POY: partially oriented yarn.
RT: regular tenacity.
IT: intermediate tenacity.

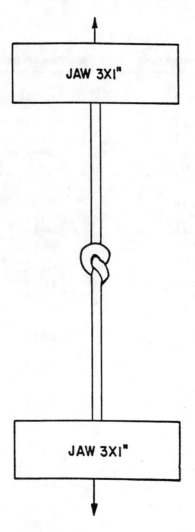

FIGURE 17.13 Simple knot test.

The same mechanisms described above for fibers in a yarn, or strands in a rope, exist with respect to individual filaments. Here the cells or fibrils of a natural fiber, or the molecular chains or crystals in natural or man-made fibers, may be considered to be similar, with respect to the fiber, as are fibers with respect to yarns. Individual fibrils are strained non-uniformly and failure ultimately results. The lower the elongation capacity of the units making up the fiber, the less the opportunity for elongation balance adjustment, and the lower the knot and loop strength efficiency.

15. FRICTION

In textile processing, fibers or yarns move relative to each other, or over metal or ceramic parts and frictional forces are generated. Carding, drafting, combing, spinning, winding, weaving, and finishing are all influenced by friction. Spun yarn strength and elongation, fabric smoothness and "hand," abrasion resistance, tendency to generate static dimensional stability, and seam slippage are also dependent upon this important property.

When one body slides over the surface of another, a resistance to the relative motion is developed which is called "frictional force," or more simply, "friction." Static or starting friction is the force which opposes the tendency of a body at rest to start to slide over another. Kinetic friction is the force which opposes the motion of two surfaces sliding by each other. When the frictional force F is divided by the normal force N which develops between the two sliding surfaces, the resulting quotient is called the friction coefficient (Figure 17.14):

$$\mu = \frac{F}{N} \qquad (17.3)$$

where

μ = coefficient of friction
F = frictional force
N = normal force

The friction which develops in textile processing or use can be fiber-to-fiber or fiber-to-other material surfaces. Therefore, in listing friction coefficients it is necessary to specify the two materials involved. The direction of rubbing of fibers or yarns, whether longitudinally or transversely, will influence friction coefficients. Even the directions of longitudinal motion along a fiber can and do produce significant and useful differences in frictional effects. For example, the ability of wool to felt is partly attributable to a greater resistance to sliding motion in the tip-to-root direction of the fiber, rather than from root-to-tip. The difference is called the differential friction effect (DFE). The

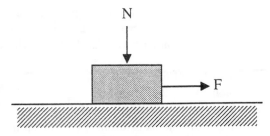

FIGURE 17.14 Frictional forces between two sliding surfaces.

scale structure of wool and other fur fibers is considered to act as a ratchet which allows motion in one direction, but resists it in the other, thus producing DFE (Figure 17.15).

16. ABRASION AND WEAR RESISTANCE

It is necessary to define these two terms at the outset. Textile technologists generally agree that the "abrasion" is confined to the pure action of rubbing, as exemplified by most of the standard abrasion machines (Chapter 20.0, Textile Testing) which function on the basis of a repetitive rubbing or abrasive action. The term *wear* is employed in a much broader sense and encompasses all of the interacting factors which may cause fabrics to fail; for example, rubbing, flexing, snagging, impacting, stretching, twisting, and exposure to chemicals, wind, sunlight, weather, wetting, drying, heating, and the like.

Much time and effort have been expended; the technical literature is replete with articles, and the industry is constantly faced with the problems of predicting the abrasion and wear resistance of new fibers and fabrics. The problems are complex. They include the selection of laboratory abrasion test methods and specifications, determination of the intrinsic abrasion resistance of new candidate fibers, the influence of yarn and fabric geometry, and the degree to which laboratory abrasion and other physical

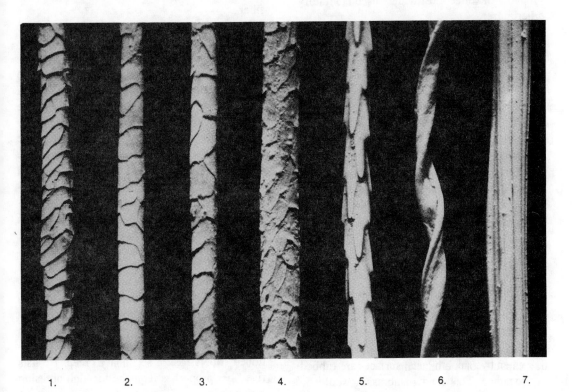

FIGURE 17.15 Surface of various fibers: 1. wool, 2. cashmere, 3. mohair, 4. camel, 5. mink, 6. cotton, 7. rayon (courtesy of Albany International).

tests correlate with service performance. It is beyond the scope of this book to discuss in great detail and review the work of the many people who have engaged in research in this important area of textile technology. However, the following sections discuss the concepts which various investigators have proposed with respect to the intrinsic abrasion resistance of textile fibers and other materials. The influence of yarn and fabric structure on abrasion resistance and a description of laboratory abrasion testers is given in Chapter 19.0, Fabric Properties and Technology.

The criteria selected to judge abrasion or wear resistance must be based upon the fabric's function. For example, for a tarpaulin, maintenance of breaking and tear strength is required. Changes in fabric thickness, weight, water permeability, electrical resistivity, surface luster, color, air permeability, the initial appearance of broken threads, or formation of a hole, are other criteria that are used or have been considered. Probably the three most widely used laboratory criteria are:

(1) Direct visual comparison of fabric appearance against a known standard, after both the test sample and the standard have been abraded for a selected number of cycles

(2) Determination of the number of abrasion cycles required to form a hole or for the fabric to fail

(3) Determination of the strength loss caused by a selected number of abrasion cycles or, more properly, a graph of abrasion cycles versus strength loss

Abrasion Mechanisms

Backer [5] proposed three mechanisms which may develop during the abrasion of a textile fiber:

(1) Smooth frictional wear such as occurs when two smooth metal surfaces are rubbed together—This kind of abrasion would be typified by sliding a fabric over a polished wooden table.

(2) Surface cutting or grinding of the fiber by a

sharp abradant of small particle size relative to the fiber diameter—This action would be typified by rubbing the fabric with a fine piece of emery cloth, and is representative of the way many laboratory abrasion testers abrade fabrics.

(3) Fiber rupture or slippage caused by rubbing the fiber with an abradant where the particle size is large compared with the fiber diameter—The fiber is then caught and plucked like a violin string. The fiber may be broken, or partially or completely pulled away from its neighbors in the yarn and fabric, thus producing a disrupted appearance.

In addition to fiber degradation or attrition, these mechanisms particularly the third, can significantly alter the positions of the yarns in the fabric, cause the fabric to distort, and otherwise change the fabric structure. Backer also pointed out that transverse rubbing action across a fiber can produce tension within the fiber.

Hamburger [6] examined the influence of the tensional repeated stress properties of fibers on abrasion resistance. He states that abrasion is a repeated stress application caused by the cyclical application and removal of stresses of a low order of magnitude. For the fiber to be abrasion resistant, it must be capable of repeatedly absorbing the energy imparted to it by the abrasive action. Therefore, it must have the ability to elongate under load application, and recover to its original dimension upon load removal. This ability to deform and recover completely is governed by the relationship between instantaneous elastic deflection, primary creep, and secondary creep. Since secondary creep is removed during the first (or second or few) loading (abrasion) cycles, the energy represented by secondary creep contributes little to abrasion resistance. Hamburger proposes that intrinsic abrasion resistance is correlative with the energy absorption of "mechanically conditioned" fibers free of secondary creep. He lists five fiber properties requisite for high abrasion resistance:

- low elastic modulus
- large immediate elastic deflection

- high ratio of primary to secondary creep
- high magnitude of primary creep
- high rate of primary creep

Hamburger determined an "energy coefficient" which is a measure of the fiber's residual energy absorption after secondary creep has been removed, and a "durability coefficient" which is a laboratory measure of the rate of loss of fiber strength with abrasion cycles. A linear correlation between these two factors appears to exist, and demonstrates that apparently abrasion resistance is dependent upon the repeated capacity of the fiber to absorb energy.

Susich [7] observed that staple yarns were always more severely abraded than corresponding multifilament yarns. He stated that although high elastic energy of fibers is the main factor necessary to prevent inherent abrasion damage, yarn extensibility, surface, and friction must also be taken into account.

In a general assessment of fiber abrasion resistance, it appears reasonable to state that types 6,6 and 6 nylon are most outstanding in their abrasion resistance. Polyester fiber is often considered as second only to nylon. Cotton has good abrasion resistance, and probably follows behind nylon and polyester. Since cotton is stronger when wet than dry, this is an added advantage which most other fibers do not have. Its wet strength and toughness give cotton excellent wet as well as dry abrasion resistance, and explains why cotton fabrics are still used in industrial applications. Wool, because of its high elongation and recovery properties, has remarkably good abrasion resistance. The acrylic fibers, viscose, acetate and regenerated protein fibers probably follow about in that order. Polyvinyl alcohol, polypropylene, and polyethylene all appear to be in the range approaching nylon for abrasion resistance.

17. MOISTURE ABSORPTION PROPERTIES

All of the natural fibers, as well as the regenerated cellulosic and protein fibers, are

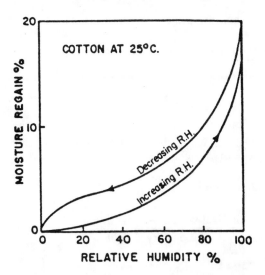

FIGURE 17.16 Relative humidity versus percent moisture regain of cotton fiber.

hygroscopic; that is, they have the ability to absorb water in the form of vapor or liquid. Hygroscopic material will absorb moisture from, or release moisture to, the surrounding air until an equilibrium moisture content is attained. Figure 17.16 shows a plot for cotton of percent moisture regain, defined below, as a function of progressively increasing relative humidity. If a dry fiber is brought to equilibrium in an increasingly humid atmosphere the water content at any relative humidity will be less than that which would result from bringing the fiber to equilibrium from a wet condition. Thus, Figure 17.16 shows moisture absorption and desorption curves which superimpose only at the 0 and 100% rh values, an area of hysteresis resulting when fabrics are conditioned from the dry versus the wet side. The moisture content of a textile material will influence its weight, its general dimensions, and many of its physical properties.

The amount of moisture is usually determined by weighing a specimen on a balance of proper sensitivity, bone drying in an oven at about 220°F and reweighing the bone dry specimen (Chapter 20.0, Textile Testing).

Percent moisture content is defined as the weight of water calculated as a percentage of the original weight of sample:

% moisture content

$$= \frac{\text{original weight} - \text{bone dry weight}}{\text{original weight}} \times 100$$

$$(17.4)$$

Percent moisture regain is defined as the weight of water calculated as a percentage of the bone dry weight:

% moisture regain

$$= \frac{\text{original weight} - \text{bone dry weight}}{\text{bone dry weight}} \times 100$$

$$(17.5)$$

17.1 Standard Moisture Regain of Fibers

Because many textile materials vary in physical dimensions, weight and properties as a function of their moisture content, it is necessary to describe precisely the water content of the air with which the textile is in equilibrium if a property is to have any meaning. In the United States, standard atmospheric conditions are 65% relative humidity and 70°F (21°C). Table 17.7 lists moisture regains of the various fibers when brought to equilibrium under these conditions of humidity and temperature. All other physical properties of fibers, yarns, or fabrics given in this book are the values obtained at standard conditions unless otherwise designated.

17.2 Commercial Moisture Regains for Fibers

Depending upon conditions of harvesting or manufacturing, hygroscope fibers can vary widely in moisture regain with a concomitant variation in true fiber weight. Therefore, for economic reasons the purchase or sale of cotton, wool, rayon, etc. must be based upon a standard amount of moisture. The normal practice is for a laboratory, often an independent organization agreeable to both buyer and seller, to measure the average moisture regain of the entire fiber lot. The actual weight of the lot is then

corrected to a standard regain agreed upon by the trading parties. Table 17.8 lists accepted commercial regains established by the separate industries for their fibers. Because of the fiber price per pound, and the size of the lots, the regain must be determined accurately. Proper selection of the test sample so that it is in fact truly representative of the lot is of equal importance.

The "as received" weight of a lot may be corrected to the accepted commercial regain weight by the following formula

$$W_c = W_l \frac{100 + R_c}{100 + R_l} \qquad (17.6)$$

where

W_c = lot weight calculated to accepted commercial regain
W_l = lot weight "as received"
R_c = percent accepted commercial regain
R_l = percent regain of "as received" lot

The bone dry weight of a lot may be corrected to the accepted commercial regain weight as follows:

$$W_c = W_b \frac{100 + R_c}{100} \qquad (17.7)$$

where W_b is bone dry weight of lot.

18. FIBER SWELLING PROPERTIES

Table 17.9 lists the swelling properties of fibers when immersed in water. Upon wetting, most hygroscope textile fibers exhibit a slight increase in length and a considerable increase in diameter and cross-sectional area. The influence of all of these interacting effects on fabric air and liquid permeability and on shrinkage are discussed in Chapter 19.0, Fabric Properties and Technology on "Cover Factor," "Fabric Density and Packing Factor," "Air Permeability," and "Water Repellency." Since hygroscopic fibers themselves do not shorten upon wetting, fabric shrinkage must be caused by other mechanisms.

TABLE 17.7 Moisture Regain of Fibers [2].

Fiber	Moisture Regain (%)	
	70°F, 65% rh	70°F, 95% rh
Polyester		
A.C.E® and Compet®		
Filament—HT	0.4	
Dacron®		
Staple and tow	0.4	
POF		
Filament—RT	0.4	
Filament—HT	0.4	
Tairilin®, staple	0.4	0.6
Trevira®		
Staple	0.4	0.6
POF	0.4	0.6
Filament—HT	0.4	0.6
Nylon		
Nylon 6		
Staple	2.8–5.0	3.5–8.5
Monofil. and filament—RT	2.5–5.0	3.5–8.5
Filament—HT	2.8–5.0	3.5–8.5
Nylon 6,6		
Staple and tow	4.0–4.5	6.1–8.0
Monofil. and filament—RT	4.0–4.5	6.1–8.0
Filament—HT	4.0–4.5	6.1–8.0
Tencel® lyocell—HT	11.5	
Rayon		
Fibro®		
RT—Multi-lobed	11	27
IT—0.9 den or 0.25 in and up	11	27
Cuprammonium, filament	11	
Rayon		
Filament—RT	11	
Filament—HT	11	
Saran®, monofilament	0	N/A
Acetate, filament and staple	6.3–6.5	14
Ryton® sulfar, staple	0.6	
Acrylic		
Acrilan®, staple and tow	1.5	3.5
Creslan®, staple and tow	1.0–1.5	2.0–2.5
MicroSupreme, staple	1.0–1.5	2.0–2.5
SEF® modacrylic, staple	2.5	

(continued)

591

TABLE 17.7 (continued).

Fiber	Moisture Regain (%)	
	70°F, 65% rh	70°F, 95% rh
Olefin		
Polyethylene		
Monofilament, low density	Negligible	Negligible
Monofilament, high density	Negligible	Negligible
Herculon®		
Staple	0.01	
Bulk filament	0.01	
Marvess® and Alpha®		
Staple and tow	−0.1	
Multifilament	−0.1	
Essera®, Marquesa Lana®, Pation® III		
BCF	0.01–0.1	
Staple	0.01–0.1	
Spectra® 900	Negligible	Negligible
Spectra® 1000	Negligible	Negligible
Fibrilon®		
Staple	0.01	0.01
Spandex		
Glospan®/Cleerspan, S-85, multifilament	Less than 1	
Lycra®		
Type 126, 127	1.3	
Glass		
Single filament−E-glass	None	Up to 0.3
Single filament−S-glass	None	Up to 0.3
Multifilament	None	Up to 0.3
Aramid		
Kevlar®		
Kevlar 29/Kevlar 49	4.3	6.5/6.0
Kevlar 149	1.2	2.3
Kevlar 68	4.3	
Kevlar HT (129)	4.3	
Nomex®, staple and filament	4.5	7.5
Fluorocarbon		
Gore-Tex®	0	0
Teflon®		
TFE multifil., staple, tow, flock	0	0
FEP, PFA monofilament	0	0
PBI, staple	15	20
Asbestos	1.0	3.0
Cotton	8.5	15
Silk	11	25
Wool	16	29

HT: high tenacity.
POF: partially oriented filament.
POY: partially oriented yarn.
RT: regular tenacity.
IT: intermediate tenacity.

TABLE 17.8 Commercial Moisture Regain
Values of Fibers [8].

Fiber	Moisture Regain (%)
Acetate (secondary)	6.5
Acrylic	1.5
Aramid, for	
Plastic reinforcement	3.5
Filtration fabrics and safety apparel	4.5
Reinforcement of rubber goods	7.0
Azlon	10.0
Cotton	
Natural cotton yarn	7.0
Dyed cotton yarn	8.0
Mercerized cotton yarn	8.5
Flax (raw)	12.0
Fluorocarbon	0.0
Glass	0.0
Hemp	12.0
Jute	13.75
Linen	8.75
Metallic	0.0
Modacrylic	
Class I	0.4
Class II	2.0
Class III	3.0
Nylon (polyamide)	4.5
Olefin	0.0
Polyester	0.4
Ramie	
Raw	7.6
Scoured	7.8
Rayon (regenerated cellulose)	11.0
Rubber	0.0
Saran	0.0
Silk	11.0
Spandex	1.3
Triacetate (primary)	3.5
Vinal	4.5
Vinyon	0.0
Wools (all forms)	13.6

19. WET FIBER STRENGTHS

Table 17.4 lists wet and dry fiber tenacities. Natural cellulosic fibers such as cotton, are 10 to 15% stronger when wet than when "dry" under standard conditions. This is of obvious advantage in end uses where fabrics are apt to become wet.

The natural cellulosics are the only fibers presently known whose wet strengths exceed dry strengths. The natural animal fibers, including silk and wool, have wet strengths which are 75 to 90% of dry strengths. Regenerated cellulose fibers such as viscose and cuprammonium rayons have wet strengths in the order of 45 to 60% of dry. The wet strengths of high modulus rayons range from 60 to 75% of dry. The modified cellulosics, such as acetate, also have wet strengths which are 60 to 80% of dry. Because many of the hydrophilic (water absorbing) hydroxyl groups in these fibers have been replaced by more hydrophobic (water resisting) acetate groups, the tendency for such fibers to absorb water is less, and so they are somewhat less sensitive to strength losses when wet.

As a general rule it can be stated that, for man-made fibers, the lower their moisture regains and liquid water absorption capacity, the less sensitive they are to wet strength losses. Thus nylon 6,6, with a moisture regain of 4.5% (Table 17.7), has a wet strength of about 85–90% of dry, while Dacron® polyester, with a regain of 0.4% has a wet strength just about equal to its dry strength. While the general rule is valid, care should be exercised in its employment, for the nature of the polymer also influences the strength loss upon wetting.

Wet Yarn Strengths

Because cotton has higher strength when wet than dry, it follows that yarns made from cotton will also be stronger at higher moisture regains as well as when wet. This is in fact the case, the increase for wet yarns being on the order of 25 percent. In addition to the intrinsic wet fiber strength increase, wet cotton fibers swell radially, thus causing a denser packing of fiber against fiber. The total frictional force increase reduces fiber slippage and so a stronger yarn results. For natural animal, regenerated cellulosic and synthetic fibers, generally the ratio of wet-to-dry yarn strengths is about the same as their respective wet-to-dry fiber strengths.

20. WET FIBER BREAKING ELONGATIONS

Table 17.5 lists wet rupture elongations. As a general rule, for both the natural and re-

TABLE 17.9 Fiber Swelling Properties.

Fiber	Length Increase (%)	Diameter Increase (%)	Cross-Sectional Area Increase (%)
Glass	–	–	0
Nylon 6,6 high tenacity	1.2	1.9–2.6	1.6–3.2
Polyethylene, low density	–	–	0
Dacron polyester, regular	0.1	0.3	0.6
Lycra spandex	Approx. 1.0		Approx. 1.0
Viscose rayon, regular	3–5	25–52	50–113
Viscose rayon, high tenacity	–	–	50
Cuprammonium rayon	4	32–53	56–62
Acrilan acrylic	–	–	Approx. 5
Orlon acrylic	0.2	2.4	4.8
Acetate	0.14	9–14	–
Cotton, raw	1.2	14–30	40–42
Cotton, mercerized	0.1	17	24–46
Flax	–	–	47
Jute	0.37	20	40
Ramie	–	–	37 (bleached)
Silk	1.7	16–20	19
Wool	1.2	16	25

generated hydrophilic fibers, the water absorbed may be considered as a plasticizer. The fiber's modulus is reduced and the total elongation-to-rupture increases. The increase for the natural cellulosic fibers, e.g., cotton, is small.

The regenerated cellulosic fibers normally have appreciably greater wet elongations than dry. Usually the greater their wet strength loss, the greater will be their rupture-elongation gain. For example, the high modulus rayons have higher wet strength than do the regular rayons, their wet elongations are concomitantly lower, and therefore they have better wet dimensional stability. The synthetic fibers are more hydrophobic and are therefore less sensitive to water. Thus there is little difference in the wet and dry strength-elongation properties of the acrylics, the polyesters, the polyolefins, and the like. Nylon, being somewhat hydrophilic, is proportionately sensitive to water.

21. EFFECT OF RELATIVE HUMIDITY ON TENACITY-ELONGATION DIAGRAMS

For all fibers except cotton, tenacities and moduli are lower and elongations are higher as relative humidity increases. Obviously the greater elongation resulting from the plasticity of wet fibers can either by recoverable or non-recoverable. For natural fibers the wet recovery properties often equal or surpass the dry properties, having high instantaneous elastic recovery, high primary creep recovery, and low permanent set. The regenerated hygroscopic fibers usually have poorer recovery properties. The hydrophobic synthetic fibers, being insensitive to water, show little or no change.

22. THERMAL PROPERTIES OF FIBERS

The influence of heat on the properties of fibers is of paramount importance with respect to textile processing as well as use. All of the natural fibers and the regenerated cellulosic and protein fibers are non-thermoplastic. They do not soften with heat, but instead char and decompose. Most synthetic fibers and the modified cellulosics such as acetate, are thermoplastic and have specific melting points.

Thermoplasticity in a fiber can be advantageous or disadvantageous, depending upon processing or requirements. Because of their cross-linked molecular structure, cotton and wool are essentially temperature insensitive and so do not soften in the 200–300°F range, nor stiffen at temperatures below freezing. Thus they can be ironed with relative safety, and do not significantly shrink, stretch, or lose strength at elevated temperatures. For processing or end uses where plasticity is necessary, water and/or steam are used rather than heat as the plasticizing means.

The thermoplastic fibers are temperature sensitive. As temperature increases, strength normally is lowered, rupture elongation may increase, and the tensional modulus thus is reduced. Unless the fiber is permanently damaged by heat, the effect often is reversible. As the fiber cools it may reassume its original properties. However, the heat exposure may produce a permanent thermal shrinkage with an accompanying increase in rupture elongation and a decrease in both strength and tenacity. This shrinkage is removable only by restretching the fiber at an elevated temperature. Thermoplasticity is an attribute when it is desired to heatset the fibers within the yarn and fabric, i.e., to make them conform permanently to the position and form into which they are placed, or where the fabric itself must then be molded or reshaped. Thermoplasticity is a disadvantage when a fabric becomes excessively limp or tacky or shrinks or distorts at high temperatures, or becomes stiff and brittle at low temperatures.

Heat may influence the properties and functions of textiles in many ways. Important criteria to consider are (1) softening, melting, or decomposition temperatures; (2) tendency of the fiber and fabric to shrink when heat-relaxed, or stretch when heated and tensioned; (3) ability of the fabric to be heat set; (4) ability of the fabric to function properly at elevated temperatures in one time or in repeated use; and (5) ability of the fabric to function properly at room temperature (or some other lower temperature) after exposure at high temperature for a given period of time.

Table 17.10 shows the effect of heat on fiber properties of various fibers. Normally for a textile fiber to be industrially useful it should not become tacky or soften below about 300°F. Upper temperature limits depend upon end-use requirements, and upon the ability of the finishing plant to provide a temperature which can

heat set the fabric in order to stabilize it and to remove wrinkles and distortions.

Table 13.2 (Chapter 13.0) lists the commercially available high temperature fibers and Table 13.3 gives their properties.

23. EFFECT OF HEAT ON THE TENSILE PROPERTIES OF FIBERS

Two subjects must clearly be separated for the influence of heat upon fiber properties:

(1) Tensile properties of fibers tested at elevated temperatures
(2) Tensile properties of fibers tested at room temperature after exposure to elevated temperatures for selected time periods

The former indicates the capability of the fiber to perform at the required elevated temperature. The latter is often used as a criterion of heat degradation resistance.

Fabric degradation by heat normally is a function of temperature, time, relative humidity, and air circulation. Since many heat degradation processes involve oxidation, the greater the air circulation the greater will be the amount of oxygen which comes in intimate contact with the fiber, and the more rapid will be the degradation. Of course, for those fibers which are insensitive to oxygen, this factor is of no consequence.

For those industrial textile uses where the item will be subjected to a tension at an elevated temperature, it is important that the fiber does not become soft and extensible at that temperature. Glass is generally used for heat insulation and heat resisting applications. While it does not begin to lose strength until 600°F, its abrasion resistance is poor, and it cannot be used where prolonged flexing or rubbing is required. Protective silicone finishes improve the abrasion resistance of glass fabrics, making them useful in certain applications, for example, high temperature resisting gas filtration fume bags. Nylon is strong and tough, but it degrades rather quickly at elevated temperatures in the presence of air, and this limits its high temperature use. For example, when exposed at 350°F for 6 hours, 210 denier type 6,6 nylon yarn loses

TABLE 17.10 Effect of Heat on Fiber Properties [2].

Fiber	Effect of Heat
Polyester	
A.C.E®	Softens at 464°F. Melts at 500°F.
Compet®	Melts at 488°F.
Dacron®	Sticks at 440 to 445°F. Melts at 482°F.
Fortrel®	Melts at 478 to 490°F.
Tairilin®	Sticks at 440–450°F. Melts at 485–495°F.
Trevira®	Softens at 445 to 465°F. Melts at 495°F.
Thornel® carbon	Does not melt. Oxidizes very slowly in air at temperatures above 600°F.
Nylon	
Nylon 6	Melts at 419 to 430°F. Slight discoloration at 300°F when held for 5 hr. Decomposes at 600 to 730°F.
Nylon 6,6	Sticks at 445°F. Melts at 480 to 500°F. Yellows slightly at 300°F when held for 5 hr.
Tencel® lyocell	Does not melt. Loses strength at about 300°F and begins to decompose at about 350°F under extended periods of exposure.
Rayon	
Fibro®	Does not melt. Loses strength at 300°F and decomposes at 350 to 464 under extended periods of exposure.
Cuprammonium	Does not soften, stick or melt. Decomposes at 350°F under extended exposure.
Rayon	Does not melt. Loses strength at 300°F and begins to decompose at about 325°F.
Saran	
Saran®	Melts at 240–280°F; will not support combustion; self extinguishing.
Acetate	Sticks at 350 to 375°F. Softens at 400 to 445°F. Melts at 500°F.
Sulfar	
Ryton®	Outstanding resistance to heat (melts at 285°C). Excellent resistance to aerial oxidation, most chemicals. Retains 70% + original strength to 400°F for 5,000 hr.
Acrylic	
Acrilan®	Does not melt.
Creslan®	Sticks at 420 to 450°F. Safe ironing at 300°F.
MicroSupreme	Does not melt.
SEF® modacrylic	Does not melt. Boiling water shrinkage equals 1%. Excellent resistance to shrinkage in dry heat. At 390°F, 5% shrinkage.
Olefin	
Polyethylene	Low density softens at 225 to 230°F, melts at 230 to 250°F. High density softens at 240–260°F, melts at 225 to 280°F. Low density shrinks 5–8% at 165°F, 50–60% at 212°F. High density shrinks 3–5% at 165°F, 8–12% at 212°F.
Herculon®	Softens at 285 to 330°F. Melts at 320 to 350°F. Decomposes at 550°F. Zero to 5% shrinkage at 212°F; 5 to 12% at 265°F.
Marvess® and Alpha®	Fiber softens at 300–320°F. Melts at 320–340°F. Decomposes at 550°F.
Essera®, Marquesa Lana®, and Pation® III	Softens at 285–340°F. Melts at 325–340°F. 0–5% shrinkage at 212°F. 5–12% shrinkage at 265°F.
Spectra®	Melts about 300°F.
Fibrilon®	Fiber softens at 290–310°F, melts at 320°F, and decomposes at 550°F.

TABLE 17.10 (continued).

Fiber	Effect of Heat
Spandex	
Glospan®/Cleerspan, S-85	Good dimensional stability. Sticks at 420°F.
Lycra®	Good dimensional stability. Can be heat set. Sticks at 347–392°F. Melts at 446°F.
Glass	None burn. At 650°F, E holds 75% tensile strength; S 80%; multifilament 50%. E softens at 1,350 to 1,611°F; melts at 2,050 to 2,160°F. S softens at 1,560 to 1,778°F; melts at 2,720°F.
Aramid	
Kevlar®	Difficult to ignite. Does not propagate flame. Does not melt. Decomposes at about 900°F.
Nomex®	Does not melt. Decomposes at 700°F.
Fluorocarbon	
Gore-Tex®	Extremely heat resistant. Can be safely used from −350°F to +550°F. Melts at 620°F.
Teflon®	
TFE	Extremely heat resistant. Safely used at −350°F to +550°F. Melts at 620°F.
FEP,PFA	Extremely heat resistant. Continuous use temperature: FEP, 400°F, PFA 500°F.
PBI	Will not ignite. Does not melt. Decomposes in air at 860°F. Retains fiber integrity and suppleness upon flame exposure. High char yield.
Asbestos	Weight loss 0.5 to 1% after exposure at 1,400°F for 2 hours. Stable up to 1,490°F.
Cotton	Yellows at 250°F, decomposes at 300°F.
Silk	Decomposes at 300°F. Rapid disintegration above 340°F.
Wool	Decomposes at 275°F, chars at 570°F.

54% of its original strength. Dacron® polyester has better heat aging resistance, losing only 8% when exposed under the same conditions.

24. FIBER FLAMMABILITY

Several terms related to the burning behavior of textiles are defined by the ASTM as follows [9]:

- combustible textiles: "a textile that will ignite and burn or that will give off vapors that will ignite and burn when subjected to external sources of ignition"
- flame resistance: "the property of a material whereby flaming combustion is prevented, terminated, or inhibited following application of a flaming or nonflaming source of ignition, with or without subsequent removal of the ignition source"
- flammability: "those characteristics of a material that pertain to its relative ease of ignition and relative ability to sustain combustion"

- flammable textile: "any combustible textile that burns with a flame"
- heat resistance: "the extent to which a material retains useful properties as measured during exposure of the material to a specified temperature and environment for a specified time"
- inherent flame-resistance: "flame resistance that derives from an essential characteristic of the fiber from which the textile is made"
- non-combustible textile: "a textile that will neither ignite nor give off vapors that will ignite when subjected to external sources of ignition"
- non-flammable textile: "any combustible textile that burns without a flame"

Cotton and other cellulosics, as well as cellulose acetate, will ignite readily and are not self-extinguishing. Once they are ignited, they will continue to burn, even when the igniting flame is removed. Silk, wool, and all the other animal fibers will burn, but are self-extinguishing. Synthetic organic fibers have varying degrees of flammability depending upon their compositions. Nylon 6,6 is stated to be moderately difficult to ignite and self-extinguishing. Glass does not burn. Carbon does not melt but oxidizes very slowly in air at temperatures above 600°F. Kevlar® aramid is difficult to ignite, does not propagate flame and does not melt. It decomposes at about 900°F. Nomex® aramid does not melt and decomposes at 700°F. Gore-Tex® and Teflon® are extremely heat resistant. They can be safely used from −350°F to +550°F. PBI will not ignite nor melt but decomposes in air at 860°F. It retains fiber integrity and suppleness upon flame exposure.

Because burning is an oxidation process, for those fibers which will burn, the type of yarns and fabrics into which they are manufactured will directly influence the burning rate. Dense, heavy fabrics will not burn as readily as will lofty, open, thin, or brushed pile fabrics. For these latter types, the oxygen in the air can surround each fiber and accelerate the burning rate. Here, if the fibers are combustible and finely dispersed, a rapid flash burning can develop. In addition to the intrinsic flammability of the fiber, and the textile into which it is manufactured, precaution must be taken to insure that a fabric finish or coating applied to impart a certain property do not produce a dangerously flammable fabric.

24.1 Flammable Fabrics Act

In 1953, the U.S. Congress passed the Flammable Fabrics Act to ban easily ignitable fabrics and wearing apparel from the market. The Act stated that "no article of wearing apparel or fabric subject to the Act and Regulations shall be marketed or handled if such article or fabric, when tested according to the procedures prescribed . . . is so highly flammable as to be dangerous when worn by 'individuals.'" These were mainly brushed or high pile cellulosic fabrics used occasionally in such garments as sweaters and children's costumes.

In 1967, the Act was amended to cover interior furnishings as well as fabrics and all wearing apparel which present an unreasonable risk from fire. In 1972, an Act of Congress established a new Consumer Product Safety Agency. In 1973, responsibility for both issuing and enforcing flammability regulations was transferred from the Commerce Department to the Agency.

The flammability standard described in the Act states, "Any fabric or article of wearing apparel shall be deemed so highly flammable within the mean . . . of this Act as to be dangerous when worn by individuals if such fabric or any uncovered or exposed part of such article of wearing apparel exhibits rapid and intense burning when tested under the conditions and in the manner prescribed in . . . 'Flammability of Clothing Textiles, Commercial Standard 191-53.'" This standard specified the methods for testing fabric flammability and classifies fabrics as follows.

Class 1. Normal Flammability

These textiles are generally accepted by the trade as having no unusual burning character.

- textiles that do not have a raised fiber surface, but have a time of flame spread in the test of 3.5 seconds or more
- textiles having a raised fiber surface that have a time of flame spread in the test of more than 7 seconds or that burn with a surface flash (time of flame spread less than 7 seconds) provided the intensity of the flame is insufficient to ignite or fuse the base fabric

Class 2. Intermediate Flammability

These textiles are recognized by the trade as having flammability characteristics between normal and intense burning. Textiles having a raised fiber surface that have a time of flame spread in the test of 4 to 7 seconds, inclusive, and the base fabric is ignited or fused.

Class 3. Rapid and Intense Burning

These textiles are considered dangerously flammable and are recognized by the trade as being unsuitable for clothing because of their rapid and intense burning.

- textiles that do not have a raised fiber surface, but that have a time of flame spread in the test of less than 3.5 seconds
- textiles having a raised fiber surface that have a time of flame spread in the test of less than 4 seconds, and the base fabric is ignited or fused

The chemical textile industries have expended large amounts of money and effort in developing flame resisting finishes for apparel, decorative and industrial fabrics. By law, curtains, draperies and other textiles in public buildings must be flameproof. Fire and flame resistance test

methods are explained in Chapter 20.0, Textile Testing. Flame retardant finishes are included in Chapter 5.0, Section 5.1.

25. RESISTANCE OF FIBERS TO MILDEW, AGING, SUNLIGHT AND ABRASION

Table 17.11 lists mildew, aging, sunlight and abrasion properties of various fibers. Wool and cotton tend to turn yellow with age. Where necessary, bleaching the cotton with hypochlorite

TABLE 17.11 Resistance of Fibers to Mildew, Aging, Sunlight and Abrasion [2].

Fiber	Resistance to Mildew, Aging, Sunlight and Abrasion
Polyester	
A.C.E®	Excellent resistance to mildew; good to abrasion, aging and sunlight with some deterioration after prolonged exposure to sunlight.
Compet®	Excellent
Dacron®	Not weakened by mildew. Excellent resistance to aging and abrasion. Prolonged exposure to sunlight causes some strength loss.
Fortrel®	Excellent resistance to mildew, aging and abrasion. Prolonged exposure to sunlight causes some strength loss.
Tairilin®	Excellent resistance to mildew, aging, abrasion. Excellent resistance to sunlight, but prolonged exposure causes some strength loss.
Trevira®	Not attacked by mildew. Excellent resistance to aging and abrasion. Excellent resistance to sunlight, but prolonged exposure causes some strength loss.
Thornel® carbon	Excellent resistance to mildew, aging, and sunlight. Poor abrasion resistance.
Nylon	
Nylon 6	Excellent resistance to mildew, aging and abrasion. Prolonged exposure to sunlight causes some degradation.
Nylon 6,6	Excellent resistance to mildew, aging and abrasion. Prolonged exposure to sunlight causes some deterioration.
Tencel® lyocell	Attacked by mildew. Good resistance to aging, sunlight and abrasion.
Rayon	
Fibro®	Attacked by mildew. Resistance to aging, sunlight and abrasion is good. Long exposure to sun yellows some intermediate rayons; aging deteriorates them.
Cuprammonium	Attacked by mildew. Stable to aging. Not affected by sunlight. Good resistance to abrasion.
Rayon	Attacked by mildew. Good resistance to sunlight, abrasion and aging.
Saran	
Saran®	Not attacked by mildew. Good resistance to aging, sunlight and abrasion.
Acetate	Impervious to aging. Good resistance to mildew discoloration and sunlight (some lose strength from long exposure). Fair abrasion resistance.
Sulfar	
Ryton®	Excellent resistance to mildew, aging and sunlight. Fair abrasion resistance.
Acrylic	
Acrilan®	Excellent resistance to mildew, aging and sunlight. Good resistance to abrasion.
Creslan® and MicroSupreme	Excellent.
SEF® modacrylic	Excellent resistance to mildew, aging and sunlight. Good resistance to abrasion.

(continued)

TABLE 17.11 (continued).

Fiber	Resistance to Mildew, Aging, Sunlight and Abrasion
Olefin	
Polyethylene	Not attacked by mildew. Good resistance to aging, sunlight and abrasion.
Herculon®	Not attacked by mildew. Good resistance to aging, indirect sunlight and abrasion. Can be stabilized to give good resistance to direct sunlight.
Marvess® and Alpha®	Not attacked by mildew. Good resistance to aging and abrasion. Stabilizers provide good resistance to sunlight fading.
Spectra®	Excellent resistance to mildew, aging and abrasion. Some loss of strength with long exposure to sunlight.
Fibrilon®	Not attacked by mildew. Good resistance to aging and abrasion. Stabilizers provide good resistance to sunlight fading.
Spandex	
Glospan®/Cleerspan, S-85	Excellent resistance to aging and abrasion. Sunlight exposure causes mild discoloration but no loss of physical properties.
Lycra®	
Type 126,127	Mildew, aging: no effect. Strength loss from prolonged sunlight. Good abrasion resistance.
Type 128	Excellent abrasion and aging resistance.
Glass	Not attacked by mildew, although binder may be affected by it. Excellent resistance to aging and sunlight.
Aramid	
Kevlar®	Excellent resistance to mildew and aging. Prolonged exposure to sunlight causes deterioration, but fibers self-screening. Good abrasion resistance.
Nomex®	Excellent resistance to mildew and aging. Prolonged exposure to sunlight causes some strength loss. Good abrasion resistance.
Fluorocarbon	
Gore-Tex®	Excellent resistance to mildew, aging, sunlight and abrasion.
Teflon®	Not weakened by mildew. Excellent resistance to aging and sunlight. Good abrasion resistance.
PBI	Good resistance to mildew and aging. Prolonged exposure to sunlight will cause darkening and some loss of tensile strength. Good abrasion resistance.

or peroxide, and the wool with sulfur dioxide or peroxide will remove the yellow color. Exclusive of any microbial or mildew degradation, weathering and sunlight cause a slow but progressive loss in strength and serviceability for cotton and rayon. Acetate is probably more resistant than rayon. Wool generally has good resistance to sunlight, weathering, and aging. This appears quite logical when one considers the natural environment of sheep.

25.1 Mildew and Rot Damage

The terms mildew, rot, and decay are variously used to indicate growth upon or damage to textiles by microorganisms such as fungi and bacteria. The major types of fungi affecting textiles are the molds such as penicillium and aspergillus, and the soil molds such as actinomycetes. Bacteria which cause textile damage

are most frequently of the bacillus or rod type. The terms *decay* and *mildew* are usually associated with damage by fungi, and *rot* with bacterial damage.

Since microorganisms require water and warmth to flourish and spread, all of the natural and regenerated hydrophilic vegetable and animal fibers are susceptible to microbial damage. Cotton, linen, viscose and cuprammonium rayons, and other cellulose based fibers, are subject to breakdown where, via a process called glycolysis, the cellulose is converted by the enzyme cellulase to cellobiose. This is then converted to glucose which serves as the food for the microorganism. Mildew is often manifested in cotton and other cellulosics by a musty odor, the formation of discolored and black areas, fabric tendering, and ultimately complete decay. Some mildews produce brightly colored spots, and often fluoresce strongly in ultraviolet light. Therefore, an inexpensive ultraviolet or "black light" lamp is a useful tool for mildew identification.

Wool, other animal fibers, and regenerated proteins are usually not quite so sensitive but they, too, can be decomposed by specific bacteria, actinomycetes, and molds which attack keratin through the production of proteolytic enzymes by these organisms. The intact cuticle and epicuticle of the wool fiber are relatively inert to such enzymes, hence such attack is more rapid on wool which previously has been mechanically or chemically damaged.

Mildew most often develops when textiles are folded and stored for lengthy periods in a warm moist atmosphere. A relative humidity of about 82% for cotton and 85% for wool is required for mildew and mold growth. Bacterial requirements are higher. Wool must ordinarily be exposed to a relative humidity of approximately 95%, or must have a moisture content on the order of 20% to support active bacteria growth. Temperatures of 75 to 100°F are optimal for most microorganisms. Local condensation can start mildew growth, occasionally at a relative humidity which is lower than expected. Once started, mildew can absorb water hygroscopically from the air or from local condensation and continue its growth. This is why packaged or folded fabrics, without adequate ventilation, will sometimes mildew in relatively "dry" rooms.

Synthetic fibers have chemical structures not found in nature. Therefore, microorganisms do not attack them. Most synthetic fibers have little or no tendency to absorb water, a necessary component required by all microorganisms if they are to survive. Clearly, this is another reason why synthetic fibers are mildew resistant. Even a slight modification in the chemical structure of a natural or regenerated fiber may be sufficient to preclude microbial attack. The nylons, polyesters, acrylics, and all other completely synthetic fibers are completely impervious to mildew growth and attendant damage. A word of caution is necessary. Often synthetic yarns may be sized, or fabrics finished, with natural gums, starches, waxes or the like that are susceptible to mildew growth. This may make it appear that the fibers themselves are being attacked. Tests usually show that while the fiber may be stained, it suffers no actual deterioration and strength loss; only the finish is attacked.

25.2 Insect and Rodent Damage

Basically the same comments made above for mildew and rot damage hold for insect damage. The natural fibers are susceptible to damage by carpet beetles, cockroaches, ants, silverfish, and moth larvae. The synthetic fibers usually are not attacked. Many insects will eat natural fibers. They may cut through synthetic or other fibers which are not attractive as food in order to liberate themselves or reach an attractive food source, or they may eat an unattractive fiber which is coated or contaminated with an attractive material. Thus, while synthetic fibers normally are not subject to insect damage, they can be damaged under specific environmental conditions. Rodents, in cases of extreme hunger, may attempt to eat cellulose or keratin base fibers. They probably will not eat synthetic fibers, but might damage them by gnawing.

26. CHEMICAL RESISTANCE OF FIBERS

Table 17.12 lists the effect of acids, alkali solvents, and special chemicals on various fibers. Because of the many fibers and many chemicals, it is impractical to list all of the effects. The "state of aggregation" (i.e., position and orientation) of fibers in yarn, and yarn in fabric, the chemical concentration, temperature, and length of exposure time also influence reaction rates and extents of degradation.

Cotton and rayon have fairly good resistance to weak acids, for example, but are severely damaged by even dilute solutions of strong acids such as sulfuric and hydrochloric. These cellulosic fibers have good resistance to strong alkalis. Mercerizing strength caustic (20% sodium hydroxide) will swell the fibers and can ultimately degrade them.

Wool shows high resistance to weak and strong acids, but is disintegrated and dissolved by alkalis. Bleaching must be accomplished on

TABLE 17.12 *Chemical Resistance of Fibers* [2].

Fiber	Effects of Acids and Alkalis	Effects of Bleaches and Solvents
Polyester A.C.E®	Good resistance to organic acids. Good resistance to inorganic acids at room temperature; moderate resistance at 212°F. Disintegrates in concentrated hot alkalis.	Excellent resistance to bleaches and other oxidizing agents.
Compet®	Excellent resistance to acids.	
Dacron®	Good resistance to most mineral acids. Dissolves with partial decomposition in concentrated solution of sulfuric acids. Good resistance to weak alkalis. Moderate resistance to strong alkalis at room temperature. Disintegrates in strong alkalis at boil.	Excellent resistance to bleaches and other oxidizing agents. Generally insoluble except in some phenolic compounds.
Fortrel®	Good resistance to most mineral acids. Dissolves with partial decomposition by concentrated solutions of sulfuric acid. Good resistance to weak alkalis and moderate resistance to strong alkalis at room temperature. Disintegrates in strong alkalis at boil.	Excellent resistance to bleaches and other oxidizing agents. Generally insoluble except in some phenolic compounds.
Tairilin®	Good resistance to most mineral acids; fair resistance to concentrated sulfuric acid. Good resistance to weak alkalis; moderate resistance to strong alkalis at room temperature. Disintegrates in strong alkalis at boil.	Excellent resistance to bleaches and other oxidizing agents.
Trevira®	Excellent resistance to mineral and organic acids at room temperature. Good resistance to weak alkalis at room temperature. Good resistance to weak alkalis at room temperature. Moderate resistance to strong alkalis, depending on concentration, temperature and time.	Excellent resistance to bleaches and oxidizing agents.
Thornel® carbon	Excellent resistance to acids and alkalis, even at high concentrations and temperatures. Strong oxidizers will degrade fiber.	Inert to all known solvents. Poor resistance to hypochlorite.
Nylon Nylon 6	Strong oxidizing agents and mineral acids cause degradation. Others cause loss in tenacity and elongation. Resists weak acids. Soluble in formic and sulfuric acids. Hydrolyzed by strong acids at elevated temperatures. Substantially inert in alkalis.	Can be bleached in most bleaching solutions. Generally insoluble in organic solvents. Soluble in some phenolic compounds.
Nylon 6,6	Unaffected by most mineral acids, except hot mineral acids. Dissolves with partial decomposition in concentrated solutions of hydrochloric, sulfuric and nitric acids. Soluble in formic acid. Substantially inert in alkalis.	Can be bleached in most bleaching solutions. Generally insoluble in most organic solvents. Soluble in some phenolic compounds.

TABLE 17.12 (continued).

Fiber	Effects of Acids and Alkalis	Effects of Bleaches and Solvents
Tencel® lyocell	Similar to cotton. Hot dilute or cold concentrated acids disintegrate fiber. Strong alkaline solutions cause swelling and reduce strength. Can be mercerized.	Attacked by strong oxidizing agents. Not damaged by bleaches. Generally insoluble in common organic solvents.
Rayon Fibro®	Similar to cotton. Hot dilute or cold concentrated acids disintegrate fiber. Strong alkaline solutions cause swelling and reduce strength.	Attacked by strong oxidizing agents. Not damaged by hypochlorite or peroxide. Generally insoluble except in cuprammonium and a few complex compounds.
Cuprammonium	Hot acids or high concentrations tend to disintegrate fiber. Strong caustic solutions tend to swell fiber and reduce strength.	Not affected by solvents. Insoluble in common organic solvents. Damaged by normal concentrations of hypochlorite or peroxide bleaches.
Rayon	Similar to cotton. Hot dilute or cold concentrated acids disintegrate fibers. Strong alkaline solutions cause swelling and reduce fiber strength.	Attacked by strong oxidizing agents. Not damaged by bleaches. Generally insoluble in common organic solvents.
Saran Saran®	Generally good resistance to acids and most alkalis. Sodium hydroxide and ammonium derivatives, in conjunction with heat, cause discoloration.	Generally good resistance to bleaches and solvents. Esters and ethers may be detrimental in varying degrees.
Acetate	Deteriorates in concentrated solutions of strong acids. Unaffected by weak acids. Strong alkalis saponify into regenerated cellulose.	Attacked by strong oxidizing agents. Not damaged by mild hypochlorite or peroxide bleaching conditions. Soluble in phenol, acetone, concentrated and glacial acetic acid.
Sulfar Ryton®	Outstanding resistance to acids and alkalis except hot concentrated sulfuric acids and concentrated nitric acid.	Excellent resistance to all solvents at high temperatures.
Acrylic Acrilan®	Good to excellent resistance to mineral acids. Fair to good resistance to weak alkalis; moderate resistance to strong, cold solutions.	Good resistance to bleaches and common solvents.
Creslan® and MicroSupreme	Generally good resistance to mineral acids and weak alkalis. Moderate resistance to strong alkalis at room temperature.	Unaffected by dry cleaning solvents. Can be bleached with sodium chlorite.
SEF® modacrylic	Resistant to most acids. Good resistance to weak alkalis; moderate resistance to strong, cold alkalis.	Good resistance to bleaches, dry cleaning fluids and most common solvents.

(continued)

TABLE 17.12 (continued).

Fiber	Effects of Acids and Alkalis	Effects of Bleaches and Solvents
Olefin Polyethylene	Low density: excellent resistance to acids and alkalis, with the exception of oxidizing agents. High density: excellent resistance to acids and alkalis, with the exception of oxidizing agents.	Low density: resists well below 150°F. Swells in chlorinated hydrocarbons at room temperature. Soluble at 160 to 175°F. High density performs generally the same but not as pronounced.
Herculon®	Excellent resistance to most acids and alkalis with the exception of elevated temperature exposure to chlorosulfonic acid, concentrated nitric acid and certain oxidizing agents.	Resistant to bleaches and most solvents. Some swells in chlorinated hydrocarbons at room temperature and dissolves at 160°F and higher. Others swell only at elevated temperatures.
Marvess® and Alpha®	Excellent resistance to concentrated acids and alkalis.	Resistant to bleaches and most solvents. Chlorinated hydrocarbons cause swelling at room temperature and dissolve fibers at 160°F and higher.
Essera®, Marquesa Lana®, Patlon® III	Excellent resistance to most acids and alkalis with the exception of elevated temperature exposure to some acids.	Resistant to bleaches and most solvents. Some hydrocarbons will cause swelling at room temperature. Some solvents will dissolve above 200°F.
Spectra®	Excellent resistance to most acids and alkalis, with the exception of elevated temperature exposure to some acids.	Resistant to most solvents, good resistance to bleaches. Some hydrocarbons cause swelling, especially at elevated temperatures.
Fibrilon®	Excellent resistance to concentrated acids and alkalis.	Resistant to bleaches and most solvents. Chlorinated hydrocarbons cause swelling at room temperature and dissolve fiber at 160°F and higher.
Spandex Glospan®/ Cleerspan, S-85	Resistant to mild acids and alkalis. Is degraded by strong acids and alkalis at high temperatures.	Good resistance to deterioration by bleaches. Discolored slightly by hypochlorite bleach. Resistant to solvents and oils except glycols.
Lycra® Type 126,127	Good resistance to most acids. Slightly yellowed by dilute hydrochloric, sulfuric acids. Good resistance to most alkalis.	Hypochlorite bleaches discolor. Can use perborate. Good resistance to most solvents, oils.
Type 128	Resistant to mild acids and alkalis. Degraded by strong acids and alkalis at high temperatures.	Good resistance to discoloring. Strength loss by hypochlorite bleaches.

TABLE 17.12 (continued).

Fiber	Effects of Acids and Alkalis	Effects of Bleaches and Solvents
Glass	Resists most acids and alkalis.	Unaffected.
Aramid		
Kevlar®	Good resistance to dilute acids and bases. Degraded by strong mineral acids and, to lesser extent, by strong mineral bases. Best chemical resistance from Kevlar 149.	Should not be bleached. Excellent solvent resistance.
Nomex®	Unaffected by most acids, except some strength loss after long exposure to hydrochloric, nitric and sulfuric. Generally good resistance to alkalis.	Unaffected by most bleaches and solvents except for slight strength loss from exposure to sodium chlorite.
Fluorocarbon		
Gore-Tex®	Essentially inert to acids and alkalis.	Essentially inert to bleaches and solvents except for alkali metals at high temperature and/or pressure.
Teflon®	Essentially inert to acids and alkalis.	Most chemical-resistant fiber known. The only known solvents are alkali metals and certain perfluorinated organic liquids at temperatures about 570°F.
PBI	Excellent resistance to most acids and alkalis. Some strength loss with strong alkali at elevated temperatures. Excellent organic solvent resistance.	Unaffected by most bleaches and solvents.

the neutral or acid side. Certain reducing agents are employed to modify the wool by rupturing sulfur-sulfur bonds. The wool is then more "plastic" and can be set into a desired shape. Reoxidation then reforms the bonds and the fiber then maintains its new configuration.

All of the natural fibers are unaffected by the common hydrocarbon and aromatic solvents.

The polyesters have good resistance to most mineral acids and to weak alkalis, but are disintegrated by strong hot alkali. They have good resistance to oxidizing agents and to most organic solvents except phenols. Generally speaking, the polyesters are considered to have better chemical resistance than the polyamides. The acrylics and modacrylics generally have excellent chemical resistance. They have excellent acid resistance, fair alkali resistance and are impervious to many solvents. The modacrylics are softened or dissolved by certain ketones. Saran has excellent acid and alkali resistance although it is sensitive to ammonia. It is unaffected by most common solvents, but is degraded by certain esters and ethers. Polyethylene and polypropylene have excellent acid and alkali resistance, but are sensitive to alkaline oxidizing

agents. Their general chemical and solvent resistance is high. They are normally soluble in hot chlorinated hydrocarbons.

27. ELECTROSTATIC PROPERTIES OF FIBERS

Both in processing and in use, the generation of static electricity on textile materials is at least a nuisance, and can be expensive or dangerous. During conversion of fiber to yarn and yarn to fabric, the large fiber surface areas are conducive to generating and holding electrostatic charges. These charges develop as the result of fiber to fiber, fiber to machinery, or even fiber to air friction. Most textile fibers are good or excellent electrical insulators; hence, electric charges cannot be conducted away. So the intensity of charge increases until fibers or yarns, all with the same charge, repel each other with accompanying "wildness" of fibers which then will not process properly. If the charge on the fibers is opposite to that on card clothing, drafting rolls, or other machinery parts, the fibers will stick to those parts of the machine oppositely charged, and processing may be slowed down or stopped. Static generation by conveyor belts and the like operating in areas where there are solvent fumes may be dangerous because a discharge spark may cause explosion and fire.

Truly distilled water is a relatively poor electrical conductor, but natural waters which contain even small quantities of soluble mineral salts are excellent conductors. Those fibers which have high moisture regains usually do not present quite as many static problems as do those which are impervious to water. If the relative humidity is sufficiently high, almost all textile fibers can be processed without electrostatic difficulty. As the humidity decreases, processing problems increase, usually geometrically with the synthetic hydrophobic fibers being the first to cause trouble. The polyesters, acrylics, olefins, and the polyamides having the lowest moisture regain are most likely to develop static charges. Rayon, silk, wool, and cotton can absorb water and hence can dissipate charges more easily. However, at low relative humidity all fibers may show processing difficulty. The textile industry, therefore, makes wide use of humidification during carding, combing, spinning, and weaving. Combing or spinning oils, warp sizes, and other processing chemicals usually contain anti-stats to eliminate the static problem. Static also may be eliminated by running the material over alpha particle radioactive sources which ionize the surrounding air.

Chemical anti-stats usually contain polar groups within their molecule. These act as conductors to prevent charge buildup or to conduct away any charges which do develop. Anti-stats also may be nothing more than humectants or water absorbing materials. The ionized water present on the textile will dissipate the static electricity. While generally an inverse relation exists between moisture regain and charge buildup, the intrinsic nature of the fiber polymer substance may also have a great bearing on static forming characteristics.

28. ELECTRICAL RESISTIVITY

Proper evaluation of the electrostatic properties of a fiber or fabric requires that all of the mechanisms for charge generation and dissipation be considered.

To pass a steady flow of electric charge, or current through any material requires a voltage. Ohm's law states that the ratio of applied voltage to current flow is called the electrical resistance. The lower the resistance, the more easily the current will flow. Flow can take place through either the volume of the material or along its surface, and the resulting resistances are called "volume resistance" and "surface resistance," respectively. Materials with high resistance will build up and hold electrostatic charges because the current cannot flow easily and allow the charge to leak off. Thus, a direct relationship normally exists between a fabric's surface or volume resistance and its tendency to build up a bothersome electrostatic charge.

Since water (other than distilled) is a good conductor, surface and volume resistances for all fibers are essentially inversely functional with moisture regain. Large changes in resistance result from small changes in regain. Many hydrophobic fibers may have high volume resistances because they are impervious to water. Their surface resistances may be proportionately lower, however, because they have thin or even molecular layers of water adhering to their surfaces. Methods for measuring resistivities are given in Chapter 20.0, Textile Testing.

References and Review Questions

1. REFERENCES

1. *ASTM D 276, Annual Book of ASTM Standards,* Vol. 7.01, 1993.

2. Textile Wold Manmade Fiber Chart 1994, McLean Hunter Publishing, Co.

3. *ASTM D 629, Annual Book of ASTM Standards,* Vol. 7.01, 1993.

4. *ASTM D 4848, Annual Book of ASTM Standards,* Vol. 7.02, 1993.

5. Backer, S., "The Relationship between the Structural Geometry of a Textile Fabric and Its Physical Properties. Part II: The Mechanics of Fabric Abrasion," *TRJ,* Vol. 21, 1951.

6. Hamburger, W. J., "Mechanics of Abrasion of Textile Materials," *TRJ,* Vol. 15, 1945.

7. Susich, G., "Abrasion Damage of Textile Fibers," *TRJ,* Vol. 24, 1954.

8. *ASTM D1909, Annual Book of ASTM Standards,* Vol. 7.01, 1993.

9. *ASTM D4391, Annual Book of ASTM Standards,* Vol. 7.02, 1993.

2. REVIEW QUESTIONS

1. Define refractive index. What is the significance of it?

2. What is tenacity? List all the fibers based on their tenacity values. Which fiber is the highest and which fiber is the lowest?

3. Is there any relation between fiber breaking elongation and its tenacity? Explain.

4. Explain viscoelasticity.

5. Define modulus. What are the different modulus definitions? Explain the significance of each. Give examples of fibers that should be tested with different modulus methods because of their properties.

6. What is toughness? Explain the significance of it for industrial textiles.

7. Discuss the significance of fiber elongation-recovery properties.

8. What methods are used to measure the loop and knot strength? How do loop and knot strength affect the performance of industrial textiles?

9. Does the fiber friction depend on fiber surface properties? Explain.

10. Why are abrasion and wear resistance critical for industrial textiles? Explain.

11. Explain how percent moisture content and percent moisture regain are determined. Which fiber has the highest moisture regain?

12. What is the effect of moisture and humidity on fiber strength?

13. Explain the effect of heat on tensile properties of fibers.

14. Why is resistance of fibers to mildew, aging and sunlight important? Explain.

15. What is the significance of electrical properties of fibers?

YARN NOMENCLATURE, PROPERTIES AND TECHNOLOGY

Yarn Nomenclature, Properties and Technology

S. ADANUR

Yarn nomenclature and the factors that influence yarn properties are discussed in this section. The textile industry has built a terminology that is dependent upon the nature of the fiber, yarn, or fabric, the country where the nomenclature developed, and the period in history during which common usage was established. Unfortunately the industry has not yet adopted one single standard terminology for defining yarn weight, twist and twist direction, number and direction of plies. The "Tex System" is a universal international system for establishing yarn weight per length and is explained below. The English system of length in inches, feet and yards, and weight in grains, ounces and pounds, is difficult and illogical not only for textile use, but in all industries. However, this system is still used by the textile industry, and so it is described in the following sections.

1. YARN NUMBER SYSTEMS

The yarn "number" or "count" defines the relationship between yarn weight and length. Two different systems are widely used in specific areas of the textile industry:

(1) The direct yarn number, or weight per unit length system, so called because the *heavier* the yarn the *greater* is the weight per unit length

(2) The indirect yarn number, or length per unit weight system, so called because the *lighter* (in weight) the yarn, the greater is the length per unit weight

The direct system is conducive to metric measure where the number of grams of weight for a given number of meters of yarn can be measured and recorded as a whole number. The indirect system is conducive to English measure where the number of yards per pound of yarn can be measured and recorded as a whole number.

"Yarn Number" by either system is a number of arbitrarily selected weights of yarn per unit length, or a number of arbitrarily selected lengths of yarn per unit weight. Examples of each system are as follows.

1.1 Direct Yarn Number

Tex: The tex is defined as the weight in grams of one kilometer (1,000 meters) of yarn. A kilotex equals 1,000 tex, a decitex equals 0.1 tex; and a millitex equals .001 tex. For example, a 300 tex yarn weighs 300 grams per kilometer.

Denier: The weight in grams of 9,000 meters of yarn. Commonly used in the man-made fiber industry, a 200 denier nylon yarn is one wherein 9,000 meters of yarn weigh 200 grams.

One tex equals one-ninth of the denier, and so these two systems are easily convertible.

1.2 Indirect Yarn Number

Cotton Hank: The number of 840 yard lengths in one pound of cotton yarn is called a 1s (ones) count yarn. A two hank cotton yarn contains $2 \times 840 = 1,680$ yards per pound and is called a 2s (twos) count yarn.

The 840 yards per pound nomenclature is used primarily for cotton. Yarns composed of other fibers use different yardages to describe their hanks. Thus a 1s worsted yarn count yarn

TABLE 18.1 Yarn Numbering and Count Systems (ASTM D 2260).

Direct or Weight per Unit Length Systems	
Tex	: the mass in grams of 1 km of yarn, filament, fiber or strand
Denier	: the mass in grams per 9,000 m of yarn, filament, fiber or other textile strand
American grain count	: the mass in grains per 120 yd of sliver or roving
Indirect or Length per Unit Weight Systems	
Cotton count	: the number of 840 yd lengths of yarn per lb
Cut (in glass yarns)	: the number of 100 yd lengths of yarn per lb
Cut (in wool yarns)	: the number of 300 yd lengths of yarn per lb
Linen lea	: the number of 300 yd lengths of yarn per lb
Metric count	: the number of meters of yarn per gram
American woolen run	: the number of 1,600 yd lengths of yarn per lb
Worsted count	: the number of 560 yd lengths of yarn per lb

contains 560 yards per pound, a 1s linen yarn contains 300 yards per pound.

Table 18.1 lists yarn numbering or count systems as defined by the ASTM D 2260. Table 18.2 gives factors for converting yarn numbers from one system to another.

2. YARN TWIST DIRECTION

Twist is necessary in order to manufacture staple fiber yarns and to give integrity, compactness, snag and abrasion resistance to filament yarns as well as to cords and ropes, whether they be made from spun or continuous filament singles. Ideally, the twist path of a single element in the yarn is a perfect cylindrical helix. It is necessary to describe the nature of the helical path with respect to whether the twist is "left" or "right." Thus when the yarn is held in a vertical (north and south) position and the twist helix rises to the right, the yarn is stated to contain "right-hand" twist; when the helix rises to the left, the yarn contains "left-hand" twist. Methods for determining the number of turns per inch will be found in Chapter 20.0, Textile Testing. The letters S and Z are used as designating letters for left and right twist, respectively. Figure 18.1 gives the usual form of illustration that accompanies these definitions.

3. SINGLES AND PRODUCER'S YARN NOMENCLATURE

Trade practice uses a relatively simple code for identifying count, yarn twist, and twist

direction. For spun cotton yarns, only the count and ply are usually given, e.g., 8s/10, meaning that ten plies of 8s singles are twisted together. Twist and twist direction are not normally listed. For continuous filament man-made producer's yarn the nomenclature on the package may read "nylon 210-34-1Z-300" which means that the yarn is 210 denier, contains 34 individual filaments, has 1 turn per inch Z twist, and the manufacturer's code type is 300. The letters B, D or SD may be given to identify bright, dull or semi-dull yarn. The fiber producer also identifies each lot of yarn with a "merge number." Since no two lots of yarn are exactly alike, in some uses care must be taken not to mix lots, otherwise streaks or bands may appear in the

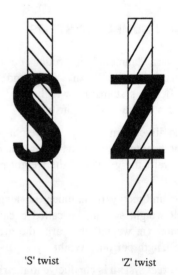

'S' twist   'Z' twist

FIGURE 18.1 Yarn twist designations.

TABLE 18.2 Conversion Factors for Converting from One Yarn Numbering System to Another[a] (copyright ASTM).

System for Which Yarn Number Is Needed	System for Which Yarn Number Is Known								
	Tex[b]	Denier	American Grain Count	Cotton Count	English Worsted Count	American Woolen Run	Metric Count	Linen Lea	yd/lb
Tex[b] (g/km)	$tex = \cdots$	$\dfrac{den}{9}$	$0.590\,541 \times gr$	$\dfrac{590.541}{cc}$	$\dfrac{885.812}{wc}$	$\dfrac{310.034}{wr}$	$\dfrac{1\,000}{mc}$	$\dfrac{1\,653.52}{lea}$	$\dfrac{496\,055}{y}$
Denier (g/9,000 m)	$den = 9 \times tex$	$\cdots$	$5.314\,87 \times gr$	$\dfrac{5\,314.87}{cc}$	$\dfrac{7\,972.31}{wc}$	$\dfrac{2\,790.31}{wr}$	$\dfrac{9\,000}{mc}$	$\dfrac{14\,881.6}{lea}$	$\dfrac{4\,464\,492}{y}$
American grain count (grains/120 yd)	$gr = \dfrac{tex}{0.590\,541}$	$\dfrac{den}{5.314\,87}$	$\cdots$	$\dfrac{1\,000}{cc}$	$\dfrac{1\,500}{wc}$	$\dfrac{525}{wr}$	$\dfrac{1\,693.36}{mc}$	$\dfrac{2\,800}{lea}$	$\dfrac{840\,000}{y}$
Cotton count (840 yd lengths/lb)	$cc = \dfrac{590.541}{tex}$	$\dfrac{5\,314.87}{den}$	$\dfrac{1\,000}{gr}$	$\cdots$	$\dfrac{wc}{1.5}$	$\dfrac{wr}{0.525}$	$0.590\,541 \times mc$	$\dfrac{lea}{2.8}$	$\dfrac{y}{840}$
English worsted count (560 yd lengths/lb)	$wc = \dfrac{885.812}{tex}$	$\dfrac{7\,972.31}{den}$	$\dfrac{1\,500}{gr}$	$1.5 \times cc$	$\cdots$	$\dfrac{wr}{0.35}$	$0.885\,812 \times mc$	$\dfrac{lea}{1.866\,67}$	$\dfrac{y}{560}$
American woolen run (1,600 yd lengths/lb)	$wr = \dfrac{310.034}{tex}$	$\dfrac{2\,790.31}{den}$	$\dfrac{525}{gr}$	$0.525 \times cc$	$0.35 \times wc$	$\cdots$	$0.310\,034 \times mc$	$0.187\,5 \times lea$	$\dfrac{y}{1\,600}$
Metric count (1,000 m/kg)	$mc = \dfrac{1\,000}{tex}$	$\dfrac{9\,000}{den}$	$\dfrac{1\,693.36}{gr}$	$\dfrac{cc}{0.590\,541}$	$\dfrac{wc}{0.885\,812}$	$\dfrac{wr}{0.310\,034}$	$\cdots$	$\dfrac{lea}{1.653\,52}$	$\dfrac{y}{496.055}$
Linen lea (300 yd lengths/lb)	$lea = \dfrac{1\,653.52}{tex}$	$\dfrac{14\,881.6}{den}$	$\dfrac{2\,800}{gr}$	$2.8 \times cc$	$1.866\,67 \times wc$	$0.187\,5 \times wr$	$1.653\,52 \times mc$	$\cdots$	$\dfrac{y}{300}$
Yards per pound (yd/lb)	$y = \dfrac{496\,055}{tex}$	$\dfrac{4\,464\,492}{den}$	$\dfrac{840\,000}{gr}$	$840 \times cc$	$560 \times wc$	$1600 \times wr$	$496.055 \times mc$	$300 \times lea$	$\cdots$

[a]The conversion factors are based on the following relationships given in Metric Practice E 380: 1 yard = 0.9144 m, exactly, and 1 lb (avoirdupois) = 0.453 592 37 kg, exactly. The conversion factors containing fewer than six significant digits are exact values.

[b]Multiples and submultiples of this basic unit may be used as a convenience to avoid large numbers or decimal fractions. For example, decitex (dtex) or tex × 10 is suitable for fine yarns and fibers; millitex (mtex) or tex × 1,000 is suitable for fibers; while kilotex (ktex) or tex/1,000 is often used for ropes, cords, rovings, tops, and slivers.

Examples of use:
1) The English worsted count equivalent to a cotton count of 10 is 1.5 times 10, or 15 English worsted count.
2) The cotton count equivalent to 30 tex is 590.54 divided by 30, or 19.7 cotton count.

dyed fabric. Yarns of differing merge numbers may have slightly different optical properties due to small differences in molecular weight, degree of crystallinity, orientation or amount of draw during manufacture. These differences can cause small but different degrees of dye acceptance, which will be manifested as streaks or bands.

4. PLIED YARN NOMENCLATURE

Because of the many singles yarn count systems used, in describing plied yarns care must be taken to properly list the construction and weight. However, once the singles yarn count system is known, the description of the number of plies is relatively straightforward.

4.1 Indirect Count

For indirect yarn numbers, the count of the singles yarn is given, followed by the number of singles combined to make the plied yarn, and the number of plied yarns combined to make the cord.

Example 1: A cotton 10s/2 (tens-2 ply) means that two cotton yarns, each of 10s count are plied together.

Example 2: If three such 10s/2 yarns are again plied to form a cord, the cord is identified as 10s/2/3. Note that in the first example the actual weight of the plied yarn, neglecting any contraction due to twist take-up (defined later in the chapter), is a 5s. This is called the "equivalent count" of the yarn. It is often necessary to know the equivalent count for purposes of calculating fabric weight or strength. For example, if a number 5 cotton duck is composed of 7s/3 warp yarns and 7s/4 filling yarns with 36 warp ends and 24 filling picks per inch, the fabric weight and strength can be reasonably estimated, predicated on the calculation of equivalent count.

4.2 Direct Count

Here the tex or denier of the singles yarn is given, followed by the number of singles making up the ply, and the number of plies making up the cord.

Example 1: A continuous filament nylon 210/3 plied yarn means that three singles yarns, each being 210 denier, are plied together. Excluding any contraction due to "twist take-up," the equivalent yarn count is 630 denier.

Example 2: If five such 210/3 plied yarns are again plied to form a cord, the cord is identified as 210/3/5, the equivalent yarn count being 3150 denier.

Sometimes the yarn is further described by giving the singles and ply twist and direction; for example, $210/3 - 8S - 6Z$ indicates three 210 denier singles yarns, each with 8 turns per inch S twist before plying, plied together with 6 turns per inch Z twist.

5. YARN TWIST MATHEMATICS

Ideally for a twisted perfectly symmetrical yarn composed of many (i.e., more than thirty) perfectly round continuous filaments of equal diameter, each filament will assume the form of a perfect cylindrical helix as shown in Figure 18.2.

If the cross sections are not round, or the symmetry is not perfect, deviations from the perfect helix will exist. In the case of plied yarns, each composed of twisted singles, a helix may be imposed upon a helix and the final path

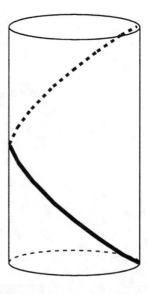

FIGURE 18.2 Ideal helix of a fiber in a yarn.

of a single filament becomes complicated. In the case of spun yarns, particularly for natural fibers, cross sections are not round, fiber diameters and lengths are not constant, and symmetry with respect to fiber position is minimum, largely because the individual fibers are not perfectly oriented longitudinally with the fiber axis but may "dart" from the center to the outer periphery of the yarn.

Certain classical mathematical relationships can be derived and are useful with respect to yarn manufacture. The relationships among filament size and shape, turns per inch, helix angle, yarn diameter and the spatial or geometric positioning of individual fibers can all theoretically be calculated. Deviations from the theoretical can be valuable in the practical phases of yarn manufacturing and in the evaluation of yarn properties. It will be apparent that in such geometrical mathematical analyses the same general principles which hold for filaments in yarns also pertain for yarns in plied yarns, and for plied yarns in cords. Several of the important factors that influence yarn properties are listed below, using the relationship of singles to plied yarns as a model. The same reasoning can be applied to the next lower (fibers into yarns), or next higher (plied yarns into cords) order of construction. A more thorough mathematical treatment is found in Appendix 5.

6. TWIST TAKE-UP

When two fixed lengths of singles yarns are twisted together to form a plied yarn, because of the helical path that each assumes with respect to the other, the resulting length of plied yarn must be less than that of each singles yarn from which the plied yarn is composed. The reduction in plied yarn length is called "twist take-up." It is defined as the change in length of a filament, yarn, or cord caused by twisting, expressed as a percent of the original untwisted length. Twist take-up is most easily determined by inserting a fixed length of yarn in a twist tester, untwisting, measuring the extended length, and calculating the percent extension as a function of the new extended length. Appendix 5 gives the mathematical analysis of twist take-up.

7. HELIX ANGLE VERSUS TURNS PER INCH

In the plied yarn the angle made by individual singles yarns with the axis of the plied yarn is called the helix angle (Appendix 5). It is a function of the turns per inch in the ply, the singles and plied yarn diameters, and a geometrical constant K, which is dependent upon the number and shape of the individual singles making up the ply.

8. YARN STABILITY

The value of the geometrical constant K, is dependent upon the shape and number of filaments making up the yarn, for these parameters will govern the manner in which the filaments pack together. Figure 18.3 shows filament packing configurations. The two and three filament (or ply) constructions are stable; four and five reasonably stable; the latter not in the form of a ring but in the flattened 2-3 form. The six develops as a central core of one, and an incomplete outer ring of five. Seven gives excellent symmetry and stability. Somewhat of a disadvantage in the seven unit construction is the problem of corkscrew (explained next) and elongation balance (Chapter 19.0, Fabric Properties and Technology). The single central filament may be slightly shorter than the peripheral filaments, and when a stress is applied the central filament may reach its limiting strain first and then break. The overall effect is somewhat lowered strength efficiency. The most stable constructions are made of 2, 3, 7 or 13 units. Of course most singles yarns, whether they be spun or continuous, usually contain more than, say, 15 filaments, and filament cross sections are not usually perfect circles. Therefore the number of filaments selected is not necessarily critical. In plied yarn or cord constructions, however, the number selected is of great importance because of the corkscrew influence.

9. CORKSCREW YARN

In plying yarns, one of the plies may run slack during the twisting while the remaining plies

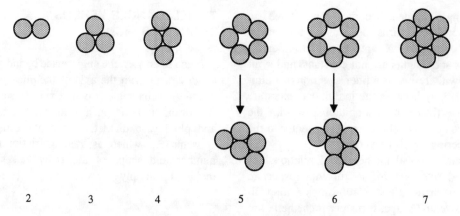

number of filaments in cross section

FIGURE 18.3 Filament packing.

are kept taut. This produces a "corkscrew" effect where the slack ply entwines about the remaining straight strands. While useful as a novelty effect, corkscrew yarns should be avoided in any industrial fabric where high strength-to-weight ratios are desired. Corkscrew yarns are weak because all plies do not break at the same time, and the yarn is stated to be "inefficient" because lack of rupture elongation balance. When tension is applied to such a yarn, the multi-yarn core, being shorter, will reach its limiting strain and will break first. Then the load is transferred to the corkscrewed yarn (or yarns). Depending upon their number, these may be too weak to support the load previously held by the straight core yarns, and so they, too, fail. Thus the strength of a corkscrew yarn must always be less than the theoretical attainable in a yarn devoid of corkscrew (see Appendix 5 for additional comments on corkscrew).

10. TWIST MULTIPLIER FOR SPUN YARNS

Because of the relationship between turns per inch and yarn size (diameter), the yarn manufacturer has found it necessary to establish a formula that will permit him to calculate the turns per inch necessary for a given size spun yarn. He uses a simple relationship between the turns per inch and the cotton count, calling the factor the "twist multiplier." This is a quantitative index of the relative steepness of the helix angle. Twist multiplier is defined as

$$TM = \frac{T}{C^{1/2}} \qquad (18.1)$$

where

T = turns per inch
C = cotton count

The twist multiplier is widely used as an empirical means for establishing proper twist in staple yarns. For cotton, for example, twist multipliers range from 2.5 for soft twisted yarns to six for very hard twisted yarns. Qualitatively, cotton singles yarns having a TM of about 4.5 have a surface helix angle of about 30°.

11. EFFECT OF YARN TWIST ON BREAKING STRENGTH

The twist that the manufacturer inserts in spun singles yarn depends upon the desired yarn character and intended use. In industrial fabrics a twist may normally be selected that will produce maximum breaking strength.

Figure 18.4 shows a graph of turns per inch versus breaking strength for a cotton yarn. The strength increases with twist until a maximum is reached. For the example shown, this is at about 22 turns per inch. Then the strength starts to decrease. Two phenomena working opposite to each other produce this type of twist-strength relationship. As the twist increases, the individual fibers assume helical paths and wrap around each other. The normal pressure and total friction of the individual fibers pressing against each other progressively increase, with accompanying yarn strength increase, the theoretical maximum being that force which will rupture the fibers.

However, as the twist increases, the helix angle becomes more acute, and the force component along the yarn axis becomes less. This produces a progressive reduction in yarn strength. Furthermore, the randomness of the position of, and points of contact between, the individual fibers in the yarn produces a distribution of fiber lengths. When these varying fiber lengths are stressed, elongation unbalance results, the fibers break one at a time instead of as a group, and a lower yarn strength results. This elongation balance phenomenon is discussed in Chapter 19.0, Fabric Properties and Technology. Thus, the factors tending to produce increased yarn strength are more effective at low twist, while the factors tending to produce decreased yarn strength are more effective at higher twist, and so a rising and then falling twist versus strength curve results.

To explain the translation of staple fiber strength into spun yarn strength is an exercise in physics and mathematics of considerable complexity. Yarn strength is based upon such intrinsic fiber properties as strength, elongation, cross-sectional shape, length, friction coefficient, and upon such geometric factors involved in yarn formation, such as yarn diameter, turns per inch, degree of fiber packing (i.e., the ratio of fiber space to void space within the yarn), and the randomness of fiber position within the yarn.

12. YARN BREAKING STRENGTH AND ELONGATION TESTS

There are mainly two types of yarn breaking strength-elongation tests: single end and skein tests.

12.1 Single End Test

The single end test consists of inserting a length of yarn between the jaws of the testing

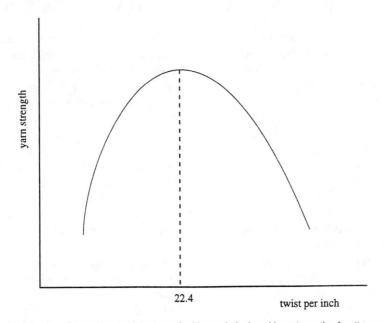

FIGURE 18.4 Effect of twist (turns per inch) on skein breaking strength of cotton yarn.

machine, separating the jaws, and recording the load-elongation curve using tensile testing machines. To avoid yarn cutting and "jaw breaks," capstan rather than flat jaws are often employed. The yarn is wound around the capstans with about ten inches of yarn in the "free gauge length," that is, the length between capstan tangent points. During the test, yarn elongation takes place not only in the free gauge area but also progressively around the capstans where the yarn is being snubbed. This snubbing elongation around the capstan must be taken into account in calculating true rupture elongation of the yarn (Chapter 20.0, Textile Testing).

12.2 Skein Test

Single end breaks give valid measures of strength and elongation under the selected test conditions, but are slow and time consuming for routine test purposes. At least five, and often fifteen or twenty tests are necessary to obtain a meaningful average. Therefore the skein test is widely used. Eighty 1.5 yard wraps of yarn are wound onto a reel, forming a skein which is then placed over two capstan jaws on the testing machine, and the breaking strength determined. ASTM D 1578 gives several options for number of turns and reel diameter. The strength per end may be calculated by dividing the recorded strength by 160 ends, or the skein strength itself may be used as an index of quality. In breaking a multiplicity of yarns in this type of test, all yarn ends do not break at the same instant of time. Because of weak spots, unequal rupture elongations and unequal tensions and lengths that develop in winding the skein, some yarns will break sooner than others. When one yarn is broken, the load that it supported is transferred to the others. This fractional additional load per yarn may exceed the capacity of another yarn, and it, in turn, will fail. The break is progressive through the skein, the phenomenon being called "serigraph effect" or lack of elongation balance (Chapter 19.0, Fabric Properties and Technology). Because of the serigraph effect, individual yarn strengths calculated from skein strength tests are lower than single end test values.

Fully automated modern yarn testing machines can test the strength of individual yarns very fast (a few seconds per yarn test), in a continuous manner, and may be used to replace skein tests.

13. COTTON YARN STRENGTH-COUNT FACTOR

For cotton yarns the higher the yarn number the lighter the yarn weight and (presumably) the lower the breaking strength. Because of this inverse relationship, the product of pounds breaking strength and yarn number tends to remain constant and it may be used as a criterion of breaking strength efficiency. The product is called the strength-count factor and is comparable to "tenacity," a term employed to describe the strength efficiency of man-made fibers, calculated as grams breaking strength per denier (or tex). The strength-count factor, of course, is not an absolute constant for singles cotton yarns, but is dependent upon fiber strength and staple length, yarn twist, whether the yarn is combed, etc. Empirical equations have been developed that relate skein strength and yarn count.

14. YARN TWIST BALANCE

Fibers in yarns assume an imposed helical shape that is not necessarily stable, there being a desire for each individual fiber to revert to its original configuration. The integrated effect sets up a torque in the yarn opposite to the direction of inserted twist, i.e., the yarn tends to untwist and therefore is stated to be "unbalanced" or "wild." Usually the greater the amount of twist, the greater is the tendency for unbalance. The term *balanced yarn* is applied to one exhibiting no tendency to double, twist or kink when held in a slack condition. To test for balance, a four or five foot length of yarn is held by the ends and allowed to hang freely in a loop while the ends are brought together. If this can be performed without the loops exhibiting any tendency to rotate or twist, the yarn is said to be balanced (Figure 18.5).

When testing for twist balance it is important

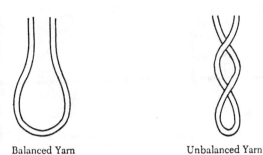

Balanced Yarn   Unbalanced Yarn

FIGURE 18.5 Twist balance.

to remove the yarn from the package without changing the twist in so doing. Upon cutting a length of yarn care must be taken not to permit one end of the yarn to rotate relative to the other. It should also be remembered that yarn twist can be changed by simply drawing the yarn over the top of the package. The yarn should always be removed from the package by the same method which is to be used for the end objective desired. Thus if the balance of a sewing thread is to be determined, and the yarn normally is paid off the spool by rotation rather than over the top of the spool, the same technique should be followed in selecting the sample for testing.

15. RELATIONSHIP OF SINGLES TO PLY TWIST

In manufacturing plied yarns or cords, the attainment of yarn balance may be desirable in order to obtain stability in the resulting fabric. If twist balance is not obtained, the yarn will be wild and difficult to manage, and the resulting fabric may pucker, curl or otherwise distort. Of course, it will be apparent that for some purposes unbalanced or wild yarns are desired – for example in the manufacture of "twist" broadloom carpets or seersucker fabrics. Presumably for most industrial uses, however, balanced yarns and cords are an advantage.

Balance can be obtained via two mechanisms. First, if the relationship of turns per inch and twist direction among singles, plies, and cords is correct, balance may result purely from the geometry of the structure. Second, independent of the geometry, balance, particularly in singles, can be obtained by relaxing the torques induced during spinning and twisting. Depending upon the nature of the fiber, this can be accomplished by immersion in water or steam for such hydrophilic fibers as cotton, rayon or wool; or by dry heat for such hydrophobic fibers as the polyesters or the polyacrylics. Aging of the yarn for prolonged periods also is useful in allowing torque stresses to decay until the yarn is "balanced." Chemical setting may also be used; in fact this is the basis of many of the so-called wash and wear finishes.

As a simple example of ply twist balance, consider an unstabilized singles yarn that contains eight turns of S twist. Referring to Figure 18.6 this would be accomplished by rotating the end of a 1 inch element through eight complete turns in a clockwise direction. If three such 8 tpi S twist singles are now plied together with eight (plus) Z turns, accomplished by rotating the three plies about the common axis in a counterclockwise direction, the original twist in the singles will be removed (approximately).

Thus, the plied structure will have eight

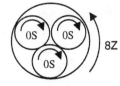
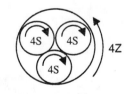

Singles
8 turns S

Plied yarn
8 turns Z in ply
produces zero turns
in singles

Plied yarn
4 turns Z in ply
causes retention of
4 turns S in singles

FIGURE 18.6 Singles-ply twist relations.

(plus) turns, while singles will theoretically be untwisted as they lie in the ply end and will possess little residual torque. The ply will contain eight turns and so a tendency for the ply to untwist will result because there is no offsetting twist in the singles. Now as a second example, if instead of eight turns Z ply twist, only four turns are inserted, the resulting structure will have approximately four turns S in singles and four Z turns in the ply, and balance is obtained. Of course this is an oversimplification of the problem. In theory, account must be taken of fiber torsional and bending rigidities, friction and stress relaxation time effects. In practice, because of these factors that cause deviations from the theoretical, the number of turns of twist in the singles and in the ply must be empirically determined in order to obtain balance. The main factor is the nature of the fiber itself. Fibers are not perfectly elastic; they do not obey Hooke's law which states that strain (elongation) is linearly proportional to applied stress (force). Fibers are said to be "viscoelastic" (Chapter 17.0, Fiber Properties and Technology); upon application of a load they initially deform approximately according to Hooke's law, but then may continue to "creep" or flow as long as a force is maintained. Upon load removal they tend to slowly "recover," i.e., revert back toward

Balanced structure: fibers parallel to plied yarn axis.

Unbalanced structure: fibers are not parallel to plied yarn axis.

FIGURE 18.7 Plied yarn balance.

their original configuration and dimension. Therefore, when a single or plied yarn is manufactured, stresses that manifest themselves as torques result, and only by empirical means can one with certainty arrive at a balanced structure. Finally it will also be obvious that balance may occur at one relative humidity and temperature, but if the fibers are sensitive to heat or moisture, a different set of ambient conditions may result in a new unbalanced state. For heavy plied structures such as cords or ropes, balance is most easily obtained if the individual fibers in the singles yarn, or the individual yarns in the cord, are parallel to the axis of the next higher structure (Figure 18.7).

18.2

Review Questions

1. REVIEW QUESTIONS

1. What is the difference between the direct and indirect yarn number systems? Which one is more logical? Why?

2. Define the following terms:
   - twist direction
   - singles yarn
   - producer's yarn
   - merge number
   - S twist
   - Z twist
   - twist take-up

3. How is the twist multiplier for spun yarns calculated?

4. Explain the effect of yarn twist on breaking strength.

5. What is the significance of having twist balance in a yarn? Explain.

FABRIC PROPERTIES AND TECHNOLOGY

Fabric Properties and Technology

S. ADANUR

1. FABRIC GEOMETRY

The visual, tactile and engineering properties of a fabric are dependent upon many intrinsic as well as geometric fiber, yarn and fabric variables. A fiber's properties are governed by the character of the polymer substance from which it is made and by such geometric factors as cross-sectional area and shape, length, amount of crimp, stiffness, and surface characteristics. Yarn properties are governed by fiber properties and by such geometric factors as yarn diameter, twist, cross-sectional shape, and degree of fiber compactness. Fabric properties are governed by yarn properties and by such geometric criteria as the fabric weave, number of threads per inch, degree of thread packing within the structure, and yarn crimp.

Peirce [1] states that there are eleven structural factors that control fabric construction. The factors, for both warp and filling are (Figure 19.1):

- total length of yarn L between yarn intersections
- reciprocal p of the number of threads per inch, i.e., horizontal projection of L
- yarn crimp, c
- maximum perpendicular displacement h of the center line of a yarn out of the fabric plane
- angle of inclination ϕ of that length of yarn which is out of the fabric plane
- sum of warp and filling yarn diameters, D

Proper utilization of these eleven interdependent variables permits the engineering design of fabrics with the desired weave, number of threads per inch, yarn crimp, degree of fabric cover (i.e., ratio of fabric surface covered by the yarns to the theoretical maximum), fabric density and thickness. These factors in turn (in addition to intrinsic properties) influence such properties as breaking and tear strength, abrasion resistance, air permeability, etc. For example, the amount of air that can flow through a fabric at a given pressure differential across the fabric is a function of the fabric weave, the number of yarns per inch, yarn crimp, and yarn cross-sectional shape, the latter influenced by yarn twist. All of these variables control the shape and number of open area interstices within and between yarns through which air flow takes place.

2. YARN COUNT AND COVER FACTOR

The "yarn count," i.e., the number of warp and filling yarns per inch in a fabric is most easily determined by a direct count with the aid of a pick counter. For a given yarn number, the larger the number of threads per inch the more dense and opaque is the fabric. If for each set of yarns the product of the thread count (per inch) and the yarn diameter (inches) is less than one, then theoretically open spaces will exist between the yarns. The ratio of fabric surface occupied by yarn to the total fabric surface is called the fabric "cover factor," and can be shown, theoretically, to be (Appendix 5):

$$C = (w \cdot d_w + f \cdot d_f - w \cdot f \cdot d_w \cdot d_f)$$

$$(19.1)$$

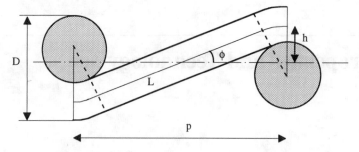

FIGURE 19.1 Schematic of a unit cell for a plain weave.

where

C = cover factor
w = warp threads per inch
f = filling threads per inch
d_w = warp yarn diameter in inches
d_f = filling yarn diameter in inches

When a condition is reached where, for either warp or filling yarns, the product of the number of yarns per inch and the diameter equals one, the yarns will touch each other, and the cover factor becomes one. When this condition exists for both sets of yarns the cover factor remains one. Often in this latter case the maximum number of warp and filling yarns possible has been woven into the fabric. Any further increase in picks or ends would result in yarns piling up on each other to give a double or multiple layer of yarns. Of course the yarns may be compressible and therefore more yarns than the theoretical maximum, calculated on the basis of uncompressed yarns, are often weavable into the fabric. Fabric cover factor is of obvious importance with respect to water, moisture vapor, or air permeability. In fabric coating the degree of openness will influence the penetration of coating into or through the fabric, and in part will affect adhesion.

3. FABRIC DENSITY, PACKING FACTOR, AND POROSITY

The density of a fabric is calculated by measuring its thickness and its weight per unit area:

$$\text{Density} = \frac{\text{weight}}{\text{unit volume}}$$

$$= \frac{\text{weight}}{\text{unit area} \times \text{thickness}} \quad (19.2)$$

A fabric is composed of both fiber substance and air space, but only the fiber substance contributes to the weight. The theoretical maximum density achievable if all yarn and fabric voids are removed by compression is that of the fiber itself. The ratio of fabric density to fiber density is a measure of the degree to which the fibers are packed together and is called the "Packing Factor."

$$\text{Packing Factor (P.F.)} = \frac{\text{fabric density}}{\text{fiber density}}$$

$$= \frac{W/V_t}{W/V_f} = \frac{V_f}{V_t} \quad (19.3)$$

where

W = fiber (as well as fabric) weight
V_t = total volume of fabric
V_f = volume of fiber

"Porosity" is the amount of open space within a fabric. Porosity may be defined as the ratio of the volume of air or void contained within the boundaries of a material to the total volume (solid material plus air or void) expressed as a percentage.

$$\text{Porosity } (P) = \frac{V_v}{V_t} \times 100 \quad (19.4)$$

where

P = porosity

V_v = volume of voids in the fabric

V_t = total volume of the fabric

Since both packing factor and porosity are measures of the ratio of fiber volume to total fabric volume, they are negatively correlated as follows:

$$P.F. = \frac{V_f}{V_t}$$

and

$$P = \frac{V_v}{V_t} = \frac{V_t - V_f}{V_t} = 1 - \frac{V_f}{V_t}$$

therefore

$$P = 1 - P.F. \qquad (19.5)$$

4. YARN CRIMP AND TAKE-UP

Since a woven fabric consists of sets of interlacing yarns, either the warp, the filling or both yarns must assume a wavy path in order to be accommodated within the fabric. This waviness is called yarn "crimp." Depending upon the tension applied to the yarns during weaving, most, or theoretically all, of the crimp can be inserted into the warp, with the filling being perfectly straight, or the crimp can be inserted into the filling, with the warp straight, or crimp can be present in both sets of yarns (Figure 19.2).

Fabric tensioning in either the warp or filling direction tends to remove the crimp and straighten the tensioned set of yarns. The straightening must inevitably produce increased crimp in the perpendicular set of yarns. This can be visualized by examining the top and middle portions of Figure 19.2. Tensioning the crimped filling produces a straight filling, but a crimped warp. This crimp transfer is called "crimp interchange." It can result in important changes in fabric gemoetry, which, in turn, can advantageously or negatively affect such properties as breaking strength and elongation, tear strength, energy absorption and crease resis-

crimped filling, straight warp

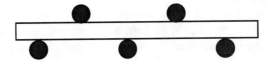

crimped warp, straight filling

crimped warp and filling

FIGURE 19.2 Fabric crimp.

tance. "Negative tension" via water or heat relaxation, with accompanying shrinkage, can also produce crimp interchange.

Fabric weave also influences the amount of yarn crimp. The greater the number of yarn intersections per unit area of fabric surface, the longer must be the path of warp over filling or filling over warp. Thus a plain weave fabric will have greater yarn crimp than a satin. In designing fabric constructions it is important that the crimp-elongation relationship be kept in mind. Thick bulky fabrics usually are designed with maximum yarn crimp. Double and triple fabrics contain high crimp yarns that travel from the face to the back of the fabric.

In addition to the crimp resulting from the passage of warp over filling, both the yarn itself and the fibers composing the yarn can also have different crimp frequencies as shown in Figure 19.3. Thus three different orders of crimp can exist within a fabric structure.

The percentage crimp which a yarn contains as it lies in a fabric is determined by placing gauge marks *on the fabric,* with crayon or ink, for example exactly ten inches apart, so that one or more of the yarns will contain the marks. Each marked yarn is then raveled out of the fabric, taking care not to strain it or disturb its twist. The flat jaws of a low capacity constant rate of extension tensile tester are adjusted to a gauge length of exactly ten inches, and the yarn inserted so that the original gauge marks on the yarn will exactly line up with the edge of each jaw. In this condition, because of the presence of yarn crimp, the yarn will be loose and floppy between the jaws. The lower jaw of the testing machine is slowly moved down, thus plotting an accurate load-elongation diagram, until the load required to straighten but not stretch, the yarn is achieved. The resulting elongation, as a per-

centage of the original length marked in the fabric is the percent crimp. Crimp is normally removed at very low loads and, therefore, meticulous techniques are required if accurate values are to be obtained.

It is important to distinguish between the terms "percent yarn crimp" and "percent yarn take-up" or "contraction." The measurement is the same, but the former is calculated on the basis of the crimped length of yarn while the latter is calculated on the extended length. Thus according to ASTM [2]:

- *Percent crimp* is the difference in distance between two points on a yarn as it lies in a fabric, and the same two points when the yarn has been removed from the fabric and straightened under specified tension, expressed as a percentage of the distance between the two points as the yarn lies in the fabric.
- *Percent take-up* is the difference in distance between two points in a yarn as it lies in a fabric, and the same two points after the yarn has been removed from the fabric and straightened under specified tension, expressed as a percentage based on the straightened out-of-fabric distance.

The weaver uses percent take-up because it tells him the length of fabric he will obtain from a given length of yarn.

5. FABRIC SKEW AND BOW

The filling yarns in a fabric may not always lie perpendicular to the warp. Furthermore, because of tensions developed during weaving, the filling yarns may assume the shape of an arc or bow as they lie in the fabric. The former condition is known as skew; the latter as bow (Figure 19.4).

Skew is defined as the distance, measured parallel to and along the selvage, between the point at which a filling yarn meets one selvage, and perpendicular from the point at which the same filling yarn meets the other selvage. It is expressed as a direct distance (AB in

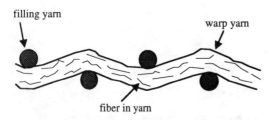

FIGURE 19.3 Three types of fiber and yarn crimp.

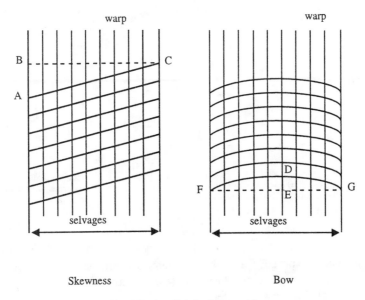

FIGURE 19.4 Fabric skew and bow.

Figure 19.4) or as a percentage of the fabric width:

$$\% \text{ Skew} = (\text{AB/BC}) \times 100 \quad (19.6)$$

Bow is defined as the greatest distance, measured parallel to the selvages, between a filling yarn and a straight line drawn between the points at which this yarn meets the two selvages. It is expressed as a direct measurement DE in Figure 19.4, or as a percentage of the fabric width:

$$\% \text{ Bow} = \text{DE/FG} \quad (19.7)$$

6. FABRIC WEIGHT

As in the case for yarns, there are both direct and indirect systems for describing fabric weights. The direct system indicates weight per area, or weight per running length for the full width of the fabric. It is unfortunate that many times when the fabric weight is given, the units are not recorded, so that one cannot know (except by experience) whether the weight per square yard or running yard is meant. Furthermore, the weight per running yard is meaningless (again except by experience) unless the width is given.

The indirect system indicates running length per weight, and is most often given as running yards per pound. Area per unit weight—for example, square yards per pound—is practically never used.

7. FABRIC THICKNESS

Because fabrics are compressible, the measurement of thickness should be made under controlled pressure of the thickness gauge. ASTM D 1777 specifies different pressures for different fabrics [3]. For dense compact fabrics, thickness is not critically influenced by pressures. For lofty, "soft," compressible fabrics it may be necessary to specify both the pressure and the area of the presser foot. Also, because the fabric might gradually compress and creep under the dead load, often the time, usually in seconds, that has elapsed between initial load application and the reading of the gauge, should be specified.

8. FABRIC COMPRESSIBILITY AND RESILIENCE

If a textile fabric is subjected to a progressively increasing compresssional load, there will be a progressive decrease in thickness.

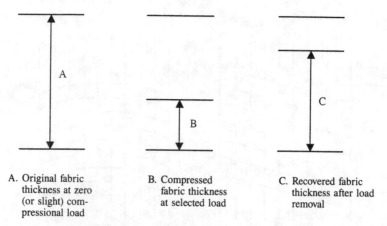

A. Original fabric thickness at zero (or slight) compressional load

B. Compressed fabric thickness at selected load

C. Recovered fabric thickness after load removal

FIGURE 19.5 Fabric compression and resilience.

When the load is gradually removed, the thickness will revert toward, but may not ever return completely to its original thickness. The tensional creep-recovery phenomena and nomenclature described in Chapter 17.0, Fiber Properties and Technology, apply equally well to fabrics in compression. Therefore the amount of compression that is instantaneously recoverable upon load removal is called instantaneous elastic recovery, the delayed recovery is called primary creep, and the non-recovery portion is called secondary creep or permanent set.

The percent of fabric compression at any selected load level is called the "softness" of the fabric (Figure 19.5). The percent of compression recovered upon load removal is often, but not always, called the "resilience." A second concept of resilience is based upon the ability of the fabric to absorb and return energy, rather than upon linear compression and recovery. Thus, referring to Figure 19.5, resilience may be defined as the energy produced when the fabric recovers from thickness B to C as a percentage of the energy required to compress the fabric from A to B. The percent of thickness not recovered is called the permanent set.

$$\text{Percent Compression} = \frac{A - B}{A} \times 100$$

(19.8)

$$\text{Percent Resilience} = \frac{C - B}{A - B} \times 100 \quad (19.9)$$
(thickness basis)

Percent Resilience
(energy basis)

$$= \frac{\text{Energy absorbed A to B}}{\text{Energy released B to C}} \times 100$$

(19.10)

$$\text{Percent Permanent Set} = \frac{A - C}{A} \times 100$$

(19.11)

A thin, hard, dense fabric will have low compressibility or softness. It will not become appreciably reduced in thickness under load. Its resilience however may be high, for upon load removal the thickness that it did lose may become almost completely restored. A thick lofty fabric under load will compress significantly and is, therefore, soft. Its resilience may be high or low depending upon its capability of recovery to its original thickness.

A fabric may be called upon to withstand repeated cycles of compressional loading and load removal. This cycling often causes the thickness to become progressively lower, approaching a limiting minimum value under maximum load. Similarly the thickness after recovery approaches a minimum value under minimum load. The ability of the material repeatedly to become compressed and to recover to its original dimension will be important where retention of thickness for maintenance of thermal insulation is required. Also, the ability to absorb energy repeatedly is a function of the ability to maintain original thickness.

9. THERMAL CONDUCTIVITY

The amount of heat (in calories or British Thermal Units) that flows through an insulating material is given by:

$$Q = \frac{KA(T_2 - T_1)t}{d} \qquad (19.12)$$

where

Q = total quantity of heat transmitted through the material

K = thermal conductivity constant

A = area through which the heat is flowing

d = thickness

T_2, T_1 = temperatures of warm and cool surfaces

t = time of heat flow

By means of thermal conductivity test equipment and procedures, the coefficient of conductivity K can be determined. The above equation presumes that K *is* a true constant independent of insulation thickness, and this is true for relatively thick insulators such as brick or glass fiber batting. However, in comparing the insulation characteristics of different fabrics on a unit thickness basis, the use of K may not be justified because it is based upon the premise that the fabric thickness can be increased or decreased without causing a change in the basic heat insulation characteristics. This is not so. Textile fabrics are relatively thin, and because of the large surface effects on both sides of the fabric, the K may be meaningless. Therefore the coefficient of "conductance" C is usually calculated for fabrics. This is a measure of the heat flow through the fabric for the thickness of the fabric as it exists, the general equation being

$$Q = CA(T_2 - T_1)t \qquad (19.13)$$

where C is the conductance. Thus

$$C = K/d \qquad (19.14)$$

For textile fabrics, a linear relationship exists between thermal conductivity and thickness. Because all fibers have higher conductivities than air, the statement that thermal insulation is largely dependent upon the entrapment of still air within the fabric is true. If the fabric is lofty and open it may be an excellent insulator in still air, but may be poor where air velocity through the fabric is high. In this case the moving air will conduct heat away and thermal insulation will fail. Tightly woven fabrics of high cover factor will not permit as much air flow, and therefore thermal insulation at high air velocity will be maintained. On the basis of thermal insulation per unit weight, open lofty structures composed of fibers of low specific gravity will be the most efficient.

Three methods are normally employed to measure thermal conductivity.

(1) The fabric is placed between two plates (usually metal) at different temperatures, and the rate of heat flow is determined.

(2) A hot surface is wrapped with the fabric and its rate of cooling is measured.

(3) A hot surface is wrapped with the fabric and the amount of heat necessary to maintain temperature is measured.

Because conductance is dependent upon thickness, it is important that fabric thickness be accurately determined and maintained throughout the test. Conductance also may be a function of the temperature differential employed in the test. For example, the conductance at a differential between 500°F and 70°F might be entirely different from that between 32°F and −40°F.

10. AIR PERMEABILITY

Air flow through a textile fabric obeys the general rules for fluid flow through an orifice. However, fabric orifices that exist because of the interlacing of warp and filling yarns are small in size, odd in shape, and large in number. These seriously complicate air flow calculations. The classical expression for the flow of an incompressible fluid through an orifice states that the volume flow is directly proportional to the product of the orifice area and the square root of the pressure differential across the orifice. The proportionality constant includes co-efficients of contraction and velocity, the approach area, the fluid density and the gravitational constant. For a fabric, air flow at any selected pressure differential is directly proportional to the product of the amount of open area within the fabric through which the air can flow, and the square root of the pressure drop across the fabric:

$$Q = K(FA)(\Delta p)^{1/2} \qquad (19.15)$$

where

Q = volume flow rate per unit fabric area

FA = free area (fractional area of the total fabric surface not occupied by yarns)

K = proportionality constant dependent upon fabric geometry and other fluid flow factors

Δp = pressure differential across the fabric

For a hypothetical fabric composed of perfectly circular filaments that do not elongate under the applied pressure differential, the air flow can be calculated from the above formula, provided K is known. However, textile fibers are viscoelastic, and deform under the tensions that develop as a result of the applied pressure. The measurement of air permeability becomes further complicated because, as the pressure increases, the yarn elongations produce changes in the shape of the interstices and the total free area increases, thus producing a concomitant increase in air flow.

It is common practice in the laboratory to measure air flow in cubic feet of air per square foot of fabric per minute at a pressure differential across the fabric equivalent to 0.5 inches of water. In actual use of fabrics, much higher pressure differentials may develop. It has been established that at pressures up to about 50 inches of water, air flow takes place primarily between adjacent yarns, and only to a minor degree between fibers within a yarn. The effects of yarn twist on air permeability are twofold. In parachute fabrics composed of multi-filament yarns, as twist increases, both the circularity and the packing density of the yarn increase. Therefore, yarn diameters are reduced, the cover factor is lowered, and the pore space of free area between yarns increases. This produces an increase in air permeability.

Any construction factor or finishing technique that changes the area, shape or length of the air flow path can appreciably change air permeability. The effect of twist has been discussed above. Yarn crimp is important because it permits the yarns to extend easily, thus opening up the fabric and increasing the free area when a pressure differential is applied. In addition, yarn crimp and fabric weave will influence the shape and length of the free area orifice. Hot calendering is often used to reduce fabric air permeability by flattening the yarns so that their cross-sectional shapes are elliptical rather than circular, thus reducing free area.

11. MOISTURE VAPOR TRANSMISSION

Water vapor will pass through a fabric provided there is a relative humidity differential across the fabric, and provided that the vapor can transpire between yarns within the fabric, between fibers within the yarns, or through the fiber substance itself. For fabrics with low cover factor where free space exists between yarns, water vapor will pass through the yarn or fabric voids. For high cover factor fabrics composed of yarns containing densely packed fibers, water vapor cannot pass through the yarn or fabric interstices. If the fiber is hydrophobic, that is, has zero moisture regain and will not absorb or release water vapor, the fabric will act as an impervious membrane. However, if the fiber is hydrophilic and can easily absorb or release water vapor, significant although limited amounts of vapor can be transmitted through the fabric. For synthetic hydrophobic fibers, when the fiber volume in the fabric exceeds about 35% of theoretical maximum, moisture transmission resistance increases rapidly, while for hydrophilic fibers such as cotton and wool, the resistance increases slowly. If the fabric construction is proper, a fabric can have low air and liquid water permeability but nominal water vapor transmission.

Moisture vapor transmission is determined by the "absorption cup" or the "evaporation" method. In the former the test fabric is placed horizontally over a cup of calcium chloride or other desiccant, the assembly is placed in a constant relative humidity atmosphere, and the absorption of water by the desiccant is measured by weighing the entire assembly at known time intervals. The second method consists of placing the test fabric horizontally over a cup of water, placing the assembly in a constant low relative humidity atmosphere, and plotting the loss in weight of water with time. For both tests transmission is calculated as grams of water transmitted per square meter of fabric per twenty-four-hour period. While these methods give valid results for impermeable materials such as plastic films, and may be useful for comparing fabrics, they do not give precise results for open fabrics of high moisture transmission. This is because as the water is absorbed into the desiccant (first method), or evaporates (second method), the relative humidity of the air adjacent to the fabric may not remain constant.

12. WATER REPELLENCY AND RESISTANCE

There are varying degrees of water repellency and resistance depending upon the nature of the fiber, the fabric construction, and whether the fabric is covered with an impermeable coating. The term *waterproof* is used to describe a fabric that is completely impermeable to the passage of water and moisture when water pressure is applied, and that will retain this property throughout its expected life under normal use.

In trade practice the canvas goods industry applies the term *waterproof* or *waterproofing* to knife coated finishes comprising such ingredients as paraffins, pigments, and resin binders with add-on weights of 50 to 90%. These types of treatments may not in fact produce fabrics that meet the strict waterproof definition. Rubber or vinyl coated fabrics that truly are waterproof are usually referred to as "coated fabrics."

The American Association of Textile Chemists and Colorists (AATCC) and the ASTM define "water resistance" as a general term denoting the ability of a fabric to resist wetting or penetration of water or both. "Water repellency" is defined as the ability of a textile fiber, yarn or fabric to resist wetting. Different tests for measuring the water resistance or repellency of fabrics are defined in Chapter 20.0, Textile Testing. The tests are of varying degrees of severity, but basically they measure (1) the resistance of the fabric to surface wetting and penetration by falling drops of water; (2) the amount of water that the fabric will absorb when immersed or manipulated under water; (3) the hydrostatic pressure required to cause fabric leakage, or the amount of water leakage at a given hydrostatic pressure; and (4) determination of fiber, yarn, or fabric wettability by contact angle measurement.

Gray unbleached cotton and linen fabrics have considerable natural water repellency because of the natural waxes present. Scoured and bleached cotton and linen fabrics, and also rayon fabrics are particularly hydrophilic unless water-repellent chemical finishes are subsequently applied. Thus they enjoy wide usage where large amounts of water are required to be absorbed. Wool initially has a fair degree of natural repellency, but when finally wet out, it has a large capacity to hold free water.

The tendency of a drop of water to spread out over a fiber, yarn, or fabric surface is dependent upon the angle of contact made by the water drop and the surface. If the fabric surface is water repellent, the drop will rest up on it and will not penetrate. The contact angle is stated to be high. If the water can easily wet the surface, spreading occurs with accompanying low contact angle. A synthetic fiber may be completely resistant to water absorption, and yet its surface may be thoroughly wetted by water. Therefore synthetic fabrics are often quickly and completely wet out, the water coating the surface of each fiber, as well as filling the voids between fibers. Because many synthetic fibers are hydrophobic, it does not follow that fabrics made from them will be water repellent. In fact, the voids that lie between fibers may act like capillaries and actually enhance the spreading and wicking of the water. Therefore many synthetic fibers must also be treated with special finishes if they are to be water repellent.

Fabric structure can play an important part in

water resistance. For example, a tightly woven cotton fabric with high cover factor that, when dry, has sufficient interstices between yarns to allow moisture vapor to dissipate, thus making the wearer comfortable. Upon wetting, the cotton fibers swell, making the fabric construction "jammed" so that water cannot get through the yarn and fabric interstices.

13. DRYING RATES AND DRYING TIME

The rate and time of drying for a textile material is a function of the state of aggregation (i.e., shape, position and distribution) of the fibers in the yarn and the yarns in the fabric, and depends as well upon the relative humidity, temperature and velocity of the air which is doing the drying via moisture absorption and subsequent dissipation. The moisture regain of the fiber itself has little influence on drying rate and time.

The drying mechanism consists of the water on the surface of the fibers evaporating into the surrounding air. Diffusion then causes water held within the fabric to come to the surface where it too can evaporate. The major part of the drying cycle from near saturation down to within 15% regain above 100% R.H. equilibrium moisture regain is characterized by a constancy in evaporation rate. This rate in grams per square centimeter of fabric per hour is equal for all fabrics of the same yarn and fabric construction, irrespective of fiber type, except that smooth-surfaced non-hairy fabrics exhibit somewhat higher rates. Total drying time under household drying conditions is dependent upon the ratio of total moisture uptake to peripheral surface area of fabric. It is generally known that identical textile structures will evaporate via identical mechanisms and at identical rates. An "Orlon" acrylic fabric and a wool fabric, both of the same yarn and fabric type, will have equal drying rates as measured by graphs of weight loss versus time. Drying times (i.e., the time required for the fabric to feel dry to the touch) will also be the same. These statements hold even though the "Orlon" has a moisture regain of 1.5% and the wool about 16%. The wool will feel dry when it has reached a regain somewhat lower than 29%, its 100% R.H. regain; but the "Orlon" must be brought below 2.5% regain, its 100% R.H. regain value before it feels dry. Thus the original regain of each fiber is of little consequence.

The general statement that synthetic fabrics are "quick drying" is fallacious. It is true that a thin, lightweight, low moisture regain synthetic filament fabric will dry more quickly than a thick, heavy, high moisture regain spun yarn fabric, but this is because the latter fabric must have a great deal more water evaporated from it before it becomes dry, not because it is made from hydrophilic fibers.

14. FABRIC BREAKING STRENGTH AND ELONGATION

The breaking strength of a fabric is popularly used both for quality control and as a performance standard. For industrial or other purposes where the fabric is in fact subjected to tension, it is proper that breaking strength be measured. In cases where the fabric is not called upon to withstand tensile forces, rightly or wrongly breaking strength is probably the most popular method for maintaining quality control. It is an easy test to make, and gives a quick and reliable indication that the fiber, yarn or fabric is up to standard. It should be remembered, however, that there are many textile applications where strength is of little consequence, and its use as a standard or quality index may be meaningless, improper, or far outweighed by other important criteria. In some cases the maximum or minimum elongation which is produced at a selected strength level below rupture is specified as a criterion.

Two types of fabric strength tests are commonly used: the "raveled strip" and "grab" methods.

14.1 Raveled Strip Strength Test

A specimen 6 inches long and 1.25 inches wide is cut in either the warp or filling direction. Edge yarns in the longer direction are raveled out and removed from the fabric until its width becomes one inch. A fringe 1/8 inch thick on each side of the fabric thus results (Figure

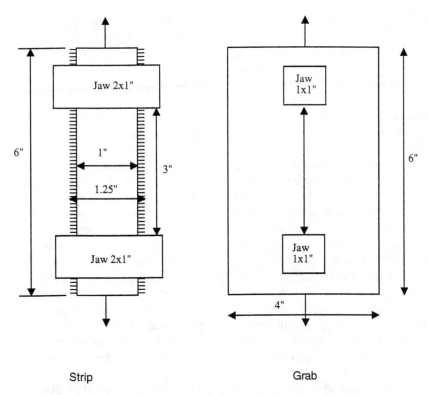

FIGURE 19.6 Raveled strip and grab breaking strength test specimens.

19.6). The test strip is clamped lengthwise in the flat jaws of the testing machine so that there are three inches between jaws. Each jaw is two inches wide so that all yarns within the fabric are held and broken during the test. If the fringe is not present, as the tensile force progressively increases, the edge yarns will pop out of the fabric, and while they are still held by both jaws, in effect a combination of yarn and fabric is tested, with the popped yarns not usually sharing their full burden of the load. The fringe prevents the yarns from popping out and hence the fabric is tested as a fabric.

Purely on a percentage basis, the need for accurately raveling coarse heavy fabrics where there may be fewer than thirty threads per inch is obviously much greater than for fine lightweight fabrics of eighty threads per inch or higher. If for any reason the yarn in the fabric is not easily raveled, it is then not necessary to ravel, for the edge yarns will not pop out during the test. In such cases the fabric is cut exactly one inch wide, and tested directly. Coated fab-

rics, resin impregnated fabrics where warp and filling yarns are bonded at yarn intersections, felts and nonwovens are cut to one inch, and are not raveled. Another advantage of raveling is that it permits the person conducting the tensile test to line up the yarns parallel with the direction of pull on the testing machine. There is less chance for a fabric to be tested slightly on the bias where yarns will not be through-going from top to bottom jaws.

The raveled or cut strip tensile test is an excellent method for studying the effects of chemical finishes, heat aging, weathering, microbial deterioration, and the like on tensile properties. Modern tensile testing machines produce a record of the load-elongation diagram from which strength elongation and energy absorption can be determined. Strip strength values are normally reported as pounds per inch. Breaking elongation is calculated as percent of original gauge length. Energy is given in inchpounds per inch length and width of specimen or gram-cm per cm².

14.2 Grab Strength Test

The ravel strip test requires slow, careful and often tedious specimen preparation and therefore is considered uneconomical by many mills. In addition, it is criticized on the basis that it does not simulate the way fabrics are used. For both of these reasons the grab test is also popular, particularly in testing heavy industrial fabrics. A specimen 4 × 6 inches is cut so that the direction of test is in the longer direction (Figure 19.6). Flat jaws one inch on a side and three inches apart at the start of the test are used to clamp the fabric. When the jaws are separated, the through-going yarns clamped in both jaws are subjected to tension. In addition, because of the interaction of warp and filling yarns with attendant frictional forces a "fabric assistance" effect develops, and yarns adjacent to those actually clamped in the jaws will help to support the load. Thus grab strength tests usually produce higher strength values than strip tests.

Because of the fabric assistance effect and the lack of uniformity in elongation between the yarns fastened within the jaws and those adjacent yarns contributing strength and energy, grab test breaking elongation and energy values may not be meaningful. Repeated stress properties are not normally measured via the grab test.

14.3 Jaw Breaks

In conducting either strip or grab tests, care must be taken to insure that fabric failure does not occur at the jaw. Jaw breaks are almost always lower in strength than breaks that occur in the "free gauge" portion of the test specimen. Particularly in heavy goods, if the breaking strength is high, considerable pressure must be applied to the flat plates of the jaw so that fabric slippage will not develop. Sometimes if the jaw pressure is too great the yarns in the fabric are crushed, deformed or otherwise damaged with resulting strength loss. If the jaw pressure is too low, fabric slippage will occur and an improper test results. Sometimes masking tape or a coated fabric will be used to line the jaws, or rosin is applied to the fabric ends to prevent slippage. The ends of the fabric specimens may

be impregnated with glue or resins such as vinyls or methacrylate dissolved in suitable solvents to permit application. In using such latter techniques care must be taken to insure that the resin or solvent does not damage the fiber.

14.4 Narrow Fabric and Ribbon Strength Tests

Fabrics, webbings, ribbons, etc., whose widths do not exceed about three inches are normally tested at full width. The test is identical with the strip test except that there is no raveling to form a fringe. Narrow fabrics usually have a woven, or resin bonded or heat fused edge so that raveling cannot develop. For heavy webbings jaw slippage or jaw breaks may occur, and so capstan jaws are often employed. The true determination of rupture elongation of a specimen tested with capstan jaws requires a correction factor because of progressive but diminishing elongation around the jaws.

14.5 Fabric Repeated Stress Properties

A raveled strip specimen can be repeatedly stressed and relaxed at load levels below rupture, and many properties can be determined similarly for the more structurally complex yarns and fabrics. However, in fibers (excluding natural or induced fiber crimp) all of the elongation, whether it be instantaneous elastic deflection, primary creep or secondary creep, is inherent in the fiber. In yarns and fabrics there can be additional elongation which is entirely geometrical, stemming from the straightening of yarn crimp. Often such geometric elongation is non-recoverable, since once the geometric deformation has occurred, there is no restoring force operating that can return the fabric to its original dimensions. Under such circumstances the fabric contains secondary creep. If, on the other hand, the geometric elongation is reversible, due to the desire of the fiber or yarn to return to its originally bent, coiled or crimped configuration, the fabric will contain primary creep. While it is not common practice in textile technology to identify them separately, there can be in fact six components of elongation in a fabric; inherent instantaneous elastic deformation, primary and secondary creep, and their geometric counterparts.

15. BIAXIAL STRESSING

The strip or grab tensile test is in general primarily concerned with the load-elongation characteristics of a specimen measured in a single direction for any given test. Tests may of course be performed in more than one fabric direction, most commonly in warp and filling directions, but each is evaluated without respect to any phenomena along any axis other than that of the test. Such tests are commonly classified as "uniaxial." That such a title is a misnomer can easily be seen by observing a simple strip tensile test, and noting the narrowing that takes place in the unrestrained region. Deformations are therefore occurring along more than one axial direction. What is really meant by a uniaxial test is not that loads or deformations take place along only one axis but that they are *applied* and *measured* along only one axis.

For many purposes a uniaxial test is adequate. It is an index of relative strength, of the degradation or improvement afforded by a treatment. Nonetheless, the practical applications of fabrics only rarely impose a truly uniaxial stress. Some examples of this are tapes and belts used for power transfer in machinery and ropes. In the more common cases stresses are simultaneously imposed in more than one direction. Obvious examples of this are the deformations at the knee or elbow of a garment, the fabric in a filter press, the walls of a fire hose, or the drying of a fabric on a tenter frame where warp and filling tensions are exerted at the same time.

Biaxial testers require control of the ratio of loading rates in the warp and filling directions and means for recording the resulting strains. Usually a cruciform specimen is employed, the two sets of jaws gripping opposite tails (Figure 19.7). In order to prevent the central portion of the fabric from skewing during the test, all four jaws must move at controlled rates. It is generally believed that the data obtained from the biaxial tests are usually not worth the effort applied. There are other empirical and easier ways of gaining an insight into the biaxial properties of fabrics, and these are the burst tests.

16. BURSTING STRENGTH

There are many industrial uses where textile fabrics are called upon to withstand bursting type forces that develop via air, fluid, or direct mechanical pressure. Pump diaphragms, screening, filter press fabrics, parachutes, hose, tarpaulins, and bags, are typical examples. The application of a pressure force perpendicular to the plane of the fabric is manifested as biaxial tensile forces along the warp and filling yarns within the fabric (Figure 19.8). The moduli of the yarns and the tensions that develop as the result of the pressure will influence the overall extension and shape of the fabric under pressure.

Two common textile structures that are called upon to withstand pressures, and that may be analyzed for resulting stresses by classical engineering methods are "thin membrane" materials in the form of (1) a hemisphere or diaphragm, and (2) a pipe. Assuming that each attains its proper respective shape upon application of pressure, the following relationships exist between the pressure applied and the tensional stresses that develop.

For a diaphragm (Figure 19.8):

$$S_1 = S_2 = \frac{PR}{2t} \qquad (19.16)$$

where

S_1 and S_2 = membrane stresses
P = pressure
R = radius of sphere
t = membrane thickness

For a pipe (Figure 19.8):

- longitudinal stress
  (lengthwise with the cylinder wall)

$$S_1 = \frac{PR}{2t} \qquad (19.17)$$

- hoop stress
  (along the wall circumference)

$$S_2 = \frac{PR}{t} \qquad (19.18)$$

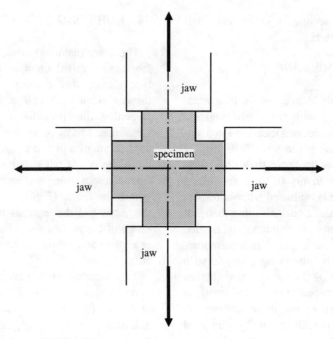

FIGURE 19.7 Biaxial stress specimen.

where

S_1 = longitudinal membrane stress
S_2 = hoop membrane stress
P = pressure
R = radius of pipe
t = membrane thickness

In the case of a pipe the longitudinal stress is always one-half the hoop stress. In manufacturing fire hose, for example, the most efficient design is one where the hoop breaking strength is twice the longitudinal breaking strength.

Maximum diaphragm bursting strength will occur when warp and filling rupture elongations are equal. Fabric manufacturers therefore attempt to produce fabrics that are "square," that is, they have equal crimps, strengths, and total rupture elongations in both warp and filling directions.

In such end uses as parachutes or filters where air or liquid passes through the fabric, the pressure causes both the warp and filling yarns to extend. Thus the fabric "opens up" and

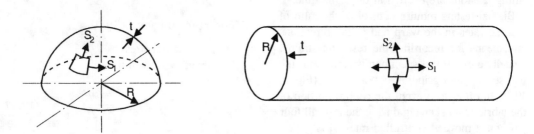

Diaphragm stresses   Pipe stresses

FIGURE 19.8 Conversions of pressure into tensile forces.

the interstices between yarns increase, i.e., the "free area" unoccupied by yarn increases and the "cover factor" is reduced. This results in more air or liquid being capable of passing through the fabric. Because textiles are viscoelastic and will creep under the load, the air or liquid permeability of a fabric may increase with time because of the continuous application of a force upon warp and filling. It is important that the fabric modulus be high enough so that the fabric will not get out of shape, distort, and open up to the point that it passes excessive air, or no longer effectively filters. On the other hand, it is equally important that the fabric not be so inextensible that it cannot absorb energy, and fails because it reaches a limiting stress instead of deforming. For repeated use, the fabric should have a high degree of creep recovery.

Coated fabrics are often subjected to bursting-type forces and for proper continued functioning without cracking or delaminating, the elongation of the coating at any load must always be greater than that of the base fabric. To prevent such types of failure, rubber and elastomer coatings are normally made with rupture elongations considerably higher than those of the fabric to which the coating is applied.

Two types of bursting load measurements are made in the laboratory. The "ball burst" test determines the force necessary to drive a one inch diameter steel ball through a 1.75 inch diameter ring clamped fabric sample. In the "diaphragm test," also commonly called the Mullen burst test, a rubber diaphragm actuated by hydraulic oil is forced against a 1.2 inch diameter sample. The oil pressure necessary to rupture the fabric is recorded. The ball burst test records total force, i.e., pounds. The diaphragm test records pressure, i.e., pounds per square inch. There is no simple constant relationship between the two set values. The rubber diaphragm in the diaphragm test accommodates to the configuration of the fabric which in turn is distorting as a function of the applied load. The area of the specimen is not a constant with load and is unknown at the time of break. Therefore, the pressure recorded on the instrument dial is the oil pressure within the machine, and is not necessarily a measure of the true pressure upon the specimen.

17. YARN AND FABRIC ELONGATION BALANCE

In Chapter 17.0, Fiber Properties and Technology, it was pointed out that corkscrew, or the wrapping of one ply around the others in a plied yarn, is undesirable because it lowers yarn strength efficiency. When tension is applied, the multi-singles core, being shorter, will reach its limiting strain and will break first. Then the load is transferred to the corkscrewed singles yarn, but it is too weak to support alone the load previously held by several singles, and it too fails. Thus the theoretical maximum strength is not attained because all singles making up the ply yarn do not break at the same time.

It is equally desirable that all yarns in a fabric break at the same limiting elongation and at the same time.

Fabric elongation usually is composed of both geometric and intrinsic components, and it is necessary that total yarn elongations be equal for all yarns if high strength efficiency is to be achieved. For example, a warp yarn of proper elongation uniformity may be woven into a fabric with a weave such that some warp yarns will be longer than others because of crimp differences. When such a fabric is put under tension, the yarns with lower crimp will fail first, the load will then be transferred to the remaining yarns of higher crimp, and they in turn will fail, but the maximum strength potentially available in the cloth will not be achieved. In weaving it is very important that warp tensions in the loom be as uniform as possible in order that equal lengths of yarn be produced in the fabric. This is particularly necessary for low elongation yarns such as glass. With more extensible yarns such as nylon, total elongations have a chance to even out, and the strength translation efficiency is higher.

18. IMPACT PROPERTIES

In a great many applications, textile materials are often called upon to withstand "high speed" or "impact" loading. The cords in automobile and particularly airplane tires, auto and airplane seat belts, parachutes, aircraft carrier arrestation cables, and body armor are typical examples of textile products that must have high impact strength. In textile processing, fibers or yarns may be impacted at low loads but at high frequencies in conjunction with picking, carding, combing, spinning, and weaving. A sewing machine causes thread to be impacted at velocities up to 180 ft/sec with short loading pulses of less than one millisecond duration.

In cases where textiles are subjected to impact forces it is required that they have the capacity to absorb large amounts of energy in an exceedingly small period of time. The area under the load-elongation diagram is a measure of the ability of the material to do work. The absorption of energy can be accomplished via various combinations of strengths and elongations. For example, the kinetic energy of a plane landing upon the deck of an aircraft carrier could be converted into strain energy by use of a solid concrete steel barrier, but the energy transfer would take place via the buildup of a high load in a very short time, and the plane would be demolished. If the deck length were no problem, the kinetic energy could be absorbed slowly by the friction of the brakes, or the reversal of the pitch of the propellers. In practice, of course, the plane is stopped by arrestation cables of both moderately high strength and elongation, thus absorbing the plane's energy over a time period sufficiently long to prevent excessive force buildup, but sufficiently short so that the plane stops within the length of the flight deck. Where impact forces are involved, the path of the load-elongation diagram, because it is a direct measure of energy absorption, must be considered with respect to the maximum stress and strain developed, and the time period during which the energy is being absorbed.

In the laboratory the load-elongation diagram of a fabric or cord is normally determined by separating the jaws of the testing machine at from about 1 to 12 inches per minute. The specimen length usually ranges from 3 to 20 inches. Therefore the maximum rate of extension of the specimen is about 12 inches per minute per 3 inches original gauge length, or 400% per minute — a relatively slow rate of deformation

when compared with service conditions, where extension rates may be in the range of 500,000% per minute. It is well known that the strengths of most viscoelastic materials are greater, and rupture elongations are less, when tested under "impact" conditions as compared with "static" or low extension rate conditions.

There are several kinds of test instruments developed to measure impact properties. Among these are freely falling weights of various masses, dropping pendulums, high speed rotating flywheels, which at a selected time will engage a jaw and elongate the test specimen, and guns of various types that fire a projectile at the test specimen. Elongation or strain rates may range from 1 to 500,000% per minute. Since the time required to rupture a specimen is usually in the thousandths or millionths of a second range, extremely sophisticated and expensive electronic devices are necessary to record load, elongation and time.

At values below the critical velocity (described below) breaking strengths normally increase and elongations decrease as impact speed is increased. A simple explanation of this phenomenon is that at slow speeds the viscoelastic fibers have time to deform, and so plastic flow rather than immediate failure takes place with accompanying greater rupture elongation. Under impact conditions the viscous components of the fiber require higher stresses to cause deformations at high speeds. Therefore the breaking strength increases. At ballistic speeds, viscoelasticity has no opportunity to function at all, and primary or secondary creep elongation cannot contribute to energy absorption. Therefore the ability of a material to withstand ballistic impact loading is probably a function of the instantaneous elastic deflection portion of its conventional stress-strain diagram, the primary and secondary creep components not being considered. It has been found, however, that at the instant of impact, stress waves develop that travel through the fabric at a speed based primarily on the velocity of sound through the fibers. In effect this defines the length of specimen which is being placed under load. The greater the speed of sound through the material the greater is the specimen length, and the greater is the total energy that can be absorbed. For each material there is a "critical velocity" at which the yarn ruptures immediately upon longitudinal impact and at the point of impact.

19. TEAR STRENGTH AND ENERGY

Tear resistance can be one of the most important properties of a fabric. For all flat sheet-like materials such as fabric, plastic films, paper and leather, the breaking strength of the material in tension is far greater than its tear resistance. While it may be difficult to induce a tear in any of these materials, usually the tear can be propagated at a relatively low load. We shall attempt briefly to analyze the stresses, and explain why tear strength is so much lower than tensile strength.

Three types of tear tests are widely, if not always properly used. The "tongue" and "trapezoid" methods employ a standard tensile testing machine for recording the force necessary to tear. The "Elmendorf" test utilizes a specially designed pendulum tester which measures tear energy.

19.1 Tongue Tear Test

In the tongue tear test a specimen 8 × 3 inches is cut so that the yarns to be ruptured during the tear lie in the shorter dimension (Figure 19.9). A cut 3.5 inches in length is made along the longer center line of the fabric (ASTM D 2261). This cut thus produces two 3 × 1.5 inch tongues which are placed in the upper and lower jaws of the testing machine. The nomenclature for identifying the direction of tear is important. A tongue sample cut so that the direction of tear is across the filling yarns, i.e., the filling yarns are ultimately broken, is called a filling tear specimen. As the jaws move apart the specimen assumes a configuration typified as shown in Figure 19.10. Each cross-yarn is subjected to progressively increasing tension that occurs in the triangular shaped distortion at the active region of tearing. The cross-yarns fail singly or often doubly, or even triply, depending upon the fabric weave, crimp distribution, yarn

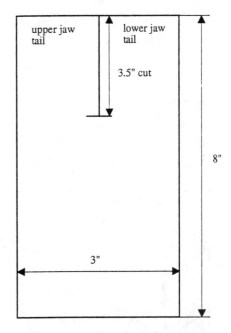

FIGURE 19.9 Tongue tear test specimen.

strength and elongation properties, and the like. It is important to note that the yarns fail individually in tension, and this is the reason that the fabric's tear strength is so much lower than its breaking strength, where all the yarns fail at the same time. The tearing action is manifested on the tensile tester recorder as a diagram of progressively increasing and then sharply decreasing loads.

The tension on a yarn (or yarns) builds up until rupture occurs at the upper peak (point P_u in Figure 19.10). Then the load drops off to P_m until the next yarn (or yarns) is brought into tension, and the procedure repeats. To obtain a proper tear diagram it is necessary to have a recorder which will respond quickly and accurately to changes in load. The electronic strain gauge tensile testers are excellent for this purpose. The average tear load may be calculated as the average of several high and low peaks.

As the tear progresses the tongues become progressively longer. They can absorb energy, which is a function of the stress-strain properties of the yarns. The yarn and tongue extensions contribute to the type of tear, and to the

$\overline{P_U}$ = AVE. UPPER LOAD LEVEL

$\overline{P_M}$ = AVE. LOWER LOAD LEVEL

$\Delta P = \overline{P_U} - \overline{P_M}$

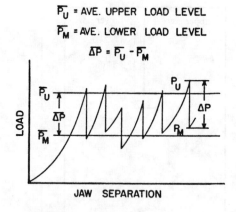

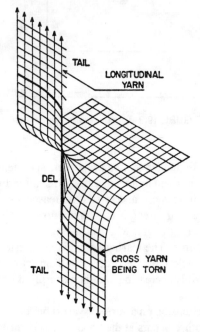

FIGURE 19.10 Failure mechanism and tear diagram in tongue tear test.

amount of tear energy or toughness. Low modulus extensible yarns have an opportunity to adjust in the tearing area. Two or more may have a chance to function together, and this produces a higher tear strength. For fabrics of the same construction at equal yarn strength levels, the higher the breaking elongation the higher will be the tear strength and energy. Often when resin finishes are applied to fabrics the tear strength is reduced severely, while the tensile strength is hardly altered or may even be increased. The answer lies with the embrittlement or increased modulus of the yarn, resulting

from the resin applications and the binding of yarns so that they cannot shift to reinforce each other. Yarn extensions are lowered, the tear formation is changed, and instead of two or three yarns breaking at the same time, they may be breaking singly because they cannot bunch up. Thus the tear strength is appreciably reduced.

It will be apparent that fabric weave can have a great influence on tear strength. All other construction factors being equal, a 2 × 2 basket weave will have more than twice the tear strength of a 1 × 1 plain weave, the reason being that the two yarns act as one and they break at the same time. A 3 × 1 twill has about 1.5 times the strength of the plain weave. For the same yarns and the same weaves, tear strength diminishes as thread count increases. A high count plain weave cotton fabric will have a lower strength than a low count fabric. It is difficult to tear a cheesecloth or gauze because the construction is so open. The tear opening is very large, and the yarns bunch up to reinforce each other, so that tearing is difficult.

The higher original tear strength of a basket weave, as compared with a plain weave, may not be retained when such fabrics are coated. The increased openness (i.e., lower cover factor) of the basket weave causes greater resin coating "strike-through" into the fabric structure. Yarns may thus become immobile and be incapable of bunching together to reinforce each other. The yarns then break individually rather than in groups, and the tear strength is thus severely lowered.

Sometimes a fabric is constructed so completely unbalanced with respect to warp and filling strengths and elongation that it is impossible to tear the stronger, tougher set of yarns. When an attempt is made to tear across the stronger set, the weaker set of yarn fails, and the tongue is torn away from the specimen. It is impossible to determine the tongue tear strength of such a fabric in the stronger direction. When this occurs the "trapezoid test" is sometimes used, but its meaning is highly questionable.

19.2 Trapezoid Tear Test

The test specimen is 6 × 3 inches with a quarter inch cut placed midway along the longer direction. The sample is inserted on the bias between the jaws of the testing machine so that the yarns are caused to fail progressively.

The increasing slack in the specimen from one side to the other causes individual yarns progressively to be brought into tension and fail. The test is entirely of a tension type, and has little meaning in terms of the practical tearing characteristics of a fabric. Therefore, this test is not used much.

19.3 Elmendorf Tear Test

The Elmendorf tester employs a pendulum to apply energy sufficient to tear through a fixed length of fabric. The tear is of the tongue type. The tester consists of a sector shaped pendulum, carrying a clamp that is in alignment with a fixed clamp when the pendulum is in the raised starting position with maximum potential energy. A rectangular specimen is fastened in the clamps and a slit is made to start the tear. The pendulum is released and the specimen is torn as the moving jaw separates from the fixed jaw. The energy required to tear is the potential energy difference at the beginning and end of the pendulum swing, as measured by the position of the pendulum at the end of the test. A graduated scale attached to the pendulum indicates the percentage of original potential energy available that was consumed in tearing the test specimen. Tear energy is recorded as inchpounds per inch of fabric torn (or gram-centimeters per centimeter). The average tearing force may be calculated by dividing the energy by twice the length of tear, because during the test the jaws separate a distance equal to twice the length of tear.

The Elmendorf test is quick and easy to use, and gives a valid reading without the necessity of making even limited calculations. The actual tearing time is in the order of one second, so that the Elmendorf can almost be considered to be an impact tear test. It does not show maximum and minimum tear strength levels, however.

20. ABRASION RESISTANCE

The abrasion resistance of a yarn is a function of both the intrinsic abrasion resistance of the fiber composing the yarn and the geometry of the yarn structure. Textiles are flexible, and depending upon the end use, repeated flexing as well as pure abrasion can contribute to product failure. Therefore to simulate service conditions, in many abrasion test methods the yarn or fabric is repeatedly flexed and abraded.

Intrinsic abrasion resistance of fibers was discussed in Chapter 17.0, Fiber Properties and Technology. The geometric arrangements of fibers in yarns and yarns in fabric are equally important in designing fabrics with high abrasion resistance. Consider a yarn composed of, say, thirty continuous filaments. The turns per inch in the yarn, the yarn crimp and the fabric weave will all influence the frequency with which a particular filament will appear on the fabric surface, such that it will come in contact with the abradant. The relationships of yarn singles to ply twist will govern yarn shape, degree of fiber packing and these in turn will also influence the amount of fiber or yarn surface in contact with the abradant, and the frequency with which a particular filament or yarn ply will become abraded. Low twist yarns tend to flatten, and a greater surface may be exposed to abrasion, although by this arrangement the stress concentration may be lowered. In addition, individual fibers in low twist yarns may not be held together well, and they may be snagged and broken by the abradant. Higher twist yarns will retain roundness and cohesiveness, presenting a smaller total surface to be abraded. However, the stress concentration may be high, and the outer surface filaments may become more severely abraded. Higher yarn twist may also reduce individual fiber mobility, and so flex resistance can be significantly lowered. For heavy plied yarns, cables, and ropes, the number and diameter of filaments, and the singles-ply-cord twist relationships can be arranged so that certain fibers are buried within the structure and bear the requisite tensile load, while other surface fibers bear the abrasive action, but are not subject to high tensile forces.

Backer and Tanenhaus [4] list the following factors as having an influence on fabric abrasion: geometric area of contact between fabric and abradant; local pressures or stress concentrations developing on specific yarn points or area; threads per inch; crown height, or the extent of yarn rise out of the fabric plane at warp and filling intersections; yarn size; fabric thickness; yarn crimp; float length; yarn cohesiveness; compressional resilience of fabric and backing; fabric tightness; and cover factor.

A number of laboratory abrasion testers are used. Independent of the design, they all probably fall into one of the following categories in terms of their actions.

(1) Plane unidirectional abrasion—The abradant travels back and forth in a straight line over the flat fabric surface.

(2) Plane multi-directional abrasion—The abradant rotates, producing a multi-directional abrasive action over the flat fabric surface. On some machines each point on the fabric surface is abraded in one direction only, although the direction may change from point to point. On other machines the directions, ideally, are entirely random.

(3) Unidirectional flex abrasion—The fabric is flexed and abraded back and forth over a cylindrical abrasive surface.

(4) Random flex abrasion—The fabric is randomly flexed and abraded in all directions.

(5) Flex without abrasion—An advantage of many flex-abrasion testers is that the abradant can be removed and replaced with a polished surface or minimum friction roller that will permit yarn or fabric flexing without surface abrasion.

Criteria of Abrasion Resistance

There are many variables that must be controlled in order to give meaning to abrasion resistance results. The abrasive surface should be uniform and not change during the test. For the planar abrasion tests, the weight of the abradant and hence the pressure on the fabric surface is significant. For the flex abrasion tests the tension on the fabric is also significant.

The three popular criteria of abrasion resistance are visual assessment, cycles to form a hole and strength loss. The fabric's end use should be known, and the type of laboratory abrasion test and performance criteria carefully selected in order to make the test meaningful. At best a laboratory abrasion test may be a hazardous way of predicting service performance.

When comparing two or more fabric samples, two basic approaches may be taken. The most direct method is to abrade all test specimens to exactly the same number of cycles, and then, evaluate their relative appearance, strength loss, thickness loss or the like. A second method is to abrade all specimens until they appear equally damaged or destroyed, and to record and then compare the number of cycles required.

The intensity of the abrasive action is of great significance. If the abrasion is too severe either inadvertently or in order to reduce testing time, so that fabric destruction occurs after a few cycles, the results may be meaningless. For example, if two lightweight fabrics are being evaluated and the abrasive action is so sharp, hard, or rough that both samples are destroyed in 25 or 30 rubs, no proper comparison can be made. If instead, a relatively mild abradant is used, requiring 1,000 or more rubs, then any differences which are observed become more meaningful. The direction of the abrasive action relative to the fabric construction can have a marked effect on abrasion resistance.

The problem of correlating laboratory abrasion tests and service performance exists because, in use, a textile may be subjected to wind, water, weather, heat, cold, bending, flexing, tension, punctures, tears, etc., all of which are grouped under the general heading of "wear." In the laboratory more idealized conditions prevail and usually only abrasion, with or without flexing, is permitted to act upon the test fabric.

21. STIFFNESS

The importance of proper fabric stiffness or flexibility should not be overlooked with respect to the proper functioning of an industrial textile fabric. Probably the prime reason textiles are useful is because of their flexibility. Two simple stiffness criteria are commonly used. The "cantilever bending length" is a measure of the tendency of the fabric to bend under its own weight. It is simply measured by sliding a strip of the fabric over the edge of a horizontal surface until gravity causes the strip to bend to a prescribed angle. "Flexural rigidity" is the fabric's resistance to bending. It is the resistance that is evidenced when the fabric is bent back and forth between the fingers. Fabric stiffness is a function of both the intrinsic stiffness of the fibers and the geometry of the yarn and fabric structure. Two fabrics of different weights can have the same bending lengths because of their intrinsic and geometrical differences, but the heavier of the two will exert a greater resistance to bending and therefore will feel stiffer. Flexural rigidity takes into account the fabric weight.

The mathematics of the bending length test have been worked out so that the length of overhanging fabric that subtends an arc of 41.5° from the horizontal is twice the bending length (Figure 19.11). Flexural rigidity G is calculated according to the formula

$$G = wc^3 \qquad (19.19)$$

where

w = weight per unit area (mg/cm²)
c = bending length (cm)

G values are expressed as gram-centimeters (or inch-pounds), and thus are units of torque. As an overall indication of fabric flexural rigidity ASTM uses the geometrical mean or square

root of the product of warp and filling values [5].

$$G_o = (G_w \cdot G_f)^{1/2} \qquad (19.20)$$

where

G_o = overall flexural rigidity
G_w = warp flexural rigidity
G_f = filling flexural rigidity

22. DRAPE AND HAND

For industrial textiles neither fabric "drape" nor "hand" are normally important. Drape is generally considered to be a visual rather than tactile property observed as the tendency of the fabric to bend under its own weight into pleasing "folds." Both warp and filling bending lengths and their interactions will influence drape. Hand or "handle" is a tactile property most often described by subjective reaction rather than objective physical measurement. The Kawabata testing system has become popular in recent years in evaluating the hand properties of fabrics.

23. CREASE RETENTION AND WRINKLE RESISTANCE

In certain industrial applications, these properties may have aesthetic as well as functional importance. Heavier-weight industrial fabrics

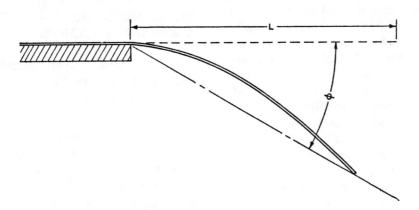

FIGURE 19.11 Cantilever bending length test.

are so thick that major permanent creases usually do not develop. However, lighter-weight goods may present more difficulty. If a fabric composed of continuous filament parallel yarns is creased so that the angle of bend approaches 180°, the fibers on the inner side of the bend are placed in compression and those on the outer side are placed in tension. There is a neutral plane where theoretically one fiber is neither compressed nor strained. If the intrinsic properties of the fibers, and the geometry of the yarn and fabric system are such that non-recoverable deformations result from the creasing action, a permanent crease will result. This can stem from two mechanisms. First, the outer fibers may be strained and the inner fibers compressed beyond their respective yield points, thus producing an objectionable crease which is called a wrinkle. Most often it is the tensional strain which is non-recoverable. Second, fibers in the yarn or yarns in the fabric may slide by each other as the result of the creasing force, but there is no restoring force which causes them to return to their original configuration.

Fibers with high extensibility coupled with high instantaneous elastic recovery and primary creep usually exhibit good-to-excellent wrinkle resistance. Fibers where high secondary creep occurs at low load and strain levels have poor wrinkle resistance. Low elongation fibers such as glass or linen have poor wrinkle resistance, since their yield points are quickly reached under the bending strain, and they cannot recover. In the case of glass fiber the creasing action often causes a strain in excess of the breaking elongation of the fiber, and so fiber failure occurs. The temperature and relative humidity of the air, as it influences the thermal or moisture sensitivity of the fiber, will also influence wrinkle resistance. Moisture sensitive fibers, particularly the rayons, have high wet secondary creep and this contributes to their poor wet wrinkle resistance. Crosslinked rayons with improved wet load elongation properties, and rayons resin-finished with "wrinkle resisting" finishes, exhibit better strain recovery properties and better wrinkle resistance. "Wash and wear" fabrics must have wet as well as dry wrinkle resistance and crease recovery.

Tightly woven fabrics generally will restrict fiber movement. When such fabrics are creased, the fibers in the yarns and the yarns in the fabric have insufficient space to enable them to move and slide relative to each other. Instead, the fibers become permanently strained or elongated beyond their yield points, consequently impairing the fabric's wrinkle recovery ability. Loosely woven fabrics allow more fiber redistribution and motion during creasing, and so less intrinsic fiber strain develops. However, the geometric movement of fibers in yarns and yarns in fabric may be large, with no inherent ability for the fabric to recover, and so any wrinkle which does develop may be more permanent.

For a given creasing force, a stiffer fabric will become creased to a lesser degree than a more flexible one. The resulting total strain will be less, and the amount of fabric creasing will be less. The crease recovery will depend upon the ability of the fibers and yarns to return to their original configuration. Thicker fabrics are generally stiffer than thin ones, and any crease which is imparted to the thicker ones is usually less permanent. For example, a crease in a tightly woven worsted fabric is more durable and permanent than an identical crease in a woolen fabric.

At equal degrees of creasing, a thicker fabric is more irrevocably deformed than a thin one, and a more permanent crease results. However, it is usually more difficult to produce equal degrees of creasing in a thick fabric, and this is why thin fabrics appear to be more easily wrinkled.

24. BLENDED AND COMBINATION FABRICS

In a strict sense, fabrics composed of more than one fiber are called blended fiber fabrics. Several types are common:

(1) Fabrics in which the spun yarns are composed of two or more types of staple fiber intimately mixed together, for example, a cotton-nylon blend. These are often called intimate blend fabrics.

(2) Combination yarn fabrics which contain plied yarns composed of two or more types of single yarns of different compositions.

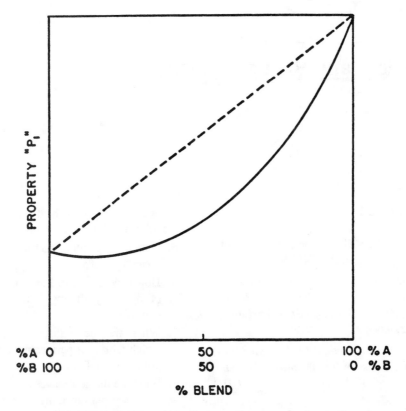

FIGURE 19.12 Effect of blend composition on yarn property *P*.

Examples are a rayon/nylon filament yarn, or spun and filament nylon yarns twisted together, or a glass core yarn wrapped with a cotton yarn cover.

(3) Combination fabrics composed of two or more types of yarns of different fiber compositions, for example, a cotton warp and a nylon filling

The objective of manufacturing a blend may stem from the utilization of a cheaper fiber as a diluent with a more expensive fiber or more properly, the attainment of certain yarn properties via the unique contribution that each component fiber gives to the blend.

It is well to consider, however, the influence of blend composition on yarn properties. Yarn strength and other properties do not necessarily follow a linear relationship with blend proportion. Figure 19.12 is a hypothetical graph that plots percent blend of a two-component mixture versus a selected property *P*.

The dotted line plots the linear relationship that would exist if each component fiber made its contribution on a direct percentage basis. This rarely happens, and the full curved line shows the kind of relationship that may develop. For example, when small amounts of a strong low-extension fiber are added to large amounts of a weaker high-extension fiber an initial reduction in yarn strength can develop so that the blended yarn is actually weaker. This phenomenon is again explained by differences in breaking elongation of the two fibers, i.e., elongation balance. The two fibers do not add the full components of their respective strengths when the less extensible one fails. Therefore the maximum strength theoretically attainable is not achieved.

In order to achieve the optimal blend at the most economical cost it is necessary that the property versus percentage composition curve be properly established. This may be done via theoretical calculation but usually experimental means are employed. Often two or more properties must be considered here, and obviously the selection of an optimum blend becomes more complicated.

References and Review Questions

1. REFERENCES

1. Peirce, F. T., "The Geometry of Cloth Structure," *JTI*, Vol. 28, T45, 1937.

2. ASTM D 3883, "Standard Test Method for Yarn Crimp or Yarn Take-Up in Woven Fabrics," *Annual Book of ASTM Standards*, Vol. 7.02, 1993.

3. ASTM D 1777, "Standard Method for Measuring Thickness of Textile Materials," *Annual Book of ASTM Standards*, Vol. 7.02, 1993.

4. Backer, S. and Tanenhaus, S. J., "The Relationship between the Structural Geometry of a Textile Fabric and Its Physical Properties," *TRJ*, Vol. 21, 1951.

5. ASTM D 1388, "Standard Test Methods for Stiffness of Fabrics," *Annual Book of ASTM Standards*, Vol. 7.01, 1993.

2. REVIEW QUESTIONS

1. Derive the fabric cover factor equation (19.1).

2. Define packing factor. How is it calculated?

3. Define fabric porosity. What is the significance of fabric porosity in industrial applications?

4. What is "crimp interchange"? Explain the effects of a typical heat setting process on warp and filling crimp.

5. Define "resilience." What is the physical meaning of it?

6. How can you control thermal conductivity of fabrics by altering fabric structure? Explain.

7. What are the industrial applications in which fabric air permeability plays an important role? Explain.

8. What are the advantages and disadvantages of waterproof functions in protective clothing?

9. What kind of information can one get from a biaxial stress test? What is the significance of this test?

10. Is bursting strength critical for geotextiles? Explain.

11. Explain the importance of impact properties in ballistic fabrics and textile composites.

12. How can you control fabric tear by fabric design? Explain.

13. What is the significance of abrasion resistance for industrial fabric applications?

14. What are the factors that affect fabric stiffness? Give application examples of industrial textiles where stiffness provides advantages and disadvantages.

15. Explain how one can improve the properties of fabrics by blended or combination structures.

TEXTILE TESTING

Textile Testing

S. ADANUR
L. B. SLATEN
D. M. HALL

1. INTRODUCTION

The purpose of this chapter is to give an overview of common textile test methods which can be used to test industrial fibers, yarns and fabrics. For an in-depth description of the test methods, the reader is referred to the references at the end of this chapter.

There are four main reasons for testing:

(1) Control of product
(2) Control of raw materials
(3) Process control
(4) Analytical information

Testing is actually a two-way street, in which the incoming raw materials that will be needed to manufacture the company's products will be scrutinized to insure that they meet the specifications. That is, any manufacturing problems will be minimized while also insuring that the textile item thus made will not result in problems for the customers; namely, that the item being manufactured is a quality product.

The term quality is somewhat intangible in its definition since it will mean different things to different people or even to different companies involved in the manufacture or use of the textile item. One way to assess "quality" is to determine its fitness for the "end use" for which the textile item is being made. That is, will the material correctly "perform" in all respects in the manner for which it was designed. One way to ascertain that the goods being inputted into the company's process or the ones being shipped to the customer will perform as they should, is to test them under conditions that will insure that a quality product is produced.

Just as any company, the customer will most likely have a set of minimum performance standards that textile products will have to meet in order to be used without problems within their process. Thus, in order to keep the customer happy, assurances of the quality of the product will be given. The company will develop Quality Assurance programs or activities that will insure that internal standard operating procedures (SOPs) are developed that will produce precise and comparable data. In this way the product is routinely checked by sampling and testing through various standard procedures through a process called Quality Control.

For industrial textiles, many of the tests that would be routinely run to assess the quality for wearing apparel or household textiles may not be appropriate. For example, tests that may be run to insure the comfort, safety or aesthetic appearance may not be required of an industrial textile. Nonethless, the need for testing to insure proper performance standards of the goods will generally not change significantly from the traditional types of tests that may be run on apparel or household-type goods. For example, the methodology required to determine the strength, abrasion resistance, blend content, etc., will require one to follow the same type of standard tests regardless of the type of textile product being tested.

When testing for quality, one has to look not only at the final product but also at each component part in order to insure that the performance expected will be obtained. For example, if a blend of cotton and polyester is specified, there are many variables that could result in a quality failure. For example, the cotton could vary in average length, maturity, tensile properties, fineness/coarseness among others. The polyester could vary in diameter (denier), cross-sectional shape, excessive surface roughness due to the presence of oligomers (unreacted monomers), fiber crimp variations, etc. Thus, any yarn made from such fibers could be of poorer quality. Even if the fibers and blend content were perfect, quality defects in the yarn such as thick/thin places, neps, twist variations, weight (count) or other yarn construction variations would again adversely impact the quality of the final product. Further, the yarn produced will ultimately be constructed into a fabric (woven, knitted, nonwoven, braided, laminated, bonded, etc.) where other variables such as thread count variations, fabrication problems (machine abrasion, barre) could occur that would produce a variable product. The final fabric will often need to be finished in order to remove size (if the fabric is woven) or scoured to remove lubricants (if knitted). The fabric may be mercerized, sanforized, singed, heat set, among others, any of which will need to be done properly in order to provide the proper end-use performance expected by the customer. Thus, a quality failure in anything from one of the subassemblies to the final fabric form could result in the entire product being classified as being a poor quality product. For this reason testing of each component at each stage of the manufacturing process is required in order to arrive at a high quality product, containing zero defects.

2. STANDARD TEST METHODS AND STANDARDS ORGANIZATIONS

Any testing which is to be done on a product will need to be done by employing standard test methods. In this way, every possible variable within the test method will be precisely controlled. The reason for this is that reproducibility must be absolutely assured. That is, the test results in a plant or laboratory will need to be the same as those obtained within the customer's laboratory; otherwise, lawsuits would abound. For example, it is a known fact that goods containing 100% cotton will be stronger when they are wet, while most other fiber genera will be weaker in strength. Thus, if the moisture content of all fabrics was not precisely fixed in the standard testing procedures, then the results could vary depending upon each laboratory or plant's test conditions. For this reason virtually all physical tests done on textile goods are done at the standard conditions of $70°F$ and 65% relative humidity (RH). Other variables such as time, reagent concentrations, speeds, number of cycles, number of tests, test apparatus employed, etc., will be precisely fixed for each test so that the test will be both accurate and reproducible from one laboratory to another. In this way the results of the tests will be the same regardless of what laboratory performs the test.

A great deal of time and effort is put into the test standardization. There are two major organizations which write standard tests for textiles. The American Society for Testing and Materials (ASTM) writes standard tests not only for textiles but also for virtually every other product such as steel, plastics, lumber, etc. [1]. The society has over 135 technical committees and over 2,000 subcommittees staffed with over 20,000 technical experts drawn from producers, users, general interest universities and consumer participants. There are presently over forty volumes of standard texts written by this organization. Just the textile tests alone require two of these volumes. For textiles, ASTM writes primarily physical type tests such as methods for testing the tensile strength, abrasion resistance, twist determination, fiber maturity, denier and yarn count among many others. Each test method will have a committee or subcommittee of experts on that subject drawn from academia, industry users, general interest personnel, consumer and government participants that will meet periodically to review the standard test to determine if any research, technique, method or system may have been developed and published which could make the test method either more accurate or reproducible. It is not unusual for a test method to require developments by a consortia of research committees through extensive investigations and inter-laboratory comparisons, often requiring years of work to finalize the test method. Each standard test will be reviewed at least once every five years (yearly for the first three years when first introduced) to determine that the test is still the state of the art in methodology, consistency, accuracy and reproducibility.

Another organization which writes standard tests for textiles is the American Association of Textile Chemists and Colorists (AATCC) [2]. This organization works very closely with ASTM but writes chemical type tests. Therefore, the two organizations are complementary and not in direct competition with each other. Thus, tests that deal with fastness properties (light, laundering, color, perspiration, chlorine, etc.) for example, will be written by this organization.

In some cases standard tests written by both organizations will need to be run to determine the whole quality story. For example, the test method that deals with fastness to chlorine bleach (a chemical reaction and thus a chemical type phenomena) will be written by AATCC. One of the properties required to be evaluated in the AATCC test method is the strength of the fabric before and after the chlorine treatment. The strength measurements would be done by the appropriate ASTM standard test method which would be referenced within the AATCC test method. Thus, each organization does not try to reinvent the wheel when there is already a recognized standard test method in existence.

Another organization having requirements for specific tests methods is the U.S. military [3]. The U.S. Army has the primary responsibility for the procurement and testing of several thousand different textile items used by the various military organizations. The specific end-use requirements for military textiles may be radically different than those that are used in the civilian sector. Thus, the military establishment will write and publish their own test methods which will cater to the specific quality needs of that military item. In most cases, the test method will follow the equivalent ASTM or AATCC test method quite closely but deviate when specific military needs dictate.

There may be times when the buyer and seller may wish to deviate from published standard procedures or tests and develop one employing their own test methodology to meet very specific requirements. For example, a nylon fishing line breaking at ten pounds when tested under standard conditions of 65% RH and 70°F will be weaker when used under actual conditions in the wet state (nylon loses about 10% strength when wet). Thus, the buyer may wish to have the fishing line tested under conditions that will more accurately reflect the end-use requirements of the textile item. In such cases, the buyer and seller may agree on different testing procedures to meet their own requirements.

Many professional organizations (e.g., INDA, TAPPI, American Filtration Society) and large industrial corporations of the United States who purchase large quantities of all sorts of materials and end items have established their own test methods and standards. The electrical, rubber, automotive, abrasive, and plastic laminating industries as well as many department store chains have established their own test methods and standards in order to insure uniformity of product and the attainment of proper quality.

Using one of the above types of standard tests will generally be the only ones that will ever be employed by most textile mills. There is, however, a significant amount of effort toward standardizing tests worldwide. The United States and most of the industrialized nations of the world have a continuing program that attempts to standardize textile testing so that one single standard test for each specific subject (abrasion, strength, perspiration, etc.) can be employed worldwide. Thus, textile products that are imported or exported can employ standard test methods under conditions that will be universally accepted worldwide. The international organization under whose umbrella the various standardization efforts are being negotiated is the International Organization for Standardization (ISO). For the negotiations, the United States has established national tests under the American National Standards Institute (ANSI). Thus, most of the AATCC and ASTM standards have been adopted as the ANSI standards which form the basis for negotiations with the other countries.

Current federal policy states that where national consensus standard test methods exist, these standards will be incorporated in Federal Standard Method 191A. This is an ongoing process and will gradually shift the responsibility for updating standard methods to the various national consensus organizations.

Evaluation Procedures

Most standard tests will give specific information such as the actual breaking strength, number of abrasion cycles to failure, the number of S or Z twist/inch, etc. Other tests will give a rating between 1 (very poor)–5 (excellent). For example, when rating the crock fastness (fastness to rubbing of a dyed fabric) of a sample, the international gray scale is used to access the amount of color that is transferred to a white test swatch. In this manner the amount of color transferred can be evaluated on the 1–5 scale. It is possible to have a rating between one of these numbers. Using the gray scale a number of 3.5 could be obtained. This would imply that the rating was not as good as a 4.0 but not as bad as a 3.0, and that the rater felt that the color transferred was about halfway between the two rating numbers.

One should be aware that all the testing procedure will do is to supply an actual property value or a rating number. The number itself will not necessarily tell if the customer will be satisfied with the rating once the fabric is put into use.

3. ENVIRONMENTAL TEST CONDITIONS

Many textile materials are sensitive to temperature and relative humidity. When testing them, it is necessary that the environment be stipulated and held constant. When all of our fibers were natural or of a regenerated animal or cellulose base, it was mandatory that relative humidity and temperature be controlled, and fabrics had to be brought to equilibrium in a standard atmosphere or "conditioned" before testing. Synthetic hydrophobic fibers are not sensitive to moisture, and yet the surface moisture characteristics of some of these fibers can

also cause differences in processing and in yarn and fabric mechanical properties. Therefore, it is equally proper to test these materials under constant temperature and humidity conditions. By general agreement, "standard conditions" for the U.S. textile industry are defined as 65% ± 2% relative humidity (RH) and 70° ± 2°F. When testing a textile material, it should be placed in a conditioning room under these conditions for several hours before testing. The length of time required for the textile to attain moisture equilibrium in the conditioning room will depend upon its preconditioned regain, and its "state of aggregation"; that is, the distribution and position of fibers in the yarns, and yarns in the fabric.

Densely packed materials originally at very low or very high moisture regain may require several hours, or even days, to come to true moisture equilibrium at standard conditions. Thin, lightweight, lofty textiles can be conditioned in a few hours or even minutes. Figure 17.16 shows that moisture absorption-desorption curves do not follow the same regain path, a hysteresis area existing. Therefore, the final moisture regain equilibrium which a material assumes when conditioned from the dry side usually is different from that which will result by conditioning from the wet side. While for most tests this difference will not be great and is usually ignored, it can influence physical properties. Therefore, for meticulous work, or in the event of a dispute, it is necessary to establish whether the material is to be conditioned from the dry or the wet side prior to testing.

The ASTM states that moisture equilibrium shall be approached from the dry side, but not from a moisture-free condition. It is an exacting task to maintain a conditioning room at the proper humidity and temperature level of 65% ± 2% RH and 70° ± 2°F. Humidistats and thermostats that record and control are available to accomplish these standards. The ASTM manual contains a psychometric chart so that relative humidity can be calculated from wet and dry bulb thermometer temperatures.

There are many test chambers on the market to aid in the testing of properties under specified conditions of high or low relative humidities and temperatures. It is often desirable to determine

TABLE 20.1 Standard Test Methods for Environmental Testing Conditions.

ASTM E 337	Measuring Humidity with a Psychrometer
ASTM D 1776	Conditioning Textile for Testing
ASTM D 1909	Commercial Moisture Regains for Textile Fibers
ASTM D 2118	Moisture Content, Commercial, for Wool and Its Products, Assigning a Standard
ASTM D 2462	Moisture in Wool by Distillation with Toluene
ASTM D 2525	Sampling Wool for Moisture
ASTM D 2654	Moisture in Textiles
ASTM D 4920	Terminology Relating to Moisture in Textiles

the strength-elongation properties of textiles under such conditions. Table 20.1 lists several ASTM standard methods relating to environmental conditions and moisture. These methods are guidelines for specific uses. Other guidelines may be different for other related products such as plastics, glass fiber items, etc.

4. SAMPLING AND QUALITY CONTROL

The importance of properly selecting fibers, yarns or fabrics for quality control or specification testing cannot be overemphasized, if the time, effort, and money expended for physical or chemical testing is to be worthwhile. In the mill, the sampling of the product from each machine must be done in a statistically sound manner so that the samples taken are truly representative of production. Obviously, the greater the number of samples tested, the more nearly the resulting average will represent the true average of the entire lot, but this increases the amount of work and cost. Therefore, it becomes necessary to establish the statistical probability one is willing to accept so that the values obtained on the samples tested in fact represent the entire lot.

The ASTM states that

the number of test specimens depends on the desired precision and the desired probability level, both of which can be selected at will. It depends also on the variability of individual test results; this is a property of the test method and of the material which is being tested and must be determined experimentally. When a certain number of tests are performed, the mean of the test results forms an estimate of the true average of the property under test.

The larger the number of tests and the lower the variability of the results, the nearer will the mean of the test results be likely to come to the true average. Precision is the difference between the mean of the test results and the true average, which is unlikely to be exceeded. Precision can be expressed in percent (equation 20.1) or in units of the property under test (equation 20.2).

As an example, if a precision of $\pm 4\%$ at a probability level of 95% is specified, then there is a 95% probability that the mean of the test results lies within $\pm 4\%$ of the true average of the property under test. Note that the expression "true average" is used here to signify the overall average of an infinite number of tests performed by the test method in use.

The general formula for number of tests are:

$$n = \frac{t^2 \nu^2}{E\%^2} \qquad (20.1)$$

$$n = \frac{t^2 \sigma^2}{E^2} \qquad (20.2)$$

where

n = number of test specimens
$E\%$ = desired precision of the mean of the test results expressed as a percentage
E = desired precision of the mean of the test results expressed in units of the property under test
ν = coefficient of variation of individual test results, determined from extensive past records on similar material (in %)
σ = standard deviation of individual test results, determined from extensive past records on similar material (in units of property under test)

t = constant depending on the probability level:

Probability Level	t
90%	1.645
95%	1.960

For other probability levels, the values of t to be used in the above formulas can be found in statistical tables.

One of the most important rules in the selection of test samples is the use of common sense. If five specimens are to be selected for warp breaking strength, it is logical to cut them so that there are entirely different warp and filling yarns in each specimen. If there is insufficient fabric to make such a completely random selection, then the next most reasonable selection is to cut the five specimens with different warp yarns but with the same filling yarns. In addition to the mathematical and statistical requirements, an understanding of the reason for conducting the test and carefully planning the utilization of the test data will be most helpful. As another example of proper sampling, consider the evaluation in the laboratory of the tear strength of a fabric before and after resin finishing to establish whether the treatment has caused a deleterious effect. Here it is proper to mark out the fabric specimen pattern so that one piece of the fabric is treated and the adjacent piece is retained as an untreated control. Then when the samples are cut for testing, the same yarns in each piece will be tested so that a direct comparison of the effect of the treatment will be established.

The proper employment of statistics for quality control and research is of great value. It is a comprehensive subject unto itself. Several excellent books are available on statistics and quality control.

A list of selected standard test method and guidelines for sampling are given in Table 20.2.

This chapter gives information and discusses the significance of the multitude of physical and chemical test methods and instrumentation for fibers, yarns and fabrics which are available to the textile scientist. ASTM D 123 (Standard Terminology Relating to Textiles) gives the definitions of textile terms.

5. FIBER TESTING

Fiber analysis includes identification, physical properties, quality of natural fibers as well as other characteristics. A selection of relevant standard test methods is listed in Table 20.3.

5.1 Fiber Identification

Identification systems are based upon differences in optical, physical, and chemical properties of fibers. The following criteria can be employed qualitatively or, if necessary, quantitatively: longitudinal and surface characteristics; cross section shape; specific gravity; refractive index; polarized light characteristics including birefringence; melting point; moisture regain; swelling and solubility characteristics in water, organic solvents, or other chemicals; burning characteristics; dyeing and staining characteristics; ultraviolet and infrared spectra; and X-ray diagrams. Several of these tests are normally employed in order to make the identification certain.

ASTM Method D 276 gives detailed procedures for identifying the following textile fibers:

TABLE 20.2 Standard Test Methods Used for Sampling.

ASTM E 105	Probability Sampling of Materials
ASTM D 1060	Core Sampling of Raw Wool in Packages for Determination of Percentage of Clean Wool Fiber Present
ASTM D 1441	Sampling Cotton Fibers for Testing
ASTM D 2258	Sampling Yarn for Testing
ASTM D 2525	Sampling Wool for Moisture
ASTM D 3333	Sampling Man-Made Staple Fibers
ASTM D 4271	Writing Statements on Sampling in Test Methods for Textiles

TABLE 20.3 *Standard Test Methods Used for Fiber Testing.*

ASTM D 276	Identification of Fibers in Textiles
ASTM D 519	Length of Fiber in Wool Top
ASTM D 629	Quantitative Analysis of Textiles
ASTM D 861	Tex System, Use of, to Designate Linear Density of Fibers
ASTM D 1060	Core Sampling of Raw Wool in Packages for Determination for Percentage of Clean Wool Fiber Present
ASTM D 1234	Staple Length of Grease Wool
ASTM D 1336	Standard Test Method for Distortion of Yarn in Woven Fabrics
ASTM D 1440	Length and Length Distribution of Cotton Fibers (Array Method)
ASTM D 1441	Sampling Cotton Fibers for Testing
ASTM D 1442	Maturity of Cotton Fibers
ASTM D 1445	Breaking Strength and Elongation of Cotton Fibers
ASTM D 1574	Extractable Matter in Wool and Other Fibers
ASTM D 1575	Length, Fiber of Wool, in Scoured Wool and Card Sliver (Flat Bundle Method)
ASTM D 1577	Linear Density of Textile Fibers
ASTM D 1909	Moisture Regains for Textile Fibers, Table of Commercial
ASTM D 2101	Tensile Properties of Single Man-Made Fibers Taken from Filament Yarns and Tow
ASTM D 2102	Shrinkage of Textile Fibers
ASTM D 2252	Specification for Fineness of Types of Alpaca
ASTM D 2480	Maturity Index and Linear Density of Cotton Fibers by Causticaire Method
ASTM D 2612	Fiber Cohesion in Sliver and Top in Static Tests
ASTM D 2494	Commercial Mass of a Shipment of Yarn or Man-Made Staple Fiber or Tow
ASTM D 3217	Breaking Tenacity of Man-Made Textile Fibers in Loop and Knot Configurations
ASTM D 3333	Sampling Man-Made Staple Fibers
ASTM D 3513	Over Length Fiber Content of Man-Made Staple Fibers
ASTM D 3660	Staple Length of Man-Made Fibers Average and Distribution (Fiber Array Method)
ASTM D 3661	Staple Length of Man-Made Fibers Average and Distribution (Single-Fiber Length Machine Method)
ASTM D 3817	Maturity Index of Cotton Fibers by Fibrograph Yarn Intermediates, and Yarns
ASTM D 3818	Linear Density and Maturity Index for Cotton Fibers
ASTM D 3822	Tensile Properties of Single Textile Fibers
ASTM D 3937	Standard Test Method for Crimp Frequency of Man-Made Staple Fibers
ASTM D 3991	Fineness of Wool or Mohair and Assignment of Grade
ASTM D 3992	Fineness of Wool Top or Mohair Top and Assignment of Grade
ASTM D 4120	Fiber Cohesion in Roving, Sliver and Top (Dynamic Tests)
ASTM D 4524	Composition of Plumage
ASTM D 5103	Length and Length Distribution of Man-Made Staple Fibers (Single-Fiber Test)
ASTM D 5104	Shrinkage of Textile Fibers (Single-Fiber Test)
ASTM D xxxx	Neps in Cotton Fibers (AFIS-N Instrument)
ASTM D yyyy	Measurement of Physical Properties of Cotton Fibers by High Volume Instruments
AATCC 20	Fiber Analysis Qualitative
AATCC 20a	Fiber Analysis Quantitative

acetate (secondary), acrylic, anidex, aramid, asbestos, cotton, cuprammonium rayon, flax, fluorocarbon, glass, hemp, jute, modacrylic, novoloid, nylon, nytril, olefin, polycarbonate, polyester, ramie, rayon (viscose), saran, silk, spandex, triacetate, vinal, vinyon, and wool.

5.2 Quantitative Analyses of Fiber Composition

Several quantitative test methods are used to determine the percent fiber composition in a yarn or fabric. All methods of fiber analyses are based upon the premise that the fiber types in the mixture to be analyzed are already known by previous qualitative analysis techniques. The design of a particular scheme for separation must be based upon this knowledge. The scheme should be as simple as possible.

The ASTM and AATCC have a jointly approved test procedure where the test methods are grouped under three headings:

(1) Mechanical Separation or Dissection

(2) Chemical Analysis

(3) Microscopic Analysis

The methods are intended primarily for the separation of binary mixtures, but by intelligent selection of the best combination of methods, multi-fiber mixtures can be analyzed. Fiber composition is generally expressed either on the oven-dry weight of the textile "as received" or on the oven-dry weight of the clean fiber after non-fibrous materials, such as finishes, are removed. In making quantitative separations, it is vital that fiber weights are precisely measured. Analytical balances of proper sensitivity are useful and often mandatory, depending upon the original size of sample and the weight of fiber remaining after a component has been removed mechanically or by dissolution. The reader is urged to examine ASTM Method D 629 or AATCC Methods 20-1990 and 20A-1989 for comprehensive information on test details.

If two or more components can easily be separated by raveling, for example a nylon warp and a cotton filling, the easiest way to ascertain percentage composition is to separate the fibers or yarns mechanically and weigh each group. Plied or cored yarns composed of two different fibers can also be handled in this fashion. Chemical separation of fibers normally consists of selecting a solvent in which one constituent is completely and easily soluble and the other constituent is completely insoluble. Tables 17.12 and 20.4 are particularly useful for selecting proper solvents. Table 20.5 lists keys to methods for the separation and determination of fiber content.

5.3 Stereoscopic, Compound, and Polarizing Microscopes

As an aid in observing the fine structure and character of fibers, as well as for examining yarns and fabrics, a range of microscopes is available to the textile scientist. These instruments, in some ways, are unique for textile applications, but most of them were developed by microscopists with broad interests in all types of natural and synthetic materials. The various microscopes are of inestimable help in the identification and examination of fibers, the study of fabric defects and the effect of chemical or resin treatments on textile fibers and yarns. There are several excellent standard books on textile microscopy, as well as several texts on general microscopy, techniques and equipment involved.

The lower power stereoscopic microscope, producing magnifications of about 6 to 60 times, permits the observer to use both eyes to see yarn or fabric structures in three dimensions, that is, depth as well as length and width. An important point often overlooked is that much information can be obtained by the use of relatively low magnification instruments. The lower the magnification, the greater is the "field of view," i.e., the diameter of the circle viewed through the stereoscopic microscope. For example, a warp streak may be very apparent to the naked eye when the fabric is viewed from a distance. Under excessive magnification, the defect may disappear because one cannot observe a sufficient amount of acceptable fabric adjacent to the defective streak to establish differences.

The compound microscope consists of a combination of a magnifying "objective," an "ocular" and a "condenser." The latter gathers and focuses the light so materials can be examined in detail at magnifications ranging from about 100 to 800 times. For fiber studies including identification, surface structure, damage, and the deposition of resins, pigments and dyestuffs, the compound microscope is a valuable tool. Generally, it is useless, as well as improper, to attempt to examine large yarns or fabrics with a compound microscope with even the lowest order of magnification. The field of view is so small, and the depth of focus so critical that only a tiny area of the yarn or fabric can be viewed, with only a part of this in focus. It is also important to point out that the objective lens of the microscope has the capability of "resolving" small distances, and thus completely controls the ability of the microscope to separate structural detail. The ocular magnifies what the objective sees. Usually, it does no good to employ high ocular magnification, without the proper previous resolution. "Empty magnification," as it is called, gives no greater

TABLE 20.4 Solubilities of Fibers in Solvents Used in Chemical Methods (ASTM D 629).[a]

Method No.:	(1) 80% Acetone	(2) Butyrolactone (A) RT	(2) Butyrolactone (B) 75°C	(3) and (9) 90% Formic Acid	(4) 59.5% H_2SO_4	(5) 70% H_2SO_4	(6) NaOCl Solution	(7) Cuprammonia Solution	(8) Hot Xylene	(10) N,N-dimethyl acetamide
Acetate	S	S	PS	S	S	S	I	I	…	…
Triacetate	I	PS	PS	S	I	I		I	…	S
Acrylic	I	S	S	I	I	I	I	I	I	S
Aramid	I									
Cellulosic, natural	I	I	I	I	SS	S	I	S	…	I
Modacrylic	I	S	S	I	I	I	I	I	…	I
Nylon	I	I	I	S	S	S	I	I	I	I
Olefin	I	I	I	I	I	I	I	I	S	I
Polyester	I	I	I	I	S	I	I	I	I	I
Rayon	I	I	I	I	S	S	I	S	I	I
Silk	…	I	I	PS	S	S	S	…	…	S
Spandex	I	I	I	PS	I	I	I	…	…	S
Wool and hair fibers	I	I	I	I	I	I[b]	S	…	…	…

[a]Key to symbols:
S = soluble
PS = partially soluble
SS = slightly soluble (a correction factor may be applied)
I = insoluble
[b]Reworked wools are soluble in 70% H_2SO_4 depending upon their previous history.

TABLE 20.5 *Chemical Methods for Analysis of Fiber Mixtures (ASTM D 629).*

	Wool	Spandex	Silk	Rayon	Polyester	Olefin	Nylon	Modacrylic	Cellulosic, Natural	Aramid	Acrylic	Triacetate
Acetate	1		1	1	1	1	1	1	1		1	1
Triacetate	3		(5)	(7^5)[b]	9	3^9	(5)	9	(5)		3	
Acrylic	(6)		(6)	(7^5)	10	(8)	10	10	(57)			
Aramid	(6)											
Cellulosic, natural	$(6)^5$	(10)	(6)	(4)	5	5	(3)	(2)				
Modacrylic	2	2	2	2	2	2	2					
Nylon	3	(10)	(6)	3	9	(8)						
Olefin	(6)		(6)	(7^5)	8							
Polyester	(6)	(10)	(6)	(7)								
Rayon	5		(6)									
Silk		6										
Spandex	(6)											

[a]Key to methods and reagents:
Method No. 1 – 80% acetone (cold)
Method No. 2 – *N*-butyrolactone
Method No. 3 – 90% formic acid
Method No. 4 – 59.5% sulfuric acid
Method No. 5 – 70% sulfuric acid
Method No. 6 – Sodium hypochlorite solution
Method No. 7 – Cuprammonia solution
Method No. 8 – Hot xylene
Method No. 9 – 90% formic acid
Method No. 10 – N,N-dimethylacetamide

[b]Each analytical method is identified by a number and where possible, two methods of analysis are provided for each binary mixture of fibers. The number or numbers inside parentheses refers to the method that dissolves the fiber shown at the top of the diagram. The number or numbers outside the parentheses indicates the method that dissolves the fiber listed at the left side of the diagram. Where two methods are listed for a specific binary mixture, the non-superscript method number represents the method of choice.

clarity of detail, and would be analogous to placing a magnifying lens in front of a television screen to make the picture bigger. This adds nothing to the clarity of the picture nor to the ability of the observer to see finer detail.

By means of precise optical and mechanical attachments, the compound microscope can be used for making quantitative measurements such as fiber length, diameter, coating thickness, cross section area, etc. Micrometer scales placed on the stage and in the eye piece (ocular) of the microscope are calibrated and then used for quantitative measurements. Glass slide holders with precise screw threads permit a specimen to be moved a fixed distance east-to-west or north-to-south on the horizontal stage. A circular specimen stage marked with the degrees of rotation may be used to measure twist angles in fine yarns or to make quantitative measurements under polarized light.

A polarizing lens converts ordinary light vibrating in many planes into polarized light which vibrates in one plane. When viewed in such light by a second polarizing lens the optical rotation exhibited by a specimen is useful to characterize molecular order, for example, a fiber's orientation, and its birefringence. The polarizing microscope is also used to determine the degree of cotton maturity.

Projection microscopes are useful for projecting images onto screens for large scale viewing or for easy quantitative measurement. In some systems, the microscope, camera, computer, monitor, video recorder and printer are all connected. The picture can be viewed on the monitor, stored on the computer disk, recorded on the videotape or printed instantly (Figure 20.1).

A wide range of photographic equipment has been developed for photographing specimens under the microscope. Such photos are properly called photomicrographs and can be taken in color, often with polarized light. Other instruments useful in microscopy laboratories are textile specimen cross-sectioning devices or microtomes. Properly prepared cross sections are useful for studying internal structure.

Fiber optic technologies developed for the computer and communications industries have been applied to imaging systems. Remote video microscopic systems are available (Figure 20.2). These systems can provide continuous viewing at magnifications greater than $\times 1,000$.

FIGURE 20.1 A modern vision system including microscope, computer, monitor and video recorder for fiber, yarn and fabric analysis (courtesy of TexTest).

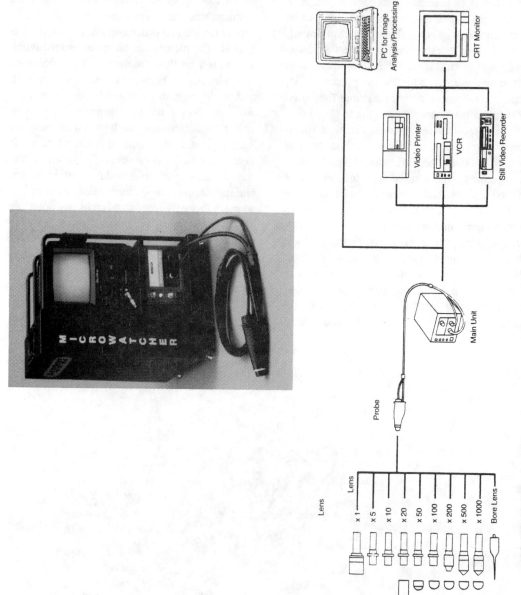

FIGURE 20.2 Remote video microscope system and its schematic (courtesy of Lawson-Hemphill Sales, Inc.).

Interfacing with computer accessories allows for image storage, analysis and printing.

5.4 Scanning Electron Microscopy

The scanning electron microscope (SEM) consists of two major parts. An electronics console provides the necessary photography. The other section of the apparatus contains the electron column where an electron gun generates an electron beam that is focused on a small spot. This small spot is where specimens are placed. The spot is scanned with the electron beam and secondary electrons generated in the specimen provide the signal, which is viewed on the screen. The electron beam is generated from a metal filament and accelerated down an evacuated column. The vacuum is necessary since electrons cannot travel very far in air.

The specimens must be conductive, so for most textile or polymeric materials they are coated in a vacuum with a conductive substance such as gold-palladium or carbon. This technique allows the observation and photographing of material surfaces. Photographs of fiber surfaces at different magnifications allow one to view the external makeup of the fibers, and in some cases, can be used for identification of fiber type or to look at changes in fiber surfaces due to external events. The advantage of SEM is the large depth of field at high magnification. The analysis of secondary X-rays excited by the primary electron beam allows the qualitative analysis of elements and their distribution in the specimen.

5.5 Fiber Denier or Tex

Many of the tests on individual fibers are difficult to execute because of a fiber's size and the extreme care needed in its manipulation. In calculating the yarn number it is relatively simple to wind off a skein of yarn, weigh on a properly sensitive balance, and determine the weight per length. Even the somewhat more meticulous task of marking out specific lengths of yarns and weighing them on an analytical balance is not overly difficult. Today analytical balances with seven decimal places are available. However, when it is necessary to determine the weight

and denier of an individual filament, the problem becomes greatly expanded. In determining the tenacity-elongation properties of individual filaments, because of large or even minute variations in fiber diameter and weight, it is necessary to establish the denier or tex of each individual breaking strength test specimen. Each fiber's strength is then related to its own weight, and a precise tenacity (grams breaking strength per tex or denier) is obtained.

The Vibrascope

One method for determining denier is by measuring the fiber's average cross-sectional area. Then, knowing its specific gravity, the weight per unit length can be calculated. This is a time-consuming task, requiring meticulous microscopical measurements of fiber cross sections. A more rapid method is based upon the principle that the weight per unit length of a completely flexible string can be calculated from its lateral frequency vibration characteristics, its length, and the applied tension. The basis of the calculation is the well-known physical law that, when a flexible string is vibrated transversely, its resonant frequency is a known function of the tension, the length and the mass per unit length.

Linear density or weight per unit length m is calculated as follows:

$$m = \frac{T}{4l^2 f_1^2} \tag{20.3}$$

where

m = linear density in grams per centimeter
T = fiber tension in dynes
l = effective fiber length in centimeters
f_1 = fundamental resonant frequency in cycles per second

The instrument, called a vibrascope, determines quickly and accurately the denier of individual fibers 2 to 4 centimeters in length. The technique consists of holding a fiber in one end, dead loading it with a small freely hanging weight, and clipping the other end of the fiber in a jaw connected to a source of sinusoidally alternating energy. The fiber is then caused to vibrate transversely setting up fundamentals or harmonics of vibrational nodes. A variable oscillator is used to adjust the frequency so the harmonic nodes are established. Knowing the frequency, the length, and the tension, the denier can be calculated. The new vibrascopes are completely automatic with respect to their operation.

Vibromat ME

The Vibromat is a fiber testing instrument for automatically determining the fineness of single fibers (Figure 20.3). This instrument uses the vibration method (ASTM D 1577). The resonance frequency of the sample is measured initially at a constant length and at known pre-tensioning weight. The fineness is then calculated. Assuming a uniform mass distribution and a circular cross section, and disregarding the influence of the bending rigidity (elastic modulus), the following formula is valid:

$$T_t = F_v/(4f^2L^2) \qquad (20.4)$$

where

T_t = fineness in dtex
F_v = pre-tensioning weight in mg
f = resonance frequency in Hz
L = test section length in mm

The Vibromat can measure the resonance frequency by frequency interval or by impulse methods.

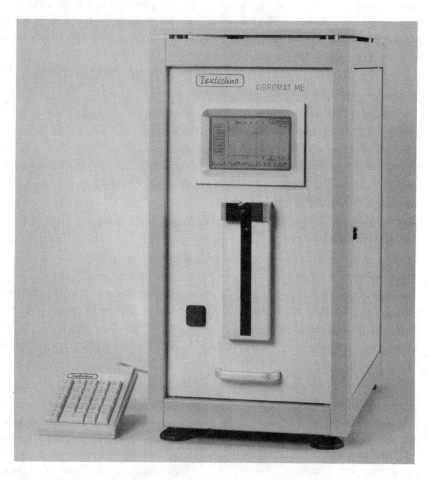

FIGURE 20.3 Vibromat single fiber fineness tester (courtesy of Textechno).

5.6 Cotton Staple Length or Array

In the laboratory procedure, a small amount of properly sampled fiber is selected. Care must be taken to insure that the sample selected, in fact, represents truly the lot being tested. This small amount of fiber is then placed across the teeth of a series of uniformly spaced parallel combs. After a series of repeated fiber combing and paralleling operations, the fibers of the longest length are withdrawn from the combs by clamping forceps and laid on a velvet board. Then one comb in the group is dropped out of the way, and fibers of next longest length are grasped by the forceps, drawn out, and laid out beside the original longest fibers. The process is repeated until all combs are dropped and the remaining shortest fibers are placed on the board.

Originally, an outline diagram of the array was drawn from which the mean staple length and other measurements were determined. Now, however, a weight versus length distribution method has been adopted because it more closely correlates with the results obtained by the cotton classers. Furthermore, by this method it is not necessary, actually, to lay down an array. Instead, the weight of fiber in each percent-inch interval length group is determined. A tabulation of weight versus fiber length is made. From this the following calculations are determined:

(1) Upper quartile length – the length that is exceeded by 25% of the fibers by weight in the sample. The ASTM states that the upper quartile length is associated with the grader staple length but does not agree exactly, because it does not adjust for character defects. It is usually slightly longer than the Fibrograph upper half mean length as defined in ASTM designation D 1447 for measuring the length of cotton fibers with the Fibrograph.

(2) The mean length – is the average length of all the fibers in the sample based upon weight-length data.

(3) The coefficient of length variation – is a measure of the variability of fiber length. It represents the standard deviation of the weight-length frequency expressed as a percentage of the mean length. The smaller the number, the more uniform are the fiber lengths.

Fibrograph Method for Determining Cotton Fiber Length

The Fibrograph is a photoelectric optical instrument which scans a combed portion of an appropriately prepared specimen of fiber and draws a length-frequency distribution curve from which are calculated the "upper half mean" and the mean length of the cotton. The upper half mean is defined as the average length of the longest half of the fibers by weight in the specimen. The mean is calculated as the average length of all fibers in the specimen that are longer than 1/4 inch. The length uniformity index is the value obtained when the mean is divided by the upper half mean and the figure multiplied by 100. The Fibrograph is of great advantage because the time required to make a test is on the order of 6 or 7 minutes as compared with the considerably longer time required to determine the length by the standard array method. The ASTM states the results obtained with the Fibrograph may not coincide with those found by a cotton grader, particularly if the specimen contains cut or broken fibers. The Fibrograph and cotton grader's assessment of length usually are reasonably close.

Classifiber Method for Determining Cotton Fiber Length

This method measures the amount of light passed through a sample clamped on combs using a one-millimeter width optical system. The Fibrogram can be drawn from the value of light at a certain length of sample and several useful pieces of information can be derived from this curve. Mean length, uniformity index short fiber content, Fibrogram and histogram are some of the data obtainable. A microprocessor allows for calculation as well as report printing. Figure 20.4 shows Classifiber sampler and measuring unit.

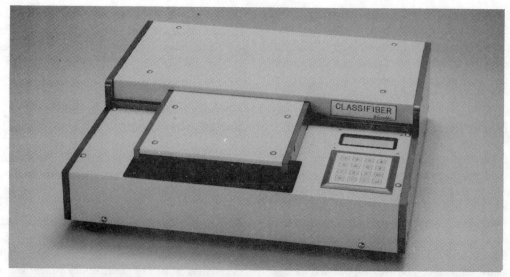

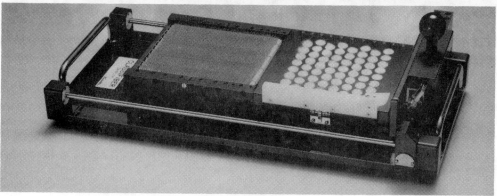

FIGURE 20.4 Classifiber sampler and measuring unit (courtesy of Keisokki).

5.7 Cotton Fiber Fineness: The Micronaire

In addition to fiber staple length, fiber diameter or fineness is an important property which influences the functional characteristics of yarns and fabrics. The shapes range from plump kidney bean-types to flat thin ribbon-types, depending upon the amount of secondary cellulose that is built up on the inside of the cell wall. The degree of fullness is called the maturity of the cotton fiber and is correlative with fineness. One measure of cotton fineness is obtained with the Micronaire, an instrument which measures, under exacting conditions, the air-flow resistance of a plug of cotton of specified dimensions (Figure 20.5).

A predetermined weight of fiber is placed in a specimen holder and compressed to a fixed shape and volume. Then air at a known pressure is forced through the plug and the rate of flow is indicated by a metering device. The exact pressure is specified for different fibers and ranges from 3 to 5 psi across the test plugs. Flow rate is recorded as Micronaire units.

5.8 Pressley Cotton Fiber Strength Test

ASTM lists several methods for testing the strength of cotton fibers by breaking fiber bundles (ASTM D 1445). The Pressley test is the most popular. A small sample is selected, the fibers are made somewhat parallel by hand drafting and lapping, and then they are combed to remove short fibers and make the remaining long ones completely parallel. A bundle of individual fibers about 1/4-inch wide results. This is inserted between two clamps, of either zero or 1/8-inch gauge length. A knife cuts off the fibers protruding from the outer sides of the clamps. The clamp and fiber assembly is inserted in the Pressley Tester, an instrument based upon the application of a force generated by a carriage rolling down an inclined plane. As the carriage rolls down the plane a lever arm exerts a progressively increasing load on the fibers until rupture occurs, at which time the carriage stops automatically, and the breaking strength is recorded. The broken fiber bundle is

removed and weighed. The breaking tenacity is calculated as follows:

Breaking Tenacity (grams/tex)

$$= \frac{5.36 \times \text{lbs. breaking load}}{\text{bundle weight in mg.}} \quad (20.5)$$

Tensile Strength (psi in thousands)

$$= \frac{10.81 \times \text{lbs. breaking load}}{\text{bundle weight in mg.} - 0.12} \quad (20.6)$$

Note: Tenacity, grams per tex = (tensile strength, 1,000 psi + 0.12) × 0.4960; tensile strength, 1,000 psi = (breaking tenacity, grams per tex × 2.016) − 0.12; tex = denier/9.

Fiber bundle tests are widely used to characterize cotton, but the man-made staple fiber industry depends largely upon single fiber strength tests in conjunction with the vibrascope in order to establish quality.

5.9 Cotton Fiber Maturity

ASTM defines maturity as "the degree to which the lumen (central open tube or channel running the length of the fiber) of the cotton fiber has been obliterated by the cellulose laid down in the walls of the fiber."

Four methods are employed to establish the extent of maturity: (1) cross-sectional shape method, (2) sodium hydroxide swelling method, (3) polarized light method, and (4) differential dyeing method.

Cross-Sectional Characteristics

The distribution of cross-sectional shapes is quantitatively determined by measuring average cross-sectional diameters, wall thicknesses, net cross-sectional areas exclusive of the lumen, and fiber shape factor or circularity. The latter is a dimensionless number indicating the roundness or shape of the fiber. It is expressed as either the "area ratio," i.e., the ratio of the actual cross-sectional area of the fiber to the greatest cross section it could have within its actual perimeter if the cross section were a perfect circle;

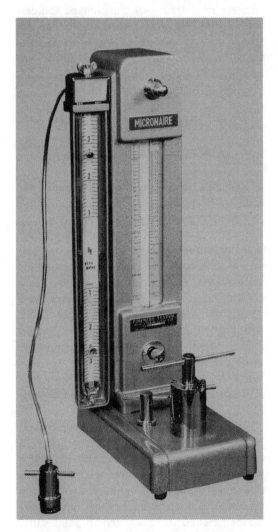

FIGURE 20.5 Micronaire (courtesy of Keisokki).

or as the "diameter ratio," i.e., the ratio of the major to the minor axis of the cross section.

Caustic Method

The fibers are immersed in cold 18% sodium hydroxide which causes them to swell. The swollen fibers are individually examined under the microscope and are considered to be mature if the wall thickness is greater than one-half the diameter of the lumen. Thus, if the combined wall thicknesses are greater than the diameter of the lumen, the fiber is classed mature. If the combined wall thicknesses are less than the diameter of the lumen, the fiber is classed immature (ASTM D 1442).

Polarized Light Method

When cotton fibers are longitudinally examined with a polarizing microscope, they exhibit different "interference" colors which are largely dependent upon the thickness of the cell walls. As a result of their different degrees of maturity, the fibers appear in different colors when observed under polarized light using a first order red selenite plate. The fibers are classified as follows: yellow fibers are fully mature, green fibers are partially mature, and blue and purple fibers are immature.

Differential Dyeing Method

It has been observed that mature and immature fibers will dye differently when treated with specific dyestuffs. Thus, when a test specimen of cotton is immersed in a mixture of Diphenyl Fast Red 5 BL (Supra I) and Chlorantine Fast Green BBL, the red dye is retained primarily by the thick-walled mature fibers, and the green dye by the thin-walled immature fibers. Thus, the degree of immaturity can be determined from the dyeing behavior. Of additional importance, these dyes can be applied to yarns or fabrics to ascertain the amount and uniformity of distribution of mature and immature fibers. Thus, the technique is helpful in determining defects which may stem from fiber nonuniformity (ASTM D 1464).

The Shirley fiber maturity tester uses Micronaire values at two compression loadings to estimate maturity. Near Infra-Red (NIR) analysis is also used to determine fiber maturity.

5.10 Wool "Grade" or Fiber Fineness

All wool fibers have cross sections which are round to oval in shape. There is no comparable property of degree of maturity as there is in cotton. However, since a fiber's stiffness is proportional to the fourth power of its diameter, the mean and distribution of wool fiber diameters are used to grade wool in terms of its softness and luxuriousness. Several hundred very short lengths of fiber, on the order of about 150 microns long, are cut by a microtome and dispersed in mineral oil on a microscopic slide. By means of a projection microscope and calibrated scale, diameters of the fibers are measured. From the calculated average and distribution of micron diameters, the fineness or grade is established. Specifications for grades 36 to 80 are shown in ASTM D 2130 and D 3991.

5.11 Wool Fiber Staple Lengths

Wool fiber staple lengths are determined basically in the same manner as for cotton. However, because the fibers are considerably longer, different techniques for hand stapling and different types of combs for quantitative stapling are required. ASTM D 1575 gives a method for quantitatively measuring staple length, using wool sorting combs. ASTM D 1234 describes a method for removing bundles of fibers from bags of wool, in the grease, laying the fibers on a velvet board, and calculating the mean staple length from the resulting array.

5.12 Determination of Moisture Regain

The most direct and obvious way to determine the moisture regain of a textile fiber sample is to weigh accurately a properly sampled amount of the fiber, then dry it in an oven at a temperature above the boiling point of water, cool it in a moisture-free atmosphere, reweigh, and calculate the weight loss. This method presumes that no materials other than water are volatilized out of the fiber by the drying method. In the event that the product itself cannot withstand the drying temperature, usually about 105°C, vacuum drying ovens can be employed. Here, at reduced pressure, the water will evaporate at a lower temperature. For textiles, however, this procedure is normally not necessary.

The nature of the fiber, yarn, or fabric, the size of sample, and the equipment available will all govern the specific techniques selected to determine regain. ASTM D 2654 suggests use of a 10-gram sample accurately weighed.

The 10-gram sample is weighed and placed in a previously accurately tarred weighing bottle or can, equipped with a moisture-tight cover. With the cover off, the bottle and its contents are dried in an oven for an hour at 105°C. While still in the oven, the cover is placed on the bottle, which is then removed and placed in a moisture-free desiccator containing calcium chloride or other drying agent. After cooling, the cover is loosened for an instant to equalize pressure, is replaced, and the bottle and contents are reweighed. The procedure is repeated, if necessary, until constant weight is obtained. The weight loss as a percentage of the bone-dry weight of the sample is calculated as the moisture regain. The textile industry has developed special test equipment for determining the regain of textiles.

5.13 Electric Moisture Meters

Electric moisture meters generally operate on the basis of a quantitative relationship that exists between a material's water content, and the effect that the water content has upon such electrical properties as resistance, capacitance, dielectric constant or power factor.

Electrical resistances of textiles vary widely with moisture content. For example, a change in regain from 5 to 10% decreases cotton's electrical resistance 1,000 times. A major disadvantage of the resistance method is that readings are not obtained on materials that themselves have very high electrical resistances—for example, acetate, and the synthetic fibers.

Capacitance and dielectric meters require that a stipulated weight of material be placed between the condenser plates in order to obtain valid calibrated readings. The dielectric constant of most textile and related organic products is about five, while that of water is seventy-eight. Therefore, the dielectric constant method is applicable for the determination of the water content in textiles and similar products.

All electrical meters must be calibrated empirically against known moisture determinations made via standard oven "bone-dry" techniques on the actual product being tested. Each instrument must be calibrated for the particular type of fiber, yarn, or fabric under test. Furthermore, the type of package or "form" which the product has, must be kept constant. Thus, if bulk rayon is to be tested on a dielectric-constant-type tester, the same approximate sample weight, shape, and bulk density must be used. The instrument must be similarly and specifically calibrated for balls of worsted top or cones of cotton yarn of approximately the same size, weight and degree-of-tightness of wind.

Modern moisture balances automatically determine moisture content and dry weight (Figure 20.6).

FIGURE 20.6 Moisture balance (courtesy of Ohaus Corporation).

5.14 Chemical Methods for Determining Moisture

Karl Fischer Titration Method

This well-known method consists of dissolving the test sample's moisture in methyl alcohol and determining the amount of water by titrating with the Karl Fischer reagent, composed of iodine, sulphur dioxide and pyridine (C_5H_5N). The water combines quantitatively with these components, and with the methyl alcohol, forming pyridine iodide ($C_5H_5N \cdot HI$) and pyridinium methyl sulfate ($C_5H_5 \cdot HI \cdot SO_4CH_3$). This method is based upon the premise that all of the water is extracted out of the sample so that it can react with the reagent and the alcohol. An electrometric titration is necessary to carry out the test in order to properly establish the end point at which the addition of the reagent is complete.

Cobalt Chloride Method

When cobalt chloride is dissolved in absolute (moisture-free) ethyl alcohol, a blue solution is obtained. Upon the addition of water, a pink color results, the reaction being reversible. By quantifying the reaction and using a spectrophotometer or other calorimetric method, the amount of water in a test sample can be determined.

5.15 The Shirley Analyzer

The Shirley Analyzer is a machine which separates lint cotton fiber from non-lint and trash. Developed by the Shirley Institute in England, the instrument and method are useful in determining the yield of spinnable cotton fiber, as well as the efficiency of ginning, opening, cleaning, picking and carding. The cotton sample is opened, and then via air-stream conveyance and separation, the clean fibers are deposited on the surface of a cage condenser, while the heavier trash and other non-lint material fall into a pan. Ultimately, the lint fiber is collected from the condenser and deposited in a delivery box. The weights of lint cotton and non-lint material, respectively, are calculated as percentages of the input weight.

5.16 High Volume Instrumentation (HVI)

HVI is a system or console that contains instruments that can evaluate cotton fiber properties (Figure 20.7). Typically, the system can measure bundle properties such as length, length uniformity, strength, elongation, Micronaire, color grade and optical trash content, short fiber content, and spinning consistency index (SCI). In 1991 nearly 100% of the U.S. cotton crop was evaluated with these types of systems. This high level of usage was a result of a requirement imposed by the United States Department of Agriculture. Cotton exported from the United States and cotton sold in international commerce may be evaluated by the previously cited standard methods or by other international standards. A standard test method using these types of instruments is currently being balloted by the ASTM.

5.17 AFIS-N, -L&D, -T Instrument

This instrument (Figure 20.8) can be used with several different test modules. These modules provide the capability to determine the following: nep count and nep size, fiber length and diameter, number and size of particles and foreign matter, dust and trash. A standard test method using this instrument for the determination of neps is currently being balloted by the ASTM.

5.18 Blend Testing

Many textile mills run blends of fibers such as polyester/cotton, polyester/rayon, cotton/rayon, etc. Most of the blended stock is either a 50/50 or a 65/35 blend. The primary method of testing blended stock is to dissolve the cotton, or cellulose fibers, in a 70% H_2SO_4 solution. NIR is another method of testing for blends.

5.19 Tests for Fiber Damage

The textile scientist has many techniques for determining whether textile fibers have been damaged, and the nature of such damage.

Purely physical damage, which may result from tension, abrasion, flexing, or cutting, can

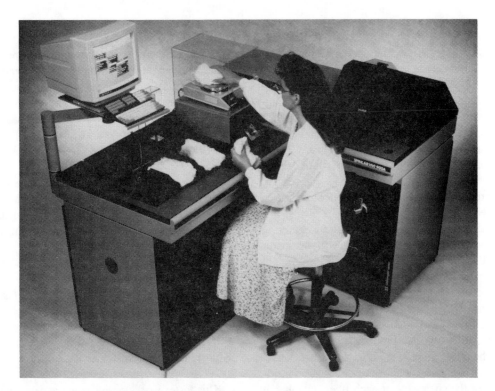

FIGURE 20.7 Uster spinlab HVI 900 fiber testing instrument.

FIGURE 20.8 Uster AFIS-N instrument for nep count.

often be observed via microscopical examination. The many texts on fiber microscopy give much information on the techniques for detecting abraded or disrupted fiber surfaces, damaged animal fiber scales, the presence of split or "broomed" fibers, etc. The structure and chemical composition of the surfaces of many fibers are different from their inner portions. Often these surface and inner components will react at different rates and to different degrees when treated with dye stains or other chemicals.

Chemical microbial, sunlight and weathering degradation of fibers normally will be manifested by a reduction in the molecular weight and chain length of the polymer. This in turn will produce more chemically reactive "end groups" which often can be qualitatively detected and quantitatively measured. Fiber degradation can also produce other changes in the chemical nature of the fiber which are detectable by specific chemical reactions. A knowledge of the chemistry of each fiber, whether it be natural or man-made is mandatory in order to ascertain the type and extent of fiber damage.

6. SLIVER, ROVING, AND YARN EVENNESS TESTS

The manufacture of good quality spun yarns is largely dependent upon uniformity of the sliver and roving—the intermediate forms produced prior to yarn spinning. Textile mills use several well-known testing instruments for checking product uniformity. Most are based upon measurement of weight, mass or thickness of the materials that pass continuously through the tester. Many instruments have graphical recorders so that point-to-point variations in properties are plotted.

6.1 Saco-Lowell Sliver Graphic Tester

The sliver is continuously tested for thickness and mass uniformity by traveling through a pair of mated tongue and groove rolls. The lower grooved roll carries the sliver, while the upper tongued roll presses down on the sliver as it moves through the groove. The thickness varia-

tion causes the pressing tongue roll to be displaced upwards or downwards, and this motion is mechanically amplified and continuously recorded on a chart.

6.2 Pacific Evenness Tester for Sliver, Roving, and Yarn

This tester works on the same principle as the Saco-Lowell sliver tester. Sliver thickness is continuously measured via the displacement of a convex pressing roll. However, by means of a magnetic displacement gauge, the roll movement is converted to an electrical signal which is continuously recorded on a chart. Mating roll sets of different groove sizes permits the testing of sliver, roving and yarn.

6.3 Brush Uniformity Analyzer for Sliver, Roving, and Yarn

This tester works on a capacitor principle. It measures variation in the mass or weight of the product which continuously travels between condenser plates. The capacitance signal is fed into an electronic recorder.

6.4 Uster Evenness Tester for Sliver, Roving, and Yarn

This instrument also works on the electrical capacitance principle (Figure 20.9). Because the shape of the sliver must be kept constant in order to obtain valid data, a "Rotofil" attachment puts false twist into the sliver to maintain a uniformly round cross section as it passes through the tester.

6.5 Beta Gauge Uniformity Tester

A method for continuously recording and controlling point-to-point thickness or weight uniformity of textile laps, slivers, etc., as well as paper, nonwovens, coatings, and plastic films is based upon the quantitative emission and detection of radioactive energy rays. The instrument is called a beta (β) ray gauge. It consists of a β particle emitter of radioactive energy, and an accompanying receiver and recorder. The test material continuously passes between the

FIGURE 20.9 Uster strand evenness tester (courtesy of Wellington Sears Company).

emitter and receiver and the signal received is inversely proportional to the mass of product passing through the instrument. While the use of such equipment is scientifically feasible, costs have not yet been brought down to the point where they are being used generally in the textile manufacturing industry.

6.6 Belger Roving Tester

This instrument indicates the ability of roving to be drawn uniformly, a prerequisite for subsequent proper yarn spinning. The roving is continuously passed over two sets of rolls running at different surface speeds, thus producing a draft on the order of 6%. The back rolls are balanced so that an arbitrarily selected median force, at the 6% draft, will maintain an indicator attached to the rolls at "zero." Then as the force required to draft the roving becomes greater or less, it is indicated on a recorder as a displacement from the zero point. Thus, the drafting force variability of the roving is continuously indicated.

7. YARN TEST METHODS

A selection of standard test methods for yarn properties are listed in Table 20.6.

7.1 Determination of Yarn Weight or Number

Yarn number or count is normally determined by winding 120 yards or some other reasonable length of yarn on a reel. The resulting skein is weighed on a sensitive balance, and the weight per unit length, or the inverse is calculated (Figure 20.10). If the yarn has a large amount of crimp, such as might be present in a bulked yarn, or if it is highly extensible or rubbery at low loads, it may be necessary to pretension it at a selected load before winding off the desired length. Then the yarn number should be reported together with the tension employed during measurement. ASTM states that the winding tension is usually 0.055 grams per denier (0.5 grams per tex) and should never exceed 0.11 grams per denier (1 gram per tex).

TABLE 20.6 Standard Test Methods Used for Yarn Testing.

ASTM D 861	Tex System, Use of, to Designate Linear Density of Fibers, Yarn Intermediates, and Yarns
ASTM D 1059	Yarn Number Based on Short-Length Specimens
ASTM D 1422	Twist in Single Spun Yarns by the Untwist-Retwist Method
ASTM D 1423	Twist in Yarns by the Direct-Counting Method
ASTM D 1907	Yarn Number by the Skein Method
ASTM D 2256	Standard Test Method for Tensile Properties of Yarns by the Single Strand Method
ASTM D 2258	Sampling Yarn for Testing
ASTM D 2259	Shrinkage of Yarns in Boiling Water or Dry Heat
ASTM D 4031	Test Method for Bulk Properties of Textile Yarns
ASTM D 4849	Yarn and Related Terms
AATCC 84-1989	Electrical Resistivity of Yarns

Selection of the tension load is always a problem. A standard method is to use the "load required to straighten but not stretch the yarn." A suggested procedure is as follows. Cut a 14-inch length of yarn and lay it flat and relaxed on a table. Make two precise ink marks on the yarn exactly ten inches apart. Place the yarn between the jaws of a low capacity, sensitive, constant rate of extension, recording tensile testing machine, with the jaws exactly ten inches apart. Separate the jaws of the machine at a slow constant speed, thus plotting an accurate load-elongation diagram (Figure 20.11). A tangent line is drawn from the linear low-load portion of the curve back to zero load (point A). The extension from 0 to A is the crimp or elongation required to straighten the specimen. Line AB parallel to the load axis is drawn. The intersection at point B defines the load required to straighten but not stretch.

An alternate to the skein weight method, useful for heavy ropes or expensive filaments, or where yarn supply is short, is to cut five or ten precisely marked short lengths of yarn (e.g., ten inches), weigh on a properly sensitive balance, and calculate.

Some new balances provide the direct read for yarn count (Figure 20.12).

7.2 Twist Testers

A twist tester normally consists of two jaws, one fixed and the other rotatable, about 10 inches apart into which a selected length of yarn is fastened for untwisting. The rotating jaw has a counter so that the number of turns required to untwist the yarn can be recorded. The other jaw is usually attached to a pulley tensioning mechanism so that a fixed length of yarn under a

FIGURE 20.10 Universal count analyzer (courtesy of James H. Heal & Co.).

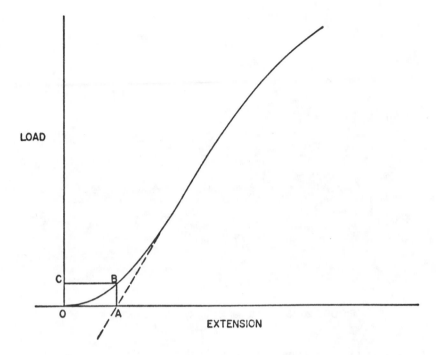

FIGURE 20.11 Determination of load required to straighten but not stretch a yarn.

selected tension can be clamped between the jaws. Figure 20.13 shows a twist tester.

The twist in a filament singles, or in a plied yarn or cord, is determined by damping a fixed length of the specimen between the jaws of the tester and untwisting until its components are completely parallel. The resulting number of turns, when divided by the original length in inches, gives the turns per inch.

When twist exists in the singles, ply and cord, the untwisting action to achieve parallelism in the largest component will produce a change in

FIGURE 20.12 Digital yarn count balance (courtesy of J. A. King & Company Inc.).

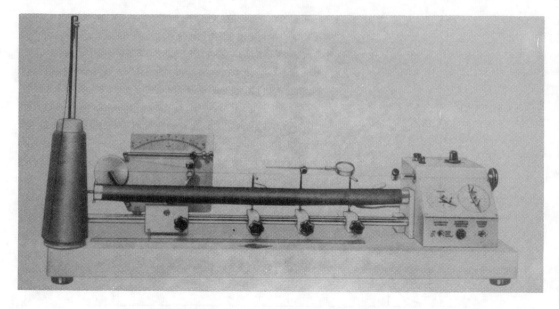

FIGURE 20.13 Twist tester (courtesy of Brance Idealair).

the twist of the remaining smaller components. Ply twisting will insert or remove singles twist, depending upon whether the ply twists and singles twists are in the same or opposite directions. Similarly, when ply turns per inch are removed by untwisting, if the ply and singles twist are in the same direction, the singles will become untwisted as the plies are untwisted. If the twist directions are opposite, untwisting the ply will add twist to the singles. In either case, when the ply is untwisted, the singles will have approximately the same number of turns that it had before plying, i.e., the turns inserted during its manufacture.

It is often desirable to determine the twist in a singles as it lies in the plied yarn. Here it is necessary to remove all of the singles yarns except the one which is to be tested, while being careful not to disturb its twist. The technique consists of clamping the plied yarn in the twist tester in the usual manner. A fine scissors or knife is used to cut away all of the singles except the one to be tested. Some twist testers have cluth mechanisms whereby both jaws can be rotated at the same time and at the same speed, so that the cut discarded yarns can be raveled away from the test specimen without altering its twist. After the extraneous yarns have been removed, the clutch is disengaged, and the twist

in the remaining singles is determined by untwisting in the usual fashion.

For spun staple yarns it is obviously impossible to untwist the yarn until all fibers are parallel, since yarn failure will occur at one point while twist still remains at other points in the yarn. Instead, an untwist-twist method is used. A length of yarn is inserted in the twist tester at a selected tension. The yarn is then untwisted and twisted in the opposite direction until the original length and yarn tension is restored. It is assumed that the same amount of twist has been reinserted as was originally present in the yarn. The turns per inch are calculated, first by dividing by two the total number of turns necessary to untwist and retwist, and then dividing by the original yarn length:

$$\text{T.P.I.} = \frac{T/2}{L} = \frac{T}{2L} \qquad (20.7)$$

where

T.P.I. = turns per inch
T = turns required to untwist and retwist
L = original yarn length

The ASTM points out that the method is only

an approximation because torsional effects may cause the fibers to resist retwisting in the opposite direction after being untwisted, and fiber slippage may occur during the test.

Various devices are available for establishing and controlling the tension during untwisting and retwisting. In the photoelectric tension control instrument, the yarn is placed under a fixed tension in the twist tester, is untwisted, and retwisted in the opposite direction until the original yarn tension is reached, at which point the twist tester automatically shuts off. The number of turns is thus automatically recorded.

A commonly used modification of the twist-untwist method is the "twist-break" method. Instead of untwisting and retwisting until the selected original tension is reobtained, additional twist, in the same direction as was used in manufacturing the yarn, is inserted until the yarn breaks. A second specimen is untwisted and retwisted in the opposite direction until the yarn breaks again. The difference in twist, divided by twice the yarn length, gives the turns per inch. This is basically the same as the tension method except that the limiting breaking tension is used. It is not as accurate as the tension method however, since the number of turns per inch, instead of being solely a function of the yarn's tension, is also a function of the yarn's breaking strength.

Twist can also be determined optically via a microscope with a rotating stage for measuring helix angles. Knowing the helix angle and the yarn diameter, the number of turns per inch can be calculated from simple mathematics. However, this method is more often used as a research tool rather than for quality control purposes.

7.3 Classification of Yarn Defects

Yarn defects can be classified using a capacitance method. This is a very common test in spun yarn mills that is important in checking yarn quality. This test classifies the defects by change in mass and length. Instruments such as Keisokki Classifault II or Uster Classimat may be used for this type of test. A standard method is currently being written by a subcommittee of ASTM for this test.

8. TENSILE TESTING OF YARNS AND FABRICS

Breaking strength-elongation testing machines are useful for determining the load-deformation characteristics of materials, as well as yarn crimp, energy absorption, tensional modulus, repeated stress properties, burst and tear strengths. These testers, when of proper sensitivity, are also useful in measuring the viscosity and surface tension of liquids, and adhesion properties of laminates.

Tensile testers operate on one of the following three principles:

(1) Constant rate of traverse
(2) Constant rate of specimen loading
(3) Constant rate of specimen extension

The ASTM test methods related to tensile testing include:

- ASTM D 76 Standard Specification for Tensile Testing Machines for Textiles
- ASTM D 1578 Breaking Strength of Yarn in Skein Form
- ASTM D 1775 Tension and Elongation of Wide Elastic Fabrics (Constant Rate-of-Load-Type Tensile Testing Machine)
- ASTM D 2256 Tensile Properties of Yarns by the Single-Strand Method
- ASTM D 4964 Tension and Elongation of Elastic Fabrics (Constant Rate-of-Extension-Type Tensile Testing Machine)
- ASTM D 5034 Breaking Force and Elongation of Textile Fabrics (Grab Test)
- ASTM D 5035 Breaking Force and Elongation of Textile Fabrics (Strip Test)

8.1 Pendulum Testers

In these testers, using heavy pendulums and bobs that produce a large mechanical moment, loads up to about 2,000 pounds can be applied to cordage, ropes and fabrics, without the testing machine itself becoming excessively large or heavy.

Pendulum testers are called "constant rate of traverse" machines. While the lower pulling jaw does move downward at a constant speed, the instrument itself neither loads nor extends the specimen at a constant rate. This is because as the load is applied to the specimen, the sprocket-chain arrangement causes the upper jaw to feed downward as the pendulum extends and rises. When the test specimen is highly deformable and considerable elongation results at low loads, the upper jaw does not feed downward as rapidly as it does when the specimen has lower extensibility. The amount and rate at which the upper jaw feeds down varies, depending upon the load-elongation properties of the test specimen. Thus, the specimen is neither loaded nor extended at a constant rate, nor does failure occur within a constant time period.

Although these type machines are still used they have been replaced in many laboratories by electronic "constant rate of extension" machines.

8.2 Electronic Tensile Testers

Bonded wire resistance strain gauges have been particularly successfully applied as load measuring units in electronic tensile testing machines. The upper jaw of the tester is attached to a steel bar of proper capacity and design so that at full load its deflection is completely elastic, and negligible, on the order of a few thousandths of an inch. The strain gauge is bonded to the bar. When a force is applied, the resulting proportional voltage change produced by the gauge is amplified and recorded as a direct measure of specimen load. Because the upper jaw moves an almost imperceptible amount even at maximum load, the pulling jaw actually elongates the test specimen at a constant rate, and such instruments, in fact, are constant rate of extension testers.

In electronic tensile testers, the various capacity load recording "cells" can be installed so that specimens with a wide range of breaking strengths can be tested on the same machine. Figure 20.14 shows an automatic tensile testing machine which operates on the principle of constant rate of extension. For different types of textile materials, different clamps are used (Figure 20.15).

The chart recorder of an electronic tensile

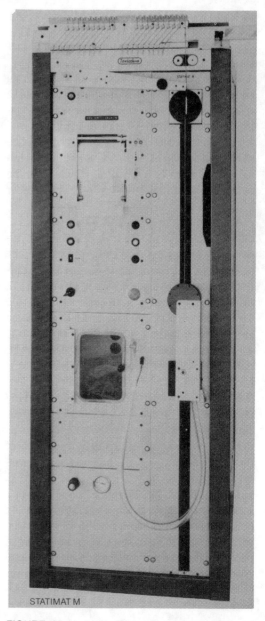

STATIMAT M

FIGURE 20.14 Automatic tensile tester (courtesy of Textechno).

tester runs at a constant speed and records force or "load" in one direction and elapsed time in the other. The elapsed time is directly proportional to the jaw separation and so a load-versus-jaw separation diagram for the tensile test specimen results. This can then be converted to a load-elongation diagram. For very brittle or low extensibility materials, the recorder chart speed can be increased so that the elongation axis is amplified. This is of significant help when measuring the properties of paper, glass, metals or other materials with low breaking elongations. An "extensometer" or strain transducer can also be attached to the specimen to measure elongation.

Modern tensile testers provide the capability of converting the load cell or transducer response into a digital output that can then be stored or processed directly using a personal computer. The various equipment manufacturers can also provide computer software that can produce reports, plot data and do statistical analyses (Figure 20.16).

8.3 Testing Speeds

The rate of loading or extension can influence the ultimate breaking strength. Therefore,

ASTM specifies the rate, and more properly, the time to break. For constant rate of traverse pendulum machines, the specified pulling jaw speed is 12 ± 0.5 inches per minute. An alternate suggestion is that the machine be operated at such a rate that the yarn will break within 20 ± 3 seconds from the start of the test. For a discussion of tensile testing machines the reader is referred to ASTM D 76.

8.4 Determination of Breaking Elongation

For most tensile testing machines, the "elongation" axis of the load-elongation recorder runs at a constant speed so a direct measure of jaw separation results. In the case of electronic constant rate of extension testers, the chart is

CLAMPS FOR HIGH-TENACITY YARNS   FABRIC CLAMPS   CLAMPS FOR SLIVERS

FIGURE 20.15 Clamps for various textile materials (courtesy of Textechno).

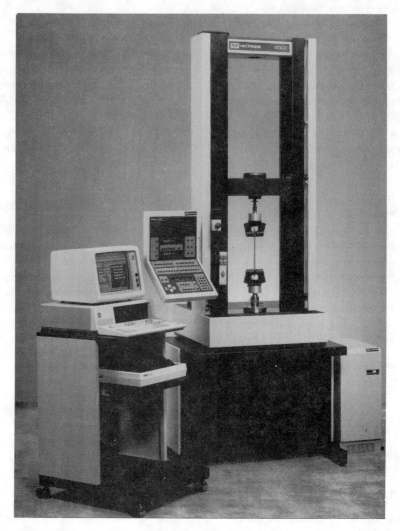

FIGURE 20.16 Computer controlled tensile testing machine (courtesy of Instron).

driven electrically at a constant speed, and is independent of the jaw motion. Usually, it runs at the same speed as the pulling jaw, or at a multiple speed, and so the time axis can be converted to jaw separation, and ultimately to specimen elongation.

With flat jaws, the recorded jaw separation can be taken as a direct measure of the specimen's elongation. For example, in a raveled strip tensile test, the gauge length (jaw separation) at the start of the test is three inches. At the instant of rupture, if the jaw has moved 0.6 inches, the breaking percent elongation can be calculated as $(0.6/3) \times 100 = 20\%$. There may be slight elongation of the ends of the specimen

held in each jaw, called jaw penetration, that may make the gauge length slightly greater than three inches, and the true elongation correspondingly slightly less, but this error is usually negligible and is ignored in the calculation.

When testing thick and heavy cords, tapes, braids or webbings, flaw jaws may not be satisfactory. Slippage occurs if the jaws are not tightened sufficiently, while cutting and "jaw breaks" result if the jaw is tightened too much. To avoid such difficulties capstan jaws are employed. The cord is wound around the capstans with about ten inches of yarn placed in the "free gauge length," that is, the length between capstan tangent points. During the test, yarn elongation takes place not only in the free gauge length, but also progressively around the capstans where the yarn is being snubbed. Thus, the yarn is gradually held and a jaw break is avoided. Because the yarn around the capstans is elongating, the jaw separation cannot be used as a direct measure of true elongation, and other means must be utilized. An approximate method is to place two marks on the yarn in the free gauge length. A ruler is placed beside the specimen. Keeping one mark at zero, the displacement of the other mark is noted at the time of rupture. This technique only gives the breaking elongation, and does not permit the plotting of a load-elongation diagram. When used on materials with high strength and high elastic recovery, a danger to the operator exists. Some materials, particularly nylon, stretch a considerable amount with accompanying storage of energy. At the instant of rupture, this energy is released with considerable "snapback" of the broken specimen. The unrestrained movement of the broken end can act like a whip and break a finger or arm of the person making the measurement. Therefore, this technique should be avoided when working with elastic materials above about 50 pounds breaking strength.

An alternate method is to employ calibrated telescopes or cathetometers to follow the gauge marks on the specimen. This is expensive because two observers are necessary. Movie cameras, multi-flash photography and electronic extensometers are also useful.

9. FABRIC TEST METHODS

Selected standard fabric test methods are presented in Table 20.7. This table does not include company methods or ISO methods. The conversion of many company methods into standard methods is an ongoing process and in the area of nonwovens many are being processed by the consensus organizations, particulary the ASTM.

9.1 Threads per Inch

A "pick counter" or pick glass is a convenient means of determining the threads per inch in a fabric. A low-power magnifier is mounted on a traveling screw with a moving reference point

and fixed scale being visible in the field of view. The pointer traverses the scale, and by means of the magnifying glass, the number of threads per inch or other unit distance can be counted. Transmitted light through the fabric or incident light upon the fabric is usually helpful in counting the threads.

A lighting table with top incident illumination or bottom transmitted illumination through a translucent glass plate is a useful piece of laboratory equipment for such work.

There are fabrics of very high sley, those which have been felted, and those where yarns overlap and are superimposed one upon the other so as to make direct visual counting impossible. It is necessary to cut a small swatch of such fabrics; for example, one-half to one inch on a side, and to ravel and count the yarns by hand.

Using optical/electronic technology, fabric threads per unit length can be measured rapidly with extreme accuracy and repeatability. Figure 20.17 shows a thread measuring device.

9.2 Fabric Weight

Fabric weight is determined by weighing the complete piece, roll, cut, or bolt, or by selecting and weighing a full width sample, 1/4-yard in length therefrom. The weight per square yard, weight per linear yard, or the number of linear yards per pound may be calculated as follows:

Full Piece, Roll, Cut, or Bolt

Weight per square yard (ounces)

$$= \frac{\text{fabric weight in lbs.} \times 16 \times 36}{\text{fabric length in yards} \times \text{width in inches}}$$

$$(20.8)$$

Weight per linear yard (ounces)

$$= \frac{\text{fabric weight in lbs.} \times 16}{\text{fabric length in yards}} \quad (20.9)$$

Number of linear yards (per pound)

$$= \frac{\text{fabric length in yards}}{\text{fabric weight in pounds}} \quad (20.10)$$

TABLE 20.7 Standard Test Methods Used for Fabric Testing.

ASTM D 434	Resistance to Slippage of Yarns in Woven Fabrics Using a Standard Seam
ASTM D 737	Air Permeability of Textile Fabrics
ASTM D 885	Testing of Tire Cords, Tire Cord Fabrics, and Industrial Filament Yarns Made from Man-Made Organic-Base Fibers (D 885 M is in metric units)
ASTM D 1117	Testing Nonwoven Fabrics
ASTM D 1230	Test Method for Flammability of Clothing Textiles
ASTM D 1388	Stiffness of Fabrics
ASTM D 1424	Tear Resistance of Woven Fabrics by Falling Pendulum (Elmendorf Apparatus)
ASTM D 1683	Failure in Sewn Seams of Woven Fabrics
ASTM D 1777	Thickness of Textile Materials
ASTM D 1908	Needle-Related Damage due to Sewing in Woven Fabrics
ASTM D 2261	Tearing Strength of Woven Fabrics by the Tongue (Single Rip) Method (CRE Tensile Testing Machine)
ASTM D 2262	Tearing Strength of Woven Fabrics by the Tongue (Single Rip) Method (CRT Tensile Testing Machine)
ASTM D 2475	Standard Specification for Wool Felt
ASTM D 2594	Stretch Properties of Knitted Fabrics Having Low Power
ASTM D 2646	Backing Fabrics
ASTM D 3107	Stretch Properties of Fabrics Woven from Stretch Yarns
ASTM D 3511	Pilling Resistance and Other Related Surface Changes of Textile Fabrics; Brush Pilling Tester Method
ASTM D 3512	Pilling Resistance and Other Related Surface Changes of Textile Fabrics; Random Tumble Pilling Tester Method
ASTM D 3514	Resistance of Apparel Fabrics to Pilling (Elastomeric Pad Method)
ASTM D 3597	Woven Upholstery Fabrics—Plain, Tufted or Flocked
ASTM D 3773	Length of Woven Fabrics
ASTM D 3774	Width of Woven Fabrics
ASTM D 3775	Fabric Count of Woven Fabric
ASTM D 3776	Mass per Unit Area (Weight) of Woven Fabric
ASTM D 3786	Hydraulic Bursting Strength of Knitted Goods and Nonwoven Fabrics—Diaphragm Bursting Strength Tester Method
ASTM D 3787	Bursting Strength of Knitted Goods—Constant-Rate-of-Traverse (CRT) Ball Burst Test
ASTM D 3884	Abrasion Resistance of Textile Fabrics (Rotary Platform, Double-Head Method)
ASTM D 3885	Abrasion Resistance of Textile Fabrics (Flexing and Abrasion Method)
ASTM D 3886	Abrasion Resistance of Textile Fabrics (Inflated Diaphragm Method)
ASTM D 3887	Standard Specification for Knitted Fabrics
ASTM D 3936	Delamination Strength of Secondary Backing of Pile Floor Coverings
ASTM D 3940	Bursting Strength (Load) and Elongation of Sewn Seams of Knit or Woven Stretch Textile Fabrics
ASTM D 4032	Stiffness of Fabric by the Circular Bend Procedure
ASTM D 4033	Resistance to Yarn Slippage at the Sewn Seam in Upholstery Fabrics (Dynamic Fatigue Method)
ASTM D 4034	Resistance to Yarn Slippage at the Sewn Seam in Woven Upholstery Fabrics
ASTM D 4157	Abrasion Resistance of Textile Fabrics (Oscillatory Cylinder Method)
ASTM D 4158	Abrasion Resistance of Textile Fabrics (Uniform Abrasion Method)
ASTM D 4848	Standard Terminology of Force, Deformation and Related Properties of Textiles
ASTM D 4852	Coated and Laminated Fabrics for Architectural Use
ASTM D 4966	Abrasion Resistance of Textile Fabrics (Martindale Abrasion Tester Method)
ASTM D 4970	Pilling Resistance and Other Related Surface Changes of Textile Fabrics (Martindale Pressure Tester Method)
AATCC 22	Water Repellency: Spray Test
AATCC 35	Water Resistance: Rain Test
AATCC 42	Water Resistance: Impact Penetration Test
AATCC 66	Wrinkle Recovery of Fabrics: Recovery Angle Method
AATCC 70	Water Repellency: Tumble Jar Dynamic Absorption Test
AATCC 76	Electrical Resistivity of Fabrics
AATCC 79	Absorbency of Bleached Textiles
AATCC 93	Abrasion Resistance of Fabrics: Accelerator Method
AATCC 115	Electrostatic Clinging of Fabrics: Fabric-to-Metal Test
AATCC 127	Water Resistance: Hydrostatic Pressure Test
AATCC 136	Bond Strength of Bonded and Laminated Fabrics

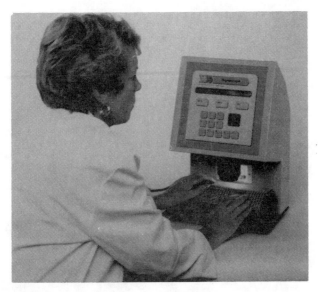

FIGURE 20.17 Thread pitch measuring device (courtesy of James H. Heal & Co.).

Quarter Yard Sample

Weight per square yard (ounces)

$$= \frac{\text{sample weight in grams} \times 45.72}{\text{sample area in square inches}} \quad (20.11)$$

Weight per linear yard (ounces)

$$= \frac{\begin{array}{c}\text{sample weight in grams} \times 36 \\ \times \text{ fabric width in inches}\end{array}}{\text{sample area in square inches} \times 28.35}$$

$$(20.12)$$

Number of linear yards (per pound)

$$= \frac{16}{\text{ounces per linear yard}} \quad (20.13)$$

For small samples a single area of at least 20 square inches, or a number of smaller die-cut specimens containing as far as possible different warp and filling threads, and having a total area of at least 20 square inches, is weighed on a properly sensitive scale. The resulting weight is converted into ounces per square yard.

9.3 Fabric Thickness

Fabric thickness is measured according to ASTM D 1777. Figure 20.18 shows an electronic thickness tester.

9.4 Tear Tests

The technology of the various tear tests and a description of the tongue, trapezoid, and the Elmendorf test methods are given in Chapter 19.0. Standard test methods such as ASTM Methods 1424 and 2261 are used for tear tests. Figure 20.19 shows a heavy-duty Elmendorf tear tester for heavy fabrics such as canvas, soil cloth, and other sheet materials beyond the capacity of the standard Elmendorf. Augmenting weights increase the testing capacity up to 25,600 grams.

9.5 Ball and Mullen Diaphragm Testers

The ball burst test determines the force necessary to drive a one-inch diameter steel ball through a 1.75-inch diameter ring-clamped fabric sample (ASTM D 3787). A sliding yoke apparatus permits the use of conventional pendulum or electronic testers to record the bursting force.

The Mullen burst tester records the oil pressure required to burst a fabric specimen where the circular area under test is 1.2 inches in diameter (ASTM D 3786).

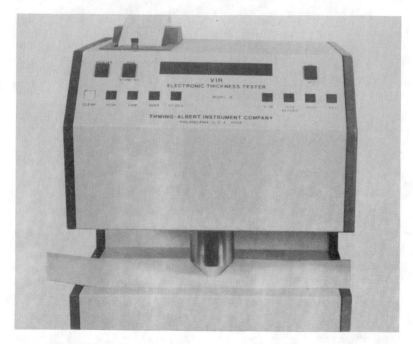

FIGURE 20.18 Electronic thickness tester (courtesy of Thwing-Albert).

FIGURE 20.19 Heavy-duty Elmendorf tear tester (courtesy of Thwing-Albert).

9.6 Compressibility and Resilience Testers

Tensile testers are easily converted to compressional resilience testers. The load-recording cell is mounted below the lower traveling jaw. The test specimen is placed on a flat plate attached to the cell and is thus compressed as the lower jaw moves downward. Plots of load versus specimen compression, i.e., jaw travel, are recorded.

A sliding cage is also useful for measuring compressional characteristics. This permits the use of tension testers in a conventional fashion. The sliding yoke arrangement puts the specimen in compression as the lower jaw travels downward.

9.7 Air Permeability Testers

ASTM D 737 describes standard test method for air permeability of textile fabrics.

Frazier Differential Pressure Air Permeability

The most popular air permeability tester is the Frazier type (Figure 20.20). The air flow in cubic feet per minute passing through one square foot of fabric under a pressure drop equal to one-half inch of water is normally measured.

The apparatus contains a suction fan for drawing air through a known area of fabric that is clamped and sealed in place by a pressure ring. A rheostat controls the speed of the fan, so that the pressure drop across the fabric, indicated by a manometer, can be precisely controlled, usually at a standard 0.5 inches of water. The resulting volume of air must then be measured. This is accomplished by utilizing a calibrated orifice through which all the air that originally passed through the fabric must also pass. A second manometer measures the pressure drop across the calibrated orifice, the readings being convertible to air flow in cubic

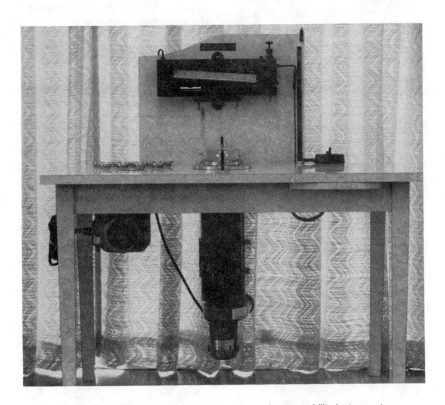

FIGURE 20.20 Frazier differential pressure air permeability instrument.

feet per minute per square foot of fabric. A set of nine calibrated orifices ranging in diameter from 1 to 16 millimeters is furnished for insertion into the tester. They cover a wide permeability range. The selection of the proper orifice is by experience and by trial and error, and, of course, depends upon the air permeability characteristics of the fabric being tested. While most test methods stipulate a pressure of 0.5 inches of water across the fabric, the Frazier permeometer can measure air flow at pressures ranging from 0.1 to 1.0 inches. With special manometers the pressure drop across the fabric can be increased to about 20 inches for low permeability fabrics.

Textest Air Permeability Tester

A different type of tester employs a mufflered suction pump drawing air through an interchangeable test head with a circular opening. The test head appropriate for the selected test standard is mounted in the instrument. The specimen is clamped over the test head which starts the suction pump. The preselected test pressure is automatically set and maintained; after a few seconds the air permeability of the test specimen is digitally displayed. The instrument is equipped with an asynchronous data port allowing computer data acquisition and analysis.

TRI air permeability tester is also commercially available.

9.8 Water Repellency and Permeability Tests

Water resistance tests measure resistance to (1) surface wetting, (2) water absorption and internal wetting, and (3) water penetration and transmission. To determine the extent of surface wetting, the spray test is used. Water absorption is determined by the "drop penetration," "dynamic absorption" and "static immersion" tests. Water penetration and transmission are measured by the rain penetration, drop penetration, impact penetration, and hydrostatic pressure tests. All of these tests are described below.

Spray Test (AATCC 22)

A fixed quantity of water is sprayed on a 6-inch diameter circle of fabric mounted in an embroidery hoop at an angle of 45° with the horizontal. After spraying, tile fabric in the hoop is tapped smartly against a solid object to remove excess water. The surface condition is then graded.

Drop Penetration Test (AATCC 79)

A drop of water is allowed to fall from a fixed height onto the taut surface of a test specimen. The time required for the reflection of the water to disappear is measured and recorded as wetting time.

Dynamic Absorption Test (AATCC 70)

A preweighed specimen is tumbled in water for a fixed period of time and is then reweighed after the excess water has been removed by blotting and wringing through squeeze rolls. The percent weight increase is taken as a measure of the absorption or resistance to internal wetting.

Static (Immersion) Absorption Test

This method measures the amount of water absorbed when the fabric is immersed but not tumbled or otherwise agitated. Therefore, it is not as severe a test as the dynamic absorption method. The 3 × 3-inch test specimens are immersed so that their plane surfaces are vertical, presumably to permit the discharge of air bubbles. By means of hooked sinkers, the fabric swatches are kept under water for twenty minutes, following which they are blotted and sent through squeeze rolls. The percent weight increase is taken as a measure of absorption.

Rain Penetration Test (AATCC 35)

To simulate the effect of rain, a horizontal water spray of fixed and constant intensity is allowed to impinge against the vertically mounted test specimen. The water penetration during a five minute spray period is determined by measuring the increase in weight of a blotter placed behind the fabric.

Impact Penetration Test (AATCC 42)

With an apparatus similar to the spray test, 500 ml of water are sprayed from a height of two feet against the taut surface of the test specimen backed by a weighed blotter. After the test, the blotter is reweighed and the percent weight increase is recorded as the water transmitted.

Low Range Hydrostatic Pressure Test (AATCC 127)

Method 1—This test measures the resistance of fabrics to wear penetration under static pressure. It is suitable for testing heavy, closely woven fabrics that are expected to be used in contact with water, such as ducks and tarpaulins. The test is not suitable for proofed or coated fabrics. A test specimen of fixed area, mounted over a well of water, is subjected to water pressure that increases at a constant rate until leakage appears on the fabric surface. The pressure is increased by means of a hydrostatic bead which is raised at the rate of one centimeter per second. The height reached when leakage occurs is recorded as the measure of water resistance.

Method II—A modification to the above technique consists of applying a preselected hydrostatic head to the fabric and measuring the amount of water leakage in milliliters which passes through the fabric in ten minutes or other specified time.

High Range Hydrostatic Pressure Test for Coated Cloth

This method is useful for determining the resistance of coated cloth to the passage of water under high pressure. A modified Mullen-type diaphragm burst tester is used to make the test. The rubber diaphragm is removed and is replaced by the test specimen. The oil or glycerine commonly used is replaced with water. In conducting the test, the water level in the pressure chamber is brought flush with the surface of the lower clamp ring by adding water through the open orifice. The specimen is then clamped into place and pressure is applied until leakage occurs. The pressure, in psi, at the first sign of water leakage through the fabric is recorded.

9.9 Abrasion Resistance Testers

Abrasion is wearing away of material through some type of frictional interaction between materials. This may occur between fibers, yarns, fabrics or between textile materials and other surfaces. The nature of the expected wear a particular textile product would be exposed to should dictate the type of abrsion test that is most appropriate. A description of seven widely used laboratory abrasion testers are described below.

Inflated Diaphragm Test (ASTM D 3886)

This method is intended for determining the abrasion resistance of woven and knitted fabrics when the specimen is mounted over an inflated rubber diaphragm and is then rubbed either unidirectionally or multi-directionally by an emery cloth-type abradant under controlled pressure conditions. The abradant is mounted upon a plate in a manner so that it can rest upon the specimen under a selected load of 0 to 5 pounds. The abradant moves back and forth unidirectionally over the specimen surface. The specimen remains in a fixed orientation or can be continuously rotated, depending upon whether unidirectional or multi-directional abrasion is desired.

Flexing and Abrasion Method (ASTM D 3885)

This method is used for determining the resistance of woven fabrics to flexing and abrasion when the specimen is subjected to unidirectional reciprocal folding and rubbing over a bar having specified characteristics under known conditions of pressure and tension. The same instrument as described in the inflated diaphragm test is used but with necessary modifications and attachments. A specimen 8 inches long and 1.25 inches wide, raveled to 1 inch, is looped around either a hard carbide bar or a blade which acts as the abradant. The bar or blade is attached to a pulley loaded with weights so a constant tension can be applied to the test specimen. An upper movable reciprocating plate contains a jaw for clamping the top edge of the specimen. A lower fixed plate also contains a jaw for clamping the lower edge of the specimen. The upper plate, which can be loaded in order to apply a specified pressure to the specimen, moves back and forth, thus causing the specimen to be both abraded and flexed over the edge of the blade. The greater the headweight on the upper plate, and the greater the tension applied by the bar, the more severe will be the abrading action. The bars and blades are edged with smooth synthetic sapphire, silicon carbide, or other highly wear-resisting surfaces, and must be precisely ground so that the abradant will remain constant for long time periods. Bars and blades are calibrated by the manufacturer but should be rechecked regularly against a standard cloth which should be set aside by each laboratory for calibration purposes.

Oscillatory Cylinder Method (ASTM D 4157)

This method is used for determining abrasion resistance when the specimen is subjected to unidirectional rubbing action under known conditions of pressure, tension and abrasive action. The instrument is composed of a 6-inch diameter oscillating cylinder upon which is mounted a sheet of emery cloth or other abrasive material. The specimens, each 9 × 1.875 inches, are mounted between clamps under a selected tension and are placed against the cylinder which oscillates through an arc 3 inches long at the rate of 90 cycles (double rubs) per minute. The unidirectional abrasive action is parallel with the long direction of the specimens, which are also subjected to mild flexing action during the test.

Rotary Platform Double Head Method (ASTM D 3884)

This method is used in determining abrasion

resistance when the specimen is subjected to rotary rubbing action under controlled conditions of pressure and abrasive action. The tester consists of a 4-inch diameter record player-type turntable which rotates in a horizontal plane, and upon which the test specimen is clamped. Above the turntable there are mounted two separate halves of a weighted frame, each containing a rubber emery composition abrading wheel similar in dimensions to a small grinding wheel. These are free to rotate in a vertical plane. At the start of the test, the wheels are lowered so they rest on the turntable and thus on the fabric. The weight of the frame can be varied, and in this manner the pressure on the sample is controlled. The turntable is driven; the abrasive wheels are not. The lines of abrasion are in the form of two arcs which crisscross each other, forming a circular track on the specimen. The specimen is not flexed during abrasion. Figure 20.21 shows single and double head abrasers.

Uniform Abrasion Method (ASTM D 4158)

This instrument multi-directionally abrades a flat surface of fabric (Figure 20.22). The essential parts of the tester consist of two flat parallel plates that rotate horizontally and can be placed in contact with each other. The lower plate supports the specimen, the upper the abradant. The plates rotate in the same direction at very nearly the same angular velocity (250 rpm) on two parallel, but not coaxial shafts. The abradant is sufficiently larger than the specimen so that all points on the specimen's surface are abraded equally and uniformly in all directions.

The Accelerotor (AATCC 93)

This instrument, developed by the American Association of Textile Chemists and Colorists, flex abrades specimens in a completely random manner. An unfettered fabric specimen is driven by a rotor in a random path so that it repeatedly impinges the walls and abradant liner of the abrading chamber. The zigzag orbit taken by the specimen is generally circular while it is being continuously subjected to extremely rapid, high-velocity impacts. The specimen is sub-

FIGURE 20.21 Single and double head abrasers (courtesy of Taber).

FIGURE 20.22 Frazier precision Schiefer uniform abrasion instrument.

jected to flexing, rubbing, shock, compression, stretching and other forces. Frictional abrasion is produced by the rubbing of surface against surface, and surface against abradant. The results of the abrasive action may be evaluated by visual comparisons, determination of strength loss or by determination of weight loss. Usually, a new fabric subjected to the Accelerotor's action will look well-worn and threadbare after a relatively short time, for example, 2 to 5 minutes.

Martindale Abrasion Tester (ASTM D 4966)

This instrument creates the abrasive wear by subjecting the specimen to rubbing motion in the form of a geometric (Lissajous) figure. Resistance to abrasion may be evaluated visually after a certain number of rubs, until an end point is reached, i.e., the presence of two broken threads; or the objective way is to remove the specimens at intervals and weigh them, so as to determine the rate of weight loss. The Martindale tester is shown in Figure 20.23.

Special Webbing Abrader

The webbing industry commonly uses a flex abrasion tester which is schematically shown in Figure 20.24. A 16-inch diameter drum oscillates back and forth pulling a weighted 30-inch

length of webbing over a 1/4-inch maximum diameter rigid hexagonal steel bar of specified hardness. A 13-inch traverse of webbing is abraded over the bar at a rate of 60 cycles per minute. The dead load tension applied to the webbing during the test is a function of its original strength. Visual appearance or strength loss after a selected number of cycles are normally used as abrasion resistance criteria. If it is desirable to study the effect of flexing without abrasion, a smooth, freely rotating rod can be substituted for the rigid hexagonal bar. A new abrasion tester developed by the University of Georgia is also becoming popular.

9.10 Pilling Tests

This test is used mostly on apparel fabrics. One result of abrasion on textile fabrics, particularly wool knit fabrics and synthetics such as nylon and polyester, is the formation of "pills." These are small fuzzy balls of fiber which form on the fabric surface. Pilling usually develops on soft, low-twist spun yarn fabrics. The mild abrasion causes protruding fiber ends to be pulled out from the fabric surface. If the fibers are brittle, for example glass or cotton, they break away from the fabric and no pills form. If the fibers are extensible and tough, they roll up on themselves, tangle with adjacent fibers, and eventually make little balls which remain attached to the fabric. Several mild abrasion tech-

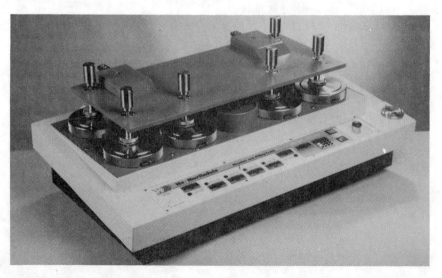

FIGURE 20.23 Nu-Martindale abrasion and pilling tester (courtesy of James H. Heal & Co.).

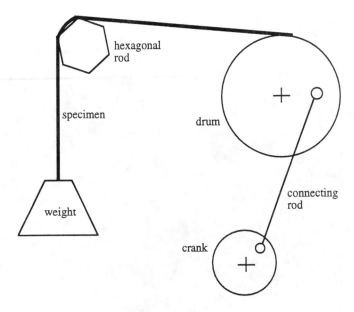

FIGURE 20.24 Webbing abrasion tester.

niques are used to evaluate pilling (Figure 20.25).

Brush Pilling Tester (ASTM D 3511)

Fabrics are brushed to form free fiber ends. Two of the specimens are then rubbed together in a circular motion. This rubbing action produces the pills. The wear as evidenced by pills is evaluated by comparison to photograph standards showing different degrees of pilling.

Random Tumble Pilling Tester (ASTM D 3512)

The wear conditions that cause pilling are simulated by tumbling the specimens in a cylindrical test chamber lined with a mildly abrasive material. The pills formed are produced from short cotton fiber added to simulate the presence of broken fibers that may be present in actual wear situations.

Elastomeric Pad Method (ASTM D 3514)

The Stoll Quartermaster Universal Wear Tester with the Frosting attachment is used for this method. The fabric specimens are washed or dry-cleaned and are then subjected to con-

trolled rubbing against an elastomeric pad possessing specific mechanical properties. The evaluation of resulting pills is accomplished by comparison with photographic standards.

Martindale Pressure Tester

This method evaluates the tendency of textile fabrics to form pills under light pressure. Fabrics are in the tester in a manner such that the face of the specimen is rubbed against another portion of the same specimen. The rubbing is done in the form of a geometric figure in an ever-widening ellipse. The pilling is evaluated against visual standards, either actual fabrics or photographs of pilled fabrics.

9.11 Fabric Stiffness Tests

The following test methods are used for fabric stiffness.

Cantilever Test

A 1 × 6-inch specimen is extended out from the horizontal until an angle of 41.5° is achieved (see Figure 19.11). One-half of the resulting overhanging length is the bending length (ASTM D 1388). The stiffness or flexural rigid-

FIGURE 20.25 ICI pilling tester (courtesy of James H. Heal & Co.).

ity is obtained by multiplying the cube of the bending length by the fabric weight per unit area.

Heart Loop Tester

The heart loop test for fabric stiffness is described in ASTM Method D 1388. No commercial tester is available or necessary.

Circular Bend

This method (ASTM D 4032) uses a metal plate with specified diameter orifice through which a fabric specimen is pushed with a plunger and the required force is recorded. This procedure gives a force value related to fabric stiffness, simultaneously averaging stiffness in all directions. Figure 20.26 shows a fabric stiffness tester based on ASTM D 4032.

9.12 Drape Testers

In addition to measuring fabric stiffness in either the warp or the filling direction, or at some specific angle to the warp or filling, certain instruments are used to measure the "draping" characteristics of fabrics. Here the interacting effects of warp and filling yarns on fabric stiffness in all directions are plotted.

9.13 Dimensional Change and Shrinkage Tests

A number of standard tests exist for the determination of the change in fabric dimension when fabrics are subjected to heat or refurbishment (Table 20.8). Tests are also available for fiber and yarn.

The methods listed in Table 20.9 are used to determine the dimensional changes which occur

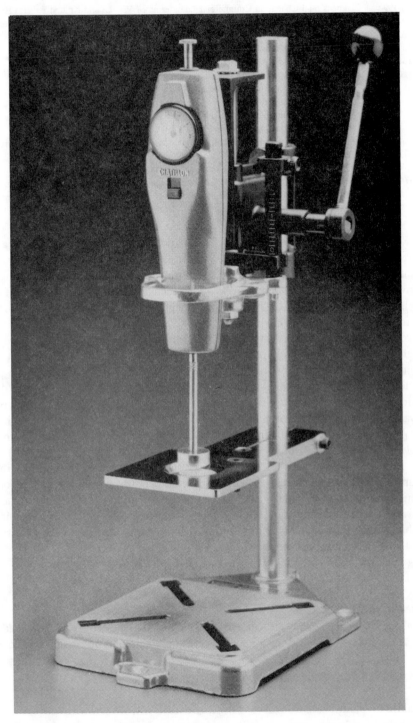

FIGURE 20.26 Manual fabric stiffness tester (courtesy of J. A. King & Company, Inc.).

TABLE 20.8 Standard Test Methods for Dimensional Change.

AATCC 96-1992	Dimensional Change in Commercial Laundering of Woven and Knitted Fabrics Except Wool
AATCC 99-1993	Dimensional Changes of Woven and Knitted Textiles: Relaxation, Consolidation and Felting
AATCC 135-1992	Dimensional Changes in Automatic Home Laundering of Woven or Knit Fabrics
AATCC 150-1992	Dimensional Changes in Automatic Home Laundering of Garments
AATCC 158-1990	Dimensional Changes on Dry-Cleaning in Perchloroethylene: Machine Method
AATCC 160-1992	Dimensional Restoration of Knitted and Woven Fabrics after Laundering
ASTM 204	Methods of Testing Sewing Threads
ASTM 461	Test Methods for Felt
ASTM 885	Methods of Testing Tire Cords, Tire Cord Fabrics, and Industrial Filament Yarns Made from Man-Made Organic-Base Fibers
ASTM 1284	Relaxation and Consolidation Dimensional Changes of Stabilized Knit Wool Fabrics
ASTM 2102	Test Method for Shrinkage of Textile Fibers
ASTM 2259	Shrinkage of Yarns in Boiling Water or Dry Heat
ASTM 4974	Thermal Shrinkage of Yarn and Cord Using the Testrite Thermal Shrinkage Oven
ASTM 5104	Shrinkage of Textile Fibers (Single Fiber Test)

in fabrics when washed in commercial laundries, as well as in the home. Because wool fabrics can undergo felting shrinkage, they are excluded from these tests. Five washing procedures ranging from mild hand washing to severe commercial laundering, five drying procedures encompassing methods used in homes and commercial laundries, and three methods for determining the dimensional restorability characteristics, are listed for those textiles which require ironing or wearing after laundering. The washing, drying and restoration procedures may be used in various combinations. For example, Test III-A-3 refers to a fabric which has been washed at 145°F for 45 minutes, tumble dried and subjected to hand ironing. This type of code is widely used to describe various wash test combinations. Selection of washing temperatures, and type of soap or detergent, depends upon the nature of the fabric being evaluated.

The test fabric is washed, as part of a full load, in a cylinder reversing wash wheel, dried and subjected to restorative forces. Normally, specimens about 22 inches square are cut and precisely marked with 18 or 20-inch gauge

marks. These are measured before and after the washing, drying and restoration treatment. The shrinkage is calculated as the change in length in each direction, as a percentage of the original length.

The tension presser is used when, upon flat-bed pressing, the shrinkage exceeds 2% in either direction. The fabric is then rewet, subjected to tensions by means of the pulleys and weights, and dried by means of a drying plate and iron. The gauge marks are then remeasured.

Other methods for determining the shrinkage of synthetic fabrics consist of immersing a 20 × 20-inch marked sample (1) in boiling water for 15 minutes, drying it flat in an oven at a temperature not above 212°F; or (2) in an autoclave under stream pressure at 290°F for 15 minutes. The percent changes in warp and filling lengths are calculated as shrinkages.

9.14 Thermal Property Tests

Transmittance Test

The ASTM Method D 1518 determines the overall thermal transmission coefficient of a test

TABLE 20.9 Washing, Drying and Restoration Procedures (AATCC 96).

Washing Procedure	Drying Procedure	Restoration Procedure
I. 105 ± 5°F for 30 min.	A. Tumble dry	1. Tension presser
II. 125 ± 5°F for 45 min.	B. Line dry	2. Knit shrinkage gauge
III. 145 ± 5°F for 45 min.	C. Drip dry	3. Hand iron
IV. 165 ± 5°F for 60 min.	D. Screen dry	
V. 207 ± 5°F for 60 min.	E. Flat-bed press	

specimen which results from the combined effects of heat conduction, convection and radiation. The time rate of heat transfer from a dry, constant temperature flat plate positioned horizontally, up through a layer of the test fabric to a calm, cool atmosphere, is measured by means of thermocouples. The test method is restricted to fabrics and battings whose thermal conductance (C) lies within a range of 0.125 to 2.5 BTU per hour per square foot of fabric per degree Fahrenheit, as defined and used by ASTM, the American Society of Heating and Air Conditioning Engineers, and the American Society of Refrigeration Engineers.

Fire and Flame Resistance Tests

There are numerous Standard Test methods that deal with fire and flammability. A compilation of over one hundred ASTM standards is given in the book *Fire Test Standards* published by the ASTM. The textile standards listed by the ASTM are shown in Table 20.10.

Vertical Flammability Test

This test is useful in evaluating the fire resistance of a wide variety of fabrics including industrial fabrics such as duck, tentage, awnings, canvas theater scenery and treated textile fabrics used for decorative or other purposes on the inside of buildings. A 3 × 12-inch test specimen, vertically mounted on a frame in the test chamber, is subjected to a standard burner flame for 12 seconds in such a manner that the lower end of the specimen is 0.75 inch above the top of the gas burner. The flame is then removed and the

specimen is evaluated for afterflaming, afterglowing and length of char. This latter is measured by inserting small hooks in the specimen, one on each side of the charred area, 0.25 inch from the adjacent outside edge and 0.25 inch from the lower end of the specimen. Weights are attached to the hook equal to about 10% of the equivalent strength of the unburned cloth. This causes the charred portion to be torn or broken away. The length of char is the distance from the end of the specimen that was exposed to the flame, to the end of the tear made lengthwise on the specimen through the center of the charred area. The acceptable levels of flame resistance depend on the end use of the fabric being tested. Various government and industry associations have different criteria for meeting their specifications or minimum performance levels (Fed. Std. 191A 5903).

Flammability of Clothing Textiles

This test is the one specified in the Flammable Fabrics Act under Commercial Standard CS 191-53 and later redesignated under the auspices of the Consumer Product Safety Commission (CPSC) as 16 CFR 1610 and is similar to ASTM D 1230. The test is designed to indicate textiles which ignite easily, and once ignited, burn with sufficient intensity and rapidity to be hazardous when worn. For purposes of complying with the CPSC the fabrics are placed in classes based on their test results.

Specimens are dried in an oven at 105°C for 30 minutes before testing, in order to put them in their most flammable condition. They are cooled in a desiccator and then tested in the

TABLE 20.10 ASTM Textile Fire and Flammability Standards.

ASTM D 461	Test Method for Felt
ASTM D 1230	Test Method for Flammability of Apparel Textiles
ASTM D 2859	Test Method for Flammability of Finished Textile Floor Covering Materials
ASTM D 3523	Test Method for Spontaneous Heating Values Liquids and Solids
ASTM D 3659	Test Method for Flammability of Apparel Fabrics by Semi-Restraint Method
ASTM D 4108	Test Method for Thermal Protective Performance of Materials for Clothing by Open Flame Method
ASTM D 4151	Test Method for Flammability of Blankets
ASTM D 4372	Specifications for Flame-Resistant Materials Used in Camping Tentage
ASTM D 4391	Terminology Relating to the Burning Behavior of Textiles
ASTM D 4723	Index of and Descriptions of Textile Heat and Flammability Test Methods and Performance Specifications

TABLE 20.11 Flammability Standards Consumer Product Safety Commission (courtesy of CPSC).

16 CFR 1610	Standard for the Flammability of Clothing Textiles (General Wearing Apparel)
16 CFR 1611	Standard for the Flammability of Vinyl Plastic Films (General Wearing Apparel)
16 CFR 1615	Standard for the Flammability of Children's Sleepwear: Sizes 0 through 6X (FF 3-71)
16 CFR 1616	Standard for the Flammability of Children's Sleepwear: Sizes 7 through 14 (FF 5-74)
16 CFR 1630	Standard for the Surface Flammability of Carpets and Rugs (FF 1-70)
16 CFR 1631	Standard for the Surface Flammability of Small Carpets and Rugs (FF 2-70)
16 CFR 1632	Standard for the Flammability of Mattresses (and Mattress Pads) (FF 4-27)
16 CFR 1633	DRAFT Proposed Standard for the Flammability (Cigarette Ignition Resistance) of Upholstered Furniture (PFF 6-81)

"flammability tester," an instrument consisting of a draft-proof ventilated chamber enclosing a standard ignition medium, a specimen holder, and an automatic timing device. The specimen is held at an angle of 45°, a standardized flame is applied to the surface near the lower end for 1 second, and the time required for the flame to travel up the fabric a distance of 5 inches is recorded. A No. 50 mercerized cotton sewing thread is attached at the 5-inch mark of the specimen and thence to a stop-watch mechanism. When the 5-inch point on the specimen is reached, the thread burns away and the stop-watch shuts off, thus precisely recording the time.

Other Flammability Standards

Other flame or fire resistance tests and standards are listed in Tables 20.11 and 20.12 and also in ASTM D 4723, an index of related test methods specified in Fed. Std. 191A, CPSC and various government and industry organizations.

9.15 Outdoor Weathering Tests

It is important to determine a textile fabric's response to the natural climatic environment. The primary elements that contribute to a material's deterioration, the end result of weathering, are: solar radiation (particularly the ultra-violet wavelength), moisture (as dew, rain or humidity), temperature (primarily the time-averaged temperature of the exposed surface) and wind. The most common method of assessing weathering durability is to expose the specimen to the natural elements of weather. The objective of outdoor real-time exposure testing is the determination of the weathering performance of materials for one or more of the following reasons:

- to provide data regarding the prediction of the influence of weather on a material's properties
- as a quality control method
- to ascertain the effective lifetime of a material

TABLE 20.12 Flammability Standards Miscellaneous Trade and Industrial Associations (courtesy of UFAC, BIFMA, and IFAI).

X.1 CPAI-84	Specification for Flame-Resistant Camping Tentage Materials
X.2 NFPA 701	Standard Methods of Fire Tests for Flame-Resistant Textiles and Films (as Applied to Tents, Tarpaulin, and Other Protective Coverings)
UFAC	Six Individual Tests:
	Fabric Classification Test Method
	Interior Fabrics Test Method
	Barrier Test Method
	Filling/Padding Component Test Method
	Welt Core Test Method
	Decking Materials Test Method (UFAC)
BIFMA F-1-78	First Generation Voluntary Upholstery Furniture
	Flammability Standard for Business and Institutional Markets: A. Small Flame Ignition; B. Cigarette Ignition

- to ascertain the stability of the material as compared to a control material

Outdoor weathering consists of several different types of exposures, with the most common being Direct Weathering and Indirect (Underglass) Weathering. Direct weathering means the material is exposed to direct sunlight and other elements of weather throughout the exposure term. This is accomplished by mounting the material on an exposure rack at a fixed angle according to the site orientation and the end-use of the material. The most common exposure angles are 45°, 5°, 90° to the horizontal, or the latitude tilt angle of the exposure site (for example, 34°S for Phoenix, Arizona and 26°S for southern Florida). In the northern hemisphere, the exposure rack should be facing south. Exposure sites are typically selected for the type of material effect exerted. Both Phoenix, Arizona and southern Florida are widely utilized testing sites because of their climates. Phoenix possesses high annual UV radiation and ambient temepratures with low humidity. Southern Florida possesses high annual UV radiation with high rainfall and humidity. These two areas have become United States and international reference climates for gauging the durability of materials. Other sites may include northern coastal or inland areas, or polluted industrial areas, if that is the material's end-use outdoor environment.

Indirect weathering is the exposure of materials under glass. This method of testing materials is used to determine the durability characteristics of materials destined for use inside the household (i.e., drapes, carpeting, upholstery, etc.) or in the interior of an automobile. This method was developed by AATCC for testing textiles and subsequently adopted by the automotive industry.

During exposure, the measurement of various environmental parameters such as total and UV radiation and temperature is important. The pyrheliometer is useful for measuring radiation within a narrow field of view for the purpose of measuring the direct-beam component of solar radiation. Direct-beam radiation is that which is received directly from the sun without scattering. The Eppley model pyrheliometer, using de-

tectors and thermocouples, measures the direct-beam irradiance incident upon a specimen exposed outdoors. Irradiance is typically measured in a power per area unit, usually Watt/m². Radiation is measured in MJ/m². Historically, accumulated radiation is measured in Langleys (g cal/cm²). To convert from MJ to Langleys, the formula used is $MJ/M^2 = Langleys \times 0.04184$. The accumulated UV and/or total radiant dosage is often correlated to the degradation of a material, as measured by physical property changes such as color changes, breaking strength and elongation, tear strength, etc. Because outdoor conditions are not standardizable, it is sometimes useful during an exposure test of a new material to include a control material with known performance for comparison purposes. Table 20.13 lists the standard weathering tests.

9.16 Accelerated Aging Tests

In addition to ourdoor weathering exposures, laboratory accelerated aging tests of many different types are widely used. They are rapid, reproducible, usually quantitative, and conveniently carried out in the laboratory. Their major disadvantage is the risk that they may not completely and properly correlate with service performance. Therefore, accelerated aging tests, as is the case with all laboratory tests, must be designed, executed, and evaluated with cautious intelligence by the experimenter.

Artificial Accelerated Weathering Chambers

Artificial weathering is the exposure of materials to laboratory controlled conditions of light, heat and moisture. The goal is to achieve accelerated degradation using radiation that is equal to or more intense than natural solar radiation. The spectrum produced by an artificial light source is extremely important in assuring that results correlate to real-time outdoor exposures. Since the development of artificial tests in the early 1900s, many types of artificial light sources have been introduced. The three most common are the carbon arc, fluores-

TABLE 20.13 Weathering Tests.

ASTM G 7	Recommended Practice for Atmospheric Environmental Exposure Testing of Nonmetallic Materials
ASTM G 23	Practice for Operating Light- and Water-Exposure Apparatus (Carbon-Arc-Type) with or without Water for Exposure of Non-metallic Materials
ASTM G 24	Standard Practice for Conducting Exposures to Daylight Filtered through Glass
ASTM G 26	Standard Practice for Operating Light-Exposure Apparatus (Xenon-arc-type) with and without Water for Exposure of Non-metallic Materials
ASTM G 53	Recommended Practice for Operating Light- and Water-Exposure Apparatus (Fluorescent UV-Condensation Type) for Exposure of Non-metallic Materials
ASTM D 4909	Color Stability of Vinyl-Coated Glass Textiles to Accelerated Weathering
AATCC 16	Colorfastness to Light
AATCC 111	General Information
AATCC 111A	Sunshine Arc Lamp Exposure with Wetting
AATCC 111C	Sunshine Arc Lamp Exposure without Wetting
AATCC 111B	Exposure to Natural Light and Weather
AATCC 111D	Exposure to Natural Light and Weather through Glass
AATCC 169	Weather Resistance of Textiles: Xenon Lamp Exposure
AATCC 177	Colorfastness to Light at Elevated Temperature and Humidity
SAE J2212	Accelerated Exposure to Automotive Interior Trim Components Using a Controlled Irradiance Air-Cooled Xenon-Arc Apparatus

cent, and xenon-arc lamps. In the past ten years, xenon-arc lamps have gained favor in the textile industry because of their superior spectral match to sunlight throughout the solar spectrum, and especially in the 300–400 nm UV range, the wavelength range which exerts the most deleterious effect upon materials.

The weathering chamber includes one to three lamps which emit elevated levels of radiation to the samples. Typical weathering chambers also can control chamber temperature and can periodically spray samples with water or create a high-humidity atmosphere in the chamber to simulate outdoor moisture conditions the material might encounter. Samples are vertically mounted in holders on a cylindrical rack which rotates around the central light source for a specified test duration, typically measured in accumulated Langleys or in accumulated MJ/m² UV. Weathering chambers also are able to be programmed for light and dark phases, again to simulate the outdoor environment to which a material is actually exposed. Specific testing standards published by various organizations such as ASTM, AATCC and SAE give specific instructions on what test cycle should be run for specific materials. As with outdoor real-time exposure testing, physical property changes that are typically assessed include color change, breaking or tear strength loss, flexibility, etc.

Accelerated Aging by Heat

Heat aging at elevated temperatures with circulating air is a useful method for evaluating treated and coated cloths. A circulating air oven, thermostatically controlled and of large capacity so that many specimens may be hung without touching each other, is all that is necessary. The temperature and time of exposure depends upon the nature of the material being tested. For example, impregnated fire-weather-water-resistant duck is aged at 200°F for five days and then tested for flexibility change (Fed. Std. 191A 5850).

Accelerated Aging by Oxygen Bonds Methods

An accelerated oxygen aging test is useful in assessing the resistance of many organic materials to oxidation degradation. The test apparatus, popularly called an "oxygen bomb," is a vessel capable of maintaining and withstanding an internal oxygen gas pressure of 300 psi for ten days at 160°F (Fed. Std. 191A 5852). The test specimens are inserted into the preheated chamber, which is then securely closed and fastened. Oxygen is introduced until the desired pressure is reached. The selected pressure and temperature must be maintained for the duration of the test. At its conclusion, the oxygen is released and the specimens are removed for examination and testing. Because of the danger of fire and explosion, the oxygen bomb test should be carried out with extreme caution.

9.17 Color Measurements and the Spectrophotometer

There are two basic systems for identifying and defining colors. One is the subjective method wherein the observer makes use of a previously assembled comprehensive collection of standard color samples and visually matches his "unknown" specimen against the nearest known standard. He depends upon his eye to judge and decide when a match is achieved. The second is an objective method wherein the color is quantitatively determined and characterized by certain physical measurements based upon the wavelength of light. The instrument which makes these measurements is called a spectrophotometer. Many textile mills use spectrophotometers together with subjective methods for matching against a "standard for shade" submitted for duplication.

The two best known systems for subjective matchings are the Munsell system and Pantone Professional Color System. These and all other subjective visual classification systems are based upon three criteria of color to which the human eye responds:

- Hue indicates the color itself, e.g., red, orange, blue, etc.
- Value or lightness is the lightness or darkness of a color and depends upon its black or white content.
- Chroma or saturation is the strength or intensity of the "pure" color.

The disadvantages of visual color classification systems are that they are qualitative and based upon the personal reaction of the observer. Objective quantitative reproducible

systems are usually preferable. Spectrophotometric systems are based upon the fact that white light is composed of a visible spectrum of colors of continuously changing wavelength ranging from 400 millimicrons (1 mμ = 1/25,400,000 inches) for violet, to 700 millimicrons for red. The subdivision of light into the six standard colors as shown in Table 20.14 is one of convenient identity. The color of the spectrum varies continuously throughout its length, the physical difference being the continuous change in wavelength.

The spectrophotometer is an instrument which can produce light of any desired wavelength (actually a narrow band of wavelengths) within the 400–700 millimicron range. The colored specimen to be measured is viewed under these progressively changing wavelengths (colors) and the percent light reflected back is plotted. For example, a "green" fabric will exhibit high reflectance when viewed with a green light (500–570 millimicrons), but low reflectance when viewed in violet, orange and red light.

The depth of shade influences the percent reflectance, and so governs the position of the curve along the reflectance axis. A perfect white specimen, as exemplified by a freshly scraped surface of calcium carbonate, will show 100% reflectance, and a perfect black will show 0% reflectance at all wavelengths. Therefore, the extent of sample grayness is also measurable from the spectrophotometric curve.

The percent transmission of clear dyestuff solutions can be determined by utilizing an optical chamber with absolutely parallel sides into which the dyestuff solution is poured. A graph of wavelength versus light transmitted through

TABLE 20.14 Wavelength Bands for Visible Colors.

Color	Wavelength (millimicrons)
Violet	400–450
Blue	450–500
Green	500–570
Yellow	570–590
Orange	590–610
Red	610–700

TABLE 20.15 Standard Test Methods Relating to Color.

AATCC EP1	Gray Scale for Color Change
AATCC EP2	Gray Scale for Staining
AATCC EP3	Chromatic Transference Scale
AATCC 8-1989	Crocking: AATCC Crockmeter Method
AATCC 16-1993	Light
AATCC 23	Burnt Gas Fumes
AATCC 61-1993	Laundering, Home and Commercial Accelerated
AATCC 116-1989	Crocking: Rotary Vertical Crockmeter Method
AATCC 125-1991	Water and Light Alternate Exposure
AATCC 126-1991	Water (High Humidity) and Light: Alternate Exposure
AATCC 148-1989	Light Blocking Effect of Curtain Materials
AATCC 164-1992	Oxides of Nitrogen in the Atmosphere under High Humidities
AATCC 177-1993	Light at Elevated Temperatures and Humidity: Water Cooled Xenon Lamp Apparatus

the solution is plotted. The concentration and thickness of a dye solution is related to its transmission according to Beer's law:

$$I = I_0 e^{-\alpha c x} \qquad (20.14)$$

where

I = intensity of light transmitted
I_0 = intensity of incident light
e = natural logarithm base
c = concentration of dyestuff
x = thickness of solution
α = absorption coefficient

One use of color measure is for fibers which is explained in ASTM D 2253 Color of Raw Cotton using the Nickerson-Hunter Cotton Colorimeter.

Colorfastness Tests

A number of standard test methods deal with the fate of colored textiles when exposed to different use conditions. A number of these methods are given in Table 20.15.

Colorfastness to Light:
the Fade-ometer

The AATCC developed test methods and grading scales for evaluating colorfastness. Fastness to light is one of the most important properties.

A test specimen and a "standard" dyeing of known light colorfastness are exposed simulta-

neously to the intense light from a calibrated light source for a time sufficient to produce "just appreciable fading" in the standard or the test specimen. Colorfastness is rated in terms of the number of AATCC Fading Units (AFU) of exposure required to produce this amount of fading. The common practice is to mount a 2.75 × 10.5-inch fabric specimen in a frame especially prepared for insertion into the Fade-ometer. The specimen is held and covered by a protective frame in which there is a rectangular opening 1.75 × 5 inches. The arc light impinges on that part of the specimen which is exposed in the frame opening. The surrounding area is protected and does not change color during the test. The specimen is examined periodically in order to note when fading develops.

While formerly the Fade-ometer and Weather-ometer were entirely different instruments, they were later combined using the same cabinet for both testers. The ultraviolet carbon arc or xenon lamp components are interchangeable so that one cabinet is useful for conducting any desired Fade-ometer or Weather-ometer test. The Weather-ometer is also used for determining the effect of a carbon arc light and water spray on colorfastness (Fed. Std. 191A 5660 and 5671).

In addition to these two accelerated methods, lightfastness may be assessed by natural daylight and by sunlight exposures. In the daylight method (AATCC 16; Fed. Std. 191A 5662.1), a specimen, together with the standard is exposed to daylight continuously under prescribed conditions while being protected from the rain. The fastness is assessed by the color change of the test specimen with that of the standard. The sunlight test is the original AATCC lightfastness test (AATCC 16B). The test specimen and blue wool lightfastness standards are exposed between 9:00 A.M. and 3:00 P.M. when the sun is at its highest intensity. Comparison of the test specimen and the standard is made directly.

Colorfastness to Washing: the Launder-ometer

The colorfastness to laundering test consists of laundering the test specimen which is attached to a standard white test cloth and observing the color change in both the test fabric and the standard. Two 2 × 4-inch test specimens are sewn against (1) a swatch of white cotton cloth and (2) a swatch of white "multi-fiber test cloth" containing yarns of several different fiber types: acetate, cotton, nylon, silk, viscose, and wool. The faces of the specimen and each test cloth are thus in contact with each other, affording the maximum opportunity for the dye to bleed and transfer from one fabric to the other. Each swatch assembly is inserted into a stainless steel can containing several 0.25-inch diameter stainless steel balls for agitation. Depending upon test requirements, a selected washing formulation of soap, bleach, detergent, etc., is added. The can with its contents is inserted into the rotating rack of a constant temperature bath, where it is rotated for a selected period of time. The swatch assembly is then removed, and examined for color change in the specimen and staining of the white test standards by means of special color charts. The can-steel ball-rotating rack-constant temperature system is called a Launder-ometer. It enjoys wide use in evaluating the wet colorfastness and dry cleaner's solvent colorfastness properties of fabrics. The Launder-ometer is also used to evaluate the permanence to repeated laundering of textile finishes such as water repellents, softeners, wrinkle resisting finishes, etc.

Gas Fading, Colorfastness to Oxides of Nitrogen in the Atmosphere

A serious colorfastness problem in the past, and one which, in part, still prevails is that of "gas fading" of dyed acetate and other synthetics. These dyes fade or change color when exposed to oxides of nitrogen produced in industrial gases and smoke in urban areas. Gas fading can also occur in some sensitive resin bonded pigments.

The laboratory test for evaluating gas fading consists of exposing small swatches of the test specimen, together with a dyed control sample, to oxides of nitrogen fumes until the control fades a prescribed standard amount. The test specimen is then examined to ascertain if fading has occurred. If it has not, the cycle is repeated until fading is obvious. The specimen is then ranked from Class 1 (poor) to Class 5 (excellent) depending upon the number of exposures required to produce an appreciable shade change. The test apparatus consists of a laboratory, oven-type, thermostatically controlled chamber containing a slowly rotating fan. Conventional illuminating gas is burned in a burner, and the combustion products conveyed into the exposure chamber. In this test, an exposure cycle to cause fading of the standard may take 5 to 10 hours. This time can be reduced in an "accelerated" test by introducing nitrous oxide into the gas flame, thus increasing its concentration in the chamber. By this technique a test cycle can be completed in about 1 hour. This accelerated method should be carried out with care. The test should not be so severe as to cause specimens to fade quickly, for then no distinction among samples can be made.

Colorfastness to Crocking

The crocking test determines the extent to which color may be transferred from the surface of a dyed fabric to another surface by rubbing. The test specimen is fastened to the base of a Crockmeter, and is rubbed under a controlled pressure by a standard white cotton print cloth for twenty rubbing cycles. The white standard cloth is then examined for the presence of color. The test may be conducted dry or wet. The extent of crocking is classified according to an AATCC color chart based upon Munsell color chips, with rankings ranging from five for no color transfer to one for large color transfer or the gray scale for staining.

Colorfastness to Other Conditions

There are many other colorfastness requirements which textiles may be called upon to meet. Many of these involve apparel and decorative fabric needs, rather than industrial textiles. These tests can be found in the AATCC manuals.

9.18 Electrical Resistivity

The electrical resistance of a material is defined by Ohm's Law as the ratio between the voltage impressed across the material and the resultant current flow through it. This observed resistance to current flow is a function of the intrinsic material properties, the geometrical shape of the material, and the size, shape, and location of the measuring electrodes.

For example, consider two electrodes placed upon the surface of a fabric specimen and a voltage applied between them. Because the electrodes span the full width of the fabric sample, the current flow along the fabric is in parallel lines as shown in Figure 20.27 by the arrows. From the values of current and voltage indicated by the two meters the effective electrical resistance applicable to this situation can be computed.

Now, if we use the same electrode configuration and electrical system on a larger piece of the same fabric, the current flow will not be along straight lines between the two conductors but will have a very complicated electrical field pattern. In this case, the current flow is appreciably higher and the effective surface resistance lower than that observed previously, although there has been no change in the fabric material properties. Therefore in any resistance measurement the test conditions must be considered in the determination of meaningful values. To do this, it is necessary to establish the quantitative relationship which exists between the geometry of the testing device, the measured resistance, and the specific resistance of the material under test.

Two types of electrical resistance can be defined, one of which relates to the resistance to current flow through the volume of bulk of the material, called volume resistivity ϱ; the other is the resistance to current flow along the surface of a piece of material and is called surface resistivity σ.

Volume Resistivity

Mathematically, volume resistivity is defined as

$$\varrho = \frac{R_v A}{L} \text{ ohm-centimeters}$$

where R_v is the volume resistance in ohms, L is the average thickness of the material, A is the ef-

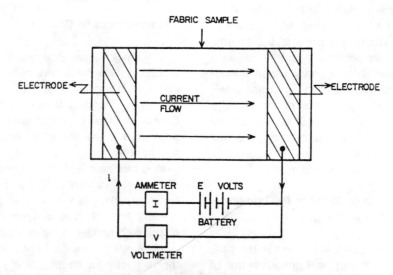

FIGURE 20.27 Idealized surface resistivity.

fective electrode area and ϱ is the resistivity of the material under test.

By Ohm's Law

$$R_v = \frac{E \text{ volts}}{I \text{ amps}}$$

Hence the resistivity of the material may be computed.

$$\varrho = \frac{E \text{ volts} \times A \text{ cm}^2}{I \text{ amps} \times L \text{ centimeters}} = \text{ohm-centimeters}$$

$$(20.15)$$

In certain test procedures, a simple ammeter cannot be used due to the very high value of R_v, and consequent difficulty in measuring the very small current flow through the specimen. Usually an electrometer would be used in this instance using calibrated shunt resistances.

Surface Resistivity

Surface resistivity may be measured using the same apparatus used for volume resistivity studies, except that the electrodes are rearranged to induce current flow along the surface of the material rather than through the mass. In mathematical terms, the surface resistivity

$$\sigma = \frac{R_s W}{L} \text{ ohm-centimeters per centimeter width}$$

$$(20.16)$$

where

σ = surface resistivity in ohm-centimeters per centimeter width

R_s = observed surface resistance along a path of length L centimeters and of width W centimeters

9.19 Fabric Hand-Kawabata System

An analysis of fabric hand has been described by the ASTM as being composed of eight components: compressibility, flexibility, extensibility, density, resilience, surface contour, surface friction, and thermal properties. The measurement of these properties does not give one an evaluation of hand. Sueo Kawabata of Japan approached the task of providing a single value for hand by starting with the development of instruments that would be capable of evaluating the desired fabric properties under low-load conditions. He believed that this would more closely relate to the human concept of hand. His instruments were designed to measure the hand-related properties: tensile and shear behavior, bending behavior, compressive behavior, and surface roughness and friction (Figure 20.28). These properties are similar to those listed by the ASTM with the exception of thermal characteristics. Kawabata developed an equation that gives a weighing to each of the measured properties and called the resultant summation Total Hand Value. The weighing factors were developed through extensive human subjective evaluations of a range of fabric types and the ranking of characteristics. The weighing factors are believed to be appropriate for the population within which the data was taken but there is some question as to the application of the same weighing factors in a different culture. The instruments of the KES system can be used for determining the listed fabric properties and are useful in providing relative data for fabric comparisons.

10. AATCC TEST METHODS

10.1 Acidity and Alkalinity

An acid is defined as any hydrogen containing compound which yields hydrogen ions $(H+)$ in water, and an alkali or base is defined as any hydroxyl compound which yields hydroxyl ions $(OH-)$ in water. At equal concentrations of $(H+)$ and $(OH-)$ ions, the solution is neutral.

Acids and alkalis which dissociate or ionize completely to produce 100% of their available $(H+)$ or $(OH-)$ ions in solution are called strong acids or alkalis. Those which do not dissociate completely are called weak acids or alkalis.

10.2 Hydrogen Ion Concentration, pH

An acidity-alkalinity scale from 0 to 14, commonly called the pH scale is used to measure

FIGURE 20.28 Kawabata surface tester (courtesy of Kato Tech Co., Ltd.).

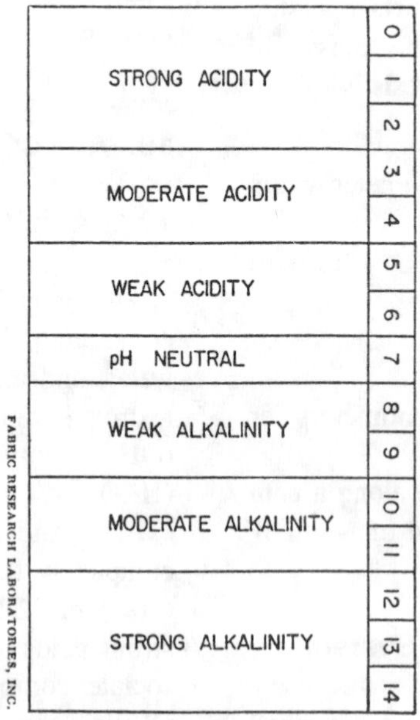

STRONG ACIDITY	0
	1
	2
MODERATE ACIDITY	3
	4
WEAK ACIDITY	5
	6
pH NEUTRAL	7
WEAK ALKALINITY	8
	9
MODERATE ALKALINITY	10
	11
STRONG ALKALINITY	12
	13
	14

FABRIC RESEARCH LABORATORIES, INC.

FIGURE 20.29 pH scale.

Test Title	Method Number
Abrasion Resistance of Fabrics: Accelerotor Method	93-1989
Absorbency of Bleached Textiles	79-1992
Aging of Sulfur-Dyed Textiles: Accelerated	26-1989
Alkali in Bleach Baths Containing Hydrogen Peroxide	98-1989
Alkali in Wet Processed Textiles: Total	144-1992
Analysis of Textiles: Finishes, Identification of	94-1992
Antibacterial Activity Assessment of Textile Materials: Parallel Streak Method	147-1993
Antibacterial Finishes on Textile Materials, Assessment of	100-1993
Antifungal Activity, Assessment on Textile Materials: Mildew and Rot Resistance of Textile Materials	30-1993
Antimicrobial Activity Assessment of Carpets	174-1993
Appearance of Apparel and Other Textile End Products after Repeated Home Laundering	143-1992
Appearance of Fabrics after Repeated Home Launderings	124-1992
Appearance of Flocked Fabrics after Repeated Home Laundering and/or Coin-Op Dry-Cleaning	142-1989
Ash Content of Bleached Cellulosic Textiles	78-1989
Bacterial Alpha-Amylase Enzymes Used in Desizing, Assay of	103-1989
Barré: Visual Assessment and Grading	178-1993
Bond Strength of Bonded and Laminated Fabrics	136-1989
CMC: Calculation of Small Color Differences for Acceptability	173-1992
Carpets: Cleaning of; Hot Water (Steam) Extraction Method	171-1989
Carpet Soiling:	
Accelerated Soiling Method	123-1989
Service Soiling Method	122-1989
Visual Rating Method	121-1989
Chelating Agents: Active Ingredients Content of Polyaminopolycarboxylic Acids and Their Salts; Copper PAN Method	168-1992
Chelating Agents: Disperse Dye Shade Change Caused by Metals; Control of	161-1992
Chelation Agents: Chelation Value of Aminopolycarboxylic Acids and Their Salts; Calcium Oxalate Method	149-1992
Chlorine, Retained, Tensile Loss: Multiple Sample Method	114-1989
Chlorine, Retained, Tensile Loss: Single Sample Method	92-1989
Chromatic Transference Scale	EP3
Color Change due to Flat Abrasion (Frosting):	
Emery Method	120-1989
Screen Wire Method	119-1989
Color Measurement of Textiles: Instrumental	153-1985
Color Measurement of the Blue Wool Lightfastness Standards: Instrumental	145-1985
Colorfastness to:	
Acids and Alkalis	6-1989
Bleaching with Chlorine	3-1989
Bleaching with Peroxide	101-1989
Burnt Gas Fumes	23-1989
Carbonizing	11-1989
Crocking: Carpets – AATCC Crockmeter Method	165-1993
Crocking: AATCC Crockmeter Method	8-1989
Crocking: Rotary Vertical Crockmeter Method	116-1989
Degumming	7-1989
Dry-cleaning	132-1993
Dye Transfer in Storage: Fabric-to-Fabric	163-1992
Fulling	2-1989
Heat: Dry (Excluding Pressing)	117-1989
Heat: Hot Pressing	133-1989

(continued)

TABLE 20.16 (continued).

Test Title	Method Number
Colorfastness to *(continued)*	
Laundering, Home and Commercial: Accelerated	61-1993
Light	16-1993
Light at Elevated Temperatures and Humidity: Water Cooled Xenon Lamp Apparatus	177-1993
Light: Detection of Photochromism	139-1989
Non-Chlorine Bleach in Home Laundering	172-1990
Oxides of Nitrogen in the Atmosphere under High Humidities	164-1992
Ozone in the Atmosphere under Low Humidities	109-1992
Ozone in the Atmosphere under High Humidities	129-1990
Perspiration	15-1989
Pleating, steam	131-1990
Solvent Spotting: Perchloroethylene	157-1990
Stoving	9-1989
Water	107-1991
Water: Chlorinated Pool	162-1991
Water: Sea	106-1991
Water Spotting	104-1989
Water and Light: Alternate Exposure	125-1991
Water (High Humidity) and Light: Alternate Exposure	126-1991
Compatibility of Basic Dyes for Acrylic Fibers	141-1989
Creases; in Fabrics, Retention of, after Repeated Home Laundering	88C-1992
Dimensional Changes in Automatic Home Laundering of Woven or Knit Fabrics	135-1992
Dimensional Changes in Automatic Home Laundering of Garments	150-1992
Dimensional Changes in Commercial Laundering of Woven and Knitted Fabrics Except Wool	96-1993
Dimensional Changes on Dry-cleaning in Perchloroethylene: Machine Method	158-1990
Dimensional Changes of Woven or Knitted Textiles: Relaxation, Consolidation and Felting	99-1993
Dimensional Restoration of Knitted and Woven Fabrics after Laundering	160-1992
Disperse and Vat Dye Migration: Evaluation of	140-1992
Dispersibility of Disperse Dyes: Filter Test	146-1989
Dispersion Stability of Disperse Dyes at High Temperature	166-1993
Dry-cleaning: Durability of Applied Designs and Finishes	86-1989
Dusting Propensity of Powder Dyes: Evaluation of	170-1989
Electrical Resistivity of Fabrics	76-1989
Electrical Resistivity of Yarns	84-1989
Electrostatic Clinging of Fabrics: Fabric-to-Metal Test	115-1989
Electrostatic Propensity of Carpets	134-1991
Extractable Content of Greige and/or Prepared Textiles	97-1989
Fabric Hand: Subjective Evaluation of	EP5
Fabrics; Appearance of, after Repeated Home Laundering	124-1992
Fiber Analysis: Qualitative	20-1990
Fiber Analysis: Quantitative	20A-1989
Finishes in Textiles: Identification	94-1992
Fluidity of Dispersions of Cellulose from Bleached Cotton Cloth	82-1989
Foaming Propensity of Disperse Dyes	167-1993
Formaldehyde Release from Fabric, Determination of: Sealed Jar Method	112-1993
Frosting (Color Change due to Flat Abrasion)	
Emery Method	120-1989
Screen Method	119-1989
Gray Scale for Color Change	EP1
Gray Scale for Staining	EP2

TABLE 20.16 (continued).

Test Title	Method Number
Hydrogen Peroxide: by Potassium Titration: Determination of	102-1992
Insect Pest Deterrents on Textiles	28-1989
Insects, Resistance of Textiles to	24-1989
Light Blocking Effects of Curtain Materials	148-1989
Mercerization in Cotton	89-1989
Migration: Disperse and Vat Dye: Evaluation of	140-1992
Mildew and Rot Resistance of Textiles: Fungicides	30-1993
Oil Repellency: Hydrocarbon Resistance Test	118-1992
Oils, Wool: Oxidation in Storage	62-1989
pH of the Water-Extract from Bleached Textiles	81-1989
Photochromism, Detection of	139-1989
Retention of Creases in Fabrics after Repeated Home Laundering	88C-1992
Rug Back Staining on Vinyl Tile	137-1989
Seams; in Fabrics; Smoothness of, after Repeated Home Laundering	88B-1992
Shampooing: Washing of Textile Floor Coverings	138-1992
Soil Redeposition, Resistance to: Launder-Ometer Method	151-1990
Soil Redeposition, Resistance to: Terg-O-Tometer Method	152-1990
Soil Release: Oily Stain Release Method	130-1990
Speckiness of Liquid Colorant Dispersions: Evaluation of	176-1993
Stain Resistance: Pile Floor Coverings	175-1993
Standard Depth Scales for Depth Determination	EP4
Thermal Fixation Properties of Disperse Dyes	154-1991
Transfer of Acid and Premetallized Acid Dyes on Nylon	159-1989
Transfer of Basic Dyes on Acrylics	156-1991
Transfer of Disperse Dyes on Polyester	155-1991
Water Repellency: Spray Test	22-1989
Water Repellency: Tumble Jar Dynamic Absorption Test	70-1989
Water Resistance: Hydrostatic Pressure Test	127-1989
Water Resistance: Impact Penetration Test	42-1989
Water Resistance: Rain Test	35-1989
Weather Resistance:	
General Information	111-1990
Sunshine Arc Lamp Exposure with Wetting	111A-1990
Sunshine Arc Lamp Exposure without Wetting	111C-1990
Exposure to Natural Light and Weather	111B-1990
Exposure to Natural Light and Weather through Glass	111D-1990
Weather Resistance of Textiles: Xenon Lamp Exposure	169-1990
Wetting Agents, Evaluation of	17-1989
Wetting Agents: Evaluation of Rewetting Agents	27-1989
Wetting Agents for Mercerization	43-1989
Whiteness of Textiles	110-1989
Wrinkle Recovery of Fabrics: Appearance Method	128-1989
Wrinkle Recovery of Fabrics: Recovery Angle Method	66-1990

the intensity of (H+) or (OH−) ions. pH is defined as the logarithm, to the base 10, of the reciprocal of the hydrogen ion concentration. Because of the conversion of the power scale to the reciprocal-log scale the pH range is 7 to 0 for progressively stronger acids and 7 to 14 for progressively stronger alkalis (Figure 20.29).

It is reiterated that the pH of a solution does not indicate the total amount of acidity or alkalinity. It only indicates the hydrogen or hydroxyl ion concentration.

pH measurement and control methods enjoy wide usage in the textile industry, for example in fiber or fabric scouring, bleaching, dyeing, and resin finishing. Selected dyestuffs which are color sensitive in specific pH ranges can be used to identify those ranges.

The pH of a solution is directly related to the electrical voltage potential which develops between the solution and a reference electrode. Electronic instruments capable of measuring these potentials have been developed and are called pH meters. The meters are usually calibrated to read direction in units.

Table 20.16 shows the major AATCC test methods and procedures for textile materials.

References and Review Questions

1. REFERENCES

1. American Society for Testing and Materials, *Annual Book of ASTM Standards* (Volumes 7.01 and 7.02), published yearly by the ASTM, Philadelphia, PA.
2. American Association of Textile Chemists and Colorists, *Technical Manual of the American Association of Textile Chemists and Colorists*, published yearly by AATCC, Research Triangle Park, NC.
3. Federal Test Methods (FTM) 191A, U.S. Printing Office, Washington, D.C.

2. REVIEW QUESTIONS

1. What is the purpose of testing? Explain.
2. List the major organizations in the United States that develop test methods and standards for textiles.
3. Why is "reproducibility" important? Is there any relationship between "textile product quality" and "reproducibility"?
4. What are the standard test conditions? What is their significance?
5. What do you understand by "quality control"? Explain.
6. What methods are used to identify fibers? Explain.
7. Find out how the scanning electron microscope works.
8. What devices are used to measure fiber denier?
9. What are the methods used to measure cotton fiber maturity?
10. Define "moisture regain." How is it measured?
11. What is HVI and AFIS? Explain.
12. Explain the major tear tests.
13. How is air permeability of fabrics measured? Can air permeability data be used to estimate water permeability of fabrics? Why?
14. What are the major abrasion resistance tests? Explain.
15. How is "drape" measured? What is the significance of it in industrial textiles?
16. What is "vertical flammability test"? Explain.
17. What properties of industrial textiles are measured with outdoor weathering tests?

TEXTILE WASTE MANAGEMENT

Textile Waste Management

S. ADANUR

1. INTRODUCTION

The textile industry suffers a great deal of material waste which is inevitable due to the variable nature of fibers, fiber-machine interaction, and the lack of purity of the raw material.

According to a survey of 348 mills done by the American Textile Manufacturers Association (ATMI), the U.S. textile companies recycled 43% of their waste in 1992. Thirty-nine percent of the waste was sent to landfills and the rest was incinerated or disposed of by other means [1]. The U.S. textile industry has invested about $1.3 billion on environmental controls during the last ten years.

Textile waste can be classified as pre-consumer and post-consumer. Pre-consumer textile waste includes by-product materials from the fiber and textile industries. Each year a great deal of this waste is reclaimed and used in new raw materials for the automotive, furniture, mattress, coarse yarn, home furnishings, paper and other industries. Post-consumer textile waste consists of any product that the owner no longer needs and decides to discard because of wear, damage, etc. Table 21.1 shows breakdown of recycled post-consumer textile waste for 1992.

Creation of new markets for waste products encourages recycling in textiles. Public awareness of environmental concerns is another driving force for recycling. In March 1992, the ATMI launched the Encouraging Environmental Excellence (E3) program. It is a voluntary program in which stringent guidelines were set for environmental protection. The ten guidelines of the E3 program are

- environmental policy

- senior management's commitment to encouraging greater environmental awareness
- environmental audit program
- employee education program
- company's dealings with outside environmental concerns
- goals and targeted achievement dates
- written emergency response plan
- relay of environmental interests and concerns to the community
- environmental assistance to citizens, interest groups, and other companies
- interaction with federal, state and local policymakers

Companies who achieve the necessary credentials to participate, are certified to use the E3 logo (Figure 21.1) on hangtags on their products, at their mill locations and in their printed materials.

The ATMI's Solid Waste Subcommittee in cooperation with the Ecological and Toxicological Association of the Dyestuffs Manufacturing Industry (ETAD) aimed at reducing waste of their member companies.

2. WASTE MANAGEMENT

In general, there are four ways of handling waste. In order of priority, they are:

- waste (source) reduction
- recycling
- incineration
- landfills

TABLE 21.1 Breakdown of Recycled Post-Consumer Textile Product Waste in 1992 [2].

	Domestic (%)	Exports (%)	Total (%)
Used clothing	–	35	35
Fiber for reprocessing	7	26	33
Recycled products	25	–	25
Landfill	7	–	7
Total	39	61	100

Factors essential to the effective implementation of any method are:

- technical feasibility
- economic viability
- marketable products
- availability of waste products

For effective waste management and reclamation purposes, clear distinctions must be made between textile wastes such as [3]:

- new production wastes
- new production wastes with a non-textile content
- discarded textiles from domestic sources
- discarded textiles from municipal refuse sources
- discarded textiles from industry
- discarded textiles from technical commodities

FIGURE 21.1 "E3" logo by the ATMI.

Discarded textiles are complex as far as material characteristics are concerned. They usually consist of different fibers and their quality is difficult to define (nature of manufacture, product characteristics, extent of wear). Due to laundering, exposure to light, etc., fiber strength can drop to 25% of its original value.

Cadieux and Godfrey suggest the following steps for waste management [4]:

- Avoid producing waste.
- When possible reuse it internally in a first quality product—preferably the same type of product originally produced.
- If not the same product, then try to find where it can be used profitably at a sister plant.
- Where it cannot be used internally, sell it to others to be used in various production programs.
- Extract combustible matter to be used as fuel.
- When further processing is impossible it is sent to a landfill.

Few textile companies process their own waste while most of the companies sell their waste to textile waste processors. A textile mill that sells its waste has the following objectives [5]:

- Sell as much waste as possible and dump as little waste as possible to avoid landfill costs and future liability exposure.
- Minimize the time, effort and hassle required to dispose of the waste.
- Sell the waste at the highest possible price.

3. SOURCE REDUCTION

Recycling is not free of cost. Capital investment, collection and handling costs and costs of treatment operations may reduce or eliminate the return on recycling investments. Therefore, source reduction (i.e., to have little or even "zero" waste) is generally the first step that should be considered in an integrated waste management system.

Efficient use of raw materials may reduce the waste generated. Improved product quality may help in reducing the textile waste. There have been many successful case histories in waste source reduction [6].

The main goals of source reduction are to reduce the amount of waste and to improve waste characteristics such as persistence, treatability, toxicity and dispersibility.

Source reduction techniques can be applied at various levels during the operations. Raw materials can be prescreened and better quality controlled. Existing processes can be modified or alternative processes can be found to minimize waste. Optimization, automation and advanced process control can also help.

4. RECYCLING

Recycling is the most viable approach to reducing solid waste stream after source reduction. The main goal of textile recycling efforts is to reprocess the textile and fiber by-products so that they can be recycled back into the original stream or into useful end products. Putting the waste back into the original system has the greatest economic value. It might go back into the original synthetic fiber, usually at a low percentage versus the virgin material. Thermoplastic waste could be diverted to less critical end uses such as a fiberfill, melt blown products or spunbonded products. Injection molding is also another possibility for textile waste [7].

Recycling methods should be technically feasible, ecologically friendly and economically viable. To achieve this for textiles, the recycling aspect must rank as high as development and production. To determine the ecological advantage, comparison should be made between the recycling method in question and what would happen if the product in question were burned in a modern incinerator or deposited in landfills. Recycling for the sake of it without any ecological and economical advantage would not serve any purpose.

4.1 Recycled Textiles

Recycled textiles are used in various products as shown in Table 21.2. These products range

TABLE 21.2 Recycled Textiles
(courtesy of ATMI).

- Mattress pads/covers
- Carpet underlay
- Decorative pillows
- Toys
- Punching bags
- Q-tips
- Cotton balls
- Blankets
- Diapers
- Package trays in autos
- Mops
- Insulation for homes
- Firemen's suits
- Quilts
- Sleeping bag liners
- Furniture decking pads
- Sound deadening pads in cars
- Casket liners
- Sanitary products
- Jewelry wrap/packaging
- Baby wipes
- Wiping clothes
- Sponges
- Ski jacket insulation
- Air filters
- Geotextiles
- Ironing board pads
- Plastic wood (used to make furniture)

from very technically sophisticated items used in automobiles and aircraft to very mundane items such as a flower pot.

4.2 Textile Waste Industry Operations

Traditional textile wastes can be in fiber, thread or fabric form. Textile waste may include card waste, motes, thread waste, selvages, substandard fibers and fabrics, filament and tow waste. There are two main processes for recycling textile wastes: tearing process and cleaning process [5,8].

Tearing Process

The tearing process consists of the following:

- grade incoming material
- collect together mixes to ensure proper consistency
- prepare the mix for feeding onto a conveyor

- blend the material while feeding the conveyor
- sort out trash, also while feeding
- cut the material to manageable size, typically 2 inch to 6 inch
- tear the material with tearing machines to return it to opened fiber form
- bale the end product

In this way, polyester/cotton threads are machined into open fiber. These fibers can be used for stuffing and for some industrial applications such as filters and caskets. In general, these fibers may be spun into very coarse count yarns such as mop yarns (a mop cord is made of four to eight plied yarns). Friction spinning is suitable for this process. Polyester filaments are cut and opened so that the end product can be used as polyester stuffing. Carpet waste can be opened up so that it may be used to produce carpet underlay. Apparel cuttings are processed into the raw materials for sound deadening pads used in automobiles. Cotton threads are opened to fibers and are used in the absorbent cotton industry. They may be bleached and purified for sanitary applications. Coloration of wasted fibers is also possible.

Recycled absorbent pads are used for oil drips and spills, walkways, packaging and cushioning. Mattress and furniture insulator pads are also made from recycled fibers (Figure 21.2). Knobs are mixed with staple fibers to produce novelty yarns.

Cleaning Process

The cleaning process handles the cleaning of soft cotton or polyester/cotton card waste and gin motes. The raw materials are blended and processed through step cleaners and lint cleaners using equipment similar to that used in cotton gins. Through this process, card waste can be cleaned up until it can be substituted for a low grade of cotton usable in mop yarns and other course count yarns. Gin motes are handled in the same way.

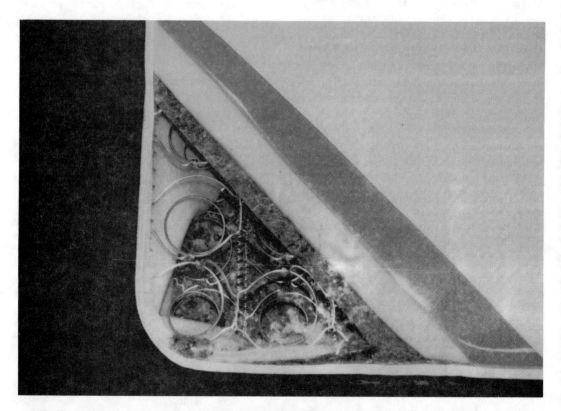

FIGURE 21.2 Nonwoven mattress pad made of recycled fibers (courtesy of Wellington Sears Utilization Plant).

The mote waste, the by-product of cotton waste cleaning operation, has been used in farming fields by mixing with the soil. Compost made of recycling waste is an excellent fertilizer with good carbon and nitrogen components.

Lehner lists the major challenges in recycling textile and fiber products as follows [5]:

- poorly segregated by-products: This occurs when different types of wastes are collected and mixed together in a way that reduces the overall value of the waste.
- contaminated by-products: When possible, textile waste should not be mixed with other waste such as metal, wood or cardboard.
- poorly packaged material: Poorly packaged fiber materials in bad bales, flimsy cartons, or unmanageable bags increase the cost of handling and storage.

5. RECYCLING OF INDUSTRIAL TEXTILES

Recycling of industrial textiles is more difficult than traditional consumer textiles because they are built as durable, high performance products. For example, coatings are designed to adhere inseparably to their substrates. Lamination adhesives are supposed to have long-term bonding properties. Therefore, engineered textile structures are difficult to disassemble and reuse.

Most of the time, technical textiles can be a solution to environmental concerns. However, sometimes they may be the cause. For example, medical textiles, whether they are disposable or reusable, must ultimately be destroyed because they are often considered hazardous. Infectious medical waste is usually disposed of through incineration. Incineration is an effective way to eliminate the infectious medical waste while reducing the size of the waste material. However, incineration results in a by-product of fly ash and emissions that may cause health problems if not filtered.

Another example of the industrial textiles that may hurt and help the environment is the industrial filters. They are used to reduce pollution, which helps the environment. When they are replaced, they are disposed of in landfills or confined if they contain noxious or radioactive substances. This will diminish the landfills.

Polymers can be classified as thermoset and thermoplastic. A thermoplastic material can be repeatedly heated to its softening point, shaped and then cooled to preserve the remolded shape. In a thermoset polymer the heating process cannot be repeated due to charring and degradation. The reason for this behavior is the cross-linking which does not exist in thermoplastic materials. Examples of thermoplastic materials are polyester, polyethylene and polypropylene; examples of thermoset materials are nitrile and butyl. Technical textiles that are made of thermoset polymers cannot be remelted for recycling purposes.

5.1 Recycling Technology

There are several techniques that can be used to recycle industrial products as discussed below [9]. It should be noted that some of the recycling techniques mentioned in this chapter may still be under development for textile applications. Some of the techniques have been used in other industries successfully and may be adapted to the textile industry.

Solvent Extraction

Solvent extraction has been used for carpet recycling. In this process, a consecutive chain of solvents is used to remove polymers of interest. For example, acetone and hexane are used to remove oils; ethylene dichloride is used to dissolve and remove the ABS and PVC plastics; xylene is used to extract polypropylene and polyethylene. PVCs can be removed with esters, ketones or chlorinated hydrocarbons. Nylons can be extracted with phenols.

Cyrogenic Fracture

In this method (with or without mechanical or ultrasonic vibration) the temperature of polymers is reduced to below glass transition temperature with liquid nitrogen or other cold temperature materials which makes the coatings or film brittle. Polymers are then broken and separated. To increase separation efficiency, mechanical or ultrasonic vibration is used.

Pyrolysis Kiln

Pyrolysis is the thermal decomposition of organic material in an oxygen-deficient environment. This technique has been used for the production of fuels and chemicals from organic feedstocks such as waste tires. The thermal organic reactor uses molten lead in a sealed chamber to reduce the organic materials to their organic residues, without harmful emissions resulting from incomplete combustion. A conveyor and filter system separates and removes the residue, allowing a continual feed process.

Powdering (Solid-State Shear Extrusion)

This method, being developed at the Illinois Institute of Technology, uses high pressure at low temperatures to grind the material for further processing.

Other Recycling Technologies

Potential future methods for recycling technology include hydrolysis to monomers, alcoholysis to monomers, catalysis and biological separation.

Hydrolysis and alcoholysis are chemical processes used to reduce polymers to their constituent monomers. Catalysis uses additives or accelerating agents that do not get used up during the chemical reaction. Biological organisms can break down certain synthetic chemicals.

Other longer term potential methods may include thermal processing, microwave heating, surface heating, gamma irradiation, ice blast, and remanufacturing.

In thermal processing, heat is applied to melt the polymers into their constituent parts. Infrared heating of the fabric surfaces (surface heating) to loosen the coatings or adhesives may prove effective. Gamma irradiation (non-ionizing radiation) may induce favorable

changes in the structure of some coated and laminated fabrics. It is known that some types of irradiation increase crosslinking in certain polymers. Ice blast is a low-impact surface scrubbing process that uses phase changes in rapidly melting ice. When small ice particles, fired from a nozzle at high speed, hit the surface, they melt and transfer energy to the surface. When the surface coating rebounds, it explodes outward and the surface is gently cleaned.

5.2 Recycling of Multi-Component Structures

Industrial textile end products are quite often multi-component structures. Typical products include tarpaulins, container curtain materials, conveyor belting, foam-laminated textiles, rubber reinforced textiles, membranes, decorative textiles on rigid moldings, nonwoven products and flocked substrates. For example, textile composites are made of reinforcement fibers and matrix materials. Nylon yarns are inserted in the filling direction in polyester papermaking fabrics for abrasion resistance. A film with excellent chemical resistance is laminated to a substrate having good dimensional stability, but allows liquid penetration. This pairing usually requires that different polymers are used for substrate and top layer. The more components are contained in a textile product, the more difficult it is to recycle it. The success of a recycling process depends on two criteria [10]:

- reduction of material vareity: Fewer fiber materials and finishing agents in a textile product make it easier to separate the individual components from each other and recycle them in a good quality process.
- reversible manufacturing processes: It is important that manufacturing processes, such as joining and bonding, are reversible for successful recycling.

Some of the current and possible future recycling methods for multi-component textile products were reviewed by Ehrler et al. They summarize the major steps in recycling of multi-component textile products as follows [11]:

(1) Separation—Separation or splitting is the treatment by which a bond between two or more materials is removed. During the separation process, the components should not be destroyed. Mechanical and chemical separation processes are possible.

(2) Isolation/fractionation—Following the separation, the "batch" produced by separation is isolated or fractionated. Criteria influencing isolation include source material, size and specific gravity.

(3) Reclamation
   - material reclamation (mechanical processing, melt-down)
   - chemical reclamation (alcoholysis, etc.)
   - "return to root" reclamation (hydrogenation, gasification)
   - heat recovery

In mechanical separation, the interface of the textile composite is overstressed and destroyed by mechanical forces (tensile, shear, bending, impact, compression-decompression and torsion). A pretreatment process can be applied before the release process to facilitate the separation. Pretreatment techniques can be low temperature treatment (inducing temporary brittleness), high temperature treatment (reducing adhesion), steam infusion and swelling of one of the components. Ehrler et al. suggested the following techniques.

For sheet materials (larger than 0.3 m²):

(1) Friction calendering
   - principle: shear forces
   - technique: calender
   - appraisal: only suitable for planar items

(2) Brushing technique
   - principle: abrasion, shear and tensile forces (ripping apart)
   - technique: brushing machine
   - appraisal: highly flexible, unsuitable for three-dimensional items

(3) Hydraulic technique
   - principle: hydromechanical forces applied by high-pressure water jets
   - technique: pump and jet array, pressure range below 100 bar
   - appraisal: highly flexible, effluent problem due to scouring action

For small dimension items:

(1) Tearing
   - principle: tensile force (in association with local tearing out), shear forces
   - technique: rag tearing machines
   - appraisal: widely used method also used for items of medium size; mechanism is unsatisfactory in numerous special cases due to inadequate clamping action; unsuitable for textiles containing elastomers

(2) Dry milling
   - principle: impact, shear and bending forces depending on the requirement
   - technique: usually an impact-type (hammer mill or pinned disc mill) cutting mill
   - appraisal: metal or thermoplastic items are troublesome depending on the type of mill

(3) Abrasion/peeling
   - principle: shear and bending forces
   - technique: peeling machine as used in the foodstuffs industry
   - appraisal: selective comminution possible, depending on product

(4) Wet milling
   - principle: shear and bending forces assisted by swelling
   - technique: beater, pulper
   - appraisal: drying necessary, cleaning usually occurs simultaneously, effluent problems

In physicochemical and chemical methods, the interface of the multi-component is stressed and destroyed by causing mutual interaction between the components of the material. The physicochemical processes vary from artificial aging of the interface through to its actual integration. Different methods can be used for separation.

Destruction of the Interface

   - principle: infiltration of the bond by agents that destroy the adhesion (swelling/dissolving agents)
   - technique: impregnation, mechanical separation

- appraisal: used for separating fibers from rubberized hair products
- principle: decomposition of polyurethane adhesives by alcoholysis/aminolysis
- technique: chemical reaction
- appraisal: used for separating membrane/substrate structures, any conversion products that may be formed must be disposed of

Temporary Change in the State of Aggregation of a Component

- principle: melting of thermoplastic components
- technique: melt filtration
- appraisal: used in the plastics and fiber production sectors to separate non-fusible melt constituents; these are generally only present as contaminants; special continuous filters are required for separating major amounts; the thermal stability of the associated material in the composite is a limiting factor
- principle: selective dissolving and separation of polyamide by means of acids (formic acid, hydrochloric acid)
- technique: extractor
- appraisal: economically viable only for polyamide contents > 30% and high throughputs
- principle: selective dissolving and isolation of polyamide by means of methanol at high pressure
- technique: autoclave
- appraisal: the polymer is dissolved and precipitated by variation of pressure and temperature
- principle: selective dissolving and separation of elastomeric filament yarns from stretch knit goods by means of cyclohexanone derivatives
- technique: modified drum scouring machine
- appraisal: knits free from elastomerics produced suitable for the opening process; solvent is combustible; process

control necessary similar to a dry-cleaning process

- principle: selective dissolving of styrene/maleic acid anhydride copolymers used as adhesives in composites for car interior trim and separation of the components of the composite
- technique: extractor
- appraisal: solution/precipitant system; additional solvent separation and processing necessary; sludge distillation product occurs as toxic waste
- principle: selective dissolving and isolation of polyethylene terephthalate from textile blends
- technique: pressure extractor
- appraisal: pure non-degraded polyester terephthalate powder; non-degraded textile waste; chlorinated hydrocarbon solvent; no precipitant used; simple preparation of the solvent; tarry toxic waste; economically viable only with polyester terephthalate contents > 30% and high throughputs

Permanent Change in the State of Aggregation of a Component

- principle: selective degradation-piece carbonization of cellulosic constituents
- technique: carbonizing machine
- appraisal: effluent problems with high contents of converted cellulose; not suitable for all blends
- principle: selective degradation/hydrolysis of the cotton in fiber mixtures
- technique: chemical reaction
- appraisal: only suitable for composites containing cellulose; effluent problems
- principle: selective reaction; esterification or etherification of cellulosic components in blends; conversion of polyamide
- technique: chemical reaction
- appraisal: procedure can only be used for specific waste materials; 100% conversion difficult in actual blends (due to contaminants); high cleaning cost for components that are not converted

5.3 Examples of Industrial Textile Recycling

Coated and Laminated Fabrics

Manufacturers of coated and laminated fabrics are interested in recycling and reclaiming of fabric scrap which is known as "edge trim." Edge trim consists of cut selvedge ends, rollends, reject lots, setup yardage and other waste materials. Disposal of this material is costly.

In 1993, the total of coated and laminated fabric scrap was 6,187 tons. At an average East Coast landfill tipping cost of $35 per ton, the cost of this problem to the industry is almost a quarter million dollars per year. An example of this is the fabric used for the Denver International Airport roof where a good deal of fabric scrap was produced. There is a need of edge-trim recycling especially for vinyl coated and vinyl laminated polyester fabrics [9].

The disassembly of laminates back into pure components is a reversal of the manufacturing process itself. The adhesive must be dissolved without destroying the textile component.

Carpets

Discarded vinyl-backed carpeting is used to produce a broad range of usable products such as park benches, birdhouses and picnic tables.

Fabrics from PET Soda Bottles

Recently work has been done to use post-consumer PET (polyethylene terephthalate) soda bottles to make fibers and fabrics. It was reported that fleece fabrics and geotextiles have been successfully made from PET soda bottles [12].

Nonwovens

A considerable amount of reclaimed fibers is used in nonwovens. Recycled nonwoven products range from middle and lower grade to functional and industrial grades. The variations in type, fineness, length and color of recycled fibers may cause high variability in the properties of the finished product. The principle parameters affecting the processing or reclaimed fibers in nonwovens are [13]:

- the degree of opening which is characterized by the percentages of fiber material and content of yarn and fabric remainders
- the fiber length which is characterized by the average fiber length and the percentage content fibers shorter than 5 mm

Reclaimed fibers are increasingly used in needled geotextiles for separation and protective applications as well as for hydraulic functions involving filtration and drainage.

Textile Structural Composites

As the use of composites spreads into more areas, recycling of composite materials is becoming economically viable and ecologically required. In the case of thermosetting composites, recycling is difficult. At best, thermosets can be reused as filler materials in limited quantities. Conditions for recycling are more favorable in the case of thermoplastic materials. This is an advantage for thermoplastic composites to increase their market share.

Carbon fiber waste from cutting of reinforcing fabric during manufacture of composites can be reused to make chopped fiber or nonwoven composites. Carbon fiber is brittle and difficult to recover from a fabric with substantial reduction in fiber length. This degradation of the fiber length generally renders the fiber unsuitable for dispersion back into the original product. The brittleness also limits the kinds of textile processing that can be used. The manufacture of nonwovens requires only that the fiber be opened, and does not require the drafting and spinning steps, both of which are mechanically strenuous on the fiber. Mixtures of waste, high modulus, thermally stable materials (like carbon) with thermoplastics waste (like polypropylene carpet backing) in a nonwoven structure might be heated and compressed to make relatively flat, moderate performance, fiber reinforced plastic panels.

The Deutsche Forschungsanstalt fur Luft und Raunfahrt (DLR) of Germany has recently introduced biodegradable construction materials called biocomposites that can be recycled as well. It was reported that the material can be used in many cases where glass fiber reinforced plastics have been used in the past. The biocomposites developed by DLR are based on natural fibers (hemp, flax or ramie) and biodegradable matrix material. The matrix material can be polyester, modified cellulose, starches and starch-containing products. It was reported that ramie or flax products achieves 50% of the rigidity of glass fiber reinforced plastics, and hemp materials as much as 100% [14].

5.4 Thermoplastic Reclamation Equipment and Techniques

Thermoplastic waste recovery systems include the following [7]:

(1) Chemical depolymerization—In chemical depolymerization, the waste is converted back to pure basic raw materials. PET bottles are recycled with this process.

(2) Discontinuous shredder: aggregators—This type of system generally consists of guillotine, rotary cutter, granulator, pelletizer and extrusion screw. The guillotine cuts off a slab of material by its moving blade acting against its fixed blade. The rotary cutter reduces the tangled web of fibers to much shorter lengths. The granulator keeps cutting the fiber until it is small enough to fall through the screen. The particles can be changed to a greater and uniform bulk density if it is composted by a pelletizer. To get the material into a pellet it is necessary to feed it into an extruder for remelting and cooling. It should be noted that the quality of the thermoplastic suffers from the multiple reheating.

(3) Continuous aggregators or friction-disk compactors—This system consists of primary granulator, feeding system, compac-

tor, secondary granulator and fines separator. The primary granulator reduces the material size that will pass through a screen. The particles are then passed over an agitator bar and advanced to the compaction chamber. An auger discharges the material between a rotating and a stationary disc where the material is forced through the stationary disk by pressure. The material out of the compactor is air-veyed to the secondary granulator where it is reduced to its final desired particle size. From here it is air-veyed to the fines separation section.

(4) Aggregator-extruder combination—This system consists of shredder-compactor, extrusion screw, melt filter and pelletizer (or granulator). The shredder cuts the material into small lengths. The material is thermally compacted. Then the material is melted at the extruder. The molten polymer produced by the extruder is forwarded through a melt filter. The filtered melt is converted to strands and granulated for very low viscosity materials like nylon and polyester. For materials like PE and PP, hot granulation is preferred. The hot granules are immediately cooled by a water spray and taken away to a vibration sieve for drying.

6. ENERGY RECOVERY

It is possible to burn solid waste in an environmentally acceptable manner in state-of-the-art energy incinerations. After source reduction and recovery of economically recyclable materials, solid waste incineration of textile waste is a vital and attractive alternative.

Textile materials have excellent waste BTU values. Materials that do not contain PVC give good energy value when used in an efficient furnace. For example, polypropylene, a basic hydrocarbon, has the same heat value as gasoline. After it gets hot enough, it burns at the rate of approximately 20,000 BTU/lb of material [4].

Textile waste products can be used in steam generation for certain production processes. Short, shredded fibers can be incorporated into a pelletized fuel product for industrial fuel users such as forest products manufacturers. Loose fibers also have potential as a fuel substitute in cement kilns.

7. LANDFILLS

Sanitary landfills should be the last alternative in an integrated waste management system after source reduction, recycling and incineration. The landfills in the United States are being filled and therefore closed rather quickly.

21.2

References and Review Questions

1. REFERENCES

1. "US Textile Mills Recycle 43% of Waste," *Textile World*, October 1993.

2. Textile Recycling Fact Sheet, Council for Textile Recycling, 1993.

3. Bottcher, P., "Textile Recycling: Current Status and Development Trends," *International Textile Bulletin*, Vol. 4, 1994.

4. Cadieux, R. D. and Godfrey, J. F., "Waste Prevention and Reclamation," *International Fiber Journal*, June 1990.

5. Lehner, C. P., "Recycling Landfill and Textile Wastes at Leigh Fibers," *Textile Recycling and Environmental Conference*, Clemson University, August 1994.

6. Hamouda, H. and Smith, B., "Source Reduction: An Alternate to Waste Treatment," *Textile Recycling and Environmental Conference*, Clemson University, August 9–10, 1994.

7. Hill, R., "Thermoplastic Reclamation Equipment and Techniques," *Recycling and Waste Minimization Conference*, TW/ITA, February 1991.

8. Strength, T. B. and Crutchfield, B. G., Wellington Sears Utilization Plant, Valley, AL, private communication.

9. Ravnitzky, M., "Tackling the Edge-Trim Problem," *Industrial Fabric Products Review*, September 1994.

10. Leckenwalter, R., "Recycling of Technical Textiles," *Canadian Textile Journal*, January/February 1993.

11. Ehrler, P. et al., "Study of Separation Methods for Textile Composites," *ITB Nonwovens*, Vol. 4, 1994.

12. Lynn, E., "Bottled Fabric," *American Dyestuff Reporter*, February 1994.

13. Bottcher, P. and Schilde, W. "Using Reclaimed Fibers in Nonwovens," *International Textile Bulletin*, Vol. 1, 1994.

14. "Biodegradable Composites to Be Employed as Environment-Friendly Construction Material," *Techtextil-Telegramm*, 28 November 1994, No. 35, Issue E.

1.1 General References

Leckenwalter, R., "Recycling of Technical Textiles," *Canadian Textile Journal*, January/February 1993.

America's Textiles: Encouraging Environmental Excellence, American Textile Manufacturers Institute (ATMI), 1993.

Strzetelski, A., "Technical Textiles: Good or Bad for the Environment?" *Technical Textiles International*, March 1993.

Proceedings of The Recycling and Waste Minimization Conference, TW/ITA, February 27–28, 1991, Charlotte, NC.

Kalogeridis, C., "Recycling Should Be Your Last Resort, Says EPA," *Textile World*, June 1992.

"Auxiliaries Tout Conservation, Recycling," *Textile World*, March 1993.

McCurrey, J. W., "Encouraging Environmental Excellence," *Textile World*, November 1993.

Riggle, D., "Tapping Textile Recycling," *Biocycle*, February 1992.

Glenn, J. M., "Erema: Recycling Units for Pelletizing Manmade Fiber and Film Materials," *Textile World*, April 1985.

"Closed Loop Recycling of Textile Sizing/Desizing Effluents," *American Dyestuff Reporter*, October 1988.

2. REVIEW QUESTIONS

1. What are the choices in waste management? Explain.

2. Explain source reduction. Is "zero waste" possible in textiles? Why?

3. Explain the current methods to recycle textiles.

4. Why is recycling industrial textiles more difficult than recycling traditional textiles? Explain with examples.

5. Which application area(s) of industrial textiles has more potential to use recycled textiles? Give examples.

COMPUTERS AND AUTOMATION
IN TEXTILES

Computers and Automation in Textiles

S. ADANUR

1. INTRODUCTION

If the motto of manufacturing in the 1940s and 1950s was "If a job can be done by machine do it by machine," the motto of the 1980s and 1990s has been "If a job can be done by computer, do it by computer." Within the last couple of decades, the textile industry certainly has not escaped from the computer revolution. Automated process control systems, management information systems, and material handling systems have found application areas throughout the textile industry. Concurrently with the computer revolution, computers have certainly found place in both design and manufacturing of industrial textiles.

In today's business environment, fast communication and electronic data interchange (EDI) are necessary elements of success. The days of massive record keeping are passing. Computer databases now store extensive information about machine settings, product parameters and manufacturing variables and retrieve them in milliseconds. Even more advantageous is the ability of the computer to compare and analyze performance of industrial fabrics of similar designs in similar applications.

Computer aided design and manufacture allows in-depth performance analyses. Problems in industrial fabric performance can be studied by the computer, engineer and scientist. As a result, more complete decisions can be made faster. On-line connections can be established between manufacturing systems and research, and design and other manufacturing databases to study, sort, correlate and analyze before orders are entered in the system.

The modern computer and communications technologies provide internal (inside the company) and external (statewide, national, and international) networking of all systems in a company. Internally a large number of personal computers (PCs) can be connected together with a master processor in one mill. Satellite plants with different hardware systems can be connected to the headquarters through computers. With this type of networking it is possible to communicate via e-mail (Internet, Bitnet) and World Wide Web (WWW), follow the status of an order throughout the manufacturing stages, etc. Some companies even have started ordering their supplies by computers. Having the ability to access the inventory of a supplier through the network makes planning easy and enables a company to meet deadlines. Companies with high speed flow of information capabilities will have an advantage of being flexible and competitive in planning, production, control and purchasing.

With the programmable logical controls (PLCs), most of the mechanisms of modern textile machines are electronically controlled and operated. With the PLCs, it is also possible to monitor the performance of machines with measurements such as picks per minute of a loom, stops, starts, etc.

The U.S. textile industry reportedly spends over $1 billion per year on capital investments. For the past ten years, a considerable amount of this equipment has been computers and microelectronic equipment. Automation by computers has established a very good outlook for itself.

2. COMPUTER TECHNOLOGY

The computer technology is changing so fast that any detailed and specific discussion of the current computer hardware or software would be meaningless in a couple of years. Computers are becoming faster and more powerful in terms of memory, versatility, adaptability, performance and supporting software programs. Besides, computer technology is not the purpose of this chapter. Therefore, only definitions of some basic computer terminology will be given.

2.1 Computer Terminology

- Bitnet: A worldwide electronic communication network
- Central Processor Unit (CPU): This is the brain of the computer. It is a chip inside the computer that runs all the other software and operations in the computer.
- compact disc (CD): a high capacity, digital disk for data storage
- cyberspace: the general name for the domain that includes e-mail, WWW, and communication networks, etc.
- database: a collection of information on a subject
- desk top publishing: publication of brochures, newsletters and even books using a specially designed word processor in a computer
- digitizer: a device that converts analog signals to digital signals which are easier to manipulate by the computer
- disk drive: a magnetic system that can read the disks or write on them
- DPI: dots per inch; an indication of the monitor resolution as well as printing quality
- EDI: Electronic Data Interchange
- e-mail: electronic mail. To be able to send electronic mail, both the sender and receiver should be connected to a network such as Internet or Bitnet.
- floppy disk: a portable magnetic computer disk with a dimension of 3.5″ or 5.25″ (the latter is becoming obsolete). The current capacity of these disks is 1.44 MB (megabyte).
- hard disk: a permanent disk inside the computer for data and program storage
- hardware: all of the physical

components (circuits, boards, cables, etc.) in a computer
- home page: information page about a user connected to the WWW (World Wide Web)
- host computer: a mother computer to which several other smaller computers are attached.
- IBM compatible: a computer that can run the same software written for the original IBM PC computers. These computers are produced by many companies such as Compaq, Dell, Gateway, Toshiba.
- interface: The interaction of the computer with outside devices such as printer, scanner, plotter, modem or machines is called interface. The communication between the computer and outside devices is done through interface cards and data converters.
- Internet: a worldwide electronic communication network
- monitor: computer screen
- MS-DOS: MicroSoft (a software development company) Disk Operating System. A communication system in the computer (CPU) that handles the file operations and establishes communications with other software programs. MS-DOS system is used by IBM and IBM compatible computers. Recently, computers have been developed that can run both IBM and Macintosh software. Examples of other operating systems are Unix and PS2.
- plotter: a device that can plot complicated designs and drawings from the computer screen. Usually a plotter has different color pans and can reduce or enlarge the drawing size.
- printer: a device used to print computer files, drawings and programs. It usually uses A4 size paper. There are different technologies such as laser printer (gives the highest printing quality), dot matrix printer and ink jet printer.
- RAM: Random Access Memory. This memory is used by the CPU to store some data that is necessary to run the

program. CPU used RAM according to the need and size of the software program that is being run. The information can be written to and retrieved from the RAM memory very fast. If RAM is small, then large software programs cannot be run. Therefore it is always good to have large RAM memory. When the computer is turned off, everything in RAM is erased.

- resolution: the number of pixels per unit area on the computer screen. It is an indication of the quality of the picture on the monitor. The higher the pixels or dpi (dots per inch), the better the resolution.
- ROM: Read Only Memory. This is a place where the information to run the computer is stored permanently. When the computer is turned on, the CPU gets the commands from ROM on how to start the computer and make it ready for the user's use. CPU can only read from ROM, i.e., it cannot write to it. The user cannot access to ROM. Information in ROM cannot be erased at any time.
- scanning: It is the reverse process of printing from a computer. In printing, the text or picture on the screen is printed on the printer. In scanning, the picture or text on paper is digitized and loaded to the computer. The device for this purpose is called scanner.
- software: any program that is used by a computer. Different software programs are available for different purposes such as drawing software, word processing software, etc.
- spreadsheet: a data table created by a software program where data manipulation is easy to do
- terminal: a monitor with a keyboard that is used to connect to a main server computer
- user friendliness: the degree of ease of using a computer
- user interface: the programs that provide communication between the user and computer

- work station: a computer that is connected to a large central computer's CPU and is run by that CPU (a work station may also have its own CPU)
- WWW: World Wide Web. A computer network system where data, voice and pictures can be exchanged easily all over the world. For example, one can connect to the White House Home Page, listen to the welcome messages from the President and Vice President, sign the White House Guest Book and leave a message, take a tour of the White House and Congress with pictures and sound, look at the Inauguration Ceremonies pictures, among many other things.

2.2 Programming Languages

These are specially developed commands to instruct the computer (CPU) to perform a specific job such as to draw a rectangle, a fabric design or to solve a math problem. There are three levels of programming languages: machine language, assembly language and high level language.

High Level Language

This is the level that people generally use to write computer programs (high level: human level; a level that a human can understand and communicate easily). The commands are in English and easy to understand by a person, e.g., WRITE, READ, INPUT, etc. Examples of high level languages are FORTRAN, Basic, C and Pascal. A program written in a high level language is converted to assembly language first by a compiler in the computer (compilation). Then the assembly language is converted to machine language before the program is executed by the computer. That is why high level language is the slowest of all.

Assembly Language

A computer program can also be written in assembly language which is the next lower language level (away from human and closer to computer's CPU). A program written in assembly language still consists of English letters, although the words are abbreviated and may not mean anything in English, e.g., CMP for "compare," INC for "increase." It is more difficult and takes longer to write a program in assembly language compared to writing it in a high level language. However, the programmer has more control over the operation of the computer. There is no need for "compilation." This also makes assembly language a lot faster than the corresponding high level language. That is why video games, where speed is critical, are written in assembly language. Each CPU type has its own assembly language.

Machine Language

This is the language that the CPU can understand, i.e., lowest level from the human level. It is made of 0's and 1's. Although a program can be written in machine language, it is extremely difficult and time consuming for a human to write a program in machine language. However, machine language gives the full control to the programmer and therefore it is very powerful. In fact one can create new computer commands by using machine langauge.

2.3 Classification of Computers

Personal Computer (PC)

A computer that can operate by itself, i.e., without being connected to a larger computer system or network (of course a PC can be connected to a larger computer system or network). A personal computer has the necessary operating system software to run itself, therefore it is self-sufficient. All that is needed to run the computer is electrical power, a monitor and a keyboard. The two largest personal computer manufacturers in the United States and in the world are IBM (International Business Machines) and Apple Computer, Inc. A laptop and a desktop are considered personal computers.

Laptop

A compact computer that can fit on one's lap. Laptops are so small compared to other com-

puters that they can fit in a briefcase. The computer (CPU), monitor and keyboard are all consolidated into one piece. Due to size limitation, the monitor screen is LCD type, i.e., similar to calculator screens.

Desktop

A computer that can physically fit on top of a desk. Most of the PCs that are used in offices and at homes are desktop size.

Mini Frame, Midi Frame, Main Frame

These terms are used to indicate successively larger host computer systems. Depending on the size, many personal computers or workstations are connected to these host computer systems or "frames." In simplistic terms, a Mini Frame may be used to run several computers in an office environment. A Midi Frame can be used to run computers in small- to medium-size companies. A Main Frame is usually used in large corporations, R&D centers and universities where large numbers of computers are used.

Super Computers

These are extremely fast and powerful main frames specially built for research purposes. The processors in the computers are made of special type semiconductors. These systems are very expensive and can only be justifiable for institutions such as NASA and big universities.

3. COMPUTER AIDED DESIGN (CAD)

Computer Aided Design is fast becoming one of the most used software packages in the industry. CAD refers to any activity that utilizes a computer to assist in the creation, modification, presentation and analysis of a design. Typical components of a CAD system are computer, monitor, scanner, plotter, printer, and digitizer as shown in Figure 22.1.

The three major functions are synthesis, analysis and presentation. Synthesis involves mental processes such as decision making and creativity. After this process, the object is analyzed, evaluated and modified. It is then prepared for

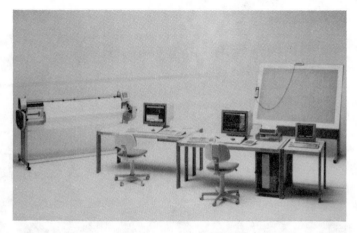

FIGURE 22.1 Typical computer aided design system (courtesy of Assyst Automation Software and Systems).

presentation and approval. Experience has shown that the CAD speeds up the engineering process. This process takes away the paperwork that inhibits productivity and creativity.

CAD's success relies on building blocks that have been previously put into the computer. The computer programs have been set up to draw on its experience and programs already used. This allows the computers to expand the knowledge given them and become more useful in future endeavors.

Computer aided design and manufacturing is more essential for industrial textiles compared to consumer textiles. Raw materials that go into industrial textiles are more expensive which requires the minimum waste possible. In general, the structures of technical textiles are also more complicated than consumer textiles which makes it a lot more difficult to visualize the structure without a prototype. With computer aided design, the correct geometry of the product can be modeled on the computer screen which helps the designer to modify or change the design. Complex computer programs (e.g., AutoCAD®, VersaCAD®) have been developed in different computer programming languages (e.g., Basic, FORTRAN, C) for this purpose. Producers of textile CAD software include Modacad, AVL, CDL and others.

3.1 Product Design

The most important and beneficial use of

CAD systems is in the design of industrial textile products. CAD programs can be used to design and characterize polymer molecules, fibers, yarns, fabrics and composites (Figure 22.2). In addition to creation of a design on the computer using a software package, designs and patterns can also be digitized or scanned into the computer. Design, development and testing of high performance textile structures with CAD shorten the development time of new, innovative products. Figure 22.3 shows a plain weave structure developed using a computer aided design system and Figure 22.4 shows a fabric pattern developed with a computer.

Powerful CAD systems allow modeling of textile structures using engineering principles. Current technology allows the combination of the power of today's computer capabilities with fundamental research in fibers, yarns and fabrics in order to model structure of textile goods and be able to predict their engineering properties and performance. CAD of industrial textiles may lead to development of new, innovative textile products. Figure 7.38 (Chapter 7.0) shows a typical composite design production cycle.

Computer modeling of textile structures is needed especially in the industrial high-tech fabric markets. Industrial textile structures are subject to significant static and dynamic forces during their end use. There are some mechanical models in technical literature to model various textile structures and determine their

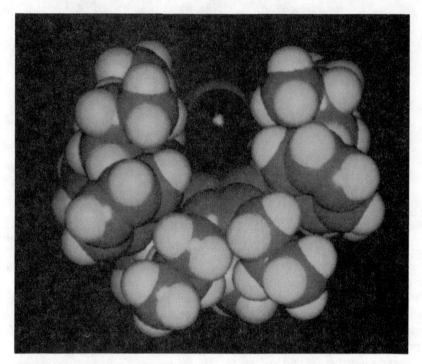

FIGURE 22.2 Synthetic polymer molecules designed with computer modeling (courtesy of Sandia National Labs).

FIGURE 22.3 Plain weave structure developed with CAD (courtesy of Adanur).

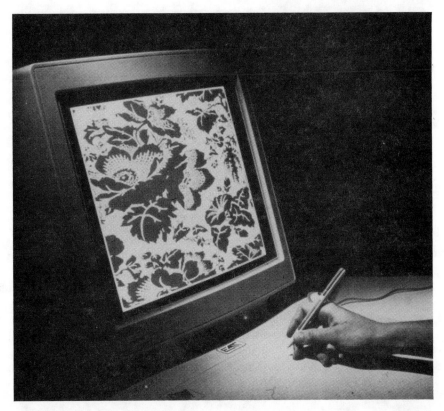

FIGURE 22.4 Fabric pattern developed by a CAD system (courtesy of Johnston Industries).

properties and performance. Modeling of high performance structures must be based on theoretical and empirical data available in the literature or through experience. An extensive data base should be established for this purpose.

CAD of industrial textile products allows the manufacturer to visualize the final product even before a decision to produce it is made. As a result, the cost of manufacturing trials and prototype development could be reduced. CAD helps the companies with quick response to market needs by reducing the development and manufacturing time of new, innovative high-tech textile products. The capabilities offered by the CAD systems will help the decision makers (management and technical personnel) with modification and improvement of existing designs and development of new products.

Various software and hardware packages are commercially available for CAD. Most of the software uses finite element analysis (FEA) to model and analyze the structures. However, these software packages are written for general use and not specifically for textile materials. Textile materials show non-linear viscoelastic behavior which may not be handled by some software packages. Moreover, some industrial textile structures are highly specialized and complicated. Therefore, industrial textile product manufacturers are usually left with the choice of developing their own software programs for their specific needs. Intense competition is another factor that justifies this choice.

3.2 Product Analysis

Various computerized vision systems are commercially available to analyze the structure of fibers, yarns and fabrics. The magnification and resolution of these systems depend on the sophistication of the hardware. Figure 22.5 shows a microvision system.

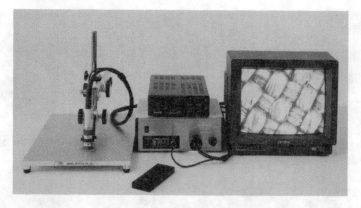

FIGURE 22.5 Microvision system (courtesy of Hirano Tecseed Co., Ltd.).

3.3 Process Design

Computers reduce the scrap or edge trim considerably during cutting operations. Computerized cutting systems provide the optimum orientation of pieces on the fabric roll by sorting the pieces automatically. They also find the shortest route for the blade which increases the throughput of the cutter. Figure 22.6 shows a computerized cutting system.

CAD and Computer Aided Manufacturing (CAM) systems are used in designing plant layouts (Figure 22.7). Plant layout is the physical arrangement of production equipment, administration equipment, utility equipment, storage and transportation equipment, raw material,

FIGURE 22.6 Computerized cutting operation (courtesy of Lectra Systems).

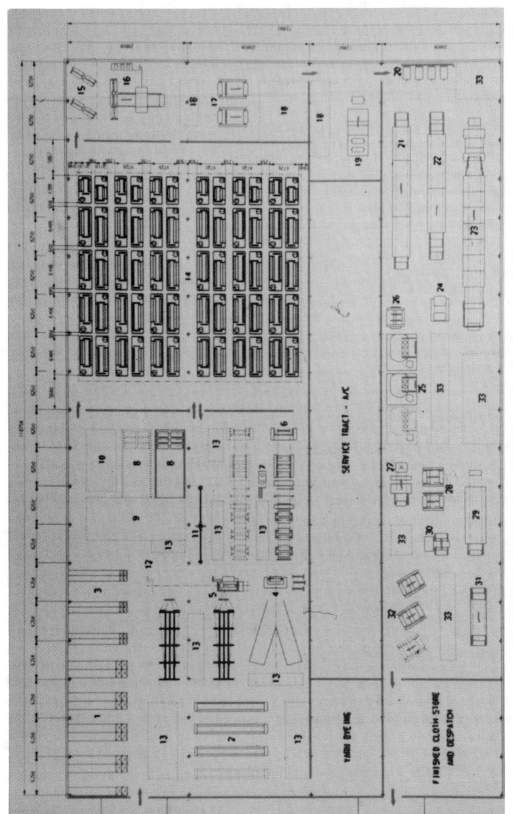

FIGURE 22.7 Plant layout and installation drawings using CAD (courtesy of Sulzer Ruti).

labor and service facilities used in a factory to produce and ship a salable item. A good plant layout is an arrangement that permits the product to be produced with minimum unit cost in the shortest time span. CAD/CAM can be used in this capacity to design an appropriate layout that will benefit the plant. This will lead to better space utilization and greater profit for all involved. It will also allow working with floor plans to let the user get familiar and experiment with the placement of the machines.

Process layout includes areas that have one kind of production equipment. CAD is used in this area to put the equipment around on the floor to get the most out of its use. The second type of layout, production layout, is the arrangement of equipment in sequence to make the product. CAD used in this way allows for the most efficient use of an entire plant.

Although the use of CAD and CAM started earlier, the commercial introduction of CAD/CAM systems for textiles took place at ITMA 87 in Paris.

3.4 Purchasing a CAD System

Purchasing a CAD system is a strategic decision that should be carefully and fully evaluated. INFO DESIGN suggests ten points to consider before purchasing a CAD system [1].

(1) Simplicity – CAD system should be assessed for ease of use which should include "user friendliness." Some of the other indicators to determine use are the variety of naturally interactive input devices, communication between different operation formats and easy installation.

(2) Customer support – It should cover two consecutive and connected periods: 1) from installation through the integration and 2) technical and commercial maintenance and development thereafter.

(3) User groups – It would be to the advantage of a purchaser if a user group is already established either by the vendor or by users of the textile CAD system. A user group meeting is a natural forum to change ideas and help each other to discover the potentials of the CAD system.

Opinions of a user group about their CAD system would be very valuable to make a decision about that particular system.

(4) Documentation (user manuals) – Considering the high employee turnover in the textile industry, good documentation is essential to train the new staff members. Documentation should be understandable, easy to follow and complete. It should explain the different functions and tools available within the program using real examples.

(5) Development – Software is a living thing and must be continuously developed and updated in order to be viable. A prospective buyer should investigate the vendor's software development plans and revenue allocation for that purpose.

(6) Focus – It would be wiser to purchase a CAD system from a vendor that focuses solely on the development of textile design and manufacturing software. Relatively recent introduction of computer aided design systems to an industry that is well established and traditional, necessitates such a need for specialization.

(7) Flexibility – Since CAD/CAM software for textile design applications is not an off-the-shelf item, a vendor should tailor the software programs to the client's specific needs. As the need for new applications develops, software modules should be added to the system.

(8) The user pool – The user pool of a CAD system should be as large as possible.

(9) Service – Before purchasing a CAD system, most companies believe that they will be doing the work they were doing before, only faster and probably with fewer employees. The reality is that, once the CAD system is integrated, the companies put out far more work than before. Therefore, the vendor should be able to offer assistance and service to their client who may need more resources in terms of hardware or software.

(10) Resources – It would be wise to purchase one of the most successful and widespread

programs available in the field. Regional and global market share of the vendor is important for several factors, including a larger user pool to interact with, a greater diversity of experience, and a larger number of likely available operators.

4. COMPUTER INTEGRATED MANUFACTURING (CIM)

CIM is a computerized total production system. CIM is not a single system but an integration of different systems into a single package. CIM improves the operation and product quality control by connecting each process (textile machinery) to a computer. The information from the network can be used to save labor, save energy, improve product quality, decrease cost and appropriate maintenance.

The manufacturing industry has been vigorously pursuing the implementation of factory automation the last several years. Plant management based on experience and intuition has only allowed the plant to obtain and evaluate data for each process individually. However, CIM makes it possible to accumulate data in real time and eliminate waste and bottlenecks between processes by controlling each process while monitoring the data. This makes it possible to improve the operation and effectively control the quality throughout the plant.

Cahill lists six subsystems for a CIM system [2]:

- high-output processing machinery
- on-machine microprocessors
- on-machine automation
- materials transport automation
- mill communication network
- machine "intelligent" software

In general, three basic technology building blocks are used to establish a CIM plant:

- process technology: production machinery and automation to minimize manufacturing labor
- control technology: microprocessors, computers, programmable controllers for self-regulatory manufacturing

process control. Microprocessors are essential parts of modern textile machinery. They monitor, control and protect mechanical and electronic functions; collect, store and retrieve machine productivity data. Microprocessors make bidirectional communication with host computer and production control systems possible. Figure 22.8 shows the console of a microprocessor on a weaving machine that is used for bidirectional communication with a host computer. Computers have made machine interface easier. Automatic control of textile machinery by the design engineer is possible using numerical control (NC).

- knowledge technology: computers capable of interpreting information and attaining real-time decision making

According to McAllister, "When a manufacturing system by itself can detect what is happening (monitors), understand the meaning of what is happening (diagnostic information), reach an appropriate decision (expert systems) and take self-action (automation), then the textile plant has entered into machine-intelligent manufacturing and that is the foundation of a next-generation textile plant" [3].

Decision science, expert systems, artificial intelligence and knowledge engineering are the tools to construct a CIM environment in a plant. A plant's overall efficiency is greatly affected and increased by using computers. Although CIM plants are highly automated, they are not "peopleless" plants. People are still the most important ingredient of a CIM plant because of their ability to learn, adapt and develop new things.

ITMA 91 is considered to be the "CIM" ITMA where automation has made significant improvements in textile processing. "Lights out" manufacturing got closer to reality. CIM plants usually have a degree of automation around 70%.

4.1 Process and Machine Data Logging

Process data includes operator data, product and material data, i.e., data that are not asso-

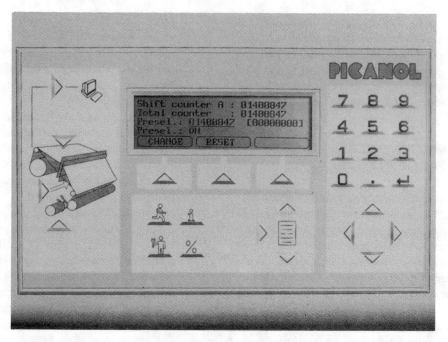

FIGURE 22.8 Microprocessor console on a weaving machine (courtesy of Picanol).

ciated with machine parameters. Machine data includes machine running times, details of defects, malfunction and waiting times, etc. The process and machine data can be stored and accessed via computers built in the machines. The data can be continuously monitored and/or transferred to a master computer for analysis which can be used to optimize machine capacity planning and reduce downtime.

4.2 Production Planning and Control Systems

Production planning and control systems are designed to help the managers and planners. They are usually part of the Management Information Systems (MIS). MIS focuses on macro-level performance and valuation monitoring. Production planning and control systems are used to integrate planning, control and execution of manufacturing processes. Production planning and control systems allow planning of quantities, checking delivery dates, monitoring order progress, materials management, capacity planning and scheduling. Process and machine data provide the necessary input for production planning and control systems.

Stocks and inventories are easily managed using computer assisted stock control systems. Stock control systems can be interfaced with production planning and control systems.

Production planning and stock control systems are very valuable for the success of Just in Time (JIT) and Quick Response (QR) systems. The JIT system requires having the right quantity of product, at the right time and in the right place. The purpose of the QR system is to provide fast flow of information from raw material supplier to the consumer. Reduction of cycle (delivery) time of a product is shortened with the QR system.

Figure 22.9 shows schematics of an integrated production control system to control the entire process in a weaving mill. Figure 22.10 shows the central control room of a modern spinning mill with CIM. A central computer directly monitors state-of-the-art equipment.

5. EXPERT SYSTEMS

Expert systems are computer programs based on scientific information systems in which the empirical and theoretical knowledge of experts

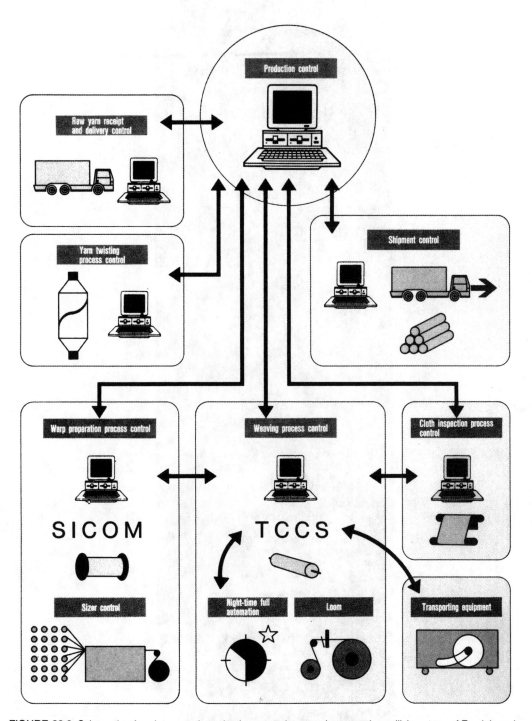

FIGURE 22.9 Schematic of an integrated production control system in a weaving mill (courtesy of Tsudakoma).

FIGURE 22.10 Central control room of a CIM spinning mill (courtesy of Muratec).

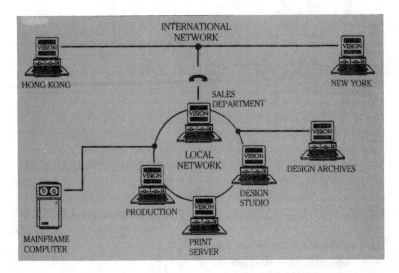

FIGURE 22.11 Schematic of a typical network system (courtesy of Info Design).

is stored and processed. Expert systems involve knowledge on the subject, relationships between various modules and utilization of events. Expert systems are used to solve problems. They are also called "intelligent systems" because they are used for perception, drawing conclusions and learning.

6. NETWORK SYSTEMS

Networking connects different information systems together. There are already several worldwide network systems that can be accessed from most of the countries in the world. Examples of these worldwide networks are Internet and Bitnet. Users of the network systems can communicate via electronic mail and electronic data interchange. Figure 22.11 shows a schematic of a typical network system.

7. FUTURE

Software incompatibility and computer dependence are the two factors that are affecting the networking and other computer applications negatively. Computer applications should be transferable among the computers. Today, manufacturers of computers use different operating systems. An optimized, standard operating system would make communications and information exchange a lot easier. Interactive communication is already becoming a reality.

Imaging technology is gaining importance. With this technology, documents such as forms, drawings, pictures and charts can be scanned and digitized so that they can be manipulated electronically. The user can combine all the product information on a single set of forms, including text and graphics. New forms can be created, stored and retrieved easily. Such systems are already commercially available.

References and Review Questions

1. REFERENCES

1. INFO DESIGN, A Vision Guide: When Researching CAD/CAM, 104 West 40th Street, 8th Floor, New York, NY 10018.
2. Cahill, N., "Textiles: Where Will the Industry be in 2018?" *Textile World,* September 1993.
3. McAllister, I. III, "Electronics Will Drive Textiles Again in the 90's," *Textile World,* August 1990.

1.1 General References

Stupperich, W., "Information Technology in the Textile Industry," *International Textile Bulletin,* Vol. 1, 1993.

"ITMA 91 Gives Boost to Textiles' Automation," *Textile World,* December 1991.

"What Automation at ITMA Means to Textiles," *Textile World,* December 1991.

CIM in Spinning Mills, Murata Machinery, Ltd.

2. REVIEW QUESTIONS

1. What is the difference between CAD, CAM and CIM? Explain.
2. What are the advantages of using PLCs in a manufacturing environment?
3. Explain the differences between high level, assembly and machine languages.
4. What are the other potential application areas of computers in textiles? Discuss how the computers could be used in those applications.

STANDARDS AND REGULATIONS

Standards and Regulations

S. ADANUR

1. INTRODUCTION

Standards and regulations are an integral part of industrial textiles. In fact, due to their nature, industrial textiles may need to be regulated more than traditional consumer textiles. There are several reasons for this. One of the most important characteristics of technical textiles is that they involve the safety element, as part of their performance, more than traditional textiles. A failure of an industrial textile may cause more severe consequences and in general may affect a group of people rather than an individual.

In recent years ISO 9000 standards have gained considerable attention in the textile industry, again more in industrial textiles than traditional textiles. ISO is important for international trade as well as domestic business. The ISO standards are reviewed in this chapter.

This chapter also includes some national and international bodies for regulations, safety and standards.

2. ISO 9000 QUALITY STANDARDS

ISO stands for the equivalent of International Organization for Standardization in French, headquartered in Geneva, Switzerland. Its purpose is to develop and promote common standards worldwide. ISO has about a hundred member countries, one of which is the United States, represented by the American National Standards Institute (ANSI). ISO has hundreds of technical committees and thousands of subcommittees and working groups to prepare standards applicable worldwide.

The ISO 9000 series is a set of international standards for quality assurance and management. It is a written set of good practice procedures which, when followed, will result in consistency for production, design, or provision of services. There are five standards in the set: ISO 9000, ISO 9001, ISO 9002, ISO 9003 and ISO 9004. These standards are generic and adaptable to quality systems already in place.

It should be emphasized that ISO 9000 is not a set of "product standards." ISO 9000 applies to all industries and is not product specific. It does not address the technological issues; i.e., it does not specify or teach any existing or new technology to be used in order to produce or improve the quality of a particular product. ISO 9000 complements the product standards and refers to quality system elements that should be implemented to have consistent and therefore high quality products and services.

ISO 9000 provides a uniform approach for registering quality systems and is acceptable worldwide. The standards evolved from the 1979 British Standards, BS 5750. Numerous standardization bodies have adopted the contents of these standards. In the United States, the ISO 9000 series of standards was adopted jointly by the ANSI and ASQC (American Society for Quality Control) as Q90 series of standards. As a result, ISO 9000–9004 series corresponds to Q90–94 in the United States. The European equivalent of these standards is the EN 29000 series.

ISO 9000 standards are designed to be used for establishing and maintaining quality management and systems for company use and to satisfy outside contracts. ISO 9000 requires extensive documentation by mandating management to record all changes and decisions. The unwritten rule concerning the application of ISO 9000 is that if all personnel were suddenly replaced, the new people, properly trained, could use the documentation to continue making the product or providing the services as before.

ISO 9000 standards were developed by the ISO Technical Committee 176 (ISO TC 176). The committee was formed in 1980 and issued the first ISO 9000–9004 series of standards in 1987. The standards are reviewed and updated if necessary, every five years.

2.1 Contents of ISO 9000 Series

As mentioned earlier, the ISO series is made up of five standards. ISO 9000 (Q90), the first standard in the series, is the road map for understanding the rest of the series and provides key quality definitions. This helps an organization to choose which contractual standard (9001, 9002 or 9003) is best suited for their business. Each element of the ISO 9000 series is suitable for a different type of business:

ISO 9000:   Quality management and quality assurance standards
            Guidelines for selection and use
ISO 9001:   Quality Systems – Model for qual-ity assurance in design/development, production, installation and servicing
ISO 9002:   Quality Systems – Model for qual-ity assurance in production and installation
ISO 9003:   Quality Systems – Model for qual-ity assurance in final inspection and testing
ISO 9004:   Quality management and quality system elements
            Guidelines for interpretation of the standards

Figure 23.1 shows the relationships among the ISO standards. ISO 9001 contains all the requirements in ISO 9002 which, in turn, contains all the requirements in ISO 9003.

ISO 9001 is a quality system model for quality assurance in design, development, production, installation and servicing. It is like a contract between supplier and customer that demonstrates the supplier's capabilities in the above areas. This is the highest ISO qualification an organization can have. It addresses all of the operations that affect the quality or fitness for use of a good or service. The elements of ISO 9001 are shown in Table 23.1 along with 9002 and 9003.

ISO 9002 is a quality system model for qual-

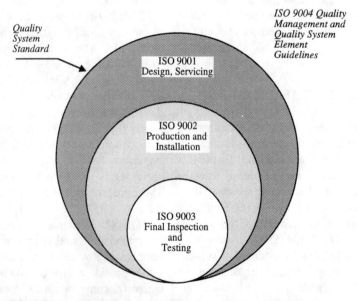

FIGURE 23.1 ISO 9000 standards (courtesy of DuPont ISO 9000 Services).

TABLE 23.1 The Elements of ISO 9001–9003 Systems (courtesy of DuPont ISO 9000 Services).

	ISO 9001	ISO 9002	ISO 9003
1. Management responsibility	X	X	X
2. Quality system	X	X	X
3. Contract review	X	X	
4. Design control	X		
5. Document control	X	X	X
6. Process control	X	X	
7. Purchasing	X	X	
8. Purchaser supplied product	X	X	
9. Product identification traceability	X	X	X
10. Inspection and testing	X	X	X
11. Inspection, measuring and test equipment	X	X	X
12. Inspection and test status	X	X	X
13. Control of non-conforming product	X	X	X
14. Corrective action	X	X	
15. Handling, storage, packaging and delivery	X	X	X
16. Quality records	X	X	X
17. Internal quality audits	X	X	
18. Training	X	X	X
19. After-sales servicing	X		
20. Statistical techniques	X	X	X

ity assurance in production and installation. It differs from ISO 9001 in that it does not cover Design Control (i.e., R&D) and Servicing after sales. ISO 9002 assures that a company has the capability to supply acceptable products and services to its customers. It also certifies that the company is able to document and control all of the processes and procedures from manufacture to installation.

ISO 9003 is a quality system model for quality assurance in final inspection and testing. It is the least strict of all the series that mainly deals with detecting, documenting, controlling nonconforming material in the final inspection and test of a product or service. In addition to design control and after-sales servicing, ISO 9003 does not cover contract review, corrective action, internal quality audit, purchasing, purchaser supplier product and process control.

ISO 9004 describes guidelines for quality management and quality system elements. It explains a basic set of elements by which quality management systems can be developed and implemented.

The elements included in ISO 9004 are:

- quality in marketing, production, specification, and design
- product safety and liability

- control in production
- product verification
- personnel
- poor quality costs

Benefits of ISO Registration

ISO registration is required in order to do business with the European Community (EC). Figure 23.2 shows the global impact of ISO 9000. ISO 9000 certification may provide a company the following benefits:

- product consistency
- advantage in government contracts
- improved quality of goods and services
- advantage for sole supplier relationship with a customer that is already ISO 9000 certified and who may require certification from its suppliers
- improved internal operations
- reduced waste and scrap
- marketing advantages

2.2 Third Party Quality System Registration

ISO 9000 was originally designed with a traditional two party audit system in mind where

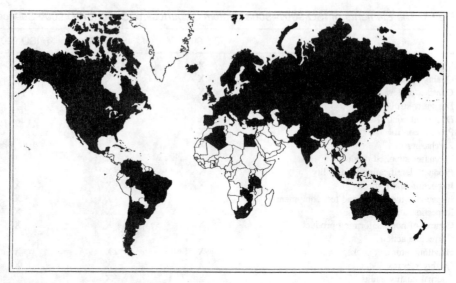

FIGURE 23.2 Global impact of ISO 9000 (Sources: DuPont, Robert Peach & Associates, 1994).

the customer directly audits the supplier. Experience showed that an impartial third party audit was the best. Just as companies require an impartial auditor to examine financial systems, a third party auditor should assess quality systems. Therefore, today, a company who wants to get registered for ISO 9000 must be certified by an accredited registrar agency. This is called registration by a third party and is shown in Figure 23.3.

Each registrar is accredited by a national accreditation body. In the United States, the "American National Accreditation Program for Registrars of Quality Systems" is established by the ANSI and the Registrar Accreditation Board (RAB, an affiliate of the American Society for Quality Control) for accrediting registrars of quality systems. The operation of the program is administered by the RAB. The RAB examines a registrar's procedures and performance against

certain criteria, and if appropriate, accredits that registrar to audit companies against the ISO 9000 standards. The RAB publishes a directory of accredited registrars and a directory of supplier companies registered by accredited registrars.

2.3 ISO 9000 Registration Process

There are several action steps that need to be completed before the ISO registration. These can be summarized as follows:

- Implementation must start at the top management level with complete commitment.
- Obtain all of the information possible on the ISO requirements from standards organization, and others who have already been through the process.

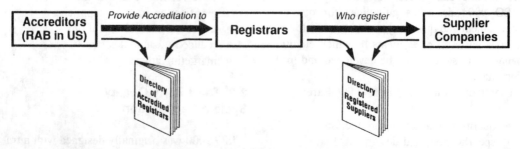

FIGURE 23.3 Third party registration (courtesy of DuPont ISO 9000 Services).

- Define the scope and choose which ISO certification best suits your organization.
- Assemble a steering committee and develop action plans.
- Select an ISO registrar to register the organization with an application.
- Confirm scope and chosen ISO qualification with registrar.
- Train employees and spell out exactly what is expected from them.
- Assess the organization's current quality programs against the chosen ISO quality system.
- Record and update the existing quality system in a quality manual. Assemble all other documentation.
- Send a copy of the quality manual to the registrar for inspection.
- Be audited by the registrar; take corrective measures based on findings by the registrar's audit team.
- Final visit by the registrar and official certification.

These steps may differ slightly from registrar to registrar. DuPont Quality Management and Technology Center (QM&TC) identifies nine basic milestones and related activities that must be completed as shown in Figure 23.4.

The registration process will have a duration of about ten to fourteen months. It is estimated that one mean-year of preparation time is required per 100 employees. The average cost of registration is $20,000–25,000, plus preparation costs that depend on how well the existing quality system is developed. Training costs are not included in this estimation. After registration, there will be surveillance twice a year and re-certification every three years, each of which requires additional expenses. The first pass rate for companies is around 30%. Figure 23.5 shows typical audit results for deficiencies.

3. ASSOCIATIONS FOR STANDARDS, REGULATIONS, AND SPECIFICATIONS

The following organizations can provide copies, information and/or assistance on revised/draft/current/new standards, guides, catalogs, indices, publications and related documents.

- American Association of Textile
  Chemists and Colorists (AATCC)
  P.O. Box 12215
  Research Triangle Park, NC 27709
  USA

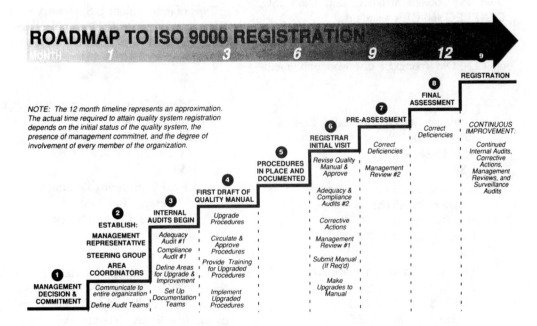

FIGURE 23.4 Roadmap to ISO 9000 registration (courtesy of DuPont ISO 9000 Services).

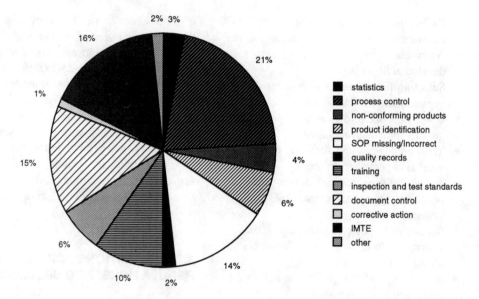

FIGURE 23.5 Typical audit results for deficiencies.

- American National Standards Institute (ANSI) (the U.S. member body for ISO and IEC)
  11 West 42nd Street, 13th Floor
  New York, NY 10036
  USA

Type of information: ANSI and ANSI approved U.S. Industry Standards. International (ISO and IEC) and Foreign Standards and Draft ISO, CENELEC and CEN standards.

All standards published by the American National Standards Institute (ANSI) are available on CD-ROM (compact disc-read only memory).

- American Society for Quality Control
  P.O. Box 3005
  Milwaukee, WI 53201
  USA

- American Society of Testing and Materials (ASTM)
  1916 Race Street
  Philadelphia, PA 19103
  USA

- American Textile Manufacturers Institute
  1801 K Street, Suite 900
  Washington, D.C. 20006
  USA

- Defense Personnel Support Center
  2800 South 20th Street
  Philadelphia, PA 19145
  USA

- Document Center
  1504 Industrial Way, Unit 9
  Belmont, CA 94002
  USA

Type of information: U.S. industry standards. Federal, military and DOD specifications. Standards and handbooks. International and foreign standards.

- Defense Technical Information Center
  Attn: DTIC-RSO
  Building B, Cameron Station
  Alexandria, VA 22304-6145
  USA

- Global Engineering Documents
  15 Inverness Way East
  Englewood, CO 80112-5704
  USA

Type of Information: U.S. Industry Standards Federal (FED), Military (MIL) and Department of Defense (DOD). Specifications and standards, handbooks, international, foreign and European standards. Historical standards.

- Industrial Fabrics Association
  International (IFAI)
  345 Cedar St., Suite 800
  St. Paul, MN 55101
  USA
- International Standards Organization
  (ISO)
  Rue de Varembe 1
  CH-1211 Geneva 20
  Switzerland
- National Institute of Standards and
  Technology (NIST)
  U.S. Department of Commerce
  Administration Building, Room A629
  Gaithersburg, MD 20899
  USA
- National Standards Association
  1200 Quince Orchard Boulevard
  Gaithersburg, MD 20878
  USA
- National Technical Information Service
  5285 Port Royal Road
  Springfield, VA 22161
  USA

- Occupational Safety and Health
  Administration (OSHA)
  Birmingham Area Office
  US Department of Labor – OSHA
  Todd Mall, 2047 Canyon Road
  Birmingham, AL 35216
  USA

- Environmental Protection Agency (EPA)
  National Headquarters
  401 M-Street, S.W.
  Washington, DC 20460
  USA

- Standards Sales Group (SSG)
  20025 Highway 18
  Apple Valley, CA 92307-2639
  USA

Type of information: International and foreign standards, publications and other reference materials. Select U.S. and foreign regulatory compliance information. Translation services.

References and Review Questions

1. REFERENCES

1.1 General References

ISO 9000 Is Here, E. I du Pont de Nemours and Co., June 1992.

Underwriters Laboratories, Inc., Questions and Answers about the UL ISO Registration Program, 1989.

Marquardt, D. W., "ISO 9000: A Universal Standard of Quality," *Management Review,* 1992.

Herzog, S. L. "Tomorrow's Stairway to Total Quality Management . . . Today," *Processing,* Vol. 5, No. 3, March 1992.

International Standards Organization, ISO 9004: 1987 "Quality Management and Quality System Elements—Guidelines."

2. REVIEW QUESTIONS

1. What is ISO 9000? Explain.
2. Does implementation of ISO 9000 provide a quality product? Why?
3. What are the advantages of third party registration?
4. What are the major steps in the ISO 9000 registration process?

FUTURE OF INDUSTRIAL TEXTILES

Future of Industrial Textiles

S. ADANUR

1. INTRODUCTION

The market share of industrial textiles has increased tremendously within the last thirty years. The major factors for this increase include new developments in synthetic fibers, new manufacturing processes, new developments in non-textile technologies and increase in human activities. Table 1.1 shows the percent share of technical textiles' production in total textile production for selected countries between 1980 and 1990. Table 24.1 shows the increase in industrial market segments within the last five years.

The table shows that the relatively new market segments had the highest annual growth rate within the last five years. The airbag market had the highest growth rate followed by geosynthetics, medical products, and safety and protective clothing. One reason for the continued growth for these markets is the standardized regulations and specifications by government and industry.

Technological innovation in polymers, fibers, fabrics and manufacturing processes and equipment will continue to be the driving factor for the growth of industrial textiles. Another factor for growth will be the ability of industrial textiles to find solutions to the new problems of non-textile industries. As an example, the human immunodeficiency virus (HIV) presented a challenge to the medical textile industry to protect medical personnel. Further advances in fiber and textile material analysis will create new application areas for textiles. For example, forensic textile science has been a helpful source for the FBI to solve some of the difficult crime cases [2].

Research will continue to be an important factor to solve new textile or non-textile related problems. Applied research will help to find new and improved applications for industrial textiles. Companies that will search for new and innovative products, processes and applications will continue to stay on top. In the increasingly competitive, rapid communication age, the fastest and the best will have the best chance to survive and accelerate.

Textile technology can combine several characteristics in a product that probably no other industry could provide: flexibility or stiffness, light weight and strength. A properly designed and developed fiber or fabric can have excellent drapeability and flexibility yet it can provide excellent strength against various loads. In fact the strength of some industrial fiber materials is a lot higher than steel and other metals. Nevertheless, in the future, the strength of textile fibers will be increased further, creating "hyperstrong" fibers. If stiffness rather than flexibility is required in a product, then the textile product can be impregnated with resin resulting in lightweight and strong textile structural composites. Another equally important characteristic of a fabric made of fibers is the porosity that can be controlled by fabric structure as well as by coating and laminating. The porosity of textile materials can be changed from zero (completely sealed structure) to very high percent open areas, such as a net structure.

Barrier fabrics for safety and protection of humans as well as products from virtually every type of hazard have great potential to growth. With the advancing and more sophisticated technologies in non-textile markets, other industries may look at industrial textiles for solutions to their specific problems.

TABLE 24.1 Fabric Usage for Some Industrial Market Segments
(estimated; units in millions of square yards, except where indicated) [1].

Industrial Market Segment	1991	1992	1993	1994	1995
Airbag	4	7	12	20	32
Architectural fabrics	3	3	3.25	3.5	3.7
Automotive	285	290	300	292	347
Awning and canopies	19	19.5	19.5	20	22
Banners and flags	9	10	11	13	15
Casual furniture	49	50	50	50	52
Geosynthetics	360	380	400	435	466
Marine products	19	18	18.5	19.2	17
Medical (in billions of sq. yards)	2.1	2.3	2.4	2.6	2.8
Safety and protective	225	250	275	300	322
Single-ply roofing	115	110	110	115	117
Tarpaulin	105	110	110	112	113
Tent and tent rental	10	10	10	10.5	11

Another growth area in industrial textiles will be "environmentally friendly" materials. Currently, there is extensive research in this area to develop high performance "biodegradable" fibers. Regulations to protect the environment will continue to play a major role for the efforts in this area. In fact, some believe that in the future, products that cannot be recycled or are non-biodegradable may not be allowed for production through regulation.

2. SMART MATERIALS

If the patent literature is any indication, one potential future growth area is "smart materials" or "intelligent materials." A smart polymer or material can be described as a material that will change its characteristics according to outside conditions or a stimulus [3]. This is a relatively new area and deserves some attention.

Mitsubishi developed a Shape Memory Polymer (SMP) which is a thermoplastic polyurethane with "elastic memory." When the polymer part is heated above its glass transition-temperature (T_g) but below its melting temperature (T_m), it changes from a rigid, glassy state to a rubbery state at which it is highly flexible and pliant. This allows it to be deformed to a new shape that can be "frozen" in by cooling below the T_g. If the part is subsequently heated above the T_g, the polymer will "remember" its original shape and spontaneously return to that shape. The polymer's shape memory properties can be tailored for various applications by formulating the polymer for a specific T_g (to within 1°F) over a wide range ($-30°F$ to $+158°F$).

Mitsubishi combined its shape memory polymer with Komatsu Seiren's polymer coating technology to make "dream cloth." It is claimed that the dream cloth polymer molecule stretches and shrinks in response to temperature; similar to the body skin, the cloth opens when hot and closes when cold. Therefore the breathability of the fabric changes. Possible application areas include skiwear and other sportswear.

Research is being done on photochromic materials. A material whose absorption spectrum changes when it is irradiated with light and reverts to its initial state upon cessation of irradiation is called a photochromic material. These types of polymers are already available. A possible application of photochromic polymers is the fabrics that change their color according to the weather and intensity of light. This, in turn, changes the reflectance properties of the fabric.

Resistively shunted piezoceramic fibers used as reinforcement in a structural composite material offer the potential to significantly increase vibration damping capability.

In general, polymers do not conduct electricity, which has been an advantage in many applications. However, within the last twenty years, research has been done to develop a new class of organic polymers with the ability to conduct electrical current. If these conductive polymers would be used in fibers and fabrics, the electrical and mechanical properties of the fabric could be changed. For example, the color of fabric may be changed by use of a small battery that would be invaluable for military applications.

Another research area for military fabrics is "adaptive camouflage" fabrics. Color control in a fabric could be achieved by using hollow fibers using a principle called electrophoresis. The fibers, with the diameter of human hair, would be filled with electrically charged particles of paint pigment suspended in a different colored dye. Different combinations of pigments and dyes would be used in the same uniform. The color combination would be controlled by moving the solid pigment toward and away from the outer surface of the fabric so that a viewer would see either the paint color or the dye color at any one time.

Research is also being done to develop fabrics that would allow medical teams to pinpoint wounded soldiers on the battlefield. The fabric, woven with metallic fibers, would emit low-intensity signals when ripped. These signals would be strong enough to be detected by a friend but not strong enough to give away the soldier's position to the enemy. The size and location of a wound could also be determined from a distance. The timing is critical to save a wounded soldier's life.

Polymers are suitable to develop biologically active materials. Biologically active textile materials are antibacterial, antifungal, anticancer, X-ray active, insecticidal, anti-inflammatory, anesthetic, antivirus, antioxidant, fermentative active, hemadilutative, immunodepressant, etc. Extensive research was done in the former Soviet Union in these areas.

3. CONCLUSIONS

In the United States alone there are approximately 30,000 patents granted for polymers. Combined with virtually endless possibilities of fiber and fabric structures and ever-advancing manufacturing technologies, development of products with custom tailored properties for any type of application will be limited only by human imagination. Industrial textiles will continue their growth and penetrate in almost every other non-textile industry. Industrial textiles can only increase their presence in mankind's journey in the years to come.

24.2

References and Review Questions

1. REFERENCES

1. Strzetelski, A., "Here Comes the Future," *Industrial Fabric Products Review*, January 1995.

2. Hall, D. M., private communication.

3. Hongu, T. and Phillips, G. O., *New Fibers*, Ellis Harwood, 1990.

2. REVIEW QUESTIONS

1. What is the definition of a "smart material"? What are its characteristics?

2. A densely woven cotton fabric swells in the rain and becomes an even more dense or tight structure. This, in turn, causes the fabric to resist further water or rain penetration. Should, therefore, a fabric made of cotton be considered a smart structure? Why?

3. What are the possible development areas of industrial textiles in the future? Discuss your predictions.

Appendix 1

Air, Tent, and Tensile Structures—
Fabric Specification Table

This table was originally published in *Fabrics and Architecture, Specguide 1995,* November–December, 1994 by the Industrial Fabrics Association International (IFAI).

Disclaimer statement:

"All information included in the table was submitted by the IFAI Awning Division member firms. The specifications were submitted voluntarily and their accuracy is the responsibility of the manufacturer. This table is not an endorsement of a company or product by *Fabrics & Architecture,* the Awning Division, or IFAI. The reader is encouraged to contact the companies directly for further information."

Footnotes:

NA   Information not available or not provided.

1. Distributed by the Astrup Co.
2. The Awntop line includes multiple products. When a range of values is given, it reflects the values of the entire product line.
3. FED-STD-191 (5041).
4. Distributed by MacKellar Associates.
5. Overlapping weld after grinding.
6. Regional, national and international distribution.
7. The material is ultraviolet protected, anti-mildew, wicking resistant, and topside acrylic coated for reinforced abrasion and easy cleaning.
8. Tenara fabric is woven by C. Cramer & Co.
9. The polyester material is Hoechst Trevira, a registered trademark of Hoechst A.G.
10. FED-STD-191 (5134) modified.

Trade Name:	8028	9032	Air Tite 1528	Architent	Architent Blackout	Architent Wideside
Trade name owner	Seaman Corp.	Seaman Corp.	Erez Thermoplastic	Herculite Products	Herculite Products[1]	Herculite Products
Base fabric						
Material	Polyester	Polyester	Polyester	Polyester	Polyester	Polyester
Weight	7.5 oz/yd^2	10 oz/yd^2	8.9 oz/yd^2	NA	NA	NA
Weave style	NA	NA	Weft insertion	NA	NA	NA
Yarn count (warp, fill)	NA	NA	120, 188 tpi	NA	NA	NA
Coating						
Material	Vinyl	Vinyl	Vinyl	Vinyl laminate	Vinyl laminate	Vinyl laminate
Weight (top, bottom)	NA	NA	10.9, 9.4 oz/yd^2	NA	NA	NA
UV topcoat material	NA	NA	Acrylic or PVFD	NA	NA	NA
UV topcoat weight	NA	NA	NA	NA	NA	NA
Finished fabric						
Test method	NA	NA	ASTM D 3776	NA	NA	NA
Thickness	NA	NA	NA	NA	NA	NA
Weight	NA	NA	28–29 oz/yd^2	14 oz/yd^2	16 oz/yd^2	10 oz/yd^2
Roll width, usable	56 in	56 in	71–73 in	61 in	61 in	90 in
Tongue tear (warp, fill)	275, 275 lb/in	300, 300 lb/in	315, 280 lb/in	115, 125 lb/in	140, 165 lb/in	30, 35 lb/in
Test method	FED-STD-191 (5134)	FED-STD-191 (5134)	ASTM D 2261	FED-STD-191 (5134)	FED STD-191 (5134)	FED-STD-191 (5134)
Trapezoidal tear (warp, fill)	85, 85/lb/in	140, 140 lb/in	80, 60 lb/in	NA	NA	NA
Test method	FED-STD-191 (5136)	FED-STD-191 (5136)	ASTM D1117	NA	NA	NA
Grab tensile (warp, fill)	700, 700 lb/in	840, 840 lb/in	670, 620 lb/in	240, 215 lb/in	250, 220 lb/in	115, 115 lb/in
Test method	FED-STD-191 (5100)	FED-STD-191 (5100)	ASTM D 751	FED-STD-191 (5100)	FED-STD-191 (5100)	FED-STD-191 (5100)
Strip tensile (warp, fill)	515, 515 lb/in	650, 650 lb/in	440, 405 lb/in	NA	NA	NA
Test method	FED-STD-191 (5102)	FED-STD-191 (5102)	ASTM D 751	NA	NA	NA

Trade Name:	8028	9032	Air Tite 1528	Architent	Architent Blackout	Architent Wideside
Adhesion	10 lb/in (min)	10 lb/in (min)	15 lb/in	NA	NA	NA
Test method	FED-STD-191 (5970)	FED-STD-191 (5970)	ASTM D 751	NA	NA	NA
Hydrostatic resistance	500 psi	500 psi	NA	340 psi	360 psi	165 psi
Test method	FED-STD-191 (5512)	FED-STD-191 (5512)	NA	FED-STD-191 (5512)	FED-STD-191 (5512)	FED-STD-191 (5512)
Cold crack	−40°F, −67°F	−40°F	−40°F	−40°F	−40°F	−40°F
Test method	Mil-C20696C, P4.4.6	Mil-C20696C, P4.4.6	Mil-C20696	FED-STD-191 (5874)	FED-STD-191 (5874)	FED-STD-191 (5874)
Burning characteristics						
Test method(s)	Meets CSFM, UL-214, NFPA-701, FED-STD-191 (5903), passes 2-s flame-out	Meets CSFM, UL-214, NFPA-701, FED-STD-191 (5903), passes 2-s flame-out	UL-214, NFPA-701, CSFM	CSFM Reg F-122.03	CSFM Reg F-122.03	CSFM Reg F-122.03
Light values						
Test method	NA	NA	Spectra radiometer	NA	NA	NA
Transmission, refl., absorp.	NA	NA	8.5–10.5%, NA, NA	NA	NA	NA
Acoustical properties						
Test method	NA	NA	NA	NA	NA	NA
Sound transmission	NA	NA	NA	NA	NA	NA
Sound absorption	NA	NA	NA	NA	NA	NA
Coefficient of thermal expansion/contraction	NA	NA	NA	NA	NA	NA
Test method	NA	NA	NA	NA	NA	NA

(continued)

Trade Name:	8028	9032	Air Tite 1528	Architent	Architent Blackout	Architent Wideside
Seams (recommended style)	Lap or butt	Lap or butt	NA	NA	NA	NA
Construction method	Heat sealed	Heat sealed	NA	Heat sealed or sewn	Heat sealed or sewn	Heat sealed or sewn
Useful temperature range	NA	NA	−10–150°F	NA	NA	NA
Dead load, seams, warp/fill						
Inch seam @ F	2 in @ room temp.	2 in @ room temp.	NA	NA	NA	NA
lb/in warp/lb/in fill	266 lbs	266 lbs	NA	NA	NA	NA
Inch seam @ F	2 in @ 160°F	2 in @ 160°F	NA	NA	NA	NA
lb/in warp/lb/in fill	133 lbs	133 lbs	NA	NA	NA	NA
Structural properties						
Effect. modulus, warp, fill	NA	NA	NA	NA	NA	NA
Poisson's ratio, warp/fill	NA	NA	NA	NA	NA	NA
Poisson's ratio, fill/warp	NA	NA	NA	NA	NA	NA
Shear modulus	NA	NA	NA	NA	NA	NA
Stretch properties	NA	NA	NA	NA	NA	NA
Uniaxial, warp	NA	NA	NA	NA	NA	NA
Uniaxial, fill	NA	NA	NA	NA	NA	NA
Biaxial, warp (1:1)	NA	NA	NA	NA	NA	NA
Biaxial, fill (1:1)	NA	NA	NA	NA	NA	NA
Mil spec/Fed spec #	NA	NA	NA	NA	NA	NA
Recommended uses	Air and tension structures	Air and tension structures	Architectural structures	NA	NA	NA

Trade Name:	Awntop[2]	Calliope	Gala Tentage	KWB-22 (Blackout)	KWN-18 (Non-Blackout)
Trade name owner	Duracote Corp.	Graniteville Co.	Graniteville Co.[1]	Super Tex	Super Tex
Base fabric					
Material	Polyester	Polyester	Cotton	Polyester	Polyester
Weight	2.7 oz/yd^2	NA	NA	NA	NA
Weave style	Weft inserted	2 ply plain weave	Army duck	Weft inserted	Weft inserted
Yarn count (warp, fill)	9, 9 tpi	NA	NA	18, 18 tpi	18, 18 tpi
Coating					
Material	Vinyl	Pigmented vinyl resin	Flame-retardant vinyl	Vinyl coated	Vinyl coated
Weight (top, bottom)	4–7, 5–7 oz/yd^2	NA	NA	NA	NA
UV topcoat material	None	NA	NA	NA	NA
UV topcoat weight	NA	NA	NA	NA	NA
Finished fabric					
Test method	FED-STD-191 (5041)	FED-STD-191	FED-STD-191	NA	NA
Thickness	NA	NA	NA	NA	NA
Weight	13–18 oz/yd^2	12.5 oz/yd^2 [3]	12.5 oz/yd^2	22 oz/yd^2	22 oz/yd^2
Roll width, usable	61 in	60 in	31 in	61–81 in	61–81 in
Tongue tear					
(warp, fill)	100, 105 lb/in	22, 19 lb/in	5.0, 7.0 lb/in	200, 180 lb/in	200, 180 lb/in
Test method	ASTM D 2262 (8 in)	FED-STD-191 (5134)	FED-STD-191 (5134)	FED-STD-191 (5134)	FED-STD-191 (5134)
Trapezoidal tear					
(warp, fill)	40, 30 lb/in	NA	NA	NA	NA
Test method	ASTM D 1117	NA	NA	NA	NA
Grab tensile					
(warp, fill)	250–265, 225–240 lb/in	500, 391 lb/in	160, 145 lb/in	280, 260 lb/in	280, 260 lb/in
Test method	ASTM D 5034	FED-STD-191 (5100)	FED-STD-191 (5100-1)	FED-STD-191 (5102)	FED-STD-191 (5102)
Strip tensile					
(warp, fill)	160–180, 140–155 lb/in	NA	NA	NA	NA
Test method	ASTM D 5304	NA	NA	NA	NA

(continued)

Trade Name:	Awntop	Calliope	Gala Tentage	KWB-22 (Blackout)	KWN-18 (Non-Blackout)
Adhesion	4–7 lb/in	NA	NA	10 lb/in	10 lb/in
Test method	ASTM D 903	NA	NA	FED-STD-191 (5970)	FED-STD-191 (5970)
Hydrostatic resistance	360–400 psi	80+ cm	55 cm	NA	NA
Test method	FED-STD-191 (5512)	FED-STD-191 (5514)	FED-STD-191 (5514)	NA	NA
Cold crack	−30°F	NA	NA	Pass −40°F	Pass −40°F
Test method	ASTM D 2136	NA	NA	MIL-C-20696	MIL-C-20696
Burning characteristics					
Test method(s)	Registered, CSFM	CSFM Reg. 76.06 UL-214 as to flammability only	CSFM Reg. 76.02 UL-214 as to flammability only	CSFM 2-s flame-out	CSFM 2-s flame-out
Light values					
Test method	NA	NA	NA	NA	NA
Transmission, refl., absorp.	Translucent to opaque	NA	NA	NA	NA
Acoustical properties					
Test method	NA	NA	NA	NA	NA
Sound transmission	NA	NA	NA	NA	NA
Sound absorption	NA	NA	NA	NA	NA
Coefficient of thermal expansion/ contraction	NA	NA	NA	NA	NA
Test method	NA	NA	NA	NA	NA
Seams (recommended style)	Lap	NA	NA	NA	NA
Construction method	Heat sealed	Heat sealed, sewn, welded	Sewn	NA	NA
Useful temperature range	0–140°F	NA	NA	NA	NA

Trade Name:	Awntop	Calliope	Gala Tentage	KWB-22 (Blackout)	KWN-18 (Non-Blackout)
Dead load, seams, warp/fill					
Inch seam @ F	NA	NA	NA	NA	NA
lb/in warp/lb/in fill	NA	NA	NA	NA	NA
Inch seam @ F	NA	NA	NA	NA	NA
lb/in warp/lb/in fill	NA	NA	NA	NA	NA
Structural properties					
Effective modulus, warp, fill	NA	NA	NA	NA	NA
Poisson's ratio, warp/fill	NA	NA	NA	NA	NA
Poisson's ratio, fill/warp	NA	NA	NA	NA	NA
Shear modulus	NA	NA	NA	NA	NA
Stretch properties	NA	NA	NA	NA	NA
Uniaxial, warp	NA	NA	NA	NA	NA
Uniaxial, fill	NA	NA	NA	NA	NA
Biaxial, warp (1:1)	NA	NA	NA	NA	NA
Biaxial, fill (1:1)	NA	NA	NA	NA	NA
Mil spec/Fed spec #	NA	NA	NA	NA	NA
Recommended uses	Tents	Tents, canopies, awnings	Circus, public assembly, and party tents	Tension structures, blackout	Tension structures, non-blackout

(continued)

Trade Name	KWN-22 (Non-Blackout)	Polaris Tentage	PRV 1310Q Weatherspan	PRV 1444K FR Klearspan	PRV 16100 Weatherspan	PRV 1613W 61" Gloss
Trade name owner	Super Tex	Graniteville Co.[1]	Snyder Mfg. Inc.	Snyder Mfg. Inc.	Snyder Mfg. Inc.	Snyder Mfg. Inc.
Base fabric						
Material	Polyester	Polyester/cotton	Polyester	Polyester	Polyester	Polyester
Weight	NA	NA	NA	NA	NA	NA
Weave style	Weft inserted	Plain weave	NA	NA	NA	NA
Yarn count (warp, fill)	18, 18 tpi	NA	NA	NA	NA	NA
Coating						
Material	Vinyl coated	Pigmented vinyl coat.	Vinyl	Vinyl	Vinyl	Vinyl
Weight (top, bottom)	NA	NA	NA	NA	NA	NA
UV topcoat material	NA	NA	NA	NA	NA	NA
UV topcoat weight	NA	NA	NA	NA	NA	NA
Finished fabric						
Test method	NA	FED-STD-191	FED-STD-191 (5041)	FED-STD-191 (5041)	FED-STD-191 (5041)	FED-STD-191 (5041)
Thickness	NA	NA	NA	NA	NA	NA
Weight	22 oz/yd^2	11 oz/yd^2	13 oz/yd^2	14 oz/yd^2	16 oz/yd^2	16 oz/yd^2
Roll width, usable	61–81 in	60 in	61 in	61 in	61 in	61 in
Tongue tear (warp, fill)	200, 180 lb/in	8.5, 7.5 lb/in	50, 60 lb/in	30, 30 lb/in	55, 63 lb/in	85, 85 lb/in
Test method	FED-STD-191 (5134)	FED-STD-191 (5134)	FED-STD-191 (5134)	FED-STD-191 (5134)	FED-STD-191 (5134)	FED-STD-191 (5134)
Trapezoidal tear (warp, fill)	NA	NA	NA	NA	NA	NA
Test method	NA	NA	NA	NA	NA	NA
Grab tensile (warp, fill)	280, 260 lb/in	145, 145 lb/in	220, 200 lb/in	140, 125 lb/in	230, 205 lb/in	320, 300 lb/in
Test method	FED-STD-191 (5102)	FED-STD-191 (5100-1)	FED-STD-191 (5100-1)	FED-STD-191 (5100-1)	FED-STD-191 (5100-1)	FED-STD-191 (5100)

Trade Name:	KWN-22 (Non-Blackout)	Polaris Tentage	PRV 1310Q Weatherspan	PRV 1444K FR Klearspan	PRV 16100 Weatherspan	PRV 1613W 61" Gloss
Strip tensile (warp, fill)	NA	NA	160, 140 lb/in	85, 85 lb/in	164, 145 lb/in	225, 210 lb/in
Test method	NA	NA	FED-STD-191 (5304.1)	FED-STD-191 (5304-1)	FED-STD-191 (5304-1)	FED-STD-191 (5304)
Adhesion	10 lb/in	NA	26 lb/in	20 lb/in	26 lb/in	25 lb/in
Test method	FED-STD-191 (5970)	NA	FED-STD-191 (5970)	FED-STD-191 (5970)	FED-STD-191 (5970)	FED-STD-191 (5970)
Hydrostatic resistance	NA	35 cm	345 psi	200 psi	360 psi	425 psi
Test method	NA	FED-STD-191 (5514)	FED-STD-191 (5512)	FED-STD-191 (5512)	FED-STD-191 (5512)	FED-STD-191 (5512)
Cold crack	Pass −40°F	NA	Pass −40°F	Pass −30°F	Pass −40°F	Pass −40°F
Test method	Mil-C-20696	NA	FED-STD-191 (5874)	FED-STD-191 (5874)	FED-STD-191 (5874)	FED-STD-191 (5874)
Burning characteristics						
Test method(s)	CSFM 2-s flame-out	CSFM, CPAI-84	NA	NA	NA	NA
Light values						
Test method	NA	NA	NA	NA	NA	NA
Transmission, refl., absorp.	NA	NA	NA	NA	NA	NA
Acoustical properties						
Test method	NA	NA	NA	NA	NA	NA
Sound transmission	NA	NA	NA	NA	NA	NA
Sound absorption	NA	NA	NA	NA	NA	NA
Coefficient of thermal expansion/contraction	NA	NA	NA	NA	NA	NA
Test method	NA	NA	NA	NA	NA	NA

(continued)

Trade Name:	KWN-22 (Non-Blackout)	Polaris Tentage	PRV 1310Q Weatherspan	PRV 1444K FR Klearspan	PRV 16100 Weatherspan	PRV 1613W 61" Gloss
Seams (recommended style)						
Construction method	NA	NA	NA	NA	NA	NA
	NA	Sewn	NA	NA	NA	NA
Useful temperature range	NA	NA	NA	NA	NA	NA
Dead load, seams, warp/fill						
Inch seam @ F	NA	NA	NA	NA	NA	NA
lb/in warp/lb/in fill	NA	NA	NA	NA	NA	NA
Inch seam @ F	NA	NA	NA	NA	NA	NA
lb/in warp/lb/in fill	NA	NA	NA	NA	NA	NA
Structural properties						
Effect. modulus, warp, fill	NA	NA	NA	NA	NA	NA
Poisson's ratio, warp/fill	NA	NA	NA	NA	NA	NA
Poisson's ratio, fill/warp	NA	NA	NA	NA	NA	NA
Shear modulus	NA	NA	NA	NA	NA	NA
Stretch properties	NA	NA	NA	NA	NA	NA
Uniaxial, warp	NA	NA	NA	NA	NA	NA
Uniaxial, fill	NA	NA	NA	NA	NA	NA
Biaxial, warp (1:1)	NA	NA	NA	NA	NA	NA
Biaxial, fill (1:1)	NA	NA	NA	NA	NA	NA
Mil spec/Fed spec #	NA	NA	NA	NA	NA	NA
Recommended uses	Tension structures, non-blackout	Sidewalls, tops on family cabin, scouting tent				

Trade Name:	PRV 2213W Twin Color	Precontraint 392	Precontraint 702 Opaque	Precontraint 1002 Fluotop S	Precontraint 1202 Fluotop T	Precontraint 1502 Fluotop T
Trade name owner	Snyder Mfg. Inc.	Ferrari[4]	Ferrari[4]	Ferrari[4]	Ferrari[4]	Ferrari[4]
Base fabric						
Material	Polyester	Hi. ten. polyester	Hi. ten. polyester	Hi. ten. polyester	Hi. ten. polyester	Hi. ten. polyester
Weight	NA	NA	NA	NA	NA	NA
Weave style	NA	Mesh fabric	NA	NA	NA	NA
Yarn count (warp, fill)	NA	NA	NA	NA	NA	NA
Coating						
Material	Vinyl	Vinyl	Vinyl	Vinyl	Vinyl	Vinyl
Weight (top, bottom)	NA	NA	NA	NA	NA	NA
UV topcoat material	NA	NA	NA	Fluotop S	Fluotop T	Fluotop T
UV topcoat weight	NA	NA	NA	NA	NA	NA
Finished fabric						
Test method	FED-STD-191 (5041)	NA	NA	NA	NA	NA
Thickness	NA	NA	240-micron top coating	350-micron top coating	NA	NA
Weight	22 oz/yd^2	25 oz/yd^2	22 oz/yd^2	31 oz/yd^2	37 oz/yd^2	43 oz/yd^2
Roll width, usable	61 in	71 in	71 in	71 in	NA	NA
Tongue tear (warp, fill)	95, 95 lb/in	NA	NA	NA	NA	NA
Test method	FED-STD-191 (5134)	NA	NA	NA	NA	NA
Trapezoidal tear (warp, fill)	NA	86, 84 lb/in	53, 38 lb/in	74, 73 lb/in	153, 131 lb/in	157, 151 lb/in
Test method	NA	FED-STD-191 (5136)	FED-STD-191 (5136)	FED-STD-191 (5136)	FED-STD-191 (5136)	ASTM D 1117
Grab tensile (warp, fill)	315, 305 lb/in	NA	445, 440 lb/in	735, 705 lb/in	945, 930 lb/in	1300, 1200 lb/in
Test method	FED-STD-191 (5100.1)	NA	FED-STD-191 (5100)	FED-STD-191 (5100)	FED-STD-191 (5100)	FED-STD-191 (5100)

(continued)

Trade Name:	PRV 2213W Twin Color	Precontraint 392	Precontraint 702 Opaque	Precontraint 1002 Fluotop S	Precontraint 1202 Fluotop T	Precontraint 1502 Fluotop T
Strip tensile (warp, fill)	235, 200 lb/in	310, 360 lb/in	320, 285 lb/in	440, 435 lb/in	600, 565 lb/in	917, 831 lb/in
Test method	FED-STD-191 (5304.1)	FED-STD-191 (5102)	FED-STD-191 (5102)	FED-STD-191 (5102)	FED-STD-191 (5102)	FED-STD-191 (5102)
Adhesion	20 lb/in	18 lb/2 in	22 lb/2 in	26 lb/2 in	26 lb/2 in	NA
Test method	FED-STD-191 (5970)	NFG 31107 (France)	NFG 31107 (France)	NFG 31107 (France)	NFG 31107 (France)	NA
Hydrostatic resistance	525	NA	NA	NA	NA	NA
Test method	FED-STD-191 (5512)	NA	NA	NA	NA	NA
Cold crack	Pass −40°F	NA	NA	NA	NA	NA
Test method	FED-STD-191 (5874)	NA	NA	NA	NA	NA
Burning characteristics Test method(s)	NA	NFPA-701	NFPA-701, CSFM	NFPA-701, CSFM	NFPA-701, CSFM	NFPA-701
Light values Test method	NA	ASHRAE 74-73	NA	NA	NA	NA
Transmission, refl., absorp.	NA	26%, 59%, 15%	NA	NA	NA	NA
Acoustical properties Test method	NA	NA	NA	NA	NA	NA
Sound transmission	NA	NA	NA	NA	NA	NA
Sound absorption	NA	NA	NA	NA	NA	NA
Coefficient of thermal expansion/contraction	NA	NA	NA	NA	NA	NA
Test method	NA	NA	NA	NA	NA	NA
Seams (recommended style)	NA	NA	Overlapping weld	Overlapping weld	NA	Overlapping weld[5]
Construction method	NA	NA	HF welded	HF welded	NA	HF welded
Useful temperature range	NA	NA	NA	NA	NA	NA

Trade Name:	PRV 2213W Twin Color	Precontraint 392	Precontraint 702 Opaque	Precontraint 1002 Fluotop S	Precontraint 1202 Fluotop T	Precontraint 1502 Fluotop T
Dead load, seams, warp/fill						
Inch seam @ F	NA	NA	NA	NA	NA	NA
lb/in warp/lb/in fill	NA	NA	NA	NA	NA	NA
Inch seam @ F	NA	NA	NA	NA	NA	NA
lb/in warp/lb/in fill	NA	NA	NA	NA	NA	NA
Structural properties						
Effect. modulus, warp, fill	NA	NA	NA	NA	NA	NA
Poisson's ratio, warp/fill	NA	NA	NA	NA	NA	NA
Poisson's ratio, fill/warp	NA	NA	NA	NA	NA	NA
Shear modulus	NA	NA	NA	NA	NA	NA
Stretch properties	NA	NA	NA	NA	NA	NA
Uniaxial, warp	NA	NA	NA	NA	NA	NA
Uniaxial, fill	NA	NA	NA	NA	NA	NA
Biaxial, warp (1:1)	NA	NA	NA	NA	NA	NA
Biaxial, fill (1:1)	NA	NA	NA	NA	NA	NA
Mil spec/Fed spec #	NA	NA	NA	NA	NA	NA
Recommended uses	NA	Shade tension structure	Tension structures, blackout effect on steel aluminum frame structures	Tension structures, air domes	Tension structures with dirt-resistant topcoat	Tension structures with dirt-resistant topcoat

(continued)

Trade Name:	SoftglassII, 200	SoftglassII, 500	SoftglassII, 600	Starfire	Sunbrella	Sunbrella Firesist
Trade name owner	DCI Inc.	DCI Inc.	DCI Inc.	The Astrup Co.	Glen Raven Mills[6]	Glen Raven Mills[6]
Base fabric						
Material	Fiberglass fabric	Fiberglass fabric	Fiberglass fabric	Polyester/cotton	Solution-dyed acrylic	Modacrylic
Weight	7 oz/yd²	12 oz/yd²	18 oz/yd²	8 oz/yd²	9.25 oz/yd²	9.25 oz/yd²
Weave style	Plain	Plain	ZENDPlain	Army duck	Plain weave	Plain weave
Yarn count (warp, fill)	44, 22 tpi	28, 20 tpi	27, 14 tpi	51, 40 tpi	75, 35 tpi	75, 35 tpi
Coating						
Material	Silicon	Silicon	Silicon	Vinyl impregnated	Resin and water repell.	Resin and water repell.
Weight (top, bottom)	4, 4 oz/yd²	6, 6 oz/yd²	9, 9 oz/yd²		NA	NA
UV topcoat material	NA	NA	NA	NA	NA	NA
UV topcoat weight	NA	NA	NA	NA	NA	NA
Finished fabric						
Test method	FED-STD-191 (5041)	FED-STD-191 (5041)	FED-STD-191 (5041)	FED-STD-191	Hydr. press. AATCC127	Hydr. press. AATCC127
Thickness		12 mil	17 mil	35 mil	NA	NA
Weight	15 oz/yd²	24 oz/yd²	35 oz/yd²	NA	9.25 oz/yd²	9.25 oz/yd²
Roll width, usable	60 in	96, 60 in	96, 60 in	NA	46, 60 in	60 in
Tongue tear (warp, fill)	NA	NA	55, 70 lb/in	NA	13.5, 8.7 lb/in	15, 9 lb/in
Test method	NA	NA	NA	FED-STD-191 (5134)	ASTM D 2262-83	ASTM D 2262-83
Trapezoidal tear (warp, fill)	25, 30 lb/in	40, 45 lb/in	55, 70 lb/in	NA	NA	NA
Test method	FED-STD-191 (5136)	FED-STD-191 (5136)	FED-STD-191 (5136)	NA	NA	NA
Grab tensile (warp, fill)	NA	NA	NA	NA	250, 151 lb/in	280, 160 lb/in
Test method	NA	NA	NA	FED-STD-191 (5100)	ASTM D 1682-64	ASTM D 1682-64

Trade Name:	SoftglassII, 200	SoftglassII, 500	SoftglassII, 600	Starfire	Sunbrella	Sunbrella Firesist
Strip tensile (warp, fill)	200, 200 lb/in (min)	540, 420 lb/in	650, 575 lb/in	NA	NA	NA
Test method	FED-STD-191 (5102)	FED-STD-191 (5102)	FED-STD-191 (5136)	NA	NA	NA
Adhesion	NA	NA	NA	NA	NA	NA
Test method	NA	NA	NA	NA	NA	NA
Hydrostatic resistance	NA	NA	NA	FED-S-191 (5514/5516)	NA	NA
Test method	NA	NA	NA	NA	NA	NA
Cold crack	−30°F	−30°F	−30°F	NA	NA	NA
Test method	Fabricator	Fabricator	Fabricator	NA	NA	NA
Burning characteristics						
Test method(s)	ASTM E 84, E 108, E 136, UL-94, NFPA-701	ASTM E 84, E 108, E 136, UL-94, NFPA-701	ASTM E 84, E 108, E 136, UL-94, NFPA-701	FED-STD-191 (5903), CSFM, MVSS-302, NFPA-701 (small sc.)	Not flame retardant	NFPA-701
Light values						
Test method	ASTM E 124	ASTM E 124	ASTM E 124	NA	NA	NA
Transmission, refl., absorp.	32%, 55%, NA	25%, 60%, NA	18%, 65%, NA	NA	NA	NA
Acoustical properties						
Test method	NA	NA	NA	NA	NA	NA
Sound transmission	NA	NA	NA	NA	NA	NA
Sound absorption	NA	NA	NA	NA	NA	NA
Coefficient of thermal expansion/contraction	NA	NA	NA	NA	NA	NA
Test method	NA	NA	NA	NA	NA	NA
Seams (recommended style)	Lap joint	Lap joint	Lap joint	NA	Lap	Lap
Construction method	Chemically bonded	Chemically bonded	Chemically bonded	Sewn	Heat sealed or sewn	Heat sealed or sewn

(continued)

Trade Name:	SoftglassII, 200	SoftglassII, 500	SoftglassII, 600	Starfire	Sunbrella	Sunbrella Firesist
Seams (recommended style) *(continued)*						
Useful temperature range	80–150°F	80–150°F	80–150°F	NA	NA	NA
Dead load, seams, warp/fill						
Inch seam @ F	3 in seam @ 180°F	3 in seam @ 180°F	3 in seam @ 180°F	NA	NA	NA
lb/in warp/lb/in fill	150 lb/in/130 lb/in	180 lb/in/130 lb/in	200 lb/in/180 lb/in	NA	NA	NA
Inch seam @ F	3 in seam @ 70°F	3 in seam @ 70°F	3 in seam @ 70°F	NA	NA	NA
lb/in warp/lb/in fill	150 lb/in/130 lb/in	240 lb/in/220 lb/in	270 lb/in/240 lb/in	NA	NA	NA
Structural properties						
Effect. modulus, warp, fill	NA	NA	NA	NA	NA	NA
Poisson's ratio, warp/fill	NA	NA	NA	NA	NA	NA
Poisson's ratio, fill/warp	NA	NA	NA	NA	NA	NA
Shear modulus	NA	NA	NA	NA	NA	NA
Stretch properties						
Uniaxial, warp	0.5% @ 20 lb/in	NA	NA	NA	NA	NA
Uniaxial, fill	1–2% @ 20 lb/in	NA	NA	NA	NA	NA
Biaxial, warp (1:1)	−0.25% @ 20 lb/in	−0.3% @ 20 lb/in	−0.2% @ 20 lb/in	NA	1% @ 40 lb/in	NA
Biaxial, fill (1:1)	2% @ 20 lb/in	2% @ 20 lb/in	1.8% @ 20 lb/in	NA	3.5% @ 40 lb/in	NA
Mil Spec/Fed spec #	NA	NA	NA	NA	NA	NA
Recommended uses	Skylights, greenhouses, acoustic liners, awning	Skylights, greenhouses, tensile structures	Tensile structures, cable domes	Tents, canopies, awning, banners, boat covers	NA	NA

Trade Name:	Supertent 14	Supertent 18	Tenara	Weblon Holiday	Weblon Tent Wall
Trade name owner	Super Tex	Super Tex	W. L. Gore & Associates[8]	Weblon Inc.[1]	Weblon Inc.
Base fabric					
Material	Polyester	Polyester	Expanded PTFE	Polyester[9]	Polyester
Weight	NA	NA	15.4 oz/yd²	1.6 oz/yd²	1.74 oz/yd²
Weave style	Woven	Woven	3 × 1 twill	Woven	Weft insertion
Yarn count (warp, fill)	16, 16 tpi	18, 18 tpi	66, 66 tpi	24, 24 tpi	18, 9 tpi
Coating					
Material	Vinyl coated	Vinyl coated	None	Vinyl laminated	Vinyl laminated
Weight (top, bottom)	NA	NA	NA	6.4 oz/yd²	4.4 oz/yd²
UV topcoat material	NA[7]	NA	None	Rainkleen	NA
UV topcoat weight	NA	NA	NA	NA	NA
Finished fabric					
Test method	NA	NA	ASTM D 1777	NA	NA
Thickness	NA	NA	0.014 in	NA	NA
Weight	14 oz/yd²	18 oz/yd²	15.4 oz/yd²	12 oz/yd²	10 oz/yd²
Roll width, usable	61–81 in	61–81 in	157 in	62 in	62 in
Tongue tear (warp, fill)	110, 90 lb/in	100, 10 lb/in	NA	18, 18 lb/in	25, 30 lb/in
Test method	FED-STD-191 (5314)	FED-STD-191 (5134)	NA	FED-STD-191 (5134)[10]	FED-STD-191 (5134)[10]
Trapezoidal tear (warp, fill)	NA	NA	NA	NA	NA
Test method	NA	NA	NA	NA	NA
Grab tensile (warp, fill)	200–170 lb/in	300, 300 lb/in	378, 385 lb/in	120, 120 lb/in	115, 115 lb/in
Test method	FED-STD-191 (5102)	FED-STD-191 (5102)	DIN 53857 (Germany)	FED-STD-191 (5100)	FED-STD-191 (5100)

(continued)

Trade Name:	Supertent 14	Supertent 18	Tenara	Weblon Holiday	Weblon Tent Wall
Strip tensile (warp, fill)	NA	NA	NA	NA	NA
Test method	NA	NA	NA	NA	NA
Adhesion	NA	10 lb/in	NA	NA	NA
Test method	FED-STD-191 (5970)	FED-STD-191 (5970)	NA	NA	NA
Hydrostatic resistance	NA	NA	NA	NA	NA
Test method	NA	NA	NA	NA	NA
Cold crack	Pass −40°F	Pass −40°F	−350°F	−20°F	−20°F
Test method	MIL-C-20696	MIL-C-20696	NA	FED-STD-191 (5874)	FED-STD-191 (5874)
Burning characteristics					
Test method(s)	CSFM 2-s flame-out	CSFM 2-s flame-out	NFPA-701 (small scale)	CSFM Reg. F-69 UL classif. as to flamm. only	CSFM F-69 UL classif. as to flamm. only
Light values					
Test method	NA	NA	DIN 67507 (Germany)	NA	NA
Transmission, refl., absorp.	NA	NA	20%, 76%, 4%	NA	NA
Acoustical properties					
Test method	NA	NA	NA	NA	NA
Sound transmission	NA	NA	NA	NA	NA
Sound absorption	NA	NA	NA	NA	NA
Coefficient of thermal expansion/contraction	NA	NA	NA	NA	NA
Test method	NA	NA	NA	NA	NA
Seams (recommended style)	NA	NA	Varies	NA	NA
Construction method	NA	NA	Sewn	Dielectrically welded, sewn	Dielectrically welded, sewn
Useful temperature range	NA	NA	−350–500°F	NA	NA

Trade Name:	Supertent 14	Supertent 18	Tenara	Weblon Holiday	Weblon Tent Wall
Dead load, seams, warp/fill					
Inch seam @ F	NA	NA	NA	NA	NA
lb/in warp/lb/in fill	NA	NA	NA	NA	NA
Inch seam @ F	NA	NA	NA	NA	NA
lb/in warp/lb/in fill	NA	NA	NA	NA	NA
Structural properties					
Effect. modulus, warp, fill	NA	NA	NA	NA	NA
Poisson's ratio, warp/fill	NA	NA	NA	NA	NA
Poisson's ratio, fill/warp	NA	NA	NA	NA	NA
Shear modulus	NA	NA	NA	NA	NA
Stretch properties					
Uniaxial, warp	NA	NA	NA	NA	NA
Uniaxial, fill	NA	NA	NA	NA	NA
Biaxial, warp (1:1)	NA	NA	NA	NA	NA
Biaxial, fill (1:1)	NA	NA	NA	NA	NA
Mil spec/Fed spec #	NA	NA	NA	NA	NA
Recommended uses	Party tents, canopies, covers, banners	Party tents, canopies, covers, banners	NA	NA	NA

Appendix 2

Awnings and Canopies—Fabric Specification Table

This table was originally published in *Fabrics & Architecture, Specguide 1995,* November–December, 1994 by the Industrial Fabrics Association International (IFAI).

Disclaimer statement:

"All specifications included in the table were submitted voluntarily by the firms and their accuracy is the responsibility of the manufacturer. This table is not an endorsement of a company or product by *Fabrics & Architecture* or the Industrial Fabrics Association International (IFAI). The reader is encouraged to contact the companies directly for further information."

Footnotes:

1. Test method ASTM 1204 (sic.).
2. Distributed by Western Rim & Baer Fabrics.
3. Test method not specified.
4. Also passes UL-214.
5. 5-Year adhesion warranty with 3M's Scotchcal.
6. Also passes 2,000 hrs. xenon test.
7. In-house test methods used.
8. Also passes NFPA-701.
9. Also passes CPAI-84.
10. Also passes ASTM E-84, Class A.
11. Also passes UL-C-109.
12. Test method FED-STD-191A (5803).
13. Also passes UL-48.
14. Test method FED-STD-191A (5803).
15. Distributed by The Astrup Co.
16. Test method ASTM D 3107.
17. Test method FED-STD-191A (5803).
18. Test method AATCC 30.
19. Passes VFAC Class 1.
20. Test method SAE J1960 "modified" xenon Weatherometer.
21. Test method AATCC 30, Parts 3 and 4.
22. Distributed by the Astrup Co., John Boyle Co., Unitex and others.
23. Test method GM Automotive 260-15 min.
24. Test method ASTM D 2594 modified.
25. Not recommended for exterior use.
26. Distributed by Unitex.

Fabric Type: Vinyl Laminated or Coated Polyester (Backlighting)

Trade Name:	Awnlit 8000	Awnlit Plus	Awnmax	Canopy	Canopy Plus	Cool Glo
Trademark owner	Duracote Corp.[26]	Duracote Corp.[26]	Hiraoka & Co.[2]	Meridian Mfg. Corp.	Meridian Mfg. Corp.	Unitex
Construction						
Base fabric	Polyester	Polyester	Polyester	Polyester	Polyester	Polyester
Finish/coating	Vinyl lam./acrylic	Vinyl lam./acrylic	Vinyl/NZ-I acrylic	Vinyl	PVDF/Fluorex-P	Vinyl coated
Put up (per roll)	50 yds	50 yds	32 or 45 yds	35 yds	35 yds	40 yds
Finished width(s)	61 in	61 in	61 in or 80 in	61 in	60 in	61 in
Finished weight (oz/sq yd)	19 oz	16 oz	19.1 oz	17–18 oz	17–18 oz	13 oz
Aesthetics						
Number of colors, top	12 solids	16 solids	27 solids	16 solids	16 solids	22 solids
Number of colors, underside	Clear	Clear	White	White	White	Same as top
Opaque	No	No	No	No	No	By color
Translucent, inherent	Yes	Yes	NA	Yes	Yes	Yes, by color
Translucent, backlighting	Yes	Yes	Yes	Yes	Yes	No
Durability						
Life expectancy*	8–10 yrs	5–8 yrs	10+ yrs	5–8 yrs	7–10 yrs	5–8 yrs
Warranty/duration	Yes/8 yrs	Yes/5 yrs	Yes/5–8 yrs	Yes/5 and 3 yrs	Yes/7 and 10 yrs	Yes/5 yrs
UV resistance**	2,000 hrs	2,000 hrs	2,000 + hrs[3]	2,000 hrs[6]	2,000 hrs[6]	1,000 hrs
Water repellent	Yes	Yes	Yes	Yes	Yes	Yes
Hydrostat†	NA	NA	NA	NA	NA	NA
Flame resistant (meets CSFM)	Yes[8]	Yes[8]	Yes[4]	Yes	Yes	Yes[8]
Mildew resistant**	Yes, ASTM G21	Yes, ASTM G21	Yes, ASTM G21	Yes, ASTM G21	Yes, ASTM G21	Yes, ASTM G21
Wick resistant**	Yes, DTM-36	Yes, DTM-36	Yes, JIS L 1096	Yes[7]	Yes[7]	NA

(continued)

781

Fabric Type: Vinyl Laminated or Coated Polyester (Backlighting)

Trade Name:	Awnlit 8000	Awnlit Plus	Awnmax	Canopy	Canopy Plus	Cool Glo
Handling						
Heat sealable (w/o seam tape)	Yes	Yes	Yes	Yes	Yes, with primer	Yes
Heat sealable (w/seam tape)	Yes	Yes	Yes	Yes	Yes	Yes
Shrinkage (warp, fill)**	0.5%, 0.2%	0.5%, 0.2%[1]	0.9%, 0.1%[3]	1.4%, 1.5%[7]	1.4%, 1.5%[7]	NA
Stretch factor (warp, fill)**	NA	NA	1.6%, 1.53%[3]	1.6%, 1.7%[7]	1.6%, 1.7%[7]	NA
Cleaning agents available	Yes	Yes	Yes	No	No	Yes
Graphics						
Heat transfer films	Yes	Yes	Yes	Yes	Yes	Yes
Heat sealed inset fabric	Yes	Yes	Yes	Yes	Yes	Yes
Sewn-in inset fabric	Yes	Yes	Yes	Yes	Yes	Yes
Eradication	No	No	No	No	No	No
Pressure-sensitive vinyl	Yes	Yes	Yes[5]	No	Yes	Yes
Silk-screening	Yes	Yes	Yes	Yes	Yes	Yes
Hand painting	Yes	Yes	Yes	Yes	Yes	Yes
Other	NA	NA	NA	NA	NA	NA
Support material						
Fabrication or technical manual	No	No	Yes	No	No	No
Specification sheet available	Yes	Yes	Yes	Yes	Yes	Yes

*Depends on care and climate conditions.
**Results depend on test method used.
†Result cited to ASTM D 583.

Fabric Type:

	Vinyl Laminated or Coated Polyester (Backlighting)					
Trade Name:	**Diko-Lit**	**Millennium**	**Portico**	**Premier**	**Premier 2000**	**Reflections**
Trademark owner	Dickson Elberton	Duracote Corp.[26]	Meridian Mfg. Corp.	Astrup	Astrup	Astrup
Construction						
Base fabric	Polyester	Polyester	Polyester	Polyester	Polyester	Polyester
Finish/coating	Vinyl/acrylic	Vinyl/Tedlar	PVDF/Fluorex-P	Vinyl/acrylic	Vinyl/urethane	Vinyl/acrylic
Put up (per roll)	38 yds	50 yds	35 yds	25 yds	25 yds	50 yds
Finished width(s)	61 in	61 in	60 in	76 in	76 in	61 in
Finished weight (oz/sq yd)	16 oz	20 oz	17 oz	16 oz	16.5 oz	15 oz
Aesthetics						
Number of colors, top	18 solids/ 5 mat. fin.	12 solids	2 solids	10 solids	4 solids	26 solids
Number of colors, underside	White	Clear	2	White	White	White
Opaque	No	No	Yes	No	No	No, except black
Translucent, inherent	No	Yes	No	No	No	No
Translucent, backlighting	Yes	Yes	No	Yes	Yes	Yes
Durability						
Life expectancy*	5 yrs	10–12 yrs	7–10 yrs	5+ yrs	8+ yrs	5+ yrs
Warranty/duration	Yes/5 yrs	Yes/10 yrs	Yes/7 and 5 yrs	Yes/5 yrs	Yes/8 yrs	Yes/5 yrs
UV resistance**	2,000 hrs	2,000 hrs	2,000 hrs[6]	3000 hrs[3]	3,000 hrs[3]	3,000 hrs[3]
Water repellent	Yes	Yes	Yes	Yes	Yes	Yes
Hydrostat†	NA	NA	NA	NA	NA	NA
Flame resistance (meets CSFM)	Yes	Yes[8]	Yes	Yes	Yes	Yes
Mildew resistant**	NA	Yes, ASTM G21	Yes, ASTM G21	Yes[3]	Yes[3]	Yes[3]
Wick resistant**	Yes[3]	Yes, DTM 36	Yes[7]	Yes[3]	Yes[3]	Yes[3]

(continued)

Fabric Type:

Vinyl Laminated or Coated Polyester (Backlighting)

Trade Name:	Diko-Lit	Millennium	Portico	Premier	Premier 2000	Reflections
Handling						
Heat sealable (w/o seam tape)	Yes	Yes	Yes, with primer	Yes	Yes	Yes
Heat sealable (w/seam tape)	Yes	Yes	Yes	Yes	Yes	Yes
Shrinkage (warp, fill)**	NA	0.5%, 0.2%[1]	1.6%, 1.8%[7]	1%, 0% max[1]	1%, 0% max[1]	1%, 0% max[1]
Stretch factor (warp, fill)**	NA	NA	1.8%, 1.9%[7]	4% max[1]	4% max[1]	4% max[1]
Cleaning agents available	Yes	Yes	No	Yes	Yes	Yes
Graphics						
Heat transfer films	Yes	No	Yes	Yes	Yes	Yes
Heat sealed inset fabric	Yes	Yes	Yes	Yes	Yes	Yes
Sewn-in inset fabric	Yes	Yes	Yes	Yes	Yes	Yes
Eradication	No	No	No	No	No	No
Pressure-sensitive vinyl	Yes	Yes	Yes	Yes	Yes	Yes
Silk-screening	Yes	Yes	Yes	Yes	Yes	Yes
Hand painting	Yes	Yes	Yes	Yes	Yes	Yes
Other	NA	NA	NA	NA	NA	NA
Support material						
Fabrication or technical manual	No	Yes	No	Yes	Yes	Yes
Specification sheet available	Yes	Yes	Yes	Yes	Yes	Yes

*Depends on care and climate conditions.
**Results depend on test method used.
†Result cited to ASTM D 583.

Fabric Type:	Vinyl Laminated or Coated Polyester (Backlighting)		Vinyl Laminated Polyester			
Trade Name:	**Signmaster Supreme**	**Starlit**	**Patio 500**	**Unishade**	**Weblon Coastline+**	**Weblon Vanguard**
Trademark owner	John Boyle & Co.	John Boyle & Co.	John Boyle & Co.	Unitex	Weblon[15]	Weblon[15]
Construction						
Base fabric	Polyester	Polyester	Polyester	Polyester	Polyester	Polyester
Finish/coating	Vinyl	Vinyl/Tedlar	Vinyl	Vinyl	Vinyl/urethane	Vinyl/urethane
Put up (per roll)	50 yds	As specified	50 yds	50 yds	50 yds	50 yds
Finished width(s)	61 in	60 in	61 in	61 in	62 in	62 in
Finished weight (oz/sq yd)	16 oz	18 oz	17 oz	16 oz	15 oz	17 oz
Aesthetics						
Number of colors, top	30	30	30 solids/6 stripes	22	48 solids/36 stripes	15 solids
Number of colors, underside	White	White	Same or white/tan	Same as top	Solids/print	Same as top
Opaque	No	No, except black	Yes	Depends on color	NA	NA
Translucent, inherent	Yes	Yes	No, exc. clear and white	Yes	Yes (dep. on color)	Yes (dep. on color)
Translucent, back-lighting	Yes	Yes	No	No	No	No
Durability						
Life expectancy*	5–8 yrs	8–10 yrs	5–8 yrs	5–8 yrs	5+ yrs	8+ yrs
Warranty/duration	Yes/5 yrs	Yes/8 yrs	Yes/5 yrs	Yes/5 yrs	Yes/5 yrs	Yes/8 yrs
UV resistance**	2,000 hrs	2,000 hrs	2,000 hrs	1,000 hrs	2,000 hrs[3]	10,000 hrs[3]
Water repellent	Yes	Yes	Yes	Yes	Yes	Yes
Hydrostat†	NA	NA	NA	NA	NA	NA
Flame resistant (meets CSFM)	Yes[4,8-11]	Yes[10,13]	Yes[4,8-11]	Yes[8]	Yes	Yes
Mildew resistant**	Yes[12]	Yes[12]	Yes[12]	Yes, ASTM G 21	Yes[3]	Yes[3]
Wick resistant**	Yes[7]	Yes[7]	Yes[7]	NA	Yes, light colors[3]	Yes[3]

(continued)

Fabric Type:	Vinyl Laminated or Coated Polyester (Backlighting)			Vinyl Laminated Polyester		
Trade Name:	Signmaster Supreme	Starlit	Patio 500	Unishade	Weblon Coastline+	Weblon Vanguard
Handling						
Heat sealable (w/o seam tape)	Yes	No	Yes	Yes	Yes	Yes
Heat sealable (w/seam tape)	Yes	Yes	Yes	Yes	Yes	Yes
Shrinkage (warp, fill)**	NA	NA	NA	NA	$<1\%$[3]	$<1\%$[3]
Stretch factor (warp fill)**	NA	NA	NA	NA	$<1\%$[3]	$<1\%$[3]
Cleaning agents available	Yes	Yes	Yes	Yes	Yes	Yes
Graphics						
Heat transfer films	Yes	No	Yes	Yes	Yes	Yes
Heat sealed inset fabric	Yes	Yes	Yes	Yes	Yes	Yes
Sewn-in inset fabric	Yes	Yes	Yes	Yes	Yes	Yes
Eradication	No	No	No	No	No	No
Pressure-sensitive vinyl	No	Yes	No	No	Yes	Yes
Silk-screening	Yes	No	Yes	Yes	Yes	Yes
Hand painting	Yes	No	Yes	Yes	Yes	Yes
Other	NA	NA	NA	NA	NA	NA
Support material						
Fabrication or technical manual	No	Yes	No	No	Yes	Yes
Specification sheet available	Yes	Yes	Yes	Yes	Yes	Yes

*Depends on care and climate conditions.
**Results depend on test method used.
†Result cited to ASTM D 583.

Fabric Type:	Vinyl Laminated or Coated Mesh		Acrylic or Resin Coated Polyester or Polyester/Cotton			
Trade Name:	Awntex 70	Awntex 90	Calliope	Fyrecote	Mirage	Pyrotone
Trademark owner	Astrup	Astrup	Graniteville[15]	John Boyle & Co.	Unitex	Unitex
Construction						
Base fabric	Polyester	Polyester	Polyester	Polyester/cotton	Polyester/cotton	Polyester/cotton
Finish/coating	Vinyl	Vinyl	Vinyl	Vinyl	Vinyl	Vinyl
Put up (per roll)	30 yds	30 yds	70 yds	30 yds	Special order	50 or 60 yds
Finished width(s)	60 in	60 in	60 in	60 in/61 in	58 in	60 in
Finished weight (oz/sq yd)	9 oz	17 oz	12.5 oz	13 oz	13.5–14.5 oz	13.5–14.5 oz
Aesthetics						
Number of colors, top	12 solids	12 solids	25 solids	19	Unlimited	30 solids
Number of colors, underside	Same as top	Same as top	Same as top	Same	30 solids	Same as top
Opaque	No	No	Yes	Yes	Yes	Yes
Translucent, inherent	Yes	Yes	No	No	No	No
Translucent, backlighting	No	No	No	No	No	No
Durability						
Life expectancy*	5 + yrs	5 + yrs	5 + yrs	5 yrs	5 yrs	5–8 yrs
Warranty/duration	Yes/5 yrs	Yes/5 yrs	Yes/5 yrs	No	Yes/5 yrs	Yes/5 yrs
UV resistance**	NA	NA	800 hrs[17]	1,000 hrs	1,000 hrs	1,000 hrs
Water repellent	No	No	Yes	Yes	Yes	Yes
Hydrostat†	NA	NA	NA	NA	NA	NA
Flame resistant (meets CSFM)	Yes	Yes	Yes	Yes[9]	Yes[8]	Yes[8]
Mildew resistant**	Yes[3]	Yes[3]	Yes, AATCC30-1988	Yes[12]	Yes, ASTM G 21	Yes, ASTM G 21
Wick resistant**	Yes[3]	Yes[3]	Yes[3]	Yes[7]	Yes[7]	Yes[7]

(continued)

Fabric Type:	Vinyl Laminated or Coated Mesh		Acrylic or Resin Coated Polyester or Polyester/Cotton			
Trade Name:	Awntex 70	Awntex 90	Calliope	Fyrecote	Mirage	Pyrotone
Handling						
Heat sealable (w/o seam tape)	Yes	Yes	No	No	No	No
Heat sealable (w/seam tape)	Yes	Yes	Yes	Yes	Yes	Yes
Shrinkage (warp, fill)	NA	NA	1–2%[3]	2%, 2%[7]	2%, 2%[7]	2%, 2%[7]
Stretch factor (warp, fill)**	1%, 2%[3]	0.5%, 4.0%[3]	1–2%[3]	0[7]	0[7]	0[7]
Cleaning agents available	Yes	Yes	Yes	Yes	Yes	Yes
Graphics						
Heat transfer film	No	No	Yes	No	No	No
Heat sealed inset fabric	Yes	Yes	Yes	No	No	No
Sewn-in inset fabric	Yes	Yes	Yes	Yes	Yes	Yes
Eradication	No	No	No	No	No	No
Pressure-sensitive vinyl	No	No	No	No	No	No
Silk-screening	Yes	Yes	Yes	Yes	Yes	Yes
Hand painting	Yes	Yes	Yes	Yes	Yes	Yes
Other	NA	NA	NA	NA	NA	NA
Support materials						
Fabrication or technical manual	Yes	Yes	Yes	No	No	No
Specification sheet available	Yes	Yes	Yes	Yes	Yes	Yes

*Depends on care and climate conditions.
**Results depend on test method used.
†Result cited to ASTM D 583.

Fabric Type:	Acrylic or Resin Coated PET or PET/Cotton		Vinyl Coated Cotton, Polyester or Polyester/Cotton			
Trade Name:	Starfire	Ultrafab	Calabana	Calabana F/R	Fabri-Awn	Gulfstream
Trademark owner	Astrup	John Boyle & Co.	Graniteville[15]	Graniteville[15]	John Boyle & Co.	John Boyle & Co.
Construction						
Base fabric	Polyester/cotton	Polyester	Cotton	Cotton	Polyester	Polyester/cotton
Finish/coating	Vinyl/acrylic	Acrylic	Vinyl	Vinyl	Vinyl	Acrylic
Put up (per roll)	45 yds	50 yds	50 yds	50 yds	35 yds	50 yds
Finished width(s)	60 in	S.61 in, str31/36/61	30–31 in	30–31 in	61 in	31 in
Finished weight (oz/sq yd)	15 oz	9.6 oz	15 oz	15 oz	15 oz	13 oz
Aesthetics						
Number of colors, top	20	25 solids/15 stripes	43 solids/26 stripes	19 solids/3 stripes	20	30 solids/26 stripes
Number of colors, underside	Same as top	Sol. same/str. nat.	Dyed pearl gray	Dyed pearl gray	Pearl gray	Gray, tan, white, floral
Opaque	Yes	Yes	Yes	Yes	Yes	Yes
Translucent, inherent	No	Yes (dep. on color)	No	No	No	No
Translucent, backlighting	No	No	No	No	No	No
Durability						
Life expectancy*	5+ yrs	5–8 yrs	5+ yrs	5+ yrs	5–8 yrs	5–8 yrs
Warranty/duration	Yes/5 yrs	Yes/5 yrs	Yes/5 yrs	Yes/5 yrs	Yes/5 yrs	Yes/5 yrs
UV resistance**	1,000 hrs	2,000+ hrs	800 hrs[17]	800 hrs[17]	2,000+ hrs	2,000+ hrs
Water repellent	Yes	Yes	Yes	Yes	Yes	Yes
Hydrostat†	NA	NA	NA	NA	NA	NA
Flame resistant (meets CSFM)	Yes	Yes	No	Yes[10]	No	No
Mildew resistant**	Yes[3]	Yes[12]	Yes[18]	Yes[18]	Yes[12]	Yes[12]
Wick resistant**	Yes[3]	Yes[7]	No	No	Yes[7]	Yes[7]

(continued)

Fabric Type:	Acrylic or Resin Coated PET or PET/Cotton		Vinyl Coated Cotton, Polyester or Polyester/Cotton			
Trade Name:	Starfire	Ultrafab	Calabana	Calabana F/R	Fabri-Awn	Gulfstream
Handling						
Heat sealable (w/o seam tape)	No	No	No	No	No	No
Heat sealable (w/seam tape)	Yes	Yes	Yes	Yes	Yes	Yes
Shrinkage (warp, fill)	2.5%[3]	NA	NA	3–5%[3]	NA	<2%/<2%[7]
Stretch factor (warp fill)	0[3]	NA	3–5%[3]	3%[3]	NA	0[7]
Cleaning agents available	Yes	Yes	Yes	Yes	Yes	Yes
Graphics						
Heat transfer films	No	No	No	No	Yes	No
Heat sealed inset fabric	No	No	No	No	Yes	No
Sewn-in inset fabric	Yes	Yes	Yes	Yes	Yes	Yes
Eradication	No	No	No	No	No	No
Pressure-sensitive vinyl	Yes	No	Yes	Yes	No	No
Silk-screening	Yes	Yes	Yes	Yes	Yes	Yes
Hand painting	Yes	Yes	Yes	Yes	Yes	Yes
Other	NA	NA	NA	NA	NA	NA
Support materials						
Fabrication or technical manual	Yes	No	Yes	Yes	No	No
Specification sheet available	Yes	Yes	Yes	Yes	Yes	Yes

*Depends on care and climate conditions.

**Results depend on test method used.

†Result cited to ASTM D 583.

Fabric Type:	Acrylic p. Cot, or P/C		100% Woven Acrylic				
Trade Name:	Pee Gee	Diklon 32	Sunbrella	Sunbrella Plus	Sunbrella Firesist	Tempotest
Trademark owner	Graniteville[15]	Dickson Elberton[15]	Glen Raven Mills[22]	Glen Raven Mills[22]	Glen Raven Mills[22]	Alutex
Construction						
Base fabric	Polyester/cotton	Sol.-dyed acrylic	Sol.-dyed acrylic	Sol.-dyed acrylic	Sol.-dyed modacr.	Sol.-dyed acrylic
Finish/coating	Acrylic	Resin/fluorocarbon	Resin/ fluorochem.	Back polyure-thane	Resin/fluorochem.	Teflon impregnated
Put up (per roll)	50 yds	65 yds	55 yds	55 yds	50 yds	65 yds
Finished width(s)	31 in	47 in/61 in	46 and 60 in (some)	60 in	60 in	47.60 in
Finished weight (oz/sq yd)	12 oz	10 oz	9.25 oz	10 oz	9.25 oz	8.85 oz
Aesthetics						
Number of colors, top	27 solids/ 23 stripes	37 solids, 66 stripes/ 13 solids	47 solids/ 57 stripes	14 solids	21 solids/7 stripes	50 solids/68 stripes
Number of colors, underside	Pearl gray	Same as top	Same	Same	Same	Same
Opaque	Yes	Yes	Yes, darker colors	Yes, darker colors	Yes, darker colors	Yes
Translucent, inherent	No	Yes	Yes, most colors	Yes, light colors	Yes, most colors	No
Translucent, backlighting	No	Yes	Yes	Yes, light colors	Yes	No
Durability						
Life expectancy*	5+ yrs	5 yrs	5+ yrs	5+ yrs	5+ yrs	5 yrs
Warranty/duration	Yes/5 yrs	Yes/5 yrs	Yes/5 yrs	Yes/5 yrs	Yes/5 yrs	Yes/5 yrs
UV resistance**	800 hrs[17]	2,000 hrs[3]	1,500 hrs[20]	1,500 hrs[20]	1,500 hrs[20]	Yes
Water repellent	Yes	Yes	Yes	Yes	Yes	Yes
Hydrostat†	NA	NA	NA	NA	NA	32 cm
Flame resistant (meets CSFM)	No	NA[19]	NA	NA	Yes[8-10]	–
Mildew resistant**	Yes[18]	Yes[18]	Yes[21]	Yes[21]	Yes[21]	Yes
Wick resistant**	No	Yes[3]	No	No	No	Yes

(continued)

Fabric Type:	Acrylic p. Cot, or P/C		100% Woven Acrylic			
Trade Name:	Pee Gee	Diklon 32	Sunbrella	Sunbrella Plus	Sunbrella Firesist	Tempotest
Handling						
Heat sealable (w/o seam tape)	No	No	No	No	No	No
Heat sealable (w/seam tape)	No	Yes	Yes	Yes	Yes	Yes
Shrinkage (warp, fill)**	2–3%[3]	NA	3%, 1%[23]	3%, 1%[23]	3%, 1%[23]	0.5%/0.5%
Stretch factor (warp, fill)**	1%[3]	NA	1%, 3%[24]	1%, 3%[24]	1%, 3%[24]	0.5%/0.5%
Cleaning agents available	Yes	Yes	Yes	Yes	Yes	Yes
Graphics						
Heat transfer films	No	No	Yes	Yes	Yes	Yes
Heat sealed inset fabric	No	Yes	Yes	Yes	Yes	Yes
Sewn-in inset fabric	Yes	Yes	Yes	Yes	Yes	Yes
Eradication	No	No	No	No	No	No
Pressure-sensitive vinyl	Yes	Yes	Yes[25]	Yes[25]	Yes[25]	No
Silk-screening	Yes	Yes	Yes	Yes	Yes	Yes
Hand painting	Yes	Yes	Yes	Yes	Yes	Yes
Other	NA	NA	NA	NA	NA	NA
Support materials						
Fabrication or technical manual	Yes	No	Yes	Yes	Yes	No
Specification sheet available	Yes	Yes	Yes	Yes	Yes	Yes

*Depends on care and climate conditions.
**Results depend on test method used.
†Result cited to ASTM D 583.

Appendix 3

Geotextile Fabric Properties

This table was originally published in *Geotechnical Fabrics Report (GFR) 1995 Specifier's Guide,* December 1994, by the Industrial Fabrics Association International.

Disclaimer Statement:

"All information included in this Appendix which has been excerpted from the *1995 Geosynthetic Fabrics Report (GFR) Specifier's Guide* was compiled from data submitted by firms in the geosynthetics industry. Specifications were submitted voluntarily and their accuracy is the responsibility of the manufacturer. The appearance of a listing herein is not an endorsement of the company or product by GFR or the Industrial Fabrics Association International (IFAI). This information is intended only as a guide, and readers are encouraged to contact the manufacturers directly for additional information. The opinions expressed and the technical data provided do not necessarily represent the opinion of the Industrial Fabrics Association International. IFAI makes no representation or warranty, either expressed or implied, as to (1) the fitness of any particular purpose of any of the information, designs or standards contained in this book or any products manufactured or constructed in accordance therewith; or (2) the merchantability of any such information, designs, standards or products. The use of any individual or entity of any such information, designs, standards or products constitutes an acknowledgement and agreement by such individual or entity that the Industrial Fabrics Association International made no representation or warranty with respect to the fitness, merchantability or quality of such information, designs, standards or products."

Advanced Drainage Systems Inc.

Product Name [Mass per unit g/m² (oz/yd²)]	Physical Properties			Filtration/Hydraulic Properties		Mechanical Properties				Wide Width Tensile Elongation[4] ASTM D 4595-96 kN/m (lb/in)		Manufacturer's Suggested Applications[5]
	Structure[1]	Polymer Composition[2]	Percent Open Area (wovens only) CWO-22125 %	Apparent Opening Size[3] ASTM D 4751-87 mm (U.S. sieve)	Permittivity ASTM D 4491-92 sec⁻¹/Flow Rate l/min/m² (gal/min/ft²)	Puncture ASTM D 4833-88 kN (lb)	Mullen Burst ASTM D 3786-87 kPa (psi)	Trapezoid Tear Strength ASTM D 4533-91 kN (lb)	Grab/Tensile Elongation ASTM D 4632-91 kN (lb)%	MD	XD	
1020	NW	PP	NA	0.150 (100)	(70)	(149)	(440)	(100)	(285)	NP	NP	P, E, S/S
1220	NW	PP	NA	0.150 (100)	(60)	(170)	(550)	(114)	(325)	NP	NP	R, P, E, S/S
4000	NW	PP	NA	0.212 (70)	(150)	(55)	(195)	(40)	(95)	NP	NP	F, D, S/S
4420	NW	PP	NA	0.212 (70)	(140)	(60)	(210)	(45)	(105)	NP	NP	F, D, S/S
6600	NW	PP	NA	0.180 (80)	(110)	(90)	(305)	(65)	(165)	NP	NP	F, D, E, P, R, S/S
8800	NW	PP	NA	0.180 (80)	(110)	(115)	(355)	(85)	(210)	NP	NP	F, D, E, P, R, S/S

Amoco Fabrics & Fibers Co.

Product Name [Mass per unit g/m² (oz/yd²)]	Structure[1]	Polymer Composition[2]	Percent Open Area (wovens only) CWO-22125 %	Apparent Opening Size[3] ASTM D 4751-87 mm (U.S. sieve)	Permittivity ASTM D 4491-92 sec⁻¹/Flow Rate l/min/m² (gal/min/ft²)	Puncture ASTM D 4833-88 kN (lb)	Mullen Burst ASTM D 3786-87 kPa (psi)	Trapezoid Tear Strength ASTM D 4533-91 kN (lb)	Grab/Tensile Elongation ASTM D 4632-91 kN (lb)%	MD	XD	Manufacturer's Suggested Applications[5]
Amoco 1198	W	PP	NA	0.425 (40)	0.05/2032 (50)	0.53 (120)	3100 (450)	0.29 (65)	1.34 x 0.89/15 (300 x 200)/15	NP	NP	E, D, F
Amoco 1199	W	PP	NA	0.212 (70)	0.28/731 (18)	0.60 (135)	3307 (480)	0.42 (95) warp 0.24 (55) fill	1.56 x 1.11/15 (350 x 250)/15	NP	NP	E, D, F
Amoco 1380	W	PP	NA	0.600 (30)	0.40/1219 (30)	0.36 (80)	2067 (300)	0.22 (50)	0.78 (175)/25	NP	NP	S/F
Amoco 2000	W	PP	NA	0.600 (30)	0.04/163 (4)	0.29 (65)	2239 (325)	0.20 (45)	0.62 (140)/15	NP	NP	S/S
Amoco 2002	W	PP	NA	0.300 (50)	0.04/163 (4)	0.40 (90)	2756 (400)	0.33 (75)	0.89 (200)/15	NP	NP	S/S
Amoco 2006	W	PP	NA	0.425 (40)	0.02/81 (2)	0.53 (120)	4134 (600)	0.53 (120)	1.34 (300)/15	21 (120)/10	21 (120)/6	S/S, R
Amoco 2016	W	PP	NA	0.425 (40)	0.55/1625 (40)	0.53 (120)	5512 (800)	0.53 (120)	1.34 (300)/20	31 (175)/15	31 (175)/15	E, S/S
Amoco 2019	W	PP	NA	0.212 (70)	0.04/203 (5)	0.62 (140)	3514 (510)	0.29 (65)	1.56 x 1.11/15 (350 x 250)/15	35 (200)/15	35 (200)/7	E
Amoco 2044	W	PP	NA	0.600 (30)	0.15/406 (10)	0.62 (140)	9302 (1350)	1.11 (250)	2.67 x 2.23/15 (600 x 500)/15	70 (400)/18	70 (400)/8	S/S, R

All values were requested to be minimum average roll values and all claims are the responsibility of the manufacturer. All product data are intended as a guide and are not all-inclusive. *Geotechnical Fabrics Report* recommends you contact manufacturers before making any specifying/purchasing decisions.

[1]
W = Woven
NW = Nonwoven
O/C = Other or combination
[2]
PET = Polyester
PP = Polypropylene
PE = Polyethylene
O/C = Other or combination
[3] Maximum opening size of tested rolls

[4] Reinforcement applications only
[5]
S/S = Separation/stabilization
R = Reinforcement
F = Filtration
D = Drainage
P = Protection
E = Erosion Control
A/O = Asphalt Overlay

S/F = Silt Fence
O/C = Other or combination
NA = Manufacturer determined that this data was not applicable to the product or the data was unavailable
NP = Data not provided by manufacturer

Amoco Fabrics & Fibers Co. (cont.)

Product Name — Mass per unit g/m² (oz/yd²)	Physical Properties			Filtration/Hydraulic Properties				Mechanical Properties			Wide Width Tensile Elongation[4] ASTM D 4595-86 kN/m (lb/in)		Manufacturer's Suggested Applications[5]
	Structure[1]	Polymer Composition[2]	Percent Open Area (wovens only) CWO-22125 %	Apparent Opening Size[3] ASTM D 4751-87 mm (U.S. sieve)	Permittivity ASTM D 4491-92 sec⁻¹/Flow Rate l/min/m² (gal/min/ft²)	Puncture ASTM D 4833-88 kN (lb)	Mullen Burst ASTM D 3786-87 kPa (psi)	Trapezoid Tear Strength ASTM D 4533-91 kN (lb)	Grab/Tensile Elongation ASTM D 4632-91 kN (lb)/%		MD	XD	
Amoco 2122	W	PP	NA	0.600 (30)	NA/3657 (90)	0.29 (65)	1895 (275)	0.22 (50)	0.53 x 0.45/15 (120 x 100)/15		NP	NP	S/F
Amoco 2125	W	PP	NA	0.850 (20)	NA/609 (15)	0.27 (60)	1895 (275)	0.22 (50)	0.45 (100)/15		NP	NP	S/F
Amoco 2127	W	PP	NA	2.00 (10)	NA/1219 (30)	0.13 (30)	1723 (250)	0.27 (60)	0.42 x 0.36/15 (95 x 80)/15		NP	NP	S/F
Amoco 2130	W	PP	NA	0.600 (30)	NA/406 (10)	0.27 (60)	2343 (340)	0.36 (80)	0.53 x 0.53/15 x 20 (120 x 120)/15 x 20		NP	NP	S/F
Amoco 2155	W	PP	NA	0.250 (20)	NA/609 (15)	0.27 (60)	1895 (275)	0.22 (50)	0.45 (100)/15		NP	NP	S/F
Amoco 7630	W	PP	NA	NA	NA	NA	NA	NA	0.13 (30) 15		NP	NP	barrier fence
Amoco 4504	NW	PP	NA	0.212 (70)	2.0/4064 (100)	0.245 (55)	1550 (225)	0.16 (35)	0.42 (95)/50		NP	NP	F (landfill)
Amoco 4506	NW	PP	NA	0.212 (70)	1.7/3657 (90)	0.401 (90)	2412 (350)	0.29 (65)	0.67 (150)/50		NP	NP	F (landfill)
Amoco 4508	NW	PP	NA	0.150 (100)	1.5/3251 (80)	0.579 (130)	3101 (450)	0.36 (80)	0.89 (200)/50		NP	NP	F (landfill)
Amoco 4510	NW	PP	NA	0.150 (100)	1.1/2544 (70)	0.734 (165)	3790 (550)	0.42 (95)	1.05 (235)/50		NP	NP	F (landfill)
Amoco 4512	NW	PP	NA	0.150 (100)	0.9/2438 (60)	0.823 (185)	4479 (650)	0.51 (115)	1.22 (275)/50		NP	NP	F (landfill) P (geomembrane)
Amoco 4516	NW	PP	NA	0.150 (100)	0.7/2032 (50)	0.979 (220)	5168 (750)	0.58 (130)	1.56 (350)/50		NP	NP	P (geomembrane)
Amoco 4535	NW	PP	NA	0.212 (70)	2.2/6298 (155)	0.200 (45)	1102 (160)	0.16 (35)	0.36 (80)/50		NP	NP	F, D
Amoco 4545	NW	PP	NA	0.212 (70)	2.1/6095 (150)	0.245 (55)	1275 (185)	0.18 (40)	0.40 (90)/50		NP	NP	F, D
Amoco 4546	NW	PP	NA	0.212 (70)	2.0/5689 (140)	0.299 (65)	1550 (225)	0.20 (45)	0.45 (100)/50		NP	NP	F, D, S/S
Amoco 4547	NW	PP	NA	0.212 (70)	1.8/4876 (120)	0.311 (70)	1654 (240)	0.22 (50)	0.53 (120)/50		NP	NP	F, D, S/S, E
Amoco 4550	NW	PP	NA	0.212 (70)	1.7/4673 (115)	0.336 (75)	1826 (265)	0.25 (55)	0.58 (130)/50		NP	NP	F, D, S/S, E

[1] W = Woven
NW = Nonwoven
O/C = Other or combination
[2] PET = Polyester
PP = Polypropylene
PE = Polyethylene
O/C = Other or combination
[3] Maximum opening size of tested rolls

[4] Reinforcement applications only
[5] S/S = Separation/stabilization
R = Reinforcement
F = Filtration
D = Drainage
P = Protection
E = Erosion Control
A/O = Asphalt Overlay

S/F = Silt Fence
O/C = Other or combination
NA = Manufacturer determined that this data was not applicable to the product or the data was unavailable
NP = Data not provided by manufacturer

All values were requested to be minimum average roll values and all claims are the responsibility of the manufacturer. All product data are intended as a guide and are not all-inclusive. *Geotechnical Fabrics Report* recommends you contact manufacturers before making any specifying/purchasing decisions.

795

Amoco Fabrics & Fibers Co. (cont.)

Product Name [Mass per unit g/m² (oz/yd²)]	Physical Properties		Filtration/Hydraulic Properties			Mechanical Properties						Manufacturer's Suggested Applications[4]
	Structure[1]	Polymer Composition[2]	Percent Open Area (wovens only) CWO-22125 %	Apparent Opening Size[3] ASTM D 4751-87 mm (U.S. sieve)	Permittivity ASTM D 4491-92 sec⁻¹/Flow Rate l/min/m² (gal/min/ft²)	Puncture ASTM D 4833-88 kN (lb)	Mullen Burst ASTM D 3786-87 kPa (psi)	Trapezoid Tear Strength ASTM D 4533-91 kN (lb)	Grab/Tensile Elongation ASTM D 4632-91 kN (lb)%	Wide Width Tensile Elongation ASTM D 4595-86 kN/m (lb/in) MD	XD %	
Amoco 4551	NW	PP	NA	0.150 (100)	1.6/4470 (110)	0.401 (90)	2170 (315)	0.29 (65)	0.67 (150)/50	NP	NP	F, D, S/S, E
Amoco 4552	NW	PP	NA	0.150 (100)	1.5/4267 (105)	0.467 (105)	2412 (350)	0.31 (70)	0.80 (180)/50	NP	NP	F, D, S/S, E
Amoco 4553	NW	PP	NA	0.150 (100)	1.4/3251 (80)	0.579 (130)	2756 (400)	0.36 (80)	0.089 (200)/50	NP	NP	F, D, S/S, E
Amoco 4555	NW	PP	NA	0.150 (100)	1.3/2844 (70)	0.668 (150)	3445 (500)	0.42 (95)	1.20 (270)/50	NP	NP	F, D, S/S, E
Amoco 4557	NW	PP	NA	0.150 (100)	1.1/2438 (60)	0.779 (175)	4134 (600)	0.51 (115)	1.22 (275)/50	NP	NP	railroad stabilization F, D, S/S, E
Amoco 4561	NW	PP	NA	0.150 (100)	0.7/2032 (50)	1.024 (230)	5168 (750)	0.58 (130)	1.45 (325)/50	NP	NP	railroad stabilization F, D, S/S, E
Petrotac 4591	NW	PP	NA	NA	NA	0.890 (200) ASTM E 154	NA	NA	0.89 (200)/NA	NP	NP	A/O, pavement crack repair, bridge membrane
Pro-Guard 459	NW	PP	NA	NA	NA	NA	NA	NA	NA	64.7 (370)/100[A]	64.7 (370)/100[A]	A/O, pavement crack repair, bridge membrane
Petromat 4596	NW	PP	NA	NA	NA	NA	1137 (165)	NA	0.35 (80)/50	NP	NP	A/O
Petromat 4597	NW	PP	NA	NA	NA	NA	1585 (230)	NA	0.53 (120)/50	NP	NP	A/O
Petromat 4597	NW	PP	NA	NA	NA	NA	1378 (200)	NA	0.40 (90)/50	NP	NP	A/O
Petromat 4597	NW	PP	NA	NA	NA	NA	1241 (180)	NA	0.40 (90)/50	NP	NP	A/O

Belton Industries Inc.

Product Name	Structure[1]	Polymer Composition[2]	Percent Open Area %	Apparent Opening Size mm (U.S. sieve)	Permittivity/Flow Rate	Puncture kN (lb)	Mullen Burst kPa (psi)	Trapezoid Tear Strength kN (lb)	Grab/Tensile kN (lb)%	MD	XD	Applications
304	W	PP	NA	0.425 (40)	0.10/305 (7.5)	0.20 (45)	2068 (300)	0.22 (50)	0.67 (150)/15 MD 0.44 (100)/10 XMD	NA	NA	S/S
776	W	PP	NA	0.425 (40)	0.10/326 (8.0)	0.36 (80)	2758 (400)	0.24 (55) MD 0.40 (90) XMD	0.89 (200)/14	NA	NA	S/S

[1] W = Woven
NW = Nonwoven
O/C = Other or combination
[2] PET = Polyester
PP = Polypropylene
PE = Polyethylene
O/C = Other or combination
[3] Maximum opening size of tested rolls

[4] Reinforcement applications only
[5] S/S = Separation/stabilization
R = Reinforcement
F = Filtration
D = Drainage
P = Protection
E = Erosion Control
A/O = Asphalt Overlay

S/F = Silt Fence
O/C = Other or combination
[A] ASTM D 882
NA = Manufacturer determined that this data was not applicable to the product or the data was unavailable
NP = Data not provided by manufacturer

All values were requested to be minimum average roll values and all claims are the responsibility of the manufacturer. All product data are intended as a guide and are not all-inclusive. *Geotechnical Fabrics Report* **recommends you contact manufacturers before making any specifying/purchasing decisions.**

Belton Industries Inc. (cont.)

Product Name [Mass per unit g/m² (oz/yd²)]	Structure[1]	Polymer Composition[2]	Percent Open Area (wovens only) CWO-22125 %	Apparent Opening Size[3] ASTM D 4751-87 mm (U.S. sieve)	Permittivity ASTM D 4491-92 sec⁻¹/Flow Rate l/min/m² (gal/min/m²)	Puncture ASTM D 4833-88 kN (lb)	Mullen Burst ASTM D 3786-87 kPa (psi)	Trapezoid Tear Strength ASTM D 4533-91 kN (lb)	Grab/Tensile Elongation ASTM D 4632-91 kN (lb)/%	Wide Width Tensile Elongation ASTM D 4595-96 kN/m (lb/in) MD	Wide Width Tensile Elongation ASTM D 4595-96 kN/m (lb/in) XD	Manufacturer's Suggested Application[4]
777	W	PP	NA	0.425 (40)	0.05/143 (3.5)	0.62 (140)	4137 (600)	0.44 (100)	1.33 (300)/15	NA	NA	S/S
113	W	PP	NA	0.425 (40)	NA	0.82 (185)	4826 (700)	0.44 (100) MD 0.33 (75) XMD	1.78 (400)/15 MD 1.51 (340)/15 XMD	NA	NA	S/S
806	W	PP	NA	0.850 (20)	0.45/1222 (30)	0.09 (20)	1379 (200)	0.13 (30)	0.31 (70)/10	NA	NA	S/F
751	W	PP	NA	0.600 (30)	0.23/610 (15)	0.20 (45)	2068 (300)	0.22 (50)	0.53 (120)/10 MD 0.42 (95)/10 XMD	NA	NA	S/F
751	W	PP	NA	1.30 (15)	0.35/978 (24)	0.09 (20)	1034 (150)	0.09 (20)	0.31 (70)/10 MD 0.18 (40)/10 XMD	NA	NA	S/F

Bradley Industrial Textiles Inc.

Product Name	Structure[1]	Polymer Composition[2]	Percent Open Area %	Apparent Opening Size[3] mm (sieve)	Permittivity sec⁻¹/Flow Rate	Puncture kN (lb)	Mullen Burst kPa (psi)	Trapezoid Tear kN (lb)	Grab/Tensile Elongation kN (lb)/%	WWT MD	WWT XD	Suggested Application
Phoenix SCS-1	NW	PET	NA	0.210 (70)	1.4 (110)	(90)	(350)	(70)	(180) 60	N/P	N/A	D, E, F, S/S, R
Phoenix M-288-D	NW	PET	NA	0.300 (50)	1.9 (145)	(45)	(180)	(35)	(90) 60	N/A	N/A	D, E, F, S/S
Phoenix M-288-E	NW	PET	NA	0.210 (70)	1.09 (90)	(95)	(370)	(75)	(200) 60	N/A	N/A	D, E, F, S/S, R
Phoenix M-288-S	NW	PET	N/A	0.210 (70)	1.6 (120)	(55)	(190)	(45)	(130) 60	N/A	N/A	D, E, F, S/S
Phoenix[A] (4-60)	NW	PET	NA	NA	NA	N/A	N/A	N/A	N/A	N/A	N/A	R,D, E, F, S/S

Carthage Mills

Product Name	Structure[1]	Polymer Composition[2]	Percent Open Area %	Apparent Opening Size[3] mm (sieve)	Permittivity sec⁻¹/Flow Rate	Puncture kN (lb)	Mullen Burst kPa (psi)	Trapezoid Tear kN (lb)	Grab/Tensile Elongation kN (lb)/%	WWT MD	WWT XD	Suggested Application
Carthage 6%	W (calendered)	PP	5-6%	0.212 (70)	0.29/773 (19)	0.664 (145)	3376 (490)	0.449 x 0.289 (110 x 65)	1.69 x 1.16 (380 x 260)/15	40.28 (230)/17	26.27 (150)/13	E, F, D, S/F, S/S, R
Carthage 10%	W (calendered)	PP	10-12%	0.425-.30 (40-50)	1.2/3621 (89)	0.578 (130)	3514 (510)	0.356 x 0.333 (80 x 75)	1.22 x 1.16 (275 x 260)/16	27.15 (155)/19	28.9 (165)/17	E, F, D, S/F, S/S, R
Carthage 15%	W	PP	12-15%	0.425-.30 (40-50)	1.5/4678 (115)	0.533 (120)	3514 (510)	0.511 x 0.356 (115 x 80)	1.64 x 0.91 (370 x 205)/10	35.9 (205)/19	24.52 (140)/8	E, F, D, S/F, S/S, R

[1] W = Woven
NW = Nonwoven
O/C = Other or combination
[2] PET = Polyester
PP = Polypropylene
PE = Polyethylene
O/C = Other or combination
[3] Maximum opening size of tested rolls
[4] Reinforcement applications only
[5] S/S = Separation/stabilization
R = Reinforcement
F = Filtration
D = Drainage
E = Erosion Control
A/O = Asphalt Overlay

S/F = Silt Fence
O/C = Other or combination
[A] Custom manufactured to meet design specifications
NA = Manufacturer determined that this data was not applicable to the product or the data was unavailable
N/P = Data not provided by manufacturer

All values were requested to be minimum average roll values and all claims are the responsibility of the manufacturer. All product data are intended as a guide and are not all-inclusive. *Geotechnical Fabrics Report* recommends you contact manufacturers before making any specifying/purchasing decisions.

Carthage Mills (cont.)

Product Name Mass per unit g/m² (oz/yd²)	Physical Properties — Structure[1]	Polymer Composition[2]	Filtration/Hydraulic Properties — Percent Open Area (wovens only) CWO-22125 %	Apparent Opening Size[3] ASTM D 4751-87 mm (U.S. sieve)	Permittivity ASTM D 4491-92 sec^{-1}/Flow Rate l/min/m² (gal/min/ft²)	Puncture ASTM D 4833-88 kN (lb)	Mullen Burst ASTM D 3786-87 kPa (psi)	Mechanical Properties — Trapezoid Tear Strength ASTM D 4533-91 kN (lb)	Grab/Tensile Elongation ASTM D 4632-91 kN (lb)%	Wide Width Tensile Elongation[4] ASTM D 4595-86 kN/m (lb/in) MD	XD	Manufacturer's Suggested Applications[5]
Carthage 20%	W	PP	15-20%	0.212 (70)	0.55/1505 (37)	0.667 (150)	4100 (595)	0.467 x 0.644 (105 x 145)	1.24 x 1.73 (280 x 390)/16	32.4 (185)/10	32.4 (185)/27	E, F, D, S/F, S/S, R
Carthage 30%	W (palmered)	PP	25-30%	0.425-30 (40-50)	2.1/6143 (151)	0.533 (120)	2859 (415)	0.422 x 0.244 (95 x 55)	1.47 x 0.93 (330 x 210)/15	31.52 (180)/20	20.14 (115)/15	E, F, D, S/F, S/S, R
Carthage 8%-HD	W (calendered)	PP	6-8%	0.30-25 (50-60)	0.824/2441 (60)	0.933 (210)	5925 (860)	0.578 x 0.667 (130 x 150)	1.78 x 2.0 (400 x 450)/17	N/P	N/P	E, F, D, S/F, S/S, R
FX-400MF	W	PP	NP	0.60 (30)	NA	0.733 (165)	8371 (1215)	0.822 x 0.822 (185 x 185)	1.82 x 1.8 (410 x 405)/9	70.05 (400)/13	70.05 (400)/9	S/S, R
FX-11	W	PP	<1%	0.85-60 (20-30)	0.20/1139 (28)	0.267 (60)	1860 (270)	0.244 (55)	0.533 (120)/15	NA	NA	S/F
FX-33	W	PP	<1%	0.30 (50)	NA	0.289 (65)	1860 (270)	0.222 (50)	0.533 (120)/20	NA	NA	S/S
FX-44	W	PP	<1%	0.425 (40)	NA	0.311 (70)	2067 (300)	0.267 (60)	0.644 (145)/20	NA	NA	S/S
FX-55	W	PP	<1%	0.425 (40)	NA	0.467 (105)	3169 (460)	0.378 (85)	0.889 (200)/15	21.02 (120)	21.02 (120)	S/S, R
FX-66	W	PP	3.8, 4.5, 5.3	<1% (40-50)	0.425-0.30	NA	0.556 (125)	4134 (600) (120)	0.533 (300)/15	1.33 (175)	30.65 (175)	S/S, R
FX-40NS	NW	PP	NA	0.212 (70)	2.2/6712 (115)	0.289 (65)	1585 (230)	0.20 (45)	0.489 (110)/50	NA	NA	D
FX-60NS	NW	PP	NA	0.212 (70)	1.54/678 (115)	0.422 (95)	2343 (340)	0.311 (70)	0.733 (165)/50	NA	NA	D
FX-80NS	NW	PP	NA	0.212 (70)	1.1/3661 (90)	0.60 (135)	3032 (440)	0.422 (95)	1.0 (225)/50	NA	NA	S/S, D
FX-38OL	NW (calendered)	PP	NA	NA	NA	0.289 (65)	1481 (215)	0.20 (45)	0.422 (95)/50	NA	NA	A/O
FX-35HS	NW (heatbonded)	PP	NA	0.212 (70)	2.2/6102 (150)	0.244 (55)	1378 (200)	0.178 (40)	0.422 (95)/50	NA	NA	D
FX-40HS	NW (heatbonded)	PP	NA	0.212 (70)	2.0/5695 (140)	0.267 (60)	1481 (215)	0.222 (50)	0.489 (110)/50	NA	NA	D
FX-60HS	NW (heatbonded)	PP	NA	0.180 (80)	1.5/4475 (110)	0.40 (90)	2136 (310)	0.289 (65)	0.756 (170)/50	NA	NA	S/S, D
FX-80HS	NW (heatbonded)	PP	NA	0.180 (80)	1.2/3661 (90)	0.511 (115)	2480 (360)	0.378 (85)	0.978 (220)/50	NA	NA	S/S, D

[1] W = Woven
NW = Nonwoven
O/C = Other or combination
[2] PET = Polyester
PP = Polypropylene
PE = Polyethylene
O/C = Other or combination
[3] Maximum opening size of tested rolls

[4] Reinforcement applications only
[5] S/S = Separation/stabilization
R = Reinforcement
F = Filtration
D = Drainage
E = Erosion Control
A/O = Asphalt Overlay

S/F = Silt Fence
O/C = Other or combination
NA = Manufacturer determined that this data was not applicable to the product or the data was unavailable
NP = Data not provided by manufacturer

All values were requested to be minimum average roll values and all claims are the responsibility of the manufacturer. All product data are intended as a guide and are not all-inclusive. *Geotechnical Fabrics Report* recommends you contact manufacturers before making any specifying/purchasing decisions.

Carthage Mills (cont.)

Product Name Mass per unit g/m² (oz/yd²)	Physical Properties			Filtration/Hydraulic Properties		Mechanical Properties					Wide Width Tensile Elongation[4] ASTM D 4595-86 kN/m (lb/in)		Manufacturer's Suggested Applications[5]
	Structure[1]	Polymer Composition[2]	Percent Open Area (wovens only) CWO-22125 %	Apparent Opening Size[3] ASTM D 4751-87 mm (U.S. sieve)	Permittivity ASTM D 4491-92 sec⁻¹/Flow Rate l/min/m² (gal/min/ft²)	Puncture ASTM D 4833-88 kN (lb)	Mullen Burst ASTM D 3786-87 kPa (psi)	Trapezoid Tear Strength ASTM D 4533-91 kN (lb)	Grab/Tensile Elongation ASTM D 4632-91 kN (lb)%		MD	XD	
FX-100HS	NW (heatbonded)	PP	NA	0.150 (100)	1.1/ 3051 (75)	0.667 (150)	3101 (450)	0.444 (100)	1,267 (285)/50		NA	NA	S/S, P, D
FX-120HS	NW (heatbonded)	PP	NA	0.150 (100)	0.8/ 2644 (65)	0.756 (170)	3790 (550)	0.511 (115)	1,444 (325)		NA	NA	S/S, P, D
FX-160HS	NW (heatbonded)	PP	NA	0.150 (100)	0.6/ 1831 (45)	1.111 (250)	5168 (750)	0.733 (165)	1,889 (425)		NA	NA	S/S, P, D

Clem Environmental   James Clem Corp.

Product Name	Structure	Polymer	% Open	AOS	Permittivity	Puncture	Mullen	Trapezoid	Grab/Tensile	MD	XD	Applications
GeoCushion 24R 815 (24)[B]	NW	O/C	NA	NA	NA	NA	NA	NA	0.445 (100)[A]	NA	NA	P (geomembrane)
GeoCushion32R 1225 (32)[B]	NW	O/C	NA	NA	NA	NA	NA	NA	0.445 (100)[A]	NA	NA	P (geomembrane)

Contech Construction Products Inc.

Product Name	Structure	Polymer	% Open	AOS	Permittivity	Puncture	Mullen	Trapezoid	Grab/Tensile	MD	XD	Applications
Trewira 011/120 112 (3.3)	NW	PET	NA	0.300 (50)	2.07/105 (155)	0.200 (45)	1173 (170)	0.133 (30)	0.400 (90) 60	9.0 (51.1) 76.1[A]	6.4 (36.6) 82.2[A]	F, D, A/O
Trewira 011/140 136 (4.0)	NW	PET	NA	0.300 (50)	2.01/102 (150)	0.222 (50)	1311 (190)	0.178 (40)	0.489 (110) 60	10.9 (62.1) 69.6[A]	7.4 (42) 70.4[A]	F, D, E, A/O
Trewira 011/200 193 (5.7)	NW	PET	NA	0.210 (70)	1.74/88 (130)	0.356 (80)	1965 (285)	0.267 (60)	0.711 (160) 60	14.4 (82.5) 70.4[A]	7.4 (42) 70.4[A]	F, D, E, S/S
Trewira 011/250 241 (7.1)	NW	PET	NA	0.210 (70)	1.47/75 (110)	0.423 (95)	2484 (360)	0.334 (75)	0.933 (210) 60	17.1 (97.6) 65.0[A]	14.8 (84.6) 75.0[A]	F, D, E, S/S
Trewira 011/280 271 (8.0)	NW	PET	NA	0.210 (70)	1.20/61 (90)	0.444 (100)	2622 (380)	0.356 (80)	1.023 (230) 60	18.3 (104.6) 68.1[A]	18.8 (107.5) 72.1[A]	F, D, E, S/S, R
Trewira 011/350 339 (10.0)	NW	PET	NA	0.210 (70)	1.07/54 (80)	0.578 (130)	3519 (510)	0.444 (100)	1.356 (305) 60	26.8 (152.9) 74.2	25.5 (134.1) 72.1	R, F, P, E, S/S
Trewira 011/420 408 (12.0)	NW	PET	NA	0.210 (70)	0.87/44 (65)	0.667 (130)	3795 (550)	0.534 (120)	1.556 (350) 60	NA	NA	R, P, S/S
Trewira 011/450 441 (13.0)	NW	PET	NA	0.149 (100)	0.53/27 (40)	0.867 (195)	5382 (780)	0.667 (150)	2.224 (500) 70	36.0 (205.5)	28.8 (164.7)	R, P, S/S

[1] W = Woven
NW = Nonwoven
O/C = Other or combination
[2] PET = Polyester
PP = Polypropylene
PE = Polyethylene
O/C = Other or combination
[3] Maximum opening size of tested rolls

[4] Reinforcement applications only
[5] S/S = Separation /stabilization
R = Reinforcement
F = Filtration
D = Drainage
E = Erosion Control
A/O = Asphalt Overlay

S/F = Silt Fence
O/C = Other or combination
NA = Manufacturer determined that this data was not applicable to the product or the data was unavailable
NP = Data not provided by manufacturer
[A] Independent test lab values—not minimum average roll values
[B] Average values

All values were requested to be minimum average roll values and all claims are the responsibility of the manufacturer. All product data are intended as a guide and are not all-inclusive. *Geotechnical Fabrics Report* recommends you contact manufacturers before making any specifying/purchasing decisions.

Contech Construction Products Inc. (cont.)

Product Name [Mass per unit g/m² (oz/yd²)]	Physical Properties		Percent Open Area (wovens only) CWO-22125 %	Filtration/Hydraulic Properties		Mechanical Properties							Manufacturer's Suggested Applications[5]
	Structure[1]	Polymer Composition[2]		Apparent Opening Size[3] ASTM D 4751-87 mm (U.S. sieve)	Permittivity ASTM D 4491-92 sec⁻¹/Flow Rate l/min/m² (gal/min/ft²)	Puncture ASTM D 4833-88 kN (lb)	Mullen Burst ASTM D 3786-87 kPa (psi)	Trapezoid Tear Strength ASTM D 4533-91 kN (lb)	Grab/Tensile Elongation ASTM D 4632-91 kN (lb)%	Wide Width Tensile Elongation[4] ASTM D 4595-86 kN/m (lb/in)			
										MD	XD		
Trevira 011/550 542 (16.0)	NW	PET	NA	0.149 (100)	0.53/27(40)	0.689 (155)	4416 (640)	0.578 (130)	1.734 (390) 65	32.1 (183.5)	27.6 (157.6)	R, P, S/S
C200	W	PP	NA	0.60 (30)	0.07/244 (6)	0.450 (100)	3101 (450)	0.38 x 0.40 (85 x 90)	0.89 x 0.89/15 (200 x 200/15)	21.0 (120)	21.0 (120)	SS, O/C
C300	W	PP	NA	0.60 (30)	0.06/203 (5)	0.533 (120)	4134 (600)	0.53 x 0.53 (120 x 120)	1.33 x 1.33/15 (300 x 300/15)	30.6 (175)	30.6 (175)	SS, O/C, R
C70/06	W monofilament	PP	4.00	0.212 (70)	0.28/732 (18)	0.667 (150)	3307 (480)	0.42 x 0.24 (95 x 55)	1.76 x 1.16/24 (395 x 260/24)	NA	NA	SS, O/C, F, D, E

Hoechst Celanese Corp.

Product Name [Mass per unit g/m² (oz/yd²)]	Structure[1]	Polymer Composition[2]	Percent Open Area %	Apparent Opening Size[3] mm (U.S. sieve)	Permittivity / Flow Rate	Puncture kN (lb)	Mullen Burst kPa (psi)	Trapezoid Tear kN (lb)	Grab/Tensile Elongation kN (lb)%	Wide Width Tensile MD	Wide Width Tensile XD	Manufacturer's Suggested Applications[5]
Trevira 011/120 112 (3.3)	NW	PET	NA	0.300 (50)	2.07/105 (155)	0.200 (45)	1173 (170)	0.133 (30)	0.400 (90) 60	9.0 (51.1) 76.1[A]	6.4 (36.6) 82.2[A]	F, D, A/O
Trevira 011/140 136 (4.0)	NW	PET	NA	0.300 (50)	2.01/102 (150)	0.222 (50)	1311 (190)	0.178 (40)	0.489 (110) 60	10.9 (62.1) 69.6[A]	7.4 (42) 70.4[A]	F, D, E, A/O
Trevira 011/200 193 (5.7)	NW	PET	NA	0.210 (70)	1.74/88 (130)	0.356 (80)	1965 (285)	0.267 (60)	0.711 (160) 60	14.4 (82.5) 68.6	7.4 (42) 70.4[A]	F, D, E, S/S
Trevira 011/250 241 (7.1)	NW	PET	NA	0.210 (70)	1.47/75 (110)	0.423 (95)	2484 (360)	0.334 (75)	0.933 (210) 60	17.1 (97.6) 65.0[A]	14.8 (84.6) 75.0[A]	F, D, E, S/S
Trevira 011/280 271 (8.0)	NW	PET	NA	0.210 (70)	1.20/61 (90)	0.444 (100)	2622 (380)	0.356 (80)	1.023 (230) 60	18.3 (104.6) 68.1[A]	18.8 (107.5) 72.1[A]	F, D, E, S/S, R
Trevira 011/350 339 (10)	NW	PET	NA	0.210 (70)	1.07/54 (80)	0.578 (130)	3519 (510)	0.444 (100)	1.356 (305) 60	26.8 (152.9) 74.2	25.5 (134.1) 73.8	R, F, P, E, S/S
Trevira 011/420 406 (12)	NW	PET	NA	0.210 (70)	0.87/44 (65)	0.667 (130)	3795 (550)	0.534 (120)	1.556 (350) 60	NA	NA	R, P, S/S
Trevira 011/450 441 (13)	NW	PET	NA	0.149 (100)	0.80/40 (60)	0.689 (155)	4416 (640)	0.578 (130)	1.734 (390) 65	32.1 (183.5) 77.6[A]	27.6 (157.6) 80.4[A]	R, P, S/S
Trevira 011/550 541 (16)	NW	PET	NA	0.149 (100)	0.53/27 (40)	0.867 (195)	5382 (780)	0.667 (150)	2.224 (500) 70	36.0 (205.5) 81.6[A]	28.8 (164.7) 77.2[A]	R, P, S/S
Trevira ProEarth 140-550 (4-16)	NW	PET	NA	NA	NA	NA	NA	NA	NA	NA	NA	F, P, D, E, S/S

[1] W = Woven
NW = Nonwoven
O/C = Other or combination
[2] PET = Polyester
PP = Polypropylene
PE = Polyethylene
O/C = Other or combination
[3] Maximum opening size of tested rolls

[4] Reinforcement applications only
[5] S/S = Separation /stabilization
R = Reinforcement
F = Filtration
D = Drainage
E = Erosion Control
A/O = Asphalt Overlay

S/F = Silt Fence
O/C = Other or combination
NA = Manufacturer determined that this data was not applicable to the product or the data was unavailable
NP = Data not provided by manufacturer
[A] = Independent test lab values—not minimum average roll values

All values were requested to be minimum average roll values and all claims are the responsibility of the manufacturer. All product data are intended as a guide and are not all-inclusive. *Geotechnical Fabrics Report* recommends you contact manufacturers before making any specifying/purchasing decisions.

Huesker Inc.

Product Name — Mass per unit g/m² (oz/yd²)	Physical Properties		Filtration/Hydraulic Properties			Mechanical Properties				Wide Width Tensile Elongation[4] ASTM D 4595-86 kN/m (lb/in)		Manufacturer's Suggested Applications[5]
	Structure[1]	Polymer Composition[2]	Percent Open Area (wovens only) CWO-22125 %	Apparent Opening Size[3] ASTM D 4751-87 mm (U.S. sieve)	Permittivity ASTM D 4491-92 sec⁻¹/Flow Rate l/min/m² (gal/min/ft²)	Puncture ASTM D 4833-88 kN (lb)	Mullen Burst ASTM D 3786-87 kPa (psi)	Trapezoid Tear Strength ASTM D 4533-91 kN (lb)	Grab/Tensile Elongation ASTM D 4632-91 kN (lb)%	MD	XD	
Huesker Inc.												
Comtrac 6G/185	W	PP	<1	(50-70)	(0.1)	0.40 (90)	2756 (400)	NA	NA	35 (200)/13	35 (200)/13	R
Comtrac 60.006	W	PP	<1	(40-70)	(0.1)	0.53 (120)	4134 (600)	NA	NA	80 (400)/13	80 (400)/13	R
Comtrac 270.270	W	PET	<1	0.4 (40)	(0.2)	2.0 (460)	10335 (1500)	(600)	NA	270 (1500)/11	270 (1500)/11	R
Comtrac 450.140	W	PET	<1	0.4 (40)	(0.2)	2.0 (460)	10335 (1500)	(600)	NA	450 (2500)/11	140 (800)/11	R
Comtrac 750.285	W	PET	<1	0.4 (40)	(0.2)	2.& (600)	11700 (1700)	(700)	NA	750 (4200)/11	285 (1600)/11	R
Linq Industrial Fabrics												
GTF180	W	PP	NP	0.84 (20)	0.05	(60)	(280)	(50)	(100)/15	NP	NP	S/F
GTF200F	W	PP	NP	0.60 (30)	0.05	(75)	(300)	(70)	(180)/15	NP	NP	S/S
GTF200	W	PP	NP	0.60 (30)	0.03	(100)	(450)	(90)	(200)/15	(150)/15	(150)/15	S/S, R
GTF300	W	PP	NP	0.60 (30)	0.02	(145)	(600)	(115)	(300)/15	(180)/15	(210)/15	S/S, R
GTF400E	W	PP	4	0.21 (70)	0.28	(135)	(480)	(95/55)	(370)/16 (250)/15	NP	NP	E
Typar 3301	NW	PP	NP	0.30 (50)	0.6	(25)	(90)	(35)	(120)/60	NP	NP	F, D
Typar 3341	NW	PP	NP	0.25 (60)	0.7	(30)	(90)	(40)	(120)/60	NP	NP	F, D
Typar 3401	NW	PP	NP	0.21 (70)	0.7	(40)	(140)	(60)	(130)/60	NP	NP	S/S, F, D
Typar 3601	NW	PP	NP	0.10 (140)	0.1	(70)	(210)	(90)	(240)/60	NP	NP	S/S, F, D, E
Typar 3631	NW	PP	NP	0.10 (140)	0.1	(75)	(210)	(90)	(250)/60	NP	NP	S/S, F, D, E

[1] W   = Woven
     NW   = Nonwoven
     O/C = Other or combination
[2] PET = Polyester
     PP   = Polypropylene
     PE   = Polyethylene
     O/C = Other or combination
[3] Maximum opening size of tested rolls

[4] Reinforcement applications only
[5] S/S = Separation /stabilization
     R   = Reinforcement
     F   = Filtration
     D   = Drainage
     P   = Protection
     E   = Erosion Control
     A/O = Asphalt Overlay

S/F = Silt Fence
O/C = Other or combination
NA   = Manufacturer determined that this data was not applicable to the product or the data was unavailable
NP   = Data not provided by manufacturer
[B] Independent test values—not MARV

All values were requested to be minimum average roll values and all claims are the responsibility of the manufacturer. All product data are intended as a guide and are not all-inclusive. *Geotechnical Fabrics Report* recommends you contact manufacturers before making any specifying/purchasing decisions.

Linq Industrial Fabrics (cont.)

Product Name [Mass per unit g/m² (oz/yd²)]	Physical Properties			Filtration/Hydraulic Properties				Mechanical Properties				Manufacturer's Suggested Applications[5]
	Structure[1]	Polymer Composition[2]	Percent Open Area (wovens only) CWO-22125 %	Apparent Opening Size[3] ASTM D 4751-87 mm (U.S. sieve)	Permittivity ASTM D 4491-92 sec⁻¹/Flow Rate l/min/m² (gal/min/ft²)	Puncture ASTM D 4833-88 kN (lb)	Mullen Burst ASTM D 3786-87 kPa (psi)	Trapezoid Tear Strength ASTM D 4533-91 kN (lb)	Grab/Tensile Elongation ASTM D 4632-91 kN (lb)%	Wide Width Tensile Elongation[4] ASTM D 4595-86 kN/m (lb/in) MD	XD	
Typar 3801	NW	PP	NP	0.07 (200)	0.08	(90)	(250)	(90)	(325)/60%	NP	NP	S/S, P
125EX	NW	PP	NP	0.21 (70)	2.2	(55)	(200)	(40)	(95)/50%	NP	NP	F, D
130EX	NW	PP	NP	0.21 (70)	2.0	(60)	(215)	(45)	(105)/50%	NP	NP	F, D
150EX	NW	PP	NP	0.18 (80)	1.3	(90)	(310)	(65)	(165)/50%	NP	NP	S/S, F, D, E
180EX	NW	PP	NP	0.18 (80)	1.2	(100)	(330)	(75)	(200)/50%	NP	NP	S/S, F, D, E, P
225EX	NW	PP	NP	0.18 (80)	1.0	(115)	(360)	(85)	(215)/50%	NP	NP	S/S, F, D, E, P
250EX	NP	PP	NP	0.15 (100)	0.8	(150)	(450)	(100)	(285)/50%	NP	NP	S/S, E, P
275EX	NP	PP	NP	0.15 (100)	0.7	(250)	(750)	(165)	(425)/50%	NP	NP	S/S, E, P
350EX	NW	PP	NP	0.15 (100)	0.5	(170)	(550)	(115)	(325)/50%	NP	NP	S/S, E, P
AOL	NW	PP	NP	0.21 (70)	1.8	(55)	(180)	(45)	(90)/50%	NP	NP	A/O
AOH	NW	PP	NP	0.21 (70)	1.8	(55)	(200)	(50)	(90)/50%	NP	NP	A/O
GTF570	W	PP	NP	0.60 (30)	0.4	(160)	(1200)	(180)	(415)/15% (410)/15%	(400)/13%	(400)/8%	R, E
GTF550T	W	PET	NP	0.43 (40)	0.1	(140)	(750)	(380)	NP	(500)/10%	(500)/10%	R
GTF1000T	W	PET	NP	0.43 (40)	0.1	(150)	(1000+)	(400)	NP	(1000)/10%	(800)/10%	R

[1] W = Woven
NW = Nonwoven
O/C = Other or combination

[2] PET = Polyester
PP = Polypropylene
PE = Polyethylene
O/C = Other or combination

[3] Maximum opening size of tested rolls

[4] Reinforcement applications only

[5] S/S = Separation/stabilization
R = Reinforcement
F = Filtration
D = Drainage
P = Protection
E = Erosion Control
A/O = Asphalt Overlay

S/F = Silt Fence
O/C = Other or combination
NA = Manufacturer determined that this data was not applicable to the product or the data was unavailable
NP = Data not provided by manufacturer

All values were requested to be minimum average roll values and all claims are the responsibility of the manufacturer. All product data are intended as a guide and are not all-inclusive. *Geotechnical Fabrics Report* recommends you contact manufacturers before making any specifying/purchasing decisions.

Nicolon/Mirafi Group

Product Name [Mass per unit g/m² (oz/yd²)]	Physical Properties		Filtration/Hydraulic Properties			Mechanical Properties				Wide Width Tensile Elongation[4] ASTM D 4595-86 kN/m (lb/in)		Manufacturer's Suggested Applications[5]
	Structure[1]	Polymer Composition[2]	Percent Open Area (wovens only) CWO-22125 %	Apparent Opening Size[3] ASTM D 4751-87 mm (U.S. sieve)	Permittivity ASTM D 4491-92 sec⁻¹/Flow Rate l/min/m² (gal/min/ft²)	Puncture ASTM D 4833-88 kN (lb)	Mullen Burst ASTM D 3786-87 kPa (psi)	Trapezoid Tear Strength ASTM D 4533-91 kN (lb)	Grab/Tensile Elongation ASTM D 4632-91 kN (lb)%	MD	XD	
Nicolon HS600	W	PET	NA	0.85 (20)	0.32 1019(25)	NA	NA	NA	NA	105(600)	88(500)	R
Nicolon HS1150	W	PET	NA	0.59 (30)	0.84 2648(65)	NA	NA	NA	NA	201(1150)	123(700)	R
Nicolon HS1715	W	PET	NA	0.59 (30)	0.65 2037(50)	NA	NA	NA	NA	298(1700)	123(700)	R
Stabilenka 2400	W	PET	NA	NA	NA	NA	NA	NA	NA	420(2400)	123(700)	R
Stabilenka 3000	W	PET	NA	NA	NA	NA	NA	NA	NA	526(3000)	123(700)	R
Stabilenka 3600	W	PET	NA	NA	NA	NA	NA	NA	NA	631(3600)	123(700)	R
Nicolon S400 (4.0)	NW	PP	NA	0.212 (70)	2.28 7333(180)	0.289 (65)	1585 (230)	0.200 (45)	0.467 (105)/60	NA	NA	F, D, E
Nicolon S600 (6.0)	NW	PP	NA	0.212 (70)	1.63 5296(130)	0.422 (95)	2412 (350)	0.29 (65)	0.712 (160)/50	NA	NA	S/S, D, E
Nicolon S700 (7.0)	NW	PP	NA	0.212 (70)	1.41 4481(110)	0.511 (115)	2756 (400)	0.356 (80)	0.89 (200)/50	NA	NA	S/S, D, E
Nicolon S800 (8.0)	NW	PP	NA	0.212 (70)	1.26 4074(100)	0.58 (130)	3101 (450)	0.4 (90)	1.001 (225)/50	NA	NA	R, P, D, F, S/S
Nicolon S1000 (10)	NW	PP	NA	0.15 (100)	0.936 3056(75)	0.73 (165)	3858 (560)	0.47 (105)	1.246 (280)/50	NA	NA	R, P, D, F, S/S
Nicolon S1200 (12)	NW	PP	NA	0.15 (100)	0.751 2444(60)	0.85 (190)	4479 (650)	0.58 (130)	1.6 (360)/60	NA	NA	R, P, S/S
Nicolon S1400 (14)	NW	PP	NA	0.15 (100)	0.642 2037(50)	0.98 (220)	4495 (725)	0.62 (140)	1.89 (425)/60	NA	NA	R, P, S/S
Nicolon S1600 (16)	NW	PP	NA	0.15 (100)	0.571 1833(45)	1.07 (240)	5512 (800)	0.69 (155)	2.225 (500)/60	NA	NA	R, P, S/S
Mirafi 100X	W	PP	1	0.841 (20)	0.2 815(20)	0.27 (60)	2067 (300)	0.27 (60)	0.53 x 0.45/10 (120 x 100/10)	NA	NA	NP
Mirafi 500XL	W	PP	1	0.595 (30)	0.04 163 (4)	0.31 (70)	2412 (350)	0.20 (45)	0.62 (140)/15	NA	NA	NP
Mirafi 500X	W	PP	1	0.595 (30)	0.03 163 (4)	0.40 (90)	2756 (400)	0.33 (75)	0.89 x 0.89/15 (200 x 200/15)	18 (100)	21 (120)	NP

All values were requested to be minimum average roll values and all claims are the responsibility of the manufacturer. All product data are intended as a guide and are not all-inclusive. *Geotechnical Fabrics Report* recommends you contact manufacturers before making any specifying/purchasing decisions.

[1]
W = Woven
NW = Nonwoven
O/C = Other or combination
[2]
PET = Polyester
PP = Polypropylene
PE = Polyethylene
O/C = Other or combination
[3] Maximum opening size of tested rolls

[4] Reinforcement applications only
[5]
S/S = Separation/stabilization
R = Reinforcement
F = Filtration
D = Drainage
P = Protection
E = Erosion Control
A/O = Asphalt Overlay

S/F = Silt Fence
O/C = Other or combination
NA = Manufacturer determined that this data was not applicable to the product or the data was unavailable
NP = Data not provided by manufacturer

Nicolon/Mirafi Group (cont.)

Product Name [Mass per unit g/m² (oz/yd²)]	Physical Properties			Filtration/Hydraulic Properties				Mechanical Properties				Manufacturer's Suggested Application[5]
	Structure[1]	Polymer Composition[2]	Percent Open Area (wovens only) CWO-22125 %	Apparent Opening Size[3] ASTM D 4751-87 mm (U.S. sieve)	Permittivity ASTM D 4491-92 sec⁻¹/Flow Rate l/min/m² (gal/min/ft²)	Puncture ASTM D 4833-88 kN (lb)	Mullen Burst ASTM D 3786-87 kPa (psi)	Trapezoid Tear Strength ASTM D 4533-91 kN (lb)	Grab/Tensile Elongation ASTM D 4632-91 kN (lb)%	Wide Width Tensile Elongation[4] ASTM D 4565-86 kN/m (lb/in) MD	XD	
Mirafi 600X	W	PP	1	0.420 (40)	0.02 81 (2)	0.53 (120)	4134 (600)	0.53 (120)	1.34 x 1.34/15 (300 x 300/15)	29 (165)	31 (175)	NP
Mirafi 700X	W	PP	4	0.210 (70)	0.28 733 (18)	0.60 (135)	3307 (480)	0.45 x 0.27 (100 x 60)	1.65 x 1.11/15 (370 x 250/15)	39 (225)	25 (145)	NP
Mirafi 700XG	W	PP	10	0.420 (40)	1.36 4074 (100)	0.51 (115)	3445 (500)	0.51 x 0.33 (115 x 75)	1.62 x 0.89/10 (365 x 200/10)	35 (200)	24 (135)	NP
Filterweave 70/20	W	PP	10	0.250 (60)	0.506 1426 (35)	0.65 (145)	4065 (590)	0.47 x 0.62 (105 x 140)	1.22 x 1.74/15 (275 x 390/15)	31.5 (180)	32 (185)	NP
Filterweave 40/10	W	PP	10	0.420 (40)	0.95 2852 (70)	0.56 (125)	3445 (500)	0.36 x 0.31 (80 x 70)	1.18 x 1.13/15 (265 x 255/15)	26 (150)	29 (165)	NP
Filterweave 40/30A	W	PP	20	0.420 (40)	2.14 5907 (145)	0.51 (115)	2756 (400)	0.40 x 0.22 (90 x 50)	1.45 x 0.89/15 (325 x 200/15)	31 (175)	19 (110)	NP
Mirafi HP800X	W	PP	1	0.210 (70)	0.1 326 (8)	0.80 (180)	4823 (700)	0.53 x 0.62 (120 x 140)	1.49 x 1.89/17 (335 x 425/17)	NA	NA	NP
Nicolon HP500	W	PP	8	0.595 (30)	1.5 4685 (115)	0.65 (145)	4995 (725)	0.65 x 0.56 (145 x 125)	1.78 x 1.49/15 (400 x 335/15)	40 (230)	39 (225)	NP
Nicolon HP550	W	PP	6	0.420 (40)	0.96 2852 (70)	0.73 (165)	4789 (695)	0.65 x 0.56 (145 x 125)	1.89 x 1.56/21 (425 x 350/21)	47 (270)	39 (225)	NP
Nicolon HP565	W	PP	NA	0.420 (40)	0.07 204 (5)	0.89 (200)	8268 (1200)	0.78 x 0.96 (175 x 215)	2.23 x 2.23/10 (500 x 500/10)	61 (350)	66 (375)	NP
Nicolon HP570	W	PP	NA	0.595 (30)	0.4 1222 (30)	0.71 (160)	8268 (1200)	0.80 x 0.80 (180 x 180)	1.85 x 1.82/8 (415 x 410/8)	70 (400)	70 (400)	NP
Mirafi 140NL	NW	PP	NA	0.210 (70)	2.0 4481 (110)	0.24 (55)	1275 (185)	0.16 (35)	0.40 (90)/50	NA	NA	NP
Mirafi 140NC	NW	PP	NA	0.210 (70)	1.5 4889 (120)	0.31 (70)	1654 (240)	0.22 (50)	0.49 x 0.58/50 (110 x 130/50)	NA	NA	NP
Mirafi 140N	NW	PP	NA	0.210 (70)	1.5 4889 (120)	0.31 (70)	1654 (240)	0.22 (50)	0.53 (120)/50	NA	NA	NP
Mirafi 160N	NW	PP	NA	0.210 (70)	1.3 4481 (110)	0.42 (95)	2239 (325)	0.27 (60)	0.67 (150)/50	NA	NA	NP
Mirafi 180N	NW	PP	NA	0.177 (80)	1.5 4481 (110)	0.58 (130)	2756 (400)	0.38 (85)	0.89 (200)/50	NA	NA	NP
Mirafi 1100N	NW	PP	NA	0.149 (100)	1.2 3463 (85)	0.71 (160)	3617 (525)	0.45 (100)	1.11 (250)/50	NA	NA	NP

[1] W = Woven
NW = Nonwoven
O/C = Other or combination
[2] PET = Polyester
PP = Polypropylene
PE = Polyethylene
O/C = Other or combination
[3] Maximum opening size of tested rolls

[4] Reinforcement applications only
[5] S/S = Separation/stabilization
R = Reinforcement
F = Filtration
D = Drainage
P = Protection
E = Erosion Control
A/O = Asphalt Overlay

S/F = Silt Fence
O/C = Other or combination
NA = Manufacturer determined that this data was not applicable to the product or the data was unavailable
NP = Data not provided by manufacturer

All values were requested to be minimum average roll values and all claims are the responsibility of the manufacturer. All product data are intended as a guide and are not all-inclusive. *Geotechnical Fabrics Report* recommends you contact manufacturers before making any specifying/purchasing decisions.

Nicolon/Mirafi Group (cont.)

Polyfelt Americas Inc.

Product Name / Mass per unit g/m² (oz/yd²)	Structure[1]	Polymer Composition[2]	Percent Open Area (wovens only) CWO-22125 %	Apparent Opening Size[5] ASTM D 4751-87 mm (U.S. sieve)	Permittivity ASTM D 4491-92 sec⁻¹/Flow Rate l/min/m² (gal/min/ft²)	Puncture ASTM D 4833-88 kN (lb)	Mullen Burst ASTM D 3786-87 kPa (psi)	Trapezoid Tear Strength ASTM D 4533-91 kN (lb)	Grab/Tensile Elongation ASTM D 4632-91 kN (lb)%	Wide Width Tensile Elongation[4] ASTM D 4595-86 kN/m (lb/in) MD	XD	Manufacturer's Suggested Applications[5]
Nicolon/Mirafi Group (cont.)												
Mirafi 1120N	NW	PP	NA	0.149 (100)	1.0 / 3056 (75)	0.80 (180)	4134 (600)	0.51 (115)	1.34 (300)/50	NA	NA	NP
Mirafi 1160N	NW	PP	NA	0.149 (100)	0.7 / 2037 (50)	1.07 (240)	5512 (800)	0.65 (145)	1.69 (380)/50	NA	NA	NP
Mirapave (3)	NW	PP	NA	0.210 (70)	NA	0.27 (60)	1378 (200)	0.16 (35)	0.40 (90)/55	NA	NA	NP
MCF 1212 (typical values)	W (coated)	PE	NA	NA	NA	0.45 (100)	2963 (430)	0.22 x 0.22 (50 x 50)	0.89 x 0.89/15 (200 x200/15)	NA	NA	NP
Polyfelt Americas Inc.												
TS420 120 (3.5)	NW	PP	NA	0.425 (40)	2.4 / 7330 (180)	0.220 (50)	1205 (175)	0.200 (45)	0.400 (90) 50	5.3[A] (30) 80	6.1[A] (35) 30	S/S, R, F, D, P, E, S/F
TS500 135 (4)	NW	PP	NA	0.300 (50)	2.2 / 6720 (165)	0.265 (60)	1380 (200)	0.220 (50)	0.510 (115) 50	8.8[A] (50) 80	8.8[A] (50) 30	S/S, R, F, D, P, E
TS550 170 (5)	NW	PP	NA	0.300 (50)	1.9 / 5700 (140)	0.310 (70)	1655 (240)	0.265 (60)	0.580 (130) 80	9.6[A] (55) 30	9.6[A] (55) 30	S/S, R, F, D, P, E, S/F
TS600 205 (6)	NW	PP	NA	0.250 (60)	1.8 / 5490 (135)	0.335 (75)	2000 (290)	0.310 (70)	0.710 (160) 50	10.5[A] (60) 80	10.5[A] (60) 30	S/S, R, F, D, P, E
TS650 240 (7)	NW	PP	NA	0.250 (60)	1.6 / 4880 (120)	0.400 (90)	2415 (350)	0.335 (75)	0.800 (180) 50	12.3[A] (70) 80	14.0[A] (80) 30	S/S, R, F, D, P, E
TS700 270 (8)	NW	PP	NA	0.212 (70)	1.3 / 4070 (100)	0.445 (100)	2760 (400)	0.380 (85)	0.935 (210) 50	13.1[A] (75) 80	15.8[A] (90) 30	S/S, R, F, D, P, E
TS750 340 (10)	NW	PP	NA	0.180 (80)	1.1 / 3260 (80)	0.555 (125)	3105 (450)	0.445 (100)	1.11 (250) 50	15.8[A] (90) 80	17.5[A] (100) 40	S/S, R, F, D, P, E
TS800 410 (12)	NW	PP	NA	0.15 (100)	1.0 / 3050 (75)	0.600 (135)	3450 (500)	0.490 (110)	1.33 (300) 50	19.3[A] (110) 80	21.0[A] (120) 40	S/S, R, F, D, P, E
TS1000 540 (16)	NW	PP	NA	0.150 (100)	0.6 / 1830 (45)	0.710 (160)	3795 (550)	0.535 (120)	1.51 (340) 50	24.5[A] (140) 80	24.5[A] (140) 40	S/S, R, F, D, P, E
PGM13 120 (3.5)	NW	PP	NA	NA	NA	0.200 (45)	NA	0.200 (45)	0.400 (90) 50	NA	NA	A/O
PGM15 142 (4.2)	NW	PP	NA	NA	NA	0.200 (45)	NA	0.220 (50)	0.445 (100) 50	NA	NA	A/O

[1] W = Woven
NW = Nonwoven
O/C = Other or combination
[2] PET = Polyester
PP = Polypropylene
PE = Polyethylene
O/C = Other or combination
[3] Maximum opening size of tested rolls

[4] Reinforcement applications only
[5] S/S = Separation /stabilization
R = Reinforcement
F = Filtration
D = Drainage
P = Protection
E = Erosion Control
A/O = Asphalt Overlay

SF = Silt Fence
O/C = Other or combination
NA = Manufacturer determined that this data was not applicable to the product or the data was unavailable
NP = Data not provided by manufacturer
[A] = Typical values

All values were requested to be minimum average roll values and all claims are the responsibility of the manufacturer. All product data are intended as a guide and are not all-inclusive. *Geotechnical Fabrics Report* recommends you contact manufacturers before making any specifying/purchasing decisions.

PolyfeltAmericas Inc. (cont.)

The Reinforced Earth Co.

Spartan Technologies

Product Name [Mass per unit g/m² (oz/yd²)]	Physical Properties			Filtration/Hydraulic Properties		Mechanical Properties				Wide Width Tensile Elongation[4] ASTM D 4595-86 kN/m (lb/in)		Manufacturer's Suggested Applications[5]
	Structure[1]	Polymer Composition[2]	Percent Open Area (wovens only) CWO-22125 %	Apparent Opening Size[3] ASTM D 4751-87 mm (U.S. sieve)	Permittivity ASTM D 4491-92 sec⁻¹/Flow Rate l/min/m² (gal/min/ft²)	Puncture ASTM D 4833-88 kN (lb)	Mullen Burst ASTM D 3786-87 kPa (psi)	Trapezoid Tear Strength ASTM D 4533-91 kN (lb)	Grab/Tensile Elongation ASTM D 4632-91 kN (lb)%	MD	XD	
PolyfeltAmericas Inc. (cont.)												
PGM20 193 (5.7)	NW	PP	NA	NA	NA	0.310 (70)	NA	0.290 (65)	0.625 (140) 50	NA	NA	A/O
The Reinforced Earth Co.												
MX4	W	PP	10	0.425/40	1.36	(115)	(500)	(115 x 75)	(365/24 x 220/10)	(200)	(160)	F, R
MX5	W	PP	8	0.600/30	1.5	(145)	(725)	(145 x 125)	(400/20 x 335/15)	(230)	(225)	F, R
MX5 5	W	PP	6	0.425/40	0.96	(165)	(695)	(145 x 125)	(425/21 x 350/21)	(270)	(225)	F, R
Spartan Technologies												
ST-OL	NW	PP	NA	NA	NA	NA	1379 (200)	NA	0.40 (90) 55	NA	NA	A/O
ST35	NW	PP	NA	0.212 (70)	2.2/6110 (150)	0.24 (55)	1379 (200)	0.18 (40)	0.42 (95) 50	NA	NA	S/S, F, D
ST40	NW	PP	NA	0.212 (70)	2.0/5700 (140)	0.26 (60)	1482 (215)	0.20 (45)	0.47 (105) 50	NA	NA	S/S, F, D
ST45	NW	PP	NA	0.212 (70)	1.8/5290 (130)	0.31 (70)	1793 (260)	0.22 (50)	0.53 (120) 50	NA	NA	S/S, F, D
ST60	NW	PP	NA	0.180 (80)	1.5/4070 (110)	0.40 (90)	2140 (310)	0.29 (65)	0.73 (165) 50	NA	NA	S/S, F, D, R, P, E
ST70	NW	PP	NA	0.180 (80)	1.3/4480 (110)	0.44 (100)	2276 (330)	0.33 (75)	0.89 (200) 50	NA	NA	S/S, F, D, R, P, E
ST80	NW	PP	NA	0.180 (80)	1.2/3255 (80)	0.51 (115)	2482 (360)	0.39 (85)	0.96 (215) 50	NA	NA	S/S, F, D, R, P, E
ST100	NW	PP	NA	0.150 (100)	1.1/2850 (70)	0.67 (150)	3103 (450)	0.44 (100)	1.27 (285) 50	NA	NA	S/S, R, P, E
ST120	NW	PP	NA	0.150 (100)	0.8/2445 (60)	0.76 (170)	3793 (550)	0.51 (115)	1.45 (325) 50	NA	NA	S/S, R, P, E

[1] W = Woven
NW = Nonwoven
OIC = Other or combination
[2] PET = Polyester
PP = Polypropylene
PE = Polyethylene
OIC = Other or combination
[3] Maximum opening size of tested rolls

[4] Reinforcement applications only
[5] S/S = Separation /stabilization
R = Reinforcement
F = Filtration
D = Drainage
P = Protection
E = Erosion Control
A/O = Asphalt Overlay

S/F = Silt Fence
O/C = Other or combination
NA = Manufacturer determined that this data was not applicable to the product or the data was unavailable
NP = Data not provided by manufacturer
[A] Typical values

All values were requested to be minimum average roll values and all claims are the responsibility of the manufacturer. All product data are intended as a guide and are not all-inclusive. *Geotechnical Fabrics Report* recommends you contact manufacturers before making any specifying/purchasing decisions.

Product Name [Mass per unit g/m² (oz/yd²)]	Physical Properties		Filtration/Hydraulic Properties			Mechanical Properties						Manufacturer's Suggested Applications[5]
	Structure[1]	Polymer Composition[2]	Percent Open Area (wovens only) CWO-22125 %	Apparent Opening Size[3] ASTM D 4751-87 mm (U.S. sieve)	Permittivity ASTM D 4491-92 sec^{-1}/Flow Rate l/min/m² (gal/min/ft²)	Puncture ASTM D 4833-88 kN (lb)	Mullen Burst ASTM D 3786-87 kPa (psi)	Trapezoid Tear Strength ASTM D 4533-91 kN (lb)	Grab/Tensile Elongation ASTM D 4632-91 kN (lb)%	Wide Width Tensile Elongation[4] ASTM D 4595-86 kN/m (lb/in) MD	Wide Width Tensile Elongation[4] ASTM D 4595-86 kN/m (lb/in) XD	
Spartan Technologies (cont.)												**Spartan Technologies (cont.)**
ST160	NW	PP	NA	0.125 (120)	0.6/1830 (45)	1.1 (250)	5172 (750)	0.73 (165)	1.89 (425) 50	NA	NA	S/S, R, P, E
Synthetic Industries												**Synthetic Industries**
901SC	W/S[E]	PP	<1%[B]	1.180 (16)	0.70 2037 (50)	0.11 (25)	1171 (170)	0.20 x 0.09 (45 x 20)	0.38 x 0.22 (85 x 50)/13 x 13	NA	NA	S/F
910SC[D]	W/S[E]	PP	<1%[B]	0.850 (20)	0.20 611 (15)	0.26 (58)	1826 (265)	0.22 x 0.22 (50 x 50)	0.45 x 0.45 (100 x 100)/15 x 15	NA	NA	S/F
913SC[D]	W/S[E]	PP	<1%[B]	0.850 (20)	0.05 122 (3)	0.16 (35)	1826 (265)	0.22 x 0.18 (50 x 40)	0.45 x 0.45 (100 x 100)/9 x 13	NA	NA	S/F
914SC	W/C[G]	PP	4.4%[A]	0.600 (30)	0.90 3055 (75)	0.27 (60)	2411 (350)	0.27 x 0.27 (60 x 60)	0.67 x 0.45 (150 x 100)/15 x 15	NA	NA	S/F
994	W/S[E]	PP	<1%[B]	0.600 (30)	0.04 81 (2)	0.31 (70)[A]	1860 (270)	0.22 x 0.18 (50 x 40)	0.62 x 0.42 (140 x 95)/15 x 15	NA	NA	S/F
200ST	W/S[E]	PP	<1%[B]	0.600 (30)	0.07 244 (6)	0.45 (100)	3101 (450)	0.38 x 0.40 (85 x 90)	0.89 x 0.89 (200 x 200)/15 x 15	21.0 (120)	21.0 (120)	S/S, R, E
300ST	W/S[E]	PP	<1%[B]	0.600 (30)	0.06 203 (5)	0.53 (120)	4134 (600)	0.53 x 0.53 (120 x 120)	1.33 x 1.33 (300 x 300)/15 x 15	30.6 (175)	30.6 (175)	S/S, R, E
400R	W/C[G]	PP	<1%[B]	0.600 (30)	0.60 1830 (45)	0.62 (140)	8270 (1200)	0.82 x 0.85 (185 x 190)	NA	70.0 (400)	70.0 (400)	S/S, R
Erosion I	W/M[F]	PP	4%[B]	0.212 (70)	0.28 732 (18)	0.67 (150)	3307 (480)	0.42 x 0.24 (95 x 55)	1.76 x 1.16 (395 x 260)/24 x 24	NA	NA	S/S, E, F, D
Erosion III	W/M[F]	PP	5%[B]	0.425 (40)	0.30 1220 (30)	0.62 (140)	3548 (515)	0.44 x 0.27 (100 x 60)	1.60 x 1.16 (360 x 260)/20 x 20	NA	NA	S/S, E, F, D
Erosion V	W/M[F]	PP	11%[B]	0.600 (30)	1.10 4475 (110)	0.58 (130)	3548 (515)	0.44 x 0.36 (100 x 80)	1.60 x 1.16 (360 x 260)/27 x 15	NA	NA	S/S, E, F, D
Erosion X	W/M[F]	PP	17%[B]	0.850 (20)	1.50 8136 (200)	0.60 (135)	2894 (420)	0.18 x 0.22 (40 x 50)	1.13 x 1.22 (255 x 275/20 x 15	NA	NA	E, F, D
Erosion XV	W/M[F]	PP	3%[B]	0.425 (40)	0.55 1630 (40)	0.47 (165)	5546 (805)	0.53 x 0.53 (120 x 120)	1.47 x 1.94 (330 x 435/15 x 20	28.0 (160)/ 12	30.6 (175)/ 12	E, F, D, S/S, R
381[C]	NW	PP	NA	NA	NA	0.27 (60)	1447 (210)	0.18 (40)	0.40 (90)/50	NA	NA	A/O

[1] W = Woven
NW = Nonwoven
O/C = Other or combination
[2] PET = Polyester
PP = Polypropylene
PE = Polyethylene
O/C = Other or combination
[3] Maximum opening size of tested rolls
[4] Reinforcement applications only
[5] S/S = Separation /stabilization
R = Reinforcement
F = Filtration
D = Drainage
P = Protection
E = Erosion Control
A/O = Asphalt Overlay
S/F = Silt Fence
O/C = Other or combination
NA = Manufacturer determined that this data was not applicable to the product or the data was unavailable
NP = Data not provided by manufacturer
[A] Typical or average value reported in absence of MARV data
[B] Area of openings ÷ total area determined from a light projection test
[C] Product calendered for pavement overlay application
[D] Indicates products available either with or without drawstring function
[E] Woven slit film
[F] Woven monofilament
[G] Woven combination

All values were requested to be minimum average roll values and all claims are the responsibility of the manufacturer. All product data are intended as a guide and are not all-inclusive. *Geotechnical Fabrics Report* recommends you contact manufacturers before making any specifying/purchasing decisions.

Synthetic Industries (cont.)

Product Name [Mass per unit g/m² (oz/yd²)]	Physical Properties			Filtration/Hydraulic Properties		Mechanical Properties				Wide Width Tensile Elongation[4] ASTM D 4595-86 kN/m (lb/in)		Manufacturer's Suggested Applications[5]
	Structure[1]	Polymer Composition[2]	Percent Open Area (wovens only) CWO-22125 %	Apparent Opening Size[3] ASTM D 4751-87 mm (U.S. sieve)	Permittivity ASTM D 4491-92 sec⁻¹/Flow Rate l/min/m² (gal/min/ft²)	Puncture ASTM D 4833-88 kN (lb)	Mullen Burst ASTM D 3786-87 kPa (psi)	Trapezoid Tear Strength ASTM D 4533-91 kN (lb)	Grab/Tensile Elongation ASTM D 4632-91 kN (lb)/%	MD	XD	
311	NW	PP	NA	0.212 (70)	2 4482 (110)	0.22 (56)	1137 (165)	0.13 (30)	0.36 (80)/45	NA	NA	F, D
351	NW	PP	NA	0.212 (70)	2 4482 (110)	0.24 (55)	1275 (185)	0.16 (35)	0.4 (90)/50	NA	NA	F, D, E
401	NW	PP	NA	0.212 (70)	2 5704 (140)	0.29 (65)	1550 (225)	0.2 (45)	0.44 (100)/50	NA	NA	F, D, E
451	NW	PP	NA	0.212 (70)	1.5 4889 (120)	0.31 (70)	1654 (240)	0.22 (50)	0.53 (120)/50	NA	NA	F, D, E
501	NW	PP	NA	0.212 (70)	1.4 4686 (115)	0.38 (85)	1894 (275)	0.25 (57)	0.60 (135)/50	NA	NA	F, D, E, S/S
601	NW	PP	NA	0.212 (70)	1.3 4482 (110)	0.42 (95)	2239 (325)	0.27 (60)	0.67 (150)/50	NA	NA	F, D, E, S/S
701	NW	PP	NA	0.212 (70)	1.5 4482 (110)	0.49 (110)	2411 (350)	0.33 (75)	0.8 (180)/50	NA	NA	F, D, E, S/S
801	NW	PP	NA	0.18 (80)	1.5 4482 (110)	0.56 (130)	2756 (400)	0.38 (85)	0.89 (200)/50	NA	NA	F, D, E, S/S
1001	NW	PP	NA	0.15 (100)	1.2 3463 (85)	0.71 (160)	3514 (510)	0.44 (100)	1.11 (250)/50	NA	NA	F, D, P, S/S
1201	NW	PP	NA	0.15 (100)	1.0 3056 (75)	0.8 (180)	4134 (600)	0.51 (115)	1.33 (300)/50	NA	NA	F, D, P, S/S
1601	NW	PP	NA	0.15 (100)	0.7 2037 (50)	1.07 (240)	5512 (800)	0.64 (145)	1.69 (380)/50	NA	NA	F, D, P, S/S

Texel Inc.

Product Name	Structure[1]	Polymer Composition[2]	Percent Open Area %	Apparent Opening Size mm (U.S. sieve)	Permittivity/Flow Rate	Puncture	Mullen Burst	Trapezoid Tear Strength	Grab/Tensile Elongation	MD	XD	Applications
Texel 7643 1460	NW	PET, PP	NA	0.06-0.1 (230-140)	0.22	NP	10200 (1479)	1.3 (293)	3.2 (720)/65-95	NA	NA	E
Texel GEO-9 310	NW reinforced	PP	NA	0.06-0.125 (200-120)	1.72	NP	2400 (348)	0.275 (62)	NP	14 (80)/21.6	NA	R, S/S
Texel 918 475	NW	PET, PP	NA	0.065-0.09 (230-170)	0.68	NP	3500 (508)	0.6 (135)	1.350 (304)/50-70	NA	NA	E, P (liner)
F-200 200	NW	PET	NA	0.04-0.05 (325-270)	0.62	NP	1265 (184)	0.2 (46)	0.4 (90)/70-100	NA	NA	F

[1] W = Woven
NW = Nonwoven
O/C = Other or combination
[2] PET = Polyester
PP = Polypropylene
PE = Polyethylene
O/C = Other or combination
[3] Maximum opening size of tested rolls
[4] Reinforcement applications only
[5] S/S = Separation /stabilization

R = Reinforcement
F = Filtration
D = Drainage
P = Protection
A/O = Erosion Control
S/F = Silt Fence
O/C = Other or combination
NA = Manufacturer determined that this data was not applicable to the product or the data was unavailable

NP = Data not provided by manufacturer

All values were requested to be minimum average roll values and all claims are the responsibility of the manufacturer. All product data are intended as a guide and are not all-inclusive. *Geotechnical Fabrics Report* **recommends you contact manufacturers before making any specifying/purchasing decisions.**

Texel Inc. (cont.)

TNS Mills Inc.

Webtec Inc.

Product Name — Mass per unit g/m² (oz/yd²)	Structure[1]	Polymer Composition[2]	Percent Open Area (wovens only) CWO-22125 %	Apparent Opening Size[3] ASTM D 4751-87 mm (U.S. sieve)	Permittivity ASTM D 4491-92 sec⁻¹/Flow Rate l/min/m² (gal/min/ft²)	Puncture ASTM D 4833-88 kN (lb)	Mullen Burst ASTM D 3786-87 kPa (psi)	Trapezoid Tear Strength ASTM D 4533-91 kN (lb)	Grab/Tensile Elongation ASTM D 4632-91 kN (lb)%	Wide Width Tensile Elongation[4] ASTM D 4595-86 kN/m (lb/in) MD	Wide Width Tensile Elongation[4] XD	Manufacturer's Suggested Applications[5]
Texel Inc. (cont.)												
F-500 / 500	NW	PET	NA	0.04-0.05 (325-270)	0.14	NP	3500 (508)	0.475 (107)	1.1 (248)/ 70-100	NA	NA	F
TNS Mills Inc.												
TNS A040 136 (4.0)	NW	PP	NP	212 (70)	1.70 (160)	(70)	(230)	(45)	(100)/50	NP	NP	F,D,E
TNS A045 153 (4.5)	NW	PP	NP	212 (70)	1.60 (140)	(75)	(250)	(50)	(120)/50	NP	NP	F,D,E
TNS A050 170 (5.0)	NW	PP	NP	180 (80)	1.50 (125)	(80)	(285)	(55)	(130)/50	NP	NP	F,D,E
TNS A060 203(6.0)	NW	PP	NP	125 (120)	1.20 (100)	(90)	(325)	(70)	(155)/50	NP	NP	S/S,D,E,P
TNS A070 237(7.0)	NW	PP	NP	125 (120)	1.10 (95)	(120)	(375)	(80)	(200)/50	NP	NP	S/S,D,E,P,R
TNS A080 271(8.0)	NW	PP	NP	125 (120)	0.95 (85)	(140)	(425)	(95)	(225)/50	NP	NP	S/S,D,E,P,R
TNS A090 305(9.0)	NW	PP	NP	125 (120)	0.93 (80)	(150)	(475)	(100)	(235)/50	NP	NP	S/S,D,E,P,R
TNS A100 339(10.0)	NW	PP	NP	125 (120)	0.90 (75)	(160)	(500)	(110)	(250)/50	NP	NP	S/S,D,E,P,R
TNS A120 407(12.0)	NW	PP	NP	106 (140)	0.80 (50)	(180)	(575)	(130)	(300)/50	NP	NP	S/S,D,E,P,R
TNS A160 543(16.0)	NW	PP	NP	106 (140)	0.70 (50)	(220)	(750)	(150)	(400)/50	NP	NP	S/S,D,P,R
Webtec Inc.												
TerraTex EP	W	PP	4	0.21 (70)	0.28 (18)	(135)	(480)	(95 x 95)	(350 x 250)/15	NA	NA	F, P, D, E
TerraTex HD	W	PP	<1	0.21 (70)	0.02 (2)	(120)	(600)	(120)	(300)/15	(175)/10	(175)/10	S/S, P, E

[1] W = Woven
NW = Nonwoven
O/C = Other or combination
[2] PET = Polyester
PP = Polypropylene
PE = Polyethylene
O/C = Other or combination
[3] Maximum opening size of tested rolls

[4] Reinforcement applications only
[5] S/S = Separation/stabilization
R = Reinforcement
F = Filtration
D = Drainage
P = Protection
E = Erosion Control
A/O = Asphalt Overlay

SF = Silt Fence
O/C = Other or combination
NA = Manufacturer determined that this data was not applicable to the product or the data was unavailable
NP = Data not provided by manufacturer

All values were requested to be minimum average roll values and all claims are the responsibility of the manufacturer. All product data are intended as a guide and are not all-inclusive. *Geotechnical Fabrics Report* recommends you contact manufacturers before making any specifying/purchasing decisions.

Product Name $\begin{bmatrix} \text{Mass per unit} \\ \text{g/m}^2 \text{ (oz/yd}^2) \end{bmatrix}$	Physical Properties		Filtration/Hydraulic Properties			Mechanical Properties						Manufacturer's Suggested Applications[5]
	Structure[1]	Polymer Composition[2]	Percent Open Area (wovens only) CWO-22125 %	Apparent Opening Size[3] ASTM D 4751-87 mm (U.S. sieve)	Permittivity ASTM D 4491-92 sec⁻¹/Flow Rate l/min/m² (gal/min/ft²)	Puncture ASTM D 4833-88 kN (lb)	Mullen Burst ASTM D 3786-87 kPa (psi)	Trapezoid Tear Strength ASTM D 4533-91 kN (lb)	Grab/Tensile Elongation ASTM D 4632-91 kN (lb)%	Wide Width Tensile Elongation[4] ASTM D 4595-86 kN/m (lb/in) MD	XD	
TerraTex GS	W	PP	<1	0.30 (50)	0.04 (4)	(90)	(400)	(75)	(200)/15	(120)/10	(175)/10	S/S, P, E
TerraTex GS-150	W	PP	<1	0.60 (30)	0.04 (4)	(65)	(325)	(45)	(140)/15	NA	NA	S/S, E
TerraTex GS-110	W	PP	<1	0.30 (50)	NA	NA	(220)	(35)	(80)/15	NA	NA	S/S
TerraTex SC	W	PP	<1	0.60 (30)	1.3 (15)	(60)	(340)	(80)	(120)/20	NA	NA	S/F
TerraTex S02	NW	PP	NA	0.60 (30)	1.0 (110)	(18)	(65)	(25)	(60)/60	NA	NA	D, F
TerraTex N03	NW	PP	NA	0.59 (30)	2.0 (110)	(50)	(170)	(25)	(80)/50	NA	NA	D, F
TerraTex N04	NW	PP	NA	0.21 (70)	2.2 (150)	(55)	(200)	(40)	(95)/50	NA	NA	D, F
TerraTex S04	NW	PP	NA	0.21 (70)	0.7 (55)	(40)	(140)	(60)	(130)/60	NA	NA	P, F, S/S
TerraTex OL	NW	PP	NA	NA	NA	(60)	(200)	(35)	(90)/50	NA	NA	D, F, A/O
TerraTex N05 (5.0)	NW	PP	NA	0.21 (70)	1.6 (110)	(80)	(270)	(55)	(135)/50	NA	NA	D, F, S/S
TerraTex N06 (6.0)	NW	PP	NA	0.18 (80)	1.5 (100)	(90)	(310)	(65)	(165)/50	NA	NA	F/D, S/S
TerraTex N07 (7.0)	NW	PP	NA	0.18 (80)	1.3 (90)	(100)	(330)	(75)	(200)/50	NA	NA	P, F, E
TerraTex N08 (8.0)	NW	PP	NA	0.18 (80)	1.2 (80)	(115)	(360)	(85)	(215)/50	NA	NA	P, F, E
TerraTex N010 (10.0)	NW	PP	NA	0.15 (100)	1.1 (70)	(150)	(450)	(100)	(285)/50	NA	NA	P, F, S/S
TerraTex N012 (12.0)	NW	PP	NA	0.15 (100)	0.8 (60)	(170)	(550)	(115)	(325)/50	NA	NA	P, F, S/S
TerraTex N016 (16.0)	NW	PP	NA	0.12 (120)	0.6 (45)	(250)	(750)	(165)	(425)/50	NA	NA	P, F, S/S

All values were requested to be minimum average roll values and all claims are the responsibility of the manufacturer. All product data are intended as a guide and are not all-inclusive. *Geotechnical Fabrics Report* recommends you contact manufacturers before making any specifying/purchasing decisions.

[1] W = Woven
NW = Nonwoven
O/C = Other or combination
[2] PET = Polyester
PP = Polypropylene
PE = Polyethylene
O/C = Other or combination
[3] Maximum opening size of tested rolls
[4] Reinforcement applications only
[5] S/S = Separation/stabilization
R = Reinforcement
F = Filtration
D = Drainage
P = Protection
E = Erosion Control
A/O = Asphalt Overlay

S/F = Silt Fence
O/C = Other or combination
NA = Manufacturer determined that this data was not applicable to the product or the data was unavailable
NP = Data not provided by manufacturer

Wellman Inc.—Nonwovens Division

Product Name / Mass per unit g/m² (oz/yd²)	Physical Properties			Filtration/Hydraulic Properties		Mechanical Properties						Manufacturer's Suggested Applications[5]
	Structure[1]	Polymer Composition[2]	Percent Open Area (wovens only) CWO-22125 %	Apparent Opening Size[3] ASTM D 4751-87 mm (U.S. sieve)	Permittivity ASTM D 4491-92 sec⁻¹/Flow Rate l/min/m² (gal/min/ft²)	Puncture ASTM D 4833-88 kN (lb)	Mullen Burst ASTM D 3786-87 kPa (psi)	Trapezoid Tear Strength ASTM D 4533-91 kN (lb)	Grab/Tensile Elongation ASTM D 4632-91 kN (lb)%	Wide Width Tensile Elongation[4] ASTM D 4595-86 kN/m (lb/in) MD	XD	
WQ150 150 (4.2)	NW	PET	NA	70	1.70 (105)	(65)	(175)	(55)	85	(35)	NP	F, A/O
WQ170 170 (4.5)	NW	PET	NA	70	1.65 (100)	(70)	(200)	(60)	85	(40)	NP	F, A/O
WQ200 200 (6)	NW	PET	NA	70	1.44 (90)	(75)	(235)	(70)	85	(45)	NP	F, D, S/S, E
WQ275 275 (8)	NW	PET	NA	70	1.26 (80)	(85)	(305)	(80)	85	(60)	NP	F, D, S/S, E
WQ350 350 (10)	NW	PET	NA	100	1.44 (75)	(105)	(385)	(90)	85	(85)	NP	F, R, P, S/S, E
WQ400 400 (12)	NW	PET	NA	100	1.04 (70)	(130)	(425)	(110)	85	(120)	NP	R, P, S/S, E
WQ475 475 (14)	NW	PET	NA	100	0.97 (65)	(165)	(525)	(120)	85	(155)	NP	R, P, S/S
WQ550 550 (16)	NW	PET	NA	100	0.90 (60)	(200)	(590)	(135)	85	(185)	NP	R, P, S/S
PN040 135 (4)	NW	PP	NA	70	2.0 (170)	(70)	(230)	(50)	50	(50)	NP	F
PN060 200 (6)	NW	PP	NA	70	1.8 (150)	(90)	(325)	(65)	50	(75)	NP	F, D, S/S, E
PN080 275 (8)	NW	PP	NA	70	1.5 (140)	(120)	(400)	(80)	50	(100)	NP	F, D, S/S, E
PN100 350 (10)	NW	PP	NA	100	1.1 (110)	(150)	(500)	(105)	50	(125)	NP	F, R, S/S, E, P
PN120 400 (12)	NW	PP	NA	100	0.9 (80)	(175)	(600)	(130)	50	(150)	NP	R, S/S, E, P
PN140 475 (14)	NW	PP	NA	100	0.7 (70)	(210)	(660)	(145)	50	(175)	NP	R, S/S, P
PN160 550 (16)	NW	PP	NA	100	0.5 (50)	(230)	(770)	(160)	50	(200)	NP	R, S/S, P

[1]
W = Woven
NW = Nonwoven
O/C = Other or combination
[2]
PET = Polyester
PP = Polypropylene
PE = Polyethylene
O/C = Other or combination
[3] Maximum opening size of sealed rolls

[4] Reinforcement applications only
[5]
S/S = Separation/stabilization
R = Reinforcement
F = Filtration
D = Drainage
E = Erosion Control
A/O = Asphalt Overlay

S/F = Silt Fence
O/C = Other or combination
NA = Manufacturer determined that this data was not applicable to the product or the data was unavailable
NP = Data not provided by manufacturer

All values were requested to be minimum average roll values and all claims are the responsibility of the manufacturer. All product data are intended as a guide and are not all-inclusive. *Geotechnical Fabrics Report* recommends you contact manufacturers before making any specifying/purchasing decisions.

Appendix 4

Characteristics, Industrial Uses, and Manufacturers of Major Generic Fibers and Trade Names*

*Sources: American Fiber Manufacturers Association, Textile World Man-Made Fiber Chart, McLean Hunter Publication, Fiber Manufacturers.

Fiber	Manufacturer	Major Characteristics	Industrial and Home Furnishing Uses
ACETATE		Luxurious feel and appearance; wide range of colors and lusters; excellent drapeability and softness; relatively fast drying; shrink-, moth- and mildew-resistant	Cigarette filters, draperies, upholstery, fiberfill for pillows, quilted products
Celebrate!	Hoechst Celanese Corporation		
Chromspun	Eastman Chemical Products		
Estron	Eastman Chemical Products		
Celanese	Hoechst Celanese Corporation		
ACRYLIC		Soft and warm, wool-like, lightweight; retains shape; resilient; quick drying; resistant to moths, sunlight, oil and chemicals	Auto tops, awnings, geotextiles, various industrial fabrics, blankets, carpets, draperies, upholstery
Acrilan	Monsanto Chemical Company		
Creslan	Cyctec Industries		
Duraspun	Monsanto Chemical Company		
Du-Rel	Monsanto Chemical Company		
Fi-lana	Monsanto Chemical Company		
Orlon	E. I. du Pont de Nemours & Company, Inc.		
Pil-Trol	Monsanto Chemical Company		
Zefkrome	Mann Industries, Inc.		
Zefran	Mann Industries, Inc.		
ARAMID		No melting point; highly flame and temperature resistant; high strength; high modulus	Hot-gas filtration fabrics, protective clothing, military helmets, protective vests, structural composites for aircraft and boats, sailcloth, tires, ropes and cables, mechanical rubber goods, marine and sporting goods
Kevlar	E. I. du Pont de Nemours & Company, Inc.		
Nomex	E. I. du Pont de Nemours & Company, Inc.		
Twaron	Akzo Industrial Fibers		
CARBON		Does not melt; high tensile strength	Composites; conductive fabrics
Thornel	Amoco Fabrics and Fibers Co.		
AS-4	Hercules, Inc.		
AS-6	Hercules, Inc.		
T40	Toray		
T300	Union Carbide/Toray		
Celion	Hoechst Celanese Corporation/ToHo		
HMS-4	Hercules, Inc.		
PAN 50	Toray		
GY-70	Hoechst Celanese Corporation		
P-55	Union Carbide		

(continued)

Fiber	Manufacturer	Major Characteristics	Industrial and Home Furnishing Uses
GLASS		Does not burn; resists alkalis and acids	Composites, insulation, sporting goods, marine products
Glass	Owens/Corning		
	PPG Industries		
LYOCELL		Good resistance to aging, sunlight and abrasion	Similar to cotton
Tencel	Courtaulds Fibers Inc.		
MODACRYLIC		Soft, resilient, abrasion and flame resistant; quick drying; resists acids and alkalis; retains shape	Industrial fabrics, filters, paint rollers, stuffed toys, nonwoven fabrics, awnings, blankets, carpets, flame-resistant draperies and curtains, knit pile fabric backings
SEF	Monsanto Chemical Company		
NYLON		Exceptionally strong; supple, abrasion resistant; lustrous; easy to wash; resists damage from oil and many chemicals; resilient	Air hoses, conveyor belts, seat belts, parachutes, racket strings, ropes and nets, sleeping bags, tarpaulins, tents, thread, tire cord, geotextiles, raincoats, ski and snow apparel, windbreakers, bedspreads, carpets, draperies, curtains, upholstery
A.C.E.	AlliedSignal Inc.		
Anso	AlliedSignal Inc.		
Antron	E. I. du Pont de Nemours & Company, Inc.		
Cantrece	E. I. du Pont de Nemours & Company, Inc.		
Capima	AlliedSignal Inc.		
Caplana	AlliedSignal Inc.		
Caprolan	AlliedSignal Inc.		
Captiva	AlliedSignal Inc.		
Compet	AlliedSignal Inc		
Cordura	E. I. du Pont de Nemours & Company, Inc.		
Crepeset	BASF Corporation		
Hydrofil	AlliedSignal Inc.		
Patina	AlliedSignal Inc.		
Resistant	BASF Corporation		
Shimmereen	BASF Corporation		
Softglo	BASF Corporation		
Tolaram	Tolaram Fibers, Inc.		
Viyana	BASF Corporation		
WorryFree	AlliedSignal Inc.		
Zefsport	BASF Corporation		
Zeftron	BASF Corporation		

Fiber	Manufacturer	Major Characteristics	Industrial and Home Furnishing Uses
OLEFIN		..... Unique wicking properties that make it very comfortable; abrasion resistant; quick drying; resistant to deterioration from chemicals; mildew, perspiration, rot and weather resistance; sensitive to heat; soil resistant; strong; very lightweight; excellent colorfastness	..... Geotextiles, filter fabrics, dye nets, automotive interiors, laundry and sandbags, cordage, doll hair, industrial sewing thread, carpet and carpet backing, slipcovers, upholstery
Alpha	Amoco Fabrics and Fibers Company		
Essera	Amoco Fabrics and Fibers Company		
Fibrilon	Synthetic Industries		
Genesis	Amoco Fabrics and Fibers Company		
Herculon	Hercules		
Marquesa Lana	Amoco Fabrics and Fibers Company		
Marvess	Amoco Fabrics and Fibers Company		
Patlon	Amoco Fabrics and Fibers Company		
Spectra	AlliedSignal Inc.		
Synera	Amoco Fabrics and Fibers Company		
Trace	Amoco Fabrics and Fibers Company		
Tolaram	Tolaram Fibers, Inc.		
POLYESTER		..... Strong; resistant to stretching and shrinking; resistant to most chemicals; quick drying; crisp and resilient when wet or dry; wrinkle and abrasion resistant; retains heat set pleats and creases; easy to wash	..... Fire hose, power belting, ropes and nets, tire cord, sail, V-belts, fiberfill for various products, carpets, curtains, draperies, sheets and pillow cases, insulated garments
A.C.E.	AlliedSignal Inc.		
Ceylon	Hoechst Celanese Corporation		
Comfort Fiber	Hoechst Celanese Corporation		
Compet	AlliedSignal Inc.		
Dacron	E. I. du Pont de Nemours & Company, Inc.		
E.S.P.	Hoechst Celanese Corporation		
Fortrel	Wellman, Inc.		
Golden Glow	BASF Corporation		
Golden Touch	BASF Corporation		
Hollofil	E. I. du Pont de Nemours & Company, Inc.		
Kodaire	Eastman Chemical		
Kodel	Eastman Chemical		
KodOfill	Eastman Chemical		
KodOsoff	Eastman Chemical		
Pentron	Hoechst Celanese Corporation		
Polarguard	Hoechst Celanese Corporation		

(continued)

Fiber	Manufacturer	Major Characteristics	Industrial and Home Furnishing Uses
POLYESTER			
Strialine	BASF Corporation		
Tolaram	Tolaram Fibers, Inc.		
Trevira	Hoechst Celanese Corporation		
Trevira Finesse	Hoechst Celanese Corporation		
Untra Touch	BASF Corporation		
PBI		Highly flame resistant; outstanding comfort factor combined with thermal and chemical stability properties; will not burn or melt; low shrinkage when exposed to flame	Suitable for high performance protective apparel such as firemens' turnout coats, astronaut space suits and applications
PBI logo	Hoechst Celanese Corporation		
RAYON		Highly absorbent; soft and comfortable; easy to dye; versatile; good drapeability	Medical/surgical products, tire, cord, industrial products, bedspreads, blankets, carpets, curtains, draperies, sheets, tablecloths, upholstery
Beau-Grip	North American Rayon		
Fibro	Courtaulds Fibers Inc.		
SPANDEX		Can be stretched 500 percent without breaking; can be stretched repeatedly and recover original length; lightweight; stronger, more durable than rubber; resistant to body oils	Surgical hose, ski pants, athletic apparel
Glospan/ Clearspan, S-85	Globe		
Lycra	E. I. du Pont de Nemours & Company, Inc.		
SULFAR		High performance fibers with excellent resistance to harsh chemicals and high temperatures; excellent strength retention in adverse environments; flame retardant; non-conductive	Filter fabric for coal-fired boiler bag houses, paper machine clothing, electrical insulation, electrolysis membranes, filter fabrics for liquid and gas filtration, high performance composites, gaskets and packings
Ryton	Amoco Fibers		
VINYON		Softens at low temperature; high resistance to chemical; non-toxic	Used in industrial applications as a bonding agent for nonwoven fabrics and products such as tea bags and automotive headliners
Vinyon	Hoechst Celanese Corporation		

Appendix 5

Derivation of Mathematical Equations

1. TWIST TAKE-UP CALCULATIONS

In Figure A5.1, if the helical path l' of each ply is equal to l before plying, then L', the plied yarn length, must be less than L, the original untwisted yarn length. The contraction or take-up can be calculated from the simple geometry of the structure:

$$\cos Q = L'/l'$$

and

$$L' = l' \cos Q$$

where

Q = helix angle of singles yarn with respect to the ply axis
D = plied yarn diameter
l' = original length of each of the singles
L' = length of the twisted plied yarn

Percent contraction due to ply twist is

$$\% \, C = \frac{(l' - L')}{l'} \times 100$$

but

$$L' = l' \cos Q$$

therefore

$$\% \, C = \frac{(l' - l' \cos Q)}{l'} 100 = (1 - \cos Q)\,100$$

If the length of the plied yarn L' is known, the length of the singles composing the ply l' can be calculated as

$$l' = L'/\cos Q$$

2. HELIX ANGLE VERSUS TURNS PER INCH

The relationship among turns per inch, singles and plied yarn diameters for perfect cylindrical helices can be shown to be

$$\tan Q = \pi T D K$$

where

Q = helix angle
T = turns per inch
D = plied yarn diameter
K = the ratio of the actual helix diameter to the plied yarn diameter. It is a geometrical constant depending upon the number and shape of the individual singles making up the ply.

K can be shown to be equal to $(D - d)/D$

where

d = singles yarn diameter
D = plied yarn diameter

For a two-ply yarn

$$K = (D - d)/D$$

but

$$D = 2d$$

therefore

$$(2d - d)/2d = 0.5$$

For a three-ply yarn $D/2$ (Figure A5.2) is slightly larger than d, and K is slightly larger than 0.5, the theoretical value being 0.54.

It was shown that K ranges from 0.80 to 0.85 as the

817

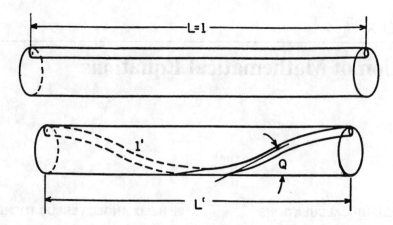

FIGURE A5.1 Helix diagram.

plies increase from 16 to 30. Theoretically K will approach unity as the number of plies increases to infinity. Obviously in multi-filament singles yarns the number of filaments becomes large and K will range from 0.8 to 0.99. When filament or singles yarn cross sections are not circular, instead of using diameters, radial distances from the plied yarn center to the center of gravity of the singles should be employed.

3. CORKSCREW YARN ANALYSIS

The strength of a corkscrew yarn must always be less than the theoretical attainable if the yarn were free of corkscrew. This may be demonstrated by the examples in Figures A5.3 and A5.4.

In Figure A5.3 where a single ply is wrapped about a central core of six, core yarn lengths l_c are always

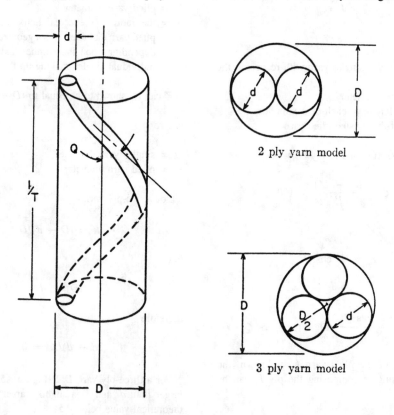

2 ply yarn model

3 ply yarn model

FIGURE A5.2 Helix diagram and yarn models.

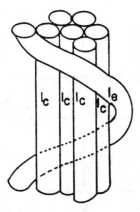

FIGURE A5.3 One yarn "corkscrewed" about a central core of six.

less than entwined yarn length l_e. Upon the application of a force to this yarn, l_c cords will reach their breaking strains (elongation) and breaking load (force) before l_e cord, since l_e is in the form of a helix and has the opportunity to assume a geometric adjustment (straightening). At the instant of break of the l_c cords, the force will be transferred to l_e, but it cannot support a load equal to six times its own strength, and so it too must fail. Thus the strength of this corkscrew structure is approximately 6/7 of theoretical, plus a contributing load assumed by l_e while l_c cords are being loaded.

In Figure A5.4 where six corkscrew yarns entwine about a central core, l_c, which is shorter, will reach its limiting elongation and load first, and will break. At the instant of break the force will be transferred to the six l_e entwining yarns. These six yarns, depending upon the load they are then already carrying, may

or may not be able to support the additional load. Here the order of magnitude of strength that the corkscrew yarn can support again remains at 6/7 of theoretical (plus whatever load is assumed by l_c while l_e cords are also being loaded).

One consequence of corkscrew is poor "elongation balance" with resulting lowered strength efficiency. Maximum strength translation can only be obtained when all supporting members reach their limiting elongation and breaking load at the same time.

4. TWIST MULTIPLIER FOR SPUN YARNS

Because of the relationship between turns per inch and yarn size (diameter), the yarn manufacturers established a formula to calculate the turns per inch necessary for a given size spun yarn. Instead of the equation $\tan Q = \pi TKD$, a simpler relationship between the turns per inch and the cotton count can be used which is called "twist multiplier." This is a quantitative index of the relative steepness of the helix angle. Twist multiplier is defined as

$$TM = T/C^{1/2}$$

where

T = turns per inch
C = cotton count

The similarity between TM and $\tan Q$ is apparent when it is realized that for equivalent yarn density, yarn diameter is proportional to the inverse of the square root of the cotton count. Thus

$$TM = \alpha TD$$

where

T = turns per inch
D = yarn diameter
α = constant relating D and $1/C^{1/2}$

In more general terms, taking into account yarn density and the number of fibers, it can be shown that

$$TM = \tan Q \frac{(P_y)^{1/2}}{\varphi K}$$

where

P_y = yarn density
φ = dimensional constant
$K = (D - d)/d$

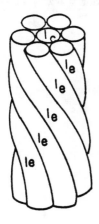

FIGURE A5.4 Six yarns "corkscrewed" about a central core of yarn.

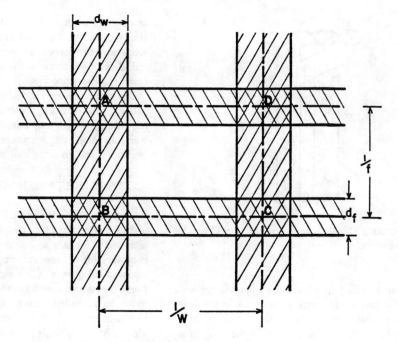

FIGURE A5.5 Cover factor diagram.

5. CALCULATION OF COVER FACTOR

Referring to Figure A5.5, if w and f are warp and filling threads per inch, and d_w and d_f are warp and filling yarn diameters,

distance between two warp yarns $= 1/w$

distance between two filling yarns $= 1/f$

total area per fabric unit $= (1/w)(1/f) = 1/(wf)$

open area per fabric unit $= (1/w - d_w)(1/f - d_f)$

closed area per fabric unit

$$= (1/wf) - (1/w - d_w)(1/f - d_f)$$

fractional closed area (cover factor)

$$= \frac{(1/wf) - (1/w - d_w)(1/f - d_f)}{1/wf}$$

or

cover factor $= wd_w + fd_f - wfd_wd_f$

Appendix 6

Measurement Units

Sources:

- ASTM
- Mark's Standard Handbook for Mechanical Engineers, McGraw-Hill Book Company, 1978.
- Standard Mathematical Tables, CRC Press, 1982.

1. U.S. CUSTOMERY UNITS

Length

    1 foot = 12 inches
    1 yard = 3 feet
    1 mile = 1,760 yards = 5,280 feet

Area

    1 square foot = 144 square inches
    1 square yard = 9 square feet
    1 acre = 43,560 square feet
    1 square mile = 640 acres

Volume

    1 cubic foot = 1,728 cubic inches
    1 gallon = 231 cubic inches
    1 cubic yard = 27 cubic feet

    liquid or fluid measures

        1 pint = 4 gills
        1 quart = 2 pints
        1 gallon = 4 quarts
        1 cubic foot = 7.4805 gallon

    dry measures

        1 quart = 2 pints
        1 peck = 8 quarts
        1 bushel = 4 pecks

Weights

    1 ounce = 16 drams = 437.5 grains
    1 pound = 16 ounces = 7,000 grains

2. SI PREFIXES

multiplication factors	prefix	SI symbol
$1\ 000\ 000\ 000\ 000\ 000\ 000 = 10^{18}$	exa	E
$1\ 000\ 000\ 000\ 000\ 000 = 10^{15}$	pecta	P
$1\ 000\ 000\ 000\ 000 = 10^{12}$	tera	T
$1\ 000\ 000\ 000 = 10^{9}$	giga	G
$1\ 000\ 000 = 10^{6}$	mega	M
$1\ 000 = 10^{3}$	kilo	k
$100 = 10^{2}$	hecto	h
$10 = 10^{1}$	deka	da
$0.1 = 10^{-1}$	deci	d
$0.01 = 10^{-2}$	centi	c
$0.001 = 10^{-3}$	milli	m
$0.000\ 001 = 10^{-6}$	micro	μ
$0.000\ 000\ 001 = 10^{-9}$	nano	n
$0.000\ 000\ 000\ 001 = 10^{-12}$	pico	p
$0.000\ 000\ 000\ 000\ 001 = 10^{-15}$	femto	f
$0.000\ 000\ 000\ 000\ 000\ 001 = 10^{-18}$	atto	a

3. SI UNITS

quantity	unit	SI symbol	Formula
Base Units			
length	meter	m	
mass	kilogram	kg	
time	second	s	
electric current	ampere	A	
thermodynamic temperature	kelvin	K	
amount of substance	mole	mol	
luminous intensity	candela	cd	
Derived Units			
acceleration	meter per second square		m/s^2
area	square meter		m^2
density	kilogram per cubic meter		kg/m^3
energy	joule	J	$N.m$
force	newton	N	$kg.m/s^2$
frequency	hertz	Hz	$1/s$
power	watt	W	J/s
pressure	pascal	Pa	N/m^2
velocity	meter per second		m/s
voltage	volt	V	W/A
volume	cubic meter		m^3
work	joule	J	$N.m$

1 meter = 100 cm = 1000 millimeter
1 kilogram = 1000 gram
1 angstrom = 10^{-8} centimeter

4. CONVERSION BETWEEN SI AND U.S. CUSTOMERY UNITS

1 yard = 0.9144 m
1 inch = 25.4 mm
1 lb = 0.453 59 kg = 453.59 g

Metric to US Customery Conversion Factors

to get	multiply	by
inches	centimeters	0.3937007874
feet	meters	3.280839895
yards	meters	1.093613298
miles	kilometers	0.6213711922
ounces	grams	0.03527396195
pounds	kilograms	2.204622622
gallons (liguid)	liters	0.2641720524
fluid ounces	milliliters (cc)	0.03381402270
square inches	centimeter squares	0.1550003100
square feet	meter squares	10.76391042
square yards	meter squares	1.195990046
cubic inches	milliliters (cc)	0.06102374409
cubic feet	cubic meters	35.31466672
cubic yards	cubic meters	1.307950619

US Customery to Metric

to get	multiply	by
microns	mils	25.4
centimeters	inches	2.54
meters	feet	0.3048
meters	yards	0.9144
kilometers	miles	1.609344
grams	ounces	28.34952313
kilograms	pounds	0.45359237
liters	gallons (liquid)	3.785411784
milliliters (cc)	fluid ounces	29.57352956
square centimeters	square inches	6.4516
square meters	square feet	0.09290304
square meters	square yards	0.83612736
milliliters (cc)	cubic inches	16.387064
cubic meters	cubic feet	0.02831684659
cubic meters	cubic yards	0.764554858

5. TEMPERATURE CONVERSION

$$^{O}F = 9/5 \ (^{O}C) + 32$$

$$^{O}C = 5/9 \ [(^{O}F) - 32]$$

Index

The Industrial Fabrics Association International (IFAI) is a not-for-profit trade association whose members represent every facet of the technical fabrics industry. IFAI serves as the industry's focal point for member companies that range in size from one-person shops to multinational corporations. Their products cover the broad spectrum of the textile industry—from high-tech fibers, fabrics and end-products, to the equipment and hardware necessary to create them. IFAI publishes a range of publications including *Industrial Fabric Products Review*, *Geotechnical Fabrics Report*, *Fabrics & Architecture*, and *IFAI's Marine Fabricator* to inform the industry.

You can contact IFAI at:

INDUSTRIAL FABRICS ASSOCIATION INTERNATIONAL

345 CEDAR STREET, SUITE #800, ST. PAUL MINNESOTA 55101-1088, U.S.A.
TELEPHONE: (612) 222-2508   FAX: (612) 222-8215